AF572693

Fundamental Proof Methods in Computer Science

Fundamental Proof Methods in Computer Science

A Computer-Based Approach

Konstantine Arkoudas and David Musser

The MIT Press
Cambridge, Massachusetts
London, England

This book was typeset in LaTeX by the authors. Printed and bound in the United States of America.

Library of Congress Cataloging-in-Publication Data

Names: Arkoudas, Konstantine, author. | Musser, David R., author.
Title: Fundamental proof methods in computer science : a computer-based approach / Konstantine Arkoudas and David Musser.
Description: Cambridge, MA : MIT Press, [2017] | Includes bibliographical references and index.
Identifiers: LCCN 2016020047 | ISBN 9780262035538 (hardcover : alk. paper)
Subjects: LCSH: Computer science–Mathematics. | Proof theory.
Classification: LCC QA76.9.M35 A74 2017 | DDC 004.01/51–dc23 LC record available at https://lccn.loc.gov/2016020047

10 9 8 7 6 5 4 3 2 1

Στους γονείς μου, Βασίλη και Χρυσούλα

K. A.

To Martha, with love and gratitude

D. M.

Contents

Preface xvii

I INTRODUCTION 1

1 An Overview of Fundamental Proof Methods 3
1.1 Equality chaining 3
1.2 Induction 8
1.3 Case analysis 10
1.4 Proof by contradiction 12
1.5 Abstraction/specialization 13
1.6 The usual case: Proof methods in combination 15
1.7 Automated proof 15
1.8 Structure of the book 16

2 Introduction to Athena 19
2.1 Interacting with Athena 22
2.2 Domains and function symbols 23
2.3 Terms 27
2.4 Sentences 34
2.5 Definitions 40
2.6 Assumption bases 42
2.7 Datatypes 44
2.8 Polymorphism 50
2.8.1 Polymorphic domains and sort identity 50
2.8.2 Polymorphic function symbols 52
2.8.3 Polymorphic datatypes 57
2.8.4 Integers and reals 59
2.9 Meta-identifiers 61
2.10 Expressions and deductions 62
2.10.1 Compositions 65
2.10.2 Nested method calls 68
2.10.3 Let expressions and deductions 69
2.10.4 Conclusion-annotated deductions 70
2.10.5 Conditional expressions and deductions 71
2.10.6 Pattern-matching expressions and deductions 72
2.10.7 Backtracking expressions and deductions 73
2.10.8 Defining procedures and methods 74
2.11 More on pattern matching 78

2.12 Directives 85
2.13 Overloading 86
2.14 Programming 89
2.14.1 Characters 90
2.14.2 Strings 90
2.14.3 Cells and vectors 90
2.14.4 Tables and maps 91
2.14.5 While loops 93
2.14.6 Expression sequences 93
2.14.7 Recursion 93
2.14.8 Substitutions 94
2.15 A consequence of static scoping 98
2.16 Miscellanea 99
2.17 Summary and notational conventions 103
2.18 Exercises 105

II FUNDAMENTAL PROOF METHODS 111

3 Proving Equalities 113
3.1 Numeric equations 113
3.2 Equality chaining preview 116
3.3 Terms and sentences as trees 117
3.4 The logic behind equality chaining 120
3.5 More examples of equality chaining 128
3.6 A more substantial proof example 130
3.7 A better proof 134
3.8 The principle of mathematical induction 135
3.8.1 Different ways of understanding mathematical induction 139
3.9 List equations 140
3.9.1 Polymorphic datatypes 147
3.10 Evaluation of ground terms 150
3.11 Top-down proof development 152
3.12 ⋆ Input expansion and output transformation 158
3.12.1 Converters 158
3.12.2 Input expansion 162
3.12.3 Output transformation 165
3.12.4 Combining input expansion and output transformation with overloading 166
3.12.5 Using `declare` with auxiliary information 167

3.13 ⋆ Conjecture falsification 168
3.14 ⋆ Conditional rewriting and additional chaining features 172
3.15 ⋆ Proper function definitions 179
3.16 Summary 185
3.17 Additional exercises 186
3.18 Chapter notes 189

4 Sentential Logic 191
4.1 Working with the Boolean constants 191
4.2 Working with conjunctions 192
4.2.1 Using conjunctions 192
4.2.2 Deriving conjunctions 193
4.3 Working with conditionals 194
4.3.1 Using conditionals 194
4.3.2 Deriving conditionals: hypothetical reasoning 195
4.4 Working with disjunctions 199
4.4.1 Using disjunctions: reasoning by cases 199
4.4.2 Deriving disjunctions 201
4.5 Working with negations 202
4.5.1 Using negations 202
4.5.2 Deriving negations: reasoning by contradiction 202
4.6 Working with biconditionals 206
4.6.1 Using biconditionals 206
4.6.2 Deriving biconditionals 207
4.7 Forcing a proof 207
4.8 Putting it all together 209
4.9 A library of useful methods for sentential reasoning 211
4.10 Recursive proof methods 226
4.11 Dealing with large conjunctions and disjunctions 236
4.12 Sentential logic semantics 238
4.13 SAT solving 246
4.14 Proof heuristics for sentential logic 272
4.14.1 Backward tactics 274
4.14.2 Forward tactics 276
4.14.3 Replacement tactics 282
4.14.4 Strategies for deploying the tactics 282
4.15 ⋆ A theorem prover for sentential logic 294
4.16 Additional exercises 304

4.17 Chapter notes 315

5 First-Order Logic 319

5.1 Working with universal quantifications 323

5.1.1 Using universal quantifications 323

5.1.2 Deriving universal quantifications 326

5.2 Working with existential quantifications 331

5.2.1 Deriving existential quantifications 331

5.2.2 Using existential quantifications 332

5.3 Some examples 338

5.4 Methods for quantifier reasoning 342

5.5 Proof heuristics for first-order logic 353

5.5.1 Backward tactics for quantifiers 354

5.5.2 Forward tactics for quantifiers 355

5.5.3 Proof strategy for first-order logic 371

5.6 First-order logic semantics 372

5.7 Additional exercises 385

5.8 Chapter notes 393

6 Implication Chaining 395

6.1 Implication chains 395

6.2 Using sentences as justifiers 402

6.2.1 Nested rules 407

6.3 Implication chaining through sentential structure 409

6.4 Using chains with chain-last 411

6.5 Backward chains and chain-first 413

6.6 Equivalence chains 415

6.7 Mixing equational, implication, and equivalence steps 417

6.8 Chain nesting 421

6.9 Exercises 423

III PROOFS ABOUT FUNDAMENTAL DATATYPES 427

7 Organizing Theory Development with Athena Modules 429

7.1 Introducing a module 429

7.2 Natural numbers using modules 431

7.3 Extending a module 433

7.4 Modules for function symbols 434

7.5 Additional module features 435

7.6 Additional module procedures 436
7.7 A note on indentation 437

8 Natural Number Orderings 439
8.1 Properties of natural number ordering functions 439
8.1.1 Trichotomy properties 443
8.1.2 Transitive and asymmetric properties 444
8.1.3 Less-equal properties 447
8.1.4 Combining ordering and arithmetic 450
8.2 Natural number subtraction 451
8.3 Ordered lists 459
8.4 Binary search trees 463
8.5 Summary and a connecting theorem 469
8.6 Additional exercises 472
8.7 Chapter notes 473

9 Integer Representations and Proof Mappings 475
9.1 Declarations and axioms 475
9.2 First proofs of integer properties 477
9.3 Another integer representation 478
9.4 Mappings between the signed and pair representations 480
9.5 Additive homomorphism property 481
9.6 Associativity and commutativity of integer addition 483
9.7 Power series 484
9.8 Summary and looking ahead 487
9.9 Additional exercises 488

10 Fundamental Discrete Structures 491
10.1 Ordered pairs 493
10.1.1 Representation and notation 493
10.1.2 Results and methods 494
10.2 Options 497
10.2.1 Representation and notation 497
10.2.2 Some useful results 498
10.3 Sets, relations, and functions 499
10.3.1 Representation and notation 499
10.3.2 Set membership, the subset relation, and set identity 501
10.3.3 Set operations 508
10.3.4 Cartesian products 518

10.3.5 Relations 521
10.3.6 Set cardinality 529
10.3.7 Powersets 530
10.4 Maps 533
10.4.1 Representation and notation 533
10.4.2 Map operations and theorems 535
10.4.3 Default maps 549
10.5 Chapter notes 558

IV PROOFS ABOUT ALGORITHMS 561

11 A Binary Search Algorithm 563
11.1 Defining the algorithm 563
11.1.1 Efficiency considerations 565
11.1.2 Correspondence to definitions in other languages 566
11.1.3 Interface design 567
11.1.4 Testing with evaluation 567
11.2 First correctness properties 569
11.3 Specifying requirements on a function to be defined 574
11.4 Correctness of an optimized binary search algorithm 575
11.5 Summary and looking ahead 576
11.6 Additional exercises 577

12 A Fast Exponentiation Algorithm 579
12.1 Mathematical background 579
12.2 Strong induction 581
12.3 Properties of half 583
12.4 Properties of odd and even 587
12.5 Properties of power 589
12.6 Properties of fast-power 590
12.7 Tail recursion, a potential optimization 592
12.8 Transforming strong induction into ordinary induction 596
12.9 Measure induction 597
12.10 Summary and looking ahead 599
12.11 Additional exercises 599

13 Euclid's Algorithm for Greatest Common Divisors 603
13.1 Quotient and remainder 603
13.2 The division algorithm 605

13.3 Divisibility 608
13.3.1 A cancellation lemma 610
13.3.2 Proof of the characterization theorem 610
13.3.3 Additional properties of divisibility 611
13.4 Euclid's algorithm 616
13.5 Summary 621
13.6 Additional exercises 621
13.7 Chapter notes 623

V PROOFS AT AN ABSTRACT LEVEL 625

14 Abstract Structures 627
14.1 Group properties 627
14.2 Theory refinement 631
14.3 Writing proofs at the level of a theory 635
14.4 Abstract proof method conventions 638
14.5 Dynamic evolution of theories 641
14.6 Testing abstract proofs 642
14.7 Group theory refinements 644
14.7.1 Abelian group theory 644
14.7.2 Multiplicative theories 646
14.7.3 Ring theory 649
14.7.4 Integral domain 652
14.7.5 Algebraic theory diagram 652
14.8 ⋆ Permutations as a group 653
14.8.1 Function theory 654
14.8.2 Permutation theory 658
14.9 Ordering properties at an abstract level 664
14.9.1 Binary-Relation 664
14.9.2 Irreflexive 665
14.9.3 Transitive 665
14.9.4 Strict partial order 666
14.9.5 Nonstrict partial orders 668
14.9.6 Strict weak order 671
14.9.7 A preorder 673
14.9.8 Strict total order 674
14.9.9 Lists over a strict weak order 674
14.9.10 Relational theory diagram 678

14.10 Additional exercises 678

15 Abstract Algorithms 683

15.1 An abstract binary search algorithm 683

15.1.1 Abstract-level binary search trees 687

15.1.2 Abstract-level binary search correctness theorems 688

15.2 An abstract fast-power algorithm 690

15.2.1 Raising to a power in a monoid 690

15.2.2 A monoid version of fast-power 693

15.2.3 Multiplicative version of fast power 700

15.2.4 A nonnumeric application 700

16 Algorithms on Memory Abstractions 703

16.1 Axioms and theorems for individual memory locations 703

16.2 Iterators and ranges 709

16.2.1 Iterator and range axioms and theorems 711

16.2.2 Trivial iterator: The base of a hierarchy of iterator theories 715

16.2.3 Forward iterators 718

16.3 Range count algorithm 721

16.4 Range replace algorithm 724

16.5 Range copy algorithm 729

16.6 Range copy-backward algorithm 734

16.7 Adapters: Reverse-iterator and reverse-range 737

16.8 Implementing copy-backward 740

16.9 Random-access iterators 743

16.9.1 Relationships among iterator functions 744

16.9.2 New properties of the length function 745

16.9.3 Theorems about collecting locations 747

16.9.4 Ordered range 753

16.10 A binary search algorithm 754

16.11 Summary and suggestions for continued study 761

VI PROOFS ABOUT PROGRAMMING LANGUAGES 765

17 A Correctness Proof for a Toy Compiler 767

17.1 Interpreting and compiling numeric expressions 767

17.1.1 Representation and notation 767

17.1.2 Defining the interpreter 769

17.1.3 An instruction set and a virtual machine 771

17.1.4 Compiling numeric expressions 774
17.1.5 Correctness 775
17.2 Handling errors explicitly 783
17.2.1 Extending the compiler with error handling 784
17.3 Chapter notes 797

18 A Simple Imperative Programming Language 799
18.1 A simple imperative language 799
18.2 Semantics of expressions 808
18.3 Semantics of commands 817
18.4 ⋆ Testing the semantics 827
18.5 Some lemmas 833
18.6 Reasoning about the language 844
18.7 Chapter notes 852

A Athena Reference 857
A.1 Syntax 857
A.2 Values 858
A.3 Operational semantics 861
A.4 Pattern matching 875
A.5 Selectors 881
A.6 Prefix syntax 882

B Logic Programming and Prolog 885
B.1 Basics of logic programming 885
B.2 Examples 888
B.3 Implementing a Prolog interpreter 893
B.4 Integration with external Prolog systems 898
B.5 Automating the handling of equations 902

Bibliography 905

Glossary 911

Index 921

Preface

Logic has been called "the calculus of computer science" [66]. Much as mathematics has proven to be an indispensable tool in the natural sciences [113], logic has proven exceedingly useful in computer science. Its applications include databases; hardware; algorithm correctness and optimization; AI subjects such as planning, constraint solving, and knowledge representation; modeling and verification of digital systems in general; and all facets of programming languages, from parsing to type systems and compilation [44]. Perhaps this should not be surprising in view of the historical kinship between the two fields: Turing machines, the key theoretical innovation that gave birth to computer science, were introduced as a formal analysis of algorithms in order to settle the question of whether the satisfiability problem in predicate logic is mechanically decidable [24].[1]

However, even though the primary focus of this text is computer science, it is worth noting from the outset that logic in general and deductive proof in particular are not just useful in computer science or mathematics. Deductive proof is a tool of universal importance and applicability for a simple reason: It is the most rationally compelling species of argument known to humans, and rational argument is the foundation not just of mathematical disciplines but of all science and engineering. Chemistry and finance may have little subject matter in common, but both are underwritten by the same canons of rationality and follow largely the same approach to inquiry: formulating key concepts, articulating and establishing their properties and interrelationships, using these to issue predictions and explanations, and revising them as needed in the face of new empirical evidence or theoretical insights. All of these activities require the ability to distinguish between good and bad arguments, between sound and erroneous trains of thought. In particular, they require the ability to tell when a proposition is a logical consequence of ("follows from") a given body of information and when it is not. And they require the ability to provide correct arguments capable of demonstrating either of these cases. Logic provides a powerful set of conceptual and methodological resources for doing that.

In particular, logic provides a rich formal language in which we can unambiguously express just about anything we care to express, namely, the language of the first-order predicate calculus, which has become the lingua franca of computer science and related fields such as linguistics. It also provides a mathematically precise semantics for that language, with a rigorous definition of the key notion of logical consequence. And perhaps most importantly, it provides effective, sound, and complete mechanisms for constructing proofs, which can be viewed as arguments given to show that a conclusion follows from some premises. These mechanisms can be implemented on the computer, and are themselves amenable to mathematical analysis. Indeed, in this book almost all of our proofs are written in a computer language, Athena, and are checkable by machine. In what follows,

1 This was David Hilbert's famous *Entscheidungsproblem*.

we briefly explain why we believe proofs to be worth studying in general, and then, more narrowly, why we are advocating a computer-based approach to them.

Why proofs?

What is the value of proof? We have already mentioned a key advantage it confers: an exceptionally strong and domain-independent epistemic warrant. If we manage to prove something, whether it is a mathematical theorem, a prediction of an economics theory, or a statement of correctness for a piece of software, we can rest assured that the conclusion is valid, as long as we accept the premises underlying the proof (these might respectively be a body of mathematical axioms, the basic tenets and working assumptions of an economics theory, and the semantics of the programming language in which the software is written). In other words, if we accept the premises, then we are rationally compelled to accept the conclusion.

Proof is the primary vehicle for knowledge generation in the mathematical sciences, including theoretical computer science. But in computer science, proof has also found a more practical use with a stronger engineering flavor: verification, namely, demonstrating that a particular system (or a component, or an algorithm, or even a design sketch) works as it should, or at least that it has certain desirable properties. Verification is becoming increasingly commonplace and will continue to grow in importance as our lives become more dependent on digital artifacts and as the economic and social cost of software and hardware malfunctions becomes more prohibitive.

Verification has so far focused on automation, in an attempt to minimize the human effort necessary for analyzing large systems. However, due to fundamental theoretical limitations, human guidance will remain necessary in most cases. The modeling and verification of complex systems will continue to require the interaction of humans and machines, with people sketching out the main proof ideas and machines filling in the details. The many examples in this book will teach students how to reason about fundamental computer science structures in a style that will hold them in good stead should they ever undertake any verification projects.

Another reason to study proofs is to increase one's level of mathematical sophistication, both in generating one's own mathematical arguments and in reading those of others. Proof is central in mathematics, and if you understand its foundations you will find it easier to understand (and do!) mathematics. You will have a deeper appreciation of how and why putative mathematical proofs work when they are successful, but you will also be able to better understand what goes wrong when they fail. And this leads to a more general reason to master logic and proofs: to become better able to spot and steer clear of reasoning errors, whether in mathematics or elsewhere.

This is, of course, a key part of critical thinking in general. Now, we all have native logical intuitions, so to a certain extent we can all manage to detect logical pitfalls. Unfortunately, however, intuition can only take us so far. Psychologists have shown that our intuitions are often wrong and susceptible to systematic biases. These can only be rectified by rigorous training in normatively correct modes of reasoning, of the kind that are embodied in formal proof methods. And while detecting reasoning errors and questioning assumptions may be useful in guarding against demagogues and other assorted charlatans, it is even more important in debugging computer systems, especially software systems. Indeed, one of the best ways to analyze and to discover errors in our systems is to attempt to *prove that there are no errors*. The gradual refinement that takes place during that process, the unearthing and explicit formulation of assumptions and their consequences, is invaluable in making us understand how our systems work, and when they fail, how and why they fail. So another benefit of training in logic and proofs is cultivating a habit of correct and careful thinking that should pay handsome dividends in the course of an engineering career.

The mention of understanding in the previous paragraph points to yet another, frequently understated, strength of proofs: their potential as tools for *explanation*. The compelling epistemic warrant given by a deductive proof and a sharpened ability to detect and avoid errors are fine and good, but we would get both of these if we had, hypothetically, access to a benevolent oracle, a black box that could always tell us whether a conclusion follows from some premises. If it turned out, perhaps after a long period of time, that the oracle was never wrong, then we would be justified in accepting its verdicts. But while the oracle might always be correct in giving us "thumbs up" or "thumbs down" for a given question, and while we could rely on its answers as robust indicators of truth, we would still lack *understanding*. Suppose the oracle tells us that P is indeed not equal to NP. That's great, but *why*? That is where proofs again enter the picture. A proof is not only a means of *verifying* a claim. It is also, importantly, a means of *explaining* why a claim holds. Or at least a good proof is.

But if a proof is to serve as an explanation, it must be digestible by humans. It must be readable, it must be properly structured, it must abstract away details when they are not needed, and so on. To a large extent that is the responsibility of the proof writer, but the underlying medium—the *language*—in which the proof is written must be able to support these desiderata. It must allow the expression of proofs in a style that is not far removed from informal mathematical practice. Athena, the proof language that we use, is based on *natural deduction*, a style of proof that was explicitly designed to capture the essential aspects of mathematical reasoning as it has been practiced for thousands of years. In combination with other features, such as abstraction mechanisms and complexity management tools borrowed from modern programming languages, Athena takes us a step closer to proofs that can explain and communicate our reasoning, but that are nevertheless entirely formal and checkable by machine.

Why this textbook?

We were motivated to write this textbook primarily by need. Many students whom we have asked to prove theorems in our courses, even in upper undergraduate—and sometimes graduate—courses, have been unable to write proofs. Some have struggled even to write down anything resembling a proof. Others have written what they thought were proofs but were in fact poor proof attempts, full of reasoning errors. Many other colleagues we have talked to who teach at other universities have reported similar experiences. It is clear that neither the training nor the practice that students are currently getting in logic and proof methods in their undergraduate courses is anywhere near adequate. A major reason is that proof exercises are difficult to grade by hand, and thus only a few such exercises are assigned when in fact many are needed in order to measure students' understanding and help them improve their skills. This situation sorely needs to be remedied, but the way that logic and proof methods are presented in current computer science or logic textbooks, and the limited number of exercises that can be assigned and graded in view of the required human-intensive effort, are inadequate for the task.

The second consideration is opportunity. Due to recent research advances, programs are now available that permit proofs to be expressed in a format that is both human-readable and machine-checkable. These programs have been developed by one of the authors and used extensively in research and teaching by both authors. Although other mechanical proof assistants have been available for many years, most were developed as aids to teaching logic to students majoring in such fields as philosophy, cognitive science, or mathematics. By contrast, the newer programs—especially Athena, the program used in this book—support an approach to logic and proof methods that is much better suited to the needs of computer science students. For example, Athena fully supports proof by induction, especially for properties of data types defined by structural induction. Furthermore, Athena allows writing proof methods that are closely related to parameterized procedures for ordinary computation. This correspondence allows us to point out many analogies between proofs-as-programs and ordinary programs, thus leveraging the natural interest of computer science students in programming and computation in a way that one could not expect of students from other disciplines.

Unable to assume interest and experience in computer programming, authors of textbooks for "Introduction to Logic" and similar courses have had to find examples and exercises in artificial domains such as puzzle solving, which are often seen as insufficiently relevant and motivating by computer science students—or, for that matter, by a broader spectrum of science and engineering students. In this textbook we instead focus on practical computer science proof applications such as algorithm correctness properties, including input-output correctness, termination, and memory safety; algorithm efficiency (correctness of optimizations); and fundamental data type properties, such as inductive properties of the recursive data structures that are pervasive across all branches of computer science. Furthermore, such applications serve to strongly motivate attention to properties of the fundamental mathematical concepts on which the correctness of important algorithms depends, such as algebraic and relational axioms and theorems.

There are several other textbooks that aim to introduce logic and proofs to a computer science audience, but in almost all cases the proof formalisms they use are unsupported by software, and therefore exercise solutions must be checked by hand. By contrast, all proofs in this textbook are machine-readable, and complete code is always given, so that students can try out the examples on a computer, experiment with them, and get immediate feedback. Student solutions can be mechanically checked for correctness.

A small number of much more recent textbooks do take a computer-based approach to proofs, and their goals are fairly similar to ours, including the goal of teaching proof methods that are fundamental in computer science; the use of a mechanical proof assistant as an integral tool in their course of study; substantial material on pervasive data types such as natural numbers and lists; and many examples and exercises, which are mechanically checkable. But there are also major differences. First, their main application area is programming languages. Our book does cover that topic, but to a somewhat lesser extent, focusing more on algorithms and data structures. To the best of our knowledge, our discussion of abstract algebraic structures and techniques, which leads into our extensive treatment of abstract algorithms and data structures, is unique. Our text also has a much more extensive treatment of equality and order relations (both for natural numbers and integers, and at an abstract level), as well as more extensive coverage of proof methods for sentential and predicate logic, including numerous heuristics for discovering proofs. Finally, material such as automated testing (falsification) of conjectures, as well as other applications of formal methods, such as SAT solving, are also not found in the aforementioned texts.

But the biggest difference lies in the style of proof supported by the respective inference systems. As we have already pointed out, Athena uses a true natural-deduction style of proof (the "Fitch style" of natural deduction), which allows for structured and readable proofs that resemble in certain key respects the informal proofs one encounters in practice. Other proof systems do not place a high premium on making proofs understandable, or even recognizable, and in fact the texts that use them seem to have taken it for granted that formal proof and human understanding are incompatible. We believe that readability is not only compatible with formal proofs, it is in fact *necessary* if formal proofs are to reach their full potential. A while back, in their preface to a classic computer science textbook [1], Abelson and Sussman urged the following: *Programs must be written for people to read, and only incidentally for machines to execute.* Few would debate this maxim. It might not always be attainable, but we should always strive to attain it. It should be no different for proofs.

Intended audience and prerequisites

The book is intended primarily for students majoring in computer science at the upper undergraduate or graduate level, but it will also be of interest to professional programmers and to practitioners of formal methods and researchers in logic-related branches of computer science, either for self-study or as a reference. More detailed suggestions are made at the end of this section. In its examples and exercises, since they are written in a machine-checkable language, it exhibits and requires complete rigor, but no more than what computer science majors and other programming practitioners are accustomed to when using other computer languages.

The only prerequisite is a basic level of programming experience and mathematical knowledge, typical of a second- or third-year computer science major. In many cases, a computer science major will take a discrete mathematics course in the first three to four semesters, and that is desirable before taking a course based on this book but not strictly necessary. The instructor must decide whether to require a discrete math background.

Prior exposure to functional programming would be helpful but not essential. A programming languages concepts or survey course, usually taken later in the curriculum, is often the first place students work with a functional language, but unless a programming languages text uses a functional language such as Scheme, ML, or Haskell throughout, as some do, the amount of time spent on functional language concepts is typically not that great. We believe that functional programming can be introduced along with other concepts, as this book does, and furthermore, that doing so provides more motivation than the typical treatment in a programming languages course. In practice, the main difficulty for students who have been taught to express computations in an imperative style, using loops and assignment statements, is learning to express them using recursion instead. There are many examples of recursive function definitions in the text, and as these can be directly executed, they should provide good practice in learning how to think recursively. But a fundamental claim that is often made about functional programs is that they are easier to reason about, mostly because the evaluation mechanism is based on substitution, reflecting standard mathematical practice; and because structural induction is a straightforward mechanism for reasoning about recursive functions defined on inductive data types. This text provides ample opportunity not only to compute with recursion, but also to reason about it mechanically. In particular, the material highlights the strong connection between structural induction over algebraic data types and recursion, a connection that is central to functional programming and yet is usually glossed over in existing functional programming texts [17].

Here are some more specific capacities in which this textbook could be used in computer science departments at the upper undergraduate or early graduate level:

1. Existing *Software Verification* courses could use the book as their primary or secondary source material, depending on whether they choose to focus on approaches based on proofs or a combination of model checking and proofs. In addition, faculty who have interests in software verification but have not previously taught a course on it may be inspired by our approach and applications emphasis to introduce a new course using the book. There is more material than could be covered in a single course, but neither of the two main application areas, algorithms (Parts IV and V) and programming languages (Part VI), is dependent on the other. Instructors can thus choose depending on students' and their own background and interests. Not all students will need the depth of coverage of logic fundamentals provided in Chapters 4 and 5, so it should not be necessary to devote much class time to that material. The use of Athena as a high-bandwidth interface to SMT and SAT solvers and other ATPs (along the lines described in the "Automated Theorem Proving" appendix on the book's web site) may be of particular interest to instructors who want to introduce students to such tools.
2. *Theory of Computation* courses could use Parts I, II, and VI as secondary source material in lieu of the more abstract treatments of sentential and predicate logic often found in theory of computation textbooks.
3. *Logic* courses could use Parts I and II as their main source material, concentrating most heavily on Chapters 4 and 5. It may be useful for motivation to quickly survey the material in one or more of the later parts of the book. Related content includes:
   - syntax and semantics of sentential logic;
   - syntax and semantics of (many-sorted) first-order logic;
   - notions of interpretation, satisfiability, tautology, entailment, and monotonicity;
   - introduction and elimination proof rules;
   - heuristics for proof development;
   - equational and implicational chaining;
   - SAT solvers and theorem proving systems.

   Inclusion of the last two topics will help prepare students for writing proofs and using automated proof systems in later courses. Because of its focus on applications, the book does not cover metamathematical results such as completeness, compactness, the Löwenheim-Skolem theorems, Gödel's incompleteness, and so on. (Soundness and completeness results for Athena are stated but not proved here.) All of these are standard and can be found elsewhere.
4. In *Algorithms* courses, the book could be used as a supplementary text, using material selected from Parts I through V, with less depth of coverage of logic fundamentals and proof methods than in software verification or logic courses. Related content includes:

- data structures: lists, binary search trees, finite sets and relations, Cartesian products, maps, memory range abstractions;
- algorithms: binary search on trees and on memory range abstractions, exponentiation, greatest common divisor (Euclid's), a few generic algorithms from the C++ Standard Template Library;
- programming techniques: recursion, replacing general recursion by tail recursion, expressing complex control structures using mutually recursive procedures, memory range traversal using iterators, behavioral abstractions as in generic programming;
- specification and verification: defining functions axiomatically, writing software specifications within a logic framework and using them together with model checking and proof attempts to detect errors in algorithm implementations, necessary and sufficient conditions on ordering relations for correct searching and sorting.

5. In *Programming Language Concepts* courses, the book could be used as a supplementary text, using material selected from Parts I, II, and VI. Related content includes:

   - Athena's programming language as an example of a higher-order functional language in the tradition of Scheme: strict and lexically scoped, encouraging a programming style based on function calls and recursion, but also offering imperative features (e.g., vectors and updatable memory cells);
   - Athena's deductive language as an example of a special-purpose language that benefits from custom-made syntax and semantics (as opposed to deductive systems that are programmed in a host programming language);
   - Athena's blend of functional and logic programming techniques, particularly the programmable interface to external Prolog systems described in Appendix B;
   - Athena's logic subset as an example of a many-sorted language with Hindley-Milner style parametric polymorphism and automatic sort inference;
   - examples of using infix operator precedence and associativity to reduce notational clutter (especially compared to Lisp-style prefix notation);
   - correctness of a compiler from an interpreter-defined toy language to a stack-based machine language, first without and then with error handling;
   - formal, executable, and mechanically analyzable definition of syntax and semantics of an imperative language, which, although still very simple compared to real programming languages, contains some of the most basic and essential features of imperative languages, providing sufficient context for discussion of many important issues related to programming language syntax and semantics.

   Logic fundamentals and proof methods could be covered in less depth than in software verification or logic courses, except perhaps when the last topic is to be treated in detail.

6. Senior students could use selected portions of the text as the basis of an independent study or capstone experience.
7. An "immigration" course or self-study requirement for new graduate students coming into computer science departments from other majors could use the book as its basis or as one of several assigned texts.

Examples and exercises

More than three hundred exercises are provided, and, in keeping with the book's emphasis, almost all require writing a complete proof or filling in one or more parts of a proof that is only sketched in the text. These exercises are generally designed to continue development of ideas presented in the book's many proof examples. (Any of the proof examples can itself be used to provide additional practice: Just glance at its overall structure, then put it aside and try to reproduce it, or even come up with a different and possibly better proof!) Exercises marked with a star are generally more difficult, and a mark of two stars indicates even greater difficulty.

Learning curve

Studying proofs via a computer language such as Athena does have a drawback: a somewhat steeper learning curve. After all, one must learn a new computer language from scratch. This is true regardless of whether one chooses to use Athena or some other computer-based proof system. However, as we discussed above, we believe that the effort required upfront is a worthwhile investment, well justified by the dividends (getting immediate feedback from the computer, a wealth of problems whose solutions can be checked by machine, capitalizing on the parallels between programs and proof methods, etc.). In the case of Athena, a minimal path to fluency can be carved out as follows: Start with Sections 2.1–2.10 and Section 2.17 from Chapter 2; continue with Sections 4.1–4.11 from Chapter 4 and Sections 5.1–5.4 from Chapter 5, and conclude with Chapters 6 and 7. This material, with assigned exercises included, can be covered in a few weeks and will ensure a basic mastery of Athena as well as a solid command of sentential and first-order proof methods, clearing the way forward to interesting applications.

Finally, it may be asked whether the learning curve can be justified if readers do not go on to use Athena later in their careers. Can the book still be of value? While Athena does have certain advantages that we hope to demonstrate in due course, ultimately it is only used as an instrument for expressing the underlying techniques concretely enough that a computer can check whether they are being used properly. As an analogy, programmers who become exposed to a functional programming language end up acquiring a set of concepts and techniques and perhaps a certain way of approaching problems and a broader view of the landscape of computation. The key benefit they derive is not learning the details of a particular language, but becoming conceptually richer and enlarging their problem-solving arsenals. The same holds for Athena and proofs.

Support materials

The authors maintain a web site, `http://proofcentral.org`, whose primary role is to provide support materials for the book. It also allows downloading of Athena (for Windows, Mac OS, and Linux) and its libraries. All Athena code that appears in the book in examples and exercises is available in files on the web site, to be used in writing exercise solutions without having to type complete proof developments. In addition, solutions to many of the book's exercises are available on the site.

Acknowledgements

Many people gave us encouragement and assistance in the course of this long project. Three anonymous reviewers of an earlier draft provided many valuable comments and helped us find ways to present the material more clearly. Nicholas Kidd provided insightful technical feedback through several iterations of this material; we are indebted to him. We are also grateful to the MIT Press team for their expert editorial and production assistance: Katherine Almeida, Susan Clark, Amy Hendrickson, Kathleen Hensley, Marie Lufkin Lee, and others behind the scenes.

Konstantine would like to thank the many people at MIT who provided support and advice during the many years he spent there working on specification, proof, and program analysis, first as a graduate student in the former AI Lab (now part of CSAIL) and then as a postdoc in the Program Analysis team (then part of the Computer Architecture Group), particularly David McAllester, Martin Rinard, Olin Shivers, and Howard Shrobe. Darko Marinov and Alex Salcianu were two of Athena's earliest users and provided exceedingly useful feedback. Several MIT students became involved in Athena projects and helped to improve early versions of the language, most notably including Teodoro Arvizo, Melissa Hao, and Jared Showalter. Amy Ehntholt went through the entire manuscript meticulously and corrected many grammatical and stylistic lapses.

David would like to thank Carlos Varela for many comments and suggestions. Students at Rensselaer Polytechnic Institute who worked with David on Athena proofs include Jonathan Brandvein, Benjamin Caulfield, Ian Dunn, Michael Gram, Benjamin Levinn, Antwane Mason, and Aytekin Vargun. Stepping back further, David gratefully acknowledges the inspiration and insights gained from working on abstract (or generic) programming with Meng Lee, Douglas Gregor, Sibylle Schupp, Alexander Stepanov, and

Changqing Wang; on the Tecton concept description language with Deepak Kapur, Rüdiger Loos, Xumin Nie, Zhiqing Shao, Sibylle Schupp, Christoff Schwarzweller, and Alexander Stepanov; on the Rewrite Rule Laboratory with John Guttag, Deepak Kapur, and Hantao Zhang; and on the Affirm specification and verification system with Raymond Bates, Roddy Erickson, Susan Gerhart, John Guttag, Dallas Lankford, Ralph London, David Taylor, David Thompson, and David Wile.

I INTRODUCTION

Logic and proof methods are fundamental tools in computer science. Our goal in this textbook is to develop proof skills applicable to key problems of program correctness and optimization. The value of reading and writing proofs goes well beyond such correctness and optimization issues: It enables a deeper understanding of computations than can be obtained through other means, such as testing or static analysis.

In the first chapter, we briefly survey proof methods that will be covered in detail in the textbook, methods chosen because they are useful in many computer science applications, including input-output correctness, termination, and memory safety; inductive properties of recursive data structures; abstract algorithms; and algorithm efficiency through verifying correctness of optimizing transformations. The goal of developing good proof skills applicable to these topics also motivates reviews throughout the text of the underlying mathematical concepts, ranging from arithmetic laws and theorems about the natural numbers and integers to more abstract developments of algebraic and ordering properties. In the last part of the text, we use proofs about programming languages and compilers to examine fundamental issues in the study of those topics. Throughout, we highlight the strong relationship between proofs and programs, showing how we can develop proofs in a style that has both parallels with, and crucial distinctions from, writing conventional computer programs. In support of all of these goals, we employ a mechanized proof system, Athena, both to ensure that at all times we employ sound logical principles and to provide plenty of practice with the proof methods studied. In Chapter 2 we introduce the basics of Athena.

1 An Overview of Fundamental Proof Methods

JUST as procedures are essential tools in programming languages for expressing computations, proof methods play a similarly essential role in expressing deductions. For a simple example of a proof method, consider one that is traditionally known as *modus ponens* (Latin for "mode that affirms"):

Given a conditional $(p \Rightarrow q)$ and its antecedent p, conclude q.

In this textbook, although we will discuss this and other basic methods (in Chapters 4 and 5), we treat them as primitives upon which *higher-level* proof methods are founded, much as higher-level programming languages are ultimately based on machine language instructions. The principal higher-level methods we will study are *equality and implication chaining, induction, case analysis, proof by contradiction*, and *abstraction/specialization*. In the following sections we preview some of the key aspects of these methods and their formulations in Athena before going on to outline the overall structure of the book.

1.1 Equality chaining

Of all the proof methods covered in this textbook, equality chaining is probably the one that is most widely used in computer science—and, for that matter, in many branches of mathematics, natural science, and engineering. In fact, readers will likely have used equality chaining many times already, albeit informally and perhaps without a thorough understanding of either its technical basis or the full extent of its applicability. As a brief but not completely trivial example of equality chaining, consider proving the algebraic identity

$$(a^{-1})^{-1} = a, \tag{1.1}$$

where we are given the following equations as axioms:

Right-Identity:	$x \cdot I = x$
Left-Identity:	$I \cdot x = x$
Right-Inverse:	$x \cdot x^{-1} = I$

Here I is the *identity element* (or *neutral element*) for the operation $\cdot$. In the case where the domain over which x ranges is the set of nonzero real numbers and $\cdot$ is real-number multiplication, I would simply be 1. But these identities also hold in many other domains, for example, $n \times n$ invertible matrices over the reals, where $\cdot$ is matrix multiplication and I is the $n \times n$ matrix with 1's on its diagonal and 0's elsewhere.

Using these identities, we might write the following equality chain as a proof of (1.1):

$$\begin{aligned}(a^{-1})^{-1} &= I \cdot (a^{-1})^{-1} \\ &= (a \cdot a^{-1}) \cdot (a^{-1})^{-1} \\ &= a \cdot (a^{-1} \cdot (a^{-1})^{-1}) \\ &= a \cdot I \\ &= a.\end{aligned}$$

Given this equality chain as proof, the reader may already be convinced of the validity of (1.1), but that impression might be something of an illusion based on familiarity (or overfamiliarity) with the somewhat informal way in which such proofs are typically presented in textbooks or journal articles. In almost all of the proofs presented or assigned as exercises in this textbook, we will adopt a more explicit style that yields a higher level of clarity, but without departing drastically from the usual textbook or journal proof style.

To preview some of the key aspects of this style, let us take a closer look at what is involved in the above proof and show where we would revise it to be clearer and more complete. As a first step, let us indicate which of the given identities is used in each step of the chain by annotating the step with the identity's label.

$$\begin{aligned}(a^{-1})^{-1} &= I \cdot (a^{-1})^{-1} && [\textit{Left-Identity}] \\ &= (a \cdot a^{-1}) \cdot (a^{-1})^{-1} && [\textit{Right-Inverse}] \\ &= a \cdot (a^{-1} \cdot (a^{-1})^{-1}) && [?] \\ &= a \cdot I && [\textit{Right-Inverse}] \\ &= a && [\textit{Right-Identity}].\end{aligned}$$

We see that the third step does not correspond to any of the given identities. We might gloss over this point by saying that that step is just a "regrouping" of the multiplication operations, but in fact the validity of such a regrouping depends on yet another identity, namely

$$\textit{Associativity}\colon (x \cdot y) \cdot z = x \cdot (y \cdot z),$$

which does hold, for example, for multiplication of reals and for multiplication of square matrices. But associativity does not hold for all binary operators (e.g., consider real or integer subtraction). Assuming we are also given this identity, our annotated proof now becomes:

$$\begin{aligned}(a^{-1})^{-1} &= I \cdot (a^{-1})^{-1} && [\textit{Left-Identity}] \\ &= (a \cdot a^{-1}) \cdot (a^{-1})^{-1} && [\textit{Right-Inverse}] \\ &= a \cdot (a^{-1} \cdot (a^{-1})^{-1}) && [\textit{Associativity}] \\ &= a \cdot I && [\textit{Right-Inverse}] \\ &= a && [\textit{Right-Identity}].\end{aligned}$$

Next, if we look more closely at the first step of the chain, we see that what is used to justify it is a particular *instance* of *Left-Identity*, namely

$$I \cdot (a^{-1})^{-1} = (a^{-1})^{-1},$$

in which the variable x of *Left-Identity* is specialized to the particular term $(a^{-1})^{-1}$. The differing roles of the variable x in *Left-Identity* and of a in the term $(a^{-1})^{-1}$ in the equation being derived can be confusing and sometimes lead to errors in reasoning. (This would be especially the case if we had written (1.1) as $(x^{-1})^{-1} = x$.) To clarify the role of the variables in the given identities, let us restate them showing explicitly that x, y, and z are *universally quantified* (over D):

$$\begin{aligned}
&\textit{Right-Identity}\text{:}\ \forall x{:}D\,.\, x \cdot I = x.\\
&\textit{Left-Identity}\text{:}\quad \forall x{:}D\,.\, I \cdot x = x.\\
&\textit{Right-Inverse}\text{:}\ \forall x{:}D\,.\, x \cdot x^{-1} = I.\\
&\textit{Associativity}\text{:}\quad \forall x{:}D\ y{:}D\ z{:}D\,.\, (x \cdot y) \cdot z = x \cdot (y \cdot z).
\end{aligned}$$

Read $\forall\, x{:}D$ as "for all x in D" or "for every x in D."

So the instance of *Left-Identity* used in the first step of the equality chain is obtained by specializing its universally quantified variable x to the term $(a^{-1})^{-1}$. But what of the variable a in that term? We can clarify its role by beginning the proof with the phrase "Let a be an arbitrarily chosen element (of D)." So the full proof reads:

To prove $\forall\, a{:}D\,.\,(a^{-1})^{-1} = a$, let a be an arbitrarily chosen element of D. Then:

$$\begin{aligned}
(a^{-1})^{-1} &= I \cdot (a^{-1})^{-1} && [\textit{Left-Identity}]\\
&= (a \cdot a^{-1}) \cdot (a^{-1})^{-1} && [\textit{Right-Inverse}]\\
&= a \cdot (a^{-1} \cdot (a^{-1})^{-1}) && [\textit{Associativity}]\\
&= a \cdot I && [\textit{Right-Inverse}]\\
&= a && [\textit{Right-Identity}].
\end{aligned}$$

In the second step of the chain, there is another aspect of the use of the identity *Right-Inverse*, namely that the instance $a \cdot a^{-1} = I$ of $\forall\, x{:}D\,.\, x \cdot x^{-1} = I$ is used to replace only a *subterm* of $(a \cdot a^{-1}) \cdot (a^{-1})^{-1}$, not the whole term. What justification is there for such a subterm replacement? The answer to that question is one of the points about equality chaining that we discuss in detail in Chapter 3. For now, we might consider further annotation of equality chain steps to identify the subterm being replaced, and perhaps also the variable substitutions being made in the justifying identity. While there are established notations for doing so, requiring their use would make equality chains very tedious to write, and the extra detail could hinder readability. But once the underlying principles are understood, equality chains can remain quite readable and convincing without the extra detail.

For mechanical checking of equality chain proofs, both the identification of the subterm being replaced and the variable substitution involved can be performed quite efficiently with limited subterm-search and term-matching capabilities. The Athena system used in this textbook provides these capabilities, among many other places, in its `chain` method, so that our example proof can be expressed at virtually the same level of explicitness as before (writing $*$ for $\cdot$ and `(inv a)` for a^{-1}):

```
1 conclude (forall ?a:D . inv (inv ?a) = ?a)
2   pick-any a:D
3     (!chain [(inv (inv a))
4             = (I * inv (inv a))                 [Left-Identity]
5             = ((a * inv a) * inv (inv a))       [Right-Inverse]
6             = (a * ((inv a) * inv (inv a)))     [Associativity]
7             = (a * I)                           [Right-Inverse]
8             = a                                 [Right-Identity]])
```

Much more on using `chain` to prove equations in Athena can be found in Chapter 3, where there are many examples and related exercises. We can also use `chain` to prove implications and equivalences. We give a couple of examples of implication chaining in Section 1.3, discuss it and equivalence chaining in detail in Chapter 6, and use them both extensively in later chapters.

Note that function applications in Athena are written in prefix notation as $(f\ a_1 \cdots a_n)$, that is, the function symbol f is written first, followed by the arguments $a_1 \cdots a_n$, which are separated simply by white space, with the whole thing surrounded by an outer pair of parentheses. However, binary function symbols can be used in infix, and different precedence levels can be given to different symbols, or even to the same symbol in different contexts, and this can greatly reduce notational clutter. In addition, parentheses can be omitted when writing successive applications of unary function symbols such as `inv`. For example, we could have written `(inv inv a)` instead of `(inv (inv a))` in the preceding proof. In fact, the notation of the proof can be simplified further if we ensure that `inv` binds tighter than the operation $*$, in which case the term in line 6 could be written simply as `(a * inv a * inv inv a)`, since every function symbol is right-associative by default. However, in the final version that follows we don't go that far, since terms are sometimes more readable if a few parentheses are used even when not strictly necessary.

The following is a self-contained Athena module containing all the code needed for this example, including the introduction of the domain D and the declarations of the relevant function symbols and their precedence levels.[1] The pound sign # begins a comment.

```
module M {
  domain D
  declare I: D
  declare *: [D D] -> D [200] # We give a precedence of 200 to *
  declare inv: [D] -> D [220] # and 220 to inv
```

1 A module M, introduced with the syntax **module** M { ... }, is essentially a namespace containing sorts, function symbols, definitions, and possibly other nested submodules. We put this code into a module mostly because it allows us to declare the symbol $*$ distinctly from its predefined meaning in Athena (as numeric multiplication).

```
  # Let's define some handy abbreviations for a few variables over D:
  define [x y z] := [?x:D ?y:D ?z:D]

  # We can now assert the four given axioms:
  assert* Left-Identity  := (I * x = x)
  assert* Right-Identity := (x * I = x)
  assert* Right-Inverse  := (x * inv x = I)
  assert* Associativity  := (x * (y * z) = (x * y) * z)

  # Finally, the desired proof:
  conclude (forall x . inv inv x = x)
    pick-any a:D
      (!chain [(inv inv a) = (I * inv inv a)              [Left-Identity]
                           = ((a * inv a) * inv inv a)    [Right-Inverse]
                           = (a * (inv a * (inv inv a)))  [Associativity]
                           = (a * I)                      [Right-Inverse]
                           = a                            [Right-Identity]])
} # close module M
```

The directive **assert** is used to insert a sentence into the current *assumption base*, while at the same time possibly giving the sentence a name for future reference (the identifier preceding the symbol :=). We will have much more to say about assumption bases later on, but for now you can think of the assumption base as the set of our working assumptions (axioms), as well as any results (theorems) we have managed to derive from those assumptions. The directive **assert*** works just like **assert**, except that it first *closes* the given sentence p; or, more precisely, it inserts into the assumption base a copy of p that is universally quantified over all of p's free variables. Accordingly, **assert*** `(I * x = x)` is equivalent to **assert** `(forall x . I * x = x)`. This shorthand saves us from having to universally quantify our axioms manually.

Before we leave the topic of equality chaining, let us consider another important aspect of such proofs: Starting with the equation $s = t$ to be proved, and letting $s_0 = s$, how do we find the right sequence of terms $s_1, s_2, \ldots, s_n = t$ to link together? Of course, we are restricted in this search in the first place by the set of available identities. Even so, it is not always easy to see how to get started. In the preceding proof, for example, the first and second steps replace the current term by a larger term (i.e., $(a^{-1})^{-1}$ by $I \cdot (a^{-1})^{-1}$ in the first step, I by $a \cdot a^{-1}$ in the second), even though the goal term a is smaller than the starting term $(a^{-1})^{-1}$. Even if the given equality chain convincingly proves the desired equation, the chain by itself does not explain how its author went about finding the right terms to link together—it may seem to have required a level of creativity that goes well beyond just the mechanics of writing proofs that are logically correct.

One approach to constructing equality chains that requires less creativity is to try to find a chain of a special structure: Working from both sides, s and t, try to *reduce them to a common term*. We discuss this strategy in detail in Chapter 3 and recommend attempting it in many cases. However, this reduction strategy is not always applicable; for example, it does not work for proving $(a^{-1})^{-1} = a$ from the given identities. In Section 1.5 we preview another strategy that can be used to limit the number of occasions in which difficult searches for a proof are required.

Before leaving this section, consider the following problem involving three people: Jack, Anne, and George.

> Jack is looking at Anne, and Anne is looking at George. Jack is married, George is not. Is some married person looking at an unmarried person?

Take a moment to try to answer the question, either (A) yes, (B) no, or (C) cannot be determined from the given information. We give the answer in a later section of this chapter. (This problem has nothing to do with proving equalities; we insert it here only to give the reader a chance to try answering before seeing the solution discussed later.)

1.2 Induction

Not every valid equation can be proved just by chaining together instances of other valid equations. Proving an equation may require some form of *mathematical induction*. We say "some form" because there are many manifestations of this important proof method, including *ordinary* and *strong* induction, and several variations thereof for different data types. Mathematical induction can also be used to prove other kinds of sentences besides equations, but we begin its study in Chapter 3 with applications to equations expressing properties such as associativity of the usual addition and multiplication operators on natural numbers. The value of these proof examples and exercises is not so much that they are needed in practice—other, higher-level methods of assuring correctness of basic arithmetic are usually preferable—but as training for formulating and proving similar properties of other abstract data types, like lists or trees. In Chapter 3 we introduce a mathematical induction principle for linear lists, and in Chapter 8 we do the same for binary trees.

For natural numbers, ordinary mathematical induction takes the form of dividing a proof of $\forall\, n \,.\, P(n)$ into two cases: (i) $P(0)$ and (ii) $\forall\, n \,.\, P(n) \Rightarrow P(n+1)$. Case (i) is called the *basis case*, case (ii) is called the *induction step*, and within it, the antecedent $P(n)$ is called the *induction hypothesis*. Proof of the basis case and the induction step suffices, basically because every natural number is either 0 or can be constructed from 0 by a finite number of increments by 1. We make these statements precise in Chapter 3, but for now we can preview how we deal with induction proofs in Athena by defining natural numbers as an *algebraic datatype* `N`:

```
datatype N := zero | (S N)
```

The symbols `zero` and `S` are the so-called *constructors* of `N`. If we interpret `S` as incrementing by 1 (and thus justly call it the "successor" function), we can read this recursive datatype[2] declaration as saying in Athena what we just said in English, that every natural number is either 0 or can by constructed from 0 by a finite number of increments by 1. Now let's look at a simple example of proving a property of natural-number addition, where the proof requires induction. We declare and axiomatize the addition function, `Plus`, as follows:

```
declare Plus: [N N] -> N [+]

define [n m] := [?n:N ?m:N]

assert* Plus-zero-axiom := (n + zero = n)
assert* Plus-S-axiom := (n + S m = S (n + m))
```

In the declaration, the annotation `[+]` means that we *overload* the built-in symbol + to designate `Plus` whenever the arguments of + are terms of sort `N`. The axioms express the identities $n+0=n$ and $n+(m+1)=(n+m)+1$. Suppose now that we want to prove the identity $0+n=n$, which in Athena we express as follows:

```
define Plus-S-property := (forall n . zero + n = n)
```

Of course, if we had already proved that `Plus` is commutative (the identity $n+m=m+n$), then we could prove `Plus-S-property` with a simple equation chain:

```
pick-any n:N
  (!chain [(zero + n) = (n + zero) [Plus-commutative]
                      = n          [Plus-zero-axiom]])
```

However, we will see in Chapter 3 that to prove commutativity of `Plus` (by induction) we need to use `Plus-S-property` in the proof! So, to avoid circular reasoning, we need to prove `Plus-S-property` without using commutativity. We must use induction, but within each case of the induction the proof is just a simple equation chain using the axioms and, in the induction step, the induction hypothesis. Here is the proof:

```
by-induction Plus-S-property {
  zero => (!chain [(zero + zero) = zero [Plus-zero-axiom]])
| (n as (S m)) =>
    conclude (zero + n = n)
    # The induction-hypothesis is already in the assumption base.
    # Here we just give it a name:
      let {induction-hypothesis := (zero + m = m)}
        (!chain [(zero + S m)
                 = (S (zero + m))       [Plus-S-axiom]
                 = (S m)                [induction-hypothesis]])
}
```

2 We use the phrase "data type" when we talk about data types in general, but the single word "datatype" when we are specifically discussing an algebraic datatype, particularly in the context of an Athena **datatype** definition.

In computation, we rarely represent numbers in the *unary* form of applications of `s` to `zero`—sometimes also called a *Peano* representation after the Italian mathematician Giuseppe Peano (1858–1932), one of the founders of mathematical logic and set theory—preferring instead a binary, octal, decimal, or hexadecimal representation. But for rigorously specifying and proving correctness of certain numeric or seminumeric algorithms, the Peano representation and mathematical induction principles based on it have the advantage of simplicity. For example, in Chapter 3 we define an exponentiation operator with equations $x^0 = 1$ and $x^{n+1} = x \cdot x^n$ and use mathematical induction to prove a few of its properties. But, more importantly, in Chapter 12 we take this simple definition as the *specification* of the meaning of x^n, against which a more efficient algorithm for computing x^n is proved correct. That algorithm reduces the problem for $n > 0$ to one for $n/2$ (instead of reducing it from $n+1$ to n), so the proof is done with a form of mathematical induction known as *strong induction*: To prove $\forall\, n \,.\, P(n)$, prove

$$\forall\, n \,.\, [\forall\, k \,.\, k < n \Rightarrow P(k)] \Rightarrow P(n).$$

Here, $[\forall\, k \,.\, k < n \Rightarrow P(k)]$ is called the *strong induction hypothesis*. There are several subtle but important points to understand about strong induction and its applications (e.g., why there is no apparent need for a basis case), as discussed in Chapter 12.

Similarly, as we progress to induction principles for abstract data types like linear lists and binary trees, we must understand new twists that distinguish them from the natural number versions. In proofs of binary tree properties, for example, we must formulate and take advantage of *two* induction hypotheses, one for each of the left and right subtrees of a nonempty tree. We will see examples of this in Section 8.4.

1.3 Case analysis

What was your answer to the little problem posed at the end of Section 1.1? If it was C ("cannot be determined from the given information"), you are in good company—but wrong. The problem is from an article in *Scientific American* [92] that discussed why people often fail to find the correct solution to various logic puzzles. In this case, C was chosen more than 80 percent of the time, apparently because solvers didn't take time to go through all of the possibilities carefully. The correct answer is A ("yes"):

> Either Anne is married or she isn't. If she is married, then since she is looking at George, and George is not married, we have a married person looking at an unmarried person. If Anne is unmarried, then since Jack, a married person, is looking at Anne, we know in this case too that a married person is looking at an unmarried one.

The form of reasoning used here, known as *case analysis*, is a powerful proof method, as we will see in many examples and exercises in this textbook. However, the kinds of applications we will discuss, beginning in Chapter 4, will *not* be simple logic puzzles like the one above, since our goal in this textbook is to develop experience with logic and proof methods in applications that are most relevant in computer science.

Thus, we will not be coming back to this or other examples of logic puzzles. Nonetheless, you might be curious about how you could set up the problem and derive the answer in Athena. The following is a complete formulation; of course, it depends on features of Athena that are only covered later, such as implication chaining, but if you compare its basic structure with the English version above, you will see a strong correspondence.

```
module Marriage-Puzzle {

  domain Person

  declare married: [Person] -> Boolean
  declare looking-at: [Person Person] -> Boolean
  declare Jack, Anne, George: Person

  define [p1 p2] := [?p1:Person ?p2:Person]

  assert [(Jack looking-at Anne)
          (Anne looking-at George)
          (married Jack)
          (~ married George)]

# Is some married person looking at an unmarried person?
# Yes, as shown with a case analysis:

  conclude goal := (exists p1 p2 . married p1 &
                                    p1 looking-at p2 &
                                    ~ married p2)
    (!two-cases
      assume case1 := (married Anne)
        (!chain-> [case1
               ==> (married Anne &
                     Anne looking-at George
                     & ~ married George)         [augment]
               ==> goal                          [existence]])
      assume case2 := (~ married Anne)
        (!chain-> [case2
               ==> (married Jack &
                    Jack looking-at Anne &
                    ~ married Anne)              [augment]
               ==> goal                          [existence]]))
} # close module Marriage-Puzzle
```

(Read the symbol ==> as "implies," & as "and," and ~ as "not." These and other sentential connectives used in Athena are discussed in Section 2.4, along with their correspondence to other commonly used symbols.)

We will see many examples of case analysis in proofs about algorithms and data structures, in many variations. More general methods of breaking a proof down into two or more cases, independently of any datatype, are introduced in Chapter 4.

1.4 Proof by contradiction

When proving a negative result, often the most natural method to apply is *proof by contradiction*: Assume the opposite, and show that that assumption leads to a contradiction. As an example, consider a relation $<$ on some domain D, with the properties

$$\textit{Irreflexivity}: \forall\, x\!:\!D\,.\, \sim x < x.$$

$$\textit{Transitivity}: \forall\, x\!:\!D\ y\!:\!D\ z\!:\!D\,.\, x < y \wedge y < z \Rightarrow x < z.$$

Then the $<$ relation also has the property

$$\textit{Asymmetry}: \forall\, x\!:\!D\ y\!:\!D\,.\, x < y \Rightarrow \sim y < x.$$

PROOF: By contradiction. Choose any a and b in D such that $a < b$ and assume, contrary to the stated conclusion, that $b < a$. Then from $a < b$, $b < a$, and transitivity, we infer $a < a$. But, by irreflexivity, we have $\sim a < a$, and hence we have a contradiction. ■

The binary `by-contradiction` method is often used to carry out such proofs in Athena. The first argument to this method is the sentence p we are trying to prove, and the second argument is a conditional of the form ($\overline{p} \Rightarrow$ `false`), where $\overline{p}$ is the *complement* of p; that is, q if p is a negation of the form ($\sim q$), and ($\sim p$) otherwise. If this conditional is in the assumption base, then the result of this method application will be the goal p. And because deductions derive their conclusions and enter them into the assumption base, this second argument is typically expressed as a deduction. The most common way to derive `false` is via the `absurd` method, which takes two arguments, both of which must be in the assumption base, and the second of which is the negation of the first. For this example, we could set up the axioms and prove the theorem as follows:

```
module Asymmetry {
  domain D
  declare <: [D D] -> Boolean

  define [x y z] := [?x:D ?y:D ?z:D]

  assert* irreflexivity := (~ x < x)
  assert* transitivity := (x < y & y < z ==> x < z)
```

```
  conclude asymmetry := (forall x y . x < y ==> ~ y < x)
    pick-any a:D b:D
      assume (a < b)
        (!by-contradiction (~ b < a)
          assume (b < a)
            let {less := (!chain-> [(b < a)
                                   ==> (a < b & b < a)     [augment]
                                   ==> (a < a)             [transitivity]]);
                 not-less := (!chain-> [true
                                       ==> (~ a < a)       [irreflexivity]])}
              (!absurd less not-less))
} # close module Asymmetry
```

All of the details of this and similar proofs about a strict inequality relation are developed in Chapter 8 for the natural numbers, and in greater generality in Chapter 14.

Proof by contradiction can also be used to prove a positive result, provided that combining its negation with other known properties leads to a contradiction. Either way, the steps toward the contradiction may use any other proof method, including equality or implication chaining (as in the preceding example), case analysis, or even other proofs by contradiction. (Even a mathematical induction proof could be used along the way, but that would be best handled as a separate proof whose result is then used as a lemma in working toward the contradiction.)

As the asymmetry proof shows, a proof by contradiction can be done as a step in a larger proof. We will see many examples in which one or more proofs by contradiction are used within larger proofs, including cases where one is nested inside a larger proof by contradiction.

1.5 Abstraction/specialization

At the end of Section 1.1, we mentioned one strategy for proving equalities that required less creativity, namely "reduction to a common term," but we noted that it does not always work. There is another strategy that we advocate to limit the number of difficult proofs one has to write, not only when constructing equality chains but in many other cases as well: *Work at an abstract level*, proving theorems in the most general setting, then specializing them to concrete instances as needed. We already pointed out that the identities on which the proof of $(a^{-1})^{-1} = a$ depends are valid in a couple of important domains, $D \equiv$ nonzero real numbers and $D \equiv$ invertible $n \times n$ matrices. Going further, we may regard these identities as *axioms of an abstract structure* T, with which we are able to derive, using the constructed equality chain, the identity $(a^{-1})^{-1} = a$ as a theorem of T. Then for *any* concrete domain D in which the identities hold, the same proof can be reused to prove the corresponding concrete specialization of $(a^{-1})^{-1} = a$. The binary operator $\cdot$ might even be specialized to an addition operator rather than a multiplication operator in a concrete domain. For example, let D be the integer domain Z, and specialize $\cdot$ to integer addition, I to its neutral element, and the inverse operator to integer negation, which we write as $+$, 0, and (unary) $-$, respectively. Then the given identities become

$$
\begin{aligned}
&\textit{Right-Identity}: \forall\, x:Z\,.\,x+0=x.\\
&\textit{Left-Identity}: \quad \forall\, x:Z\,.\,0+x=x.\\
&\textit{Right-Inverse}: \forall\, x:Z\,.\,x+(-x)=0.\\
&\textit{Associativity}: \quad \forall\, x:Z\; y:Z\; z:Z\,.\,x+(y+z)=(x+y)+z.
\end{aligned}
$$

And the proof of $(--a=a)$ becomes: Let a be an arbitrarily chosen element of Z. Then

$$
\begin{aligned}
(--a) &= (0+--a) && [\textit{Left-Identity}]\\
&= ((a+-a)+--a) && [\textit{Right-Inverse}]\\
&= (a+(-a+--a)) && [\textit{Associativity}]\\
&= (a+0) && [\textit{Right-Inverse}]\\
&= a && [\textit{Right-Identity}].
\end{aligned}
$$

But we don't have to construct this proof from scratch! Once we have found the abstract-level proof, we can store it along with the abstract theorem it proves, to be simply reused, with appropriate specialization, in any concrete domain (like Z) in which the axioms hold. We will see that Athena fully supports this strategy.

In mathematics, the advantages of working at an abstract level have long been known (they were conclusively demonstrated by mathematicians who were developing the field of *abstract algebra* over one hundred years ago). In computer science, the same benefits are seen in some textbooks and journal articles in which a theory is developed at an abstract level and then specialized in different ways. But readers with less mathematical training might be more familiar with how similar principles have also been applied in *computer programming*, not for constructing proofs, but for creating—automatically, during compilation—concrete algorithms or data structures as instances of abstract algorithms or data structures (often called *generic* algorithms or data structures).[3] In this book we will demonstrate the advantages of working at an abstract level by doing proofs in abstract theories and specializing them to various concrete domains, including examples very similar to the ones introduced in this section, starting in Chapter 14. Moreover, we will also apply this strategy to prove correctness properties of abstract algorithms in Chapters 15 and 16, including ones that are very similar to abstract algorithms in the C++ Standard Template Library.

---

3 Major languages that support abstract programming (or "generic programming"), to one degree or another, include Ada, C++, C#, and Java.

1.6 The usual case: Proof methods in combination

In this introductory overview, we have sampled several of the most important proof methods: equality chaining, induction, case analysis, proof by contradiction, and abstraction and specialization. In simple cases, a proof can be done with only one of these methods, but most proofs require combinations of them. A key goal of this textbook is to develop skills in choosing and properly applying a suitable combination of proof methods, particularly when proving a correctness or optimization property of an algorithm or data structure. We will see that, in many cases, a method that can be successfully applied to advance a proof is strongly suggested by the syntactic structure of the current proof goal and/or the current set of premises. Chapter 4 especially explores this line of attack. In other cases, it is the semantic content of the proof goal, or of available axioms and theorems, that provides important clues as to how to proceed. In such cases, it can be much more difficult to discern the right proof strategy than in the syntactically driven cases. We begin discussing how one can build up one's experience with such strategies in Section 3.11, and we return to this important topic repeatedly in the rest of the textbook. But indeed, in all cases, whether syntactically or semantically driven, there is no substitute for experience. Accordingly, throughout the book we have provided an abundance of proof examples and exercises.

1.7 Automated proof

Athena is integrated with a number of powerful automated theorem-proving systems (ATPs) and model-building systems such as SMT and SAT solvers.[4] Using these systems, all of the example proofs in this chapter could be performed automatically, in an entirely "push-button" manner. Having this level of automation available can be handy when tackling large verification projects, with Athena used to specify the system, test the specification, and outline the high-level structure of the proof, while ATPs fill in some of the more tedious proof details.

In this book, however, we make no use of ATPs. That decision was made for pedagogical reasons. We believe that the only way to gain sufficient experience in developing structured proofs is to write them from the ground up, without resorting to black-box oracles to cut corners or potentially get around tricky aspects of the required reasoning. Of course, building a proof "from the ground up" does not mean that the proof should be expressed exclusively in terms of low-level primitives such as modus ponens, in the same way that developing a computer program from the ground up does not mean that we should write the program in machine language. Abstraction mechanisms are indispensable in both cases.

4 The integration is seamless in that these systems can be invoked as primitive methods similar to Athena's built-in inference methods, or as primitive procedures, and the interaction occurs at the level of Athena's polymorphic sort system; the details of the underlying translations are hidden.

But, for pedagogical purposes, we believe that the internal workings of those abstraction mechanisms should be understandable in terms of the semantics of the host language, and that is not the case for external ATPs. That said, readers are encouraged to experiment with automated proof. To that end, we describe how ATPs can be used in Athena and provide a number of examples in the "Automated Theorem Proving" appendix on the book's web site.

1.8 Structure of the book

This part continues with an introduction to the basics of Athena in Chapter 2. As noted in the Preface, not all of this chapter needs to be studied before going on; much of it can be treated as a reference while continuing through the rest of the book.

Part II (Fundamental Proof Methods) begins with an in-depth look at equality chaining and induction applied to natural numbers and lists, including natural-number addition, multiplication, and exponentiation, and list concatenation and reversal. Chapter 3 also introduces term evaluation and other tools that can be used before attempting proof, such as automated conjecture testing.

We then continue with sentential and first-order logic (Chapter 4 and Chapter 5, respectively), and generalize chaining to encompass implications and equivalences (Chapter 6).

In Part III (Proofs about Fundamental Datatypes), we begin by describing Athena's `module` mechanism for managing namespaces and for packaging large numbers of Athena proofs and programs into properly organized components (Chapter 7). In Chapter 8 we introduce and prove many simple properties of ordering relations over natural numbers. With these results at hand, we are able to formulate and prove properties of natural-number subtraction, which in later chapters are applied in defining integer arithmetic and natural-number division. The final sections of the chapter develop ordering properties of lists and binary search trees (over natural numbers).

In Chapter 9 we show how to develop the integer domain, including introduction of the important tool of homomorphic mappings between different representations to simplify proofs. The chapter concludes with a brief treatment of power series arithmetic.

Chapter 10 introduces some fundamental discrete structures, including ordered pairs, sets, relations, functions, and maps, and proves a number of useful theorems about them.

In Part IV (Proofs about Algorithms), the main emphasis is on correctness and optimization of algorithms, focusing again on operations on natural numbers. In Chapter 11, a binary-search algorithm provides the setting for discussion of whether initial correctness requirements placed on the behavior of the algorithm are sufficient to rule out unsatisfactory candidates. In Chapter 12, several variants of an exponentiation algorithm are formulated and proved correct. While each variant does some computation that is unnecessary, it is noted that this will be corrected in Chapter 15, in an abstract version that is also much more generally applicable. For the third example in this part, we extend the study of fundamental numeric properties to natural-number division, with proofs about Euclid's algorithm motivating continued development of this basic mathematical theory (Chapter 13).

In Part V (Proofs at an Abstract Level), we turn to the study of abstraction and specialization, starting in Chapter 14 with fundamental algebraic abstractions: monoid, group, ring, and so forth. Such abstractions can be expressed and organized in Athena using a *structured theory* data structure that allows both refinement (defining a new theory as a combination of existing theories and additional axioms) and evolution (adding theorem/proof pairs to a theory). A central focus is how theory refinement and evolution can be combined to greatly reduce the number of proofs that have to be written. The chapter continues with examples of fundamental relational abstractions and their expression in Athena: reflexivity, transitivity, partial ordering, and so on.

Both algebraic and relational abstractions play an important role in the lifting of concrete algorithms to an abstract level, as illustrated in the next two chapters. Chapter 15 revisits the binary-search and exponentiation algorithms studied in Chapters 11 and 12, obtaining abstract versions and correctness-property proofs that are much more generally applicable, via specialization. In Chapter 16, these abstraction and specialization methods are brought to bear on algorithms drawn from the C++ Standard Template Library (STL) that operate on memory ranges. Though only a few algorithms are presented, the discussion deals fully with issues of memory updating and the role of the STL's iterator hierarchy.

The last part of the book, Part VI (Proofs about Programming Languages), presents techniques for representing and reasoning about programming languages. Chapter 17 introduces a simple functional language for numeric expressions, a machine language for manipulating stacks of numbers, and a toy compiler from the expression language to the machine language. We go on to prove the correctness of this simple compiler and then to extend both the compiler and its correctness proof with explicit error handling. We use evaluation throughout to test our function definitions, while also employing the simple form of automated testing introduced in Section 3.13 to try to falsify conjectures.

The last chapter introduces a more realistic imperative programming language that is still simple enough to be rigorously modeled and investigated. We start by representing the abstract grammar of the language, and we build a parser that can map concrete strings to trees of that abstract grammar. We go on to formalize the semantics of expressions and commands by means of an inference system (a widely used technique in computer science), and to prove that the semantics of the language are deterministic.

Finally, the book has the following appendices:

- Appendix A is a compact description of the syntax and operational semantics of Athena's core constructs.

- Appendix B is a brief discussion of logic programming and Prolog, including an Athena implementation of a Prolog interpreter for definite clauses. Apart from its own intrinsic interest (in view of the subject's intimate connection to logic and computation), this appendix is included because the treatment of programming language semantics in the last chapter uses logic programming for evaluation purposes.

Three additional appendices are available on the book's website:

- Appendix C is a short tutorial on the intersection of functional programming and proofs in Athena.
- Appendix D is a discussion of automated theorem proving, briefly surveying some of the mechanisms available in Athena for automated reasoning through external theorem provers.
- Appendix E contains solutions to many of the book's exercises.

2 Introduction to Athena

ATHENA is a language for expressing proofs and computations. We begin with a few remarks on computation. As a programming language, Athena is similar to Scheme. It is higher-order, meaning that procedures are first-class values that can be passed as (possibly anonymous) arguments or returned as results; it is dynamically typed, meaning that type checking is performed at run-time; it has side effects, featuring updatable cells and vectors; it relies heavily on lists for structuring data, and those lists can be heterogeneous (containing elements of arbitrary types); and the main tool for control flow is the procedure call.

Nonetheless, Athena's programming language differs from Scheme in several respects. At the somewhat superficial level of concrete syntax, it differs in allowing infix notation in addition to the prefix (s-expression) notation of Scheme. A more significant difference lies in Athena's formal semantics, which are a function of not just the usual semantic abstractions, such as a lexical environment and a store, but also of a novel semantic item called an *assumption base*, which represents, unsurprisingly, a finite set of assumptions. The fact that this new semantic abstraction is built into the formal semantics of the core language is enabled by (and in turn requires) another point of divergence of Athena from more conventional programming languages: differences in the underlying data values, the most salient one being that the concept of a logical proposition, expressed by a formal sentence, takes center stage in Athena.

The basic data values of Scheme are more or less like those of other programming languages: numbers, strings, and so on. From these basic values, Scheme can build other, more complex values, such as lists and procedures. But the fundamental computational mechanisms of Scheme (which are essentially those of the λ-calculus, i.e., mainly procedural abstraction and application) can be deployed on any domain (type) of primitive values, not just the customary ones. The computational mechanisms remain the same, though the universe of primitive values can vary. Athena deploys these same computational mechanisms but on a largely different set of primitive values.

Athena's primitive data values include characters and other such standard fare. But these are not the fundamental data values of the language. The fundamental data values are *terms* and *sentences*. A term is a symbolic structure, essentially a tree whose every node contains either a *function symbol* or a *variable*. Terms should be familiar to you from previous logic courses, or even from elementary math. Expressions such as $3+5$, $x/2$, 78, etc., are all terms. So are names such as *Joe*, expressions such as *Joe's father*, and so on. The job of terms is to *denote* or represent individual objects in some domain of interest. For instance, the term $3+5$ denotes the integer 8, while the term *Joe* presumably designates some individual by that name, and *Joe's father* represents the father of that individual. A *sentence* is essentially a formula of first-order logic: either an atomic formula, or a Boolean combination of formulas, or a quantification. We will describe these in greater detail later in this chapter. Sentences are used to express propositions about various domains of interest. They serve as the conclusions of proofs.

Computations in Athena usually involve terms and sentences. Procedures, for instance, typically take terms or sentences as inputs, manipulate them in some way or other, and eventually produce some term or sentence as output. Athena does provide numbers (though these are just special kinds of terms), characters, strings, rudimentary I/O facilities, etc., so it can be used as a general-purpose programming language. But the focus is on terms and sentences, as these are the fundamental ingredients for writing proofs. So the programming language of Athena can be viewed as a Scheme-like version of the λ-calculus designed specifically to compute with terms and sentences.[1]

To repeat, there are two uses for Athena: (a) writing programs, usually, though not necessarily, intended to compute with terms and sentences; and (b) writing proofs. Accordingly, there are two fundamental syntactic categories in the language: *deductions*, which represent proofs, and *expressions*, which represent computations (programs).[2] We typically use the letters D and E to range over the sets of all deductions and expressions, respectively. Deductions and expressions are distinct categories, each with its own syntax and semantics, but the two are intertwined: Deductions always contain expressions, and expressions may contain deductions. A *phrase* F is either an expression or a deduction. Thus, if we were to write a BNF grammar describing the syntax of the language, it would look like this:

$$
\begin{aligned}
E &:= \cdots && \text{(Expressions, for computing)} \\
D &:= \cdots && \text{(Deductions, for proving)} \\
F &:= E \mid D && \text{(Phrases)}
\end{aligned}
\tag{2.1}
$$

Intuitively, the difference between an expression and a deduction is simple. A deduction D represents a logical argument (a proof), and so, if successful, it can only result in one type of value as output: a *sentence*, such as a conjunction or negation, that expresses some proposition or other, say, that all prime numbers greater than 2 are odd. That sentence represents the *conclusion* of the proof. An expression E, by contrast, is not likewise constrained. An expression represents an arbitrary computation, so its output could be a value of any type, such as a numeric term, or an ASCII character, or a list of values, and so on. It may even be a sentence. But there is a crucial difference between sentences produced by expressions versus sentences produced by deductions. A sentence that results from a deduction is guaranteed to be a logical consequence of whatever assumptions were in effect at the time when the deduction was evaluated. No such guarantee is provided for a sentence produced by a computation. A computation can generate any sentence it wishes, including an outright contradiction, without any restrictions whatsoever. But a deduction has to play

1 While both terms and sentences could be represented by appropriate data structures in any general-purpose programming language, having them as built-in primitives woven into the syntactic and semantic kernel of the language carries a number of advantages.

2 In this book, we use the terms "deduction" and "proof" interchangeably.

within a sharply delimited sandbox: It can only result in sentences that are entailed by the assumptions that are currently operative. So the moves that a deduction can make at any given time are much more curtailed.

As with most functionally oriented languages, Athena's mode of operation can be pictured as follows: On one side we have phrases (i.e., expressions or deductions), which are the syntactically well-formed strings of the language; and on the other side we have a class of *values*, such as characters, terms, sentences, and the like. Athena works by evaluating input phrases and displaying the results. This process is called *evaluation*, because it produces the value that corresponds to a given phrase. Evaluation might not always succeed in yielding a value. It might fail to terminate (by getting into an infinite loop), or it might generate an error (such as a division by zero). We write $\mathbf{V}$ for the set of all values, and we use the letter V as a variable ranging over $\mathbf{V}$.

A phrase F cannot be evaluated in a vacuum. For one thing, the phrase may contain various names, and we need to know what those names mean in order to extract the value of F. Consider, for example, a procedure application such as `(plus x 2)`, where `plus` is a procedure that takes two numeric terms and produces their sum. To compute the value of this expression, we need to know the value of `x`. In general, we need to know the lexical *environment* in which the evaluation is taking place. An environment is just a mapping from names (identifiers) to values. For instance, if we are working in the context of an environment in which the name `x` is bound to `7`, then we can compute the output `9` as the value of `(plus x 2)`. Second, since a phrase may be a proof, and every proof relies on some working assumptions, also known as *premises*, we also need to be given a set of premises—a so-called *assumption base*. Third, since Athena has imperative features such as updatable memory locations ("cells") and vectors, we also need to be given a *store*, which is essentially a function that tells us whether a given memory location is occupied or empty, and if occupied, the value residing there. Finally, we need a *symbol set*, which is a finite collection of sorts and function symbols along with their signatures (we will explain what these are shortly).

In summary, the evaluation of a phrase F always occurs *relative to*, or *with respect to* these four semantic parameters: (a) a lexical environment ρ; (b) an assumption base β; (c) a store σ; and (d) a symbol set γ. Schematically:

$$\text{Input: Phrase } F \xrightarrow[\text{(w.r.t. given } \rho, \beta, \sigma, \gamma)]{\textit{Evaluation}} \text{Output: Value } V$$

A rigorous description of the language would require a precise specification of the syntax of phrases, via a grammar of the form (2.1); a precise specification of the set of values $\mathbf{V}$; and a precise description of exactly what value V, if any, corresponds to any phrase F, relative to a given environment, assumption base, store, and symbol set. Such a description is given in Appendix A. In this chapter we take a less formal approach. We will be chiefly concerned with expressions, that is, with the computational aspects of Athena. We will describe most available kinds of expressions and explain what values they produce, usually without explicit references to the given environment, assumption base, store, or symbol set. We start with terms and sentences in Sections 2.3 and 2.4, respectively. After that we move on to other features of Athena. Deductions will be briefly addressed in general terms in Section 2.10, but the syntax and semantics of most deductions will be introduced in later chapters as needed. Before we can get to terms and sentences, we need to cover some preliminary material on domains and function symbols, in Section 2.2. And before that still, we need to say a few words on how to interact with Athena.

Bear in mind that it is not necessary to go through this chapter from start to finish. Only Sections 2.1–2.10 and the summary (Section 2.17) need to be read in their entirety before continuing to other chapters. (Also, at least the first five exercises are very strongly recommended.) You can return later to the remaining sections on an as-needed basis.

2.1 Interacting with Athena

Athena can be used either in batch mode or interactively. The interactive mode consists of a read-eval-print loop similar to that of other languages (Scheme, ML, Python, Prolog, etc.). The user enters some input in response to the prompt >, Athena evaluates that input, displays the result, and the process is then repeated. The user can quit at any time by entering `quit` at the input prompt.

Typing directly at the prompt all the time is tiresome. It is often more convenient to edit a chunk of code in a file, say `file.ath`, and then have Athena read the contents of that entire file, processing the inputs sequentially from the top of the file to the bottom as if they had been entered directly at the prompt. This can be done with the **load** directive. Typing

```
load "file.ath"
```

at the input prompt will process `file.ath` in that manner. The file extension `.ath` can be omitted; to load `foo.ath` it suffices to write `load "foo"`. Also, when Athena is started, it can be given a file name as a command-line argument, and the file will be loaded into the session.

Note that **load** commands can themselves appear inside files. If the command

```
load "file1"
```

is encountered while loading `file.ath`, Athena will proceed to load `file1.ath` and then resume loading the rest of `file.ath`.

Inside a file, the character # marks the beginning of a comment. All subsequent characters until the end of the line are ignored.

A word on code listings: A typical listing displays the input prompt >, followed by some input to Athena, followed by a blank line, followed by Athena's response. Additional Athena inputs, usually preceded by the prompt > and their respective outputs, may follow. If Athena's response to a certain input is immaterial, we omit it, and, in addition, we do not show the prompt before that input. So the absence of the prompt > before some Athena input indicates that the system's response to that input is irrelevant for the purposes at hand and will be omitted. We encourage you to try out all of the code that you encounter as you read this chapter and the ones that follow.

When input is directly entered at the prompt, it may consist of multiple lines. Accordingly, Athena needs to be told when the input is complete, so that it will know to evaluate what has been entered and display the result. This end-of-input condition can be signalled by typing a double semicolon ;;, or by typing `EOF`, and then pressing Enter. However, this is only necessary if the input is not syntactically balanced. The input is syntactically balanced if it either consists of a single token (a single word or number), or else it starts with a left parenthesis (or bracket) and ends with a matching right parenthesis (or bracket, respectively). For those inputs it is not necessary to explicitly type ;; or `EOF` at the end—simply press Enter (even if the input consists of multiple lines) and Athena will realize that the input is complete because it has been balanced in the sense we just described, and it will then go on to evaluate what you have typed. But if the input is not syntactically balanced in that way, then ;; or `EOF` must be typed at the end. Of course, if one is writing Athena code in a file `f.ath` to be loaded later, then these end-of-input markers do not need to appear anywhere inside `f`. In the code listings that follow we generally omit these markers.

2.2 Domains and function symbols

A domain is simply a set of objects that we want to talk about. We can introduce one with the **`domain`** keyword. For example,

```
> domain Person

New domain Person introduced.
```

introduces a domain whose elements will be (presumably) the persons in some given set.[3] The following is an equivalent way of introducing this domain:

```
(domain Person)
```

3 Whatever set we happen to be concerned with; it could be the set of students registered for some college course, the set of citizens of some country, or the set of all people who have ever lived or ever will live. In general, the *interpretation* of the domains that we introduce is up to us; this point is elaborated in Section 5.6.

Both of the above declarations are syntactically valid. In general, Athena code can be written either in prefix form, using fully parenthesized Lisp-like "s-expressions," or in (largely) infix form that requires considerably fewer parentheses. In this book we generally use infix as the default notation, because we believe that it is more readable, but s-expressions have their own advantages and, for those who prefer them, Appendix A specifies the prefix version of every Athena construct.

Multiple domains can be introduced with the **domains** keyword:

```
domains Element, Set
```

There are no syntactic restrictions on domain names. Any legal Athena identifier *I* can be used as the name of a domain.[4]

Domains are *sorts*. There are other kinds of sorts besides domains (namely *datatypes*, discussed in Section 2.7), but domains are the simplest sorts there are.

Once we have some sorts available we can go ahead and declare *function symbols*, for instance:

```
> declare father: [Person] -> Person

New symbol father declared.
```

This simply says that `father` is a symbol denoting an operation (function) that takes a person and produces another person. We refer to the expression `[Person] -> Person` as the *signature* of `father`. At this point Athena knows nothing about the symbol `father` other than its signature.

The general syntax form for a function symbol declaration is

$$\texttt{declare}\ f\texttt{:}\ \texttt{[}D_1\ \cdots\ D_n\texttt{]}\ \texttt{->}\ D$$

where f is an identifier and the D_i and D are previously introduced sorts, $n \geq 0$. We refer to $D_1 \cdots D_n$ as the *input sorts* of f, and to D as the *output sort*, or as the *range* of f. The number of input sorts, n, is the *arity* of f. Function symbols must be unique, so they cannot be redeclared at the top level (although a symbol can be freely redeclared inside different modules). There is no conventional overloading whereby one and the same function symbol can be given multiple signatures at the top level, but Athena does provide a different (and in some ways more flexible) form of overloading, described in Section 2.13.

For brevity, multiple function symbols that share the same signature can be declared in a single line by separating them with commas, for instance:

---

4 An identifier in Athena is pretty much any nonempty string of printable ASCII characters that is not a reserved word (Appendix A has a complete list of all reserved words); does not start with a character from this set: {!, ", #, $, ', (,), ,, :, ;, ?, [,], `}; and doesn't contain any characters from this set: {", #,), ,, :, ;,]}. Throughout this book, we use the letter *I* as a variable ranging over the set of identifiers.

```
declare union, intersection: [Set Set] -> Set

declare father, mother: [Person] -> Person
```

A function symbol of arity zero is called a *constant symbol*, or simply a constant. A constant symbol c of domain D can be introduced simply by writing

declare c: D

instead of **declare** c: [] -> D. The two forms are equivalent, though the first is simpler and more convenient.

```
> declare joe: Person

New symbol joe declared.

> declare null: [] -> Set

New symbol null declared.
```

Multiple constant symbols of the same sort can be introduced by separating them with commas:

```
declare peter, tom, ann, mary: Person

declare e, e1,e2: Element

declare S, S1, S2: Set
```

There are several built-in constants in Athena. Two of them are `true` and `false`, both of which are elements of the built-in sort `Boolean`. The two numeric domains `Int` (integers) and `Real` (reals) are also built-in. Every integer numeral, such as `3`, `594`, `(- 2)`, etc., is a constant symbol of sort `Int`, while every real numeral (such as `4.6`, `.217`, `(- 1.5)`, etc.) is a constant symbol of sort `Real`.

A function symbol whose range is `Boolean` is also called a *relation* (or *predicate*) symbol, or just "predicate" for short. Some examples:

```
declare in: [Element Set] -> Boolean

declare male, female: [Person] -> Boolean

declare siblings: [Person Person] -> Boolean

declare subset: [Set Set] -> Boolean
```

Here in is a binary predicate that takes an element and a set and "returns" true or false, presumably according to whether or not the given element is a member of the given set.[5] The symbols male and female are unary predicates on Person, while siblings and subset are binary predicates on Person and Set, respectively.

Some function symbols are built-in. One of them is the binary equality predicate =, which takes any two objects of the same sort S and returns a Boolean. S can be arbitrary.[6]

We will generally use the letters f, g, and h as typical function symbols; a, b, c, and d as constant symbols; and P, Q, and R as predicates. The letters f, g, and h will actually do double duty; they will also serve as variables ranging over Athena procedures. The context will always disambiguate their use.

Function symbols are first-class data values in Athena. They can be denoted by identifiers, passed to procedures as arguments, and returned as results of procedure calls.

We stress that function symbols are not procedures. They are ordinary data values that can be manipulated by procedures.[7] An example of a procedure that takes function symbols as arguments is get-precedence, which we discuss in the next section.

A procedure is the usual sort of thing written by users, a (possibly recursive) lambda abstraction designed to compute anything from the factorial function to much more complicated functions. Here is an Athena procedure for computing factorials, for example:

```
define (fact n) :=
  check {
    (less? n 1) => 1
  | else => (times n (fact (minus n 1)))
  }
```

where less?, times, and minus are primitive (built-in) procedures operating on numeric terms:

```
> (less? 7 8)

Term: true

> (times 2 3)

Term: 6

> (minus 5 1)
```

5 As with domains, the interpretation of the function symbols that we declare (what these symbols actually *mean*) is up to us. We will soon see how to write down axioms that can help to prescribe the meaning of a symbol.

6 More precisely, = is a *polymorphic* function symbol. We discuss polymorphism in more detail in Section 2.8.

7 Of course, procedures themselves are also "ordinary data values that can be manipulated by procedures," as is the case in every higher-order functional language. But there is an important difference: Procedures cannot be compared for equality, whereas function symbols can. This ostensibly minor technicality has important notational ramifications.

```
Term: 4
```

We will see a few more primitive procedures in this chapter.

A function symbol like `union` or `father`, by contrast, is just a constant data item. We cannot perform any meaningful computations by applying such function symbols to arguments; all we can get by such applications are unreduced terms, as we will see in the next section.

2.3 Terms

A term is a syntactic object that represents an element of some sort. The simplest kind of term is a constant symbol. For instance, assuming the declarations of the previous section, if we type `joe` at the Athena prompt, Athena will recognize the input as a term:

```
> joe

Term: joe
```

We can also ask Athena to print the sort of this (or any other) term:

```
1 > (println (sort-of joe))
2
3 Person
4
5 Unit: ()
```

Athena knows that `joe` denotes an individual in the domain `Person`, so it responds by printing the name of that domain on line 3. (The *unit value* `()` that appears on line 5 is the value returned by the procedure `println`. Most procedures with side effects return the unit value. We will come back to this later.)

A *variable* is also a term. It can be thought of as representing an indeterminate or unspecified individual of some sort, but that is just a vague intuition aid—what exactly is an "unspecified individual" after all? It is more precise to say that a variable is a syntactic object acting as a placeholder that can be replaced by other terms in appropriate contexts. Variables are of the form ?I:S, for any identifier I and any sort S; we refer to I as the *name* of the variable and to S as its sort. Thus, the following are all legal variables:

```
?x:Person
?S25:Set
?foo-bar:Int
?b_1:Boolean
?@sd%&:Real
```

There is no restriction on the length of a variable's name, and very few restrictions on what characters may appear in it, as illustrated by the last example (see also footnote 4). We will use $x, y, z, \ldots$ as metavariables, that is, as variables ranging over the set of Athena variables. Occasionally we might write a metavariable x as $x:S$ to emphasize that its sort is S. Thus, a metavariable $x:S$ ranges over all variables of sort S.

Constant symbols and variables are primitive or *simple* terms , with no internal structure. More complex terms can be formed by applying a function symbol f to n given terms $t_1 \cdots t_n$, where n is the arity of f. Such an application is generally written in prefix form as $(f\ t_1 \cdots t_n)$,[8] though see the remarks below about infix notation for binary and unary function symbols. We say that f is the *root symbol* (or just the "root") of the application, and that $t_1 \cdots t_n$ are its *children*. Thus, the application

```
(father (mother joe))
```

is a legal term. Its root is the symbol `father`, and it has only one child, the term `(mother joe)`. Some more examples of legal terms:

```
(in e S),
(union null S2),
(male (father joe)),
(subset null (union ?X null)).
```

There are two primitive procedures that operate on term applications, `root` and `children`, which respectively return the root symbol of an application and its children (as a list of terms, ordered from left to right). For example:

```
define t := (father (mother joe))

> (root t)

Symbol: father

> (children t)

List: [(mother joe)]
```

Note that constant symbols may also be viewed as applications with no children.[9] Thus:

```
> (root joe)

Symbol: joe
```

8 A more common syntax for such terms that you may be familiar with is $f(t_1, \ldots, t_n)$. When we use conventional mathematical notation we will write terms in that syntax.

9 Indeed, Athena treats `(joe)` and `joe` interchangeably.

```
> (children joe)

List: []
```

It is convenient to be able to use these procedures on variables as well. Applying `root` and `children` to a variable x will return x and the empty list, respectively.

We will have more to say about lists later on, but some brief remarks are in order here. A list of $n \geq 0$ values $V_1 \cdots V_n$ can be formed simply by enclosing the values inside square brackets: $[V_1 \cdots V_n]$. For instance:

```
> [tom ann]

List: [tom ann]

> []

List: []

> [tom [peter mary] ann]

List: [tom [peter mary] ann]
```

Note that Athena lists, like those of Scheme but unlike those of ML or Haskell, can be heterogeneous: They may contain elements of different types. The most convenient and elegant way to manipulate lists is via pattern matching, which we discuss later. Outside of pattern matching, there are a number of primitive procedures for working with lists, the most useful of which are `add`, `head`, `tail`, `length`, `rev`, and `join`. The first one is a binary procedure that simply takes a value V and a list L and produces the list obtained by prepending V to the front of L. The functionality of the remaining five procedures should be clear from their names. Note that the concatenation procedure `join` can take an arbitrary number of arguments (one of the few Athena procedures without a fixed arity). Some examples:

```
> (add 1 [2 3])

List: [1 2 3]

> (head [1 2 3])

Term: 1

> (tail [1 2 3])

List: [2 3]

> (rev [1 2 3])

List: [3 2 1]
```

```
> (length [1 2 3])

Term: 3

> (join [1 2] ['a 'b] [3])

List: [1 2 'a 'b 3]
```

The infix-friendly identifier `added-to` is globally defined as an alias for `add`, so we can write, for example:

```
> (rev 1 added-to [2])

List: [2 1]
```

Returning to terms, we say that a term t is *ground* iff it does not contain any variables. Alternatively, t is ground iff it only contains function symbols (including constants). Thus, `(father joe)` is ground, but `(father ?x:Person)` is not.

We generally use the letters s and t (possibly with subscripts/superscripts) to designate terms.

The *subterm* relation can be inductively defined through a number of simple rules, namely:

1. Every term t is a subterm of itself.
2. If t is an application $(f\ t_1 \cdots t_n)$, then every subterm of a child t_i is also a subterm of t.
3. A term s is a subterm of a term t if and only if it can be shown to be so by the above rules.

Inductive definitions like this one can readily be turned into recursive Athena procedures, and we will see such examples throughout the book. A *proper* subterm of t is any subterm of t other than itself. If t_3 is a proper subterm of t_1 and there is no term t_2 such that t_2 is a proper subterm of t_1 and t_3 is a proper subterm of t_2, we say that t_3 is an *immediate* subterm of t_1.

As should be expected, every legal term is of a certain sort. The sort of `(father joe)`, for example, is `Person`; the sort of

```
(in ?x:Element ?s:Set)
```

is `Boolean`; and so on. In general, the sort of a term $(f\ t_1 \cdots t_n)$ is D, where D is the range of f. Some terms, however, are *ill-sorted*, for example, `(father true)`. An attempt to construct this term would result in a "sort error" because the function symbol `father` requires an argument of sort `Person` and we are giving it a `Boolean` instead. Athena performs sort checking automatically and will detect and report any ill-sorted terms as errors:

```
> (father true)

standard input:1:2: Error: Unable to infer a sort for the term:(father true)

(Failed to unify the sorts Boolean and Person.)
```

Here `1:2` indicates line 1, column 2, the precise position of the offending term in the input stream; and `standard input` means that the term was entered directly at the Athena prompt. If the term `(father true)` had been read in batch mode from a file `foo.ath` instead, and the term was found, say, on line 279, column 35, the message would have been: `foo.ath:279:35: Error: Unable to infer ···`. Athena prints out the ill-sorted term and gives a parenthetical indication (line 5 above) of why it was unable to infer a sort for it.

Note that we speak of "ill-sorted" terms, "sort errors," and so on, rather than ill-typed terms or type errors. There is an important distinction between types and sorts in Athena. Only terms have sorts, but terms are just one type of value in Athena. There are several other types: sentences, lists, procedures, methods, the unit value, and more. So we say that the type of `null` is term, and its sort is `Set`. This crucial distinction will become clearer as you gain more experience with the language.

It is not necessary to annotate every variable occurrence inside a term with a sort. Athena can usually infer the proper sort of every variable occurrence automatically. For instance:

```
> (in ?x ?S)

Term: (in ?x:Element ?S:Set)
```

Here we typed in the term `(in ?x ?S)` without any sort annotations. As we can tell from its response, Athena inferred that in this term the sort of `?x` must be `Element` while that of `?S` is `Set`. Another example:

```
> (father ?p)
Term: (father ?p:Person)
```

Omitting explicit sort annotations for variables can save us a good deal of typing, and we will generally follow that practice, especially when the omitted sorts are easily deducible from the context. Occasionally, however, inserting such annotations can make our code more readable.

A binary function symbol can always be used in infix, that is, between its two arguments rather than in front of them. For example, instead of `(union null S)` we may write

```
(null union S).
```

Also, when we have successive applications of unary function symbols, it is not necessary to enclose each such application within parentheses. Thus, instead of writing

```
(father (mother joe))
```

we may simply write (father mother joe). Accordingly, some of the terms we have seen so far can also be written as follows:

```
(e in S),
(null union S2),
(male father joe),
(null subset ?x union null).
```

Even though users may enter terms in infix, Athena always prints output terms in full prefix form, as so-called "s-expressions":

```
> (null union ?s)

Term: (union null ?s:Set)
```

Prefix notation is generally better for output because it is easier to indent, and proper indentation can be invaluable when trying to read large terms (or sentences, as we will soon see). Note also that when Athena prints its output, all variables are fully annotated with their most general possible sorts. This is the default; it can be turned off by issuing the top-level directive **set-flag** print-var-sorts "off"; see also Section 2.16.

Every function symbol has a *precedence* level, which can be obtained with the primitive procedure get-precedence:

```
> (get-precedence subset)

Term: 100

> (get-precedence union)

Term: 110
```

By default, every binary relation symbol (i.e., binary function symbol with Boolean range) is given a precedence of 100, while other binary or unary function symbols are given a precedence of 110.

Precedence levels can be set explicitly with the directive **set-precedence**:

```
> set-precedence union 200

OK.
```

Now that union has a higher precedence than intersection (which has the default 110), we expect an expression such as

```
(?s1 union ?s2 intersection ?s3)
```

to be parsed as (intersection (union ?s1 ?s2) ?s3), and we see that this is indeed the case:

```
> (?s1 union ?s2 intersection ?s3)

Term: (intersection (union ?s1:Set ?s2:Set)
                    ?s3:Set)
```

Binary function symbols also have associativities that can be programmatically obtained and modified. By default, every binary function symbol associates to the right:

```
> (?s1 union ?s2 union ?s3)

Term: (union ?s1:Set
             (union ?s2:Set ?s3:Set))
```

The associativity of a binary function symbol can be explicitly specified with the directives **left-assoc** and **right-assoc**:

```
> left-assoc union

OK.

> (?s1 union ?s2 union ?s3)

Term: (union (union ?s1:Set ?s2:Set)
             ?s3:Set)
```

The unary procedure get-assoc returns a string representing the associativity of its input function symbol.

Infix notation can be used not just for function symbols, but for binary procedures as well. For instance:

```
> (7 less? 8)

Term: true

> (2 times 3)

Term: 6

> (5 minus 1)

Term: 4
```

Thus, the earlier factorial example could also be written as follows:

```
define (fact n) :=
  check {
    (n less? 1) => 1
  | else => (n times fact n minus 1)
  }
```

2.4 Sentences

Sentences are the bread and butter of Athena. Every successful proof derives a unique sentence. Sentences express propositions about relationships that might hold among elements of various sorts. There are three kinds of sentences:

1. Atomic sentences, or just *atoms*. These are simply terms of sort `Boolean`. Examples are

   `(siblings peter (father joe))`

   and `(subset ?s1 (union ?s1 ?s2))`.

2. Boolean combinations, obtained from other sentences through one of the five *sentential constructors*:[10] `not`, `and`, `or`, `if`, and `iff`, or their synonyms, as shown in the following table.

<table>
<tr><th>Constructor</th><th>Synonym</th><th>Prefix mode</th><th>Infix mode</th><th>Interpretation</th></tr>
<tr><td>not</td><td>~</td><td>(not p)</td><td>(~ p)</td><td>Negation</td></tr>
<tr><td>and</td><td>&</td><td>(and p q)</td><td>(p & q)</td><td>Conjunction</td></tr>
<tr><td>or</td><td>|</td><td>(or p q)</td><td>(p | q)</td><td>Disjunction</td></tr>
<tr><td>if</td><td>==></td><td>(if p q)</td><td>(p ==> q)</td><td>Conditional</td></tr>
<tr><td>iff</td><td><==></td><td>(iff p q)</td><td>(p <==> q)</td><td>Biconditional</td></tr>
</table>

   One can use any connective or its synonym in either prefix or infix mode, although, by convention, the synonyms are typically used in infix mode. As with terms, Athena always writes output sentences in prefix. In infix mode, precedence levels come into play, often allowing the omission of parenthesis pairs that would be required in prefix mode. The procedure `get-precedence` can be used to inspect the precedence levels of these connectives, but briefly, conjunction binds tighter than disjunction, and both bind tighter than conditionals and biconditionals. Negation has the highest precedence of all five sentential constructors. Therefore, for example, `(~ A & B | C)` is understood as the

10 We use the term *sentential* (or "logical") *connective* interchangeably with "sentential constructor." Note that the term "constructor" is not used here in the technical datatype sense introduced in Section 2.7.

disjunction of `((~ A) & B)` and `C`. All binary sentential constructors associate to the right by default. (This can be changed with the **`left-assoc`** directive, but such a change is not recommended.) In a conjunction $(p$ `&` $q)$, we refer to p and q as *conjuncts*, and in a disjunction $(p$ `|` $q)$, we refer to p and q as *disjuncts*. A conditional $(p$ `==>` $q)$ can also be read as "if p then q" or even as "p implies q."[11] We call p the *antecedent* and q the *consequent* of the conditional. We will see that a biconditional $(p$ `<==>` $q)$ is essentially the same as the conjunction of the two conditionals $(p$ `==>` $q)$ and $(q$ `==>` $p)$, and for that reason it is sometimes called an *equivalence*, as it states that p and q are equivalent (insofar as each implies the other).

The combination of negation with equality has its own shorthand:

Procedure	Synonym	Prefix	Infix	Interpretation
`unequal`	`=/=`	`(unequal` s t`)`	$(s$ `=/=` $t)$	s is not equal to t

Because function symbols and sentential constructors are regular data values, users can define their own abbreviations and use them in either prefix or infix mode. For instance:

```
define \/ := or
```

When written in prefix, conjunctions and disjunctions can be polyadic; that is, the sentential constructors `and` and `or` may receive an arbitrary number of sentences as arguments, not just two. We will have more to say on that point shortly.

3. Quantified sentences. There are two quantifiers in Athena, `forall` and `exists`. A quantified sentence is of the form $(Q$ x:S . $p)$ where Q is a quantifier, x:S is a variable of sort S, and p is a sentence—the *body* of the quantification, representing the *scope* of the quantified variable x:S. The period is best omitted if the body p is written in prefix, but, when present, it must be surrounded by spaces. For example:

```
(forall ?x . ?x =/= father ?x).          (2.2)
```

Intuitively, a sentence of the form (`forall` $x : S$. p) says that p holds for every element of sort S, while (`exists` $x : S$. p) says that there is some element of sort S for which p holds. In summary:

11 It is debatable whether the sentential constructor ==> has any meaningful correspondence to the informal notion of implication. Many logicians have argued that implication between two propositions requires strong relationships that cannot be captured by any truth-functional semantics (like the semantics of ==> that we present in Section 4.12). For the purposes of this book, however, such debates may be ignored, so we will freely "speak with the vulgar" and refer to conditionals as implications.

Quantifier	Prefix	Infix	Interpretation
`forall`	`(forall` $x:S$ p`)`	`(forall` $x:S$ `.` p`)`	p holds for every $x:S$
`exists`	`(exists` $x:S$ p`)`	`(exists` $x:S$ `.` p`)`	p holds for some $x:S$

We generally use the letters p, q, and r as variables ranging over sentences.

The following shows what happens when we type sentence (2.2) at the Athena prompt:

```
> (forall ?x . ?x =/= father ?x)

Sentence: (forall ?x:Person
            (not (= ?x:Person
                    (father ?x:Person))))
```

Athena acknowledges that a sentence was entered, and displays that sentence using proper indentation. Note that we did not have to explicitly indicate the sort of the quantified variable. Athena inferred that `?x` ranges over `Person`, and annotated every occurrence of `?x` with that sort. But, had we wanted, we could have explicitly annotated the quantified variable with its sort:

```
> (forall ?x:Person . ?x =/= father ?x)

Sentence: (forall ?x:Person
            (not (= ?x:Person
                    (father ?x:Person))))
```

Or we could annotate every variable occurrence with its sort, or only some such occurrences:

```
> (forall ?x:Person . ?x:Person =/= father ?x:Person)

Sentence: (forall ?x:Person
            (not (= ?x:Person
                    (father ?x:Person))))

> (forall ?x . ?x =/= father ?x:Person)

Sentence: (forall ?x:Person
            (not (= ?x:Person
                    (father ?x:Person))))
```

It is usually best to omit as many variable annotations as possible and let Athena work them out.

Because the body of (2.2) is written entirely in infix, we did not have to insert any additional parentheses in it. Had we wanted, we could have entered the sentence in a more explicit form by parenthesizing either the `(father ?x)` term or the entire body of the quantified sentence:

```
> (forall ?x:Person . ?x:Person =/= (father ?x:Person))

Sentence: (forall ?x:Person
            (not (= ?x:Person
                    (father ?x:Person))))
```

or:

```
> (forall ?x:Person . (?x:Person =/= (father ?x:Person)))

Sentence: (forall ?x:Person
            (not (= ?x:Person
                    (father ?x:Person))))
```

Typically, however, when we write sentences in infix we omit as many pairs of parentheses as possible. In full prefix notation, we would write sentence (2.2) as `(forall ?x (=/= ?x (father ?x)))`. Note that in prefix we typically omit the dot immediately before the body of the quantified sentence.

As a shorthand for iterated occurrences of the same quantifier, one can enter sentences of the form (Q $x_1 \cdots x_n$. p) for a quantifier Q and $n > 0$ variables $x_1 \cdots x_n$. For instance:

```
> (forall ?S1 ?S2 . ?S1 = ?S2 <==> ?S1 subset ?S2  & ?S2 subset ?S1)

Sentence: (forall ?S1:Set
            (forall ?S2:Set
              (iff (= ?S1:Set ?S2:Set)
                   (and (subset ?S1:Set ?S2:Set)
                        (subset ?S2:Set ?S1:Set)))))
```

Another shorthand exists for prefix applications of `and` and `or`. Both of these can take an arbitrary number $n > 0$ of sentential arguments: (`and` $p_1 \cdots p_n$) and (`or` $p_1 \cdots p_n$). In infix mode we can get somewhat similar results with (p_1 `&` $\cdots$ `&` p_n) and (p_1 `|` $\cdots$ `|` p_n), respectively. Note, however, that these two latter expressions will give nested and right-associated applications of the binary versions of `and` and `or`, whereas in the case of (`and` $p_1 \cdots p_n$) or (`or` $p_1 \cdots p_n$) we have a *single* sentential constructor applied to an arbitrarily large number of $n > 0$ sentences.

In addition, all five sentential constructors can accept a *list* of sentences as their arguments, which is quite useful for programmatic applications of these constructors:

```
> (and [A B C])

Sentence: (and A
               B
               C)

> (if [A B])
```

```
Sentence: (if A B)

> (not [A])

Sentence: (not A)
```

We inductively define the *subsentence* relation as follows:

1. Every sentence p is a subsentence of itself.
2. If p is a negation (~ q), then every subsentence of q is also a subsentence of p.
3. If p is a conjunction (or disjunction), then any subsentence of any conjunct (or disjunct) is a subsentence of p.
4. If p is a conditional, then every subsentence of the antecedent or consequent is a subsentence of p.
5. If p is a biconditional (p_1 <==> p_2), then every subsentence of p_1 or p_2 is a subsentence of p.
6. If p is a quantified sentence, then every subsentence of the body is a subsentence of p.
7. A sentence q is a subsentence of p if and only if it can be shown to be so by the preceding rules.

As was the case with subterms, this definition too can be easily turned into a recursive Athena procedure; see exercise 3.36. A *proper subsentence* of p is any subsentence of p other than itself. If r is a proper subsentence of p and there is no sentence q such that q is a proper subsentence of p and r is a proper subsentence of q, then we say that r is an *immediate* subsentence of p.

A variable occurrence x:S inside a sentence p is said to be *bound* iff it is within a quantified subsentence of p of the form (Q x:S . p'), for some quantifier Q. A variable occurrence that is not bound is said to be *free*. Free variable occurrences must have consistent sorts across a given sentence. For instance,

```
(?x = joe & male ?x)
```

is legal because both free variable occurrences of ?x are of the same sort (Person), but

```
(male ?x & ?x = 3)
```

is not, because one free occurrence of ?x has sort Person and the other has sort Int:

```
1 > (?x = joe & male ?x)
2
3 Sentence: (and (= ?x:Person joe)
4                (male ?x:Person))
5
```

```
6  > (male ?x & ?x = 3)
7
8  Error, top level, 1.1: Unable to verify that this sentence is well-sorted:
9  (and (male ?x:Person)
10      (= ?x:Int 3))
11
12 (Could not satisfy the constraint 'T189 = (Int /\ Person)
13 because Int and Person do not have a g.l.b.)
```

(You can ignore the parenthetical message on lines 12 and 13.) However, bound variable occurrences can have different sorts in the same sentence:

```
> ((forall ?x . male father ?x) & (forall ?x . ?x subset ?x))

Sentence: (and (forall ?x:Person
                 (male (father ?x:Person)))
               (forall ?x:Set
                 (subset ?x:Set ?x:Set)))
```

Intuitively, that is because the name of a bound variable is immaterial. For instance, there is no substantial difference between `(forall ?x . male father ?x)` and

`(forall ?p . male father ?p).`

Both sentences say the exact same thing, even though one of them uses `?x` as a bound variable and the other uses `?p`. We say that the two sentences are *alpha-equivalent*, or alpha-convertible, or "alphabetic" variants. This means that each can be obtained from the other by consistently renaming bound variables. Alpha-equivalent sentences are essentially identical, and indeed we will see later that Athena treats them as such for deductive purposes. Alpha-conversion, and specifically *alpha-renaming*, occurs automatically very often during the evaluation of Athena proofs. A sentence is alpha-renamed when all bound variable occurrences in it are consistently replaced by fresh (hitherto unseen) variables ranging over the same sorts.

A very common operation is the safe replacement of every free occurrence of a variable v inside a sentence p by some term t. By "safe" we mean that, if necessary, the operation alpha-renames p to avoid any *variable capture* that might come about as a result of the replacement. Variable capture is what would occur if any variables in t were to become accidentally bound by quantifiers inside p as a result of carrying out the replacement. For instance, if p is

`(exists ?x:Real . ?x:Real > ?y:Real),`

v is `?y:Real`, and t is `(2 * ?x:Real)`, then naively replacing free occurrences of v inside p will result in

`(exists ?x:Real . ?x:Real > 2 * ?x:Real),`

where the third (and boxed) occurrence of `?x:Real` has been accidentally "captured" by the leading bound occurrence of `?x:Real`. An acceptable way of carrying out this replacement is to first alpha-rename p, obtaining an alphabetic variant p' of it, say

p' = `(exists ?w:Real . ?w:Real > ?y:Real)`,

and then carry out the replacement inside p' instead, obtaining the result

```
(exists ?w:Real . ?w:Real > 2 * ?x:Real).
```

This is legitimate because p and p' are essentially identical. We write $\{v \mapsto t\}(p)$ for this operation of safely replacing all free occurrences of v inside p by t.

We close this section with a tabular summary of Athena's prefix and infix notations for first-order logic sentences vis-à-vis corresponding syntax forms often used in conventional mathematical writing:

	Athena prefix	Athena infix	Conventional
Atoms:	$(R\ t_1 \cdots t_n)$	$(t_1\ R\ t_2)$ when $n = 2$	$R(t_1, \ldots, t_n)$ $(t_1\ R\ t_2)$
Negations:	`(not` p`)`	`(~` p`)`	$(\neg p)$
Conjunctions:	`(and` $p_1\ p_2$`)`	$(p_1$ `&` $p_2)$	$(p_1 \wedge p_2)$
Disjunctions:	`(or` $p_1\ p_2$`)`	$(p_1$ `|` $p_2)$	$(p_1 \vee p_2)$
Conditionals:	`(if` $p_1\ p_2$`)`	$(p_1$ `==>` $p_2)$	$(p_1 \Rightarrow p_2)$
Biconditionals:	`(iff` $p_1\ p_2$`)`	$(p_1$ `<==>` $p_2)$	$(p_1 \Leftrightarrow p_2)$
Un. quantifications:	`(forall` $x : S\ p$`)`	`(forall` $x : S\ .\ p$`)`	$(\forall x : S\ .\ p)$
Ex. quantifications:	`(exists` $x : S\ p$`)`	`(exists` $x : S\ .\ p$`)`	$(\exists x : S\ .\ p)$

Occasionally we may use conventional notation, if we believe that it can simplify or shorten the exposition.

2.5 Definitions

Definitions let us give a name to a value and then subsequently refer to the value by that name. This is a key tool for managing complexity. Athena provides a few naming mechanisms. One of the most useful is the top-level directive **`define`**. Its general syntax form is:

`define` I `:=` F

where I is any identifier and F is a phrase denoting the value that we want to define. Once a definition has been made, the defined value can be referred to by its name:

```
> define p := (forall ?s . ?s subset ?s)

Sentence p defined.

> (p & p)

Sentence: (and (forall ?s:Set
                 (subset ?s:Set ?s:Set))
               (forall ?s:Set
                 (subset ?s:Set ?s:Set)))
```

Athena has static scoping (also known as *lexical scoping*), which means that new definitions override older ones:

```
> define t := joe

Term t defined.

> t

Term: joe

> define t := 0

Term t defined.

> t

Term: 0
```

Because function symbols are regular denotable values, we can easily define names for them and use them as alternative notations. For instance, suppose we want to use the symbol \/ as an alias for `union` in a given stretch of text. We can do that with a simple definition:

```
define \/ := union
```

We can now go ahead and use \/ to build terms and sentences and so on:

```
> \/

Symbol: union

> (e in null \/ S)

Term: (in e
          (union null S))
```

Because logical connectives and quantifiers are also regular denotable values, we can do the same thing with them. For instance, if we prefer to use `all` as a universal quantifier, we can simply say:

```
> define all := forall

Quantifier all defined.

> (all ?s . ?s subset ?s)

Sentence: (forall ?s:Set
            (subset ?s:Set ?s:Set))
```

The ability to compute with function symbols (as well as logical connectives and quantifiers) has some distinct advantages. It enables a flexible form of overloading and other powerful notational customizations. These are features that we will be using extensively in this book.

Keep in mind that **define** is a top-level directive, not a procedure. It can only appear as a direct command to Athena; it cannot appear inside a phrase.

We can use square-bracket list notation to define several values at the same time. For instance:

```
> define [x y z] := [1 2 3]

Term x defined.

Term y defined.

Term z defined.

> [z y x]

List: [3 2 1]
```

This feature is similar to the use of patterns inside **let** bindings, discussed in Section 2.16.

2.6 Assumption bases

At all times Athena maintains a global set of sentences called the *assumption base*. We can think of the elements of the assumption base as our premises—sentences that we regard (at least provisionally) as true. Initially the system starts with a small assumption base. Every time an axiom is postulated or a theorem is proved at the top level,[12] the corresponding sentence is inserted into the assumption base.

12 A theorem is simply a sentence produced by a deduction D. We say that the deduction proves or derives the corresponding theorem (which is also the deduction's *conclusion*).

The assumption base can be inspected with the procedure show-assumption-base. It is a nullary procedure (i.e., it does not take any arguments), so it is invoked as

(show-assumption-base).

A related nullary procedure, get-ab, returns a list of all the sentences in the current assumption base. Because this is a very commonly used procedure, it also goes by the simpler name ab.

When Athena is first started, the assumption base has a fairly small number of sentences in it, mostly axioms of some very fundamental built-in theories, such as pairs and options.

```
> (length (ab))

Term: 20
```

A sentence can be inserted into the assumption base with the top-level directive **assert**:

```
> assert (joe =/= father joe)

The sentence
(not (= joe
        (father joe)))
has been added to the assumption base.
```

The unary procedure holds? can be used to determine whether a sentence is in the current assumption base. We can confirm that the previous sentence was inserted into the assumption base by checking that it holds afterward:

```
> (holds? (joe =/= father joe))

Term: true
```

There is a similar procedure hold? that takes a list of sentences and returns true if all of them are in the assumption base and false otherwise.

Multiple sentences can be asserted at the same time; the general syntax form is

assert $p_1, \ldots, p_n$

Lists of sentences can also appear as arguments to **assert**. A list of sentences $[p_1 \cdots p_n]$ can be given a name with a definition, and then asserted by that name. This is often the most convenient way to make a group of assertions:

```
define facts := [(1 = 1) (joe = joe)]

assert facts
```

Alternatively, we can combine definition with assertion in one fell swoop:

```
assert facts := [(1 = 1) (joe = joe)]
```

All of these features work with **assert*** as well, which is equivalent to **assert** except that it first universally quantifies each given sentence over all of its free variables before inserting it into the assumption base (see page 7).

There are also two mechanisms for removing sentences from the global assumption base:

clear-assumption-base

and **retract**. The former will empty the assumption base, while the second will only remove the specified sentence:

```
> clear-assumption-base

Assumption base cleared.

> assert (1 = 2)

The sentence
(= 1 2)
has been added to the assumption base.

> retract (1 = 2)

The sentence
(= 1 2)
has been removed from the assumption base.
```

All three of these (**assert**, **clear-assumption-base**, and **retract**) are top-level directives; they cannot appear inside other syntax forms.

The last two constructs are powerful and should be used sparingly. A directive

retract p,

in particular, simply removes p from the assumption base without checking to see what other sentences might have been derived from p in the interim (between its assertion and its retraction), so a careless removal may well leave the assumption base in an incorrect state. This tool is meant to be used only when we assert a sentence p and then right away realize that something went wrong, for instance, that p contains certain free variables that it should not contain, in which case we can promptly remove it from the assumption base with **retract**.

2.7 Datatypes

A *datatype* is a special kind of domain. It is special in that it is *inductively generated*, meaning that every element of the domain can be built up in a finite number of steps by applying certain operations known as the *constructors* of the datatype. A datatype D is specified by giving its name, possibly followed by some sort parameters (if D is polymorphic; we discuss that in Section 2.8), and then a nonempty sequence of constructor profiles separated by the symbol `|`. A constructor profile without selectors[13] is of the form

$$(c\ S_1 \cdots S_n), \tag{2.3}$$

consisting of the name of the constructor, c, along with n sorts $S_1 \cdots S_n$, where S_i is the sort of the ith argument of c. The range of c is not explicitly mentioned—it is tacitly understood to be the datatype D. That is why, after all, c is called a "constructor" of D; because it builds or *constructs* elements of the datatype. Thus, every application of c to n arguments of the appropriate sorts produces an element of D. In terms of the **declare** directive that we have already seen, a constructor c of a datatype D with profile (2.3) can be thought of as a function symbol with the following signature declaration:

declare c: $[S_1 \cdots S_n]$ -> D.

A nullary constructor (one that takes no arguments, so that $n = 0$) is called a *constant constructor* of D, or simply a constant of D. Such a constructor represents an individual element of D. The outer parentheses in (2.3) may be (and usually are) omitted from the profile of a constant constructor.

Here is an example:

```
datatype Boolean := true | false
```

This defines a datatype by the name of `Boolean` that has two constant constructors, `true` and `false`. (This particular datatype is predefined in Athena.) Thus, this definition says that the datatype `Boolean` has two elements, `true` and `false`. The definition conveys more information than that, however. It also licenses the conclusion that `true` and `false` are *distinct* elements, and, moreover, that `true` and `false` are the *only* elements of `Boolean`.

The intended effect of this datatype definition could be approximated in terms of mechanisms with which we are already familiar as follows:

```
domain Boolean

declare true, false: Boolean

assert (true =/= false)

assert (forall ?b:Boolean . ?b = true | ?b = false)
```

Here we have made two assertions that ensure that (a) `true` and `false` refer to distinct objects; and (b) the domain `Boolean` does not contain any other elements besides those

13 Selectors are discussed in Section A.5.

denoted by `true` and `false`. For readers who are familiar with universal algebra, these axioms hold because the datatype is freely generated, meaning that it is a *free algebra*. Given a datatype D, we will collectively refer to these axioms as the *free-generation axioms* for D. Roughly, the axioms express the following propositions: (a) different constructor applications create different elements of D; and (b) every element of D is represented by some constructor application. Axioms of the first kind capture what are known as *no-confusion* conditions, while axioms of the second kind capture a limited form of a so-called *no-junk* condition.

Thus, to a certain extent, a datatype definition can be viewed as syntax sugar for a domain declaration (plus appropriate symbol declarations for the constructors), along with the relevant free-generation axioms. We say "to a certain extent" because an additional and important effect of a datatype definition is the introduction of an inductive principle for performing structural induction on the datatype (we discuss this in Section 3.8). For infinite datatypes such as the natural numbers, to which we will turn next, structural induction is an ingredient that goes above and beyond the free-generation axioms, meaning that the induction principle allows us to prove results that are not derivable from the free-generation axioms alone.

Here is a datatype for the natural numbers that we will use extensively in this book:

```
datatype N := zero | (S N)
```

This defines a datatype `N` that has one constant constructor `zero` and one unary constructor `S`. Because the argument of `S` is of the same sort as its result, `N`, we say that `S` is a *reflexive* constructor. Thus, `S` requires an element of `N` as input in order to construct another such element as output. By contrast, `zero`, as well as `true` and `false` in the `Boolean` example, are *irreflexive* constructors—trivially, since they take no arguments.

Unlike domains, which can be interpreted by arbitrary sets, the interpretation of a datatype is fixed: A datatype definition always picks out a unique set (up to isomorphism). Roughly, in the case of `N`, that set can be understood as given by the following rules:

1. `zero` is an element of `N`.
2. For all n, if n is an element of `N`, then `(S` n`)` is an element of `N`.
3. Nothing else is an element of `N`.

The last clause, in particular, ensures the minimality of the defined set, whereby the *only* elements of `N` are those that can be obtained by the first two clauses, namely, *by a finite number of constructor applications*. This is what ensures that the specified set does not have any extraneous ("junk") elements.

Essentially, every datatype definition can be understood as a recursive set definition of the preceding form. How do we know that a recursive definition of this form always succeeds in determining a unique set, and how do we know that a datatype definition can always be thus understood? We will not get into the details here, but briefly, the answer to the first question is that we can prove the existence of a unique set satisfying the recursive definition using fixed-point theory; and the answer to the second question is that there are simple syntactic constraints on datatype definitions that ensure that every datatype gives rise to a recursive set definition of the proper form.[14]

The following are the free-generation axioms for N:

```
(forall ?n . zero =/= S ?n)

(forall ?n ?m . S ?n = S ?m ==> ?n = ?m)

(forall ?n . ?n = zero | exists ?m . ?n = S ?m)
```

You might recognize these if you have ever seen Peano's axioms. The first two are no-confusion axioms: (1) `zero` is different from every application of `S` (i.e., `zero` is not the successor of any natural number), and (2) applications of `S` to different arguments produce different results (i.e., `S` is injective). The third axiom says that the constructors `zero` and `S` span the domain `N`: Every natural number is either `zero` or else the successor of some natural number.[15]

The free-generation axioms of a datatype can be obtained automatically by the unary procedure `datatype-axioms`, which takes the name of the datatype as an input string and returns a list of the free-generation axioms for that datatype:

```
> (datatype-axioms "N")

List: [
(forall ?y1:N
  (not (= zero
          (S ?y1:N))))

(forall ?x1:N
  (forall ?y1:N
    (iff (= (S ?x1:N)
            (S ?y1:N))
         (= ?x1:N ?y1:N))))

(forall ?v:N
  (or (= ?v:N zero)
      (exists ?x1:N
        (= ?v:N
           (S ?x1:N)))))
]
```

14 For instance, definitions such as **datatype** `D := (c D)` are automatically rejected.

15 Strictly speaking, the third axiom (and other axioms of a similar form, in the case of other datatypes) can be proved by induction, but it is useful to have it directly available.

```
> (datatype-axioms "Boolean")

List: [
(not (= true false))

(forall ?v:Boolean
  (or (= ?v:Boolean true)
      (= ?v:Boolean false)))
]
```

Whether or not these axioms are added to the assumption base automatically is determined by a global flag `auto-assert-dt-axioms`, off by default. If turned on (see Section 2.12), then every time a new datatype T is defined, its axioms will be automatically added to the assumption base. In addition, the identifier T`-axioms` will be automatically bound in the global environment to the list of these axioms.

Once a datatype D has been defined, it can be used as a bona fide Athena sort; for example, we can declare functions that take elements of D as arguments or return elements of D as results. We can introduce binary addition on natural numbers, for instance, as follows:

```
> declare Plus: [N N] -> N

New symbol Plus declared.
```

A number of mutually recursive datatypes can be defined with the **datatypes** keyword, separating the component datatypes with `&&`. For example:

```
> datatypes Even := zero | (s1 Odd) &&
            Odd := (s2 Even)

New datatypes Even and Odd defined.
```

The constructors of a top-level datatype (or a set of mutually recursive datatypes) must have distinct names from one another, as well as from the constructors of every other top-level datatype and from every other function symbol declared at the top level. However, the same sort, constructor, or function symbol name can be used inside different modules.

Not all inductively defined sets are freely generated. For example, in some cases it is possible for two distinct constructor terms to denote the same element. Consider, for instance, a hypothetical datatype for finite sets of integers:

```
datatype Set := null | (insert Int Set)
```

This definition says that a finite set of integers is either `null` (the empty set) or else of the form `(insert` i s`)`, obtained by inserting an integer i into the set s. Now, two sets are identical iff they have the same members, so that $\{1,3\} = \{3,1\}$. Hence,

```
(insert 3 (insert 1 null))
```

and

```
(insert 1 (insert 3 null))
```

ought to be identical. That is, we must be able to prove

```
((insert 3 (insert 1 null)) = (insert 1 (insert 3 null))).        (2.4)
```

But one of the no-confusion axioms would have us conclude that `insert` is injective:

```
(forall ?i1 ?s1 ?i2 ?s2 . (insert ?i1 ?s1) = (insert ?i2 ?s2)
                          ==> ?i1 = ?i2 & ?s1 = ?s2).
```

This is inconsistent with (2.4), as it would allow us to conclude, for example, that `(1 = 3)`.

The preferred approach here is to define `Set` as a **`structure`** rather than a **`datatype`**:

```
structure Set := null | (insert Int Set)
```

A *structure* is a datatype with a coarser identity relation. Just like a regular datatype, a structure is inductively generated by its constructors, meaning that every element of the structure is obtainable by a finite number of constructor applications. This means that structural induction (via **`by-induction`**) is available for structures. The only difference is that there may be some "confusion," for instance, the constructors might not be injective (the usual case), so that one and the same constructor applied to two distinct sequences of arguments might result in the same value. More rarely, we might even obtain the same value by applying two distinct constructors. It is the user's responsibility to assert a proper identity relation for a structure. In the set example, we would presumably assert something along the following lines:

```
(forall ?s1 ?s2 . ?s1 = ?s2 <==> ?s1 subset ?s2 & ?s2 subset ?s1),
```

where `subset` has the usual definition in terms of membership.

The unary procedure `structure-axioms` will return a list of all the inductive axioms that are usually appropriate for a structure, assuming that the only difference is that constructors are not injective. The input argument is the name of the structure:

```
> (structure-axioms "Set")

List: [
(forall ?y1:Int
  (forall ?y2:Set
    (not (= null
            (insert ?y1:Int ?y2:Set)))))

(forall ?v:Set
  (or (= ?v:Set null)
      (exists ?x1:Int
        (exists ?x2:Set
          (= ?v:Set
             (insert ?x1:Int ?x2:Set))))))
]
```

If the structure has other differences (most notably, if applying two distinct constructors might result in the same value), then it is the user's responsibility to assert the relevant axioms as needed.

2.8 Polymorphism

2.8.1 Polymorphic domains and sort identity

A domain can be polymorphic. As an example, consider sets over an arbitrary universe, call it S:

```
> domain (Set S)

New domain Set introduced.
```

The syntax for introducing polymorphic domains is the same as before, except that now the domain name is flanked by an opening parenthesis to its left and a list of identifiers to its right, followed by a closing parenthesis; say, (Set S), or in the general case, $(I\ I_1 \cdots I_n)$, where I is the domain name. The identifiers $I_1, \ldots, I_n$ are parameters that serve as *sort variables* in this context, indicating that I is a *sort constructor* that takes any n sorts $S_1, \ldots, S_n$ as arguments and produces a new sort as a result, namely $(I\ S_1 \cdots S_n)$. For instance, Set is a unary sort constructor that can be applied to an arbitrary sort, say the domain Int, to produce the sort (Set Int). For uniformity, monomorphic sorts such as Person and N can be regarded as nullary sort constructors. Polymorphic datatypes and structures, discussed in Section 2.8.3, can also serve as sort constructors.

Equipped with the notion of a sort constructor, we can define Athena sorts more precisely. Suppose we have a collection of sort constructors ***SC***, with each $sc \in$ ***SC*** having a unique arity $n \geq 0$, and a disjoint collection of sort variables ***SV***. We then define a *sort over* ***SC*** *and* ***SV*** as follows:

- Every sort variable ψ in $\boldsymbol{SV}$ is a sort over $\boldsymbol{SC}$ and $\boldsymbol{SV}$.
- Every nullary sort constructor $sc \in \boldsymbol{SC}$ is a sort over $\boldsymbol{SC}$ and $\boldsymbol{SV}$.
- If $S_1, \ldots, S_n$ are sorts over $\boldsymbol{SC}$ and $\boldsymbol{SV}$, $n > 0$, and $sc \in \boldsymbol{SC}$ is a sort constructor of arity n, then $(sc\ S_1 \cdots S_n)$ is a sort over $\boldsymbol{SC}$ and $\boldsymbol{SV}$.
- Nothing else is a sort over $\boldsymbol{SC}$ and $\boldsymbol{SV}$.

We write $Sorts(\boldsymbol{SC}, \boldsymbol{SV})$ for the set of all sorts over $\boldsymbol{SC}$ and $\boldsymbol{SV}$.

For example, suppose that $\boldsymbol{SC} = \{$`Int`, `Boolean`, `Set`$\}$ and $\boldsymbol{SV} = \{$`S1`, `S2`$\}$, where `Int` and `Boolean` are nullary sort constructors, while `Set` is unary. Then the following are all sorts over $\boldsymbol{SC}$ and $\boldsymbol{SV}$:

```
Boolean,
Int,
(Set Boolean),
S1,
(Set Int),
(Set S2),
(Set (Set Int)),
(Set (Set S1)).
```

There are infinitely many sorts over these three sort constructors and two sort variables.

Sorts of the form $(sc\ S_1 \cdots S_n)$ for $n > 0$ are called *compound*, or complex. A *ground* (or *monomorphic*) sort is one that contains no sort variables. All of the above examples are ground except `S1`, `(Set S2)`, and `(Set (Set S1))`. A sort that is not ground is said to be polymorphic.

A polymorphic domain is really a domain *template*. Substituting ground sorts for its sort variables gives rise to a specific domain, which is an *instance* of the template. Intuitively, you can think of a polymorphic domain as the collection of all its ground instances.

Formally, let us define a *sort valuation* τ as a function from sort variables to sorts. Any such τ can be extended to a function

$$\widehat{\tau} : Sorts(\boldsymbol{SC}, \boldsymbol{SV}) \to Sorts(\boldsymbol{SC}, \boldsymbol{SV})$$

(i.e., to a function $\widehat{\tau}$ from sorts over $\boldsymbol{SC}$ and $\boldsymbol{SV}$ to sorts over $\boldsymbol{SC}$ and $\boldsymbol{SV}$) as follows:

$$\widehat{\tau}(\psi) = \tau(\psi)$$
$$\widehat{\tau}((sc\ S_1 \cdots S_n)) = (sc\ \widehat{\tau}(S_1) \cdots \widehat{\tau}(S_n))$$

for any sort variable $\psi \in \boldsymbol{SV}$ and sort constructor $sc \in \boldsymbol{SC}$ of arity n. We say that a sort S_1 is an *instance of* (or *matches*) a sort S_2 iff there exists a sort valuation τ such that $\widehat{\tau}(S_2) = S_1$. And we say that two sorts S_1 and S_2 are *unifiable* iff there exists a sort valuation τ such that $\widehat{\tau}(S_1) = \widehat{\tau}(S_2)$. There are algorithms for determining whether one sort matches another, or whether two sorts are unifiable, and these algorithms are widely used in Athena's operational semantics.

Two sorts are considered identical iff they differ only in their variable names, that is, iff each can be obtained from the other by (consistently) renaming sort variables. It follows that a ground sort (such as N) is identical only to itself, since it does not contain any sort variables.

2.8.2 Polymorphic function symbols

Polymorphic sorts pave the way for polymorphic function symbols. The syntax for declaring a polymorphic function symbol f is the same as before, except that the relevant sort variables must appear listed within parentheses and separated by commas before the list of input sorts. Thus, the general syntax form is

$$\texttt{declare}\ f\texttt{:}\ (I_1,\ldots,I_n)\ \texttt{[}S_1\cdots S_n\texttt{]}\ \texttt{->}\ S \tag{2.5}$$

where $I_1,\ldots,I_n$ are distinct identifiers that will serve as sort variables in the context of this declaration, and as before, S_i is the sort of the ith argument and S is the sort of the result. Presumably, some of these sorts will involve the sort variables. For example:

```
declare in: (S) [S (Set S)] -> Boolean

declare union: (S) [(Set S) (Set S)] -> (Set S)

declare =: (S) [S S] -> Boolean
```

The first declaration introduces a polymorphic membership predicate that takes an element of an arbitrary sort S and a set over S and yields either true or false. The second declaration introduces a polymorphic union function that takes two sets over an arbitrary sort S and produces another set over the same sort. Finally, the last declaration introduces a polymorphic equality predicate; that particular symbol is built-in.

Polymorphic constants can be declared as well, say:

```
declare empty-set: (T) [] -> (Set T)
```

A function symbol declaration of the form (2.5) is admissible iff (a) f is distinct from every function symbol (including datatype constructors) introduced before it; (b) every S_i, as well as S, is a sort over $\boldsymbol{SC}$ and $\{I_1,\ldots,I_n\}$, where $\boldsymbol{SC}$ is the set of sort constructors available prior to the declaration of f; and (c) every sort variable that appears in the output sort S appears in some input sort S_i. Hence, the following are all inadmissible:

```
> declare g: (S) [(Set Int S)] -> Boolean

standard input:1:18: Error: Ill-formed sort: (Set Int S).
```

```
> declare g: (S) [(Set T) S] -> Int

standard input:1:18: Error: Ill-formed sort: (Set T).

> declare g: (S) [Int] -> S

standard input:1:25: Error: The sort variable S appears in the
resulting sort but not in any argument sort.
```

The first declaration is rejected because `(Set Int S)` is not a legal sort (we introduced `Set` as a unary sort constructor, but here we tried to apply it to *two* sorts). The second attempt is rejected because `T` is neither a previously introduced sort nor one of the sort variables listed before the input sorts. The third error message explains why the last attempt is also rejected.

Intuitively, a polymorphic function symbol f can be thought of as a collection of monomorphic function symbols, each of which can be viewed as an instance of f. The declaration of each of those instances is obtainable from the declaration of f by consistently replacing sort variables by ground sorts. For example, the foregoing declaration of the polymorphic predicate symbol `in` might be regarded as syntax sugar for infinitely many monomorphic function symbol declarations:

```
declare in_Int: [Int (Set Int)] -> Boolean

declare in_Real: [Real (Set Real)] -> Boolean

declare in_Boolean: [Boolean (Set Boolean)] -> Boolean

declare in_(Set Int): [(Set Int) (Set (Set Int))] -> Boolean
```

and so on for infinitely more ground sorts. This is elaborated further in Section 5.6.

Polymorphic function symbols are harnessed by Athena's polymorphic terms and sentences. Try typing a variable such as `?x` into Athena without any sort annotations:

```
> ?x

Term: ?x:'T175
```

Note the sort that Athena has assigned to the input variable, namely, `'T175`. This is a sort variable. Athena generally prints out sort variables in the format `'T`*n* or `'S`*n*, where *n* is some integer index. This is the most general sort that Athena could assign to the variable `?x` in this context. So this is a polymorphic variable, and hence a polymorphic term. Users can also enter explicitly polymorphic variables, that is, variables annotated with polymorphic sorts, writing sort variables in the form `'`*I*. For example:

```
> ?y:'T3

Term: ?y:'T177
```

Note that the sort that Athena assigned to ?y is 'T177, which is identical to 'T3, since each can be obtained from the other by renaming sort variables (recall our discussion of sort identity in the previous subsection). Here are some more examples of polymorphic terms:

```
> (?x in ?y)

Term: (in ?x:'T203
          ?y:(Set 'T203))

> (?a = ?b)

Term: (= ?a:'T206 ?b:'T206)

> (?x:(Set 'T) in ?y:(Set (Set 'T)))

Term: (in ?x:(Set 'T209)
          ?y:(Set (Set 'T209)))
```

In the first two examples, Athena automatically infers the most general possible polymorphic sorts for every variable occurrence. Also note that the common occurrence of 'T203 in the first example indicates that Athena has inferred a constraint on the sorts of ?x and ?y, namely, that whatever the sort of ?x is, ?y must be a set of elements of *that* sort. Likewise for the second example: The sorts of ?a and ?b can be arbitrary but they must be identical. In the third example we have explicitly provided specific polymorphic sorts for ?x and ?y: (Set 'T) for ?x and (Set (Set 'T)) for ?y. Athena verified that these were consistent with the signature of in and thus accepted them (modulo the sort-variable permutation 'T $\leftrightarrow$ 'T209). Observe, however, that it is not necessary to provide both of these sorts. We can just annotate one of them and have the sort of the other be inferred automatically:

```
> (?x:(Set 'T) in ?y)

Term: (in ?x:(Set 'T213)
          ?y:(Set (Set 'T213)))
```

We can write sort annotations not just for variables but also for constant terms:

```
> (in ?x empty-set:(Set Real))

Term: (in ?x:Real
          empty-set:(Set Real))

> empty-set:(Set (Set 'S))

Term: empty-set:(Set (Set 'T4395))
```

Any constant term can be annotated in Athena, including monomorphic ones. Athena will just ignore such annotations, as long as they are correct:

```
> joe:Person

Term: joe

> (S zero:N)

Term: (S zero)
```

There is an easy way to check whether a term t is polymorphic: Give it as input to the Athena prompt and then scan Athena's output for sort variables (prefixed by `'`). If Athena annotates at least one variable or constant symbol in t with a polymorphic sort (i.e., one containing sort variables), then t is polymorphic; otherwise it is monomorphic. But there is also a primitive unary procedure `poly?` that will take any term t and return `true` or `false` depending on whether or not t is polymorphic.

Informally, it is helpful to think of a polymorphic term as a schema or template that represents an entire set of (possibly infinitely many) monomorphic terms. For instance, you can think of the polymorphic term `empty-set` as representing infinitely many monomorphic terms, such as

```
empty-set:(Set Int),
empty-set:(Set Boolean),
empty-set:(Set (Set Int)),
```

and so on. Keep in mind that the mere presence of a polymorphic symbol in a term does not make that term polymorphic. For instance, the term `(?x:Int = 3)` contains the polymorphic symbol =, but it is not itself polymorphic. No variable or constant symbol in it has a nonground sort, as you can verify by typing the term into Athena:

```
> (?x = 3)

Term: (= ?x:Int 3)
```

or by supplying it as an argument to `poly?`:

```
> (poly? (?x = 3))

Term: false
```

The presence of polymorphic function symbols is a necessary condition for a term to be polymorphic, but it is not sufficient.

A polymorphic sentence is one that contains at least one polymorphic term, or a quantified variable with a nonground sort. Here are some examples:

```
> (forall ?x . ?x = ?x)

Sentence: (forall ?x:'S
            (= ?x:'S ?x:'S))

> (forall ?x ?y . ?x union ?y = ?y union ?x)

Sentence: (forall ?x:(Set 'S)
            (forall ?y:(Set 'S)
              (= (union ?x:(Set 'S)
                        ?y:(Set 'S))
                 (union ?y:(Set 'S)
                        ?x:(Set 'S)))))

> (~ exists ?x . ?x in empty-set)

Sentence: (not (exists ?x:'S
                 (in ?x:'S
                     empty-set:(Set 'S))))
```

Note that quantified variables can be explicitly annotated with polymorphic sorts:

```
> (exists ?x:(Set (Set 'T)) . ?x =/= empty-set)

Sentence: (exists ?x:(Set (Set 'S))
            (not (= ?x:(Set (Set 'S))
                    empty-set:(Set (Set 'S)))))
```

As with terms, a simple way to test whether an unannotated sentence is polymorphic is to give it as input to the Athena prompt and then scan the output for sort variables. If you see any, the sentence is polymorphic, otherwise it is monomorphic. But `poly?` can also be used on sentences:

```
> (poly? (forall ?x . ?x = ?x))

Term: true
```

Also as with terms, a polymorphic sentence such as `(forall ?x . ?x = ?x)` can be seen as a collection of (potentially infinitely many) monomorphic sentences, namely:

```
(forall ?x:Int . ?x = ?x),
(forall ?x:Boolean . ?x = ?x),
(forall ?x:(Set Int) . ?x = ?x),
```

and so on. This expressivity is the power of parametric polymorphism. A single polymorphic sentence can express infinitely many propositions about infinitely many sets of objects.

2.8.3 Polymorphic datatypes

Since datatypes are just special kinds of domains, they too can be polymorphic, and likewise for structures. Athena's polymorphic datatypes resemble polymorphic algebraic datatypes found in languages such as ML and Haskell. Here are two examples for polymorphic lists and ordered pairs:

```
datatype (List S) := nil | (:: S (List S))

datatype (Pair S T) := (pair S T)
```

The syntax is the same as for monomorphic datatypes, except that the datatype name is now flanked by an opening parenthesis to its left and a list of distinct identifiers to its right, followed by a closing parenthesis, as in `(List S)` or `(Pair S T)`, or, in general, $(I\ I_1 \cdots I_n)$, where I is the name of the datatype. Just as for polymorphic domains, the identifiers $I_1, \ldots, I_n$ serve as local sort variables, indicating that I is a sort constructor that takes n sorts $S_1, \ldots, S_n$ as arguments and produces a new sort as a result, $(I\ S_1 \cdots S_n)$. For instance, `List` is a unary sort constructor that can be applied to an arbitrary sort, say the domain `Int`, to produce the sort `(List Int)`; while `Pair` is a binary sort constructor and can thus be applied to any two sorts to produce a new one, say, `(Pair Boolean (List Int))`.

The profile of each constructor is the same as before: $(c\ S_1 \cdots S_k)$, where c is the name of the constructor.[16] Here, each S_i must be a sort over all previously introduced sort constructors *plus* the datatype that is being defined (thus allowing recursion), and the sort variables $I_1, \ldots, I_n$. Recursive datatype definitions are quite useful and common, the definition of lists above being a typical example.

The notion of a reflexive constructor remains unchanged: If an argument of a constructor c is of a sort that involves the name of the datatype of which c is a constructor, then c is reflexive; otherwise it is irreflexive. Every datatype must have at least one irreflexive constructor. More specifically, we say that a definition of a datatype D is admissible iff (1) the name D is distinct from all previously introduced sorts (domains or datatypes); (2) the constructors of D are distinct from one another, as well as from every function symbol introduced prior to the definition of D; (3) every argument sort of every constructor of D is a sort over $\boldsymbol{SC} \cup \{D\}$ and $\{I_1, \ldots, I_n\}$, where $\boldsymbol{SC}$ is the set of previously available sort constructors and $I_1, \ldots, I_n$ are the sort variables (if any) listed in the definition of D; and finally, (4) D has at least one irreflexive constructor.

The procedures `datatype-axioms` and `structure-axioms` work just as well with polymorphic datatypes, for example:

```
> (datatype-axioms "List")

List: [
```

16 Also as before, the outer parentheses may be dropped when $k = 0$.

```
(forall ?y1:'S
  (forall ?y2:(List 'S)
    (not (= nil:(List 'S)
            (:: ?y1:'S
                ?y2:(List 'S))))))

(forall ?x1:'S
  (forall ?x2:(List 'S)
    (forall ?y1:'S
      (forall ?y2:(List 'S)
        (if (= (:: ?x1:'S
                   ?x2:(List 'S))
               (:: ?y1:'S
                   ?y2:(List 'S)))
            (and (= ?x1:'S ?y1:'S)
                 (= ?x2:(List 'S)
                    ?y2:(List 'S))))))))

(forall ?v:(List 'S)
  (or (= ?v:(List 'S)
         nil:(List 'S))
      (exists ?x1:'S
        (exists ?x2:(List 'S)
          (= ?v:(List 'S)
             (:: ?x1:'S
                 ?x2:(List 'S)))))))
]
```

When D is a polymorphic datatype of arity n and $S_1, \ldots, S_n$ are ground sorts, then the sort $(D\ S_1 \cdots S_n)$ may be regarded as a monomorphic datatype. That datatype's definition can be retrieved from the definition of D by consistently replacing the sort variables $I_1, \ldots, I_n$ by the sorts $S_1, \ldots, S_n$, respectively. For example, `(Pair Int Boolean)` may be seen as a monomorphic datatype $\texttt{Pair}_{\texttt{Int}\times\texttt{Boolean}}$ with one binary constructor

$$\texttt{pair}_{\texttt{Int}\times\texttt{Boolean}}$$

whose first argument is `Int` and whose second argument is `Boolean`, namely, as the datatype

$$\textbf{datatype}\ \texttt{Pair}_{\texttt{Int}\times\texttt{Boolean}} := (\texttt{pair}_{\texttt{Int}\times\texttt{Boolean}}\ \texttt{Int Boolean}).$$

Likewise, `(List Int)` can be understood as a monomorphic datatype $\texttt{List}_{\texttt{Int}}$, definable as

$$\textbf{datatype}\ \texttt{List}_{\texttt{Int}} := \texttt{nil}_{\texttt{Int}} \mid (\texttt{::}_{\texttt{Int}}\ \texttt{Int}\ \texttt{List}_{\texttt{Int}}).$$

Thus, just as with polymorphic domains, a polymorphic datatype may be viewed as the collection of all its ground instances.

2.8.4 Integers and reals

Athena comes with two predefined numeric domains, Int for integers and Real for real numbers. The domain Int has infinitely many constant symbols associated with it, namely, all integer numerals, positive, negative, and zero:

```
> (?x = 47)

Term: (= ?x:Int 47)
```

Note that Athena automatically recognized the sort of ?x as Int. Negative integer numerals are written as (- *n*):

```
> (exists ?x . ?x = (- 5))

Sentence: (exists ?x:Int
            (= ?x:Int
               (- 5)))
```

Real numerals are written in the usual format, with dots embedded in them to mark the fractional part of the number:

```
> 3.14

Term: 3.14

> (?x = .158)

Term: (= ?x:Real 0.158)
```

There are five predefined binary function symbols that take numeric terms as arguments: + (addition), - (subtraction), * (multiplication), / (division), and % (remainder). They are overloaded so that they can be used both with integers and with reals, or indeed with any combination thereof:

```
> (?x + 2)

Term: (+ ?x:Int 2)

> (2.3 * ?x)

Term: (* 2.3 ?x:Real)
```

These symbols adhere to the usual precedence and associativity conventions:

```
> (2 * 7 + 35)

Term: (+ (* 2 7)
         35)
```

The subtraction symbol can be used both with one and with two arguments:

```
> (- 2)

Term: (- 2)

> (7 - 5)

Term: (- 7 5)
```

As a unary symbol it represents integer/real negation, and as a binary symbol it represents subtraction. Likewise, + can be used both as a unary and as a binary symbol.

There are also function symbols for the usual comparison operators: < (less than), > (greater than), <= (less than or equal to), and >= (greater than or equal to). They are likewise overloaded:

```
> (2 <= 5)

Term: (<= 2 5)

> (forall ?x . ?x + 1 > ?x)

Sentence: (forall ?x:Int
            (> (+ ?x:Int 1)
               ?x:Int))
```

Note that if we wanted ?x to range over Real in the last sentence, we could annotate the quantified variable:

```
> (forall ?x:Real . ?x + 1 > ?x)

Sentence: (forall ?x:Real
            (> (+ ?x:Real 1)
               ?x:Real))
```

Alternatively, we could just write the numeral 1 in real-number format, and Athena would then automatically treat ?x as ranging over Real:

```
> (forall ?x . ?x + 1.0 > ?x)

Sentence: (forall ?x:Real
            (> (+ ?x:Real 1.0)
               ?x:Real))
```

There are also predefined procedures for performing the usual computations with numbers: plus, minus, times, div, mod, less?, greater?, and equal?. We stress that these are not function symbols, that is, they do not make terms but actually perform the underlying computations:

```
> (1 plus 2 times 3)

Term: 7

> (100 div 2)

Term: 50

> (3.0 div 1.5 equal? 2)

Term: true
```

The procedure `equal?` (also defined as `equals?`) is a generic equality test that can be applied to any two values, not just numbers.

2.9 Meta-identifiers

We have already seen that several sorts are predefined in Athena. These include domains such as `Int` and `Real`, as well as datatypes such as `Boolean`. Another built-in domain is `Ide`, the domain of *meta-identifiers*. There are infinitely many constants pre-declared for `Ide`. These are all of the form `'`*I*, where *I* is a regular Athena identifier. For example:

```
'foo
'x
'233
'*
'sd8838jd@!
```

These are called meta-identifiers.[17] Examples:

```
> 'x

Term: 'x

> (println (sort-of 'x))

Ide

Unit: ()

> (exists ?x . ?x = 'foo)
```

17 These should not be confused with sort variables. Both have printed representations that start with the quote symbol `'`, but they are very different kinds of things. A meta-identifier such as `'foo` is an Athena *term*. A sort variable such as `'S35` represents a sort, and sorts are what terms have. The sort of every meta-identifier is `Ide`.

```
Sentence: (exists ?x:Ide
            (= ?x:Ide 'foo))
```

Among many other uses, meta-identifiers can represent the variables of some object language whose abstract syntax needs to be modeled by an Athena datatype. As a brief example, consider the untyped λ-calculus, where an expression is either a variable (identifier), or an abstraction, or an application. That abstract grammar could be represented by the following Athena datatype:

```
datatype Exp := (Var Ide) | (Lambda Ide Exp) | (App Exp Exp)
```

Then the term

```
(Lambda 'x (Var 'x))
```

would represent the identity function.[18]

2.10 Expressions and deductions

In this section we discuss some common kinds of expressions and deductions. The most basic kind of expression is a *procedure call*, or procedure application. The general syntax form of a procedure call is:

$$(E\ F_1 \cdots F_n) \tag{2.6}$$

for $n \geq 0$, where E is an expression whose value must be a procedure and the arguments $F_1 \cdots F_n$ are phrases whose values become the inputs to that procedure.

Similarly, the most basic kind of deduction is a *method call*, or method application. The syntax form of a method call is

$$(\texttt{apply-method}\ E\ F_1 \cdots F_n) \tag{2.7}$$

where E is an expression that must denote a *method* M and the arguments $F_1 \cdots F_n$, $n \geq 0$, are phrases whose values become the inputs to M. The exclamation mark ! is typically used as a shorthand for the reserved word **apply-method**, so the most common syntax form of a method call is this:

$$(!E\ F_1 \cdots F_n). \tag{2.8}$$

Observe that both in (2.6) and in (2.7) the arguments are *phrases* $F_1, \ldots, F_n$, which means that they can be *either* deductions *or* expressions.[19] By contrast, that which gets applied in both cases must be the value of an *expression* E, the syntactic item immediately preceding the arguments $F_1 \cdots F_n$. The reason why we have an expression E there and not

18 Note that `Lambda` is different from the keyword **`lambda`**. Athena identifiers are case-sensitive.

19 Recall that a phrase F is either an expression E or a deduction D.

a phrase F is that deductions can only produce sentences, and what we need in that position is something that can be applied to arguments—such as a procedure or a method, but not a sentence.

To evaluate a method call of the form (2.7) in an assumption base β, we first need to evaluate E in β and obtain some method M from it; then we evaluate each argument phrase F_i in β, obtaining some value V_i from it (with a small but important twist that we discuss in Section 2.10.2); and finally we apply the method M to the argument values $V_1, \ldots, V_n$. What exactly is a method? A method can be viewed as the deductive analogue of a procedure: It takes a number of input values, carries out an arbitrarily long *proof* involving those values, and eventually yields a certain *conclusion* p. Alternatively, the method might (a) halt with an error message, or (b) diverge (i.e., get into an infinite loop). These are the only three possibilities for the outcome of a method application.

The simplest available method is the nullary method `true-intro`. Its result is always the constant `true`, no matter what the assumption base is:

```
> (!true-intro)

Theorem: true
```

The next simplest method is the unary reiteration method `claim`. This method takes an arbitrary sentence p as input, and if p is in the assumption base, then it simply returns it back as the output:

```
1  assert true
2
3  > (!claim true)
4
5  Theorem: true
```

Note that the result of any deduction D is always reported as a *theorem*. That is because the result of D is guaranteed to be a logical consequence of the assumption base in which D was evaluated.

Every theorem produced by evaluating a deduction at the top level is automatically added to the global assumption base. In this case, of course, the global assumption base already contains the sentence `true` (because we asserted it on line 1), so adding the theorem `true` to it will not change its contents.

Claiming a sentence p will succeed if and *only* if p itself is in the assumption base. Suppose, for instance, that starting from an empty assumption base we **`assert`** p and then we claim $(p \mid p)$. We cannot expect that claim to succeed, despite the fact that $(p \mid p)$ is logically equivalent to p. The point here is that `claim` will check to see if *its input sentence* p is in the assumption base, not whether the assumption base contains some sentences from which p could be inferred, or some sentence that is equivalent to p. For example:

```
clear-assumption-base

assert true

> (!claim true)

Theorem: true

> (!claim (true | true))

Error, standard input, 1.9: Failed application of claim---the sentence
(or true true) is not in the assumption base.
```

More precisely, the identity required by `claim` is alpha-equivalence. That is, applying `claim` to *p* will succeed iff the assumption base contains a sentence that is alpha-equivalent to *p*:

```
assert (forall ?x . ?x = ?x)

> (!claim (forall ?y . ?y = ?y))

Theorem: (forall ?y:'S
           (= ?y:'S ?y:'S))
```

Alpha-equivalence is the relevant notion of identity for all primitive Athena methods.

Most other primitive Athena methods are either *introduction* or *elimination* methods for the various logical connectives and quantifiers. We describe their names and semantics in detail in chapters 4 and 5, but we will give a brief preview here, starting with the methods for conjunction introduction and elimination. Conjunction introduction is performed by the binary method `both`, which takes any two sentences *p* and *q*, and provided that both of them are in the assumption base (again, up to alpha-equivalence), it produces the conclusion (and *p* *q*):

```
declare A, B, C: Boolean

assert A, B

> (!both A B)

Theorem: (and A B)
```

Conjunction elimination is performed by the two unary methods `left-and` and `right-and`. The former takes a conjunction (and p_1 p_2) and returns p_1, provided that (and p_1 p_2) is in the assumption base (an error occurs otherwise). The method `right-and` behaves likewise, except that it produces the right conjunct instead of the left:

```
clear-assumption-base

assert (A & B)
```

```
> (!left-and (A & B))

Theorem: A

> (!right-and (A & B))

Theorem: B

> (!right-and (C & B))

Error, standard input, 1.2: Failed application of right-and---the sentence
(and C B) is not in the assumption base.

> (!left-and (A | B))

Error, standard input, 1.2: Failed application of left-and---the given
sentence must be a conjunction, but here it was a disjunction: (or A B).
```

Another unary primitive method is dn, which performs double-negation elimination. It takes a premise p of the form (not (not q)), and provided that p is in the assumption base, it returns q:

```
assert p := (~ ~ A)

> (!dn p)

Theorem: A
```

There are several kinds of deductions beyond method applications. In what follows we present a brief survey of the most common deductive forms other than method applications.

2.10.1 Compositions

One of the most fundamental proof mechanisms is *composition*, or sequencing: Assembling a number of deductions $D_1, \ldots, D_n$, $n > 0$, to form the compound deduction

$$\{D_1; \; \cdots \; ;D_n\}. \tag{2.9}$$

To evaluate such a deduction in an assumption base β, we first evaluate D_1 in β. If and when the evaluation of D_1 results in a conclusion p_1, we go on to evaluate D_2 in $\beta \cup \{p_1\}$, that is, in *β augmented with the conclusion of the first deduction*. When we obtain the conclusion p_2 of D_2, we go on to evaluate D_3 in

$$\beta \cup \{p_1, p_2\},$$

namely, in the initial assumption base augmented with the conclusions of the first two deductions; and so on.[20] Thus, each deduction D_{i+1} is evaluated in an assumption base that incorporates the conclusions of all preceding deductions $D_1, \ldots, D_i$. The conclusion of each intermediate deduction thereby serves as a *lemma* that is available to all subsequent deductions. The conclusion of the last deduction D_n is finally returned as the conclusion of the entire sequence. For example, suppose we evaluate the following code in the empty assumption base:

```
1  > assert (A & B)
2
3  The sentence
4  (and A B)
5  has been added to the assumption base.
6
7  > {
8      (!left-and  (A & B));                    # This gives A
9      (!right-and (A & B));                    # This gives B
10     (!both B A)                              # And finally, (B & A)
11   }
12
13 Theorem: (and B A)
```

The **assert** directive adds the conjunction `(A & B)` to the initially empty assumption base. Accordingly, the starting assumption base in which the proof sequence on lines 7–11 will be evaluated is $\beta = \{$`(A & B)`$\}$. The proof sequence itself consists of three method calls. The first one, the application of `left-and` to `(A & B)` on line 8, successfully obtains the conclusion `A`, because the required premise `(A & B)` is in the assumption base in which this application is evaluated. Then the second element of the sequence, the application of `right-and` on line 9, is evaluated in the starting assumption base β augmented with the conclusion of the first deduction, that is, in $\{$`(A & B)`$,$`A`$\}$. That yields the conclusion `B`. Finally, we go on to evaluate the last element of the sequence, the application of `both` on line 10. That application is evaluated in β augmented with the conclusions of the two previous deductions, that is, in

$$\{\texttt{(A \& B)}, \texttt{A}, \texttt{B}\}.$$

Since both of its inputs are in that assumption base, `both` happily produces the conclusion `(B & A)`, which thus becomes the conclusion of the entire composition, reported as a theorem on line 13.

At the end of the entire composite proof, only the final conclusion `(B & A)` is retained and added to the starting assumption base β. The intermediate conclusions generated

20 Strictly speaking, we should say that D_2 is evaluated *in a copy of* β augmented with p_1; D_3 is evaluated in a copy of β augmented with p_1 and p_2; and so on. We don't statefully (destructively) add each intermediate conclusion to the initial assumption base itself.

during the evaluation of the composition (namely, the sentences `A` and `B` generated by the `left-and` and `right-and` applications) are *not* retained. For instance, here is what we get if we try to see whether `A` is in the assumption base immediately after the above:

```
> (holds? A)

Term: false
```

In general, whenever we evaluate a deduction D at the top level and obtain a result p:

```
> D

Theorem: p
```

only the conclusion p is added to the global assumption base. Any auxiliary conclusions derived in the course of evaluating D are discarded.

As you might have noticed, when we talk about evaluating a deduction D we often speak of the contents of "the assumption base." For instance, we say that evaluating an application of `left-and` to a sentence `(`p `&` q`)` produces the conclusion p provided that the conjunction `(`p `&` q`)` is in "the assumption base." This does not necessarily refer to the global assumption base that Athena maintains at the top level. Rather, it refers to *the assumption base in which the deduction D is being evaluated*, which may or may not be the global assumption base. For instance, if D happens to be the third member of a composite deduction, then the assumption base in which D will be evaluated will not be the global assumption base; it will be some superset of it. We will continue to speak simply of "the assumption base" in order to keep the exposition simple, but this is an important point to keep in mind.

We refer to a deduction of the form (2.9) as an *inference block*. Each D_i is a *step* of the block. This is not reflected in (2.9), but in fact a step does not have to be a deduction D; it may be an expression E. So, in general, a step of an inference block can be an arbitrary phrase F. The very last step, however, must be a deduction.

Even more generally, a step of an inference block may be named, so that we can refer to the result of that step later on in the block by its given name. A named step is of the form I := F, where I is an identifier or the wildcard pattern (underscore), and F is a phrase. Using this feature, we could express the above proof block as follows:

```
{
  p1 := (!left-and A-and-B);
  p2 := (!right-and A-and-B);
  (!both p2 p1)
}
```

However, we encourage the use of **`let`** deductions instead of inference blocks; we will discuss **`let`** shortly (Section 2.10.3). A **`let`** proof can do anything that an inference block can do, but it is more structured and usually results in more readable code.

2.10.2 Nested method calls

Procedure calls can be nested, that is, the arguments to a procedure can themselves be procedure calls. This is a common feature of all higher-level programming languages and a style that is especially emphasized in functional languages. Similarly, the arguments to a method can themselves be method calls or other deductions. (Arguments to a method can also be arbitrary expressions.) Suppose that a deduction D appears as an argument to an application of some method M:

$$(!M \ \cdots D \cdots).$$

To evaluate such a method call in an assumption base β, we first need to evaluate every argument in β; in particular, we will need to evaluate D in β, obtaining from it some conclusion p. Now, when all the arguments have been evaluated and we are finally ready to apply M to their respective values, that application will occur in β *augmented with* p. Thus, the conclusion of D will be available as a lemma by the time we come to apply M. This is, therefore, a basic mechanism for proof composition.

For example, suppose we have defined `conj` as `(A & (B & C))` and consider the deduction

$$\texttt{(!left-and (!right-and conj))} \tag{2.10}$$

in an assumption base β that contains only the premise `conj`:

$$\beta = \{\texttt{(A \& (B \& C))}\}.$$

Here, the deduction `(!right-and conj)` appears directly as an argument to `left-and`. To evaluate (2.10) in β, we begin by evaluating the argument `(!right-and conj)` in β. That will successfully produce the conclusion `(B & C)`, since the required premise is in β. We are now ready to apply the outer method `left-and` to `(B & C)`; but this application will take place in β *augmented with* `(B & C)`, that is, in the assumption base

$$\beta' = \beta \cup \{\texttt{(B \& C)}\} = \{\texttt{(A \& (B \& C))}, \texttt{(B \& C)}\},$$

and hence it will successfully produce the conclusion `B`:

```
clear-assumption-base

assert conj := (A & (B & C))

> (!left-and (!right-and conj))

Theorem: B
```

In general, every time a deduction appears as an argument to a method call, the conclusion of that deduction will appear in the assumption base in which the method will be applied.

2.10.3 Let expressions and deductions

Composing expressions with nested procedure calls is common in the functional style of programming, but Athena also allows for a more structured style using **let** expressions, which let us *name* the results of intermediate computations. Similarly, in addition to composing deductions with nested method calls, Athena also permits **let** deductions, which likewise allow for naming intermediate results. The most common syntax form of the **let** construct is:[21]

$$\texttt{let}\ \{I_1 := F_1; \cdots ; I_n := F_n\}\ F \tag{2.11}$$

where $I_1, \ldots, I_n$ are identifiers and $F_1, \cdots, F_n$ and F are phrases. If F, which is called the *body* of the **let** construct, is an expression, then so is the whole **let** construct. And if the body F is a deduction, then the whole **let** construct is also a deduction. So whether or not (2.11) is a deduction depends entirely on whether or not the body F is a deduction.

An expression or deduction of this form is evaluated in a given assumption base β as follows: We first evaluate the phrase F_1. Now, F_1 is either a deduction or an expression. If it is an expression that produces a value V_1, then we just bind the identifier I_1 to V_1 and move on to evaluate the next phrase, F_2, in β. But if F_1 is a deduction that produces some conclusion p_1, we not only bind I_1 to p_1, but we also go on to evaluate the next phrase F_2 in β *augmented with* p_1. We then do the same thing with F_2. If it is an expression, we simply bind I_2 to the value of F_2 and proceed to evaluate F_3; but if it is a deduction, we bind I_2 to the conclusion of F_2, call it p_2, and then move on to evaluate F_3 in β augmented with p_1 *and* p_2. Thus, the conclusion of every intermediate deduction becomes available as a lemma to all subsequent deductions, including the body F when that is a deduction. Moreover, if the conclusion of an intermediate deduction happens to be a conjunction p_i, then all the conjuncts of p_i (and their conjuncts, and so on) are also inserted into the assumption base before moving on to subsequent deductions, and hence they also become available as lemmas.

Here is an example of a **let** expression:

```
> let { a := 1;
        b := (a plus a)
      }
    (b times b)

Term: 4
```

Here, `(a plus a)` is evaluated with `a` bound to `1`, producing `2`; `b` is then bound to `2`, so the body `(b times b)` evaluates to `4`. The entire **let** phrase is an expression because the body `(b times b)` is an expression (a procedure application, specifically).

An example of a **let** deduction is:

---

21 We will see later (in Appendix A) that more general *patterns* can appear in place of the identifiers $I_1, \ldots, I_n$.

```
assert hyp := (male peter & female ann)

> let { left  := (!left-and hyp);
        right := (!right-and hyp)
      }
    (!both right left)

Theorem: (and (female ann)
              (male peter))
```

Here the call to both succeeds precisely because it is evaluated in an assumption base that contains the results of the two intermediate deductions—the calls to `left-and` and `right-and`. The entire **let** phrase is a deduction because its body is a deduction (a method application, specifically).

In the expression example, all of the phrases involved are expressions, and in the deduction example all of the phrases are deductions. But any mixture of expressions and deductions is allowed. For example,

```
assert hyp := (A & B)

> let { goal := (B & A);
        _ := (print "Proving: " goal);
        _ := (!right-and hyp);          # this proof step deduces B
        _ := (!left-and hyp) }          # and this one derives A
     (!both B A)

Proving:
(and B A)

Theorem: (and B A)
```

This example also illustrates that when we do not care to give a name to the result of an intermediate phrase F_i, we can use the wildcard pattern _ as the corresponding identifier.

2.10.4 Conclusion-annotated deductions

Sometimes a proof can be made clearer if we announce its intended conclusion ahead of time. This is common in practice. Authors often say "and now we derive p as follows: $\cdots$" or "p follows by $\cdots$." In Athena such annotations can be made with the **conclude** construct, whose syntax is **conclude** p D. Here D is an arbitrary deduction and p is its intended conclusion.

To evaluate a deduction of this form in an assumption base β, we first evaluate D in β. If and when we obtain a conclusion q, we check to ensure that q is the same as the expected conclusion p (up to alpha-equivalence). If so, we simply return p as the final result. If not, we report an error to the effect that the conclusion was different from what was announced:

```
assert p := (A & B)

> conclude A
    (!left-and p)

Theorem: A

> conclude B
    (!left-and p)

standard input:1:2: Error: Failed conclusion annotation.
The expected conclusion was:
B
but the obtained result was:
A.
```

In its full generality, the syntax of this construct is **conclude** E D, where E is an arbitrary expression that denotes a sentence. This means that E may spawn an arbitrary amount of computation, as long as it eventually produces a sentence p. We then proceed normally by evaluating D to get a conclusion q and then comparing p and q for alpha-equivalence. In addition, a name I may optionally be given to the conclusion annotation, whose scope becomes the body D:

conclude I := E D.

This is often useful with top-level uses of **conclude**, when we prove a theorem and define its name at the same time:

```
conclude plus-commutativity :=
  (forall ?x ?y . ?x + ?y = ?y + ?x)
  D
```

2.10.5 Conditional expressions and deductions

In both expressions and deductions, conditional branching is performed with the **check** construct. The syntax of a **check** expression is

$$\textbf{check } \{F_1 \texttt{=>} E_1 \mid \cdots \mid F_n \texttt{=>} E_n\} \qquad (2.12)$$

where the F_i => E_i pairs are the *clauses* of (2.12), with each clause consisting of a *condition* F_i and a corresponding *body* expression E_i. A **check** deduction has the same form, but with deductions D_i as clause bodies:

$$\textbf{check } \{F_1 \texttt{=>} D_1 \mid \cdots \mid F_n \texttt{=>} D_n\}.$$

To evaluate a **check** expression or deduction, we evaluate the conditions $F_1, \ldots, F_n$, in that order. If and when the evaluation of some F_i produces `true`, we evaluate the corresponding body E_i or D_i and return its result as the result of the entire expression or deduction. The last condition, F_n, may be the keyword **else**, which is treated as though it were `true`. It is an error if no F_i produces `true` and there is no **else** clause at the end.

```
assert A

> check {(holds? false) => 1 | (holds? A) => 2 | else => 3}

Term: 2
```

2.10.6 Pattern-matching expressions and deductions

Another form of conditional branching is provided by pattern-matching expressions or deductions. A pattern-matching expression has the form

$$\texttt{match}\ F\ \{\pi_1 \texttt{=>} E_1\ \texttt{|}\ \cdots\ \texttt{|}\ \pi_n \texttt{=>} E_n\} \tag{2.13}$$

where the phrase F is called the *discriminant*, while the π_i => E_i pairs are the *clauses* of (2.13), with each clause consisting of a *pattern* π_i and a corresponding *body* expression E_i. For a description of the various forms of patterns and the details of the pattern-matching algorithm, see Appendix A.4; additional discussion and examples can be found in this chapter in Section 2.11.

The syntax of a pattern-matching deduction is the same, except that the body of each clause must be a deduction:

$$\texttt{match}\ F\ \{\pi_1 \texttt{=>} D_1\ \texttt{|}\ \cdots\ \texttt{|}\ \pi_n \texttt{=>} D_n\}.$$

A pattern-matching expression or deduction is evaluated in a given environment ρ and assumption base β as follows. We first evaluate the discriminant F, obtaining from it a value V. We then try to *match* V against the given patterns $\pi_1, \ldots, \pi_n$, in that order. If and when we succeed in matching V against some π_i, resulting in a number of bindings, we go on to evaluate the corresponding body E_i or D_i in ρ augmented with these bindings, and in β.[22] The result produced by that evaluation becomes the result of the entire pattern-matching expression or deduction. An error occurs if the discriminant value V does not match any of the patterns.

```
> match [1 2] {
    [] => 99
  | (list-of h _) => h
  }

Term: 1
```

22 If the discriminant F is a deduction that produces a conclusion p, and the body is a deduction D_i, then D_i will be evaluated in $\beta \cup \{p\}$. In that case, therefore, the conclusion of the discriminant will serve as a lemma during the evaluation of D_i.

```
> match [1 2] {
    [] => (!claim false)
  | (list-of _ _) => (!true-intro)
  }

Theorem: true
```

2.10.7 Backtracking expressions and deductions

A form of backtracking is provided by **try** expressions and deductions. A **try** expression has the form

$$\texttt{try}\ \{E_1\ \mid\ \ldots\ \mid\ E_n\} \tag{2.14}$$

where $n > 0$. Such an expression does what its name implies: It tries to evaluate each expression E_i in turn, $i = 1, \ldots, n$, until one succeeds, that is, until some E_i is found that successfully produces a value V_i. At that point V_i is returned as the result of the entire **try** expression. It is an error if all n expressions fail. For example:

```
> try { (4 div 0) | 2 }

Term: 2

> try { (4 div 0) | 25 | (head []) }

Term: 25

> try { (4 div 0) | (head []) }

standard input:1:2: Error: Try expression error; all alternatives failed.
```

A **try** deduction has the same form as (2.14), except that the alternatives are deductions rather than expressions:

$$\texttt{try}\ \{D_1\ \mid\ \ldots\ \mid\ D_n\}$$

for $n > 0$. For example:

```
> try { (!left-and false) |
        (!true-intro)      |
        (!right-and (true ==> false))}

Theorem: true

> try { (!left-and false) |
        (!right-and (true ==> false))}

standard input:1:2: Error: Try deduction error; all alternatives failed.
```

2.10.8 Defining procedures and methods

Users can define their own procedures with the **lambda** construct, and then use them as if they were primitive procedures. For instance, here is a procedure that computes the square of a given number:[23]

```
> define square := lambda (n) (n times n)

Procedure square defined.

> square

Procedure: square (defined at standard input:1:32)

> (square 4)

Term: 16
```

The general form of a **lambda** expression is

$$\texttt{lambda}\ (I_1 \cdots I_n)\ E$$

where $I_1 \cdots I_n$ are identifiers, called the *formal parameters* of the **lambda** expression, and E is an expression, called the *body* of the **lambda**.

At the top level it is not necessary to define procedures with **lambda**. An alternative notation is the following:

```
> define (square n) := (n times n)

Procedure square defined.
```

or in more traditional Lisp notation:

```
(define (square n)
   (times n n))
```

Any of these alternatives gives the name `square` to the procedure

```
lambda (n) (n times n).
```

It is also possible to use the **lambda** construct directly, without giving a name to the procedure it defines. In that case we say that the procedure is *anonymous*. This is particularly useful when we want to pass a procedure as an argument to another procedure. For example, the built-in `map` procedure takes a unary procedure f as its first argument and a list L as its second, and returns the list formed by applying f to each element of L. If $L = [V_1 \cdots V_n]$, then

$$(\texttt{map}\ f\ L) = [(f\ V_1) \cdots (f\ V_n)].$$

23 For simplicity, we often use "number" synonymously with "numeric literal" (of sort `Int` or `Real`).

For example, since we already defined square, we can write

```
> (map square [1 2 3 4 5])

List: [1 4 9 16 25]
```

But we could also pass the squaring procedure to map anonymously:

```
> (map lambda (n) (n times n)
       [1 2 3 4 5])

List: [1 4 9 16 25]
```

Defining a procedure is often a process of abstraction: By making an expression the body of a procedure, with some (not necessarily all) of the free identifiers of the expression as formal parameters, we abstract it into a general algorithm that can be applied to other inputs.

Likewise, a given deduction can often be abstracted into a general method that can be applied to other inputs. Consider, for instance, a deduction that derives (B & A) from (A & B):

```
let {_ := (!left-and (A & B));
     _ := (!right-and (A & B))}
  (!both B A)
```

It should be clear that there is nothing special here about the atoms A and B. We can replace them with any other sentences *p* and *q* and the reasoning will still go through, as long as the conjunction (*p* & *q*) is in the assumption base. So, just like a particular computation such as the squaring of 4 can be abstracted into a general squaring procedure by replacing the constant 4 by a formal parameter like n, so we can turn the preceding deduction into a general proof *method* as follows:

```
method (p q)
  let {_ := (!left-and (p & q));
       _ := (!right-and (p & q))}
    (!both q p)
```

This method can be applied to two arbitrary conjuncts *p* and *q* and will produce the conclusion

$$(q \,\&\, p),$$

provided that the premise (*p* & *q*) is in the assumption base. While the method could be applied anonymously, it is more convenient to give it a name first:

```
1 clear-assumption-base
2
3 define commute-and :=
4   method (p q)
```

```
5      let {_ := (!left-and (p & q));
6           _ := (!right-and (p & q))}
7        (!both q p)
8
9  assert (B & C)
10
11 > (!commute-and B C)
12
13 Theorem: (and C B)
14
15 > (!commute-and A B)
16
17 standard input:3:15: Error: Failed application of left-and---the sentence
18 (and A B) is not in the assumption base.
```

The last example failed because the necessary premise `(A & B)` was not in the assumption base at the time when the method was applied.

These examples illustrate an important point: When a defined method M is called, M is applied in the assumption base in which the call takes place, not the assumption base in which M was defined.[24] Here, when `commute-and` was defined (on lines 3–7), the assumption base was empty. But by the time the call on line 11 is made, the assumption base contains exactly one sentence, namely `(B & C)`, so *that* is the logical context in which that application of `commute-and` takes place. Later, when `commute-and` is called on line 15, the assumption base contains exactly two sentences, `(B & C)` as well as `(C & B)`, the theorem produced by the preceding deduction.

A stylistic note: In order to make methods composable, it is preferable, when possible, to define a method M so that its only inputs are the premises that it requires. Doing so ensures that M can take deductions as arguments, which will serve to establish the required premises prior to the application of M (by the semantics of nested method calls, as discussed in Section 2.10.2). Accordingly, the preceding method is better written so that it takes the required conjunction as its sole input, rather than the two individual conjuncts as two separate arguments:

```
define commute-and' :=
  method (premise)
    match premise {
      (p & q) => let {_ := (!left-and premise);
                      _ := (!right-and premise)}
                   (!both q p)
    }
```

This version uses pattern matching to take apart the given premise and retrieve the individual conjuncts, after which the reasoning proceeds as before. The interface and style

24 We thus say that method closures have static name scoping but *dynamic assumption scoping*.

of commute-and' would normally be preferred over that of commute-and on composability grounds. For instance, suppose the assumption base contains (~ (~ (A & B))) and we want to derive (B & A). Using the second version, we can express the proof in a single line by composing double negation and conjunction commutation:

```
assert premise :=  (~ ~ (A & B))

> (!commute-and' (!dn premise))

Theorem: (and B A)
```

Such composition is not possible with the former version.

The same alternative notation that is available for defining procedures can also be used for defining methods: Instead of writing

define M := **method** $(I_1 \cdots I_n)$ D

we can write

define $(M\ I_1 \cdots I_n)$:= D,

or, in prefix,

(**define** $(M\ I_1 \cdots I_n)$ D).

For instance:

```
1  > define (commute-and p q) :=
2      let {_ := (!left-and (p & q));
3           _ := (!right-and (p & q))}
4        (!both q p)
5
6  Method commute-and defined.
```

How does Athena know that this is a method and not a procedure (observe that it responded by acknowledging that a *method* by the name of commute-and was defined)? It can tell because the body (lines 2–4) is a deduction. And how can it tell that? In general, how can we tell whether a given phrase F is an expression or a deduction? Recall that expressions and deductions are distinct syntactic categories; there is one grammar for expressions and another for deductions. In most cases, a deduction is indicated just by the leading keyword (the reserved word with which the phrase begins):

apply-method (usually abbreviated as !)
assume
pick-any
pick-witness

```
pick-witnesses
generalize-over
with-witness
suppose-absurd
conclude
by-induction
datatype-cases
```

(Don't worry if you don't yet recognize some of these; we will explain all of them in due course.) In other cases, when the beginning keyword is `let`, `letrec`, `check`, `match`, or `try`, it is necessary to peek inside the phrase. A `let` or `letrec` construct is a deduction if and only if its body is. With a `check` or `match` phrase we simply look at its first clause body, since the clause bodies must be all deductions or all expressions. With a `try` phrase, we look at its first alternative. Finally, any phrase not covered by these rules (such as the unit `()`, or a meta-identifier) is an expression. Thus, the question of whether a given phrase is a deduction or an expression can be mechanically answered with a trivial computation, and in Athena this is done at parsing time.

It is important to become proficient in making that determination, to be able to immediately tell whether a given phrase is an expression or a deduction. Sometimes Athena beginners are asked to write a proof D of some result and end up accidentally writing an expression E instead. Readers are therefore advised to review the above guidelines and then tackle Exercise 2.1 in order to develop this skill.

2.11 More on pattern matching

Recall that the general form of a `match` deduction is

$$
\begin{array}{l}
\texttt{match } F \; \{ \\
\quad \pi_1 \;\; \texttt{=>} \;\; D_1 \\
\texttt{|} \; \pi_2 \;\; \texttt{=>} \;\; D_2 \\
\qquad \vdots \\
\texttt{|} \; \pi_n \;\; \texttt{=>} \;\; D_n \\
\}
\end{array}
$$

where $\pi_1, \ldots, \pi_n$ are *patterns* and $D_1, \ldots, D_n$ are deductions. The phrase F is the discriminant whose value will be matched against the patterns. The pattern-matching algorithm is described in detail in Section A.4, but some informal remarks and a few examples will be useful at this point. We focus on `match` deductions here, but what we say will also apply to expressions.

As we already mentioned in Section 2.10.6, to evaluate a **match** proof of the above form in an assumption base β and an environment ρ,[25] we start by evaluating the discriminant F in β and ρ, to obtain a value V_F.[26] We then start comparing the value V_F against the patterns π_i sequentially, $i = 1, \ldots, n$, to determine if it matches any of them. When we encounter the first pattern π_j that is successfully matched by V_F under some set of bindings $\rho' = \{I_1 \mapsto V_1, \ldots, I_k \mapsto V_k\}$, we evaluate the corresponding deduction D_j in the context of β and ρ *augmented with* ρ';[27] the result of that evaluation becomes the result of the entire **match** deduction.

In general, an attempt to match a value V against a pattern π will either result in failure, indicating that π does not reflect the structure of V; or else it will produce a matching environment $\rho' = \{I_1 \mapsto V_1, \ldots, I_k \mapsto V_k\}$, signifying that V successfully matches π under the bindings $I_j \mapsto V_j, j = 1, \ldots, k$. In the latter case we say that V *matches* π *under* ρ'.

The underscore _ is the wildcard pattern that is matched by any value whatsoever.

Suppose, for example, that the discriminant is the sentence `(true | ~ false)`. This sentence:

- matches the pattern `(p1 | p2)` under {`p1` $\mapsto$ `true`, `p2` $\mapsto$ `(not false)`};
- matches the pattern `p` under {`p` $\mapsto$ `(or true (not false))`};
- matches `(or true (not q))` under {`q` $\mapsto$ `false`};
- matches all three patterns `_`, `(or _ _)`, and `(or _ (not _))` under the empty environment {}.

Here are the first three examples in Athena:

```
> define discriminant := (true | ~ false)

Sentence discriminant defined.

> match discriminant {
    (p1 | p2) => (print "Successful match with p1: " p1 "\nand p2: " p2)
  }

Successful match with p1: true
and p2:
(not false)
```

25 Recall that an environment is a finite function mapping identifiers to values. Athena maintains a global environment that holds all the definitions made by the user (as well as built-in definitions). For instance, when a user issues a directive like **define** `p := (true | false)`, the global environment is extended with the binding `p` $\mapsto$ `(or true false)`. Since it is a finite function, an environment can be viewed as a finite set of identifier-value bindings, where each binding is an ordered pair consisting of an identifier I and a value V. We typically write such a binding as $I \mapsto V$.

26 We ignore the store and symbol set here because they do not play a central role in what we are discussing.

27 The result of augmenting (or "extending") an environment ρ_1 with another environment ρ_2 is the unique environment that maps an identifier I to $\rho_2(I)$, if I is in the domain of ρ_2; or to $\rho_1(I)$ otherwise.

```
Unit: ()

> match discriminant {
    p => (print "Successful match with p: " p)
  }

Successful match with p:
(or true
    (not false))

Unit: ()

> match discriminant {
    (or true (not q)) => (print "Successful match with q: " q)
  }

Successful match with q: false

Unit: ()
```

Patterns are generally written in prefix, but binary sentential constructors (as well as function symbols) can also appear inside patterns in infix, as seen in the first example above. Thus, for instance, `(and p1 (not (or p2 p3)))` and

`(p1 & (~ (p2 | p3)))`

are two equivalent patterns. However, patterns must always be fully parenthesized, so we cannot omit parentheses and rely on precedence and associativity conventions to determine the right grouping. For example, a pattern such as `(p & q ==> r)` will not have the intended effect; the pattern should be written as `((p & q) ==> r)` instead.

Sentences are not the only values on which we can perform pattern matching. We can also pattern-match terms, lists, and any combination of these. Consider, for instance, the following patterns:

1. `t`
2. `(mother t)`
3. `(mother (father person))`
4. `(father _)`

The term `(mother (father ann))` matches the first pattern under

{`t` ↦ `(mother (father ann))`};

it matches the second pattern under {`t` ↦ `(father ann)`}; it matches the third pattern under {`person` ↦ `ann`}; and it does not match the fourth pattern.

```
define discriminant := (mother father ann)

> match discriminant {
    t => (print "Matched with t: " t)
  }

Matched with t:
(mother (father ann))

Unit: ()

> match discriminant {
    (mother t) => (print "Matched with t: " t)
  }

Matched with t:
(father ann)

Unit: ()

> match discriminant {
    (mother (father t)) => (print "Matched with t: " t)
  }

Matched with t:  ann

Unit: ()

> match discriminant {
    (father _) => (print "Matched...")
  }

standard input:1:1: Error: match failed---the term
(mother (father ann))
did not match any of the given patterns.
```

Term variables such as `?x:Boolean` are themselves Athena values, and hence can become bound to pattern variables. For example, the term `(union ?s1 ?s2)` matches the pattern `(union left right)` under the bindings {`left` ↦ `?s1:Set`, `right` ↦ `?s2:Set`}. Any occurrence of a term variable inside a pattern acts as a constant—it can only be matched by that particular variable. Hence, the pattern `(S ?n)` can only be matched by one value: the term `(S ?n)`.

Quantified sentences can also be straightforwardly decomposed with patterns. Consider, for instance, the pattern `(forall x p)`. The sentence

`(forall ?human . male father ?human)`

will match this pattern under the bindings

{`x` ↦ `?human:Person`, `p` ↦ `(male (father ?human:Person))`}.

The sentence

```
(forall ?x . exists ?y . ?x subset ?y & ?x =/= ?y)
```

will also match, under

{x ↦ ?x:Set, p ↦ (exists ?y:Set . ?x subset ?y & ?x =/= ?y)}.

Any identifier inside a pattern that is not a function symbol or a sentential constructor or quantifier (such as `if`, `forall`, etc.) is interpreted as a pattern variable, and can become bound to a value during pattern matching. For instance, assuming that `joe` has *not* been declared as a function symbol, the term `(father ann)` will match the pattern `(father joe)` under {joe ↦ ann}. But if `joe` has been introduced as a function symbol, then it can no longer serve as a pattern variable, that is, it cannot become bound to any values. It can still appear inside patterns, but it can only match itself—the function symbol `joe`. Thus, the only value that will match the pattern `(mother joe)` at that point (after `joe` has been declared as a function symbol) is the term `(mother joe)` itself, and nothing else. So it is crucial to distinguish function symbols (and of course sentential constructors and quantifiers) from regular identifiers inside patterns.

Athena also supports nonlinear patterns, where multiple occurrences of the same pattern variable are constrained to refer to the same value. Consider, for instance, the pattern `(p | p)`. The sentence `(true | true)` matches this pattern under {p ↦ true}; but the sentence `(true | false)` does not. Likewise, the terms `(union null null)` and

```
(union (intersection ?x ?y) (intersection ?x ?y))
```

match the pattern `(union s s)`, but the term `(union ?foo null)` does not.

Lists are usually taken apart with two types of patterns: patterns of the form

(**list-of** π_1 π_2)

and those of the form [$\pi_1 \cdots \pi_n$]. The first type of pattern matches any nonempty list

[$V_1 \cdots V_k$]

with $k \geq 1$ and such that V_1 matches π_1 and the tail [$V_2 \cdots V_k$] matches π_2. For example, the three-element list

```
[zero ann (father peter)]
```

matches the pattern (**list-of** head tail) under the bindings

{head ↦ zero, tail ↦ [ann (father peter)]}.

The second kind of pattern, [$\pi_1 \cdots \pi_n$], matches all and only those n-element lists [$V_1 \cdots V_n$] such that V_i matches π_i, $i = 1, \ldots, n$. For instance, `[ann peter]` matches the pattern `[s t]` under $\{$`s` $\mapsto$ `ann`, `t` $\mapsto$ `peter`$\}$; but it does not match the pattern `[s s]`—a nonlinear pattern that is only matched by two-element lists with identical first and second elements.

These types of patterns can be recursively combined in arbitrarily complicated ways. For instance, the pattern

```
[((p & q) ==> (~ r)) (list-of (forall x r) _)]
```

is matched by any two-element list whose first element is a conditional of the form

```
((p & q) ==> (~ r))
```

and whose second element is any nonempty list whose first element is a universal quantification whose body is the sentence `r` that appears as the body of the negation in the consequent of the aforementioned conditional.

An arbitrary sentence of the form

$$(\circ\ p_1 \cdots p_n)$$

with $\circ \in \{$`not`, `and`, `or`, `if`, `iff`$\}$ can match a pattern of the form

$$((\texttt{some-sent-con}\ I)\ \pi_1 \cdots \pi_n),$$

provided that each p_i matches π_i in turn, $i = 1, \ldots, n$, in which case I will be bound to $\circ$:

```
> match (A & ~ B) {
    ((some-sent-con sc) left right) => [sc left right]
  }

List: [and A (not B)]

> match (A ==> B) {
    ((some-sent-con sc) left right) => [sc left right]
  }

List: [if A B]
```

A sentence of the form $(\circ\ p_1 \cdots p_n)$ can also match a pattern of the form

$$((\texttt{some-sent-con}\ I)\ \pi)$$

when π is a list pattern. Here I will be bound to $\circ$ and the list $[p_1 \cdots p_n]$ will be matched against π. For example, the following procedure takes any sentence p, and provided that p is an application of a sentential constructor sc to some subsentences $p_1 \cdots p_n$, it returns a pair consisting of sc and the sentences $p_1 \cdots p_n$ listed in reverse order:

```
> define (break-sentence p) :=
    match p {
     ((some-sent-con pc) (some-list args)) => [pc (rev args)]
    }
```

```
Procedure break-sentence defined.

> (break-sentence (~ true))

List: [not [true]]

> (break-sentence (and A B C))

List: [and [C B A]]

> (break-sentence (A | B))

List: [or [B A]]

> (break-sentence (false ==> true))

List: [if [true false]]

> (break-sentence (iff true true))

List: [iff [true true]]

> (break-sentence true)

Error, standard input, 2.5: match failed---the term true
did not match any of the given patterns.
```

Another useful type of pattern is the **where** pattern, of the form

$$(\pi \ \mathbf{where} \ E),$$

where E is an expression that may contain pattern variables from π. The idea here is that we first match a discriminant value V against π, and if the match succeeds with a set of bindings, then we proceed to evaluate E in the context of those bindings (on top of the environment in which V was obtained). The overall pattern succeeds (with that same set of bindings) if the evaluation of E produces `true` and fails otherwise. For instance, the following matches a list whose head is an even integer:

```
define (first-even? L) :=
  match L {
    ((list-of x _) where (even? x)) => true
  | _ => false
  }
```

We have only scratched the surface of Athena's pattern-matching capabilities here. The subject is covered in greater detail in Section A.4.

2.12 Directives

In addition to expressions and deductions, the user can give various *directives* as input to Athena. These are commands that direct Athena to do something, typically to enter new information or adjust some setting about how it processes its input or how it displays its output. Most such directives have been mentioned already: **load** (page 22), **set-precedence** (page 32), **left-assoc/right-assoc** (page 33), **define** (page 40), **assert** (page 43), **clear-assumption-base/retract** (page 44), and **set-flag** `auto-assert-dt-axioms` (page 48). In this section we describe a couple of **set-flag** directives for controlling output, and the next section continues with the **overload** directive for adapting function symbols to have different meanings depending on context.

- **set-flag** `print-var-sorts` *s*, where *s* is either the string `"on"` or the string `"off"`. The default value is `"on"`. When set to `"off"`, Athena will not print out the sorts of variables. Examples:

```
> set-flag print-var-sorts "off"

OK.

> (?x = ?y)

Term: (= ?x ?y)

> set-flag print-var-sorts "on"

OK.

> (?x = ?y)

Term: (= ?x:'T185 ?y:'T185)
```

- **set-flag** `print-qvar-sorts` *s*, where *s* is again either the string `"on"` or `"off"`. When `print-vars-sorts` is turned off, variable sorts in the body of a quantified sentence are not printed, but variable occurrences that immediately follow a quantifier occurrence continue to have their sorts printed. The printing of even those sorts can be disabled by turning off the flag `print-qvar-sorts`:

```
> set-flag print-qvar-sorts "off"

OK.

> (forall ?x . exists ?y . ?y > ?x)

Sentence: (forall ?x
```

```
          (exists ?y
            (> ?y ?x)))
```

Turning off variable sort printing can sometimes simplify output significantly, especially when there are several long polymorphic sorts involved, but keeping it on can often provide useful information.

2.13 Overloading

Say we have introduced Plus as a binary function symbol intended to represent addition on the natural numbers:

```
declare Plus: [N N] -> N
```

While we could go ahead and use Plus in all of our subsequent proofs, it might be preferable, for enhanced readability, if we could use + instead of Plus, since + is traditionally understood to designate addition. However, + is already used in Athena to represent addition on real numbers (i.e., it is already declared at the top level as a binary function symbol that takes two real numbers and produces a real number). This means that we cannot redeclare + to take natural numbers instead:

```
> declare +: [N N] -> N

Warning, standard input:1:9: Duplicate symbol---the name + is
already used as a function symbol.
```

We could, of course, simply *define* + to be Plus, but then we would lose the original meaning of + as a binary function symbol on the real numbers.

At the top level we can get around these difficulties by *overloading* + so that it can stand for Plus whenever that makes sense but revert to its original meaning at all other times:

```
> overload + Plus

OK.
```

We can now use + for both purposes: as an alias for Plus, to denote addition on natural numbers; and also to denote the original function, addition on real numbers. Which of these alternatives is chosen depends on the context. More specifically, it depends on the sorts of the terms that we supply as arguments to +. If the arguments to + are natural numbers, then + is understood as Plus. If, on the other hand, the arguments are not natural numbers, then

Athena infers that + is being used in its original capacity, to represent a function on real numbers:

```
1  > (?a + zero)
2
3  Term: (Plus ?a:N zero)
4
5  > (?a + 2.5)
6
7  Term: (a:Real + 2.5)
```

Here, on line 1, Athena interpreted + as `Plus`, because even though the variable `?a` was not annotated, the second argument was `zero`, a natural number. Hence, the only viable alternative was to treat + as `Plus`. On line 5, by contrast, + could not possibly be understood as `Plus`, since the second argument was `2.5`, of sort `Real`, and hence + was treated as it would have been normally treated prior to the overloading. If the sorts of the arguments to + are completely unconstrained, in which case both interpretations are plausible, then the most recently overloaded meaning takes precedence, in this case `Plus`:

```
> (?a + ?b)

Term: (Plus ?a:N ?b:N)
```

Essentially (if somewhat loosely), after a directive of the form **`overload`** f g has been issued, every time Athena encounters an application of the form (f $\cdots$) it tries to interpret it as (g $\cdots$). If that fails, then it interprets the application based on the original meaning of f.

This is not overloading in the conventional sense, because we do not directly declare + to have two distinct signatures, one of which expects two natural numbers as inputs and produces a natural number as output. Instead, we first introduce `Plus`, and then we essentially announce that + is to be used as an alias for `Plus` in any context in which it is sensible to do so. Thus, for instance, `(+ zero zero)` actually produces the term `(Plus zero zero)` as its output. So the name + here really is used simply as a synonym for `Plus`.

Note that after the overloading, + no longer denotes a function symbol. Rather, it denotes a binary procedure that does what was described above: It first tries to apply `Plus` to its two arguments, and if that fails, it tries to apply to them *whatever was previously denoted by* +, which in this case is a function symbol.[28] This process can be iterated indefinitely. For instance, after first overloading + to represent `Plus`, we might later overload it even further, say, to represent `::`, the reflexive constructor of the `List` datatype (see page 57):

28 Of course, + remains a function symbol in the current symbol set.

```
1  > overload + Plus
2
3  OK.
4
5  > (?a + ?b)
6
7  Term: (Plus ?a:N ?b:N)
8
9  > overload + ::
10
11 OK.
12
13 > (?a + ?b)
14
15 Term: (:: ?a:'T2
16           ?b:(List 'T2))
17
18 > (?a + zero)
19
20 Term: (Plus ?a:N zero)
21
22 > (?a + 3.14)
23
24 Term: (+ ?a:Real 3.14)
```

Here, after the second overloading on line 9, + denotes a binary procedure that takes two values v_1 and v_2 and tries to apply :: to them. If that fails, then it applies to v_1 and v_2 whatever was previously denoted by +, which in this case is the procedure that resulted from the first overloading, on line 1.

Multiple overloadings can be carried out in one fell swoop as follows:

```
> overload (+ Plus) (- Minus) (* Times) (/ Div)

OK.
```

The above is equivalent to the following sequence of individual directives:

```
overload + Plus
overload - Minus
overload * Times
overload / Div
```

The `overload` directive is most useful when we are working exclusively at the top level or inside a single module. Across different modules there is rarely a need for explicit overloading, since the same function symbol can be freely declared in multiple modules with different signatures. That is, two distinct modules M and N are allowed to declare a function symbol of one and the same name, f. No conflict arises because the enclosing modules serve to disambiguate the symbols: in one case we are dealing with $M.f$ and in the other with $N.f$. So we could, for instance, develop our theory of natural numbers inside a module named, say, `N`, and then directly introduce a function symbol `+` inside `N` to designate addition, which would altogether avoid the introduction of `Plus`. This is, in fact, the preferred approach when developing specifications and proofs in the large. The alternative we have described here, **`overload`**, does not involve modules and can be used in smaller-scale projects (though it can also come in handy sometimes inside modules).

We have seen that if f is a function symbol, then after a directive like

`overload` $f\ g$

is issued, f will no longer denote the symbol in question; it will instead denote a procedure (recall that function symbols and procedures are distinct types of values). This raises the question of how we can now retrieve the *symbol* f. For instance, consider the first time we overload +:

```
> overload + Plus

OK

> +

Procedure: +
```

How can we now get hold of the actual function symbol +? (We might need the symbol itself for some purpose or other.) The newly defined procedure can still make terms with the symbol + at the top, so all we need to do is grab that symbol with the `root` procedure, though simple pattern matching would also work:

```
> (root (1 + 2))

Symbol: +
```

But probably the easiest way to obtain the function symbol after the corresponding name has been redefined via **`overload`** is to use the primitive procedure `string->symbol`, which takes a string and, assuming that the current symbol set contains a function symbol f of the same name as the given string, it returns f:

```
> (string->symbol "+")

Symbol: +
```

2.14 Programming

In this section we briefly survey some Athena features that are useful for programming.

2.14.1 Characters

A literal character constant starts with ‘ and is followed by either the character itself, if the character is printable, or by *d*, where *d* is a numeral of one, two, or three digits representing the ASCII code of the character:

```
> 'A

Character: 'A

> '\68

Character: 'D
```

Standard escape and control sequences (also starting with ‘) are understood as well, for example, `'\n` indicates the newline character, `'\t` the tab, and so on.

2.14.2 Strings

A string is just a list of characters:

```
> (print (tail "hello world"))

ello world
Unit: ()
```

Built-in procedures like `rev` and `join` can therefore be directly applied to them.

2.14.3 Cells and vectors

A cell is a storage container that can hold an arbitrary value (possibly another cell). At a lower level of abstraction, it can be thought of as a constant pointer, pointing to a specific address where values can be stored. A cell containing a value *V* can be created with an expression of the form **cell** *V*. The contents of a cell *c* can be accessed by writing **ref** *c*. We can destructively modify the contents of a cell *c* by storing a value *V* in it (overwriting any previous contents) with an expression of the form **set!** *c V*.

```
> define c := cell 23

Cell c defined.

> ref c

Term: 23

> set! c "a string now..."
```

```
Unit: ()

> ref c

List: ['a '\blank 's 't 'r 'i 'n 'g '\blank 'n 'o 'w '. '. '.]
```

A vector is a fixed-length array of values. If N is an expression whose value is a nonnegative integer numeral and V is any value, then an expression of the form

make-vector N V

creates a new vector of length N, every element of which contains V. If A is a vector of size N and i is an index between 0 and $N - 1$,

vector-sub A i

returns the value stored in the ith element of A. To store a value V into the ith element of A (assuming again that A is of size N and $0 \leq i < N$), we write

vector-set! A i V.

2.14.4 Tables and maps

A *table* is a dictionary ADT (abstract data type), implemented as a hash table mapping keys to values. Keys are hashable Athena values: characters, numbers, strings, but also terms, sentences, and lists of such values (including lists of lists of such values, and so on). Tables provide constant-time insertions and lookups, on average. The functionality of tables is accessible through the module `HashTable`,[29] which includes the following procedures:

1. `HashTable.table`: A nullary procedure that creates and returns a new hash table. Optionally, it can take an integer argument specifying the table's initial size.
2. `HashTable.add`: A binary procedure that takes a hash table T as its first argument and either a single key-value binding or a list of them as its second argument; and augments T with the given binding(s). Each binding is either a pair of the form [*key val*] or else a 3-element list of the form [*key* --> *val*].[30] Multiple bindings are added from left to right. The unit value is returned.
3. `HashTable.lookup`: A binary procedure that takes a hash table T as its first argument and an arbitrary key as its second argument and returns the value associated with that key in T. It is an error if the key is not bound in T.

29 See Chapter 7 for more details on Athena's modules. For those familiar with `C++`, you can think of modules as namespaces similar to those of `C++`, with minor syntactic differences (e.g., `M.x` is used instead of `M::x`).

30 `-->` is a constant function symbol declared in Athena's library.

4. `HashTable.remove`: A binary procedure that takes a hash table T as its first argument and an arbitrary *key* as its second argument. It is an error if *key* is not bound in T. Otherwise, if *key* is bound to some *val*, then that binding is removed from T and *val* is returned.
5. `HashTable.clear`: A unary procedure that takes a hash table T and clears it (removes all bindings from it). The unit value is returned.
6. `HashTable.size`: A unary procedure that takes a hash table T and returns its size (the number of bindings in T).
7. `HashTable.table->string`: A unary procedure that takes a hash table and returns a string representation of it.
8. `HashTable.table->list`: A unary procedure that takes a hash table and returns a list of all the bindings in it (as a list of key-value pairs).

A *map* is also a dictionary ADT, but one that is implemented as a functional tree: inserting a new key-value pair into a map m creates and returns another map m', obtained from m by incorporating the new key-value binding; the old map m is left unchanged. Maps provide logarithmic-time insertions and lookups. To apply a map m to a key k, we simply write (m k). Thus, notationally, maps are applied just like procedures, although semantically they form a distinct type. The rest of the interface of maps is accessible through the module `Map`, which includes the procedures described below. Note that map keys can only come from types of values that admit of computable total orderings. These coincide with the hashable value types described above. Any value can be used as a map key, but if it's not of the right type (admitting of a computable total ordering) then its printed representation (as a string) will serve as the actual key.

- `Map.make`: A unary procedure that takes a list of key-value bindings

  $$[[key_1\ val_1]\ \cdots\ [key_n\ val_n]]$$

  and returns a new map that maps each key_i to val_i, $i = 1, \ldots, n$. The same map can also be created from scratch with the custom notation:

  $$|\{key_1 := val_1, \ldots, key_n := val_n\}|.$$

- `Map.add`: A binary procedure that takes a map m and a list of bindings (with each binding represented as a [*key* *val*] pair) and returns a new map m' that extends m with the given bindings, added from left to right.
- `Map.remove`: A binary procedure that takes a map m and a *key* k and returns the map obtained from m by removing the binding associated with k. If m has no such binding, it is returned unchanged.
- `Map.size`: A unary procedure that takes a map and returns the number of bindings in it.

- `Map.keys`: A unary procedure that takes a map *m* and returns a list of all the keys in it.
- `Map.values`: Similar, except that it returns the list of values in the map.
- `Map.key-values`: Similar, but a list of `[`*key val*`]` pairs is returned instead.
- `Map.map-to-values`: A binary procedure that takes a map *m* and a unary procedure *f* and returns the map obtained from *m* by applying *f* to the values of *m*.
- `Map.map-to-key-values`: Similar, except that the unary procedure *f* expects a pair as an argument, and *f* is applied to each key-value pair in the map.
- `Map.apply-to-key-values`: A binary procedure that takes a map *m* and a side effect-producing unary procedure *f* and applies *f* to each key-value pair in *m*. The unit value is returned.
- `Map.foldl`: A ternary procedure that takes a map *m*, a ternary procedure *f*, and an initial value *V*, and returns the value obtained by left-folding *f* over all key-value pairs of *m* (passing the key as the first argument to *f* and the value as the second; the third argument of *f* serves as the accumulator), and using *V* as the starting value.

2.14.5 While loops

A **`while`** expression is of the form **`while`** E_1 E_2. The semantics of such loops are simple: As long as E_1 evaluates to `true`, E_2 is evaluated. Such loops are rarely used in practice. Even when side effects are needed, (tail) recursion is a better alternative.

2.14.6 Expression sequences

A sequence of one or more phrases ending in an expression can be put together with an expression of this form:

(**`seq`** $F_1 \cdots F_n$ E).

The phrases $F_1, \ldots, F_n, E$ are evaluated sequentially and the value of E is finally returned as the value of the entire **`seq`** expression. This form is typically used for code with side effects.

2.14.7 Recursion

Phrases of the form

`letrec` $\{ I_1 := E_1 ; \cdots ; I_n := E_n \}$ F

allow for mutually recursive bindings: An identifier I_j might occur free in any E_i, even for $i \leq j$. The precise meaning of these expressions is given by a desugaring in terms of **`let`** and cells (via **`set!`** expressions), but an intuitive understanding will suffice here. In most cases the various E_i are **lambda** or **method** expressions. The phrase F is the *body* of the **letrec**. If F is an expression then the entire **letrec** is an expression, otherwise the **letrec** is a deduction. For example, the following code defines two mutually recursive procedures for determining the parity of an integer:[31]

```
define [even? odd?] :=
  letrec {E := lambda (n)
                 check {(n equal? 0) => true
                       | else => (O n minus 1)};
          O := lambda (n)
                 check {(n equal? 0) => false
                       | else => (E n minus 1)}}
    [lambda (n) (E (abs n))
     lambda (n) (O (abs n))]
```

where abs is a primitive procedure that returns the absolute value of a number (integer or real).

Explicit use of **letrec** is not required when defining procedures or methods at the top level. The default notation for defining procedures, for instance, essentially wraps the body of the definition into a **letrec**. So, for example, we can define a procedure to compute factorials as follows:

```
define (fact n) :=
  check {(n less? 2) => 1
        | else => (n times fact n minus 1) }
```

However, if we want to define an inner recursive procedure (introduced inside the body of a procedure defined at the top level), then we need to use **letrec**:

```
define (f ···) :=
  ...
   letrec {g := lambda (···)
                  ··· (g ···) ···}
    ...
```

2.14.8 Substitutions

Formally, a *substitution* θ is a finite function from variables to terms:

$$\theta = \{x_1 \mapsto t_1, \ldots, x_n \mapsto t_n\}, \tag{2.15}$$

where each term t_i is of the same sort as the variable x_i. We say that the set of variables $\{x_1, \ldots, x_n\}$ constitutes the *support* of θ. In Athena, substitutions form a distinct type of

31 This is purely for illustration purposes; a much better way to determine parity is to examine the remainder after division by 2 (using the mod procedure).

values. A substitution of the form (2.15) can be built with the unary procedure make-sub, by passing it as an argument the list of pairs [[x_1 t_1] $\cdots$ [x_n t_n]]. For example:

```
> (make-sub [[?n zero] [?m (S ?k)]])

Substitution:
{?n:N --> zero
?m:N --> (S ?k:N)}
```

Note the format that Athena uses to print out a substitution: {x_1 --> t_1 $\cdots$ x_n --> t_n}. The support of a substitution can include variables of different sorts, say:

```
> (make-sub [[?counter (S zero)] [?flag true]])

Substitution:
{?counter:N --> (S zero)
?flag:Boolean --> true}
```

It may also include polymorphic variables:

```
> (make-sub [[?list (:: ?head ?tail)]])

Substitution: {?list:(List 'S) --> (:: ?head:'S ?tail:(List 'S))}
```

The support of a substitution can be obtained by the unary procedure supp. The empty substitution is denoted by empty-sub. A substitution θ of the form (2.15) can be *extended* to incorporate an additional binding $x_{n+1} \mapsto t_{n+1}$ by invoking the ternary procedure extend-sub as follows:

(extend-sub θ x_{n+1} t_{n+1}).

If we call (extend-sub θ x t) with a variable x that already happens to be in the support of θ, then the new binding for x will take precedence over the old one (i.e., the resulting substitution will map x to t).

Substitutions can be *applied* to terms and sentences. In the simplest case, the result of applying a substitution θ of the form (2.15) to a term t is the term obtained from t by replacing every occurrence of x_i by t_i. The syntax for such applications is the same syntax used for procedure applications: (θ t).[32] For example:

```
> theta

Substitution:
{?counter:N --> (S zero)
?flag:Boolean --> true}

> (theta ?flag)
```

32 When we use conventional mathematical notation, we might write such an application as $\theta(t)$ instead.

```
Term: true

> (theta (?foo = S ?counter))

Term: (= ?foo:N
         (S (S zero)))
```

The result of applying a substitution θ of the form (2.15) to a sentence p, denoted by $(\theta\ p)$, is the sentence obtained from p by replacing every free occurrence of x_i by t_i.

In many applications, substitutions are obtained incrementally. For instance, first we may obtain a substitution such as θ_1 = {?a:N $\mapsto$ (S ?b)}, and then later we may obtain another one, for example,

$$\theta_2 = \{\texttt{?b:N} \mapsto \texttt{zero}\}.$$

We want to combine these two into a single substitution that captures the information provided by both. We can do that with an operation known as substitution *composition*. In this case, the composition of θ_2 with θ_1 yields the result:

$$\theta_3 = \{\texttt{?a:N} \mapsto \texttt{(S zero)},\ \texttt{?b:N} \mapsto \texttt{zero}\}.$$

We can think of θ_3 as combining the information of θ_1 and θ_2. To obtain the composition of θ_2 with θ_1 in Athena, we apply the binary procedure compose-subs to the two substitutions. In general, for any

$$\theta_1 = \{x_1 \mapsto t_1, \ldots, x_n \mapsto t_n\}$$

and

$$\theta_2 = \{y_1 \mapsto s_1, \ldots, y_m \mapsto s_m\}$$

we have:

$$\texttt{(compose-subs } \theta_2\ \theta_1\texttt{)} = \{x_1 \mapsto (\theta_2\ t_1), \ldots, x_n \mapsto (\theta_2\ t_n), y_{i_1} \mapsto s_{i_1}, \ldots, y_{i_k} \mapsto s_{i_k}\},$$

where the set $\{y_{i_1}, \ldots, y_{i_k}\} \subseteq \{y_1, \ldots, y_m\}$ contains every variable in the support of θ_2 that is "new" as far as θ_1 is concerned, that is, every y_i whose name is not the name of any x_j.

Two very useful operations on terms that return substitutions are *term matching* and *unification*. We say that a term s *matches* a term t if and only if we can obtain s from t by consistently replacing variables in t by certain terms. Thus, the term t is viewed as a template or a pattern. The variables of t act as placeholders—empty boxes to be filled by sort-respecting terms. Plugging appropriate terms into these placeholders produces the term s. We say that s is an *instance* of t.

For example, s = (zero Plus S ?a) matches the term t = (?x Plus ?y), and hence s is an instance of t, because we can obtain s from t by substituting zero for ?x and (S ?a) for ?y. A more precise way of capturing this relation is to say that s matches t iff there exists a substitution that produces s when applied to t. The binary procedure `match-terms` efficiently determines whether a term (the first argument) matches another (the second). If it does not, `false` is returned, otherwise a matching substitution is produced. Matching is a fundamental operation of central importance that arises in many areas of computer science, from programming languages and databases to artificial intelligence. In this book, matching will be widely used in dealing with equational proofs by means of *rewriting*. For example:

```
> (match-terms (null union ?x) (?s1 union ?s2))

Substitution:
{?s2:Set --> ?x:Set
?s1:Set --> null}

> (match-terms (father joe) (mother ?x))

Term: false
```

There is a corresponding primitive procedure `match-sentences` that extends this notion to sentences. Roughly, a sentence p matches a sentence q iff either both p and q are atomic sentences and p matches q just as a term matches another term; or else both are complex sentences built by the same sentential constructor or quantifier and their corresponding immediate subsentences match recursively. In particular, when p and q are quantified sentences $(Q\ x\ p')$ and $(Q\ y\ q')$, respectively (where Q is either the universal or the existential quantifier), we proceed by recursively matching $\{x \mapsto v\}(p')$[33] against $\{y \mapsto v\}(q')$, where v is some fresh variable of the same sort as x and y, with the added proviso that v cannot appear in the resulting substitution (to ensure that if any variables appear in a resulting substitution, these are among the free variables of the two sentences that were matched). Some examples:

```
> (match-sentences (~ joe siblings ann) (~ ?x siblings ?y))

Substitution:
{?y:Person --> ann
 ?x:Person --> joe}

> (match-sentences (forall ?x . ?x siblings joe)
                   (forall ?y . ?y siblings ?w))

Substitution: {?w:Person --> joe}
```

33 Recall that for any sentence p, variable v, and term t, we write $\{v \mapsto t\}(p)$ for the sentence obtained from p by replacing every free occurrence of v by t, taking care to rename bound variables as necessary to avoid variable capture.

We conclude with a brief discussion of unification. Informally, to unify two terms s and t is to find a substitution that renders them identical, that is, a substitution θ such that $(\theta\ s)$ and $(\theta\ t)$ are one and the same term. Such a substitution is called a *unifier* of s and t. Two terms are *unifiable* if and only if there exists a unifier for them. Consider, for instance, the two terms $s =$ `(S zero)` and $t =$ `(S ?x)`. They are unifiable, and in this case the unifier is unique: {`?x` $\mapsto$ `zero`}. Unifiers need not be unique. For instance, the terms `(S ?x)` and `?y:N` are unifiable under infinitely many substitutions.

The binary procedure `unify` can be used to unify two terms (i.e., to produce a unifier for them); `false` is returned if the terms cannot be unified. Note that unifiability is not the same as matching. Neither of two terms might match the other, but the two might be unifiable nevertheless. Athena's procedures for matching and unification handle polymorphic inputs as well, consistent with the intuitive understanding of a polymorphic term (or sentence) as a collection of monomorphic terms (or sentences).

2.15 A consequence of static scoping

Athena is statically scoped. Roughly speaking, this means that free identifier occurrences inside the body of a procedure (or method) get their values from the lexical environment in which the procedure (or method) was defined. It also means that consistent renaming of bound identifier occurrences does not affect the meaning of a phrase. This is in contrast to *dynamic scoping*, in which free identifier occurrences inside the body of a procedure (or method) get their values from the lexical environment in which the procedure (or method) is called, not the environment in which it was defined. Almost all modern programming languages use static scoping, because dynamic scoping is conducive to subtle errors that are difficult to recognize and debug. For our purposes in this textbook we need not get into all the complexities of static vs. dynamic scoping, but it is worth noting a consequence that could seem puzzling to the novice. First, consider the following definitions of procedures `f` and `g`:

```
1  > define (f x) := (x plus x)
2
3  Procedure f defined.
4
5  > define (g x) := ((f x) plus 3)
6
7  Procedure g defined.
8
9  > (g 5)
10
11 Term: 13
```

Suppose we now realize we should have defined `f` as squaring rather than doubling `x`, so we redefine it:

```
> define (f x) := (x times x)

Procedure f defined.

> (g 5)

Term: 13
```

Why do we still get the same result? Because in the earlier code, at the point where we defined g (line 5), the free occurrence of f in the body of g referred to the doubling procedure defined in line 1. This is a statically fixed binding, unchanged when we bind f to a new procedure in the second definition. Thus, we must also redefine g in order to have a definition that binds f to the new function, by reentering the (textually) same definition as before:

```
> define (g x) := ((f x) plus 3)

Procedure g defined.

> (g 5)

Term: 28
```

In general, if you are interactively entering a series of definitions and you then revise one or more of them, you'll also need to update the definitions of other values that refer to the ones you have redefined. Of course, if you enter all the definitions in a file that you then load, things are simpler: You can just go back and edit the text of the definitions that need changing and then reload the file.

2.16 Miscellanea

Here we describe some useful features of Athena that do not neatly fall under any of the subjects discussed in the preceding sections.

1. *Short-circuit Boolean operations*: `&&` and `||` perform the logical operations AND and OR on the two-element set {`true, false`}. They are special forms rather than primitive procedures precisely in order to allow for short-circuit evaluation.[34] In particular, to evaluate `(&&` $F_1 \cdots F_n$`)`, we first evaluate F_1 to get a value V_1. V_1 must be either `true` or `false`, otherwise an error occurs. If it is `false`, the result is `false`; if it is `true`, then we proceed to evaluate F_2, to get a value V_2. Again, an error occurs if V_2 is neither `true` nor `false`. Assuming no errors, we return `false` if V_2 is `false`; otherwise V_2 is `true`, so we

34 Because Athena is a strict (call-by-value) language, if, say, `&&` were just an ordinary procedure, then all of its arguments would have to be fully evaluated before the operation could be carried out, and likewise for `||`.

proceed with F_3, and so on. If every F_i is `true` then we finally return `true` as the result. We evaluate`(|| ` $F_1 \cdots F_n$`)` similarly, but with the roles of `true` and `false` reversed.

2. *Fresh variables*: There is a predefined procedure `fresh-var` that will return a *fresh variable*: a variable whose name is guaranteed to be different from that of every other variable previously encountered in the current Athena session. When called with zero arguments, the procedure returns a fresh variable whose sort is completely unconstrained:

```
> (fresh-var)

Term: ?v1:'T190
```

A fresh variable of a specific sort can be created by passing the desired sort as a string argument to `fresh-var`:

```
> (fresh-var "Int")

Term: ?v2:Int

> (fresh-var "(Pair 'T Boolean)")

Term: ?v3:(Pair 'T192 Boolean)

> ?v4

Term: ?v4:'T193

> (fresh-var)

Term: ?v5:'T194
```

If we want the name of the fresh variable to start with a prefix of our choosing rather than the default `v`, we can pass that prefix as a second argument, in the form of a meta-identifier:

```
> (fresh-var "Int" 'foo)

Term: ?foo253:Int
```

The ability to generate fresh variables is particularly useful when implementing theorem provers.

3. *Dynamic term construction*: Sometimes we have a function symbol f and a list of terms `[`$t_1 \cdots t_n$`]`, and we want to form the term $(f\ t_1 \cdots t_n)$, but we cannot apply f directly because the number n is not statically known (indeed, often both f and the list of terms are input parameters). For situations like that there is the binary procedure `make-term`, which takes a function symbol f and a list of terms $[t_1 \cdots t_n]$ and returns $(f\ t_1 \cdots t_n)$, provided that this term is well sorted:

```
> (make-term siblings [joe ann])

Term: (siblings joe ann)
```

4. *Free variable computation*: The primitive unary procedure `free-vars` (also defined as `fv`) takes any sentence p and returns a list of those variables that have free occurrences in p:

```
define p := (?x < ?y + 1 & forall ?x . exists ?z . ?x = ?z)

> (fv p)

List: [?x:Int ?y:Int]

> (fv (forall ?x . ?x = ?x))

List: []
```

5. *Free variable replacement*: The operation of safely replacing every free occurrence of a variable v inside a sentence p by some term t, denoted by $\{v \mapsto t\}(p)$ (see page 40), is carried out by the primitive ternary procedure `replace-var`. Specifically,

   (replace-var v t p)

   produces the sentence obtained from p by replacing every free occurrence of v by t, renaming as necessary to avoid variable capture.

6. *Dummy variables*: Sometimes we need a "dummy" variable whose name is unimportant. We can use the underscore character to generate a fresh dummy variable, that is, a variable that has not yet been seen during the current session:

```
> _

Term: ?v70:'T646

> _

Term: ?v71:'T647
```

   Observe that we get a different variable each time, first `?v70` and then `?v71`. That is what makes these variables fresh. The sorts of these variables are completely unconstrained, which makes the variables maximally flexible: They can appear in any context whatsoever and take on the locally required sorts.

```
> (father _)

Term: (father ?v73:Person)
```

```
> (_ in _)

Term: (in ?v350:'T4428
          ?v351:(Set 'T4428))
```

7. *Proof errors*: A primitive nullary method `fail` halts execution when applied and raises an error:

```
> (!fail)

standard input:1:1: Error: Proof failure.
```

A related method, `proof-error`, has similar behavior except that it takes a string as an argument (an error message of some kind), and prints that string in addition to raising an error.

8. *Patterns inside* **`let`** *phrases*: The syntax of **`let`** phrases is somewhat more flexible than indicated in (2.11). Specifically, instead of identifiers I_j, we can have patterns appearing to the left of the assignment operators :=. These can be term patterns, sentential patterns, list patterns, or a mixture thereof. For example:

```
> let {[x y] := [1 2]}
    [y x]

List: [2 1]

> let {[(siblings L (father R))] := [(siblings joe (father ann))]}
    [L R]

List: [joe ann]
```

9. *last-val*: At the beginning of each iteration of the read-eval-print loop, the identifier `last-val` denotes the value of the most recent phrase that was evaluated at the top level:

```
> (rev [1 2])

List: [2 1]

> last-val

List: [2 1]
```

10. *Primitive methods*: A primitive method in this context refers to an explicitly defined method M whose body is an *expression* E (rather than a deduction, as is the usual requirement for every nonprimitive method). M can take as many arguments as it needs,

and it can produce any sentence it wants as output—whatever the expression E produces for given arguments. Thus, M becomes part of our trusted computing base and we had better make sure that the results that it produces are justified. Such a method is introduced by the following syntax:

$$\texttt{primitive-method}\ (M\ I_1 \cdots I_n)\ \texttt{:=}\ E$$

where $I_1 \cdots I_n$ refer to the arguments of M. One can think of Athena's primitive methods—such as `mp`—as having been introduced by this mechanism. For example, one can think of modus ponens as:

```
primitive-method (mp premise-1 premise-2) :=
  match [premise-1 premise-2] {
    ([(p ==> q) p] where (hold? [premise-1 premise-2])) => q
  }
```

Normally there is no reason to use **primitive-method**, unless we need to introduce infinitely many axioms in one fell swoop, typically as instances of a single axiom schema. In that case a **primitive-method** is the right approach. An illustration (indeed, the only use of this construct in the entire book) is given in Exercise 4.38.

2.17 Summary and notational conventions

Below is a summary of the most important points to take away from this chapter, as well as the typesetting and notational conventions we have laid down:

- Expressions and deductions play fundamentally different roles. Expressions represent arbitrary computations and can result in values of any type whatsoever, whereas deductions represent logical derivations and can only result in sentences. We use the letters E and D to range over the sets of expressions and deductions, respectively.
- A *phrase* is either an expression or a deduction. We use the letter F to range over the set of phrases.
- Expressions and deductions are not just semantically but also syntactically different. Whether a phrase F is an expression or a deduction is immediately evident, often just by inspecting the leading keyword of F.
- Athena keywords (such as **assume**) are displayed in bold font.
- Athena can be used in batch mode or interactively. In interactive mode, if the input typed at the prompt is not syntactically balanced (either a single token or else starting and ending with parentheses or brackets), then it must be terminated either with a double semicolon `;;` or with `EOF`.

- The syntax of expressions and deductions is defined by mutual recursion. Expressions may contain deductions and all deductions contain expressions.
- All phrases (both expressions and deductions) are evaluated with respect to a given lexical environment ρ, assumption base β, store σ, and symbol set γ. If a deduction D is evaluated with respect to an assumption base β and produces a sentence p, then p is a logical consequence of β. That is the main soundness guarantee provided by Athena.
- Athena *identifiers* are used to give names to values (and also to sorts and modules). They are represented by the letter I (possibly with subscripts, superscripts, etc.). Identifiers can become bound to values either at the top level, with a directive such as **`define`**, or inside a phrase, with a mechanism like **`let`** or via pattern matching inside a **`match`** phrase, and so on.
- Athena *values* are divided into a number of types, enumerated below.[35] Evaluating any phrase F must produce a value of one of these types, unless the evaluation diverges or results in an error:

  1. *Terms*, such as `(father peter)` or `(+ ?x:Int 1)`, which are essentially syntax trees used to represent elements of various sorts. We use the letters s and t to represent terms. *Variables* are a special kind of terms, of the form $?I:S$, where S is a sort. They act as syntactic placeholders. We use the letters x, y, z, and v to range over variables.
  2. *Sentences*, such as `(= 1 1)` or `(not false)`. We use the letters p, q, and r to represent sentences. Evaluating a deduction can only produce a value of this type.
  3. *Lists* of values, such as `[1 2 [joe (not true)] 'foo]`. We use the letter L to range over lists of values.
  4. The *unit value* `()`.
  5. *Function symbols*, such as `true` or `+`. We use the letters f, g, and h to range over function symbols. Function symbols whose range is `Boolean` are called *relation* or *predicate* symbols; we use the letters P, Q, and R to range over those.
  6. *Sentential constructors* and *quantifiers*, namely `not`, `and`, `or`, `if`, `iff`, `forall`, and `exists`.
  7. *Procedures*, whether they are primitive (such as `plus`) or user-defined.[36]
  8. *Methods*, whether they are primitive (such as `both`) or user-defined.

---

35 The division is not a partition, as some values belong to more than one type. For instance, every Boolean term (such as `true`) is both a term and a sentence.

36 We often use the word "procedure" to refer both to syntactic objects, namely, *expressions* of the form **`lambda`** $(I_1 \cdots I_n)$ E; and to semantic objects, namely, the *denotations* of such expressions, which are proper mathematical functions. Sometimes, when we want to be explicit about the distinction, we speak of a *procedure value* to refer to such a function. But usually the context will make clear whether we are talking about an expression (syntax) or the abstract function that is the value of such an expression (semantics). Similar remarks apply to our use of the term "method."

9. ASCII *characters*.
10. *Substitutions*, which are finite mappings from term variables to terms.
11. *Cells* and *vectors*, which can be used to store and destructively modify arbitrary values or sequences thereof.
12. *Tables* and *maps*, which implement dictionary data types whose keys can be (almost) arbitrary Athena values, the former as hash tables and the latter as functional trees.

We use the letter V to range over Athena values.

- Every term t has a certain *sort*, which may be monomorphic (such as `Ide` or `Boolean`) or polymorphic (such as `'S` or `(Pair 'S1 'S2)`). Sorts are not the same as types. Types, enumerated above, are used to classify Athena values, whereas sorts are used to classify Athena terms, which are just one particular type of value among several others.

2.18 Exercises

Exercise 2.1: Determine whether each of the following phrases is an expression or a deduction. Assume that `A`, `B`, `C`, and `D` have been declared as `Boolean` constants.

```
1. (!both A B)

2. let {p := (A | B)}
     (!both p p)

3. 1700

4. (2 plus 8.75)

5. 'cat

6. let {p := (A & B)}
     match p {
       (_ & _) => (!left-and p)
     }

7. "Hello world!"

8. assume A
     assume B
       (!claim A)

9. [1 2 3]

10. (tail L)

11. lambda (x)
```

```
    (x times x times x)

12. lambda (f)
      lambda (g)
        lambda (x)
          (f (g x))

13. let {L := ['a 'b 'c]}
      (rev L)

14. let {x := 2;
         y := 0}
      try { (x div y) | (x times y) }

15. let {p := A;
         q := (B | C)}
      match (p & q) {
        (p1 & (p2 | p3)) => 'match
      | _ => 'fail
      }

16. pick-any x
      (!reflex x)

17. let {g := letrec {fact := lambda (n)
                                check {
                                  (n less? 2) => 1
                                | else => (n times fact n minus 1)
                                }}
                fact}
      (g 5)


18. pick-witness w for (exists ?x . ?x = ?x)
      (!true-intro)

19. check {
      (less? x y) => (!M1)
    | else => assume A (!M 2)
    }

20. letrec {M := method (p)
                   match p {
                     (~ (~ q)) => (!M (!dn p))
                   | _ => (!claim p)
                   }}
      assume h := (~ ~ ~ ~ A)
        (!M h)

21. let {M := method (p)
```

```
             (!both p p)}
    [M 1]
```

Explain your answer in each case. □

Exercise 2.2: Determine the type of the value of each of the following expressions.

1. `2`
2. `true`
3. `(not false)`
4. `[5]`
5. `()`
6. `'a`
7. `+`
8. `(head [father])`
9. `‘A`
10. `or`
11. `lambda (x) x`
12. `|'a := 1|`
13. `(father joe)`
14. `(+ ?x:Int 1)`
15. `"foo"`
16. `make-vector 10 ()`
17. `(match-terms 1 ?x)`
18. `method (p) (!claim (not p))`
19. `(HashTable.table 10)`

If the value is a term, also state the sort of the term. □

Exercise 2.3: Find a phrase *F* such that `(!claim` *F*`)` always succeeds (in every assumption base). □

Exercise 2.4: Find a deduction *D* that always fails (in every assumption base). □

Exercise 2.5: Implement some of Athena's primitive procedures for manipulating lists, specifically: `map`, `foldl`, `foldr`, `filter`, `filter-out`, `zip`, `take`, `drop`, `for-each`, `for-some`, from-to, and rd. These are all staples of functional programming, and most of them generalize naturally to data structures other than lists (e.g., to trees and beyond). They are specified as follows:

- map takes a unary procedure f and a list of values [$V_1 \cdots V_n$] and produces the list [(f V_1) $\cdots$ (f V_1)].
- foldl takes a binary procedure f, an identity element e (typically the left identity of f, i.e., whichever value e is such that (f e V) $= V$ for all V), and a list of values [$V_1 \cdots V_n$] and produces the result

  $$(f \cdots (f\ (f\ e\ V_1)\ V_2) \cdots V_n).$$

- foldr takes a binary procedure f, an identity element e (typically the right identity of f), and a list of values [$V_1 \cdots V_n$] and produces the result

  $$(f\ V_1\ \cdots (f\ V_{n-1}\ (f\ V_n\ e)) \cdots).$$

- filter takes a list L and a unary procedure f that always returns true or false and produces the sublist of L that contains all and only those elements x of L such that (f x) $=$ true, listed in the order in which they occur in L (with possible repetitions included).
- filter-out works like filter except that it only keeps those elements x for which (f x) $=$ false.
- zip is a binary convolution procedure that maps a pair of lists to a list of pairs. Specifically, given two lists [$V_1 \cdots V_n$] and [$V_1{}' \cdots V_m{}'$] as arguments, zip returns the list [[V_1 $V_1{}'$] $\cdots$ [V_k $V_k{}'$]], where k is the minimum of n and m.
- take is a binary procedure that takes a list L and an integer numeral n and returns the list formed by the first n elements of L, assuming that n is nonzero and L has at least n elements. If L has fewer than n elements or n is negative, L is returned unchanged. It is an error if n is not an integer numeral.
- drop takes a list L and an integer numeral n and returns the list obtained from L by "dropping" the first n elements. If n is not positive then L is returned unchanged, and if n is greater than the length of L, the empty list is returned.
- for-each takes a list L and a unary procedure f that always returns true or false; and returns true if (f x) is true for *every* element x of L, and false otherwise.
- for-some has the same interface as for-each but returns true if (f x) is true for *some* element x of L, and false otherwise.
- from-to takes two integer numerals a and b and produces the list of all and only those integers i such that $a \leq i \leq b$, listed in numeric order. If $a > b$ then the empty list is returned. This procedure is also known by the infix-friendly name to, used as follows:

```
> (5 to 10)

List: [5 6 7 8 9 10]
```

- The `rd` procedure takes a list L and produces the list L' obtained from L by removing all duplicate element occurrences from it, while preserving element order. □

Exercise 2.6: Define a unary procedure `flatten` that takes a list of lists $L_1, \ldots, L_n$ and returns a list of all elements of $L_1, \ldots, L_n$, in the same order in which they appear in the given arguments. For instance,

```
(flatten [[1 2] [3 4 5]])
```

should return `[1 2 3 4 5]`. □

Exercise 2.7: Athena's library defines a unary procedure `get-conjuncts` that takes as input a conjunction p and outputs a list of all its nonconjunctive conjuncts, in left-to-right order, where a nonconjunctive conjunct of p is either an immediate subsentence of p that is not itself a conjunction, or else it is a nonconjunctive conjunct of an immediate subsentence of p that is itself a conjunction. If p is not a conjunction, then the singleton list `[`p`]` is returned. Thus, for example, if the input p is

```
((A & (B | C)) & (D & ~ E) & F),
```

then the result should be `[A (B | C) D (~ E) F]`. Use one of the higher-order list procedures of the previous exercises to implement `get-conjuncts`. Implement a similar procedure `get-disjuncts` for disjunctions. □

Exercise 2.8: Define a ternary procedure `list-replace` that takes (i) a nonempty list of values $L = [V_1 \cdots V_n]$; (ii) a positive integer $i \in \{1, \ldots, n\}$; and (iii) a unary procedure f; and returns the list `[`$V_1 \cdots V_{i-1}$ `(`f V_i`)` $V_{i+1} \cdots V_n$`]`. That is, it returns the list obtained from the input list L by replacing its ith element, V_i, by `(`f V_i`)`. For instance,

```
(list-replace [1 5 10] 2 lambda (n) (n times n))
```

should return `[1 25 10]`. An error should occur if the index i is not in the proper range. □

II FUNDAMENTAL PROOF METHODS

In this part, we present and begin working with almost all of the proof methods that are fundamental in computer science. We begin in Chapter 3 with the method of chaining together rewriting steps to prove equations, either directly or as cases of an induction proof. Examples are drawn from properties of addition, multiplication, and exponentiation on natural numbers, as well as list concatenation and reversal. Term rewriting and mathematical induction are both explored in depth, building intuition for these critical concepts but also providing a sufficient formal foundation. An important feature of the presentation is reinforcement of the technical details by reference to their mechanized counterparts in Athena, a theme that is repeated throughout the book. We also begin discussion of proof strategies: bottom-up, top-down, or (more usually) some combination.

We have chosen to begin with reasoning about equations because most readers already have some experience with and intuition for this important class of sentences, even if they have not been particularly aware of the role of equations in logic. In going beyond equations in the next two chapters, our goal is partly to build or reinforce the reader's intuitions for working with any sentence of full first-order logic. To do this and at the same time provide a thorough formal grounding in the most basic principles of logic, we start in Chapter 4 with sentential logic, which encompasses sentences built with the logical connectives for negation, conjunction, disjunction, implication, and equivalence. We continue in Chapter 5 with (many-sorted) predicate logic, also called first-order logic, which adds universal and existential quantifiers.

By the end of these two chapters, we theoretically have at hand all we need for any proof in first-order logic with equality, but in practice proofs would be too long and tedious to develop using such primitive methods. Fortunately, as we show in later chapters, we can work at a much higher level of abstraction. Indeed, we begin to develop higher-level proof methods already in these two chapters, building up a library of useful components in Sections 4.9 and 5.4. We then go on in Chapter 6 to introduce one of the most useful higher-level proof methods, implication chaining. This technique chains together sentences with the implication or equivalence connective, analogously to chaining together terms with the equality symbol, and is expressed in Athena with the same `chain` method. Since it is implemented in Athena in terms of primitive methods of predicate logic, its soundness is guaranteed.

3 Proving Equalities

IN this chapter we concentrate on *equational proof*, one of the most common and useful types of inference. Our first examples will involve simple functions on natural numbers, like addition and multiplication. Later in the chapter we will work with list functions like concatenation and reversal. For both natural numbers and lists, the simplest examples involve only equality chaining, but for proofs of more interesting and useful properties we need some form of induction, corresponding to the way in which natural numbers and lists are defined as datatypes. Such an induction proof divides into cases, but in all of the examples and exercises in this chapter, each proof case can be completed with equality chaining.

3.1 Numeric equations

As we already saw in Section 2.7, we can define natural numbers in Athena as follows:

```
datatype N := zero | (S N)
```

This definition says that the values of sort `N` are

```
zero
(S zero)
(S (S zero))
(S (S (S zero)))
...
```

In general, 0 is represented by `zero` and to obtain the natural number $n+1$ we apply the "successor function" `S` to the natural number n. Put another way, the natural number n is obtained by an n-fold application of `S` to `zero`. Furthermore—and this is a crucial point—we define the values so obtained to be *the only values* that are natural numbers.[1] Recall from page 30 that terms without any variables, like those in the sequence above, are called *ground terms*. Terms containing variables, such as `(S ?n)`, are called *nonground* (or sometimes *general*) terms.

Suppose now we want to define and reason about the addition function `Plus` that takes two natural numbers as inputs and returns their sum. In Athena, we first **declare** the function, specifying the number and sorts of its inputs (also called *arguments*) and the sort of its output (or "return value"), as follows:

```
declare Plus: [N N] -> N [+]
```

1 Roughly speaking, we are saying that any expression that purports to denote a natural number must have an equivalent expression just in terms of `S` and `zero`. Later in the chapter we shall see how this fundamental property is formalized, in conventional mathematical terms as well as in Athena.

This declaration says that `Plus` takes two natural numbers as inputs and produces a natural number as output. The expression `[+]` at the end of the declaration overloads the built-in symbol + so that it can be used as an alias for `Plus` whenever the context allows it. We can now write terms such as `(Plus (S zero) zero)`, or

$$\texttt{((S zero) Plus zero)} \tag{3.1}$$

if we prefer infix notation. In fact, by making sure that `S` binds tighter (has greater precedence) than `Plus`, with a directive like:

```
set-precedence S 350
```

we can write term (3.1) even more simply as:

`(S zero Plus zero).`

And since the declaration above has also overloaded the predefined operator + to designate `Plus` when applied to terms of sort `N`, we can also write (3.1) as

`(S zero + zero).`

(Another way to permit the use of + for arguments of sort `N` is to declare our function using the identifier + in the first place, placing the declaration inside a module to avoid conflict with the predefined +. That's how it is done in the Athena library, and how it will be done in this book after the discussion of modules in Chapter 7.) So far these are strictly matters of syntax; they do not say anything about the *meaning* of a term such as `(S zero + zero)` other than that its sort is `N` (and is thus one of the values `zero`, `(S zero)`, `(S S zero)`, ...[2]). At this point, for all we know, it could be any natural number. To give `Plus` the intended meaning, we will *define* it by asserting some appropriate universally quantified equations. But before we do that, it will be convenient to give names to a few variables of sort `N` so that we don't have to keep typing question marks:

```
define [x y z n m k] := [?x:N ?y:N ?z:N ?n:N ?m:N ?k:N]
```

As we saw in the previous chapter, variables are regular denotable values, so we can now use these names to refer to the corresponding variables:

```
> (x Plus S y)

Term: (Plus ?x:N
            (S ?y:N))
```

2 Recall that consecutive applications of a unary function symbol such as `S` need not be separated by parentheses, so we can write `(S S S zero)` rather than `(S (S (S zero)))`. We use this convention frequently.

We can even use the defined names as arguments to quantifiers:

```
> (forall n m . n Plus m = m Plus n)

Sentence: (forall ?n:N
            (forall ?m:N
              (= (Plus ?n:N ?m:N)
                 (Plus ?m:N ?n:N))))
```

Defining variables like that is a common practice that we will follow often in this book. (In fact, x, y, and z are already predefined at the top level as the polymorphic variables ?x, ?y, and ?z, respectively.)

We now introduce the following universally quantified equations:

```
assert right-zero := (forall n . n + zero = n)

assert right-nonzero := (forall n m . n + S m = S (n + m))
```

Since we introduced them with **assert**, these equations are also entered into the global assumption base.

Now, of course, the meaning of (S zero + zero) is determined by the equation that is just the special case of right-zero with the ground term (S zero) substituted for n. One way to say this in Athena, and thereby get the special-case equation into the assumption base, is

```
(!instance right-zero [(S zero)])
```

to which Athena responds:

```
Theorem: (= (Plus (S zero)
                  zero)
            (S zero))
```

That is, $1 + 0 = 1$. Likewise:

```
> (!instance right-nonzero [zero (S zero)])

Theorem: (= (Plus zero
                  (S (S zero)))
            (S (Plus zero
                     (S zero))))
```

In general, the first argument to instance is a universally quantified sentence p in the assumption base, and the second is a list L of terms.[3] If

$$p = (\texttt{forall } v_1 \cdots v_n \texttt{ . } q)$$

3 For convenience, we can also give a term t by itself as the second argument to instance. That has the same effect as passing the one-element list [t] as the second argument.

and $L = [t_1 \cdots t_k]$, where $k \leq n$, then `instance` produces the sentence

$$\texttt{(forall } v_{k+1} \cdots v_n \texttt{ . } q'\texttt{)}$$

where q' results from substituting t_i for v_i in q, $i = 1, \ldots, k$.[4] In the first case above, $n = k = 1$, and in the second, $n = k = 2$. In the following case $n = 2$ and $k = 1$, so the result still has one quantifier:

```
> (!instance right-nonzero [zero])

Theorem: (forall ?v303:N
           (= (Plus zero
                    (S ?v303:N))
              (S (Plus zero ?v303:N))))
```

3.2 Equality chaining preview

What about the meaning of `Plus` for larger ground term inputs, like

`(S S zero + S S zero)`?

In other words, can we now deduce that $2 + 2 = 4$? Yes, and here is one way to do it:

```
(!chain [(S S zero + S S zero)
       = (S (S S zero + S zero))    [right-nonzero]
       = (S S (S S zero + zero))    [right-nonzero]
       = (S S S S zero)             [right-zero]
        ])
```

Here we have used `chain`, an Athena method[5] for proving equations by chaining together a sequence of terms connected by equalities. In general,

$$\texttt{(!chain [}t_0 \texttt{ = } t_1 \texttt{ [}p_1\texttt{] = } t_2 \texttt{ [}p_2\texttt{] = } \cdots \texttt{ = } t_n \texttt{ [}p_n\texttt{]])}$$

attempts to derive the identity $(t_0 = t_n)$, provided that each p_i is in the assumption base and each equation $(t_{i-1} = t_i)$ follows from p_i,[6] typically by virtue of one of the five fundamental axioms of equality listed in Section 3.4, for $i = 1, \ldots, n$. Here $n = 3$, the working

4 The substitutions are only for the *free* occurrences of the variables, namely, those not bound by quantifiers within q. Moreover, the substitutions are carried out in a safe manner, so as to avoid variable capture. These points are explained more fully in Section 5.1.

5 This is not a primitive method; `chain` is defined in Athena's library.

6 Section 5.6 will define more precisely what we mean by "follows from" in the case of first-order logic with equality, and it will also explicate the notion of an interpretation for first-order logic, but an intuitive understanding of these concepts will suffice for now.

assumptions used at each step are $p_1 = p_2 =$ `right-nonzero` and $p_3 =$ `right-zero`, and Athena responds with the theorem proved:

```
Theorem: (= (Plus (S (S zero))
                  (S (S zero)))
            (S (S (S (S zero)))))
```

We refer to each list `[`p_i`]` as the *justification list* for the corresponding step (from t_{i-1} to t_i), and to each p_i as a *justifier* for the step.

The interface of `chain` is actually a bit more flexible than the above description suggests. For example, multiple sentences may appear inside the square brackets (rather than a single p_i), and the structure of these sentences can be fairly complex (e.g., each sentence may be a *conditional* equation or, say, a conjunction, rather than a simple equation). But before examining how equality chaining works in general, we need to understand terms and sentences as tree structures.

3.3 Terms and sentences as trees

Terms and sentences are tree-structured objects, and in some cases it makes sense to treat them as trees explicitly. Let us start with terms. A variable or a constant symbol can be viewed as a simple one-node tree (a *leaf* node), while an application of the form $(f\ t_1 \cdots t_n)$ for $n > 0$ can be viewed as a tree with the symbol f at the root and with the trees corresponding to $t_1, \ldots, t_n$ as its immediate subtrees, arranged from left to right in that order. For instance, the tree corresponding to the term `(Plus (S zero) (Plus (S x) (S y)))` can be depicted as follows:

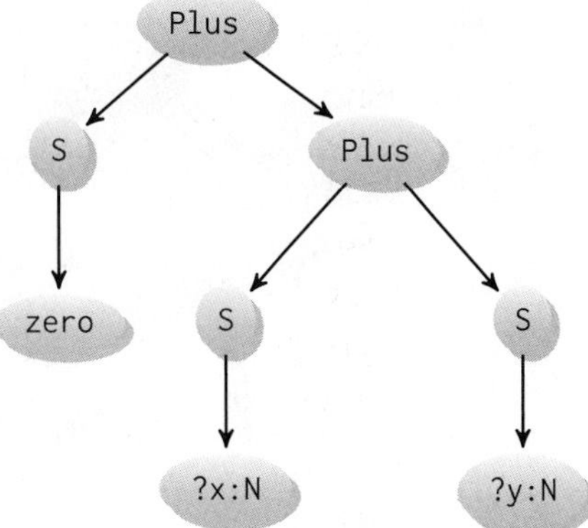

Thus, leaves stand for simple terms (variables and constants), while internal nodes represent compound terms—applications.

A two-dimensional tree representation of a compound term t depicts the essential syntactic structure of t, telling us exactly what function symbols are applied to what arguments and in what order, but without specifying how to write down t as a linear string. It does not tell us whether to use prefix, infix, or postfix notation; how to separate the arguments from one another (with commas, periods, spaces, indentation, etc.); what characters to use for grouping the arguments of a single application together (parentheses, square brackets, etc.); and so on. Such notational details are decided by choosing a particular *concrete syntax* for terms. The concrete syntax that Athena uses to output terms is the prefix notation common in Lisp dialects. The same prefix notation is available for input as well, but one can also use infix notation for binary function symbols, which often reduces notational clutter, especially in combination with precedence and associativity conventions. But tree representations, by dispensing with such details, are said to depict the *abstract syntax* of terms. We will have more to say about abstract syntax in Chapter 18.

Every node in the tree representation of a term can be assigned a unique list of positive integers $[i_1 \cdots i_m]$ indicating the path that must be traversed in order to get from the root of the tree to the node in question. That list represents the *position* of the node in the tree. As an example, Figure 3.1 shows the positions of all nodes in the tree representation of `(Plus (S zero) (Plus x y))`. As you can see, the position of the `S` node is [1], because we get to it by traveling down the first (leftmost) edge attached to the root; the position of `x` is [2 1], because we get to it by first visiting the second child of the root, and then moving to the first child of that node; and so on. The position of the root node is always the empty list [ ]. These integer sequences are sometimes called *Dewey paths* (or Dewey positions), because they are somewhat similar in their structure to the sequences of the Dewey decimal classification system used by libraries to organize book collections.

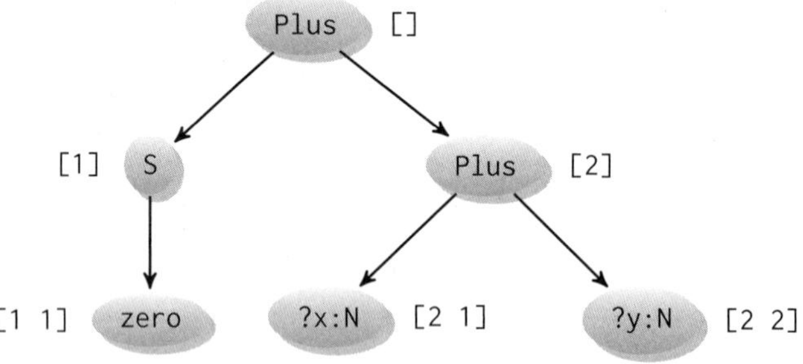

Figure 3.1
The term `(Plus (S zero) (Plus ?x ?y))` depicted as a tree structure. Nodes are annotated with their Dewey positions.

The procedure `positions-and-subterms` takes a term and produces a list of all positions in the term, each paired in a sublist with the subterm at that position. For example, for the term in Figure 3.1 we have:

```
> (positions-and-subterms (Plus (S zero) (Plus x y)))

List: [[[] (Plus (S zero) (Plus ?x ?y))]
       [[1] (S zero)]
       [[1 1] zero]
       [[2] (Plus ?x ?y)]
       [[2 1] ?x]
       [[2 2] ?y]]
```

We use this procedure in an exercise in the next section, and to help implement a basic form of equality chaining.

The following is a useful procedure that takes a term t and a position I (as a list of positive integers) and returns the subterm of t that is located at position I in the tree representation of t. An error will occur if there is no such subterm, that is, if I is not a valid position for t.

```
define (subterm t I) :=
  match I {
    [] => t
  | (list-of i rest) => (subterm (ith (children t) i)
                                 rest)
  }
```

The primitive binary procedure `ith` takes a list of $n > 0$ values $[V_1 \cdots V_n]$ and an integer $i \in \{1, \ldots, n\}$ and returns V_i.

If we only want the node at a given position, we can use the following procedure:

```
define (subterm-node t I) := (root (subterm t I))

> (subterm-node (x + S S zero) [2 1])

Symbol: S
```

Another useful procedure is `replace-subterm`, where (`replace-subterm` t I t') returns the term obtained from t by replacing the subterm at position I by t', provided that the result is well sorted. We leave the definition of this procedure as an exercise.

Similar ideas apply to sentences. An atomic sentence is just a term t, so its tree representation is that of t. A sentence of the form $(\circ\ p_1 \cdots p_n)$, for $\circ \in$ {`not`,`and`,`or`,`if`,`iff`}, can be viewed as a tree with the sentential constructor $\circ$ at the root and the trees corresponding to $p_1, \ldots, p_n$ as its immediate subtrees, listed from left to right in that order. Finally, a quantified sentence of the form $(Q\ x\ p)$ can be viewed as a tree with the quantifier Q at the root, the sole leaf x as its left subtree, and the tree corresponding to p as its right subtree.

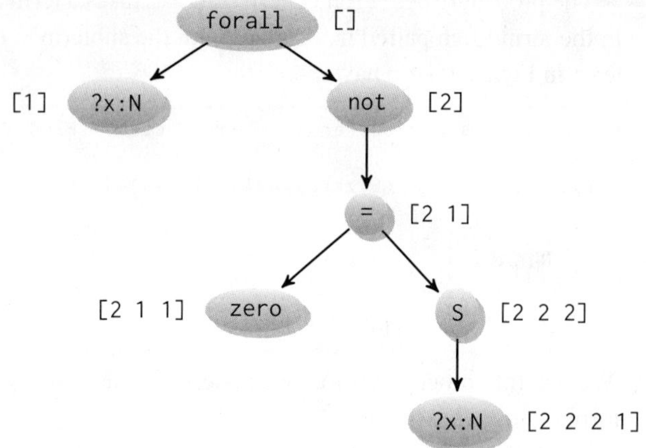

Figure 3.2
The sentence `(forall ?x (not (= zero (S ?x))))` depicted as a tree. Nodes are annotated with their respective positions.

Node positions are defined as they were for term trees. Thus, for example, the sentence `(forall x . zero =/= S x)`, which in prefix notation is written as

```
(forall x (not (= zero (S x)))),
```

is represented by the tree shown in Figure 3.2, whose nodes have been annotated with their corresponding positions. Procedures `subsentence`, `subsentence-node`, and `replace-subsentence` can be implemented analogously to `subterm`, `subterm-node`, and `replace-subterm`; see Exercise 3.37.

3.4 The logic behind equality chaining

A firm foundation for reasoning about equalities is provided by the basic *equality axioms*, which can be expressed in conventional notation as follows:

1. **Reflexivity:** $\forall\, x\, .\; x = x.$
2. **Symmetry:** $\forall\, x\, y\, .\; x = y \Rightarrow y = x.$
3. **Transitivity:** $\forall\, x\, y\, z\, .\; x = y \wedge y = z \Rightarrow x = z.$
4. **Functional Substitution:** For any function symbol f of n arguments,

$$\forall\, x_1 \cdots x_n\; y_1 \cdots y_n\, .\; x_1 = y_1 \wedge \ldots \wedge x_n = y_n \Rightarrow f(x_1, \ldots, x_n) = f(y_1, \ldots, y_n).$$

5. **Relational Substitution:** For any relation symbol R of n arguments,

$$\forall\, x_1 \cdots x_n\; y_1 \cdots y_n\, .\; x_1 = y_1 \wedge \ldots \wedge x_n = y_n \wedge R(x_1, \ldots, x_n) \Rightarrow R(y_1, \ldots, y_n).$$

Strictly speaking, the last two are *axiom schemas*, because they are indexed by f and R. Each schema has multiple axioms as instances, one for each function and relation symbol. Axiom schemas are captured in Athena with methods; we will present the relevant methods shortly. Because these are captured by methods, they are not sentences to be entered into the assumption base. Rather, any one of the first three axioms and any instance of the last two axiom schemas can be readily derived at any time by applying the corresponding method to appropriate arguments.

For simplicity, we will refer to all five of the above as "axioms" (rather than "axioms and axiom schemas"). Not all five are necessary, by the way. In fact, we can get away with reflexivity and a variation of the relational substitution axiom known as Leibniz's law. Symmetry, transitivity, and functional substitution would then become derivable as consequences. For practical purposes, however, it is more convenient to start out with all five.

Equational proofs in Athena are ultimately based on primitive methods corresponding to each of the five equality axioms, namely `reflex`, `sym`, `trans`, `fcong` ("functional congruence"), and `rcong` ("relational congruence"), as illustrated here:

```
domain D
declare a, b, c: D

> conclude (a = a)
    (!reflex a)

Theorem: (= a a)

> conclude (a = b ==> b = a)
    assume h := (a = b)
      (!sym h)

Theorem: (if (= a b)
             (= b a))

> conclude (a = b & b = c ==> a = c)
    assume (a = b & b = c)
      (!tran (a = b) (b = c))

Theorem: (if (and (= a b)
                  (= b c))
             (= a c))

declare f:[D D] -> D
declare R:[D D] -> Boolean
declare a1, a2, b1, b2: D

> conclude (a1 = b1 & a2 = b2 ==> (f a1 a2) = (f b1 b2))
    assume (a1 = b1 & a2 = b2)
      (!fcong ((f a1 a2) = (f b1 b2)))

Theorem: (if (and (= a1 b1)
```

```
             (= a2 b2))
          (= (f a1 a2)
             (f b1 b2)))

> conclude (a1 = b1 & a2 = b2 & a1 R a2 ==> b1 R b2)
    assume (a1 = b1 & a2 = b2 & a1 R a2)
      (!rcong (a1 R a2) (b1 R b2))

Theorem: (if (and (= a1 b1)
                  (and (= a2 b2)
                       (R a1 a2)))
             (R b1 b2))
```

The following is a more precise specification of these methods:

- `reflex`: A unary method that takes an arbitrary term t and produces the sentence $(t = t)$, in any assumption base.
- `sym`: A unary method that takes an equality $(s = t)$ as an argument. If the sentence $(s = t)$ is in the assumption base, then the conclusion $(t = s)$ is produced. Otherwise an error occurs.
- `tran`: A binary method that takes two equalities of the form $(t_1 = t_2)$ and $(t_2 = t_3)$ as arguments. If both of these are in the assumption base, the conclusion $(t_1 = t_3)$ is produced. Otherwise an error occurs.
- `fcong`: A unary method that takes an equality p of the form

  $$((f\ s_1 \cdots s_n) = (f\ t_1 \cdots t_n))$$

  as an argument. If the assumption base contains the n equalities $(s_i = t_i)$, $i = 1, \ldots, n$, then the input p (which represents the desired conclusion) is returned as the result. An error will occur if some $(s_i = t_i)$ is not in the assumption base, unless s_i and t_i are syntactically identical.
- `rcong`: A binary method that takes two atoms of the form $(R\ s_1 \cdots s_n)$ and $(R\ t_1 \cdots t_n)$ as arguments, where R is a relation symbol of arity n. If the assumption base contains the first sentence, $(R\ s_1 \cdots s_n)$, along with the n equalities $(s_1 = t_1), \ldots, (s_n = t_n)$, then the second sentence $(R\ t_1 \cdots t_n)$ is produced, otherwise an error is generated.

It is not difficult to see that the given axioms are *true* under the conventional interpretation of the equality symbol as the identity relation, regardless of how we interpret other symbols. Therefore, these five methods are *sound*, meaning that they can never take us from true premises to false conclusions. How about completeness? Are these methods sufficient for deriving *all* identities that follow from a given set of equations? In tandem with `instance`, the answer is affirmative. More precisely, if we have a finite set E of universally quantified equations and are presented with a new equation that follows from E, and we are asked to derive that equation from E, a proof could always be carried out using a combination of applications of `instance` to derive substitution instances of the various equations in

E, along with applications of `reflex`, `sym`, `trans`, `fcong`, and `rcong`. Although we will not prove it here (it was first proved by Birkhoff in 1935 [10]), this is an important result that ensures that the six methods in question (the five equational methods along with `instance`) constitute a complete inference system for equational logic. Any equation that follows from E can be derived from it with a—potentially very long—sequence of applications of these few methods.

However, such proofs would be operating at a very low level of abstraction, not unlike programs written in machine language, and nontrivial cases would require long and tedious proofs. Instead, from these basic ingredients we can derive results that justify *term rewriting*, which is the kind of larger step in reasoning about equalities that is routinely used by the `chain` method. We begin with:

Theorem 3.1: First Substitution Theorem

If $s = s'$ then for any function symbol f of n arguments and terms t_i of appropriate sorts,

$$f(t_1, \ldots, t_{k-1}, s, t_{k+1}, \ldots, t_n) = f(t_1, \ldots, t_{k-1}, s', t_{k+1}, \ldots, t_n).$$

PROOF: By Reflexivity, $t_i = t_i$ for i ranging from 1 to n except k. The desired equation now follows from $s = s'$ and Functional Substitution. ■

Exercise 3.1: Athena's primitive method `fcong` actually implements the First Substitution Theorem as well as the Substitution Axiom, implicitly invoking Reflexivity as needed. Thus, it accepts the desired equation (of the form that appears in the conclusion of the First Substitution Theorem) and proves it, provided that $s = s'$ is in the assumption base.

We can then apply `fcong` again, with the new equation now in the assumption base, to prove an equation between larger terms. Continue the following development:

```
declare g:[N N] -> N
declare h:[N] -> N
declare i:[N N] -> N

assert premise :=  (= (i (S zero) zero)
                      (S (i zero zero)))
```

using repeated applications of `fcong` to prove the following `goal`:

```
define goal :=
   (= (h (g zero
            (i (S zero)
                  (i (S zero) zero))))
      (h (g zero
            (i (S zero)
                  (S (i zero zero))))))
```

Note: We are using prefix notation here to make it easier to see the terms as trees. □

We now generalize this idea of repeated applications of the First Substitution Theorem:

Theorem 3.2: Second Substitution Theorem

Let s and t be terms, let $I = [i_1 \cdots i_m]$ be a position that is in both terms, and suppose s and t are identical except at I. Let s' be the subterm of s at position I, and t' the subterm of t at position I. If $s' = t'$ then $s = t$.

Remark: Understanding the following proof requires some familiarity with proof by induction, one of the important proof methods to be studied in detail in this book (beginning in Section 3.8). You may wish to skip it and come back to it later, accepting this theorem for now without proof.

PROOF: By induction on the length m of I. For the basis case, $m = 0$, we have $I = [\,]$, hence $s = s'$ and $t = t'$, so we already have $s = t$ by the assumption $s' = t'$. Otherwise, assume the result for positions of length $m - 1$ and let $s = f(s_1, \ldots, s_{i_1}, \ldots, s_n)$ and $t = g(t_1, \ldots, t_{i_1}, \ldots, t_n)$, as shown in Figure 3.3. By assumption, $f = g$ and $s_i = t_i$ for all i from 1 to n except i_1. But we also have $s_{i_1} = t_{i_1}$, by applying the induction hypothesis to these terms and position $I' = [i_2 \cdots i_m]$. Then $s = t$ follows by the First Substitution Theorem. ■

The First and Second Substitution Theorems are based on an equation $s = t$ between two terms, but we can derive an even more useful proof principle that starts with a *universally quantified* equation, implicitly using any instance of it as the basis of an application of the

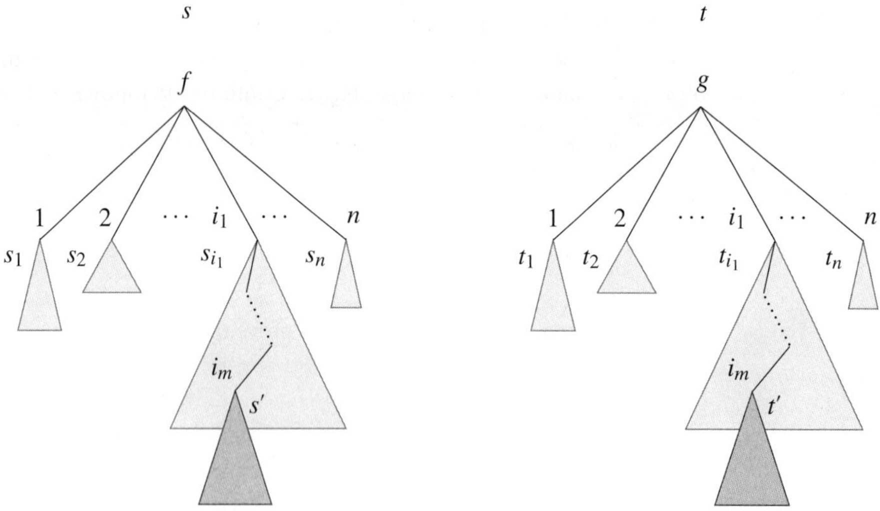

Figure 3.3
Illustration of the Second Substitution Theorem.

Second Theorem. It is best explained and justified using the notions of *rewrite rules* and *term rewriting*, which are in turn based on the notions of substitutions and matching that were described in Section 2.14.8. (It might help to review that section before continuing.)

Rewrite Rules Suppose that p is a universally quantified equation of the following form:[7]

$$p \equiv \forall\, v_1 \cdots v_n \,.\, L = R, \tag{3.2}$$

where the variables in the term L are $\{v_1, \ldots, v_n\}$ and those in R are a subset of $\{v_1, \ldots, v_n\}$. We call such a sentence p a *rewrite rule* (or, less often, a *rewriting rule*). Both of the axioms for `Plus`, namely `right-zero` and `right-nonzero`, are in the form of rewrite rules. If we were dealing with addition on integers rather than on natural numbers, then we might have an axiom like this:

```
(forall ?i . ?i + (- ?i) = 0).
```

But if this sentence were written instead as

```
(forall ?i . 0 = ?i + (- ?i)),
```

it would *not* be a rewrite rule, since the variable `?i` appears on the right-hand side of the equation but not on the left. Using such an equation for term rewriting, as defined next, would result in spurious variables being introduced into terms.

Many of the theorems that we derive, such as `left-zero` below, will also be in the form of rewrite rules and will thus be suitable for use by proof techniques that are based on rewriting, such as chaining.

In fact, most of our rewrite rules will be in a more structured form than what has been described above: They will usually be *constructor-based* rewrite rules. A constructor-based rewrite rule is of the same general form as (3.2) and adheres to the same variable restriction, but, in addition, the left-hand side L is of the form $(f\ s_1 \cdots s_n)$, where:

a. each s_i contains only variables and constructors;

b. f is not a constructor (but rather a function symbol that we are defining, such as `Plus`).

Term Rewriting Let p be a rewrite rule whose body is $L = R$. Further, suppose s and t are terms, I is a valid position for both s and t, the subterm s' of s at I matches L with substitution θ, and t is identical to s except at I, where $\theta(R)$ occurs instead. Then we say that s *rewrites* or

---

7 The case of $n = 0$, where p is an unquantified equation, is allowed. Also, as we mentioned earlier, instead of (L = R) we could have a *conditional equation* (q ==> L = R) as the body of the rule. Conditional equations are discussed in Section 3.14, and the full generality of the forms of the justifying sentences allowed in the `chain` method is described in Chapter 6.

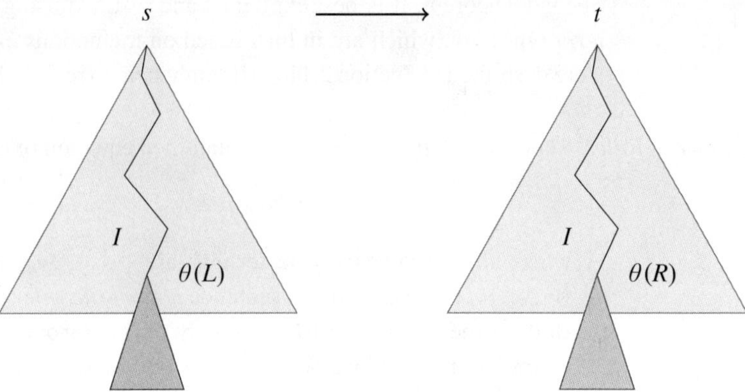

Figure 3.4
Illustration of the Rewriting Theorem.

reduces to t using p, and we call $\theta(L)$ and $\theta(R)$ the *redex* and the *contractum*, respectively. We write this relation as $s \rightarrow_\theta t$ [p], or simply $s \rightarrow t$ [p] when θ is clear from the context or immaterial. The relation is depicted in Figure 3.4. From this definition of rewriting we obtain:

Theorem 3.3: Rewriting Theorem

Let p be a rewrite rule, and $t \rightarrow u$ [p]. If p then $t = u$.

PROOF: The sentence p implies $\theta(L) = \theta(R)$, and therefore $t = u$ follows by the Second Substitution Theorem. ■

All of this machinery is brought to bear in the following finite proof principle:

Principle 3.1: Single Rewriting Step

To prove an equation $s = t$ using p, where

$$p \equiv \forall\, v_1 \cdots v_n \,.\, L = R$$

is a rewrite rule, search s for a position I with subterm s' such that s' matches L; i.e., there is a substitution θ such that s' is $\theta(L)$. If t is identical to s except at position I, where $\theta(R)$ occurs instead of s', then $s \rightarrow t$ using p, from which $s = t$ follows by the Rewriting Theorem. Otherwise (if any of these conditions fail for position I), continue to the next position. If the search of all subterm positions of s fails, then perform the same search but with the roles of s and t reversed. If successful, $t = s$ is proved, from which $s = t$ follows by Symmetry.

The basic equality chaining that the `chain` method performs is based on the above principle, although `chain` supports multiple rewrite rules in one step (and multiple redex-contractum pairs), as well as conditional rewrite rules and indeed more complex justifying sentences. If all of the individual steps of a `chain` method application

$$\texttt{(!chain [}t_0 \texttt{ = } t_1 \texttt{ [}p_1\texttt{] = } t_2 \texttt{ [}p_2\texttt{] = } \cdots \texttt{ = } t_n \texttt{ [}p_n\texttt{]])}$$

are successful, it deduces the equation $t_0 = t_n$ by combining the individually deduced equations using $n-1$ applications of Transitivity.

We close this section with a notational convention: for a list of sentences $[p_1 \ldots p_k]$, we write

$$s \rightarrow_\theta t \, [p_1 \ldots p_k]$$

(or simply $s \rightarrow t \, [p_1 \ldots p_k]$ when θ is obvious or immaterial) iff $s \rightarrow_\theta t \, [p_i]$ for *some* p_i, $i \in \{1, \ldots, k\}$. This is a natural extension of the notation $s \rightarrow_\theta t \, [p]$. We will use it whenever we want to indicate that a term rewrites to another term by virtue of *a collection* of rewrite rules (say, a set of axioms defining a function), without having to single out specifically which axiom is used for the rewrite step in question.

Exercise 3.2: This exercise develops a simple example of the Single Rewriting Step principle, with

$$s = \texttt{(Plus (S zero) (Plus (S zero) zero))}$$
$$t = \texttt{(Plus (S zero) (S (Plus zero zero)))}$$

and the rule `(forall x (= (Plus (S x) y) (S (Plus x y))))`.

(a) Show the list of all position and subterm pairs that would be produced if the procedure `positions-and-subterms` (see page 119) were applied to s. (Check your solution by actually running the procedure or with the answer given in the solutions.)

(b) Find two distinct positions I_1 and I_2 in s such that for both $k = 1, 2$, the subterm s' at I_k matches the left-hand side `(Plus (S x) y)`.

(c) For each of the two distinct positions I_1 and I_2 in s found in the previous exercise, show the substitution θ_i by which `(Plus (S x) y)` matches the corresponding subterm.

(d) For which of I_1 and I_2 in s does the Single Rewriting Step principle succeed in proving $s = t$? In that case, what term corresponds to $\theta(R)$? □

$\star$ ***Exercise 3.3:***[8] Define a method `basic-rewrite` for proving an equation between terms based on the Second Substitution Theorem. Let s and t be terms, I be a valid position in

8 Starred exercises are generally more difficult than unstarred ones. Two stars indicate even greater difficulty.

both s and t, let s' be `(subterm s I)`, let t' be `(subterm t I)`, and assume `(s' = t')` is in the assumption base. Then `(!basic-rewrite s I t)` should derive `(s = t)`. *Hint:* The solution can be expressed as a recursive method, corresponding directly to the inductive proof given for the Second Substitution Theorem. □

⋆ ***Exercise 3.4:*** Define a method `ltr-rewrite` ("left-to-right rewrite") for proving an equation based on the Rewriting Theorem. `(!ltr-rewrite s p t)` should prove $(s = t)$ if $s \rightarrow t\,[p]$. *Hint:* Base the search for a redex on procedure `positions-and-subterms`. Check whether a candidate redex matches the left-hand side of the equation in p using procedure `match-terms` (see page 97); when it does, derive the corresponding substitution instance of p using the `instance` method. Then the `basic-rewrite` procedure of the preceding exercise can be used to complete the proof, if one exists for this redex. □

Exercise 3.5: Define a method `rewrite` for proving an equation based on the Single Rewriting Step principle: `(!rewrite s p t)` should prove `(s = t)` if either $s \rightarrow t\,[p]$ or $t \rightarrow s\,[p]$. □

Exercise 3.6: Define a method `chain` that implements equality chaining as described at the beginning of Section 3.2. While Athena's `chain` method is quite a bit more powerful—it also supports implication chaining, multistep-rewriting, and other extensions described later—the basic version that you can now implement in terms of `rewrite` should be able to handle all of the examples in this chapter (except that examples of directional rewriting, introduced on page 132, would need to be modified to use `=`, with **`define`** `--> := =;` **`define`** `<-- := =`). □

3.5 More examples of equality chaining

Chaining equalities together to prove a new equality is one of the most useful proof methods in mathematics and computer science, as we shall see throughout this book. However, proving equations like $2+2=4$ provides underwhelming evidence for this method's importance, so let's move on to more interesting examples.

In preparation, we note a couple of more general properties of `Plus`. We begin with the following property:

```
define left-zero := (forall n . zero + n = n)
```

which differs from `right-zero` in that `zero` appears as the first input to `Plus` rather than the second.

Exercise 3.7: Although the `chain` method alone is inadequate to prove `left-zero` (we will see later exactly why this is so), `chain` can prove instances of it, with specific ground terms substituted for `n`. Prove each of the following equations using `chain`:

```
(zero + zero = zero),
(zero + S zero = S zero),
(zero + S S zero = S S zero).
```

Your solution may also use one or both of `right-zero` and `right-nonzero`. □

Later in the chapter we will come back and prove `left-zero`, but for now, we assert it into the assumption base (treating it like an axiom):

```
assert left-zero
```

Similarly:

```
assert left-nonzero := (forall m n . (S n) + m = S (n + m))
```

Although there's quite a bit more to say and prove about `Plus`, let's continue by introducing a multiplication function, `Times`, that takes two natural numbers as inputs and returns their product. First, the syntax:

```
declare Times: [N N] -> N [*]
```

Here, we have overloaded the symbol `*` to mean `Times` when applied to `N` arguments. Next, the semantics:

```
assert Times-zero := (forall x . x * zero = zero)
assert Times-nonzero := (forall x y . x * S y = x * y + x)
```

If we read `(S` n`)` as $n+1$, the second axiom just says

$$x \cdot (y+1) = x \cdot y + x.$$

Note that `*` has a built-in precedence greater than that of `+`, so that, for example, `(`x `*` y `+` z`)` is parsed as `((`x `*` y`)` `+` z`)`.

Let's also introduce a name one and give its meaning with an equation:

```
declare one: N

assert one-definition := (one = S zero)
```

The proof of the following property:

```
define Times-right-one := (forall x . x * one = x)
```

provides another simple illustration of equality chaining:

```
conclude Times-right-one
  pick-any x:N
    (!chain [(x * one)
```

```
           = (x * S zero)          [one-definition]
           = (x * zero + x)        [Times-nonzero]
           = (zero + x)            [Times-zero]
           = x                     [left-zero]])
```

The main difference from the previous examples is that the equalities involved are not just between ground terms; the terms include the (fresh) variable denoted by `x`, which, loosely speaking, represents an arbitrary value of sort `N`, as indicated by the way it is introduced in the **`pick-any`** step.

Exercise 3.8: How is `Times-right-one` related to `right-zero`? □

Here is one more property of `Times`, which for the moment we will treat as an axiom by asserting it into the assumption base:

```
assert Times-associative := (forall x y z . (x * y) * z = x * (y * z))
```

As a final bit of preparation for more substantial examples of equational proof, let us define an exponentiation function, `**`. We set the precedence of `**` higher than that of `*` (which is predefined to be 300).

```
declare **: [N N] -> N [310]
```

For semantics, we write:

```
assert Power-right-zero := (forall x . x ** zero = one)

assert Power-right-nonzero := (forall x n . x ** S n = x * x ** n)
```

3.6 A more substantial proof example

Recall the following result from elementary algebra:

```
define power-square-theorem := (forall n x . (x * x) ** n = x ** (n + n))
```

(In more conventional notation, $(x^2)^n = x^{2n}$.) If we define the following procedure:

```
define (power-square-property n) :=
  (forall x . (x * x) ** n = x ** (n + n))
```

we can then express `power-square-theorem` more succinctly as the proposition that every natural number has `power-square-property`:

```
(forall n . power-square-property n).
```

Athena will verify that the two formulations are identical:

```
> (power-square-theorem equals? (forall n . power-square-property n))

Term: true
```

We call power-square-property a *property procedure*; its role will become clear soon.

So how do we go about proving power-square-theorem? In this case it might be a good idea to start by writing down a few specific instances of the result, to convince ourselves that it is valid at least in those cases. For this theorem, instantiating only the variable n for a few small values yields:

```
(forall x . (x * x) ** zero = x ** (zero + zero))

(forall x . (x * x) ** S zero = x ** (S zero + S zero))

(forall x . (x * x) ** S S zero = x ** (S S zero + S S zero))

(forall x . (x * x) ** S S S zero = x ** (S S S zero + S S S zero))

⋮
```

These sentences can be automatically obtained by applying power-square-property to zero, (S zero), and so on. While we could also try instantiating x so that we could check the resulting equations with pure calculation (and indeed in Section 3.13 we discuss a technique that automates that approach), let us instead bring proofs into play. Thus, before trying to prove power-square-theorem in its most general form, let's see if we can prove some of its above instances. The proof of (power-square-property zero) is simple:

```
conclude power-zero-case := (power-square-property zero)
  pick-any x:N
    (!chain [((x * x) ** zero)
           = one                           [Power-right-zero]
           = (x ** zero)                   [Power-right-zero]
           = (x ** (zero + zero))          [right-zero]])
```

For n ≡ (S zero), first consider the following proof:

```
conclude power-one-case := (power-square-property (S zero))
  pick-any x:N
    (!combine-equations
      (!chain [((x * x) ** S zero)
             = ((x * x) * (x * x) ** zero)  [Power-right-nonzero]
             = ((x * x) * one)              [Power-right-zero]
             = (x * x)                      [Times-right-one]])
      (!chain [(x ** (S zero + S zero))
             = (x ** (S (S zero + zero)))   [right-nonzero]
             = (x ** (S S zero))            [right-zero]
             = (x * (x ** S zero))          [Power-right-nonzero]
```

```
           = (x * (x * (x ** zero)))        [Power-right-nonzero]
           = (x * x * one)                  [Power-right-zero]
           = (x * x)                        [Times-right-one]]))
```

The structure of this proof, combining *two* applications of chain, illustrates a frequently useful variation of equality chaining. Instead of a single chain of equations, we first write a chain in which we start with the left-hand side of the equation we want to prove, at each step using some axiom or other equation in a *left-to-right* direction. That is, we match the equation's left-hand side to some subterm of the current term, replacing it with the right-hand side. Then we work in the same manner starting with the right-hand side, again using axioms and other equations in a left-to-right direction. Diagrammatically:

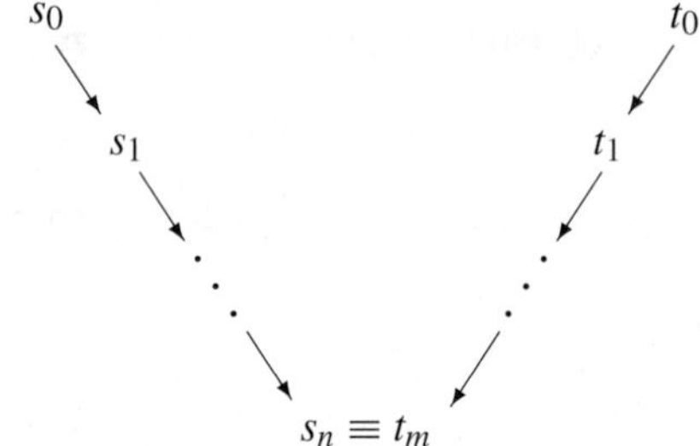

If we can rewrite each side of the equation we are trying to prove down to the very same term, then we can combine the two chain conclusions to obtain the proof. The method combine-equations does just that:

(!combine-equations (s_0 = s_n) (t_0 = t_m))

proves (s_0 = t_0) when both (s_0 = s_n) and (t_0 = t_m) are in the assumption base and s_n and t_m are identical.

In addition, chain allows the direction of rewriting to be indicated on each step: Instead of an equality symbol between two terms, either --> or <-- can be used. If --> is used, chain only attempts to rewrite left-to-right: $t_i \rightarrow t_{i+1}$; and if <-- is used, it only attempts to rewrite right-to-left: $(t_{i+1} \rightarrow t_i)$. If = is specified, chain first tries left-to-right, and if that fails, it tries right-to-left.

Note that in right-to-left rewriting, as defined here, it is *not* the equation within p_i, say $L_i = R_i$, that is reversed. That must be avoided, as there might be variables in L_i that do not occur in R_i. (Recall the requirement that the variables of the right-hand side of the equation must be a subset of those of the left.)

For example, every equality symbol in the preceding proof of power-one-case could be replaced by -->, and the proof of power-zero-case could be written as follows:

```
conclude (forall x . (x * x) ** zero = x ** (zero + zero))
  pick-any x:N
    (!chain [((x * x) ** zero)
         --> one                        [Power-right-zero]
         <-- (x ** zero)                [Power-right-zero]
         <-- (x ** (zero + zero))       [right-zero]])
```

Thus, in the original power-zero-case proof, where = was used, chain tries to do the second step using Power-right-zero left-to-right and fails, but it succeeds in using it right-to-left, and likewise for the third step.

A fair question at this point is: Do we really need combine-equations? Why not write a single application of chain? For example, instead of

```
(!combine-equations
  (!chain [s0 --> s1 [p1] --> s2 [p2] --> ··· --> sn [pn]])
  (!chain [t0 --> t1 [q1] --> t2 [q2] --> ··· --> tm [qm]]))
```

we could write a single chain in which we reverse the order of the second sequence of chain links:

```
(!chain [s0 --> s1 [p1] --> s2 [p2] --> ··· --> sn [pn]
             <-- tm-1 [qm] <-- ··· <-- t1 [q2] <-- t0 [q1]])
```

Though possible, it is often more difficult to find the correct sequence of right-to-left proof steps to get from the term s_n (which must be identical to t_m) to the desired term t_0 than it is to find the separate chains working down to a common term using only (or primarily) left-to-right rewrites. In this book, we have a goal more important than succinctness: We want to illustrate and promote good strategies for *finding* proofs and writing them down in a clear, easy-to-understand style. The strategy of rewriting both sides to a common term often leads to a successful proof more easily than by trying to construct one as a single chain of rewrites, and the proof is usually easier for someone else to read and understand.

Thus, we will usually avoid writing equality proofs as single chains, restricting that practice to cases where (1) the proof can be done just with left-to-right rewriting anyway, or (2) only a few right-to-left rewriting steps are used, and they are only applied to fairly simple terms, as in our proof above of the n = zero case. Otherwise, we will tend to write two chains and combine them, as in the n = (S zero) case.

Exercise 3.9: Modify the given proof of power-one-case to use a single application of chain. Indicate the direction of each rewrite with --> or <-- instead of just using = signs. Verify that the resulting proof works. □

3.7 A better proof

The proof given for power-one-case, whether written with one chain application or two, is longer than it needs to be. It can be shortened by taking advantage of the power-zero-case theorem, as follows:

```
conclude (forall x . (x * x) ** S zero = x ** (S zero + S zero))
  pick-any x:N
    (!combine-equations
      (!chain [((x * x) ** S zero)
           --> ((x * x) * ((x * x) ** zero))        [Power-right-nonzero]
           --> ((x * x) * (x ** (zero + zero)))     [power-zero-case]
           --> (x * x * x ** (zero + zero))         [Times-associative]])
      (!chain [(x ** (S zero + S zero))
           --> (x ** (S (S zero + zero)))           [right-nonzero]
           --> (x ** (S (S (zero + zero))))         [left-nonzero]
           --> (x * (x ** (S (zero + zero))))       [Power-right-nonzero]
           --> (x * x * x ** (zero + zero))         [Power-right-nonzero]]))
```

With this change, the proof is not just shorter than the previous one; it has a structure that can be generalized to prove the theorem for *any* value of n. To see this, note that the zero that appears in the (S zero) term in the theorem itself and throughout the proof *never plays any role in the proof.* (For example, neither right-zero nor Times-zero-axiom or Power-right-zero is ever used.) We can take advantage of this observation most easily if we encapsulate the proof in a *method*, as follows:

```
define power-square-step :=
  method (n)
    let {previous-result := (power-square-property n)}
      conclude (power-square-property (S n))
        pick-any x:N
          (!combine-equations
            (!chain [((x * x) ** S n)
                 --> ((x * x) * ((x * x) ** n))     [Power-right-nonzero]
                 --> ((x * x) * (x ** (n + n)))     [previous-result]
                 --> (x * x * (x ** (n + n)))       [Times-associative]])
            (!chain [(x ** (S n + S n))
                 --> (x ** (S (S n + n)))           [right-nonzero]
                 --> (x ** (S S (n + n)))           [left-nonzero]
                 --> (x * (x ** (S (n + n))))       [Power-right-nonzero]
                 --> (x * x * (x ** (n + n)))       [Power-right-nonzero]]))
```

This method definition does not itself prove a theorem, but a successful call of it does. In particular, if we apply this method to any specific natural number n, we will obtain the theorem

(power-square-property (S n)),

provided that (power-square-property n) *is already in the assumption base*, which is required for the appeal to previous-result in the body of the method to succeed. Hence the name, power-square-step, indicating the "stepping" of power-square-property from one natural number to the next. Let's also encapsulate the previously given proof of the n $\equiv$ zero case in a separate method, power-square-base, which needs no argument:

```
define power-square-base :=
  method ()
    conclude (power-square-property zero)
      pick-any x:N
        (!chain [((x * x) ** zero)
                   = one                    [Power-right-zero]
                   = (x ** zero)            [Power-right-zero]
                   = (x ** (zero + zero))   [right-zero]])
```

Then the following sequence of calls could be extended to obtain the proof of (power-square-property n) for *any* natural number n:

```
(!power-square-base)
(!power-square-step zero)
(!power-square-step (S zero))
        ⋮
```

Exercise 3.10: Verify in an Athena session that the above sequence of calls does prove (power-square-property n) for $n \equiv$ zero, (S zero), (S S zero), and (S S S zero). □

3.8 The principle of mathematical induction

At this point we have proved a few special cases of power-square-theorem, and we can even imagine repeatedly invoking power-square-step to eventually obtain the proof for *any* given n. In Athena we could even program an iterative or recursive proof method that carries out all of the proofs up to the desired natural number, which in fact we'll do later, mainly as an illustration of Athena's proof-programming facilities. While that is probably enough to convince us of the validity of power-square-theorem,[9] we still don't have a formal proof of it in its general form, with n universally quantified. But we do now have the main ingredients of a proof, namely our power-square-base and power-square-step methods. To complete the job, we now invoke a fundamental proof method known as *the principle of mathematical induction*. As we will see, this principle does not apply only to

9 In Section 3.13 we will see another technique that can help to convince us that a conjecture p is true without having a proof for it: the falsify procedure.

sentences about natural numbers, but the form it takes for the natural numbers is one of the most important, and is exactly what we need here:

Principle 3.2: Mathematical Induction for Natural Numbers

To prove $\forall\, n \,.\, P(n)$ where n ranges over the natural numbers, it suffices to prove:

1. *Basis case*: $P(0)$.
2. *Induction step*: $\forall\, n \,.\, P(n) \Rightarrow P(n+1)$.

In the induction step, the antecedent assumption $P(n)$ is called the *induction hypothesis*.

This principle is embodied in Athena's **by-induction** proof construct. We can use it to prove power-square-theorem as follows:

```
by-induction power-square-theorem {
  zero   => (!power-square-base)
| (S n) => (!power-square-step n)
}
```

The keyword **by-induction** is followed by the sentence to be derived, which is a goal of the form

$$\forall\, n : N \,.\, P(n),$$

followed by a number of *clauses*, enclosed in curly braces and separated by |, expressing the cases that together are sufficient to complete the proof. There are usually two clauses (there can be more): one that expresses the basis case, corresponding to $P(0)$, and the other expressing the induction step (or "inductive step"), corresponding to

$$\forall\, n \,.\, P(n) \Rightarrow P(n+1).$$

Each clause is essentially a pair consisting of a constructor pattern π_i that represents one of the cases of the inductive argument, and a corresponding subproof D_i. The arrow keyword => separates π_i from D_i. The subproof D_i will be evaluated in the original assumption base *augmented with all appropriate inductive hypotheses*. There may be zero inductive hypotheses if the pattern π_i corresponds to a basis case.

The induction-step sentence for our example can be written in Athena as follows:

```
(forall n . power-square-property n ==> power-square-property (S n))
```

If we were trying to prove this sentence from scratch, without the benefit of **by-induction**, we could do it with a proof along the following lines:

```
pick-any n:N
  assume induction-hypothesis := (power-square-property n)
    conclude (power-square-property (S n))
      (!power-square-step n)
```

But with **by-induction** it is not necessary to write this much detail. Essentially, Athena automatically takes care of the first three steps: the **pick-any**, **assume**, and **conclude**. The **pick-any** identifiers will be all and only those identifiers that occur as pattern variables in π_i. These identifiers will be bound to freshly generated variables of the appropriate sorts throughout the evaluation of D_i, obviating the need for an explicit **pick-any**. And the inductive hypotheses will be automatically constructed and inserted into the assumption base for you (temporarily, only while evaluating the subproof D_i), thus obviating the need for an explicit **assume**. So all D_i needs to do is to derive the result $P(n+1)$. Athena will then check that the produced result is of the right form, which also avoids the need for an explicit **conclude**.

Exercise 3.11: Verify that the above proof derives the induction step, and that the **by-induction** proof before that derives power-square-theorem. □

We expressed proof principle 3.2 in informal notation, writing, for example, $\forall\, n \,.\, P(n)$ instead of using Athena notation. But what exactly is property P? Perhaps the best way to think of it is as a unary procedure that takes a natural number term t and produces a sentence. We can always define such a procedure and use it to drive the entire workflow of an inductive proof, as outlined in the following schema:

```
# Start by defining a unary ''property procedure'':

define (P t) := ...

# Then use it to define a goal which says that
# every object has this property:

define goal := (forall n:N . P n)

# Finally, prove the goal by induction:

by-induction goal {
  zero    => conclude (P zero)
               (!basis-case ...)
| (n as (S m)) =>
    conclude (P n)            # Here the assumption base contains
      (!induction-step ...)   # the inductive hypothesis (P m).
}
```

where basis-case and induction-step are methods that may take any number of arguments, depending on the situation. A brief word on **as** patterns: A pattern of the form (I **as** π) works just like π, except that it also gives the name I to whatever value corresponds to π. The scope of I is the deduction that appears on the right-hand side of the arrow =>.

In this particular example, the procedure encoding the property of interest was power-square-property:

```
define (power-square-property n) :=
  (forall x . (x * x) ** n = x ** (n + n))
```

This simple procedure can be applied to an arbitrary term *t* of sort `N` and will produce a sentence essentially stating that `power-square-theorem` holds for *t*:

```
> (power-square-property zero)

Sentence: (forall ?x:N
            (= (** (Times ?x:N ?x:N)
                   zero)
               (** ?x:N
                   (Plus zero zero))))

> (power-square-property (S zero))

Sentence: (forall ?x:N
            (= (** (Times ?x:N ?x:N)
                   (S zero))
               (** ?x:N
                   (Plus (S zero)
                         (S zero)))))

> (power-square-property ?k)

Sentence: (forall ?x:N
            (= (** (Times ?x:N ?x:N)
                   ?k:N)
               (** ?x:N
                   (Plus ?k:N ?k:N))))
```

But it is not necessary to adhere to this style of defining property procedures for each inductive proof. Neither is it necessary to define proof methods like `power-square-base` and `power-square-step` in order to use **`by-induction`**. We can simply write out the proofs of the basis case and induction step *inline*. For a simpler example of this approach, let us take a property that we defined earlier in the chapter:

```
define left-zero := (forall n . zero + n = n)
```

There we asserted it into the assumption base, but now let us prove it, using **`by-induction`** and inline proofs:

```
by-induction left-zero {
  zero => conclude (zero + zero = zero)
            (!chain [(zero + zero) --> zero [right-zero]])
| (n as (S m)) =>
    conclude (zero + n = n)
      let {induction-hypothesis := (zero + m = m)}
        (!chain [(zero + S m)
                 --> (S (zero + m))      [right-nonzero]
                 --> (S m)               [induction-hypothesis]])
}
```

The principal advantage of writing the subproofs inline is that it requires less preparation. As with ordinary programming, however, when "proof code" gets larger it may be worth the extra trouble to encapsulate it in methods. Then the proof of the basis case can be tested even before attempting to write down the proof of the induction step. Similarly, the proof of the induction step can be separately tested, either with successive applications to `zero`, `(S zero)`, `(S S zero)`, and so on, or in the manner shown above for the induction step for `power-square-theorem`.[10] Similar remarks apply to defining procedures like `power-square-property`; while not necessary, they are often useful in making our proofs more structured.

3.8.1 Different ways of understanding mathematical induction

With this simple proof of `left-zero` at hand, let's return to a point we alluded to at the beginning of the chapter, namely that the *only* natural number values are those that can be obtained by starting with `zero` and applying `S` some finite number of times. *One way to understand the principle of mathematical induction for natural numbers is that it says precisely the same thing.* Why? Because:

1. For any value n that is obtained by starting with `zero` and applying `S` some finite number of times, we can prove $P(n)$ by starting with the basis case and applying the induction step the same number of times; and
2. the principle says that just proving the basis case and the induction step is sufficient to prove P for all natural numbers.

As another aid to intuition about mathematical induction, consider what would happen if we posited a natural number called, say, `unnatural`, that we assume is *not* equal to any of the values generated by `zero` and `S`. Consider then, once again, `left-zero`. Is it still valid? We proved it, but our proof, using the principle of mathematical induction, did not consider the possibility that there could be natural numbers other than those that can be generated from `zero` and `S`. And, without further information, we *cannot* conclude

```
(zero + unnatural = unnatural),
```

so the property isn't necessarily valid. We might in fact have `(zero + unnatural = zero)` or `(zero + unnatural = one)`, or any other value besides `unnatural`, and that would invalidate `left-zero`.

Exercise 3.12: Joe Certain claims that he can prove

```
(zero + unnatural = unnatural)
```

10 Another, more general approach to testing partial proofs will be introduced in Section 4.7.

using the fact that Plus satisfies a commutative law:

(forall m n . m + n = n + m).

Thus:

```
(!chain [(zero + unnatural)
    --> (unnatural + zero)         [Plus-commutative]
    --> unnatural                  [right-zero]])
```

What's wrong with Joe's "proof"? After all, one *can* prove Plus-commutative using mathematical induction; see Exercise 3.20. In answering, assume that unnatural has somehow been declared, and in such a way that we do not get a sort-checking error with any of the terms in the above proof (i.e., your answer shouldn't simply be that unnatural is an undefined identifier or that there is a sort error). □

Here is still another angle on mathematical induction, this time from the point of view of defining datatypes like N. Recall that N is introduced with the following definition:

```
datatype N := zero | (S N)
```

which can be read as "N consists of all values, and only those values, recursively generated by the *constructors* zero and S." The values in question are called *canonical terms*. We will have more to say about these in Section 3.10, but in general, a canonical term is defined as a term that contains only constructors and/or numerals of sort Int or Real and/or meta-identifier constants of the form '*I*.

It is this **datatype** definition that determines **by-induction**'s requirements for proof clauses: When the first universally quantified variable of the stated goal is of sort N, **by-induction** requires subsequent clauses that "cover" all possible values that can be generated using constructors zero and S.

Exercise 3.13: Experiment with a declaration of a datatype N' that differs from N in having an additional value, unnatural:

```
datatype N' := zero' |  unnatural | (S' N)
```

Which of the proofs in this chapter (to this point) would still work if N is replaced by N', zero by zero', and S by S'? □

3.9 List equations

We now turn to a different class of equations, those involving *lists* rather than natural numbers. We will introduce lists as a datatype with two constructors, nil and ::, which are analogous to the natural number constructors zero and S, respectively. (The :: constructor is adopted from similar usage in ML and is pronounced "cons" in the tradition of lists in Lisp and other functional programming languages.) We will see how useful functions on lists, such as concatenation and reversal, can be precisely defined with equational axioms and how further properties can be derived from these axioms. Since the axioms and additional properties are all stated as universally quantified equations, like those concerning natural numbers, we already have at hand the necessary proof methods—equality chaining and induction. We will see that equational chaining works for lists exactly as it does for natural numbers, and that induction requires only a minor adjustment to account for the difference in constructors.

To begin, we introduce lists in Athena with a **datatype** definition, first in a form in which the list elements are natural numbers:

```
datatype N-List := nil | (:: N N-List)
```

Thus, the constructor `nil` takes no arguments and the constructor `::` takes two: a natural number, and, recursively, an `N-List`. Here are a few ground terms of this sort:

```
nil
(one :: nil)
(zero :: nil)
(zero :: S zero :: S S zero :: S S S zero :: nil)
(one :: zero :: nil)
(one :: zero :: S one :: nil)
```

We interpret `nil` as the empty list (that contains no elements) and $(x :: L)$ as the list whose first element is x and whose remaining elements, if any, are those of list L. (Borrowing from Lisp terminology, we often say that $(x :: L)$ is the result of "consing" x onto L.) So the last example has `one` as its first element, `zero` as its second element, and `(S one)` as its third and final element.

This interpretation is reflected in the definition of the first list function we will study, `join`, for concatenating two lists. By concatenation we mean that the result of `(L1 join L2)` is a list that begins with the elements of list `L1`, in the same order as in `L1`, followed by the elements of `L2`, in the same order as in `L2`.

```
define [L L' L1 L2 L3] :=
       [?L:N-List ?L':N-List ?L1:N-List ?L2:N-List ?L3:N-List]

declare join: [N-List N-List] -> N-List [++]
```

We defined some handy variable names to avoid having to type variables in their fully explicit form (preceded by question marks) in what follows, and we also introduced ++ as an alias for the function symbol `join`.

We want `::` to bind tighter than ++ (to make sure, e.g., that `(x::L1 ++ L2)` is understood as the join of `x::L1` and `L2` rather than the result of consing `x` onto the join of `L1` and `L2`), so we give a higher precedence to `::` with the following directive:

```
set-precedence :: 150
```

We now introduce two axioms that define concatenation:

```
assert left-empty := (forall L . nil ++ L = L)
assert left-nonempty := (forall x L1 L2 . x::L1 ++ L2 = x::(L1 ++ L2))
```

With the previous definitions for natural-number functions such as `Plus` and `Times`, we were able to rely on a lot of experience. Lists, by contrast, are likely not as familiar, so let's make sure we understand how these axioms define `join`. The first axiom states that the list produced by joining `nil` and `L` is identical to `L`.

The second axiom can be read as follows: By consing `x` to a list `L1` and then joining the result with a list `L2`, we get the same list that we would get by first joining `L1` and `L2` and then consing `x` to the result. Figure 3.5 illustrates this relation. The left part of the diagram depicts the computation of $(x::L_1 \text{ ++ } L_2)$ and the right part that of $(x::(L_1 \text{ ++ } L_2))$. We see that the results are identical.

Besides holding up defining axioms to scrutiny and drawing diagrams, it is also helpful to use the axioms to compute the function for a few ground term inputs, such as

```
(one::zero::nil ++ one::zero::S one::nil).
```

We can reduce such a ground term to one that only involves `::` and `nil` by applying the `join` axioms in a chain of equalities:

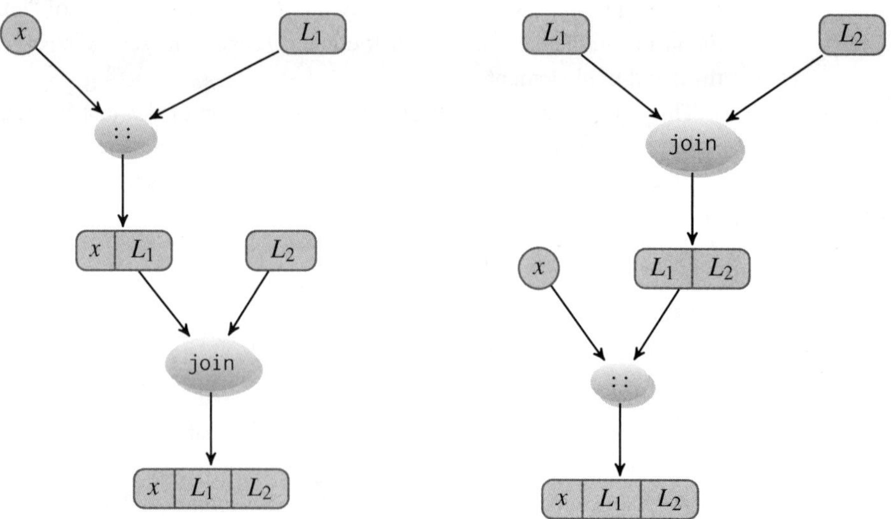

Figure 3.5
Illustration of the second axiom for list concatenation.

```
> (!chain [(one::zero::nil ++ one::zero::S one::nil)
          = (one::(zero::nil ++ one::zero::S one::nil)) [left-nonempty]
          = (one::zero::(nil ++ one::zero::S one::nil)) [left-nonempty]
          = (one::zero::one::zero::S one::nil)          [left-empty]])

Theorem: (= (join (:: one
                      (:: zero nil))
                  (:: one
                      (:: zero
                          (:: (S one)
                              nil))))
            (:: one
                (:: zero
                    (:: one
                        (:: zero
                            (:: (S one)
                                nil))))))
```

Carrying out such equality proofs and examining the results provides additional evidence that the axioms for join give it the meaning we informally prescribed for it. (A more direct computational way to obtain such evidence—without constructing proofs—will be introduced in Section 3.10.) But still another way to bolster our confidence is to state, as conjectures, more properties that we intuitively expect a list concatenation function to have. The simplest such property is one like the first axiom, but with nil as the second argument:

```
define right-empty := (forall L . L ++ nil = L)
```

Note the analogy with the natural number function Plus, with nil now playing the role of zero: We had an axiom

```
assert right-zero := (forall n . n + zero = n),
```

and we later stated and proved

```
define left-zero := (forall n . zero + n = n).
```

If we can prove right-empty we will have completed the analogy (except that our list axiom and property have nil in the first and second arguments of join, respectively, the reverse of what we had with zero for Plus).

We cannot prove right-empty just by equality chaining, since neither of our axioms can be used to rewrite one side of its equation to the other. But as we did with Plus, let's see what happens with a few ground term inputs to join when its second argument is nil. For example:

```
(!chain [(one::nil ++ nil)
       = (one::(nil ++ nil))    [left-nonempty]
       = (one::nil)             [left-empty]])

(!chain [(one::zero::nil ++ nil)
```

```
        = (one::(zero::nil ++ nil))      [left-nonempty]
        = (one::zero::(nil ++ nil))      [left-nonempty]
        = (one::zero::nil)               [left-empty]])

(!chain [(S one::one::zero::nil ++  nil)
        = (S one::(one::zero::nil ++ nil))   [left-nonempty]
        = (S one::one::(zero::nil ++ nil))   [left-nonempty]
        = (S one::one::zero::(nil ++ nil))   [left-nonempty]
        = (S one::one::zero::nil)            [left-empty]])
```

In each case, we verify that the result computed by `join` is the same as its first argument, and we strongly suspect this must be true for lists of any length. But if we continued this pattern, then to verify `right-empty` for a list of length n we would need a calculation of length n. We can do better by noting that we can use the result for length n in the proof for length $n+1$. The following method definitions and evaluations are analogous to those of `power-square-base` and `power-square-step` in Section 3.7, but they are a little more complicated due to the extra argument for the list element in the `::` constructor.

```
define (join-nil-base) :=
  conclude (nil ++ nil = nil)
    (!chain [(nil ++ nil) = nil [left-empty]])

define (join-nil-step IH x L) :=
  conclude (x::L ++ nil = x::L)
    (!chain [(x::L ++ nil)
            = (x::(L ++ nil))      [left-nonempty]
            = (x :: L)             [IH]])
```

We now have:

```
> define p0 := (!join-nil-base)

Theorem: (= (join nil nil)
            nil)

Sentence p0 defined.

> define p1 :=
    pick-any x
      (!join-nil-step p0 x nil)

Theorem: (forall ?x:N
           (= (join (:: ?x:N nil)
                    nil)
              (:: ?x:N nil)))

Sentence p1 defined.
```

```
> define p2 :=
    pick-any x:N y:N
      (!join-nil-step p1 x (y::nil))

Theorem: (forall ?x:N
           (forall ?y:N
             (= (join (:: ?x:N
                          (:: ?y:N nil))
                      nil)
                (:: ?x:N
                    (:: ?y:N nil)))))

Sentence p2 defined.

> define p3 :=
    pick-any x:N y:N z:N
      (!join-nil-step p2 x (y::z::nil))

Theorem: (forall ?x:N
           (forall ?y:N
             (forall ?z:N
               (= (join (:: ?x:N
                            (:: ?y:N
                                (:: ?z:N nil)))
                        nil)
                  (:: ?x:N
                      (:: ?y:N
                          (:: ?z:N nil)))))))

Sentence p3 defined.
```

The pattern of these proofs suggests that we could prove `right-empty` in full generality by mathematical induction, perhaps by reformulating the property in terms of the length of the first list, which is a natural number. In fact, however, we need no such reformulation; we can prove it directly. In Athena, the following proof works:

```
> by-induction right-empty {
    nil => (!join-nil-base)
  | (h::t) => let {IH := (t ++ nil = t)}
                (!join-nil-step IH h t)
  }

Theorem: (forall ?L:N-List
           (= (join ?L:N-List nil)
              ?L:N-List))
```

or, writing the proofs of the `nil` and `::` cases inline:

```
> by-induction right-empty {
    nil => (!chain [(nil ++ nil) = nil [left-empty]])
  | (h::t) =>
      let {IH := (t ++ nil = t)}
        conclude (h::t ++ nil = h::t)
          (!chain [(h::t ++ nil)
                 = (h :: (t ++ nil))    [left-nonempty]
                 = (h::t)               [IH]])
  }

Theorem: (forall ?L:N-List
           (= (join ?L:N-List nil)
              ?L:N-List))
```

The way **by-induction** works for lists is based on the following:

Principle 3.3: Mathematical Induction for Lists of Natural Numbers

To prove $\forall\, L\,.\, P(L)$ where L ranges over lists of natural numbers, it suffices to prove:

1. *Basis case*: $P(\mathtt{nil})$.
2. *Induction step*: $\forall\, L\,.\, P(L) \Rightarrow \forall\, x\,.\, P(x\texttt{::}L)$.

In the induction step, the antecedent assumption $P(L)$ is called the *induction hypothesis*, and x ranges over the natural numbers.

So we have seen two flavors of mathematical induction, for natural numbers and for lists of natural numbers. Both are supported by Athena's **by-induction** proof form. The way **by-induction** adapts to different datatypes is through the information supplied in **datatype** definitions about the given constructors. Thus, from

```
datatype N-List := nil | (:: N N-List),
```

by-induction deduces that it must expect clauses corresponding to at least two cases, one corresponding to `nil` and one corresponding to `::`. The `nil` case is called a *basis case* because `nil` is an irreflexive constructor that takes no argument of the datatype being defined; there would be more than one basis case if there were more such constructors. The `::` case, since `::` does take an argument of the datatype being defined, corresponds to an *induction step* (or "inductive step"), and **by-induction** temporarily assumes an appropriate inductive hypothesis for the duration of the proof given in that clause. The general evaluation semantics of **by-induction** are discussed in Appendix A.3.

Exercise 3.14: Reformulate the inductive proof of `right-empty` in the property-procedure style described in Section 3.8. □

3.9.1 Polymorphic datatypes

Before going on to further examples of proofs about list functions, it should be noted that none of the axioms or proofs so far have depended in any way on the sort of the list elements. In place of `N`, we could have used any sort, and everything other than the actual values in ground terms would be the same. So, for example, if we defined

```
datatype Boolean-List := nilb | (consb Boolean Boolean-List)
```

then we could repeat all of the development by replacing any `N` ground values with `true` or `false`; nothing else would need to change. But such repetition is something we should avoid if possible; writing proofs is hard enough without having to repeat them over and over again for different sorts! Fortunately, we can avoid it by issuing declarations and definitions with *sort parameters*, as discussed in Section 2.8:

```
datatype (List S) := nil | (:: S (List S))

declare join: (S) [(List S) (List S)] -> (List S) [++]
```

(Note that `(List S)` is predefined in Athena and in practice there would be no need to actually issue the above definition.)

As a convenience, we define a few general (polymorphic) variable names, as well as some specifically for polymorphic lists; and we set the precedence level of `::` to 150:

```
define [x y z x1 x2 h h1 h2] :=  [?x ?y ?z ?x1 ?x2 ?h ?h1 ?h2]

define [L L' L0 L1 L2 t t1 t2] :=
       [?L:(List 'S1) ?L':(List 'S2) ?L0:(List 'S3) ?L1:(List 'S4)
        ?L2:(List 'S5) ?t:(List 'S6) ?t1:(List 'S7) ?t2:(List 'S8)]

set-precedence :: 150
```

We can now give a polymorphic definition of `join` as follows:

```
assert left-empty    := (forall L . nil ++ L = L)
assert left-nonempty := (forall x L1 L2 . x::L1 ++ L2 =  x::(L1 ++ L2))
```

All of the proofs we previously did with `N-List` values could now be redone using `(List N)` values instead; for example, one of the proofs on page 142:

```
let {L := (one::zero::S one::nil)}
  (!chain [(one::zero::nil ++ L)
         = (one::(zero::nil ++ L))  [left-nonempty]
         = (one::zero::(nil ++ L))  [left-nonempty]
         = (one::zero::L)           [left-empty]])
```

or the proof on page 145:

```
define right-empty := (forall L . L ++ nil = L)

> by-induction right-empty {
    nil => (!chain [(nil ++ nil) = nil    [left-empty]])
  | (h::t) =>
       let {IH := (t ++ nil = t)}
         conclude (h::t ++ nil = h::t)
           (!chain [(h::t ++ nil)
                  = (h :: (t ++ nil))     [left-nonempty]
                  = (h :: t)              [IH]])
  }

Theorem: (forall ?L:(List 'S)
           (= (join ?L:(List 'S)
                    nil:(List 'S))
              ?L:(List 'S)))
```

The first of these proofs involves ground terms of the list element sort (`zero`, `one`, etc.), and so would have to be redone for lists of another sort (`Boolean`, say, with `true` and `false` elements). But the second proof contains no occurrence of ground terms, so it applies without change to (`List Boolean`) or indeed to (`List` S) for *any* sort S: Once the theorem it proves is in the assumption base, the theorem can be used with (`List S`) values for any sort `S`. The same will be true for all of the example proofs and exercises that follow.

The following more general induction principle pertains to lists of values of an arbitrary sort S:

Principle 3.4: Mathematical Induction for Lists over sort S

To prove $\forall\, L\, .\, P(L)$ where L ranges over lists of sort S, it suffices to prove:

1. *Basis case*: $P(\texttt{nil})$.
2. *Induction step*: $\forall\, L\, .\, P(L) \Rightarrow \forall\, x\, .\, P(x::L)$.

As before, the antecedent assumption $P(L)$ in the induction step is called the *induction hypothesis*; the variable x is of sort S.

Exercise 3.15: Our axioms for `join` define equations for terms in which `nil` and `::` terms occur in the first argument to `join`, and we proved `right-empty` for the case of `nil` in the second argument. Write an equation for the case of a `::` term in the second argument. (An answer is given in Exercise 3.30, which asks you to prove the given equation.) □

By now we have seen several cases where a binary function obeys one or more axioms, such as identity element axioms (`zero` is an identity element for `Plus`, `one` is for `Times`, and now, `nil` is for `join`), and associative and commutative axioms (with which of course we are familiar in the cases of `Plus` and `Times` for the natural numbers, but which exercises in

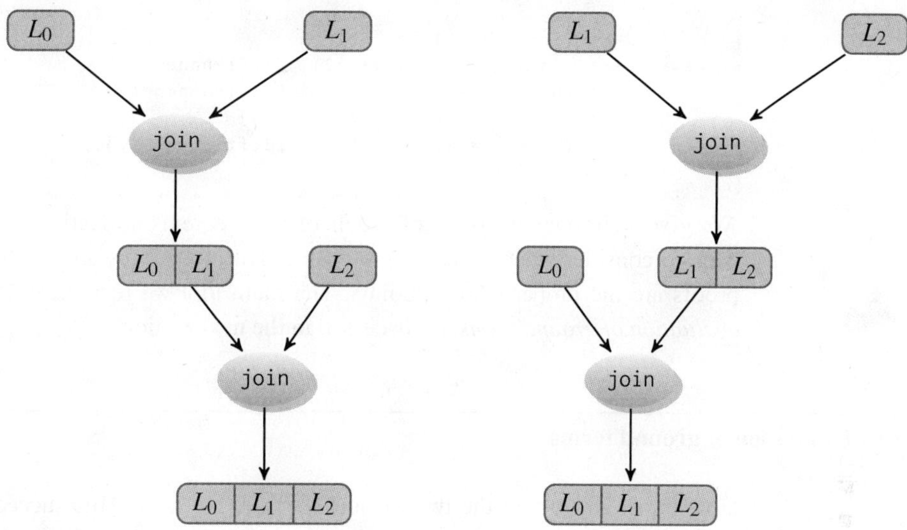

Figure 3.6
Illustration of the associativity of list concatenation.

Section 3.17 ask you to confirm with proofs). Such basic axioms are so frequently useful in reasoning about specific applications of binary functions that when we are presented with a new binary function it is a good strategy to determine whether or not they hold for it. So let's next consider whether our list `join` function is associative:

```
define join-associative :=
  (forall L0 L1 L2 . (L0 ++  L1) ++ L2 = L0 ++ (L1 ++ L2))
```

Intuitively, this relation should hold, as Figure 3.6 suggests. Note the similarity of this diagram to the one on page 142 depicting a relation between `join` and `::`, which can be regarded as a "mixed" associativity relation.

In fact, in the following proof of `join-associative` by induction, we make several uses of that same relation, `left-nonempty`, between ++ and `::`:

```
by-induction join-associative {
  nil =>
    pick-any L1 L2
      (!chain [((nil ++ L1) ++ L2)
           --> (L1 ++ L2)                [left-empty]
           <-- (nil ++ (L1 ++ L2))    [left-empty]])
| (L as (h::t)) =>
    let {IH := (forall L1 L2 . (t ++ L1) ++ L2 = t ++ (L1 ++ L2))}
      conclude (forall L1 L2 . (L ++ L1) ++ L2 = L ++ (L1 ++ L2))
        pick-any L1 L2
          (!chain
```

```
       [((h::t ++ L1) ++ L2)
    --> ((h::(t ++ L1)) ++ L2)  [left-nonempty]
    --> (h::((t ++ L1) ++ L2))  [left-nonempty]
    --> (h::(t ++ (L1 ++ L2)))  [IH]
    <-- (h::t  ++ (L1 ++ L2))   [left-nonempty]])
}
```

We give additional examples of proofs of list theorems in Section 3.11, but we first step back to consider what we can do to ensure that the axioms on which we are basing our proofs are the proper starting points. The main tool we recommend for this purpose is *evaluation of ground terms*, as discussed in the next section.

3.10 Evaluation of ground terms

How do we know that the two axioms we have given for `Plus` succeed in capturing our intentions, namely, that they characterize `Plus` as the binary addition function on the natural numbers? Admittedly, these axioms are simple and it is arguable that nothing more than a moment's reflection is needed to grasp their meaning. Moreover, we are able to prove various results about `Plus` that we know to be true of the addition function, such as commutativity, associativity, and so on. Nevertheless, it would be reassuring to obtain some more concrete evidence that the axioms are indeed successful in defining addition, for instance, that they ensure that applying `Plus` to 3 and 2 yields 5, applying it to 1 and 6 yields 7, and so forth. In other words, it would be reassuring to be able to “test” the axioms on various “inputs.” That is especially true right after we first write down some equational axioms, before we have managed to derive any expected results from them.

One way to do that is to write proofs that derive the desired identities, as we did in Section 3.2. For instance, to test the axioms on inputs 3 and 2, we may give a proof that derives the conclusion

```
(S S S zero Plus S S zero = S S S S S zero).
```

But having to write such a proof anew for each pair of inputs that we want to test is onerous. Instead, we can try to write a method that takes any two natural numbers (represented by canonical terms) and derives the corresponding identity. The following is such a method for `Plus`:

```
define (derive-plus a b) :=
  match b {
    zero => (!chain [(a + b) = a [right-zero]])
  | (S k) => match (!derive-plus a k) {
               ((a Plus k) = sum) =>
                 (!chain [(a + S k)
                          = (S (a + k))    [right-nonzero]
```

```
                           = (S sum)          [(a + k = sum)]])
          }
  }
```

The method works by checking the form of its second argument, `b`. If it is `zero`, then the axiom `right-zero` is used to derive the identity `(a + zero = a)`. Otherwise, if `b` is of the form `(S k)` for some canonical term `k`, then we recursively invoke the method on `a` and `k`.[11] Presumably that recursive call will result in a theorem of the form `(a Plus k = sum)`, where `sum` is a canonical term representing the sum of `a` and `k`. With that intermediate result, we can readily derive the appropriate identity for `a` and `(S k)` by chaining via `right-nonzero`. We can now test as many inputs as we like:

```
> (!derive-plus (S zero) (S zero))

Theorem: (= (Plus (S zero)
                  (S zero))
            (S (S zero)))

> (!derive-plus (S S S zero) (S S zero))

Theorem: (= (Plus (S (S (S zero)))
                  (S (S zero)))
            (S (S (S (S (S zero))))))
```

While this approach is an improvement over writing deductions from scratch, it has a serious drawback: It requires writing a new method (such as `derive-plus`) for each new function symbol that we axiomatize with rewrite rules and that we want to test; we would need to write similar methods for `Times`, for `join`, and so on.

As an alternative, Athena provides a built-in unary procedure `eval` that takes any term *t*, containing arbitrary function symbols, and attempts to reduce it to a canonical form by using the various rewrite rules that have been asserted as axioms. Some examples:

```
> (eval (S zero + S zero))

Term: (S (S zero))

> let {three := (S S S zero)}
    (eval (three + three))

Term: (S (S (S (S (S (S zero))))))
```

The built-in precedence of `eval` is very low, so when we apply `eval` to a complex term *t* it is not necessary to insert outer parentheses around *t*. Thus, the preceding examples could be written as follows:

11 This recursive call corresponds to the recursion that is implicit in `right-nonzero`, manifested by the occurrence of the symbol `Plus` on the right-hand side of the identity `(n Plus S m = S (n Plus m))`.

```
> (eval S zero + S zero)

Term: (S (S zero))

> let {three := (S S S zero)}
    (eval three + three)

Term: (S (S (S (S (S (S zero))))))
```

A few more examples:

```
> (eval zero::nil ++ S zero::nil)

Term: (:: zero
          (:: (S zero)
              nil:(List N)))

> let {1|0|nil := (S zero :: zero :: nil);
       3*3|nil := ((S S S zero * S S S zero) :: nil)}
   (eval 1|0|nil ++ 3*3|nil)

Term: (:: (S zero)
          (:: zero
              (:: (S (S (S (S (S (S (S (S (S zero)))))))))
                  nil:(List N))))
```

Note that `eval` is not guaranteed to terminate successfully. The failure of `eval` to produce a result does not necessarily mean that something is wrong with the underlying axioms. It could be that some of the relevant axioms were expressed in an unorthodox format that could not be handled by `eval`. Typically, however, if the mechanical evaluation of a ground term fails or diverges, then some of the given definitions warrant further scrutiny. The kinds of issues that arise with function definitions are discussed more fully in Section 3.15.

3.11 Top-down proof development

Another simple but extremely useful function on lists is `reverse`, which takes a list of any length and returns a list with the same elements but in the opposite order. We define it from the start as parameterized by the sort of the list elements:

```
declare reverse: (S) [(List S)] -> (List S) [120 [(alist->clist int->nat)]]

assert* reverse-def :=
  [(reverse nil = nil)
   (reverse h::t = (reverse t) ++ h::nil)]

define [reverse-empty reverse-nonempty] := reverse-def
```

The precedence of `reverse` is set to 120, just a little higher than that of `++` (`join`), which is 110, so that a term like

```
(reverse x::y ++ L)
```

is parsed as `((reverse x::y) ++ L)` rather than `(reverse (x::y ++ L))`. The other annotation in the `reverse` declaration,

```
[(alist->clist int->nat)],
```

is an example of *input expansion*: It allows `reverse` to accept as inputs Athena lists (in square-bracket notation), which are then converted to `::`-built terms, while converting integer numerals to natural numbers along the way, using `int->nat`. This means that we can apply `reverse` directly to an expression like `[1 2 3]`. Another mechanism that aids readability is *output transformation*, which we illustrate here by transforming the output of `eval` by converting lists built with `::` to more readable square-bracket lists:

```
transform-output eval [(clist->alist nat->int)]

> (eval reverse [1 2 3 4 5 6])

List: [6 5 4 3 2 1]
```

Input expansion and output transformation are explained and illustrated more fully in Section 3.12.

As an aid to understanding the second axiom, `reverse-nonempty`, we diagram it as we have previous equations, in Figure 3.7.

Exercise 3.16: Use the `reverse` and `join` axioms to prove the following:

```
define reverse-pair-property :=
  (forall x y . reverse x::y::nil = y::x::nil)
```

That is, for any two-element list, `reverse` produces a list with the same two elements in the opposite order. □

As a further test of understanding, let's see if we can prove the following:

```
define reverse-reverse-theorem := (forall L . reverse reverse L = L)
```

We begin a proof by induction and note that the basis case, `L = nil`, is quite easy.

```
by-induction reverse-reverse-theorem {
  nil => conclude (reverse reverse nil = nil)
           (!chain [(reverse reverse nil)
                --> (reverse nil)               [reverse-empty]
                --> nil                         [reverse-empty]])
| (L as (h::t)) => ...
}
```

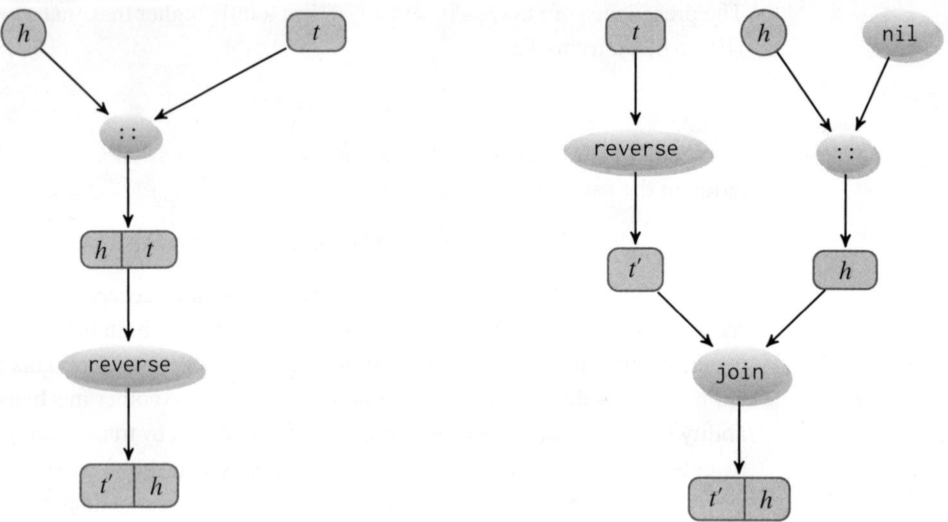

Figure 3.7
Illustrating the second `reverse` axiom (writing t' for `(reverse` t`)`).

But when we try to prove the induction step, we run into a problem:

```
by-induction reverse-reverse-theorem {
  nil =>
    conclude (reverse reverse nil = nil)
      (!chain [(reverse reverse nil)
           --> (reverse nil)           [reverse-empty]
           --> nil                     [reverse-empty]])
| (L as (h::t)) =>
    conclude (reverse reverse L = L) {
      let {IH := (reverse reverse t  = t)}
        (!chain [(reverse reverse h::t)
             --> (reverse ((reverse t) ++ h::nil)) [reverse-nonempty]
             --> ?                                 [??]
```

Neither of the `reverse` axioms and none of the `join` axioms or theorems can be applied, and the same goes for the induction hypothesis. So at this point we have to try to see if there is another property of `reverse` or `join` (or a relation between them) which, if we could prove it, could be used here as a lemma to help complete the proof of `reverse-reverse-theorem`. This situation—where we start working *top-down* from a goal property we want to prove to conjecturing and proving other new properties as subgoals and then using them as lemmas—is actually far more typical of the way proof development works in practice than the *bottom-up* type of development we have previously illustrated in this chapter.

The advantage of top-down proof development is that it is goal-driven: The new properties that you conjecture and try to prove are motivated by the need to complete the proof of some property that is itself a useful, or at least interesting, goal. With bottom-up proof development, by contrast, you must try to anticipate what properties will be needed and invest effort in proving them even before you know for certain that they will ever be used. After all, there are infinitely many possible theorems!

The advantage of a top-down strategy over a bottom-up one is not limited to proof activity; it shows up in many other disciplines. In software development, for example, it is widely accepted that a complex system should be programmed in a modular manner, with extensive use of subroutines, rather than as a monolithic program. The advantages of modularity are many, including ease of understanding, documenting, and testing the code, all of which are essential for its future maintenance. But while design and development of subroutines could proceed either top-down or bottom-up, it is usually much more productive to do it top-down, driven by the needs of the system at hand.

There are also some disadvantages of top-down proof development, which we will discuss later, but for now let's get back to what we need to finish the proof. When we left off, we were staring at

```
(reverse ((reverse t) ++ h::nil)),
```

a term that we somehow need to prove equal to `(h::t)`. Intuitively, we know that the last element of `((reverse t) ++ h::nil)` is `h`, and therefore `h` should be the first element of the reversed list. So the result is of the form `(h :: ···)`, and thus if we can fill in that ··· with `t`, we'll be done. But, in fact, without the `h` on the end, `((reverse t) ++ h::nil)` is just ++'s first argument, `(reverse t)`. And by the induction hypothesis, `(reverse reverse t)` is `t`, just what we need.

To replace this intuitive argument with a real proof, we write the key step as a formal sentence and try to prove it:

```
define reverse-join-1 :=
  (forall L x . reverse ((reverse L) ++ x::nil) = x::L)

by-induction reverse-join-1 {
  nil => pick-any x
           (!chain [(reverse ((reverse nil) ++ x::nil))
                --> (reverse (nil ++ x::nil))              [reverse-empty]
                --> (reverse x::nil)                       [left-empty]
                --> ((reverse nil) ++ x::nil)              [reverse-nonempty]
                --> (nil ++ x::nil)                        [reverse-empty]
                --> (x::nil)                               [left-empty]])
| (L as (h::t)) =>
   conclude (forall x . reverse ((reverse L) ++ x::nil) = x::L)
     let {IH := (forall x . reverse ((reverse t) ++ x::nil) = x::t)}
       pick-any x
```

```
    (!chain
      [(reverse ((reverse h::t) ++ x::nil))
    --> (reverse (((reverse t) ++ h::nil) ++ x::nil)) [reverse-nonempty]

    --> ?                                              [??]
        ...])
}
```

Alas, we get stuck again, without any axiom or theorem that can be applied, and neither can the induction hypothesis be applied (because of the extra `join` in the term we are trying to rewrite). To continue, we could repeat our top-down strategy, conjecturing another property that might help us here, but that doesn't look too promising since the term we are working with now is even more complicated than the one we started with.

There is another possibility, though, and that is to step back and see if what we are trying to prove can be simplified by *generalizing* it. Such a step may seem counterintuitive—how can it be easier to prove something even more general than the property we are having difficulty with? But in fact, generalization often does help, *especially in proofs by mathematical induction*. The reason is simple: A more general property gives us a *more general induction hypothesis*. Being more general, the induction hypothesis gives us more ammunition that we can use to advance the proof.

We can make `reverse-join-1` more general by replacing `(reverse L)` by `L` on the left-hand side of the equation and compensating by the opposite replacement on the right-hand side (which is well-motivated by our intuitive belief in `reverse-reverse-theorem`, though we haven't yet proved it!):

```
define reverse-join-1 := (forall L x . reverse (L ++ x::nil) = x::reverse L)
```

Now, we have no difficulty in proving this more general conjecture:

```
by-induction reverse-join-1 {
  nil =>
    conclude (forall x . reverse (nil ++ x::nil) = x::reverse nil)
      pick-any x
        (!chain [(reverse (nil ++ x::nil))
            --> (reverse x::nil)                        [left-empty]
            --> (reverse nil ++ x::nil)                 [reverse-nonempty]
            --> (nil ++ x::nil)                         [reverse-empty]
            --> (x::nil)                                [left-empty]
            <-- (x::reverse nil)                        [reverse-empty]])
| (L as (h::t)) =>
    conclude (forall x . reverse (L ++ x::nil) = x::reverse L)
      let {ih := (forall x . reverse (t ++ x::nil) = x::reverse t)}
        pick-any x
          (!chain [(reverse (h::t ++ x::nil))
              --> (reverse h::(t ++ x::nil))            [left-nonempty]
              --> (reverse (t ++ x::nil) ++ h::nil)     [reverse-nonempty]
```

```
                 --> (x::(reverse t) ++ h::nil)             [ih]
                 --> (x::(reverse t ++ h::nil))             [left-nonempty]
                 <-- (x::(reverse h::t))                    [reverse-nonempty]])
}
```

And, finally, with this lemma at hand, `reverse-reverse-theorem` can be easily derived:

```
by-induction reverse-reverse-theorem {
  nil => conclude (reverse reverse nil = nil)
           (!chain [(reverse reverse nil)
                  --> (reverse nil)                      [reverse-empty]
                  --> nil                                [reverse-empty]])
| (list as (h::t)) =>
      conclude (reverse reverse list = list)
        let {ih := (reverse reverse t = t)}
          (!chain [(reverse reverse h::t)
                --> (reverse (reverse t ++ h::nil)) [reverse-nonempty]
                --> (h::reverse reverse t)           [reverse-join-1]
                --> (h::t)                           [ih]])
}
```

Thus, we have succeeded with a top-down strategy, discovering and proving a lemma needed to help complete the proof of our original goal. In this case, it wasn't too difficult, but carrying out top-down proof development can be quite challenging in general—it is perhaps the most complex technique to master once the more mechanistic proof skills have been acquired. It is worth reiterating the key ideas that even this relatively simple proof exercise has illustrated.

1. Start with the main goal and try to advance the proof by applying any previously proved axiom or theorem (or, in a proof by induction, the induction hypothesis). But beware of going in circles! (See Exercise 3.31.)
2. If you run out of properties that can be applied, make one up! Form a conjecture, based not only on what is needed to move the current proof along but also on the ultimate goal of the current proof. Of course, whatever you conjecture, you must show that it is valid by separately finding a proof for it. (If the conjecture seems really particular to the current proof—and thus unlikely to be useful elsewhere—you might want to include its proof inline, as a step within the current proof. But in most cases it is better to develop its proof separately, just as it is often better in software development to code a nontrivial computation as a subroutine and call it, rather than putting it inline.)
3. If you are having trouble proving your conjecture, step back and see if you can generalize it, say by replacing a nonvariable subterm with a variable, as we did in the second version of `reverse-join-1`. That may make it easier to prove, and if you succeed, you will still be able to use this more general property as a lemma in proving the original goal.

4. Although we succeeded in proving `reverse-reverse-theorem` after conjecturing and proving only one lemma, you will occasionally have to do this several times in order to complete your original proof. The need for additional conjectures can arise serially in the original proof or by deeper subgoaling within the proof of a conjecture. But when we are successful in proving all of the needed lemmas, we not only have the satisfaction of achieving our original goal, but we have also significantly added to our arsenal of results available for future proof efforts.

In cases where several new lemmas are required, we meet some of the greatest challenges in proof development. It is easy to get lost in a forest of failed proof attempts, and in the confusion there is a danger of going in circles—conjecturing a property and "proving" it using the original goal property. This is one of the potential disadvantages of top-down proof development, but when the proofs are done with a proof system like Athena, such circularity is prevented—any sentence other than an axiom must be proved before it can enter the assumption base, and it cannot be used in a proof unless it is already in the assumption base.

Overcoming such challenges requires a good deal of creativity. And that level of creativity usually comes only with experience. Fortunately, much useful experience can be gained by working on simpler proof problems, including ones that can be done with a purely bottom-up strategy, using only previously proved properties. Most of the additional proof exercises in this chapter can be done bottom-up and contain hints about which previously proved properties are likely to be useful. The more of these exercises you do, the better prepared you will be for tackling more complex proofs, where creativity is required, such as those we will encounter in later chapters.

3.12 ⋆ Input expansion and output transformation

3.12.1 Converters

It is tedious to write large natural numbers in canonical form, solely in terms of `S` and `zero`. Moreover, when evaluating ground terms containing large natural numbers, it is difficult to determine by visual inspection whether the input terms were typed correctly or whether the output is correct. For instance, multiplying 7 by 7 should obviously result in 49, but if we type

```
> (eval S S S S S S S zero * S S S S S S S zero)
```

and obtain the answer

```
Term: (S (S (S (S (S (S (S (S (S (S (S (S (S (S (S (S
      (S (S (S (S (S (S (S (S (S (S (S (S (S (S (S (S
      (S (S (S (S (S (S (S (S (S (S (S (S (S (S (S (S
      (S zero)))))))))))))))))))))))))))))))))))))))))))))))))
```

it is not easy to verify that the answer is correct, or even that the inputs were what we intended.

We can rectify this by using Athena's built-in integer numerals as shorthands for natural numbers: 0 for `zero`, 1 for `(S zero)`, 2 for `(S S zero)`, and so on. We can easily write procedures to convert back and forth between these two representations. For instance, the following will convert any canonical natural number term to the corresponding integer numeral:

```
1 define (nat->int n) :=
2   match n {
3     zero => 0
4   | (S m) => (1 plus nat->int m)
5   | _ => n
6   }
```

Observe the last **match** clause on line 5: If the input `n` is neither `zero` nor a term with the successor symbol `S` at the top, then it is returned unchanged as the output. This is an important point, as we will explain shortly.

Likewise, we can write a procedure that converts nonnegative integer numerals into canonical natural number terms:

```
define (int->nat n) :=
  check {
    (numeral? n) => check {(n less? 1) => zero
                          | else => (S int->nat (n minus 1))}
  | else => n
  }
```

Thus:

```
> (int->nat 0)

Term: zero

> (int->nat 3)

Term: (S (S (S zero)))

> (nat->int zero)

Term: 0

> (nat->int S S zero)

Term: 2

> (nat->int S S S S S S S zero)

Term: 7
```

With conversion procedures like these, the inputs and outputs to eval can become much more readable and writable:

```
> (eval (int->nat 2) * (int->nat 3))

Term: (S (S (S (S (S (S zero))))))

> (eval (int->nat 7) * (int->nat 7))

Term: (S (S (S (S (S (S (S (S (S (S (S (S (S (S (S (S
      (S (S (S (S (S (S (S (S (S (S (S (S (S (S (S (S
      (S (S (S (S (S (S (S (S (S (S (S (S (S (S (S (S
      (S zero)))))))))))))))))))))))))))))))))))))))))))))))))
```

To make the output more readable as well, we can pass it through the nat->int filter:

```
> (nat->int eval (int->nat 7) * (int->nat 7))

Term: 49
```

We can write similar converters for other sorts. For instance, instead of writing out lists in the long form required by :: and nil, we can use bracket notation, converting Athena lists to terms built by :: and nil on the fly with procedures such as the following:

```
define (alist->clist L) :=
  match L {
    [] => nil
  | (list-of x rest) => (x :: (alist->clist rest))
  | _ => L
  }
```

This procedure converts any Athena list ("alist") of homogeneously sorted terms[12] into the corresponding ::-built list (clist for "cons list"), for example:

```
> (alist->clist [true false true])

Term: (:: true
          (:: false
              (:: true
                  nil)))

> (alist->clist [1 2])

Term: (:: 1
          (:: 2
              nil))
```

A similar conversion can be performed in the reverse direction:

12 That is, any Athena list of terms, all of which have the same sort. Clearly, we could not convert a list such as ['a zero true] into a ::-built list, since the result would not be well sorted.

```
define (clist->alist L) :=
  match L {
    nil => []
  | (x :: rest) => (add x (clist->alist rest))
  | _ => L
  }

> (clist->alist 1::nil)

List: [1]

> (clist->alist true::false::nil)

List: [true false]
```

Thus, we can now test the axioms for, say, `join`, as follows:

```
define (test-join L1 L2) :=
  (clist->alist eval (alist->clist L1) join (alist->clist L2))

> (test-join [1 2] [3 4])

List: [1 2 3 4]

> (test-join [99] [])

List: [99]
```

Now, writing something like

```
((int->nat 28) Plus (int->nat 356))
```

is much better than writing the same term using function symbols only, but having to explicitly insert the `int->nat` converters is tedious. What we would like to do is tweak `Plus` so that it *automatically* applies the `int->nat` converter to its inputs, by default. We can do that by redefining `Plus` as the following procedure:[13]

```
define (Plus s t) :=
  (make-term (string->symbol "Plus") [(int->nat s) (int->nat t)])
```

We can now directly write:

```
> (3 Plus 5)

Term: (Plus (S (S (S zero)))
            (S (S (S (S (S zero))))))
```

13 Recall the definition of `make-term` from page 100.

In fact, because int->nat leaves its argument unchanged if it is not an integer numeral, we can continue to apply Plus to natural-number terms as well, just as before:

```
> (S zero Plus S S zero)

Term: (Plus (S zero)
            (S (S zero)))
```

More than that, we can freely mix both input formats, for instance, a canonical natural-number term on the left and an integer numeral on the right, or vice versa:

```
> (3 Plus S zero)

Term: (Plus (S (S (S zero)))
            (S zero))

> (S zero Plus 3)

Term: (Plus (S zero)
            (S (S (S zero))))
```

That is why, as we mentioned earlier, it is important for input converters such as int->nat to leave their arguments unchanged when they are not of a form that we want to convert. Suppose we had instead defined int->nat like this:

```
define (int->nat n) :=
   check { (less? n 1) => zero
         | else => (S (int->nat (n minus 1)))
         }
```

and then went on to (re)define Plus as before:

```
define (Plus s t) :=
  (make-term (string->symbol "Plus") [(int->nat s) (int->nat t)])
```

Then applications like (zero Plus 1) would generate errors, as this version of int->nat cannot handle inputs like zero:

```
> (int->nat zero)

standard input:2:13: Error: Wrong sort of term given as first argument
to less?---a numeric term was expected, but the given term was of sort N.
```

3.12.2 Input expansion

However, having to explicitly redefine Plus so as to make it apply int->nat to its arguments before constructing the output term is still somewhat inconvenient. This is where Athena's **expand-input** directive enters the picture. Instead of explicitly writing a definition like this:

```
define (Plus s t) :=
  (make-term (string->symbol "Plus") [(int->nat s) (int->nat t)])
```

we can instead say simply:

```
> expand-input Plus [int->nat int->nat]

OK.

> (3 Plus 5)

Term: (Plus (S (S (S zero)))
            (S (S (S (S (S zero))))))
```

What **expand-input** Plus [int->nat int->nat] effectively says is this: Redefine Plus so that it automatically applies the input converters listed within the square brackets to the corresponding arguments. In general, if a function symbol f has arity $n > 0$, then the effect of a directive

expand-input f [$C_1 \cdots C_n$]

is this: Redefine f so that it automatically applies converter C_i to its ith argument. Each converter C_i must be an expression denoting a unary procedure. We must list n converters if f has arity n, otherwise we will get an error message:

```
> expand-input Plus [int->nat]

standard input:1:1: Error: Input expansion error: wrong number
of converters (1) given for Plus; exactly 2 are required.
```

Athena will also check that each converter is unary (expects exactly one argument as input).

What if one of the arguments is not amenable to conversion, meaning that it need not or should not be converted? We can then list the identity procedure, id,[14] as the converter for that argument position. For example,

expand-input f [C_1 id C_3]

would make f apply C_1 and C_3 to its first and third arguments, respectively, leaving the second argument unchanged.

As another example, here is how we would make join able to receive lists in bracket notation as well as cons-based (::-built) lists:

```
expand-input join [alist->clist alist->clist]
```

We can now write, for instance:

---

14 id is defined at the top level as the identity procedure: **lambda** (x) x.

```
> (eval [1 2 3] join [4 5 6])

Term: (:: 1
          (:: 2
              (:: 3
                  (:: 4
                      (:: 5
                          (:: 6
                              nil:(List Int)))))))

> (eval [true false] join [true])

Term: (:: true
          (:: false
              (:: true
                  nil:(List Boolean))))
```

Note, however, that input converters for a function symbol f can be conveniently specified when f is first declared (see Section 3.12.4), so there is rarely a need to issue **expand-input** directives explicitly.

The following exercise explores how to make converters like `alist->clist` and `clist->alist` more versatile, by making them into higher-order procedures. For example, we can make `clist->alist` into a higher-order procedure that takes a unary procedure f as an argument and returns a conventional converter that not only transforms a given `::`-built list into an Athena list but also applies f to each element of the resulting list. And we can do the same thing in the converse direction for `alist->clist`. (Indeed, the declaration of `reverse` on page 152 used this more flexible version of `alist->clist`.) Such converters for lists and other datatypes will be used in Chapters 10 and 17 and can be applied in many other contexts to simplify term inputs, as well as outputs from procedures like `eval`.

Exercise 3.17: Sometimes when we write converters intended to take a structurally complex value v and convert it to some other value, we want to dig a little deeper and transform not just the spine of v (viewing v as a tree) but also values that are hanging off v's spine. For instance, let's say we want to transform `::`-built lists of natural numbers to Athena lists. Instead of just transforming something like `(zero::(S zero)::nil)` to `[zero (S zero)]`, we might want to transform it instead to `[0 1]`. While we could always do this by applying two converters in two passes, devise a converter that can do it in one pass. Your converter should be able to perform the following conversions in one pass:

- Convert `('foo::'bar::nil)` to `["foo" "bar"]`; more generally, convert applications of `::` to applications of `add` *and also* convert meta-identifiers to strings.
- Convert `(zero::(S zero)::nil)` to `[0 1]`; more generally, convert applications of `::` to applications of `add` *and* natural-number terms to integers.

- Convert

```
((zero::nil)::((S zero)::(S S zero)::nil)::nil)
```

to `[[0] [1 2]]`; more generally, convert both outer and inner applications of :: to applications of add, and also convert natural-number terms to integers. □

Exercise 3.18: Write a second converter that works in the opposite direction. □

3.12.3 Output transformation

With **expand-input** under our belt we can write input terms succinctly and fluidly, but as the preceding examples demonstrate, the outputs that we occasionally get back, especially from `eval`, can be unwieldy—unless we take explicit pains to pass them through some sort of converter that would, for example, convert natural numbers to integers. The purpose of the directive **transform-output** is to avoid having to do this explicitly. By issuing the directive:

```
> transform-output eval [nat->int]

OK.
```

we instruct Athena to automatically redefine `eval` so that its output is passed through `nat->int`:

```
> (eval 12 Plus 8)

Term: 20
```

Multiple converters $C_1 \cdots C_n$ can be listed in a single **transform-output** directive:

transform-output f `[`$C_1 \cdots C_n$`]`

and in that case every converter C_i will be sequentially applied to the output of f in the specified order: $i = 1, \ldots, n$. For example:

```
> transform-output eval [nat->int clist->alist]

OK.

> (eval [1 2 3] join [4 5 6])

List: [1 2 3 4 5 6]
```

In this case, the output of `eval` is the term

```
(:: 1
    (:: 2
        (:: 3
```

```
(:: 4
    (:: 5
        (:: 6
            nil:(List Int))))))
```

and first the output transformer `nat->int` is applied to it, leaving the result unchanged (since it is not a natural number), after which the second given transformer `clist->alist` is applied to it, giving the final output `[1 2 3 4 5 6]`. Note that output transformers are typically the inverses of input expanders.

Input expansions and output transformations are powerful notational tools and will be used widely in the rest of the book. Input expansions, in particular, will be useful not just for testing and evaluation purposes, but in the bodies of proofs as well.

Just as was the case with **`overload`**, both the **`expand-input`** and **`transform-output`** directives can be applied to function symbols as well as procedures. However, again as with overloading, it must be kept in mind that if f is a function symbol, then after a directive like

`expand-input` $f \cdots$

takes effect, the name f will no longer denote the corresponding function symbol. Rather, it will denote a *procedure*, namely, the procedure that accepts the same number of arguments, applies the listed converters to them, and then applies whatever was previously denoted by f to the results. As before, if need be, the original function symbol named f can be retrieved through `string->symbol` (recall the discussion on page 89).

3.12.4 Combining input expansion and output transformation with overloading

Overloading can be combined with input expansions and/or output transformations. For instance, we could first input-expand `Plus` so that it accepts integer numerals, and then overload + so that it can be used as an alias for the input-expanded version of `Plus`:

```
> expand-input Plus [int->nat int->nat]

OK.

> overload + Plus

OK.

> (n + 5)

Term: (Plus ?n:N
            (S (S (S (S (S zero))))))
```

In some cases we need to pay attention to the order in which these directives are carried out. Generally, it is better practice to perform input expansion first and then overloading.

3.12.5 Using declare with auxiliary information

In the interest of brevity, it is possible to **declare** a function symbol and also to specify its precedence in one fell swoop:

```
declare f: [N N] -> N [150]
```

This has the same effect as the following sequence:

```
declare f: [N N] -> N

set-precedence f 150
```

The abbreviated form simply includes the information [150] after the range of the function symbol. In addition to precedence, we can also specify the associativity of the declared symbol:

```
declare f: [N N] -> N [150 left-assoc]
```

This is equivalent to the following sequence:

```
declare f: [N N] -> N

set-precedence f 150

left-assoc f
```

We could also specify its associativity only, without bothering to specify a precedence:

```
declare f: [N N] -> N [left-assoc]
```

In addition to precedence and/or associativity, we can also specify a symbol or procedure to be overloaded as an alias for the newly declared symbol, for example:

```
declare minus: [N N] -> N [200 -]
```

This is equivalent to:

```
declare minus: [N N] -> N
set-precedence minus 200
overload - minus
```

If there is only overloading information included but no precedence and/or associativity, then the precedence and associativity of the newly declared symbol are set to be identical to those of the symbol or procedure that is being overloaded. For example:

```
> declare minus: [N N] -> N [-]

New symbol minus declared.
```

```
> (equal? (get-precedence minus) (get-precedence -))

Term: true
```

A brand new name *I* can appear inside the square brackets representing that which is to be overloaded; if *I* is not bound to anything, then it will be introduced into the current environment as an alias for the function symbol being declared. For instance, assuming that the identifier ^ is unbound in the current environment, the directive

```
declare g: [N N] -> N [^]
```

will introduce ^ as an alias for g.

Finally, we may also specify a list of input converters, which will be used to automatically carry out an **expand-input** directive on the new symbol. For instance, the following declaration introduces a new binary function symbol `mult` on natural numbers (intended to represent multiplication); it specifies its precedence to be 300; it overloads * so that it can be used as an alias for the new `mult` symbol; and it prescribes that `int->nat` be automatically applied to each of the two arguments of the new symbol:

```
declare mult: [N N] -> N [300 * [int->nat int->nat]]
```

By and large, the order in which these auxiliary pieces of information are listed inside the square brackets is immaterial, with one exception: If a list of input converters appears at all, it must appear last.

3.13 ⋆ Conjecture falsification

Before we try to prove a conjecture *p*, it is often useful to first try to *falsify p* by finding a counterexample to it. This falls under a class of techniques collectively known as *model checking* (although in some contexts that term can have a much more narrow technical meaning), or *automated testing*. If we manage to find a counterexample, then any attempt to prove *p* would be futile,[15] and therefore this kind of automated testing can be a time saver. Moreover, the feedback provided by the testing procedure, given in the form of a specific counterexample to the conjecture, is often very valuable in helping us to debug and, in general, to better understand the theory or system we are developing.

Athena is integrated with a number of external systems that can be used for model building/checking, most notably SMT and SAT solvers, but there is also a built-in model checker that can be used when a theory is constructive, and specifically when the relevant axioms

15 That is because Athena's proofs are guaranteed to be sound, so we can never prove something that doesn't follow logically from the assumption base, and if we find a counterexample to *p*, then *p* does not follow from the assumption base.

are rewrite rules of the kind that can be handled by `eval`. That model checker is perhaps the simplest to use, and can be surprisingly effective. At its simplest, to falsify a conjecture p we call

(falsify p N),

where N is the desired *quantifier bound*, namely, the number of values of the corresponding sort that we wish to examine (in connection with the truth value of p) at each quantifier of p. To take a contrived but illustrating example, suppose that p is the conjecture that addition on the natural numbers is commutative:

`(forall n m . n + m = m + n).`

Then the call `(falsify p 10)` will try to falsify p by examining 10 natural numbers at each quantifier. In this case, the procedure would essentially check whether addition is commutative on the first 10 natural numbers, or more specifically, whether `(n + m = m + n)` evaluates to `true` when both `n` and `m` take the successive values `zero`, `(S zero)`, ..., `(S S S S S S S S S S zero)`. This would result in 100 evaluations (10×10) of the body of p, and of course in this case the falsification would fail, since commutativity does in fact hold:

```
> define p := (forall n m . n + m = m + n)

Sentence p defined.

> (falsify p 10)

Term: 'failure
```

When falsification fails within the given bound, the term `'failure` is returned.

When falsification succeeds, it returns specific values for the quantified variables that make the conjecture false. As another example, suppose we conjecture that list concatenation is commutative. Now this conjecture is clearly false, so let's see how `falsify` handles it:

```
define p := (forall L1 L2 . L1 ++ L2 = L2 ++ L1)

> (falsify p 10)

List: ['success
|{
?L1:(List Int) := (:: 1 nil:(List Int))
?L2:(List Int) := (:: 0 nil:(List Int))
}|]
```

The result is now a pair (two-element list) whose first element is the term `'success` and whose second element is a mapping from the quantified variables to appropriate values, namely, values that falsify the given conjecture. Indeed, we see that the returned values do falsify the conjecture: Concatenating `[1]` with `[0]` gives `[1 0]`, whereas concatenating `[0]` with `[1]` gives `[0 1]`, and these two lists are certainly different (using square-bracket notation for lists, instead of `::` and `nil`, for brevity). We can easily verify that this is a genuine counterexample with the aid of `eval`:

```
define [t1 t2] := [(1::nil) (0::nil)]

> (eval t1 ++ t2 = t2 ++ t1)

Term: false
```

Note that the given conjecture was polymorphic, but `falsify` automatically worked with a monomorphic variant of it. Typically `Int` is used to instantiate sort variables, but other domains or datatypes might also be used.

The `falsify` procedure can work with both integers and reals, as well as meta-identifiers. It has a native understanding of all primitive operations on these domains, such as addition, multiplication, and so on. For example:

```
define p := (forall ?x:Int ?y:Int . ?x Real.* ?y > ?x)

> (falsify p 10)

List: ['success |{?x:Int := 1, ?y:Int := 1}|]

define p := (forall ?x:Int . ?x Real.* ?x >= ?x)

> (falsify p 10)

Term: 'failure

> (falsify p 50)

Term: 'failure

> (falsify p 100)

Term: 'failure
```

The fact that a conjecture cannot be falsified within a given bound does not mean that the conjecture is valid. It could simply be that counterexamples emerge only outside the range of values that we have examined. That is, it could be that if we increase the bound we might be able to falsify the conjecture. A lot of errors, however, tend to manifest themselves on fairly small input values. So if we keep increasing the falsification bound and continue to fail to find a counterexample, that can give us a good amount of confidence in the conjecture. But that kind of evidence will remain inductive, rather than deductive, and hence it can never establish with certainty that the conjecture is true. No matter how large the bound, no matter how extensively we have tested the conjecture, it is conceivable that the conjecture is still false. Only deductive proof can ultimately assure us that the conjecture holds.

Of course, increasing the bound makes falsification more expensive, since the number of tests that need to be carried out increases polynomially with the number of quantifiers in the conjecture, and specifically with the length of the longest sequence of consecutive quantifiers. If we have three consecutive quantifiers, for instance, as we would, say, in a transitivity conjecture of the form

```
(forall x y z . x R y & y R z ==> x R z),
```

and we specify a bound of 10, then `falsify` would need to carry out 1000 (10^3) tests. If we specify a bound of 100, we would need to carry out

$$100 \times 100 \times 100 = (10^2)^3 = 10^6 = 1 \text{ million}$$

tests, and so on. Therefore, the more consecutive quantifiers that a conjecture has, the lower the bound that is feasible. The recommended strategy is to start with a low bound such as 5–10, and to gradually increase that until a counterexample is found or the search becomes too expensive.

Sometimes it makes little sense to fix the same upper bound N for *every* quantified variable of the conjecture. Different variables might instead call for different bounds. Recall, for instance, the following:

```
define power-square-theorem := (forall n x . (x * x) ** n = x ** (n + n))
```

If we were trying to discover counterexamples to this sentence, we would not want to do so with a call like

```
(falsify power-square-theorem 30),
```

as we neither want nor need to raise x = 20 to the n = 29th power. Instead, we want to explore more values for x (the base) and significantly fewer values for n (the exponent). To accommodate this, the second argument to `falsify` need not be a single positive numeral. Rather, it can be a map from variables to numerals. In this case we might want to try 10 values for x and 3 values for n:

```
> (falsify power-square-theorem |{x := 10, n := 3}|)

Term: 'failure
```

We close with some brief remarks on the algorithm for selecting the terms with which to replace each quantified variable. A fair enumeration algorithm is used for that purpose, meaning that, in the limit, every canonical term of the relevant sort is guaranteed to be generated. With structurally complex terms, however, such as abstract syntax trees (e.g., expressions of a given grammar), two fair enumeration algorithms can vary widely in their effectiveness depending on how exactly they enumerate the terms—for instance, depending on whether deeper vs. bushier trees are favored (generated earlier). The default strategy tries to strike a reasonable balance under all sets of circumstances, but some degree of customization may be necessary in some cases. There is a global cell variable named `breadth-factor` that can be used to control the exact degree by which broader (bushier) terms are preferred over deeper terms. The minimum possible value of that variable is 1, which will tend to strongly favor deep terms. A value of around 20 or 30 will usually strike a decent balance, whereas a value of 100 or 200 or more will strongly favor broader over deeper terms. For ultimate control on how exactly to enumerate the canonical terms of a given sort, the user can provide an infinite stream of such terms using the binary procedure `register-sort-stream`, which takes a sort name as its first argument (represented as a string) and an infinite stream of terms as its second argument, represented as a lazy list in the API used in the library file `dt-streams.ath`.

3.14 ⋆ Conditional rewriting and additional chaining features

All examples of equational chaining that we have seen so far have been based on the definition of rewrite rules on page 125, namely, as sentences of the form

$$\forall\, v_1 \cdots v_n \,.\, l = r$$

(where r does not contain any v_i that does not occur in l). Rewrite rules of this simple form, however, are not sufficiently expressive. Many applications of equational proof require rewriting based on *conditional rewrite rules*. Such a rule is typically of the form

$$\forall\, v_1 \cdots v_n \,.\, l_1 = r_1 \wedge \cdots \wedge l_n = r_n \Rightarrow l = r \tag{3.3}$$

for $n \geq 1$, where any variable v_j that appears in r must also appear either in l or in some l_i or r_i, $i \in \{1, \ldots, n\}$. In other words, the variables of the left-hand side l together with those of the antecedent form a superset of the variables in r. However, the antecedent may contain variables that do not occur in l. The n identities $(l_i = r_i)$ are the *conditions* or *guards* of the rule. The `chain` method actually allows for conditional rules of more complicated sentential structure. Specifically, while the consequent must always be an identity $(l = r)$, the antecedent need not be a conjunction of identities. It may be a disjunction, or a negation, or a conjunction of identities and/or negations of identities, and so on. So we may more generally write a conditional rewrite rule in the form

$$\forall\, v_1 \cdots v_n \,.\, p \Rightarrow l = r \tag{3.4}$$

where, again, the variables of l and the (free) variables of p comprise a superset of the variables of r. We say that p is the rule's condition. If p is a conjunction, we may also collectively refer to the conjuncts of p as the conditions (or guards) of the rule.

Reduction with a conditional rule R of the form (3.4) works just as it does with plain rewrite rules, but with one additional twist: We say that

$$s \to_\theta t \; [R]$$

if there is a subterm s' of s that matches l under θ and t is obtained from s by replacing s' with $\theta(r)$. We require θ to include in its support all the variables of l and p.[16] Further, we now have an additional obligation to discharge before we can allow this reduction from s to t: We must ensure that the corresponding instance of the condition p, namely $\theta(p)$, actually holds (i.e., that $\theta(p)$ is in the assumption base).[17] Using more conventional notation, we might define this relation through the following inference rule:

$$\frac{\theta(p)}{s \to_\theta t \; [R]}$$

provided there is a subterm s' of s that matches l under θ,
θ includes all variables of l and p in its support, and t is
obtained from s by replacing s' by $\theta(r)$.

It is not difficult to see that the antecedent along with the provisos spelled out in the auxiliary condition of this inference rule ensure that `(s = t)` is derivable. First we can derive a substitution instance of R by replacing each universally quantified variable v_i by $\theta(v_i)$, using `uspec` or `instance`; that will give us the conditional

$$\theta(p) \Rightarrow \theta(l = r).$$

We can then apply modus ponens to this and $\theta(p)$ to obtain the identity $\theta(l = r)$, that is to say, $\theta(l) = \theta(r)$. And from that point on $s = t$ can be derived using the five primitive equational methods of Section 3.4.

Thus, operationally, when Athena evaluates a `chain` step like

```
s = t [R]
```

where R is a conditional rewrite rule of the form (3.4), it checks, in addition to all the other conditions listed above, that $\theta(p)$ is in the assumption base.[18] However, it is fairly lenient about that condition. For instance, if p is a complex conjunction with conjuncts $p_1, \ldots, p_n$ and if each $\theta(p_i)$ is in the assumption base, then the check will succeed even if $\theta(p)$ itself—as a conjunction—is not in the assumption base. Likewise, if p is a disjunction with disjuncts $p_1, \ldots, p_n$ and at least one $\theta(p_i)$ is in the assumption base, then the check will succeed even if $\theta(p)$ itself, as a disjunction, is not in the assumption base. Essentially, the implementation of `chain` attempts to *derive* the instantiated condition $\theta(p)$ on the fly. That

16 Therefore, θ will be applicable to all the variables of r, since these are a subset of the variables of l and p.

17 See page 96 for the definition of $\theta(p)$.

18 Of course, it first needs to discover an appropriate substitution θ.

derivation might boil down to a simple application of claim, if $\theta(p)$ happens to be in the assumption base. But it might be a little more involved, as illustrated by the two examples we just cited, involving a complex conjunctive or disjunctive condition. As another example, if the instantiated condition is a negated conjunction, then chain will try to derive at least one of the complemented conjuncts, after which it will derive the desired negation by applying an instance of De Morgan's laws. Nevertheless, the attempted derivation of $\theta(p)$ is guaranteed to terminate very quickly; the theorem proving that chain does to this end is very light. Therefore, it is largely the proof writer's responsibility to ensure that when a conditional rewrite rule is used in a chain step, the relevant instantiated condition(s) are either in the assumption base at that point or else can be very easily derived from it.

As an example, consider:

```
1  declare P, Q: [N] -> Boolean
2
3  assert* CR := (P n & Q m ==> n * m = m)
4
5  > pick-any a:N b:N
6      assume (P a)
7        assume (Q b)
8          (!chain [(a * b) = b [CR]])
9
10 Theorem: (forall ?a:N
11             (forall ?b:N
12               (if (P ?a:N)
13                   (if (Q ?b:N)
14                       (= (Times ?a:N ?b:N)
15                          ?b:N)))))
```

Here, at the time when the conditional rule CR is used in a chain step, on line 8, the relevant instance of the antecedent, namely (P a & Q b), is *not* in the assumption base.[19] However, both (P a) and (Q b) are individually in the assumption base, as hypotheses of the **assume** proofs on lines 6 and 7; therefore, chain successfully derives the required conjunction of the two atoms and licenses the step.

We close this section with a brief discussion of some of the more advanced features of chain. We will illustrate each feature with the following three rules:

```
assert* R1 := (n * one = n)
assert* R2 := (n * zero = zero)
assert* comm := (n * m = m * n)
```

The implementation of chain allows for:

19 That this is the relevant instance was discovered by matching the desired chain step (a * b = b) to the consequent of the rule, (n * m = m), which resulted in the substitution {?n:N --> ?a:N, ?m:N --> ?b:N}. This matching is automatically performed by chain.

- *Multiple redexes from a single rule in one step*: A single rewrite rule can be used to rewrite multiple redexes in a starting term, all in one single chain step. For example:

```
> (!chain [(m * one + n * one)
        =  (m + n)                [R1]])

Theorem: (= (Plus (Times ?m:N one)
                  (Times ?n:N one))
            (Plus ?m:N ?n:N))
```

Here R1 was applied both to (m * one) and to (n * one) in the starting term

(m * one + n * one),

yielding (m + n) in one step.

- *Multiple redexes from multiple rules in one step*: More generally yet, the justification list for a given step may contain multiple rewrite rules, and each rule may be used to rewrite one or more subterms of the starting term. For example:

```
> (!chain [(m * zero + n * one)
          = (zero + n)               [R1 R2]])

Theorem: (= (Plus (Times ?m:N zero)
                  (Times ?n:N one))
            (Plus zero ?n:N))
```

Here R2 was used to rewrite the subterm (m * zero) and R1 was used to rewrite the subterm (n * one), resulting in (zero + n). As can be seen, it is not necessary to list the justifiers in any particular order. The implementation will determine which rule applies to which subterms.

- *Multiple successive rewrites in one step*: Not only can multiple rewrite rules be applied to disjoint subterms of the starting term in one step, but multiple rewrite steps in nondisjoint positions can also be taken in one step. For instance, consider going from the starting term

$s =$ (zero * m + one * n)

to $t =$ (zero + n). It is impossible to do so using R1 and R2 alone, because neither is in fact applicable to any subterm of s. We first need to apply the comm rule twice, in order to reverse each of the two complex subterms of s, resulting in (m * zero + n * one), and *then* we can finally apply R1 and R2 to obtain t. But it is in fact possible to take all of these actions in one step, simply by listing all three rules as justifiers:

```
> (!chain [(zero * m + one * n)
         =  (zero + n)               [R1 comm R2]])
```

```
Theorem: (= (Plus (Times zero ?m:N)
                  (Times one ?n:N))
            (Plus zero ?n:N))
```

The implementation of chain will launch an equational proof search, using various heuristics, that will attempt to derive the destination term *t* from the starting term *s* using the given rules. Because the search must always terminate very quickly, this feature cannot typically be used to find long and deep chains of equational reasoning. Nevertheless, as the example illustrates, this can be a handy feature when we want to perform a few simple rewriting steps in one fell swoop.

- *Redundant rules*: It is not necessary for every listed justifier of a given step to be used or even usable in that step. We can always list irrelevant justifiers; Athena will figure out that they are irrelevant and will ignore them. This can be helpful when we have given a single name to *several* rewrite rules, as a group, and we want to justify a chain step by appealing to these rules collectively. This is often done with lists of rewrite rules intended to define functions. For instance, instead of defining Times with two separate axioms, and giving each axiom its own name, as we did before:

```
assert Times-zero := (forall x . x * zero = zero)

assert Times-nonzero := (forall x y . x * S y = x * y + x)
```

we might define it with a single list of axioms, and only give a name to that list, say:

```
assert* Times-definition :=
  [(x * zero = zero)
   (x * S y = x * y + x)]
```

This simply inserts every sentence in the given list into the assumption base,[20] and it gives the name Times-definition to the entire list.

Now later on we might want to justify a chain step that is enabled by one of these axioms by citing the list of axioms as a whole, Times-definition, rather than by taking the time to explicitly identify the relevant axiom. As an example, consider the earlier proof of Times-right-one (from page 129):

```
1  conclude Times-right-one
2    pick-any x:N
3      (!chain
4        [(x * one)
5        = (x * S zero)      [one-definition]
6        = (x * zero + x)    [Times-nonzero]
7        = (zero + x)        [Times-zero]
```

20 After universally quantifying each sentence over all of its free variables; that is the effect of using **assert*** instead of **assert**.

```
8       = x                    [left-zero]])
```

Here, each of the steps on lines 6 and 7 is justified by appealing to the individual axiom that licenses the step, `Times-nonzero` for the first one and `Times-zero` for the other. However, it might be more convenient to write simply:

```
1  conclude Times-right-one
2    pick-any x:N
3      (!chain
4        [(x * one)
5        = (x * S zero)        [one-definition]
6        = (x * zero + x)      [Times-definition]
7        = (zero + x)          [Times-definition]
8        = x                   [left-zero]])
```

and let Athena figure out which axiom(s) exactly, from the `Times-definition` list, is needed on each step.

This brings up another related point: the justification list J of a `chain` step

$$s = t \quad J$$

need not contain individual sentences only; it may also contain *lists* of sentences. On lines 6 and 7 in the preceding proof, the justification list was `[Times-definition]`, which is a list containing a list. In general, each element of J can be either a sentence or a list of sentences. The reason for this is readability. In informal proofs, we sometimes justify an inference step by appealing to a single sentence (say an axiom, lemma, or hypothesis), other times by appealing to a whole collection of sentences, and other times by appealing to both (e.g., "$s = t$ follows from theorem X and the equations defining f"), and the flexible format of justification lists is intended to emulate that style. Underneath, the implementation of `chain` will "flatten out" any justification list J into a single collection of sentences.

As pointed out earlier, Athena processes a `chain` step of the form

$$s = t \quad J$$

by launching a search for an equational derivation of $(s = t)$ from J that could potentially consist of multiple rewrite steps. This means that we can often shorten `chain` proofs by skipping one or more steps. For instance, we could compress the steps on lines 6 and 7 into one, as follows:

```
conclude Times-right-one
  pick-any x:N
    (!chain
     [(x * one)
     = (x * S zero)        [one-definition]
     = (zero + x)          [Times-definition]
```

```
     = x                    [left-zero]])
```

In fact, in this case we could actually skip *all* steps:

```
conclude Times-right-one
  pick-any x:N
    (!chain [(x * one) = x [one-definition Times-definition left-zero]])
```

and the proof would still go through, as the rewrite steps would be automatically discovered by Athena.[21] Such radical shortening is not always possible, but even when it is, we will generally refrain from taking steps that are too large, as the proof might then fail to shed any light on the underlying reasoning.

- *No rules*: When we don't care to single out any rules for a given step (e.g., because we feel that the rules justifying the step are obvious and do not need to be explicitly specified), we may appeal to the entire assumption base, that is, the justifier may simply be (ab). Equivalently, we may omit a justifier altogether for that step. This has the same effect as citing the entire assumption base as the justifier. For example, the last two steps of the following chain do not list any justifiers, so the entire assumption base becomes the default justifier list for each:

```
conclude Times-right-one
  pick-any x:N
    (!chain
     [(x * one)
     = (x * S zero)        [one-definition]
     = (zero + x)
     = x                   ])
```

We rarely write chain steps without any justifiers, and you should do the same, as such steps tend to be inefficient and can obscure the reasoning behind a proof.

---

21 If a relatively short proof can be quickly found by rewriting, the discovered derivation can be displayed by using the ternary procedure find-eqn-proof, which takes a term s, a term t, and a list of rewrite rules L in the assumption base, and attempts to find a sequence of rewrite steps from s to t using the rules in L. If it succeeds, the output is a list of triples

$$[[s_1\ R_1\ \theta_1] \cdots [s_n\ R_n\ \theta_n]]$$

where, for $i = 2, \ldots, n$, s_i can be derived from s_{i-1} by R_i using the substitution θ_i, where R_i is an element of L, $s_1 = s$, and $s_n = t$, so that

$$s_1 \rightarrow_{\theta_2} s_2\ [R_2]\ \rightarrow_{\theta_3} s_3\ [R_3]\ \cdots \rightarrow_{\theta_n} s_n\ [R_n].$$

(The first equation R_1 is the trivial identity ($s = s$) and θ_1 is the empty substitution.) If find-eqn-proof fails, it returns the unit value (). For an illustration, apply the procedure to this example as follows:

```
(find-eqn-proof (x * one) x eqns)
```

with eqns defined as [one-definition Times-zero Times-nonzero left-zero].

3.15 ⋆ Proper function definitions

Most of the functions in this book will be defined by collections of (possibly conditional) equations. We say "collections" rather than "lists" or "sequences" because the order in which these equations are given is immaterial; we will return to this point shortly.

Specifically, a function f with signature `[`$S_1 \cdots S_n$`] ->` S will usually be defined by $k > 0$ possibly conditional equations. Since an unconditional identity can also be regarded as a conditional one, with the trivial condition `true`, we can think of all k defining equations of f, $E_1, \ldots, E_k$, as conditional, having the following form:

$$\begin{array}{llll} E_1 \equiv & (p_1 & \texttt{==>} & ((f\ t_1^1 \cdots t_n^1)\ \texttt{=}\ r_1)) \\ E_2 \equiv & (p_2 & \texttt{==>} & ((f\ t_1^2 \cdots t_n^2)\ \texttt{=}\ r_2)) \\ \vdots & \vdots & \vdots & \vdots \\ E_k \equiv & (p_k & \texttt{==>} & ((f\ t_1^k \cdots t_n^k)\ \texttt{=}\ r_k)) \end{array}$$

where each t_i^j consists exclusively of constructors and/or variables. The right-hand sides r_j will typically consist of constructors, variables, and other function symbols (possibly including f) that have been likewise defined by similar sets of equations.

We often intend the given equations to constitute a *total* specification of f, which means that they ought to determine a unique result for *every* possible sequence of inputs

$$x_1 : S_1 \cdots x_n : S_n. \tag{3.5}$$

For that reason, we usually require that there is *exactly one equation* E_j applicable to any given input (3.5). That boils down to requiring that (a) there is *at least one* E_j applicable to any such input, and (b) there is *at most* one E_j applicable to any such input. We call these the *exhaustiveness* and *disjointness* requirements, respectively. We discuss each in turn:

- *Exhaustiveness*: The requirement that there is at least one equation applicable to any input is formally captured by the following sentence:

$$\texttt{(forall}\ x_1 : S_1 \cdots x_n : S_n\ \texttt{(or}\ D_1 \cdots D_k\texttt{))} \tag{3.6}$$

  with each D_j as follows:

$$D_j \equiv \texttt{(exists}\ \overrightarrow{v_j}\ \texttt{(and}\ (x_1\ \texttt{=}\ t_1^j)\ \cdots\ (x_n\ \texttt{=}\ t_n^j)\ p_j\texttt{))} \tag{3.7}$$

  where we write $\overrightarrow{v_j}$ for a fixed listing of the union of the variables that appear in $(f\ t_1^j \cdots t_n^j)$ along with the free variables of p_j, and we assume that the variables $x_1 : S_1 \cdots x_n : S_n$ are fresh, so that they do not occur in any defining equation E_j. Intuitively, (3.6) says that no matter what input values $x_1 : S_1 \cdots x_n : S_n$ we choose, there is at least one equation E_j such that $(f\ t_1^j \cdots t_n^j)$ and $(f\ x_1 \cdots x_n)$ become identical and p_j is satisfied (by replacing the variables in the former term and in p_j by suitable values; the sentence D_j simply asserts the existence of such values). Operationally, this means that we would then be able to use E_j as a rewrite rule to reduce $(f\ x_1 \cdots x_n)$ to an appropriate instance of r_j.

- *Disjointness*: The requirement here is that if one equation is applicable, then no others are:

$$\texttt{(forall } x_1\texttt{:}S_1 \cdots x_n\texttt{:}S_n \texttt{ (and (}D_1 \texttt{ ==> } D_1{'}\texttt{)} \cdots \texttt{ (}D_k \texttt{ ==> } D_k{'}\texttt{)))} \tag{3.8}$$

where each D_j is defined as in (3.7) and

$$D_j{'} = \bigwedge_{i \in \{1,\ldots,k\} \setminus \{j\}} (\texttt{\~{}}\ D_i). \tag{3.9}$$

These two requirements should be viewed as general guidelines that should normally be followed when defining functions in this style. Both of these conditions should normally hold; indeed, both should be provable. Neither is set in stone, however. A set of equations might violate both conditions and still be a perfectly good function definition. That is probably easier to see in the case of exhaustiveness. Sometimes we do not care to specify the value of a function for certain inputs, and in such cases our definitions will be nonexhaustive by design. For instance, when we later come to define division on the natural numbers, we will not specify the value of the function when the second argument (the divisor) is zero. Our equations will therefore serve only as a partial definition; they will *underspecify* the division function. This simply means that there will be many different total functions with signature `[N N] -> N` satisfying the definition. A term such as `(S zero / zero)` (1 divided by 0) will still have a unique value under any given interpretation, but we will not be able to derive any identities of the form

$$\texttt{(S zero / zero = } n\texttt{)}$$

for a canonical natural number n (assuming a consistent assumption base), because we will not have any relevant axioms to work with. Underspecification is fairly common.

Disjointness violations occur less frequently, and when they do they are more likely to represent genuine problems in a definition. The main concern here is that if two distinct defining equations are applicable to one and the same input term $s = (f \cdots)$, then we might end up getting two different outputs $r_i \neq r_j$ for that input, r_i from one equation and r_j from the other, which could potentially lead to an inconsistency, as we would then have to conclude that $(r_i = r_j)$ (from $(s = r_i)$ and $(s = r_j)$). But disjointness violations can also be innocuous in some cases, as long as the overlapping equations are confluent, meaning that ultimately they always lead to the same result. To take a simple if contrived example, consider the following unorthodox definition of the factorial function:

```
declare fact: [N] -> N

assert* fact-def :=
```

```
[(fact zero = S zero)
 (fact S zero = S zero)
 (fact S n = S n * fact n)]
```

The disjointness condition does not hold here because the last two equations are both applicable to the term `(S zero)`. According to the second equation, the input `(S zero)` should produce the output `(S zero)`:

```
(fact S zero = S zero).                                    (3.10)
```

And according to the last equation, it should produce the output

```
(S zero * fact zero).                                      (3.11)
```

However, `(fact zero)` is itself mapped to `(S zero)`, hence (3.11) further reduces to

```
(S zero * S zero),
```

which, by the definition of multiplication, becomes `(S zero)`. We thus arrive at the same conclusion as (3.10), and this is in fact a legitimate—if unorthodox and unnecessarily complicated—definition of the factorial function.

By and large, however, we will want disjointness to hold. If and when it does not, we should be aware of that fact and either reformulate our definition so as to make the equations disjoint, or else at least examine the violations closely to ensure that they do not yield an inconsistency. Likewise for exhaustiveness: We should be aware when a definition is nonexhaustive to ensure that the omitted cases are intentional (because we want to underspecify the function at hand) rather than due to oversight.

Athena provides a unary procedure `fun-def-conds` that can help in that task. It takes any function symbol f[22] that has been defined in the style we have been discussing, and produces a list of sentences that capture the two conditions discussed above, the exhaustiveness and disjointness conditions for f. These two can also be produced separately by the procedures `fun-def-conds-e` (for exhaustiveness) and `fun-def-conds-d` (for disjointness). The output sentences may actually be somewhat simpler than the definitions given above, because these procedures first try to carry out a few elementary simplifications.

When the equations defining f are not conditional, both exhaustiveness and disjointness are automatically decidable, and moreover, when either condition is violated, appropriate counterexamples can be mechanically generated. Procedure `fun-def-conds` will perform these checks automatically whenever it realizes that all defining equations are unconditional. It will still return a list of sentences representing the exhaustiveness and disjointness conditions, but it will also print a message that will either declare that these sentences hold or else report counterexamples, namely, specific inputs that violate these conditions. For instance, here is the output for `Plus`:

22 Or overloaded procedure f.

```
> (fun-def-conds Plus)

The definition of Plus is well-formed; the following conditions hold:

List: [
(forall ?v1273:N
  (or (exists ?v1269:N
        (= ?v1273
           (S ?v1269)))
      (= ?v1273 zero)))

(forall ?v1286:N
  (if (exists ?v1282:N
        (= ?v1286
           (S ?v1282)))
      (not (= ?v1286 zero))))
]
```

Note that the output reports that the conditions are satisfied. Also note that these two conditions follow directly from the datatype axioms for the natural numbers. By contrast, here is the output we get for the previous definition of `fact`:

```
> (fun-def-conds fact)

[Warning: the equational axioms defining fact are not disjoint.
For example, at least  two distinct axioms apply to every instance
of this term: (fact (S zero)).]

List: [
(forall ?v1308:N
  (or (exists ?v1307:N
        (= ?v1308
           (S ?v1307)))
      (= ?v1308
         (S zero))
      (= ?v1308 zero)))

(forall ?v1313:N
  (if (exists ?v1312:N
        (= ?v1313
           (S ?v1312)))
      (not (= ?v1313 zero))))

(forall ?v1313:N
  (if (exists ?v1312:N
        (= ?v1313
           (S ?v1312)))
      (not (= ?v1313
              (S zero)))))
```

```
(forall ?v1313:N
  (if (= ?v1313
         (S zero))
      (not (= ?v1313 zero))))
]
```

Here, the procedure determined that disjointness does not hold and issued a warning to that effect, along with a specific input, `(S zero)`, that violates the condition. This could have also been discovered by `falsify` if we had tried to generate counterexamples to the given sentences:

```
define [c1 c2 c3 c4] := (fun-def-conds fact)

> c3

Sentence: (forall ?v1327:N
            (if (exists ?v1326:N
                  (= ?v1327:N
                     (S ?v1326:N)))
                (not (= ?v1327:N
                        (S zero)))))

> (falsify c3 5)

List: ['success |{?v1326:N := zero, ?v1327:N := (S zero)}|]
```

The assignment `?v1327:N := (S zero)` gives us a specific input value that violates the disjointness condition `c3` (which is universally quantified over `?v1327:N`). But again, when the defining equations are not conditional, this work is done automatically by `fun-def-conds`.

When some of the defining equations are conditional, both exhaustiveness and disjointness can become arbitrarily difficult to decide. In that case, one should look at each sentence in the output list of `fun-def-conds` and try to prove each separately, or at least convince oneself that each holds, or, finally, if it is violated, that the violation is innocuous. Violations, when they occur, can often be found automatically by `falsify`, and usually at low bounds. When no violations can be found, that usually means that a condition holds, in which case it should be provable. Again, these conditions can occasionally require nontrivial proof, but oftentimes a proof can be found automatically. Users could, for instance, try deriving each of the relevant conditions via the unary method `induction*`, which takes a universally quantified sentence and attempts to derive it by using (possibly nested) structural inductions along with external ATPs integrated with Athena (refer to the "Automated Theorem Proving" appendix on the book's web site). Datatype axioms typically need to be in the assumption base for such proofs to succeed.

For those who are used to programming with pattern matching in functional programming languages like ML, note that the operational semantics of such languages guarantee disjointness by default, because patterns are tried sequentially from top to bottom. It would be impossible for two clauses to be applicable to one and the same input because, by definition, only the first (topmost) clause that matched could ever be applied to the input. For instance, the foregoing definition of `fact` would be given as follows in SML:

```
fun fact(zero) = S(zero)
  | fact(S(zero)) = S(zero)
  | fact(S(n)) = Times(S(n),fact(n));
```

The last two clauses here are in principle overlapping, but this is resolved because the third one will only be applied if the second does not match, that is, only when the input is greater than 1. By contrast, in the Athena style we have been discussing, the order in which the equations are listed is irrelevant, and hence disjointness needs a little more care. (Nonexhaustive definitions in languages like ML are also automatically detected and trigger warnings.)

Apart from the conditions discussed above, any recursion present in function definitions must be well-founded, meaning that there should be no infinite chains

$$t_1 \rightarrow t_2\ [R_1] \rightarrow t_3\ [R_2] \rightarrow \cdots$$

whereby a term t_1 is reduced to t_2, t_2 to t_3, and so on *ad infinitum*. Here is a trivial example of a function definition that fails this requirement:

```
declare loop: [N] -> N
assert* loop-def := [(loop x = loop x)]
```

Clearly, for any canonical natural number n we have

$$\texttt{(loop}\ n\texttt{)} \rightarrow \texttt{(loop}\ n\texttt{)}\ \texttt{[loop-def]} \rightarrow \texttt{(loop}\ n\texttt{)}\ \texttt{[loop-def]} \rightarrow \cdots$$

ad infinitum. So, for computational purposes, this definition is unacceptable because if we tried to use it to mechanically evaluate a term, such as `(loop zero)`, we would never arrive at a result. And for logical purposes the definition is uninteresting, because the axiom is satisfied by infinitely many (indeed, by *all*) unary functions on the natural numbers. Therefore, the axiom has no content, as it fails to weed out any models, thereby failing to say anything interesting about `loop`. Care must be taken to ensure that function definitions are terminating. Formally, this can be done by introducing a function that associates an integer quantity with the arguments of a defined function, and by establishing that each recursive call decreases that quantity. An alternative approach [106, 107], applicable to many of the function definitions considered in this book, is based on constructing, for each new function definition, a corresponding "termination condition," a sentence expressing sufficient conditions for termination of the function for all inputs. The sentence is then proved by an induction that mirrors the recursive structure in the function definition and is thus very similar to inductions for proving correctness properties of the function.

Informally, in many cases testing with `eval` will reveal infinite loops, among other unintended behaviors. Ultimately, it is the user's responsibility to ensure that a defined function f determines a unique result for all inputs in its intended purview.

3.16 Summary

Here are the most important points to take away from this chapter:

- Equality chaining is a broadly useful proof method, explored here in terms of addition and multiplication functions on natural numbers and concatenation and reversal functions on lists. Athena's `chain` method allows expression of equality chains in a form that closely resembles the way in which these are commonly written.
- Understanding equality chaining is aided by considering terms and sentences as tree structures. A fundamental step involved is *term rewriting*, defined as replacement of one subterm t' of a term t by a new term t''. Rewriting (of the nonconditional variety) involves matching of t' to the left-hand side of a *rewrite rule* (a universally quantified equation $L = R$ where all variables in R also appear in L), which determines a *substitution* θ that is applied to R to yield t''.
- Not all useful equalities can be derived by equality chaining alone. Some form of mathematical induction may be necessary. Here we have considered a simple form for natural numbers and lists called *ordinary induction*. In later chapters we will consider variations such as *strong induction* and *measure induction*, which are better suited for proving properties of functions that recurse in less trivial ways.
- Mathematical induction for a given datatype T is based on how T is defined by constructors. Athena's **`by-induction`** proof form automatically extracts the necessary information from T's definition to handle the required basis and inductive-step proof cases.
- To understand function definitions by way of equational axioms, it helps to be able to evaluate ground terms using the defining axioms, to see if the result agrees with one's intuition. Athena's `eval` procedure provides this capability when the defining axioms have appropriate constructive content.
- Proofs can be developed bottom-up (working from axioms and proving lemmas and then a desired result) or top-down (from the desired result, conjecturing needed lemmas, attempting to prove them and possibly conjecturing more lemmas until axioms are reached), or some combination of the two approaches. Learning to manage such proof development is challenging, but the examples and exercises in this book will provide extensive experience with the key issues.
- The last four sections of the chapter are starred because they can be treated as reference material to be consulted when related topics come up in later chapters:

- Athena provides support for translating values expressed in familiar notation to/from datatype terms that can be directly used in proofs, term evaluations, and so on. For example, base-10 numeric literals can be translated to/from the unary notation of Peano arithmetic (`datatype N`), and lists written in square-bracket notation can be converted to/from terms of sort `(List` S`)`.
- There is also a simple form of *conjecture testing* or model checking, via the `falsify` procedure. Applied to a conjectured sentence containing function symbols that are constructively defined by (possibly conditional) equations, it can find a counterexample if the sentence is not valid and a sufficiently small counterexample exists. Besides saving time and effort spent on futile proof attempts, *falsify* can help in debugging a conjecture. The generated counterexamples can help to guide us in replacing a false conjecture with one that actually holds (and is therefore provable).
- The rewriting notion was extended to rewrite rules that have conditions, making it far more useful than the unconditional rules studied in the rest of the chapter.
- Finally, the last section discussed proper function definitions and singled out a number of conditions ensuring that a set of conditional rewrite rules define a function totally and without introducing inconsistencies. This boils down to requirements of *exhaustiveness*, *disjointness*, and *termination* on the rules.

Other aspects of complexity in proof development will arise as we move beyond equational sentences and the equation-chaining proof methods we have studied in this chapter. While equations are widely used in mathematics and the sciences (including computer science) and equation-chaining proofs are relatively easy both to construct and to understand, they are only a small part of the full range of useful logic concepts and proof methods. And even with equations, we have not yet seen significant examples of the conditional equations introduced in Section 3.14. Other crucial questions include: How do we properly deal with conjunction (`&`), disjunction (`|`), and equivalence (`<==>`)? How do quantifiers interact with negation? And so on.

To prepare to address the many aspects of proof development that we cover in later chapters, we will devote the next two chapters to a tour of the most fundamental principles of logic and proof, starting with sentential logic in Chapter 4 and continuing with first-order logic in Chapter 5.

3.17 Additional exercises

Exercise 3.19: Prove `left-nonzero` (defined in Section 3.5). □

Exercise 3.20: Prove `Plus-commutative` (defined in Exercise 3.12). *Hint:* You will need both `left-zero` and `left-nonzero`. □

Exercise 3.21: State and prove `Plus-associative`, an associative axiom for `Plus` analogous to `Times-associative` (defined on page 130). □

Exercise 3.22: Prove the following property of `Times`:

```
define Times-left-zero := (forall x . zero * x = zero)
```

Use induction. □

Exercise 3.23: Prove the following property of `Times`:

```
define Times-left-nonzero := (forall x y . (S y) * x = x + y * x)
```

Use induction. □

Exercise 3.24: Consider:

```
define Times-Right-Distributive :=
  (forall x y z . (x + y) * z = x * z + y * z)
```

Prove `Times-Right-Distributive`. □

Exercise 3.25: Prove `Times-associative` (defined on page 130). (*Hint:* You will need to use the three preceding `Times` properties.) □

Exercise 3.26: State and prove `Times-commutative`, a commutative axiom for `Times`. □

To prepare for the following exercises, insert in your solutions file Athena commands declaring `**` and defining and asserting the two axioms for `**` given on page 130. Also define `Times-right-one` and assert or prove it.

Exercise 3.27: Consider the following property of `**`:

```
(forall n . one ** n = one).
```

Prove it by induction. □

Exercise 3.28: Consider the following property of `**`:

```
define Power-Plus :=
  (forall m n x . x ** (m + n) = x ** m * x ** n)
```

Prove it by induction. □

Exercise 3.29: Prove the following property of `**`:

```
define Power-Distributive := (forall n x y . (x * y) ** n = x ** n * y ** n)
```

□

To prepare for the following exercises, insert in your solutions file Athena commands declaring `join` and defining and asserting the two `join` axioms given on page 147; do the same for the `reverse` declaration and axioms given on page 152. Also, assert or reprove `right-empty` (page 145); define `reverse-reverse-theorem` (page 153); introduce `++` as an alias for `join`; and define `L1` and `L2` as the variables `?L1` and `?L2`.

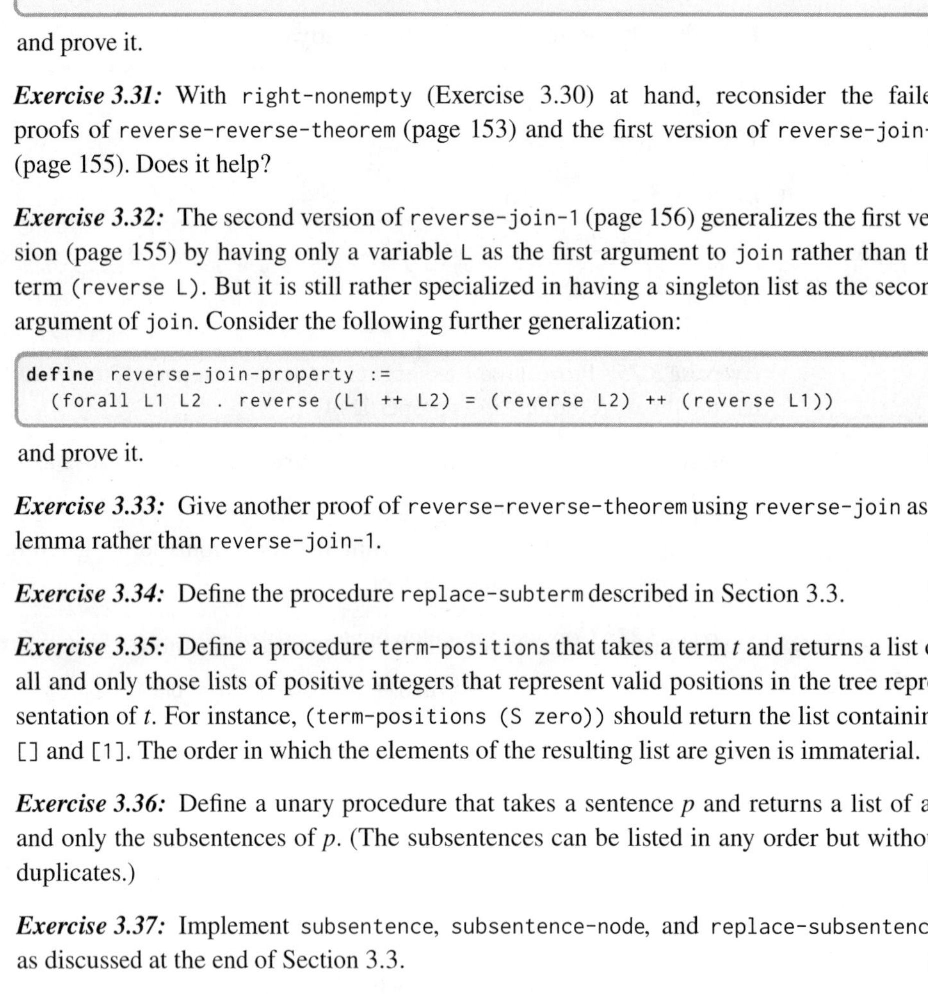

Exercise 3.30: Define the property `right-nonempty` as follows and prove it:

```
define right-nonempty :=
  (forall L1 x L2 . L1 ++ x::L2 = (L1 ++ x::nil) ++ L2)
```

and prove it. □

Exercise 3.31: With `right-nonempty` (Exercise 3.30) at hand, reconsider the failed proofs of `reverse-reverse-theorem` (page 153) and the first version of `reverse-join-1` (page 155). Does it help? □

Exercise 3.32: The second version of `reverse-join-1` (page 156) generalizes the first version (page 155) by having only a variable `L` as the first argument to `join` rather than the term `(reverse L)`. But it is still rather specialized in having a singleton list as the second argument of `join`. Consider the following further generalization:

```
define reverse-join-property :=
  (forall L1 L2 . reverse (L1 ++ L2) = (reverse L2) ++ (reverse L1))
```

and prove it. □

Exercise 3.33: Give another proof of `reverse-reverse-theorem` using `reverse-join` as a lemma rather than `reverse-join-1`. □

Exercise 3.34: Define the procedure `replace-subterm` described in Section 3.3. □

Exercise 3.35: Define a procedure `term-positions` that takes a term t and returns a list of all and only those lists of positive integers that represent valid positions in the tree representation of t. For instance, `(term-positions (S zero))` should return the list containing `[]` and `[1]`. The order in which the elements of the resulting list are given is immaterial. □

Exercise 3.36: Define a unary procedure that takes a sentence p and returns a list of all and only the subsentences of p. (The subsentences can be listed in any order but without duplicates.) □

Exercise 3.37: Implement `subsentence`, `subsentence-node`, and `replace-subsentence`, as discussed at the end of Section 3.3. □

Exercise 3.38: Define a procedure `sentence-positions` that takes a sentence p and returns a list of all and only those lists of positive integers that represent valid positions in the tree representation of p, in any order. (This is the sentential analogue of `term-positions` from Exercise 3.35.) □

Exercise 3.39: Recall the definition of `int->nat` given in Section 3.12:

```
define (int->nat n) :=
  check {
    (numeral? n) => check {(less? n 1) => zero
                          | else => (S (int->nat (minus n 1)))}
  | else => n
  }
```

Can you spot any sources of inefficiency that could be easily removed? □

3.18 Chapter notes

Birkhoff [10] laid the foundations of equational logic in 1935, based on identities in abstract algebras. In 1970, Knuth and Bendix [59] achieved a breakthrough in automating proofs of equations using term rewriting systems. Two properties that were briefly discussed in Section 3.15 for ensuring proper function definitions, *termination* and *confluence* of the defining axioms, are also crucial for fully automating proofs of equations. The Knuth-Bendix *completion procedure* achieves both properties, in many cases by modifying the original set of rules and augmenting it with new rules, but neither the original procedure nor any of its subsequent refinements always terminates. Dershowitz and Joaunnaud [30] survey much of the very extensive research on term rewriting systems and give many references. One line of this research builds on a surprising discovery by Musser [75] that under certain conditions it is possible to prove an equation in the inductive theory of the axioms without explicitly setting up and proving basis and inductive step cases. The conditions are that the axioms satisfy the *exhaustiveness* property (discussed in Section 3.15 and also known in the literature as *sufficient completeness*) and that one can generate a terminating and confluent system that includes the axioms and the new equation.

In this book our main concern is not with automation of equational proofs but with expressing proofs in a format—an equation chain or induction cases that each consist of equation chains—that is intuitive to a human reader and easily checkable by machine. An advantage of this approach is that we do not always have to have termination and confluence of the sets of equations we use to axiomatize data types: We do use the equations as rewrite rules, but we have full control over how they are applied in proofs. When each link of an equation chain is justified by a single rewrite rule, there is no issue of termination or confluence (with rare exceptions such as a rule that by itself produces an infinite sequence of rewrites). If we want to shorten proofs by using more than one axiom in a link justification, then we need to be a bit more concerned about termination, as multiple rules might be combined in unexpected ways to produce an infinite rewrite sequence. In many cases, however, termination is naturally satisfied, and we furthermore usually have disjointness, which trivially guarantees confluence, so we avoid the need to run a completion procedure.

4 Sentential Logic

THE focus of this chapter is sentential logic. Sentential logic, also known as *propositional logic*, is concerned with zero-order sentences, such a sentence being either a Boolean term, say, `(zero < S zero)`, or the result of applying one of the five sentential connectives (`not`, `and`, `or`, `if`, `iff`) to other zero-order sentences. Sentences that are Boolean terms are viewed as indivisible units having no internal structure; the only meaning they are capable of having in sentential logic is a binary indication of truth or falsity, as will become clear when we come to discuss sentential semantics in Section 4.12.

Mastering sentential logic is a crucial step in becoming able to write and understand proofs, because sentential logic forms the core of more powerful systems such as first-order logic, which we will study in subsequent chapters, and indeed lies at the heart of mathematical reasoning in general. In this chapter we cover all available mechanisms for constructing proofs in sentential logic, giving numerous examples along the way; we develop a library of useful sentential proof methods; we touch on some more advanced topics, such as recursive proof methods; and we introduce heuristics for constructing such proofs. We also give a computational treatment of the semantics of sentential logic, and we discuss some useful applications enabled by algorithms for solving the satisfiability (SAT) problem. Finally, we develop a theorem-proving method for sentential logic that will be used in the sequel. In all of the examples in this chapter we use the identifiers `A`, `B`, `C`, `D`, `E`, and so on, as Boolean atoms. We assume that appropriate declarations of these constants have already been made at the top level.

4.1 Working with the Boolean constants

The two Boolean atoms `true` and `false` are the ultimate foundations of sentential reasoning. We can derive `true` any time by applying the nullary method `true-intro`:

```
> (!true-intro)

Theorem: true
```

The constant `false` can only be derived if the assumption base is inconsistent. Specifically, if the assumption base contains two contradictory sentences of the form *p* and `(~` *p*`)`, then applying the binary method `absurd` to *p* and `(~` *p*`)` will derive `false`:

```
> assume A
    assume (~ A)
      (!absurd A (~ A))
```

```
Theorem: (if A
             (if (not A)
                 false))
```

It is an error if the two arguments to absurd are not of the form p and (~ p), respectively, or if either of them is not in the assumption base.

Finally, we can derive (~ false) at any time through the nullary method false-elim:

```
> (!false-elim)

Theorem: (not false)
```

4.2 Working with conjunctions

4.2.1 Using conjunctions

It is part and parcel of the meaning of a conjunction (p & q) that it logically implies each of its two components, p and q. If it is raining and it is windy, it follows that it is raining; and it also follows that it is windy. Accordingly, if we have a conjunction (p & q) in the assumption base, we should be able to derive p from it, as well as q. In Athena, the former is done by the unary method left-and. If the argument to left-and is a conjunction (p & q) that is present in the assumption base, then the output will be the conclusion p:

```
assert p := (A & B)

> (!left-and p)

Theorem: A
```

There is a similar unary method, right-and, that does the same thing for the right component of a conjunction: If (p & q) is in the assumption base,

(!right-and (p & q))

will produce the conclusion q. It is an error if the argument to left-and or right-and is not a conjunction, or if it is a conjunction that is not in the assumption base.

We say that left-and performs conjunction *elimination*, because it takes a conjunction as input and detaches its left component. Likewise for right-and. Most primitive deductive constructs of Athena either eliminate or else *introduce* one of the five logical connectives, or one of the quantifiers, thereby serving as introduction or elimination mechanisms for the corresponding connective or quantifier.

4.2.2 Deriving conjunctions

The introduction mechanism for conjunctions is the binary method `both`. Given any two sentences p and q in the assumption base, the method call

(!both p q)

will produce the conclusion (p & q). As an example that uses all three methods dealing with conjunctions, consider the derivation of `(D & A)` from the premises `(A & B)` and `(C & D)`:

```
assert (A & B), (C & D)

> let {_ := conclude D
              (!right-and (C & D));
       _ := conclude A
              (!left-and (A & B))}
    (!both D A)

Theorem: (and D A)
```

Alternatively:

```
assert A-and-B := (A & B)
assert C-and-D := (C & D)

> {
    conclude D
      (!right-and C-and-D);
    conclude A
      (!left-and A-and-B);
    (!both D A)
  }

Theorem: (and D A)
```

or:

```
> (!both conclude D
           (!right-and C-and-D)
         conclude A
           (!left-and A-and-B))

Theorem: (and D A)
```

or even more tersely:

```
> (!both (!right-and C-and-D)
         (!left-and  A-and-B))

Theorem: (and D A)
```

4.3 Working with conditionals

4.3.1 Using conditionals

One of the most fundamental inference rules in logic is *modus ponens*, also known as *conditional elimination*. Starting from two premises of the form (*p* ==> *q*) and *p*, modus ponens yields the conclusion *q*, thereby detaching ("eliminating") the conditional connective. In Athena, modus ponens is performed by the primitive binary method `mp`. When the first argument to `mp` is of the form (*p* ==> *q*), the second argument is *p*, and both arguments are in the assumption base, `mp` will derive *q*. For instance, assuming that `(A ==> B)` and `A` are both in the assumption base, we have:

```
> (!mp (A ==> B) A)

Theorem: B
```

Another important method involving conditionals is *modus tollens*. Given two premises of the form (*p* ==> *q*) and (~ *q*), modus tollens generates the conclusion (~ *p*). That's a valid form of reasoning. Suppose, for instance, that we know the following two sentences to be true:

1. If the keys are in the apartment, then they are in the kitchen.
2. The keys are not in the kitchen.

From these two pieces of information we can infer that the keys are not in the apartment.[1] This inference is licensed by modus tollens.

In Athena, modus tollens is performed by the binary method `mt`. When the first argument to `mt` is a conditional (*p* ==> *q*), the second is (~ *q*), and both are in the assumption base,

1 Indeed, suppose that the keys are in the apartment. Then, by the first premise and modus ponens, they would have to be in the kitchen. But the second premise tells us that, in fact, they are not in the kitchen. We have thus arrived at two inconsistent conclusions: that the keys are in the kitchen and that they are not. This contradiction shows that our starting assumption is untenable—the keys cannot be in the apartment. This is an example of reasoning by contradiction, a topic that we discuss more extensively in Section 4.5.2.
An astute reader might point out that the derivation of a contradiction from our two premises along with the hypothesis that the keys are in the apartment does not necessarily compel us to retract that hypothesis. All it shows is that the three sentences together, as a whole, form an inconsistent set, and therefore that at least one of them must be false. Which one we choose to reject is up to us. That is true. We could restore consistency by rejecting one of the two premises instead. However, we generally operate under the convention that we are not willing to give up the initial premises of an argument—the premises are taken for granted, and if any sentence *p* conflicts with them, then it is *p* that must be given up, not any of the premises. (Of course this is an otiose point if the starting premises are themselves jointly inconsistent.)

`mt` will derive `(~` p`)`. For instance, assuming that `(A ==> B)` and `(~ B)` are both in the assumption base, we have:

```
> (!mt (A ==> B) (~ B))

Theorem: (not A)
```

4.3.2 Deriving conditionals: hypothetical reasoning

The standard way of proving a conditional `(`p `==>` q`)` is to assume the antecedent p (i.e., to add p to the current assumption base) and proceed to derive the consequent q. In Athena this is done with deductions of the form **`assume`** p D, where D is a proof that derives q from the augmented assumption base. To illustrate this technique, here is a proof of `(A ==> A)`:

```
> assume A
    (!claim A)

Theorem: (if A A)
```

Deductions of the form **`assume`** p D are called *conditional* or *hypothetical* deductions. We refer to p and D as the *hypothesis* (or *assumption*) and the *body* of the conditional deduction, respectively. We also say that D represents the *scope* of the hypothesis p. So, in the above example, the hypothesis is the atom `A` and its scope is the deduction `(!claim A)`, which constitutes the body of the conditional proof.

To evaluate a deduction of the form

$$\texttt{assume}\ p\ D \tag{4.1}$$

in an assumption base β, we add p to β and go on to evaluate the body D in the augmented assumption base $\beta \cup \{p\}$. The fact that D is evaluated in $\beta \cup \{p\}$ means that *the assumption p can be freely used anywhere within its scope*, that is, anywhere inside D. If and when the evaluation of D in $\beta \cup \{p\}$ produces a conclusion q, we return the conditional `(`p `==>` q`)` as the result of (4.1). If the evaluation of D in $\beta \cup \{p\}$ diverges or results in an error, then the evaluation of (4.1) also diverges or results in that same error, respectively.

The body D is said to be a *subproof* (or *subdeduction*) of the conditional proof (4.1). Subproofs can be nested inside one another, as the following figure illustrates. Note that, strictly speaking, the scope of a hypothesis is not determined by indentation but rather by the syntax rules that define what counts as a deduction. Nevertheless, proper indentation can be very helpful in demarcating the scope of an assumption (and, more generally, in clarifying the structure of a proof). Typically, the body D of a hypothetical deduction **`assume`** p D is written on the line after the hypothesis p and indented two spaces to the right of the keyword **`assume`**.

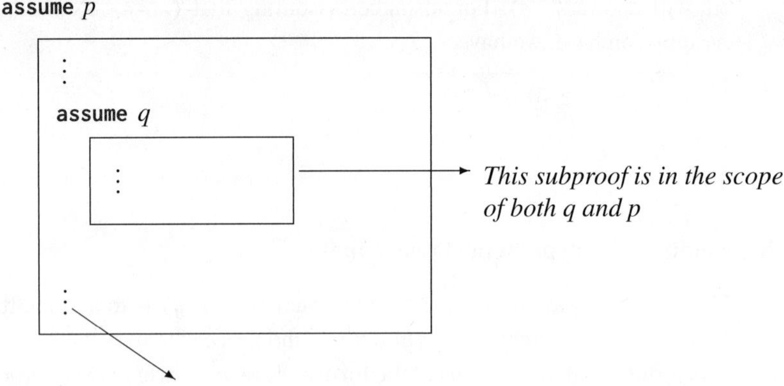

For instance, in the following proof the body `(!claim A)` is in the scope of both the inner assumption `B` and the outer assumption `A`:

```
> assume A
    assume B
      (!claim A)

Theorem: (if A
             (if B A))
```

Note that the contents of the assumption base are immaterial for the above proof. Evaluating this proof in any β whatsoever will successfully produce the result

```
(A ==> B ==> A).
```

That is the case for all and only those sentences that are *tautologies*. A tautology is precisely a sentence that can be derived from every assumption base. Given that Athena's classical logic is monotonic,[2] an alternative way of saying the same thing is this: A tautology is a sentence that can be derived from the empty assumption base. The equivalence of the two conditions can be shown as follows. In one direction, if p can be derived from every assumption base, then it can certainly be derived from the empty assumption base. Conversely, if p can be derived from the empty assumption base then, by monotonicity, it can be derived from every assumption base.[3]

---

2 Monotonicity means that if p is derivable from some β, then it is also derivable from every superset of β. Intuitively, the import of this is that *adding* new information to an assumption base does not invalidate any conclusions previously obtainable from it.

3 This is a syntactic or proof-theoretic characterization of tautologies. A semantic characterization would say that a tautology is a sentence that is true under all interpretations, or, alternatively, logically entailed by the empty assumption base; we will pursue the semantic direction later on, in Section 4.12. Since the proof system of Athena

Oftentimes it is useful to give a name to the hypothesis of a conditional deduction, and then use that name inside the body to refer to the hypothesis. This can be done with the following variant of the **assume** construct:

assume $I := p$
D

which is really a shorthand for:

let $\{I := p\}$
assume I
D

As an example, consider the proof of the tautology `(A & B ==> B & A)`:

```
> assume hyp := (A & B)
    let {left   := (!left-and hyp);
         right  := (!right-and hyp)}
      (!both right left)

Theorem: (if (and A B)
             (and B A))
```

Actually, Athena allows a simpler proof:

```
> assume (A & B)
    (!both B A)

Theorem: (if (and A B)
             (and B A))
```

This proof is possible because whenever the hypothesis of a conditional proof is a conjunction, Athena not only adds that conjunction to the assumption base, but it also adds each of the conjuncts (and the conjuncts of those conjuncts, and so forth). Thus, in the above proof, both `A` and `B` are in the assumption base when `both` is invoked. A similar automatic decomposition of conjunctions also happens during intermediate proof steps in a **let** deduction. For instance:

```
assert A=>B&C := (A ==> B & C)
assert A

let {B&C := (!mp A=>B&C A);
     ··· # We now have not just (B & C), but also B and C in the a.b.
    }
  ···
```

---

is complete (meaning that a sentence p is entailed by an assumption base β iff it is derivable from β), these two views of tautologies coincide.

Consider next this implication:

$$\texttt{((A ==> B ==> C) ==> (B ==> A ==> C))}. \tag{4.2}$$

The following proof derives (4.2):

```
1  > assume hyp := (A ==> B ==> C)
2      assume B
3        assume A
4          let {B=>C := (!mp hyp A)}
5            conclude C
6              (!mp B=>C B)
7
8  Theorem: (if (if A
9                   (if B C))
10               (if B
11                  (if A C)))
```

Observe that the structure of the proof mirrors the structure of the goal. The proof is constructed by reasoning as follows: Our top goal, (4.2), is a conditional of the form (p ==> q), hence our proof, starting on the first line, will be of the form **assume** p D, where D must be a proof capable of deriving the subgoal q. Now, in this case q is also a conditional, namely, (B ==> A ==> C), so D, starting on line 2, will be a deduction of the form

assume B D',

where D' is a deduction that will derive the subgoal (A ==> C). Since that subgoal is again a conditional, the proof D' that starts on line 3 is of the form

assume A D'',

where D'' now has to derive the subgoal C from the preceding three assumptions. This can be done by the two inferences on lines 4 and 6, first by deriving (B ==> C) and finally by deducing C; both steps use modus ponens on the appropriate premises. As we discuss in Section 4.14, most proofs are constructed through similar combinations of reasoning that is driven exclusively by the syntactic form of the goal and reasoning that is driven by the contents of the assumption base.

Of course, a conditional may be derived in many other ways. For instance, if the assumption base contains a conjunction of the form (p & (q ==> r)), then it is simpler and more direct to derive the conditional (q ==> r) by performing right conjunction elimination on that premise instead of using **assume**. The same goes for sentences of other forms, say, for biconditionals. The proof techniques discussed in this chapter are the canonical or most common ways of inferring the corresponding kinds of sentences, but, depending on the context, it may be simpler to use an alternative method.

4.4 Working with disjunctions

4.4.1 Using disjunctions: reasoning by cases

Suppose we are trying to derive some goal p. If the assumption base contains a disjunction (p_1 | p_2), we can often put that disjunction to use as follows: We know that p_1 holds or p_2 holds. If we can show that p_1 implies the goal p *and* that p_2 also implies p, then we can conclude p. For, if p_1 holds, then p follows from the implication (p_1 ==> p), p_1, and modus ponens; while, if p_2 holds, then p follows from (p_2 ==> p), p_2, and modus ponens. This type of reasoning (called "case analysis" or "reasoning by cases") is pervasive, both in mathematics and in real life.[4]

In Athena this type of reasoning is carried out by the ternary method cases. The first argument of this method must be a disjunction, say (p_1 | p_2); while the second and third arguments must be conditionals of the form (p_1 ==> p) and (p_2 ==> p), respectively. If all three sentences are in the assumption base, then the conclusion p is produced as the result. It is an error if the arguments are not of the form just described, or if one of them is not in the assumption base. For instance:

```
assert (C1 | C2), (C1 ==> B), (C2 ==> B)

> conclude B
    (!cases (C1 | C2)
            (C1 ==> B)
            (C2 ==> B))

Theorem: B
```

As another example, suppose the assumption base contains (C ==> B) and ((A & B) | C) and from these two premises we want to derive the conclusion B. We can do that with the following case analysis:

```
> (!cases (A & B | C)
    assume (A & B)
      (!claim B)
    assume C
      (!mp (C ==> B) C))

Theorem: B
```

4 For instance, suppose we know that either the transmission of a car is broken or the battery needs to be replaced. And we know further that the cost of fixing the transmission exceeds $150, while replacing the battery also costs more than $150. Then we can conclude that in either case, fixing the car will cost more than $150.

Note how the implications required by `cases` in its second and third argument positions are naturally generated as results of **assume** deductions. And in the first case, `(!claim B)` succeeds because, as noted earlier, **assume** `(A & B)` adds the conjuncts `A` and `B` to the assumption base, along with `(A & B)`.

A useful variant of reasoning by cases performs a case analysis on a sentence *p* and its negation (~ *p*). We know that for any given *p*, either *p* or (~ *p*) holds; this is the law of the *excluded middle*. Therefore, if we can show that a goal *q* follows both from *p* and from (~ *p*), we should be able to conclude *q*. This is done with the binary method `two-cases`, which takes two premises of the form (*p* ==> *q*) and (~ *p* ==> *q*) and derives *q*. For example:

```
assert (A ==> B), (~ A ==> B)

> (!two-cases
    (A ==> B)
    (~ A ==> B))

Theorem: B
```

The two premises can be passed in either order, for example, the negative conditional can be the first argument and the positive second:

```
> (!two-cases
    (~ A ==> B)
    (A ==> B))

Theorem: B
```

This is not a primitive Athena method. It is defined in terms of `cases` as follows:

```
define two-cases :=
  method (cond-1 cond-2)
    let {M := method (p q)
                (!cases conclude (p | ~ p)
                          (!ex-middle p)
                  (p ==> q)
                  (~ p ==> q))}
      match [cond-1 cond-2] {
        [(p ==> q) ((~ p) ==> q)] => (!M p q)
      | [((~ p) ==> q) (p ==> q)] => (!M p q)
      }
```

where `ex-middle` is a unary method that takes an arbitrary sentence *p* and derives the theorem (*p* | ~ *p*). (We ask you to implement this method in Exercise 4.15.)

4.4.2 Deriving disjunctions

The most straightforward way to derive a disjunction `(`p `|` q`)` is to derive the left component, p, or the right component q. If we have p in the assumption base, then `(`p `|` q`)` can be derived by applying the binary method `left-either` to p and q:

`(!left-either` p q`).`

Likewise, if q is the assumption base, then the method call

`(!right-either` p q`)`

will derive the disjunction `(`p `|` q`)`. For instance:

```
> assume A
    (!claim A)

Theorem: (if A A)

> (!left-either (A ==> A) B)

Theorem: (or (if A A)
             B)

> (!right-either B (A ==> A))

Theorem: (or B
             (if A A))
```

In addition to these two primitive methods, Athena offers a third, more versatile mechanism for disjunction introduction, the binary method `either`. If either p *or* q is in the assumption base, then `(!either` p q`)` derives the disjunction `(`p `|` q`)`. Otherwise, if neither argument is in the assumption base, `either` fails.

In realistic proofs, one is unlikely to derive a disjunction simply by deriving one of the two disjuncts. Usually there is a relationship between the two that must be taken into account, and that relationship is lost when each disjunct is treated in isolation from the other. Consider, for instance, the derivation of `(A | ~ A)` from the empty assumption base. Neither `A` nor `(~ A)` is derivable by itself, but their disjunction is.

There are a few different avenues for dealing with such cases. One of them is to reason by contradiction, that is, to negate the entire disjunction and proceed to derive a contradiction. In the foregoing example, that would mean assuming `(~ (A | ~ A))` and trying to derive a contradiction from that. (Reasoning by contradiction is discussed in Section 4.5.2.) Another useful technique is to treat a disjunction `(`p `|` q`)` as the conditional

`(~` p `==>` q`)`

and use the corresponding introduction mechanism for conditionals (Section 4.3). The conditional `(~ p ==> q)` is logically equivalent to the disjunction `(p | q)`, so if we succeed in deriving the former, we should be able to derive the latter as a matter of routine. Indeed, in Section 4.9 we develop a method that can take an arbitrary conditional of the form `(~ p ==> q)` and produce the corresponding disjunction (method `cond-def`, page 222). With that method under our belt, we can transform a disjunction-derivation problem into a conditional-derivation problem.

4.5 Working with negations

4.5.1 Using negations

The only primitive method for negation elimination is `dn`, which stands for "double negation." It is a unary method whose argument must be of the form `(~ ~ p)`. If that sentence is in the assumption base, then the call

`(!dn (~ ~ p))`

will produce the conclusion p, thereby eliminating the two leading negation signs. For instance:

```
> assume h := (~ ~ A)
    (!dn h)

Theorem: (if (not (not A))
             A)
```

There are two other primitive methods that require some of its arguments to be negations: `mt` and `absurd`. Therefore, depending on what exactly we are trying to prove, we might be able to use a negation in the assumption base as an argument to one of these methods.

4.5.2 Deriving negations: reasoning by contradiction

Proof by contradiction[5] is one of the most useful and common forms of deductive reasoning. The basic idea is to establish a negation `(~ p)` by assuming p and showing that this assumption (perhaps in tandem with other working assumptions) leads to an absurdity, namely, to `false`. That entitles us to reject the hypothesis p and conclude the desired `(~ p)`.

The binary method `by-contradiction` is one way to perform this type of reasoning in Athena. The first argument to `by-contradiction` is simply the sentence we are trying to establish, typically a negation `(~ p)`. The second argument must be the conditional

5 Also known as *reductio ad absurdum* and *indirect proof*.

(p `==> false`), essentially stating that the hypothesis p leads to an absurdity. If that conditional is in the assumption base, then the desired conclusion (~ p) will be produced.

As an example, suppose that the assumption base contains the premises `(A ==> B & C)` and `(~ B)`, and from these two pieces of information we want to derive the negation `(~ A)`. We can reason by contradiction as follows: Suppose `A` holds. Then, by the first premise and modus ponens, we would have `(B & C)`, and hence, by conjunction elimination, `B`. But this contradicts the second premise, `(~ B)`, and that contradiction allows us to reject the hypothesis `A`, inferring `(~ A)`.

```
assert premise-1 := (A ==> B & C)
assert premise-2 := (~ B)

> (!by-contradiction (~ A)
    assume A
      let {p1 := conclude (B & C)
                   (!mp premise-1 A);
           _ := conclude B
                  (!left-and p1)}
        (!absurd B premise-2))

Theorem: (not A)
```

Of course, we could have derived the conditional `(A ==> false)` first, and then, in a separate step, deduced `(~ A)` by applying `by-contradiction`:

```
> assume A
    let {_ := conclude (B & C)
                (!mp premise-1 A)}
      (!absurd B premise-2)

Theorem: (if A false)

> (!by-contradiction (~ A)
    (A ==> false))

Theorem: (not A)
```

However, we believe that the first version is more readable, so typically the requisite conditional (p `==> false`) will be established by a hypothetical deduction of the form

assume p D

that will appear directly as the second argument to `by-contradiction`. Observe that the semantics of nested deductions that we discussed in Section 2.10.2 are crucial here. A proof like

```
(!by-contradiction (~ p)
  assume p ···)
```

works precisely because the **assume**, being a deductive argument to the outer method call, will have its conditional conclusion incorporated into the assumption base by the time `by-contradiction` is applied.

Recall that the most direct way to derive `false` is to apply the binary method `absurd` to two contradictory sentences of the form q and `(~` q`)` in the assumption base. For instance, assuming that `A` and `(~ A)` are both in the assumption base, we have:

```
> (!absurd A (~ A))

Theorem: false
```

Accordingly, a proof of `(~` p`)` by contradiction often has the following logical structure:

```
(!by-contradiction (~ p)
  assume p
    let {p1 := conclude q
                 D1;
         p2 := conclude (~ q)
                 D2}
      (!absurd p1 p2))
```

In many cases, however, one of the two contradictory sentences, q or `(~` q`)`, is already in the assumption base, so we do not have to deduce it explicitly. We simply derive the other element of the contradictory pair and then move directly to the `absurd` application.

As another example, here is a proof that derives `(~ B)` from `(~ (A ==> B))`:

```
assert premise := (~ (A ==> B))

> (!by-contradiction (~ B)
    assume B
      let {A==>B := assume A
                      (!claim B)}
        (!absurd A==>B premise))

Theorem: (not B)
```

Sometimes the sentence p that we want to establish by contradiction is not a negation; it might be an atom, or a disjunction, or something else. That would seem to present a problem, since, according to what we have said so far, `by-contradiction` can only introduce negations. However, in classical logic every sentence p is equivalent to the double negation `(~ ~` p`)`, so this is not really a limitation: We can simply infer `(~ ~` p`)` by assuming `(~` p`)` and deriving a contradiction. After that, we can eliminate the double negation sign with `dn`. For example, suppose that the assumption base contains `A` and `(~ (A & ~ B))`, and we want to derive the atom `B`. Following this recipe, we have:

```
assert premise-1 := (~ (A & ~ B))
assert premise-2 := A

> let {--B := (!by-contradiction (~ ~ B)
                 assume (~ B)
                   (!absurd (!both A (~ B)) premise-1))}
    (!dn --B)

Theorem: B
```

However, this routine of inferring a double negation first and then eliminating the two leading negation signs is somewhat tedious. For that reason, `by-contradiction` offers a shortcut: When the conclusion p to be established is not in the explicit form of a negation, it suffices to establish the conditional (`~` p `==> false`). We can then apply `by-contradiction` directly to p and this conditional:

```
(!by-contradiction p
  (~p ==> false))
```

and the desired p will be obtained. Thus, for instance, the foregoing proof could be more concisely written as follows:

```
> (!by-contradiction B
    assume (~ B)
      (!absurd (!both A (~ B)) premise-1))

Theorem: B
```

Let us say that the *complement* of a negation (`~` p) is the sentence p, while the complement of a sentence p that is not a negation is the sentence (`~` p). Thus, for example, the complement of `A` is `(~ A)`, while the complement of `(~ A)` is `A`. We can define complementation with the following procedure:

```
define (complement p) :=
  match p {
    (~ q) => q
  | _ => (~ p)
  }
```

We write $\overline{p}$ to denote the complement of a sentence p. Note that $\overline{\overline{p}} = p$.

With this notion, we can succinctly specify the behavior of `by-contradiction` as follows. If the conditional ($\overline{p}$ `==> false`) is in the assumption base, then

```
(!by-contradiction p
  (p̄ ==> false))
```

will produce the conclusion p; otherwise it will fail.

There are two other auxiliary methods for reasoning by contradiction:

1. The unary method `from-false` derives any given sentence, provided that the assumption base contains `false`. That is, `(!from-false` p`)` will produce the theorem p whenever the assumption base contains `false`. This captures the principle that "everything follows from `false`."
2. The ternary method `from-complements` derives any given sentence p provided that the assumption base contains two complementary sentences q and $\overline{q}$. Specifically,

   `(!from-complements` p q $\overline{q}$`)`

   will derive p provided that both q and $\overline{q}$ are in the assumption base. Such an application can be read as: "Infer p from the complements q and $\overline{q}$."

These two methods are not primitive. Both are defined in terms of `by-contradiction`, and indeed each is definable in terms of the other; see Section 4.9 and Exercise 4.10. Nevertheless, it is convenient to have both available as separate methods.

4.6 Working with biconditionals

4.6.1 Using biconditionals

There are two elimination methods for biconditionals, `left-iff` and `right-iff`. For any given biconditional `(`p `<==>` q`)` in the assumption base, the method call

`(!left-iff (`p `<==>` q`))`

will produce the conclusion `(`p `==>` q`)`, while

`(!right-iff (`p `<==>` q`))`

will yield `(`q `==>` p`)`. For instance:

```
assert bc := (A <==> B)

> (!left-iff bc)

Theorem: (if A B)

> (!right-iff bc)

Theorem: (if B A)
```

4.6.2 Deriving biconditionals

The introduction method for biconditionals is `equiv`. Given two conditionals (p ==> q) and (q ==> p) in the assumption base, the call

```
(!equiv (p ==> q) (q ==> p))
```

will derive the biconditional (p <==> q):

```
assert (A ==> B), (B ==> A)

> (!equiv (A ==> B) (B ==> A))

Theorem: (iff A B)
```

Note that there is a similarity between the introduction and elimination methods for conjunctions and those for biconditionals, with `left-and` and `right-and` being analogous to `left-iff` and `right-iff`, respectively, and with `both` analogous to `equiv`. This similarity is no accident, as a biconditional (p <==> q) is logically equivalent to the conjunction of the two conditionals (p ==> q) and (q ==> p). Indeed, if we regard (p <==> q) as syntax sugar for the conjunction ((p ==> q) & (q ==> p)), the analogy becomes exact.

4.7 Forcing a proof

When we set out to write a proof, we often know the rough outline of the proof's structure, but we don't know ahead of time how to fill in the details. (This is a recurrent theme that will come to the forefront when we turn to proof heuristics in Section 4.14.) Suppose, for example, that our goal is some sentence p, and that we have decided to proceed by a case analysis on a disjunction (p_1 | p_2). Accordingly, we know that the general form of the proof will be

```
(!cases (p1 | p2)
  assume p1
    conclude p
      D1
  assume p2
    conclude p
      D2)
```

although we have not yet figured out what D_1 and D_2 should be. The fact that we don't yet have a complete proof should not prevent us from entering into the system the partial outline that we do have and verifying that it properly derives the goal. Later on we will make additional progress on D_1 and on D_2, and we will want to evaluate that progress as well. In general, proof headway is made incrementally, and we should have a way of testing these contributions as they are being made, gradually, instead of having to wait until we have a finished proof with all the details filled in. Yet Athena can only evaluate syntactically complete deductions. So it would seem, for example, that the only way to evaluate the preceding proof outline will be to expand D_1 and D_2 into complete deductions. How can we resolve that tension?

The answer lies in the built-in primitive method **force**. This is a quite special unary method (hence the distinct font) that takes an arbitrary sentence p as its argument and produces p right back as its result. It is, in a sense, the deductive equivalent of the identity function, and it is therefore similar to the reiteration method `claim`. Unlike `claim`, however, the **force** method does not bother to check that the given sentence is in the assumption base. In fact, **force** doesn't perform any checking at all; it simply returns as a theorem whatever sentence p it is given as input. It should be clear that this is an unsound and generally dangerous method that should not be used anywhere in a *finished*, complete proof. However, it can be a useful tool during proof development, for the reasons described. For example, using **force** we can express the foregoing proof sketch as follows:

```
(!cases (p1 | p2)
  assume p1
    conclude p
      (!force p)
  assume p2
    conclude p
      (!force p))
```

This a proof that can be evaluated perfectly well on its own and will produce the intended conclusion p—as long as the `cases` application is correct. Once we have tested this outer proof skeleton, we can go about replacing the first **force** application with another proof sketch, and so on, until eventually there are no occurrences of **force** left anywhere in the proof text and we have a complete proof that successfully derives p.[6] Using methods we could decompose and subdivide the work further in a more structured way, for example:

```
(!cases (p1 | p2)
  assume p1
    conclude p
```

6 Another alternative would be to insert dummy deductions in place of D_1 and D_2, provided we remove the "**conclude** p" annotations, for example, we might replace both D_1 and D_2 by `(!true-intro)`. This would enable us to evaluate the overall deduction. Then, as we begin to fill in the details, these bogus `(!true-intro)` applications would be removed. However, that approach has two serious drawbacks, apart from requiring the deletion of the **conclude** annotations. First, the result of the entire proof would become `true`, rather than the intended p. This means that we cannot have this partial proof embedded in a larger context (a surrounding partial proof) in which it would be expected to produce the proper conclusion, p. Second, we cannot patch up the proof in a truly incremental fashion. For instance, we would be forced to remove both `true-intro` applications simultaneously, because if we remove only the first one, replacing it by a proper deduction D_1 that derives p, while keeping the second `true-intro` application, the `cases` application would fail. The use of **force** averts both complications.

```
      (!handle-case-1 p1 p)
  assume p2
    conclude p
      (!handle-case-2 p2 p))
```

where the case handlers could be defined as independent methods, initially defined simply as applications of **force**:

```
define handle-case-1 :=
  method (premise goal)
   # Assume the premise holds, derive the goal
    (!force goal)
.
.
.
define handle-case-2 :=
  method (premise goal)
    (!force goal)
```

The removal of the **force** applications could now proceed incrementally in these separate methods.

4.8 Putting it all together

In this section we will work through a proof that illustrates several of the mechanisms discussed in this chapter. Suppose we are given the following two premises:

```
assert premise-1 := (A & B | (A ==> C))

assert premise-2 := (C <==> ~ E)
```

and our task is to write a proof D that derives the conditional `(~ B ==> A ==> ~ E)` from the two premises.

Since our goal is a conditional, we will derive it with a conditional proof that assumes the antecedent and derives the consequent, so our top-level deduction D will be of the form:

```
assume (~ B)
  D'
```

where D' must derive the subgoal `(A ==> ~ E)` from the hypothesis `(~ B)` and the two premises. Now this subgoal is itself a conditional, so D' will be a conditional proof as well. Thus, the skeleton of our top-level proof D now takes the following form:

```
assume (~ B)
  assume A
   D''
```

where D'' needs to derive the subgoal `(~ E)` from the two previous hypotheses, `A` and `(~ B)`, along with the given premises.

We can now proceed in two ways. Continuing our "backward" or "goal-driven" analysis (more on that in Section 4.14), we could look at the form of the current subgoal and use the corresponding introduction mechanism for it. The current subgoal is a negation, `(~ E)`, so we would use proof by contradiction to derive it, and D'' would take the following form:

```
(!by-contradiction (~ E)
  assume E
    D''')
```

where D''' would have to derive `false`.

The second way to proceed is to examine the information that is available to D'' and see whether we can derive the desired conclusion `(~ E)` in a more direct manner. Let us pursue this second alternative here. D'' has access to four sentences: the two starting premises and the two hypotheses `(~ B)` and `A`. How can we use these to derive `(~ E)`? Well, the first premise is a disjunction,

`(A & B | (A ==> C)),`

and we saw that the chief way to use a disjunction $(p \mid q)$ in order to infer a goal r is to perform a case analysis: Show that the goal r is implied both by p and by q. So in this case we would have to show that `(~ E)` follows both from the first disjunct `(A & B)` and from the second one, `(A ==> C)`. Accordingly, D'' takes the following form:

```
(!cases premise-1
  assume (A & B)
    conclude (~ E)
      D1
  assume (A ==> C)
    conclude (~ E)
      D2)
```

where now D_1 has to derive `(~ E)` from the four available sentences along with `(A & B)`; while D_2 has to derive `(~ E)` from the four sentences along with `(A ==> C)`. These deductions are now fairly easy to construct. Let us start with D_1. The hypothesis `(A & B)` gives us `B`, since **assume** `(A & B)` places `A` and `B` in the assumption base. But we are currently operating within the scope of the hypothesis `(~ B)`, so this contradiction allows us to conclude the desired `(~ E)` by a straightforward use of `from-complements`. D_1 thus becomes:

```
(!from-complements (~ E) B (~ B))
```

Continuing with D_2: By the second premise, `C` is equivalent to `(~ E)`, so we can detach the conditional `(C ==> ~ E)` via `left-iff`. But the innermost hypothesis `(A ==> C)` along with the preceding hypothesis `A` gives us `C`, so `(~ E)` can now be obtained simply by `mp`:

```
let {C1 := conclude (C ==> ~ E);
             (!left-iff premise-2);
     C2 := conclude C
            (!mp (A ==> C) A)}
  conclude (~ E)
    (!mp C1 C2)
```

Assembling all the pieces together, and using a little additional naming, we arrive at the following final version:

```
assert premise-1 := (A & B | (A ==> C))
assert premise-2 := (C <==> ~ E)

assume -B := (~ B)
  assume A
    conclude -E := (~ E)
      (!cases premise-1
              assume (A & B)
                (!from-complements -E B -B)
              assume A=>C := (A ==> C)
                let {C=>-E := (!left-iff premise-2);
                     C     := (!mp A=>C A)}
                  (!mp C=>-E C))
```

In Section 4.14 we present heuristics for performing this kind of proof-finding analysis more systematically.

4.9 A library of useful methods for sentential reasoning

The inference mechanisms we have covered so far constitute a complete deduction system for sentential logic, meaning that if any zero-order sentence p follows logically from a finite set of zero-order sentences β, then there exists a proof expressed in terms of these mechanisms that derives p from β. However, these mechanisms are too low-level, in that the inference steps we can take with them are very small. Each such mechanism either eliminates or else introduces a logical connective. But expressing a realistic proof exclusively in terms of elimination and introduction steps can be unduly cumbersome. Very often we would like to take a larger inferential step in one fell swoop. To take but one example, we might want to apply one of De Morgan's laws to directly infer

$$(\sim p \ \& \ \sim q) \tag{4.3}$$

from

$$(\sim (p \mid q)) \tag{4.4}$$

in one step, instead of painstakingly writing out the entire proof in terms of primitive introduction and elimination methods.

In a programmable proof system such as Athena, this issue has a straightforward solution: We can implement new methods in terms of methods that are already available. (We will see later that methods can even be recursively defined in terms of themselves.) Once a new method has been defined, it can be used as if it were a primitive method, just like modus ponens (`mp`) or conjunction introduction (`both`). For instance, once we have defined a method `de-morgan`, we can write

```
(!de-morgan (~ (p | q)))
```

and automatically derive the conclusion

```
(~ p & ~ q).
```

Method definition is very similar to procedural abstraction in programming, and carries the same well-known advantages of reusability and modularity. (Indeed, the similarity between writing proofs and programs will be a pervasive theme of this book.) For instance, a client of `de-morgan` need not be concerned with *how* the method is implemented, namely, with how exactly conclusions of the form (4.3) are generated from premises of the form (4.4). Instead, `de-morgan` can be viewed simply as a black box capable of taking premises of one form and deriving conclusions of some other form.

An important point here is that Athena guarantees the soundness of every defined method, meaning that if and when a method call manages to derive a conclusion p, we can be assured that p is a logical consequence of the assumption base in which the call occurred. This guarantee stems from the formal semantics of Athena, which will not be discussed here. The larger point to keep in mind is that defined methods "cannot go wrong," in that they can never produce a result that is not logically entailed by the assumption base.[7]

Therefore, starting with a very small but sound and complete kernel of primitive introduction and elimination methods, and armed with an array of versatile mechanisms for expressing proof recipes (including pattern matching and recursion), we can go on to define highly complex inference methods that are nevertheless guaranteed to behave soundly. In what follows we develop a collection of methods that are particularly useful for sentential reasoning. Full working code will be given for most of these, but some will be left as exercises. All of the methods that we discuss in this section are defined in Athena's library, so they are all available at the top level when Athena starts. The library definitions are similar to the ones that appear below, except that they are often written to handle a larger

---

7 Assuming that all primitive methods are sound.

range of inputs. Nevertheless, the definitions we are about to give here capture the essential behavior of these methods.

- *Contrapositive*: The contrapositive of a conditional (p ==> q) is the conditional

  (~ q ==> ~ p).

  These two sentences are equivalent, and it is useful to be able to go from each to the other. The following method does that in one direction, going from any conditional (p ==> q) to (~ q ==> ~ p). In Exercise 4.21 we ask you to write a more general two-way version.

```
define (contra-pos premise) :=
  match premise {
    (p ==> q) => assume -q := (~ q)
                   conclude (~ p)
                     (!mt premise -q)
  }
```

- *Excluded middle*: This method, whose implementation is left as an exercise, takes an arbitrary sentence p as input and derives the disjunction (p | ~ p):

```
define (ex-middle p) :=
  conclude (p | ~ p)
    ...
```

  This method is used to implement two-cases (page 200).

- *Commutation*: Both conjunction and disjunction are commutative. Accordingly, we should be able to infer (q & p) given (p & q) and (q | p) given (p | q). Biconditionals, being essentially conjunctions, are also commutative, so given (p <==> q) we should be able to derive (q <==> p). We define three such methods as follows:

```
define (and-comm premise) :=
  match premise {
    (_ & _) => (!both (!right-and premise) (!left-and premise))
  }

define (or-comm premise) :=
  match premise {
     (p | q) => (!cases premise
                        assume p
                          (!right-either q p)
                        assume q
                          (!left-either q p))
  }

define (iff-comm premise) :=
```

```
  match premise {
    (_ <==> _) => (!equiv (!right-iff premise) (!left-iff premise))
  }
```

Observe the similarity between and-comm and iff-comm.

We now define a generic commutation method that can accept conjunctions, disjunctions, or biconditionals as inputs:

```
define (comm premise) :=
  match premise {
    (_ & _)     => (!and-comm premise)
  | (_ | _)     => (!or-comm premise)
  | (_ <==> _) => (!iff-comm premise)
  }
```

- *Bidirectional double negation*: The primitive Athena method dn works only in one direction: It takes as input a premise that is at least doubly negated and removes the two leading negation signs, going from $(\sim \sim p)$ to p. It is often useful to be able to go in the other direction as well, whereby we are given an arbitrary premise and insert two negation signs in front of it. The method idn below implements that direction:

```
define (idn premise) :=
  (!by-contradiction (~ ~ premise)
     assume -premise := (~ premise)
       (!absurd premise -premise))
```

It is also useful to have a "bidirectional" double-negation method, bdn, which inserts two negation signs in front of the given premise p *only* when p is not already doubly negated; if it is, then it applies dn to it to remove the two leading negation signs:

```
define (bdn premise) :=
  match premise {
    (~ (~ _)) => (!dn premise)
  | _ => (!idn premise)
  }
```

Thus:

```
> assume h := (~ ~ A)
    (!bdn h)

Theorem: (if (not (not A))
             A)

> assume A
    (!bdn A)
```

```
Theorem: (if A
             (not (not A)))
```

- *Ex falso quodlibet*: This Latin name means "From false, everything follows." It reflects a well-known and useful principle of classical bivalent logic: Everything can be proved from a contradiction. The following method derives any given sentence p, provided that the assumption base contains `false`:

```
define (from-false goal) :=
  (!by-contradiction goal
    assume (~ goal)
      (!claim false))
```

A similar method, `from-complements`, derives any given sentence p from two complementary sentences q_1 and q_2 in the assumption base:

```
define (from-complements p q1 q2) :=
  let {M := method (goal q -q)
              let {_ := (!absurd q -q)}
                (!from-false goal)}
    match [q1 q2] {
      [q (~ q)] => (!M p q1 q2)
    | [(~ q) q] => (!M p q2 q1)
    }
```

- *Hypothetical syllogism*: This binary method takes two premises of the form (p_1 ==> p_2) and (p_2 ==> p_3) and derives the conclusion (p_1 ==> p_3):

```
define (hsyl premise-1 premise-2) :=
  match [premise-1 premise-2] {
    [(p1 ==> p2) (p2 ==> p3)] =>
      assume p1
        conclude p3
          (!mp premise-2
               conclude p2
                 (!mp premise-1 p1))
  }
```

- *Disjunctive syllogism*: This is a binary method that accepts two premises of the form (p | q) and $\overline{p}$ (recall that $\overline{p}$ is the complement of p), and derives the conclusion q. Alternatively, the second premise may be the complement of q, in which case the conclusion will be p. In summary, the first premise is always a disjunction, the second premise is the complement of one the two disjuncts, and the conclusion is the other disjunct. The validity of this reasoning pattern should be intuitively clear. For example, if we know that the murderer is either Peter or Paul and we then find out that Peter is innocent, we can conclude that the culprit is Paul.[8]

We first define a method dsyl-1 that handles only those cases where the complemented disjunct is the left one, which is then used to define a single method dsyl that meets the more general specification:

```
define (dsyl-1 premise-1 premise-2) :=
  match [premise-1 premise-2] {
    [(p | q) p'] =>
      check {
        (complements? p p') =>
          (!cases premise-1
                  assume p
                    (!from-complements q p p')
                  assume q
                    (!claim q))
      }
  }

define (dsyl premise-1 premise-2) :=
  match [premise-1 premise-2] {
    [(p | q) r] =>
      check {
        (complements? p r) => (!dsyl-1 premise-1 premise-2)
      | else => (!dsyl-1 (!comm premise-1) premise-2)
      }
  }
```

The procedure complements? takes two sentences p and q and returns true if p and q are complements and false otherwise:

```
define (complements? p q) :=
  (|| (q equal? complement p) (p equal? complement q))
```

The version of dsyl defined in Athena's library is considerably more powerful than the preceding implementation.[9] The library version has the following additional features:

1. The first argument, p_1, can be an arbitrarily long and arbitrarily structured disjunction, such as (or A B C) or ((A | B) | (C ==> D) | E).

---

8 This is actually the basis of a powerful inference method known as *resolution*. In the general case, the resolution method takes as input two premises of the form (or $p_1 \cdots p_i \cdots p_m$) and (or $q_1 \cdots q_j \cdots q_n$). Provided that p_i and q_j are complementary (negations of each another), it produces the conclusion

$$(\texttt{or}\ p_1 \cdots p_{i-1}\ p_{i+1} \cdots p_m\ q_1 \cdots q_{j-1}\ q_{j+1} \cdots q_n).$$

We implement this more general method in Exercise 4.35.

9 Nevertheless, the two are compatible insofar as any inputs that are successfully handled by the present implementation lead to the exact same output by the more powerful implementation; so the library version is a strict extension of the version defined here.

2. The second argument, p_2, can be a conjunction or even a plain list of sentences, each of which is complementary to some disjunct of the first argument. (If p_2 is a list, then all of its elements must be in the assumption base.) It can also be a single sentence that is complementary to one of these disjuncts, which is what ensures the compatibility of the two versions mentioned in footnote 9.

3. The final conclusion is the disjunction obtained from p_1 by removing all those disjuncts that have complementary occurrences in p_2. If there is only one such disjunct d left, then d is produced (rather than `(or` d`)`). And if there are no such disjuncts, then `false` is produced, as would happen, for instance, if we applied the method to $p_1 =$ `(A | B | C)` and $p_2 =$ `(~ A & ~ B & ~ C)`.

- *De Morgan*: This rule has four different parts, which we define below.

1. Given a premise of the form `(~ (`p `&` q`))`, derive the conclusion `(`$\overline{p}$ `|` $\overline{q}$`)`.
2. In the converse direction, given a premise of the form `(`p `|` q`)`, derive `(~ (`$\overline{p}$ `&` $\overline{q}$`))`.
3. Given a premise of the form `(~ (`p `|` q`))`, derive `(`$\overline{p}$ `&` $\overline{q}$`)`.
4. Conversely, given a premise of the form `(`p `&` q`)`, derive `(~ (`$\overline{p}$ `|` $\overline{q}$`))`.

These are implemented by four different methods. The definition of the first one is left as an exercise.

```
define (dm-1 premise) :=
  ...

define (dm-2 premise) :=
  match premise {
    (p | q) =>
      let {[p' q'] := [(complement p) (complement q)];
           goal := (~ (p' & q'))}
        (!by-contradiction goal
          assume -goal := (~ goal)
            let {_ := (!dn -goal)} # We now have (p' & q')
              (!cases premise
                      assume p
                        (!from-complements false p p')
                      assume q
                        (!from-complements false q q')))
  }

define (dm-3 premise) :=
  match premise {
    (~ (p | q)) =>
      let {p' := (!by-contradiction (complement p)
                    assume p
                      (!absurd (!left-either p q) premise));
```

```
        q' := (!by-contradiction (complement q)
                 assume q
                   (!absurd (!right-either p q) premise))}
    (!both p' q')
  }

define (dm-4 premise) :=
  match premise {
    (p & q) =>
      let {[p' q'] := [(complement p) (complement q)];
           goal := (~ (p' | q'))}
        (!by-contradiction goal
          assume -goal := (~ goal)
            let {p'|q' := (!dn -goal)}
              (!cases
                p'|q'
                assume p'
                  (!from-complements false (!left-and premise) p')
                assume q'
                  (!from-complements false (!right-and premise) q')))
  }
```

Finally, we define a method that takes a premise of any one of these four forms and acts accordingly:

```
define (dm premise) :=
  match premise {
    (~ (_ & _)) => (!dm-1 premise)
  | (_ | _)     => (!dm-2 premise)
  | (~ (_ | _)) => (!dm-3 premise)
  | (_ & _)     => (!dm-4 premise)
  }
```

The version of `dm` defined in Athena's library is more flexible than the above, as it can handle polyadic conjunctions and disjunctions.

As we have defined them here, De Morgan's laws are based on complementation rather than negation. The complementation version tends to be more useful in practice, as we typically want to go from something like `(~ (~ A & ~ B))` to `(A | B)` rather than `(~ ~ A | ~ ~ B)`. Nevertheless, occasionally we want the plain-negation version of De Morgan, which is defined as follows:

1. Given a premise of the form (~ (*p* & *q*)), derive the conclusion (~ *p* | ~ *q*).
2. Given a premise of the form (*p* | *q*), derive (~ (~ *p* & ~ *q*)).
3. Given a premise of the form (~ (*p* | *q*)), derive (~ *p* & ~ *q*).
4. Conversely, given a premise of the form (*p* & *q*), derive (~ (~ *p* |~ *q*)).

There is a method `dm'` defined in Athena's library that implements this formulation of De Morgan's laws.[10] A few examples illustrating the two methods:

```
> assume h :=  (~ (~ A | ~ B))
    (!dm h)

Theorem: (if (not (or (not A)
                      (not B)))
             (and A B))

> assume h :=  (~ (~ A | ~ B))
    (!dm' h)

Theorem: (if (not (or (not A)
                      (not B)))
             (and (not (not A))
                  (not (not B))))

> assume h := (~ A & ~ B)
    (!dm h)

Theorem: (if (and (not A)
                  (not B))
             (not (or A B)))

> assume h := (~ A & ~ B)
    (!dm' h)

Theorem: (if (and (not A)
                  (not B))
             (not (or (not (not A))
                      (not (not B)))))
```

- *Distributivity*: Conjunction distributes over disjunction and vice versa. Thus, from a premise of the form $(p$ `&` $(q$ `|` $r))$ we should be able to derive

  $$(p \;\&\; q \mid p \;\&\; r)$$

  and conversely. Likewise, from a premise of the form $(p \mid (q \;\&\; r))$ we should be able to infer

  $$((p \mid q) \;\&\; (p \mid r))$$

  and vice versa. Below we present four methods that can carry out such derivations, and we then use them to define a single distribution method that invokes the proper member of the quadruple depending on the form of the input:

10 This is, in fact, the more conventional formulation of De Morgan's laws.

```
define (cd-dist-1 premise) :=
  match premise {
    (p & (q | r)) =>
      conclude (p & q | p & r)
        let {_ := (!left-and premise)}
          (!cases ((q | r) by (!right-and premise))
            assume q
              (!left-either (!both p q) (p & r))
            assume r
              (!right-either (p & q) (!both p r)))
  }

define (cd-dist-2 premise) :=
  match premise {
    ((p & q) | (p & r)) =>
      conclude (p & (q | r))
        (!cases premise
          assume (p & q)
            (!both p (!left-either q r))
          assume (p & r)
            (!both p (!right-either q r)))
  }

define (dc-dist-1 premise) :=
  match premise {
    (p | (q & r)) =>
      conclude ((p | q) & (p | r))
        (!cases premise
          assume p
            (!both (!left-either p q) (!left-either p r))
          assume (q & r)
            (!both (!right-either p q)
                   (!right-either p r)))
  }

define (dc-dist-2 premise) :=
  match premise {
    ((p | q) & (p | r)) =>
      conclude (p | (q & r))
        let {_ := (!left-and premise);
             _ := (!right-and premise)}
          (!cases (p | q)
            assume p
              (!left-either p (q & r))
            assume q
              (!cases (p | r)
                assume p
                  (!left-either p (q & r))
                assume r
                  (!right-either p (!both q r))))
  }
```

```
define (dist premise) :=
  match premise {
    ((p | _) & (p | _)) => (!dc-dist-2 premise)
  | (_ & (_ | _))       => (!cd-dist-1 premise)
  | ((p & _) | (p & _)) => (!cd-dist-2 premise)
  | (p | (q & r))       => (!dc-dist-1 premise)
  }
```

The order in which the patterns appear in the definition of `dist` is important, particularly if we wish the method to be bidirectional, so that

$$(\texttt{!dist (!dist } p) = p$$

in every assumption base that contains p. That would fail, for instance, if the first two clauses were swapped, as the first pattern is an instance of the second and thus every sentence that matches the first pattern also matches the second. Hence, if the order were reversed, method `dc-dist-2` would never be invoked.

Note, finally, that `dist` is incomplete: It cannot be applied to disjunctions such as `((`p `&` q`) |` r`)` when r is not a conjunction, or to conjunctions such as `((`p `|` q`) &` r`)` when r is not a disjunction. To distribute sentences of this form, we first need to commute them in order to make them conform to the patterns recognized by `dist`.

- *Conditional definition*: One of the most useful equivalences in logic is the one between `(`p `==>` q`)` and `(~` p `|` q`)`. This amounts to a definition of conditionals in terms of disjunction.[11] It is quite handy to be able to go back and forth between these two characterizations of conditionals. We would thus like a unary method `cond-def` that can take a conditional `(`p `==>` q`)` in the assumption base and derive the disjunction `(~` p `|` q`)` from it. And conversely, given a disjunction `(`p `|` q`)` in the assumption base, `cond-def` should produce the conditional `(~` p `==>` q`)`. Accordingly, we will define two separate methods, `cond-def-1` and `cond-def-2`, each of which will handle one of these two directions: The first will convert conditionals to disjunctions, and the second disjunctions to conditionals. We can then implement `cond-def` by invoking one of these two methods, depending on the form of the input:

```
define (cond-def-1 premise) :=
  match premise {
    (p ==> q) =>
      let {goal := (~ p | q)}
        (!by-contradiction goal
          assume -goal := (~ goal)
            let {p1   := conclude (~ ~ p & ~ q)
```

11 Many older texts used to treat conditionals as syntax sugar, defining `(`p `==>` q`)` either as an abbreviation for `(~` p `|` q`)` or as an abbreviation for `(~ (`p `& ~` q`))`. (You can use De Morgan and double negation to verify that the two are equivalent.)

```
                        (!dm' -goal);
                p := (!dn (~ ~ p))}
             (!absurd (conclude q
                        (!mp premise p))
                      (~ q)))
  }

define (cond-def-2 premise) :=
  match premise {
    (p | q) => assume -p := (~ p)
                 (!cases premise
                        assume p
                          (!from-complements q p -p)
                        assume q
                          (!claim q))
  }

define (cond-def premise) :=
  match premise {
    (_ ==> _) => (!cond-def-1 premise)
  | (_ | _) =>   (!cond-def-2 premise)
  }
```

While this implementation of cond-def works as specified, we often want a slightly different behavior. First, given a conditional of the form (~ p ==> q), we typically want to derive the disjunction (p | q), instead of (~ ~ p | q). And given a disjunction (~ p | q), we usually want to derive the conditional (p ==> q) rather than (~ ~ p ==> q). Essentially, we want a version of cond-def that works with complementation rather than negation, just like the De Morgan method dm that we defined above. It would seem, therefore, that we need to write two additional methods, cond-def-1' and cond-def-2', each handling conditionals and disjunctions of the said form, and then write another method cond-def' that dispatches on cond-def-1' and cond-def-2' appropriately.

While we could do that, an alternative approach is to express the conditional-definition logic in a generic way, via a unary procedure that takes as its argument the negation procedure that we want to use (conventional negation or complementation) and produces the corresponding method. We take that approach here because it illustrates the useful technique of higher-order procedures that produce methods as outputs:

```
define (make-cond-def-1 neg) :=
  method (premise)
    match premise {
      (p ==> q) => let {p' := (neg p);
                        goal := (p' | q)}
                   (!by-contradiction goal
```

```
                       assume -goal := (~ goal)
                         let {_ := (!dm' -goal);
                              p := try { (!dn (~ ~ p)) | (!claim p) }}
                           (!absurd (!mp premise p)
                                    (~ q)))
     }

define (make-cond-def-2 neg) :=
  method (premise)
    match premise {
      (p | q) => assume p' := (neg p)
                   (!cases premise
                           assume p
                             (!from-complements q p p')
                           assume q
                             (!claim q))
    }

define (make-cond-def neg) :=
  method (premise)
     match premise {
       (_ ==> _) => (!(make-cond-def-1 neg) premise)
     | (_ | _)   => (!(make-cond-def-2 neg) premise)
     }
```

We can now define two versions, one that works with conventional negation and one that works with complementation, simply by supplying the two different versions of negation as arguments to `make-cond-def`. Because the complementation version is more useful in practice, we define that as `cond-def` and define the primed version, `cond-def'`, as the one based on negation:

```
define cond-def  := (make-cond-def complement)
define cond-def' := (make-cond-def not)
```

We illustrate the two methods with some examples:

```
> assume h := (A ==> B)
    (!cond-def h)

Theorem: (if (if A B)
             (or (not A)
                 B))

> assume h := (A ==> B)
    (!cond-def' h)

Theorem: (if (if A B)
             (or (not A)
                 B))
```

```
> assume h := (~ A ==> B)
    (!cond-def h)

Theorem: (if (if (not A)
                 B)
             (or A B))

> assume h := (~ A ==> B)
    (!cond-def' h)

Theorem: (if (if (not A)
                 B)
             (or (not (not A))
                 B))

> assume h := (A | B)
    (!cond-def h)

Theorem: (if (or A B)
             (if (not A)
                 B))

> assume h := (A | B)
    (!cond-def' h)

Theorem: (if (or A B)
             (if (not A)
                 B))

> assume h := (~ A | B)
    (!cond-def h)

Theorem: (if (or (not A)
                 B)
             (if A B))

> assume h := (~ A | B)
    (!cond-def' h)

Theorem: (if (or (not A)
                 B)
             (if (not (not A))
                 B))
```

- *Conditional negation*: To deny a conditional (p `==>` q) is to affirm the antecedent p and deny the consequent q. It is useful to be able to go from a negated conditional (`~` (p `==>` q)) to the conjunction (p `&` `~` q) and conversely. Below is a method `neg-cond` that does that. It uses three auxiliary methods: `neg-cond-ant` takes a premise of the form (~ (p ==> q)) and derives the antecedent p; neg-cond-con takes a premise of the same form and derives the negated consequent, (~ q); and finally, neg-cond-conv takes a premise of the form (p & ~ q) and derives (~ (p ==> q)).

```
define (neg-cond-ant premise) :=
  match premise {
    (~ (p ==> q)) => (!by-contradiction p
                        assume -p := (~ p)
                          let {p==>q := assume p
                                          (!from-complements q p -p)}
                            (!absurd p==>q premise))
  }

define (neg-cond-con premise) :=
  match premise {
    (~ (p ==> q)) => (!by-contradiction (~ q)
                        assume q
                          let {p==>q := assume p (!claim q)}
                            (!absurd p==>q premise))
  }

define (neg-cond-conv premise) :=
  match premise {
    (p & (~ q)) => (!by-contradiction (~ (p ==> q))
                      assume h := (p ==> q)
                        (!absurd conclude q
                                   (!mp h (!left-and premise))
                                 (!right-and premise)))
  }

define (neg-cond premise) :=
  match premise {
    (~ (_ ==> _)) => (!both (!neg-cond-ant premise)
                             (!neg-cond-con premise))
  | _ => (!neg-cond-conv premise)
  }
```

- *Biconditional definition*: As we remarked earlier, a biconditional (p <==> q) is equivalent to the conjunction of (p ==> q) and (q ==> p). The following method takes a biconditional premise (p <==> q) and derives that conjunction, and vice versa:

```
define (bicond-def premise) :=
  match premise {
    (p <==> q) => (!both (!left-iff premise)
                         (!right-iff premise))
  | ((p ==> q) & (q ==> p)) => (!equiv (!left-and premise)
                                       (!right-and premise))
  }
```

A biconditional (p <==> q) is also equivalent to the disjunction

$$(p \ \& \ q \mid \overline{p} \ \& \ \overline{q}),$$

where $\overline{p}$ and $\overline{q}$ are the complements of p and q, respectively, which states that either both p and q hold or neither does.[12] This is also a useful equivalence, and the following method can derive one from the other:

```
define (bicond-def' premise) :=
  match premise {
    (p <==> q) =>
      let {[p' q'] := [(complement p) (complement q)]}
        (!two-cases
          assume p
            let {q := (!mp (!left-iff premise) p)}
              (!left-either (!both p q) (p' & q'))
          assume p'
            let {_ := conclude q'
                        (!mt (!right-iff premise) p')}
              (!right-either (p & q) (!both p' q')))
  | ((p & q) | (p' & q')) =>
      (!cases premise
        assume (p & q)
          (!equiv assume p (!claim q)
                  assume q (!claim p))
        assume (p' & q')
          (!equiv assume p (!from-complements q p p')
                  assume q (!from-complements p q q')))
  }
```

- *Biconditional negation*: It follows from the above that to deny a biconditional (p <==> q) is to say that either p holds and q does not, or that p does not hold and q does. That is, (~ (p <==> q)) is equivalent to

$$(p \ \& \ \sim q \mid \sim p \ \& \ q),$$

and each is derivable from the other by a method `negated-bicond`, whose implementation is left as an exercise.

4.10 Recursive proof methods

Recursive methods (methods that call themselves) are useful in a wide variety of situations. One of their most prominent applications arises when the reasoning that must be carried

12 As with previous examples, it is possible to formulate this principle using straight negation rather than complementation, but the latter tends to be more useful in practice so we prefer it here. Exercise 4.25 implements a more flexible binary version of this method.

out depends on the size of the input, meaning that in order to derive the desired result the method needs to perform a number of inferences that is not a priori fixed but is instead dynamically dependent on the inputs. Consider, for instance, the commutation method that takes a premise (p & q) and derives (q & p). Here there is no need for iteration, because regardless of how big the input is, our work is always the same: an application of `left-and` and one of `right-and`, followed by an application of `both`. If we view sentences as abstract syntax trees (ASTs), we might say that the commutation method does not need to descend down the subtrees of the given conjunction; we simply detach the left subtree, detach the right subtree, and then join them together in the reverse order. We therefore never have to visit any node in the AST below the second level. In other situations, however, we do need to examine and work on various parts of the input, indeed on arbitrarily many parts, and in such cases the required iteration is typically expressed by recursion.

To take but one example, consider normalizations, where a given sentence p must be converted into some normal form or other, say negation normal form (Exercise 4.31), conjunctive normal form (Exercise 4.32), disjunctive normal form (Exercise 4.33), or some other variation. In a deductive setting, this would have to be done by a method that took p as input and derived the corresponding normal form as a theorem. For instance, we might want to take an arbitrary conjunction and flatten it out into a right-chained list of atoms, so that `((A & B) & (C & D))` is transformed into `(A & (B & (C & D)))`. The solution here is to express the desired method in terms of itself. (We ask you to write such a method in Exercise 4.30.)

As a simpler example, consider writing an iterative version of double negation: a method `dn*` that takes a premise with an arbitrary number of pairs of negation signs in front of it, and eliminates all such pairs by repeated applications of `dn`. Thus, for instance, applying `dn*` to `(~ ~ A)` will derive `A`, so in this case the effect is the same as calling `dn`. Applying `dn*` to `(~ ~ ~ A)` gives `(~ A)` (and again the effect is the same as calling `dn`); applying `dn*` to `(~ ~ ~ ~ A)` gives `A`; applying it to `(~ ~ ~ ~ ~ A)` gives `(~ A)`; and so on. The method should have no effect on an input that does not have at least two negation signs in front of it. Any such input should simply be reiterated, so that, for example, applying `dn*` to `A` should result in `A`.

The following implementation of `dn*` fits the bill:

```
1  define (dn* premise) :=
2    match premise {
3      (~ (~ p)) => let {_ := conclude p
4                              (!dn premise)}
5                    (!dn* p)
6    | _ => (!claim premise)
7    }
```

When the given premise has at least two negation signs in front of it, that is, when it is of the form (~ ~ p), `dn*` does two things. First, it derives the conclusion p by applying `dn` to the premise; and then it recursively calls itself with p as the input. The recursion bottoms out when all double negation signs have been eliminated, that is, when the input premise is not of the form (~ ~ p). At that point dn* will simply claim the given premise (line 6).

Recall that in a deduction of the form

let $\{I$:= $D_1\}$ D

the body D is evaluated in an assumption base that has been augmented with the conclusion of D_1. Here this means that the recursive call to dn* on line 5 will occur in an assumption base that contains the result of the first step, namely, the sentence p. This is important because the input argument to the recursive call is p, and the specification requires that the input to dn* must be in the assumption base at the time when dn* is applied to it. The use of a proof sequencing mechanism such as **let** ensures that this precondition is satisfied. Accordingly, successive recursive calls are made in increasingly larger assumption bases, with one new conclusion added on each iteration. When the final call to dn* occurs, the assumption base contains all the elements of the initial assumption base (when dn* was first called), along with every intermediate conclusion obtained by every intervening recursive call.

An alternative—and more succinct—way of defining dn* is the following:

```
define (dn* premise) :=
  match premise {
    (~ (~ p)) => (!dn* (!dn premise))
  | _ => (!claim premise)
  }
```

In general, assuming that D_1 derives p, a deduction of the form

let {_ := **conclude** p D_1} (!M p)

is equivalent to (!M D_1), because Athena's semantics dictate that when an argument to a method call is a deduction, the conclusion of that deduction must be incorporated in the assumption base when the method is finally applied to the values of the arguments. In our case this means that the conclusion of the deduction (!dn premise), namely, p, will be in the assumption base when dn* is finally applied to the values of its arguments (i.e., to p), thereby recursively satisfying the precondition of dn*.

As another example, consider a method that takes a conditional premise of the form

(p ==> q ==> r)

and derives (p & q ==> r). This is traditionally known in logic as *importation*. We can implement such a method as follows:

```
define (import premise) :=
  match premise {
    (p ==> (q ==> r)) => assume (p & q)
```

```
                                          let {q=>r := (!mp premise p)}
                                            (!mp q=>r q)
    }
```

Suppose now that we want a more general version of this method, one that can take an arbitrary conditional of the form (p_1 ==> p_2 ==> p_3 ==> $\cdots$) and repeatedly apply importation, obtaining, say, (((p_1 & p_2) & p_3) ==> p_4) from

(p_1 ==> p_2 ==> p_3 ==> p_4).

A recursive method is again the solution:

```
define (import* premise) :=
  match premise {
    (p1 ==> (p2 ==> p3)) => let {p1&p2=>p3 := (!import premise)}
                              (!import* p1&p2=>p3)
  | _ => (!claim premise)
  }
```

or, more succinctly:

```
define (import* premise) :=
  match premise {
    (p1 ==> (p2 ==> p3)) => (!import* (!import premise))
  | _ => (!claim premise)
  }
```

You should try to trace the behavior of `import*` on an input premise such as

```
(A ==> B ==> C ==> D)
```

to make sure that you understand how the conclusion is produced.

We now proceed to develop a powerful recursive method that implements what is traditionally known in logic as the *replacement rule*, which lets us replace an arbitrary part q of a premise p by any equivalent sentence q'. Indeed, our implementation will allow us to replace multiple parts of p (arbitrarily many) by equivalent sentences. This exercise will demonstrate the flexibility of recursive proof methods, and it will also illustrate the use of higher-order methods (methods that take other methods as inputs).

For example, suppose we have

$p =$ `(A & ~ ~ B)` (4.5)

in the assumption base. Since `(~ ~ B)` is equivalent to `B`, and specifically since we know how to derive each from the other, via the `bdn` method, we should be able to conclude

`(A & B).` (4.6)

This conclusion is obtained from p by replacing the double negation `(~ ~ B)` with the equivalent `B`. Thus, our reasoning here can be roughly expressed as follows: (4.6) follows from (4.5) by virtue of the `bdn` method. More precisely, this inference will be captured by the method call

```
(!transform (A & ~ ~ B) (A & B) [bdn]).
```

This should be read as: "Transform the first argument (the premise) into the second one by replacement, using the method `bdn`." The reason why the third argument is a list of methods (in this case a singleton list containing `bdn`) will become clear in the next paragraph.

As another example, suppose the following conditional is in the assumption base:

```
(~ ~ A ==> ~ (B & C)).
```
(4.7)

From De Morgan's laws, we know that the consequent is equivalent to `(~ B | ~ C)`, while from double negation we know that the antecedent is equivalent to `A`. The `transform` method will allow us to derive

```
(A ==> ~ B | ~ C)
```

from (4.7) simply by citing these two methods, `dm` (De Morgan) and `bdn` (two-way double negation):

```
(!transform (~ ~ A ==> ~ (B & C)) (A ==> ~ B | ~ C) [bdn dm]).
```

In general, then, calls to `transform` will be of the form

$$(\texttt{!transform}\ p\ q\ \texttt{[}M_1 \cdots M_n\texttt{]})$$

and should be read as: "Transform p into q by replacement, using the methods $M_1 \cdots M_n$." The desired conclusion q should be obtainable from p by replacing $k \geq 0$ subsentences of it, $p_1, \ldots, p_k$, by sentences $q_1, \ldots, q_k$, where each equivalence $(p_i \texttt{ <==> } q_i)$, $i = 1, \ldots, k$, is provable by some method M_j, $j \in \{1, \ldots, n\}$. The starting sentence p must be in the assumption base, and each M_i will (usually) be a bidirectional unary method, namely, such that if

$$(\texttt{!}M_i\ p) = q$$

then $(\texttt{!}M_i\ q) = p$, in all appropriate assumption bases. The order in which the various methods appear in the list should be irrelevant.

We say that each M_i will "usually" be a bidirectional unary method because we also allow M_i to be binary, as discussed below; and also because a unary M_i need not be bidirectional for *every* possible argument, as long as it is bidirectional for the arguments that will be supplied to it during a particular application of `transform`. The De Morgan method `dm`, for instance, is not fully bidirectional: If we let

$$p = \texttt{~ (~ ~ A \& ~ B)} \tag{4.8}$$

and

$$q = \texttt{(~ A | B)} \tag{4.9}$$

then we have

$$(\texttt{!dm}\ p) = q$$

(in an assumption base that contains p) but

$$(\texttt{!dm}\ q) \neq p$$

(in an assumption base that contains q). However, in the context of example (4.7), `dm` will happily derive `(~ B | ~ C)` from `(~ (B & C))` *and* will also derive `(~ (B & C))` from `(~ B | ~ C)`.

In general, however, especially when dealing with arbitrary sentences whose structure is not statically known, it is better—and safer—to perform replacement either with unary methods that are fully bidirectional (such as `bdn`) or else with binary methods. Binary methods in this context allow for greater precision; the first argument is always the premise and the second argument is the desired conclusion. The extra information that is conveyed by the additional argument, namely, the goal that must be derived from the premise, can be quite useful. For instance, a binary version of De Morgan along these lines can be much more flexible than the unary versions `dm` and `dm'`, because it can sometimes use negation and other times complementation, as required by the structure of the goal, whereas both `dm` and `dm'` always use either complementation (`dm`) or regular negation (`dm'`), unable to switch on demand. Consequently, neither `dm` nor `dm'` can derive (4.9) from (4.8) *and also* (4.8) from (4.9), as one direction requires complementation and the other negation.[13] A binary version of De Morgan that receives the goal as its second argument, by contrast, can handle such cases easily. We develop such a version in Exercise 4.24 and use it further in methods that derive NNF and CNF variants of a given sentence.

The above amounts to a specification of the replacement method. We now have to implement it. The brunt of the work will be done by a ternary method `prove-equiv`, which takes two sentences p and q as inputs and a list of methods $[M_1 \cdots M_n]$ and derives the biconditional (p <==> q), provided that q is obtainable from p by replacing $k \geq 0$ subsentences of it, $p_1, \ldots, p_k$, by sentences $q_1, \ldots, q_k$, where each equivalence (p_i <==> q_i) is provable by some M_j, $j \in \{1, \ldots, n\}$. With `prove-equiv` under our belt, we can readily implement `transform` as follows:

```
define (transform p q methods) :=
  let {E := conclude (p <==> q)
              (!prove-equiv p q methods)}
    conclude q
      (!mp (!left-iff E) p)
```

13 In fact, in some cases we might need both negation and complementation in one and the same direction, say, in going from `(~ (~ A | ~ B))` to `(~ ~ A & B)` or vice versa.

Our task has thus been reduced to implementing prove-equiv. Recall that the intended effect of an application

$$\texttt{(!prove-equiv } p\ q\ \texttt{[}M_1 \cdots M_n\texttt{])}$$

is to derive the biconditional (p <==> q) using the given methods. The algorithm for that derivation is as follows. We first check to see whether there is some method M_j that can derive q from p and vice versa. If so, we prove the equivalence of p and q by applying that method. If no such method exists, then we check to see if p and q are syntactically identical. If so, we are done, since we can readily prove (p <==> p) for any p by the following method:

```
define (ref-equiv p) :=
  (!equiv assume p (!claim p)
          assume p (!claim p))
```

Finally, if the two inputs p and q are neither syntactically identical nor interconvertible by one of the supplied methods, they must be nonatomic sentences obtained by applying the same sentential constructor, that is, they must both be negations, or they must both be conjunctions, and so on.[14] Say that they are both negations, so that p and q are of the form (~ p_1) and (~ q_1), respectively. We then recursively apply prove-equiv to p_1 and q_1 (and the same list of methods [$M_1 \cdots M_n$]), which will presumably result in the biconditional (p_1 <==> q_1). But then it is a simple matter to derive (p <==> q), since for any p_1 and q_1, (~ p_1 <==> ~ q_1) is derivable from (p_1 <==> q_1). The method not-cong[15] does precisely that:

```
define (not-cong premise) :=
  match premise {
    (p1 <==> p2) =>
      let {-p1=>-p2 := assume -p1 := (~ p1)
                         (!by-contradiction (~ p2)
                           assume p2
                             (!absurd (!mp (!right-iff premise) p2)
                                      -p1));
           -p2=>-p1 := assume -p2 := (~ p2)
                         (!by-contradiction (~ p1)
                           assume p1
                             (!absurd (!mp (!left-iff premise) p1)
```

14 If that is not the case, then the relevant precondition of prove-equiv fails, since q is then *not* "obtainable from p by replacing $k \geq 0$ subsentences of it by $\cdots$."

15 A binary relation R on a set S is *compatible* with an operator $f: S \rightarrow S$ iff $R(f(x), f(y))$ whenever $R(x,y)$. And it is compatible with an operator $g: S \times S \rightarrow S$ iff $R(g(x_1,x_2), g(y_1,y_2))$ whenever $R(x_1,y_1)$ and $R(x_2,y_2)$. A *congruence* is an equivalence relation on a set S that is compatible with a certain set of operators on S. We have called not-cong a congruence method because, viewing not as an algebraic operator on sentences and the derivability of a biconditional between two sentences as an equivalence relation, that relation is compatible with negation (and hence a congruence), and this method constructively demonstrates the compatibility. Likewise for the remaining congruence methods.

```
                                                    -p2))}
        (!equiv -p1=>-p2 -p2=>-p1)
  }
```

The reasoning for conjunctions, disjunctions, etc., is similar. We illustrate the former case, when p and q are respectively of the form (p_1 & p_2) and (q_1 & q_2). We then place two recursive calls to prove-equiv, one applied to p_1 and q_1 (and the given list of methods); and the other applied to to p_2 and q_2 (and said list). These will presumably result in the biconditionals (p_1 <==> q_1) and (p_2 <==> q_2), and from these two the biconditional (p <==> q) is deduced by the following method:

```
define (and-cong premise1 premise2) :=
  match [premise1 premise2] {
    [(p1 <==> q1) (p2 <==> q2)] =>
        let {p1&p2=>q1&q2 := assume (p1 & p2)
                                (!both (!mp (!left-iff premise1) p1)
                                       (!mp (!left-iff premise2) p2));
             q1&q2=>p1&p2 := assume (q1 & q2)
                                (!both (!mp (!right-iff premise1) q1)
                                       (!mp (!right-iff premise2) q2))}
          (!equiv p1&p2=>q1&q2 q1&q2=>p1&p2)
  }
```

The treatment of the remaining cases is analogous, and requires corresponding congruence methods for disjunctions, conditionals, and biconditionals. We present such a method for conditionals here; the other two are left as exercises.

```
define (if-cong premise1 premise2) :=
  match [premise1 premise2] {
    [(p1 <==> q1) (p2 <==> q2)] =>
      let {p1=>p2-gives-q1=>q2 :=
             assume (p1 ==> p2)
               assume q1
                 (!mp (!left-iff premise2)
                      (!mp (p1 ==> p2)
                           (!mp (!right-iff premise1) q1)));
           q1=>q2-gives-p1=>p2 :=
             assume (q1 ==> q2)
               assume p1
                 (!mp (!right-iff premise2)
                      (!mp (q1 ==> q2)
                           (!mp (!left-iff premise1) p1)))}
        (!equiv p1=>p2-gives-q1=>q2 q1=>q2-gives-p1=>p2)
  }
```

There are two auxiliary pieces of code that we need before we can present the definition of prove-equiv. First, we need a binary method find-first-element that takes a unary method M and an arbitrary list $L = [V_1 \cdots V_m]$ (of values of any type), and repeatedly applies M to each V_i in turn, $i = 1, \ldots, m$, until one such application succeeds, in which case the conclusion returned will be whatever result is produced by (!M V_i). If (!M V_i) fails for every $i = 1, \ldots, m$, then the entire call to `find-first-element` fails.

```
define (find-first-element M list) :=
  match list {
    (list-of first rest) =>
      try { (!M first) | (!find-first-element M rest) }
  }
```

Finally, it will be convenient to have a selection procedure that takes an arbitrary sentential constructor and chooses the appropriate congruence method:

```
define (choose-cong-method pc) :=
  match pc {
    &     => and-cong
  | |     => or-cong
  | ==>   => if-cong
  | <==> => iff-cong
  }
```

We can now define `prove-equiv` as follows:

```
define (prove-equiv p q methods) :=
  try {
    (!find-first-element method (M)
                             try { (!equiv assume p (!M p)
                                           assume q (!M q))
                                 | (!equiv assume p (!M p q)
                                           assume q (!M q p)) }
                          methods)
  | check {
      (p equals? q) => (!ref-equiv p)
    | else => match [p q] {
                  [(~ p1) (~ q1)] =>
                    (!not-cong (!prove-equiv p1 q1 methods))
                | [((some-sent-con pc) p1 p2) (pc q1 q2)] =>
                    (!(choose-cong-method pc)
                      (!prove-equiv p1 q1 methods)
                      (!prove-equiv p2 q2 methods))
                }
    }
  }
```

Try tracing `transform` on the aforementioned inputs and other examples of your choice to make sure you understand the implementation.

In a traditional logic system, a similar inference method for replacement would be proven sound by induction on the structure of the first argument, the premise p. Our implementation is essentially a constructive rendering of such an inductive argument, with recursive method calls serving in lieu of inductive hypotheses.

We close this section with a recursive method that illustrates a useful technique: *proof continuation-passing style*. By its very semantics, an Athena method call can only derive at most one single result. But oftentimes we want a single method M to derive a number of results, not just a single sentence. How can we write such methods? An answer is provided by proof continuations. Because Athena methods are higher-order, they can take other methods as arguments. Thus, along with its regular arguments, we can pass to M an extra argument, a method K that receives *a list* of all the intermediate results derived by M and then goes on to perform "the rest of the proof," whatever that might be. We refer to K as a *proof continuation*.

For example, sometimes we have a structurally complex conjunctive premise p and we need to derive all of its conjuncts before we can proceed with the rest of the derivation. By "structurally complex" we mean that, viewed as a tree, p could have an arbitrary shape, with the sentential constructor and appearing at some uppermost part of the tree. For instance, p could be (and p_1 p_2 p_3 p_4), or it could be

(and p_1 (and (and p_2 p_3 p_4) p_5) p_6),

where potentially some of the p_i could themselves be conjunctions, and so on. We would like to derive as many conjuncts as we possibly can from p, put them in a list L, and then call a continuation K with L as its argument. Of course, we need to make sure that K is called in an assumption base that contains every derived conjunct; that is, *every element of L must be in the assumption base when K is applied.* This is needed to ensure that the derived results can indeed serve as lemmas in the remainder of the argument; and it is typically achieved through tail recursion. The following method decomposes such arbitrarily complex conjunctions and passes the results to the given continuation:

```
1  define (decompose p K) :=
2    match p {
3      (and (some-list _)) =>
4        (!decompose (!left-and p)
5                    method (L)
6                      (!decompose (!right-and p)
7                                  method (R)
8                                    (!K (join L [p] R))))
9    | _ => (!K [p])
10   }
```

The basis case here occurs when the input premise p is not a conjunction (line 9), in which case the continuation K is called with [p] as its argument. If p is a conjunction, we first derive its left component and go on to recursively decompose that with a proof continuation that will then do the same thing to the right component of p, while finally invoking the original continuation with all intermediate results joined together (including p itself).

This can sometimes result in duplicate sentences being passed to the original continuation. One way to avoid that would be the following definition:

```
define (decompose p K) :=
  letrec {loop := method (p K)
                    match p {
                      (and (some-list _)) =>
                        (!loop (!left-and p)
                               method (L)
                                 (!loop (!right-and p)
                                        method (R)
                                          (!K (join L [p] R))))
                    | _ => (!K [p])
                    }}
    (!loop p method (results) (!K (dedup results)))
```

The library procedure dedup removes all duplicate element occurrences from a list.

4.11 Dealing with large conjunctions and disjunctions

When a conjunction has anywhere from 2 to 4 conjuncts, we typically build it by using and as a right-associative binary constructor. For example, we might write

$$(p_1 \ \& \ p_2 \ \& \ p_3) \tag{4.10}$$

for a conjunction with three components p_1, p_2, and p_3. Given that and associates to the right by default, (4.10) represents $(p_1 \ \& \ (p_2 \ \& \ p_3))$. So, if (4.10) is in the assumption base, applying `left-and` to it will yield p_1, and applying `right-and` to it will give $(p_2 \ \& \ p_3)$. Likewise for disjunctions: If we are dealing with 2 to 4 or maybe 5 disjuncts, we usually build the disjunction by using or as a binary sentential constructor associating to the right.

But sometimes we need to deal with larger conjunctions or disjunctions, and it then becomes convenient to use and and or as polyadic constructors, writing, say,

$$(\& \ p_1 \cdots p_n) \tag{4.11}$$

for arbitrarily large n. Recall, by the way, that and (as well as or) can also be used as a unary constructor that accepts a *list* of sentences. For example, we may write:

```
> (and [true (1 =/= 2) (~ false)])

Sentence: (and true
               (not (= 1 2))
               (not false))
```

This can be handy for programmatic purposes, when we don't know ahead of time the number of arguments that will be passed to these sentential constructors.

There are some special inference methods for dealing with polyadic conjunctions and disjunctions. Only one of them, `and-intro`, is primitive; the others are just variants of methods we have already discussed. To introduce (4.11), assuming that all immediate subsentences $p_1, \ldots, p_n$ are in the assumption base, we use the unary `and-intro` method, which accepts a list of sentences as its argument:

$$(\texttt{!and-intro [}p_1 \cdots p_n\texttt{]}).$$

This can be viewed as a generalization of the `both` method.[16] For elimination, the `left-and` and `right-and` methods can also be applied to polyadic conjunctions of the form (4.11), except that `right-and` will then produce the polyadic conjunction

$$(\texttt{\&}\ p_2 \cdots p_n).$$

Thus, by repeated applications of `left-and` and `right-and`, we can detach any component of a polyadic conjunction. There are library methods that automate this, such as the binary `conj-elim` method that takes two sentence inputs p and q and, assuming that p is an immediate subsentence of q and that q is a polyadic conjunction (like (4.11)) in the assumption base, it derives p.

For introducing polyadic disjunctions we can use `either` with a single *list* of sentences $[p_1 \cdots p_n]$ as its input. If at least one p_i is in the assumption base, the polyadic disjunction

$$(\texttt{or}\ p_1 \cdots p_n) \tag{4.12}$$

will be produced. For elimination we can still use `cases`, which is actually a method of flexible arity (we only discussed the ternary version earlier). Specifically, the first argument to `cases` can be an arbitrarily large polyadic disjunction of the form (4.12), while the remaining n arguments must be conditionals of the form $(p_i\ \texttt{==>}\ p)$, where p is the desired conclusion of the case analysis, for $i = 1, \ldots, n$. For example:

```
assert d := (or A B C D)

> (!cases d
          assume A (!true-intro)
          assume B (!true-intro)
          assume C (!true-intro)
          assume D (!true-intro))

Theorem: true
```

16 There is a related unary method, `conj-intro`, which takes an arbitrarily complicated conjunction as input and derives it, provided that all of its conjuncts are in the assumption base (where the "conjuncts" of a sentence p are as given by `get-conjuncts`, see Exercise 2.7). However, `conj-intro` is not a primitive method; it is defined in Athena's library in terms of other primitives.

The conditionals can be listed in any order. In fact, they can all be grouped in a single list that can be given to cases as its second argument. This is also useful for programmatic purposes.

```
assert cond1 := (A ==> E)
assert cond2 := (B ==> E)
assert cond3 := (C ==> E)
assert cond4 := (D ==> E)

> (!cases d [cond1 cond2 cond3 cond4])

Theorem: E
```

Several of the methods we defined in Section 4.9 have slightly different definitions in Athena's library in order to handle arbitrarily large conjunctions and disjunctions.

4.12 Sentential logic semantics

Our treatment of sentential logic so far has been entirely focused on proofs. While this is appropriate in view of the book's title, it is important to also say a few things about *semantics*. Semantics in general, both for natural and for artificial languages, is concerned with the specification and analysis of meaning. Usually this amounts to working out the conditions under which the sentences of a language become true. In the case of formal logics, semantics serves as a useful tool for constructing and analyzing proof systems. In particular, we want to make sure that our proof system (e.g., deductions D in the case of Athena) can never lead us from true premises to false conclusions. If that is the case, we say that our proof system is *sound*. Formal semantics makes it possible to investigate such issues rigorously.

A hallmark of semantics is that the meaning of a sentence, or any piece of syntax for that matter, is rarely specifiable in isolation. Additional information is usually required. This should be familiar from elementary algebra. Consider a formula like

$$x + 10 < 20. \tag{4.13}$$

Is this formula true or not? The answer depends, among other things, on what value we choose to assign to the variable x. If we let x be 5, then (4.13) is true. But if we give x the value 37, then (4.13) is false. And this is assuming that we understand $+$ as numeric addition, $<$ as the usual numeric comparison, and 10 and 20 as the numbers ten and twenty! So meaning in general, and truth in particular, is always determined with respect to some given *context*.

Artificial languages have a formal syntax, typically given by enumerating some primitive syntactic building blocks, along with some mechanisms for recursively building more complex syntactic objects out of simpler components. When it comes time to give the semantics of such languages, fixing the aforementioned "context" boils down to providing the meanings of the primitive syntactic blocks. Given such a context, the semantics tells us how to assign meanings to the more complex constructs.

In particular, the meaning of a complex syntactic construct is usually given as a (computable) function of the meanings of its components, an approach known as *compositional semantics*. So, to determine the meaning of a complex expression, we apply this function to the meanings of its subexpressions, which are themselves obtained by applying appropriate semantic functions to the meanings of their subexpressions, and so on until we finally reach the primitive building blocks, whose meanings are then obtained by consulting the given context.

In the case of sentential logic, the primitive building blocks are the atomic sentences—terms of sort `Boolean` in Athena. The mechanisms for building more complex sentences are the five sentential constructors: `not`, `and`, `or`, `if`, and `iff`. Thus, following the methodology outlined above, a compositional semantics for sentential logic would be a computable function $\boldsymbol{V}$ that takes a complex sentence p and some context and yields the meaning of p in terms of the meanings of its subsentences, with the context supplying the meanings of the atoms that occur in p. Since the "meaning" of a sentence will be truth or falsity, we can model $\boldsymbol{V}$ as a function that takes p and the context and produces either `true` or `false`, according to whether or not p is true with respect to the given context. The context itself is modeled as a mapping from atomic sentences to `true` or `false`. We call such a mapping an *interpretation* I.

With this background in mind, the semantic function that we are after can be precisely defined in Athena as follows:

```
define (V p I) :=
  match p {
    (|| true false) => p
  | (some-atom _) => (I p)
  | (~ q) => (negate (V q I))
  | (and (some-list args)) => (&&* args lambda (q) (V q I))
  | (or (some-list args)) =>  (||* args lambda (q) (V q I))
  | (q ==> r) => (|| (V (~ q) I) (V r I))
  | (q <==> r) => (|| (V (q & r) I) (V (~ q & ~ r) I))
  }
```

The procedures `&&*` and `||*` are generalizations of the built-in short-circuit expression forms `&&` and `||`, defined as follows:

```
1 define (&&* L f) :=
2   match L {
3     [] => true
4   | (list-of h t) => (&& (f h) (&&* t f))
5   }
```

```
6
7  define (||* L f) :=
8    match L {
9      [] => false
10   | (list-of h t) => (|| (f h) (||* t f))
11   }
```

Procedure `&&*` takes an arbitrary list $L =$ `[`$V_1 \cdots V_n$`]` as its first argument and a unary procedure f as its second argument, where f will produce either `true` or `false` when applied to any element V_i of L. It then starts going through the list one by one, ensuring that each $(f\ V_i)$ is `true`; the minute it finds some i such that $(f\ V_i)$ is `false`, it quits and returns `false`. If every $(f\ V_i)$ is `true`, then it returns `true`. The idea behind `||*` is similar, except that the roles of `true` and `false` are reversed: We stop the minute we discover some i such that $(f\ V_i)$ is `true`, while we finally return `false` if every $(f\ V_i)$ is `false`. The reason for passing the second argument f to these procedures is that we want to maintain the short-circuit behavior of `&&` and `||`. We produce the truth values that we want to check incrementally, as needed, by applying the anonymous procedure `lambda (q) (V q I)` to the immediate subsentences of p, and the semantics of the built-in forms `&&` and `||` take care of the short-circuiting on lines 4 and 10.

Here is an example of the semantics in action:

```
> define I := |{ A := true , B := false }|

Map I defined.

> (V (A & B) I)

Term: false
```

Thus, we see that the "meaning" (i.e., truth value) of the sentence `(A & B)` under `I` is `false`. By contrast, we have:

```
> (V (A | B) I)

Term: true
```

To enhance readability, we define an infix-appropriate alias for `V`, `wrt` ("with respect to"), and set its precedence lower than that of any sentential constructor:

```
define wrt := V

set-precedence wrt 5
```

We can now write:

```
> (A & B wrt I)

Term: false

> (A & B <==> B & A wrt I)

Term: true
```

The next lemma follows directly from the definition of `negate`:

Lemma 4.1

We have

$$(\texttt{\~{}}\ \texttt{\~{}}\ p\ \texttt{wrt}\ I) = (p\ \texttt{wrt}\ I)$$

for every sentence p and interpretation I.

Since an interpretation I is represented here as a finite map, and since there are always infinitely many atoms,[17] it follows that some (most!) atoms will not be in I's domain.[18] If we try to obtain the truth value of some sentence p with respect to an interpretation I that does not assign truth values to all of p's atoms,[19] we might well get a "no such key in the map" error. We say that I is an *appropriate* interpretation of a sentence p iff it assigns truth values to all atoms that occur in p. And we say that I is appropriate for a collection of sentences $\{p_1, \ldots, p_n\}$ iff it is appropriate for every p_i. When I is appropriate for p, `(V p I)` will return either `true` or `false`. We are interested only in appropriate interpretations, so for the remainder of this section, whenever we speak of an interpretation I in connection with some sentence p (or a number of sentences $p_1, \ldots, p_n$), it should be understood that we are supposing I to be appropriate for p (or for $p_1, \ldots, p_n$).

Let us say that an interpretation I *satisfies* a sentence p, written $I \models p$, iff

$$(p\ \texttt{wrt}\ I) = \texttt{true}.$$

We also then say that p is *true* under I. Likewise, we say that I satisfies a finite set of sentences β, written $I \models \beta$, iff I satisfies every sentence in β. And we say that *I falsifies p*, written $I \not\models p$, iff I does not satisfy p (i.e., iff $(p\ \texttt{wrt}\ I) = \texttt{false}$[20]). Satisfaction is a key notion that forms the basis of the following important definitions and results.

---

17 This is so even without any declarations of `Boolean` function symbols, since infinitely many variables of sort `Boolean` are always available (`?x1:Boolean`, `?x2:Boolean`, ...), and these count as atoms.

18 We could instead model interpretations as total computable procedures that map *every* atomic sentence to `true` or `false`. Athena could represent such interpretations directly, since the language is higher-order, and in fact the definition of `V` given earlier would not need to change at all. That would actually be necessary if we were to define satisfaction for infinite sets of sentences and present results such as the compactness theorem. However, the present approach is adequate for our purposes and makes for more readable code.

19 Other than `true` and `false`, which receive special treatment in the definition of `V`.

20 Recall that we are assuming that I is appropriate for p.

Definition 4.1: Tautologies

A sentence p is a *tautology* of sentential logic iff

$$I \models p$$

for every interpretation I.

Consider, for instance, the sentence:

```
(A & B ==> B & A).
```

The following demonstrates that this sentence is a tautology:

```
define p := (A & B ==> B & A)

> (p wrt |{ A := true, B := true}|)

Term: true

> (p wrt |{ A := true, B := false}|)

Term: true

> (p wrt |{ A := false, B := true}|)

Term: true

> (p wrt |{ A := false, B := false}|)

Term: true
```

We see that *no matter how we assign truth values to its atoms*, that is, no matter what (appropriate) interpretation we choose, the semantic valuation function keeps producing `true` as the meaning of

```
(A & B ==> B & A),                                    (4.14)
```

and that is precisely the definition of a tautology. A tautology is not falsified by any interpretation. Here we have only checked four different interpretations, but that is all we need to check, because interpretations that assign truth values to additional atoms other than `A` and `B` will produce the same truth value for (4.14) as one of these four interpretations, and hence will not make a difference on whether (4.14) is a tautology. In general, for any interpretations I_1 and I_2, if I_1 and I_2 agree on (assign the same truth value to) the atoms of p, then (*p* `wrt` I_1) and (*p* `wrt` I_2) will be identical. This can be shown by induction on the structure of p. Hence, when examining whether p is a tautology, we can focus on those interpretations that assign truth values to all and only the atoms of p. If p has n atoms then there are 2^n such interpretations. We will write a procedure shortly that generates all such interpretations and uses them to check whether a given p is a tautology.

Definition 4.2: Satisfiability

A sentence p is *satisfiable* iff there is some interpretation I such that $I \models p$.

We say that p is *unsatisfiable* iff it is not satisfiable, that is, iff every interpretation I falsifies p. The following list summarizes these notions:

1. *No* interpretations satisfy p: p is unsatisfiable.
2. *Some* interpretations satisfy p: p is satisfiable.
3. *All* interpretations satisfy p: p is a tautology.

Theorem 4.1

p is a tautology iff `(~ p)` is unsatisfiable.

PROOF: Assume that p is a tautology and let I be any interpretation. Then

`(V (~ p) I)`	=	`(negate (V p I))`	(definition of `V`)
	=	`(negate true)`	(since p is a tautology)
	=	`false`	(definition of `negate`).

This shows that `(V (~ p) I) = false` for every interpretation I, which is to say that every interpretation falsifies p. Hence, p is unsatisfiable. Conversely, suppose that `(~ p)` is unsatisfiable. Then:

`(V p I)`	=	`(V (~ ~ p) I)`	(Lemma 4.1)
	=	`(negate (V (~ p) I))`	(definition of `V`)
	=	`(negate false)`	(since `(~ p)` is unsatisfiable)
	=	`true`	(definition of `negate`).

We have thus shown that `(V p I) = true` for every I, so p is a tautology. ■

Definition 4.3: Entailment

A finite set of sentences β *entails* a sentence p, written $\beta \models p$, iff for all interpretations I, $I \models p$ whenever $I \models \beta$. If $\beta \models p$ we also say that p follows (logically) from—or is a (logical) consequence of—β.

Theorem 4.2

p is a tautology iff $\emptyset \models p$ iff $\{\texttt{true}\} \models p$.

PROOF: If p is a tautology then $\emptyset \models p$ and $\{\texttt{true}\} \models p$ follow trivially, because any I will satisfy p (and will also trivially satisfy $\emptyset$ and true). Conversely, suppose that $\emptyset \models p$ and pick any I. Then, trivially, $I \models \emptyset$, so, since $\emptyset \models p$, we conclude that $I \models p$. Since I was chosen arbitrarily, this shows that every I satisfies p, and hence that p is a tautology. Similar reasoning will show that if $\{\texttt{true}\} \models p$ then p is a tautology, which makes all three statements equivalent. ■

Theorem 4.3

$(p \texttt{ ==> } q)$ is a tautology iff $(p \texttt{ \& } \sim q)$ is unsatisfiable.

PROOF: In one direction, suppose that $(p \texttt{ ==> } q)$ is a tautology and pick any interpretation I. We need to show that I falsifies $(p \texttt{ \& } \sim q)$, which is the case iff I falsifies p or falsifies $(\sim q)$, that is, iff I falsifies p or satisfies q. But because $(p \texttt{ ==> } q)$ is a tautology, it is satisfied by I, and by the definition of $\vee$, that means precisely that I falsifies p or satisfies q. The converse direction is likewise shown. ■

Theorem 4.4

$\{p_1,\ldots,p_n\} \models p$ iff $(p_1 \texttt{ \& } \cdots \texttt{ \& } p_n \texttt{ ==> } p)$ is a tautology.

PROOF: Suppose that $\{p_1,\ldots,p_n\} \models p$ and pick any I. We need to show that I satisfies $(p_1 \texttt{ \& } \cdots \texttt{ \& } p_n \texttt{ ==> } p)$. By contradiction, suppose that it does not. Then it must satisfy $(p_1 \texttt{ \& } \cdots \texttt{ \& } p_n)$ and falsify p. But if it satisfies the conjunction $(p_1 \texttt{ \& } \cdots \texttt{ \& } p_n)$ then it satisfies every p_i, and that would mean that I satisfies $\{p_1,\ldots,p_n\}$ but falsifies p, contradicting the hypothesis $\{p_1,\ldots,p_n\} \models p$. Conversely, suppose that

$$(p_1 \texttt{ \& } \cdots \texttt{ \& } p_n \texttt{ ==> } p) \tag{4.15}$$

is a tautology. Now pick any I and suppose that I satisfies every p_i, so that I also satisfies the conjunction $(p_1 \texttt{ \& } \cdots \texttt{ \& } p_n)$. In that case, because we have assumed that (4.15) is a tautology and is therefore satisfied by I, we must have $I \models p$, for otherwise, by the definition of $\vee$, it would be impossible for I to satisfy (4.15), given that I satisfies $(p_1 \texttt{ \& } \cdots \texttt{ \& } p_n)$. This shows that $\{p_1,\ldots,p_n\} \models p$. ■

Theorem 4.5

For any sentence p, $\{p\} \models$ `true` and $\{$`false`$\} \models p$. Equivalently, by theorem 4.4, `(`p `==> true)` and `(false ==>` p`)` are tautologies.

PROOF: For $\{p\} \models$ `true` to fail, there must be some interpretation that satisfies p but not `true`, and that is impossible because every interpretation satisfies `true`. And for $\{$`false`$\} \models p$ to fail, there must be some interpretation that satisfies `false` but not p, which is also impossible because no interpretation satisfies `false`. ■

Theorem 4.6: Monotonicity

If $\beta \models p$ and $\beta \subseteq \beta'$ then $\beta' \models p$.

PROOF: Assume $\beta \models p$ and $\beta \subseteq \beta'$, and consider any I such that $I \models \beta'$. Since I satisfies every element of β' and β' is a superset of β, it follows that I satisfies every element of β, and therefore $I \models \beta$. Hence, by the assumption $\beta \models p$, it follows that $I \models p$. ■

Corollary 4.1

p is a tautology iff it follows from every assumption base, that is, iff $\beta \models p$ for every β.

PROOF: Suppose that p is a tautology and consider an arbitrary β. By Theorem 4.2, $\emptyset \models p$, hence, because $\emptyset \subseteq \beta$, monotonicity implies that $\beta \models p$. Conversely, if $\beta \models p$ for every β, it trivially follows that $\emptyset \models p$, hence p is a tautology by Theorem 4.2. ■

It follows from these results that all of the questions below are equivalent, in that if we can answer any one of them then we can answer the others:

- Is p satisfiable?
- Is p a tautology?
- Does p follow from $\{p_1, \ldots, p_n\}$?

The first of these questions is the so-called *satisfiability* problem (SAT for short), a problem of key importance both in theoretical computer science (where it was one of the first problems to be proven NP-complete, and remains the quintessential NP-complete problem) and in the modeling and verification of digital systems. Oftentimes when we ask whether a given p is satisfiable we do not just want a yes/no answer; if p is satisfiable, we also want a interpretation I such that $I \models p$. In the next few sections we will present a number of different techniques for solving the SAT problem, some of which are dramatically more efficient than others, exhibiting remarkably good performance even for very large and structurally complex sentences. However, given that all of these problems are paradigmatically NP-complete, we cannot expect any algorithm to have guaranteed good (polynomial-time) performance on all possible inputs—at least not unless $P = NP$! Even the most efficient SAT solver will suffer degraded performance on some inputs. We study techniques for solving the SAT problem in the next section.

Earlier in this chapter, on page 196, we wrote that a tautology is "a sentence that can be *derived* from every assumption base." That was a proof-centered or syntactic characterization of tautologies. We mentioned in footnote 3 that the syntactic characterization coincides with the semantic one that we developed in this section because Athena's proof system is *complete*. A more precise formulation of completeness is given by Theorem 4.8 below. A similar result is given for a closely related (and arguably more important) property, *soundness*. We do not prove these results here. Soundness could be established by induction on the structure of Athena deductions. Proving completeness is a bit more challenging, but it could be done either directly, using techniques that go back to Henkin [45], or by reduction to a proof system that is already known to be complete, for example, by showing that every result that can be derived in that system can also be derived in Athena's sentential logic.

Theorem 4.7: Soundness

If a deduction D produces a conclusion p in an assumption base β, then $\beta \models p$.

Theorem 4.8: Completeness

If $\beta \models p$ then there is a deduction D that derives p from β.

4.13 SAT solving

The most naive way of solving the tautology problem (and hence also SAT) for a given p is the enumeration technique mentioned in the previous section: Generate all 2^n interpretations that distribute truth values to the n atoms of p in different ways and check that p comes out `true` under each of these interpretations.

```
define (all-interpretations atoms) :=
  match atoms {
    [] => [|{}|]
  | (list-of a t) => let {L := (all-interpretations t)}
                       (join (map lambda (I) (Map.add I [[a true]]) L)
```

```
                              (map lambda (I) (Map.add I [[a false]]) L))
  }

define (taut? p) :=
  (for-each (all-interpretations (atoms p)) lambda (I) (V p I))

set-precedence taut? 5

> (taut? A & B ==> B & A)

Term: true

> (taut? (A & B <==> C | D) ==> ~ C ==> ~ D ==> ~ A | ~ B)

Term: true

> (taut? A & B)

Term: false
```

(Note: The library procedure `atoms` returns a (deduped) list of all the atoms that occur in a given sentence.) We can now define a procedure `sat?` for determining satisfiability directly in terms of `taut?` by taking advantage of Theorem 4.1 and Lemma 4.1:

```
define (sat? p) := (negate taut? ~ p)

set-precedence sat? 5

> (sat? A & B)

Term: true

> (sat? A & ~ A)

Term: false
```

However, as mentioned earlier, for satisfiable sentences we are typically not content with a "yes" answer—we also want a satisfying interpretation (and sometimes several or all satisfying interpretations). The following implementation of `sat?` returns an interpretation (as a map) if there is one, otherwise it returns `false`.

```
define (sat? p) :=
 (find-element (all-interpretations atoms p)
               lambda (I) (V p I)
               lambda (x) x
               lambda () false)
set-precedence sat? 5
```

The generic list procedure `find-element` has the following definition:

```
define (find-element L pred success failure) :=
  letrec {loop := lambda (L)
                    match L {
                      [] => (failure)
                    | (list-of h t) => check {(pred h) => (success h)
                                             | else => (loop t)}
                    }}
    (loop L)
```

We now have:

```
> (sat? A & B)

Map: |{A := true, B := true}|

> (sat? A & ~ B)

Map: |{A := true, B := false}|

> (sat? A & ~ A)

Term: false
```

While this approach works well enough for small inputs, its exponential complexity quickly becomes prohibitive.

```
declare P: [Int] -> Boolean

define (if-chain N) :=
  let {P_1 := (P 1);
       P_N := (P N);
       conditionals := (map lambda (i) (P i ==> P i plus 1)
                            (1 to N minus 1));
       antecedent := (& (P_1 added-to conditionals))}
    (antecedent ==> P_N)

define p := (if-chain 3)

> p

Sentence: (if (and (P 1)
                   (if (P 1)
                       (P 2))
                   (if (P 2)
                       (P 3)))
              (P 3))
```

Clearly, `(if-chain` n`)` is a tautology for every $n > 1$. Even for $n = 20$,

`(taut? if-chain` n`)`

takes a while to finish. After all, the algorithm has to generate and then check more than one million interpretations ($2^{20} = 1048576$)! For larger values yet, the combinatorics become hopeless. By contrast, we will see that state-of-the-art SAT solvers solve instances of `(if-chain` n`)` instantly, even for values of n exceeding 10^6. (More precisely, they solve instances of `(~ if-chain` n`)`, recognizing that such sentences are unsatisfiable, which, by Theorem 4.1, amounts to `(if-chain` n`)` being a tautology.)

Modern SAT solving is based on the *DPLL procedure*, an acronym for the "Davis-Putnam-Logemann-Loveland" algorithm invented in the late 1950s by Martin Davis, Hilary Putnam, George Logemann, and Donald Loveland.[21] In what follows we will implement the core DPLL algorithm in Athena. DPLL expects its input formula p to be in CNF (conjunctive normal form), meaning that p must be a conjunction of *clauses*, where a clause is a disjunction of *literals*, and a literal is either an atom or the negation of an atom. See Exercise 4.32 for additional discussion of this normal form and for algorithms that can transform arbitrary sentences into CNF. We will use Athena's built-in CNF conversion procedure later to test our implementation of DPLL.

We represent a disjunction d of literals as a list L_d of those literals, with the understanding that the order of the literals in the list is immaterial (in view of the commutativity and associativity of disjunction), and we represent a conjunction of such disjunctions by a list of the underlying disjunctive lists. So, for instance, the CNF formula

```
((A | B) & (B |~ C) & D)
```

might be represented by the list

```
[[A B] [B (~ C)] [D]].
```

A list of this kind will be the main input to our DPLL implementation. The output will be either the term `'unsat`, indicating that the input is unsatisfiable, or else an interpretation, represented simply as a list of literals. Observe that an empty list of clauses is trivially satisfiable (every interpretation satisfies it); and that an empty clause `[]` is tantamount to `false`. To see the latter, recall that a clause is meant to represent a disjunction of its literals. Therefore, an interpretation I satisfies a clause $c = [l_1 \cdots l_n]$ iff I satisfies *some* l_j. Clearly, no such l_j exists when $n = 0$, so an empty clause is unsatisfiable.

DPLL can be viewed as a constraint-solving algorithm that must assign appropriate values to each of a number of variables, or else announce that the input constraint is not solvable. In this case the variables are the atoms in the given CNF input and the values are `true` and `false`, so a variable assignment is essentially an interpretation of the corresponding sentence. The assigned values must satisfy (solve) the input constraint. Many problems, ranging from solving systems of equations to unification and mathematical programming, can be formulated as constraint-solving problems. And there is a general methodology for solving such problems that can be described—at a high level of abstraction—as follows:

21 For the early history behind this algorithm, refer to Davis [25].

1. If the constraint is in a trivial form that directly represents a solution (positive or negative), then stop and give the relevant output.
2. Otherwise, choose a hitherto unassigned variable x and assign to it a value v.
3. Compute and propagate some local consequences of that assignment, thereby obtaining a (hopefully simpler) constraint.
4. Repeat.

DPLL (roughly) follows the same pattern, except that the propagation occurs at the beginning of the algorithm, so it computes either consequences of a variable assignment from the previous iteration or consequences of structural properties of the input. The process of propagating consequences is known as *Boolean constraint propagation*, or BCP for short. In the case of DPLL, BCP consists of two transformations: *unit clause propagation* and *pure literal propagation*. We will describe both shortly. Using this terminology, the core algorithm for computing `(dpll` L`)` for a given list of clauses L can be described as follows:

- If the input list L is empty, return `true`.
- If the input list L contains an empty clause, return `false`.
- If BCP (Boolean constraint propagation) is applicable to L, apply it to get a new list of clauses L' and then apply `dpll` recursively to L'.
- Otherwise, pick a currently unassigned literal l and return

  `(|| (dpll (add [`l`]` L`)) (dpll (add [`$\bar{l}$`]` L`)))`,

  where $\bar{l}$ is the complement of l.

This version only returns `true` or `false`, but we will see that it is easy to instrument it so that it returns a satisfying interpretation when there is one.

We now turn to the two BCP transformations, starting with unit clause propagation. A *unit clause* is simply a clause with only one literal, such as `[A]` or `[(~ B)]`. Let us write l_u for the unique literal of any unit clause u. Thus, if u is the unit clause `[(~ B)]`, then $l_u =$ `(~ B)`.

Clearly, if our input list L contains a unit clause u then any interpretation that satisfies L *must* satisfy l_u. Accordingly, we can use u to simplify other clauses in L as follows:

1. If a clause c in L contains l_u then it must be satisfied by any interpretation that satisfies L, so we can remove c from L.
2. If a clause c contains $\overline{l_u}$ then we can remove $\overline{l_u}$ from c because, again, any interpretation I satisfying L must satisfy l_u, and therefore if I is to satisfy c, it must satisfy one of the other literals of c.

For example, suppose

$$L = [c_1\ c_2\ c_3\ c_4],$$

where

$$c_1 = \texttt{[A]}, \tag{4.16}$$
$$c_2 = \texttt{[A (\~{} B) C]}, \tag{4.17}$$
$$c_3 = \texttt{[(\~{} C) (\~{} A) D]}, \tag{4.18}$$
$$c_4 = \texttt{[(\~{} D)]}; \tag{4.19}$$

and let us focus on the unit clause $u = c_1 =$ `[A]`. (There is one more unit clause here, `[(~ D)]`, but we can ignore that one for now.) For starters, we can remove c_2 from L because any interpretation of L must satisfy `A` and therefore also $c_2 =$ `[A (~ B) C]` (recall that a clause is a disjunction of its literals). In fact we can remove u itself (because it certainly contains l_u!), although we will remove unit clauses with care—we will store l_u in the interpretation that we will be incrementally building. And finally, we can remove $\overline{l_u} =$ `(~ A)` from c_3. Thus we are left with two clauses after this round of unit clause propagation:

`[(~ C) D]` and `[(~ D)]`.

There is a unit clause in this set of clauses as well, namely `[(~ D)]`, so we can apply one more round of unit propagation. This would then leave us with a single clause:

`[(~ C)]`,

which is itself a unit clause that could then be removed, at which point we would be left with the empty list of clauses, which is trivially satisfiable. Retrieving the unit literals that we removed along the way would then give us a satisfying interpretation, consisting of `A`, `(~ D)`, and `(~ C)`.

Note that this particular problem was solved purely with BCP, and in fact only with unit clause propagation. We didn't even get a chance to get to the hard part, which is the selection of some previously unassigned literal and the subsequent case analysis. Many easy problems are indeed solved just by BCP, which can be implemented very efficiently. Another example is given by the `if-chain` tautologies above. Consider again the tautology produced by `(if-chain 3)`, which is essentially the conditional which says that if `P1` and `(P1 ==> P2)` and `(P2 ==> P3)` all hold, then `P3` also holds (where, for simplicity, we write `P1` instead of `(P 1)`, etc.). To verify that this is a tautology using DPLL, we would try to verify that its negation is unsatisfiable. If we negate this conditional and convert to CNF, we end up with the following clauses:

```
[[P1] [(~ P1) P2] [(~ P2) P3] [(~ P3)]].
```

A few repeated applications of unit propagation to this input will leave us with the clause list `[[]]`, a singleton consisting of the unsatisfiable empty clause.

Let L be a list of clauses and let l be a literal that appears in some of the clauses of L. We say that l is *pure* w.r.t. L iff the complement $\overline{l}$ does not appear anywhere in L. Thus, for instance, the earlier clause list $L =$ `[`c_1 c_2 c_3 c_4`]`, with c_1–c_4 as given by (4.16)–(4.19), has only one pure literal, `(~ B)`, because its complement `B` does not appear in any of the clauses of L. All other literals of L are impure because they have *both* negative and positive occurrences in L.

The *pure-literal transformation* removes from a clause list L all those clauses that contain pure literals $l_1, \ldots, l_m$. The resulting list L' is satisfiable iff L is. One direction of this claim is easy: If I is any interpretation that satisfies L, then it also satisfies L', since L' is a subset of L.[22] In the other direction, suppose that I is any interpretation satisfying the reduced set of clauses L'. Then we can easily extend I to a new interpretation that satisfies L by mapping each positive pure literal l_i to `true` and each negative pure literal l_j to `false`, for $i,j \in \{1, \ldots, m\}$. Since the new interpretation differs from I only in the pure literals $l_1, \ldots, l_m$, it follows that it satisfies the L' part of L. And it satisfies all the remaining clauses as well (the removed clauses containing pure literals), because it satisfies all the pure literals. For example, the pure-literal transformation entitles us to go from

```
[ [A B] [(~ A) C] [(~ C) B] ]
```

to

```
[ [(~ A) C] ]
```

by removing the two clauses containing the pure literal `B`. It is *not* the case here that any single interpretation that satisfies one set of clauses must also satisfy the other. For instance, the interpretation that assigns `A` and `B` to `false` and `C` to true satisfies the reduced set of clauses but not the original. The claim, rather, is that each set of clauses is satisfiable iff the other is, so that the transformation "preserves satisfiability."

We will implement `dpll` as a single call to an auxiliary procedure `dpll0` that will do all the hard work. `dpll0` will take two arguments, the input list of clauses L and a list of literals I representing an interpretation. Initially I will be empty; it will be incrementally extended. Both BCP transformations will also take the same two inputs as `dpll0`: the current list of clauses L and the current interpretation I. And they will both return a triple of the following form:

`[flag` L' I'`]`,

where *flag* is a bit indicating whether the transformation (unit clause or pure-literal propagation) was successful. If *flag* is `true`, then L' is the resulting list of clauses (obtained from L by applying the transformation) and I' is resulting interpretation (obtained from I by taking into account the effects of the transformation). With these signatures, we can write `dpll` and `dpll0` as follows:

---

22 We often speak as if L were a set, since it is in fact meant to represent one.

```
define (dpll clauses) := (dpll0 clauses [])

define (dpll0 clauses I) :=
  match clauses {

    [] =>  I # empty list of clauses, return I.

  | (_ where (member? [] clauses)) => 'unsat # empty clause indicates unsat.

  | _ => let {[success? clauses' I'] := (unit-propagation clauses I)}
           check
             {success? => (dpll0 clauses' I')
             | else => let {[success? clauses' I'] := (pure-literal clauses I)}
                         check
                           {success? => (dpll0 clauses' I')
                           | else => let {l := (choose-lit clauses)}
                                       match (dpll0 (add [l] clauses) I) {
                                         (some-list L) => L
                                       | _ => let {l' := (complement l)}
                                                (dpll0 (add [l'] clauses) I)
                                       }
                           }
             }
  }
```

Apart from the two BCP procedures, unit-propagation and pure-literal, we are also missing choose-lit, the procedure for selecting a literal for the case analysis. There are many different heuristics for making that selection, although none of them can guarantee an always-optimal selection. Perhaps the simplest most effective heuristic is to choose the literal that appears in the greatest number of clauses, the rationale being that assigning a value to that literal will help to simplify the largest possible number of clauses. The following is an implementation of this heuristic:

```
define (choose-lit clauses) :=
  let {counts := (HashTable.table);
       process := lambda (c)
                    (map-proc
                      lambda (l)
                        let {count := try { (HashTable.lookup counts l)
                                          | 0 }}
                          (HashTable.add counts [l --> (1 plus count)])
                      c);
       _ := (map-proc process clauses);
       L := (HashTable.table->list counts)}
    letrec {loop := lambda (pairs lit max)
                      match pairs {
                        [] => lit
                      | (list-of [l count] more) =>
                          check {(count greater? max) =>
```

```
                                  (loop more l count)
                             | else => (loop more lit max)}
                      }}
     (loop L () 0)
```

On average, this runs in linear time in the number of clauses. We could sidestep the need to traverse all the clauses in order to make a selection by passing around a data structure containing additional information about the unassigned variables, which could let us make a selection more efficiently. However, we would then need to do some additional bookkeeping in order to update that structure every time we modify the clause list.

We continue with the implementation of unit literal propagation:

```
1  define (unit-propagate l clauses) :=
2    let {l' := (complement l)}
3      (map-select lambda (c)
4                    check {(member? l c) => ()
5                          | else => (filter c (unequal-to l'))}
6                  clauses
7                  (unequal-to ()))
8
9  define (unit-propagation clauses I) :=
10   let {f := lambda (c)
11               match c {
12                 [l] => l
13               | _ => ()
14               }}
15     (find-some-element
16       clauses
17       (unequal-to ())
18       f
19       lambda (l)
20         let {clauses' := (unit-propagate l clauses)}
21           [true clauses' (l added-to I)]
22       lambda ()
23         [false clauses I])
```

We look for the first unit clause we can find in the `clauses` list. If we find one, then the success continuation on lines 19–21 takes the literal `l` of that clause and produces a new list of clauses, `clauses'`, by carrying out the actual unit propagation, through the auxiliary procedure `unit-propagate`. We then return the triple `[true clauses' (l added-to I)]`. If no unit clause is found, then the failure continuation on lines 22–23 is applied. The implementation of `(unit-propagate l clauses)` is straightforward: Every clause that contains `l` is removed, and the complement of `l` is removed from every other clause. As with `choose-lit`, we could simplify the discovery of unit literals (so that we don't have to traverse the entire clause list in the worst case, as we do now) in a number of ways. One way would be to always keep the input clauses partitioned into unit and nonunit clauses, which we could do by representing `clauses` as a pair of clause lists (unit and nonunit). When removing literals from clauses, either via unit propagation or the pure-literal transformation, we would then need to make sure that new clauses end up in the right partition, but this is straightforward. The implementation of `pure-literal` is worked out in Exercise 4.34.

To test `dpll` on a regular sentence p, we first need to convert p to a list of clauses. Athena provides a primitive procedure `cnf-core` for that purpose. It is a binary procedure whose first argument is the sentence p to be converted and whose second argument is a meta-identifier specifying the format of the output. This argument has three possible values:

- `'dimacs-list`;
- `'sentence-list`; and
- `'sentence`.

If `'dimacs-list` is specified, then the result of the conversion will be a list of clauses, where each clause is a list of integers. These integers represent literals. A negative integer is a negated atom, whereas a nonnegative integer is a positive atom. An integer atom in the output corresponds either to one of the atoms in the input p, or else it is a *Tseitin variable* introduced during the conversion (refer to Exercise 4.34 for more details). In order for the caller to know which integers correspond to which atoms in the input p, `cnf-core` also includes in its output a hash table from integers to the said atoms. More specifically, the output of `cnf-core` is a map m that should be thought of as a record with several fields. One of the fields is `'result`, which is the actual list of clauses, and another field is `'atom-table`, which is the aforementioned hash table. The remaining fields provide statistics about the conversion and the output clauses:

- `'cnf-conversion-time`;
- `'total-var-num`;
- `'tseitin-var-num`;
- `'clause-num`;
- `'table-formatting-time`; and
- `'clause-formatting-time`.

The value of `'total-var-num` represents the total number of variables in the output clauses, while `'tseitin-var-num` gives the number of Tseitin variables introduced by the conversion. The sum of `'tseitin-var-num` and

```
(HashTable.size (m 'atom-table))
```

should be equal to `'total-var-num`, as any variable in the output is either a Tseitin variable or a variable corresponding to some atom in the input. The value of `'cnf-conversion-time` is the number of seconds spent on the CNF conversion. The other two time outputs give the seconds spent on the other two tasks that are needed to prepare the output of cnf-core, namely, formatting the table and the clauses. These numbers are typically smaller than the value of 'cnf-conversion-time. The sum of these numbers should be the total time consumed by an invocation of cnf-core. An example:

```
define p := (A & B | C & D)

> define m := (cnf-core p 'dimacs-list)

Map m defined.

> m

Map:
|{
'result := [[7] [(- 3) 2] [(- 3) 1] [(- 6) 5] [(- 6) 4] [(- 7) 6 3]]
'atom-table :=
|[1 := B,
  2 := A,
  4 := D,
  5 := C
]|
'cnf-conversion-time := 0.0
'total-var-num := 7
'tseitin-var-num := 3
'clause-num := 6
'table-formatting-time := 0.0
'clause-formatting-time := 0.0
}|

> (m 'result)

List: [[7] [(- 3) 2] [(- 3) 1] [(- 6) 5] [(- 6) 4] [(- 7) 6 3]]

> (HashTable.lookup (m 'atom-table) 1)

Sentence: B
```

The 'dimacs-list option is convenient for preparing text files in DIMACS format, which is the standard input format for SAT solvers.[23] The primitive Athena procedure cnf is defined as a wrapper call around cnf-core with 'dimacs-list as its value:

```
define (cnf p) := (cnf-core p 'dimacs-list)
```

If 'sentence-list is given as the second argument of cnf-core, then the 'result field of the output map will be a list of clauses expressed as sentences, that is, as native Athena

23 See the information on DIMACS-CNF in the Wikipedia page on the *Boolean Satisfiability Problem*: http://en.wikipedia.org/wiki/Boolean_satisfiability_problem.

disjunctions of literals. A Tseitin variable such as x_2 will appear as `?X2:Boolean` in an Athena clause. For example:

```
define m := (cnf-core (A & B | C & D) 'sentence-list)

> (m 'result)

List: [
?X7:Boolean

(or (not ?X3:Boolean)
    A)

(or (not ?X3:Boolean)
    B)

(or (not ?X6:Boolean)
    C)

(or (not ?X6:Boolean)
    D)

(or (not ?X7:Boolean)
    ?X6:Boolean
    ?X3:Boolean)
]
```

Finally, if the second argument of `cnf-core` is `'sentence` then the `'result` field of the output map will be an Athena sentence in CNF (i.e., a conjunction of disjunctions of literals):

```
> ((cnf-core (A & B | C & D) 'sentence) 'result)

Sentence: (and ?X7:Boolean
               (or (not ?X3:Boolean)
                   A)
               (or (not ?X3:Boolean)
                   B)
               (or (not ?X6:Boolean)
                   C)
               (or (not ?X6:Boolean)
                   D)
               (or (not ?X7:Boolean)
                   ?X6:Boolean
                   ?X3:Boolean))
```

Note that `cnf-core` can handle arbitrarily complicated atoms, for example:

```
define p := (?x < ?y ==> ?x + 1 < ?y + 1)

> (cnf p)
```

```
Map:
|{
'result := [[2 (- 1)]]
'atom-table := |[

1 :=
(< ?x:Int ?y:Int)

2 :=
(< (+ ?x:Int 1)
   (+ ?y:Int 1))

]|
'cnf-conversion-time := 0.0
'total-var-num := 2
'tseitin-var-num := 0
'clause-num := 1
'table-formatting-time := 0.0
'clause-formatting-time := 0.0
}|
```

Even quantified sentences can serve as atoms:

```
define p := (forall ?x . ?x = ?x)

> (cnf (p & ~ p))

Map:
|{
'result := [[1] [(- 1)]]
'atom-table := |[1 :=
(forall ?x:'T3399
  (= ?x:'T3399 ?x:'T3399))
]|
'cnf-conversion-time := 0.0
'total-var-num := 1
'tseitin-var-num := 0
'clause-num := 2
'table-formatting-time := 0.0
'clause-formatting-time := 0.0
}|
```

We can now test dpll as follows:

```
define (test-dpll p) :=
  let {m := (cnf-core p 'sentence-list);
       clause-sentences := (m 'result);
       clauses := (map get-disjuncts clause-sentences)}
    (dpll clauses)

set-precedence test-dpll 10;;
```

```
> (test-dpll A ==> B)

List: [
(not A)
 B]

> (test-dpll ((A ==> B) & A & ~ B))

Term: 'unsat

> (test-dpll ~ if-chain 20)

Term: 'unsat
```

Even this rudimentary implementation of dpll is much more efficient than the brute-force approach of taut?. For example, it recognizes (~ if-chain 400) as unsatisfiable in a fraction of a second, whereas the exhaustive search procedure would never finish given the number of interpretations it would need to generate and test (2^{400}). As we pointed out earlier, all instances of this particular problem are solvable by unit clause propagation alone, but dpll has decent performance on harder problems as well. Beyond the simple improvements to which we have already alluded (slightly better organized data structures to facilitate the retrieval of unit clauses and literal selection for case analysis), there are several techniques that can be deployed on top of this basic infrastructure and that typically result in dramatic performance improvements, such as *conflict learning* and *backjumping*. For further information and implementation details, refer to *The Handbook of Propositional Satisfiability* [9].

Industrial-strength SAT solvers can be invoked in Athena through the primitive procedure sat-solve, which takes an arbitrary sentence p as input (or a list of sentences, see below), converts p to CNF as needed, prepares a DIMACS file, feeds that file to a SAT solver,[24] and then extracts the output of the SAT solver and formats it in a way that makes sense at the Athena level. The result of a call to sat-solve is a map containing various meta-identifier fields. The main output field is 'satisfiable, which is true or false depending on whether the input p was found to be satisfiable. There is also an 'assignment field. When 'satisfiable is true, the value of the 'assignment field is a hash table representing an interpretation of p, mapping atoms of p to true or false. When 'satisfiable is false, the value of 'assignment is the unit (). Other fields in the output map include:

- 'cnf-conversion-time;
- 'sat-solving-time;

24 Usually some (possibly modified) version of MiniSat.

- 'dimacs-prep-time;
- 'total-var-num;
- 'tseitin-var-num; and
- 'clause-num,

all of which should be self-explanatory in view of our foregoing discussion of cnf-core.

Here are some examples of sat-solve in action:

```
define p := (A ==> B)

> (sat-solve p)

Map:
|{
'cnf-conversion-time := 0.0
'total-var-num := 2
'tseitin-var-num := 0
'clause-num := 1
'satisfiable := true
'assignment := |[A := false, B := false]|
'sat-solving-time := 0.326
'dimacs-prep-time := 0.001
}|

> (sat-solve (p & ~ p))

Map:
|{
'cnf-conversion-time := 0.0
'total-var-num := 2
'tseitin-var-num := 0
'clause-num := 3
'satisfiable := false
'assignment := ()
'sat-solving-time := 0.288
'dimacs-prep-time := 0.001
}|
```

As with cnf-core, the input p can contain arbitrarily complicated atoms, including quantified sentences. For example:

```
define p := (forall ?x . ?x = ?x)

> (sat-solve (p & ~ p))

Map:
|{
'cnf-conversion-time := 0.0
'total-var-num := 1
```

```
'tseitin-var-num := 0
'clause-num := 2
'satisfiable := false
'assignment := ()
'sat-solving-time := 0.389
'dimacs-prep-time := 0.001
}|
```

Also, as mentioned earlier, the input to `sat-solve` can be a list of sentences L instead of just one single sentence. The output will then indicate whether all the elements of L are jointly satisfiable, and if so, the `'assignment` table will provide an appropriate interpretation:

```
> ((sat-solve [A (A ==> B) (~ A)]) 'satisfiable)

Term: false
```

The satisfiability problem is not just of theoretical interest. Because SAT is so expressive, many other difficult combinatorial problems have fairly straightforward reductions to it. Owing to the remarkable progress of SAT-solving technology in recent times, it is now feasible in some cases to reduce a tough problem to a SAT formulation and throw an off-the-shelf SAT solver at it instead of coding up a custom-made algorithm for the problem. Athena is particularly suitable for such problems because the reductions can be expressed fluidly and the integration with SAT solvers is seamless. In what follows we demonstrate this type of problem solving by reducing graph coloring—for an arbitrary number of colors—to SAT.

Graph coloring is a key problem in computer science that surfaces in many contexts under different guises. Mathematically, assuming we have k colors available $c_1, \ldots, c_k$, the problem asks whether it is possible to assign one of these colors to each vertex of an undirected graph G in such a way that adjacent vertices receive distinct colors. If so, we say that G is *k-colorable*. If the answer is affirmative, we usually want the solution to produce such a coloring.

With sufficient imagination on what counts as a color and what counts as a vertex and an edge, it is possible to express many practical problems as instances of graph coloring, ranging from job scheduling and register assignment in compiler code generation to frequency allocation in radio communications and many more. For $k = 2$ colors the problem can be solved in polynomial time, as it reduces to the question of whether the graph G is bipartite. But for $k > 2$ the problem is NP-complete. For $k > 3$ the question always has a positive answer for *planar* graphs:[25] With 4 or more colors available it is possible to color *any* planar graph so that adjacent vertices are given distinct colors. That is the content of the famous four-color theorem. (Of course, in practice we are usually interested in discovering

25 A graph is planar if it can be drawn on the plane so that its edges do not intersect anywhere (other than their endpoints).

such a coloring, not simply knowing that one exists.) For $k = 3$, by contrast, there are many simple examples of planar graphs that are not k-colorable; we will give such an example shortly. The smallest k for which a graph is k-colorable is called its *chromatic number*. In some problems we are interested in a k-coloring for the smallest possible k, as the "colors" in those cases are resources that we want to minimize. The procedure we will write takes an arbitrary graph G and number of available colors k and produces a k-coloring of G, if one exists, or else reports that there is no such coloring.

The first question we need to answer is how to represent graphs. We will opt for two representations, one of which is intuitive and convenient for input purposes but not terribly efficient; and another "internal" representation that is much more efficient. It will be easy to convert from one to the other. The first representation encodes an undirected graph simply as a list of pairs of nodes, where a node is always identified by a nonnegative integer i. Thus, for instance, the following graph

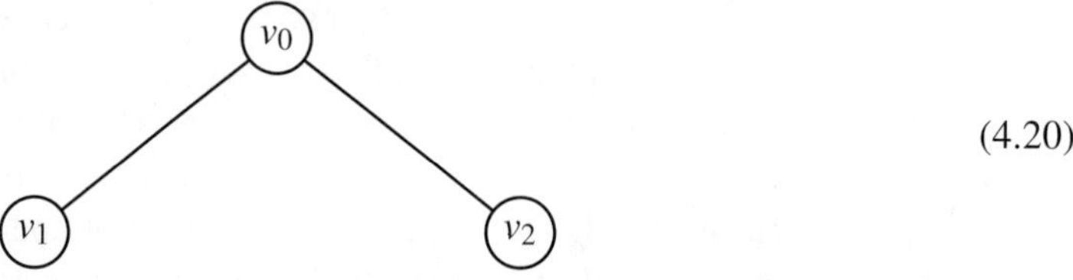

(4.20)

is represented by the list `[[0 1] [0 2]]`. The "internal" representation of a graph G with N nodes $v_0, \ldots, v_{N-1}$ will be a pair consisting of the number N and a vector V of length N, where the ith element of V contains a list of the neighbors (adjacent vertices) of v_i. The following procedure takes a list of edges and produces a graph under this canonical representation. For simplicity, the number of nodes N is assumed to be 1 plus the maximum index of any node appearing in the given edges. The procedure could be easily modified to take N as an additional parameter, instead of computing it in this way.

```
define (make-graph edges) :=
  let {N := (max* (join (map first edges) (map second edges)));
       V := make-vector (1 plus N) [];
       _ := (map-proc lambda (e)
                        match e {
                          [i j] => let {N-i := vector-sub V i;
                                        N-j := vector-sub V j;
                                          _ := vector-set! V i (add j N-i)}
                                     vector-set! V j (add i N-j)
                        }
                      edges);
       _ := (map-proc lambda (i) vector-set! V i (dedup vector-sub V i)
                      (0 to N))}
   [(1 plus N) V]
```

Note that we add each edge in both directions and dedup all neighbor lists at the end.

```
define E := [[0 1] [0 2]]

> (make-graph E)

List: [3 |[1 2] [0] [0]|]
```

We use two simple datatypes to represent colors and nodes. Neither is necessary, as both colors and nodes are essentially integers, but they make the code more readable. We also introduce a binary predicate has to represent an assignment of a color to a node. Thus,

(node i has color j)

will have the obvious meaning.

```
datatype Color := (color Int)

datatype Node := (node Int)

declare has: [Node Color] -> Boolean
```

All we now have to do is write a procedure `coloring-constraints` that takes a graph G and the number K of available colors and produces a list of sentences, using has atoms, that captures the coloring constraints. We need three groups of constraints to that end. First, we must ensure that every node is assigned *at least* one color. Second, that every node is assigned *at most* one color. (Clearly, these two sets of constraints jointly imply that every node is assigned exactly one color.) And finally, for every node v_i we must ensure that if v_i is assigned a color c, then none of its neighbors are assigned c. We then join these three groups of constraints together and return the resulting list as the output of `coloring-constraints`. The code provides a good illustration of a situation in which polyadic conjunctions and disjunctions come handy:

```
define (coloring-constraints G K) :=
  match G {
    [N neighbors] =>
      let {all-nodes := (0 to N minus 1);
           all-colors := (1 to K);
           at-least-one-color :=
             (map lambda (i)
                    (or (map lambda (c)
                               (node i has color c)
                             all-colors))
                  all-nodes);
           at-most-one-color :=
             (map lambda (i)
                    (and (map lambda (c)
                                (if (node i has color c)
                                    (and (map lambda (c')
```

```
                                    (~ node i has color c')
                                  (list-remove c all-colors))))
                  all-colors))
           all-nodes);
     distinct-colors :=
       (map lambda (i)
            (and (map lambda (j)
                   (and (map lambda (c)
                           (node i has color c ==>
                             ~ node j has color c)
                         all-colors))
                  vector-sub neighbors i))
           all-nodes)}
   (join at-least-one-color at-most-one-color distinct-colors)
}
```

Here are two of the constraints we get for graph (4.20) for $K = 2$; the first expressing that the first node (node 0) must be assigned at least one color, and the second one expressing the distinctness requirement for (node 0):

```
(or (has (node 0)
         (color 1))
    (has (node 0)
         (color 2)))

(and (and (if (has (node 0)
                   (color 1))
              (not (has (node 1)
                        (color 1))))
          (if (has (node 0)
                   (color 2))
              (not (has (node 1)
                        (color 2)))))
     (and (if (has (node 0)
                   (color 1))
              (not (has (node 2)
                        (color 1))))
          (if (has (node 0)
                   (color 2))
              (not (has (node 2)
                        (color 2))))))
```

Note that this implementation duplicates distinctness constraints for adjacent vertices. For instance, if v_2 and v_5 are adjacent and we state that if v_2 is red then v_5 should not be red, we will later also state that if v_5 is red then v_2 should not be red, which will be superfluous at that point. Generally speaking, in industrial applications of SAT solving we try to minimize the number of generated constraints,[26] and so in a case like this we would not generate the redundant constraints. This can be easily done here by keeping a hash table of all generated distinctness constraints involving an edge `[i j]`. We leave that as an exercise (Exercise 4.37).

We now write a top-level `graph-coloring` procedure that takes a list of edges and the number of available colors K and either produces a K-coloring of the graph G represented by the given edges or else outputs the unit value `()`, indicating that no such coloring exists. A third parameter is a map M assigning strings to color codes, that can be used to print the output coloring in a more readable way. For instance, we can agree that colors 1, 2, and 3 are "red," "green," and "blue," respectively. If M does not assign a string to a given color c_i, then the color is printed simply as `(color i)`, so we can pass the empty mapping `|{}|` as the value of M if we don't care to specify any color names.

```
define (show-results L color-names) :=
  let {_ :=  (map-proc lambda (p)
                        match p {
                          [((node i) has (color c)) true] =>
                            let {name := try {(color-names c)
                                             | (val->string color c)}}
                              (print "\nNode" i "-->" name)
                        | _ => ()
                        }
                      L)}
    (print "\n")


define (graph-coloring edges K m) :=
  let {G := (make-graph edges);
       constraints :=  (coloring-constraints G K);
       res := (sat-solve constraints)}
    match (res 'assignment) {
      (some-table ht) => (show-results (rev HashTable.table->list ht) m)
    | r => (print K "colors are not enough for this graph.")
    }

define color-names :=
  |{1 := "red", 2 := "blue", 3 := "green", 4 := "yellow"}|
```

Let's test the algorithm with the simple graph (4.20):

26 However, the relationship between the total number of constraints (or CNF clauses, ultimately) and the search complexity of the resulting SAT problem is not always clear. Sometimes adding constraints can actually speed up the search.

```
> (graph-coloring [[0 1] [0 2]] 2 color-names)

Node 0 --> blue
Node 1 --> red
Node 2 --> red

Unit: ()
```

The following graph makes for a more interesting example:

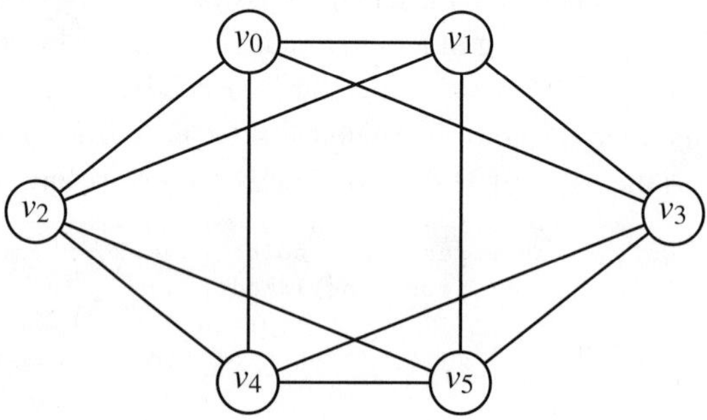

It should be clear that it cannot be colored with 2 colors alone. Can it be colored with 3? Let's define it first as a list of edges (note that we don't need to repeat symmetric edges, for example, we do not list `[1 0]` given that we have listed `[0 1]`):

```
define g := [[0 1] [0 3] [0 4] [0 2]
             [1 2] [1 3] [1 5]
             [2 4] [2 5]
             [3 4] [3 5]
             [4 5]]
```

The answer, as we see, is affirmative:

```
> (graph-coloring g 3 color-names)

Node 0 --> green
Node 1 --> blue
Node 2 --> red
Node 3 --> red
Node 4 --> blue
Node 5 --> green
```

Figure 4.1 provides a visual representation of this coloring (writing G for green, etc.).

As a negative example, and perhaps somewhat surprisingly, it can be shown that the following graph has no 3-coloring:

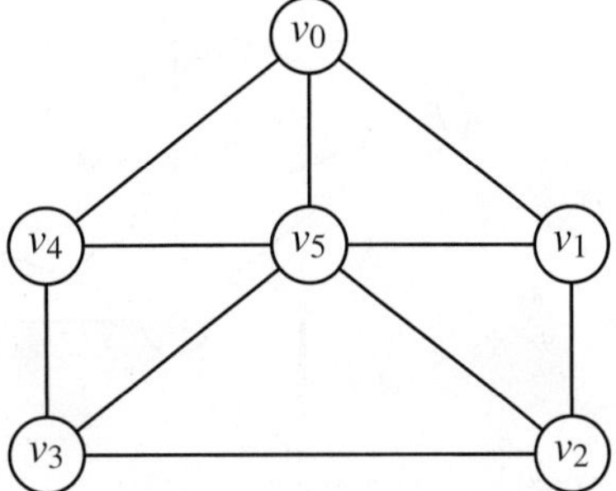

Let's verify this:

```
define g := [[0 4] [0 5] [0 1]
             [1 2] [1 5]
             [2 5] [2 3]
             [3 5] [3 4]
             [4 5]]

> (graph-coloring g 3 color-names)

3 colors are not enough for this graph.
```

In all of these examples, answers were returned instantaneously. For more extensive testing, we can write a procedure that generates random undirected graphs. One way to do that is through the *Gilbert random graph model* $G(N,p)$, determined by the number of vertices n and a real number $p \in (0,1)$, where the existence of an edge between any two vertices in the graph has independent probability p. A procedure for generating random Gilbert graphs can be coded as follows:

```
1  define (make-random-graph N p) :=
2   let {all-nodes := (0 to N minus 1);
3        all-edges := (filter-out (all-nodes X all-nodes)
4                                  lambda (p) (first p equal? second p))}
5     (filter all-edges lambda (_) (bernoulli-trial p))
```

where a Bernoulli trial with probability of success p is defined as follows:

```
define (bernoulli-trial p)  :=
  let {log := (length tail tail val->string p);
       limit := (10 raised-to log);
       r := (random-int limit)}
    (r leq? p times limit)
```

The binary procedure `X` (line 3 in the listing of `make-random-graph`) builds the Cartesian product of two lists. For example:

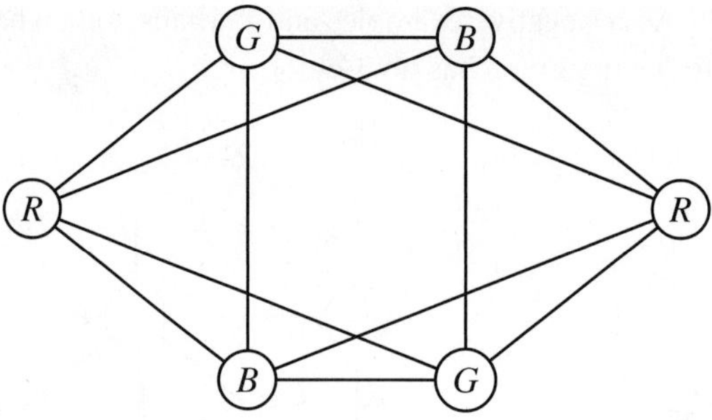

Figure 4.1
A coloring discovered by SAT solving.

```
> ([1 2 3] X ['a 'b])

List: [[1 'a] [1 'b] [2 'a] [2 'b] [3 'a] [3 'b]]
```

We can now test our code as follows:

```
define (test-gc N p K) :=
  (graph-coloring (make-random-graph N p) K |{}|)
```

Experimentation will show that, depending on how connected the graph is, the SAT solver can quickly find colorings even for hundreds or thousands of nodes, and for hundreds of thousands of constraints.

A drawback of the Gilbert model is that it examines every possible edge $e = v_i, v_j$ to decide whether to keep e. Therefore its run time is quadratic in the number of vertices, which can be nontrivial even for a few thousand vertices. An alternative technique for generating random graphs over a given set of vertices is to take as input the required number of edges, and then keep generating random pairs of vertices until we have the desired total number of (distinct) edges. We leave the implementation of that approach as an exercise.

Exercise 4.1 (Sentence Polarities): Let p be the sentence `(A & ~ B)`. We say that the occurrence of `A` in p is *positive*, or that it has a positive *polarity*, whereas the occurrence of `B` is *negative*, or that it has a negative polarity. Intuitively, this is intended to reflect the fact that `B` is embedded inside p within an odd number of negation signs,[27] while `A` is within an even—in this case zero—number of negations. Likewise, the occurrence of `B` in

```
(~ ~ B | ~ C)
```

27 Meaning that if we traverse the abstract syntax tree of p from the root until `B`, we will encounter an odd number of negations along the way.

is positive, as is the occurrence of `(~ C)`. However, the occurrence of `C` is negative, as is the occurrence of `(~ B)`. Of course, for any *p*, the unique occurrence of *p* in *p* itself is positive.

It is possible for a sentence *q* to have both positive and negative occurrences inside a given *p*. For instance, `(A | B)` has two occurrences in

$$(\underline{\texttt{(A | B)}} \texttt{ | (C \& ~ } \overline{\texttt{(A | B)}}\texttt{))},$$

the first of which (the underlined occurrence) is positive, while the second (overlined) occurrence is negative.

When it comes to conditionals and biconditionals, polarity is not just a matter of traversing an abstract syntax tree and counting negations. Consider first a conditional `(`*p* `==>` *q*`)`. Because this is essentially equivalent to

`(~` *p* `|` *q*`)`,

we regard the occurrence of *p* in the conditional to be negative and the occurrence of *q* to be positive. By the same token, since a biconditional

$$(p \texttt{ <==> } q) \tag{4.21}$$

is equivalent to the conjunction of the two conditionals `(`*p* `==>` *q*`)` and `(`*q* `==>` *p*`)`, we regard the occurrence of *p* in (4.21) as *both* positive and negative, and likewise for the occurrence of *q*. Similarly, every occurrence of a subsentence of *p* in (4.21) is both positive and negative, and likewise for every occurrence of a subsentence of *q* in (4.21). We will use the term `'pn` to denote a polarity that is both positive and negative—or "positive-negative." The terms `'p` and `'n` will denote positive and negative polarities, respectively.

We can make the above discussion more precise by defining a binary procedure `polarity` that takes two sentences *p* and *q* and returns `'p`, `'n`, or `'pn` depending on whether *p* has a positive, negative, or positive-negative occurrence in *q*, respectively. However, this simple interface would not be able to handle cases in which (a) *p* has various occurrences of different types in *q* (e.g., one positive occurrence and one negative); and (b) *p* has no occurrences in *q*. A better specification is this: Return *a list* of terms of the form `'p`, `'n`, or `'pn`, one for each occurrence of *p* in *q*. If *p* does not occur in *q*, the empty list should be returned. We thus name the procedure in the plural, `polarities`, and define it as follows:

```
define (flip pol) :=
  match pol {
    'p => 'n
  | 'n => 'p
  | 'pn => 'pn}

define (polarities p q) :=
  match q {
    (val-of p) => ['p]
  | (~ q1) => (map flip (polarities p q1))
```

```
| (q1 ==> q2) => (join (map flip (polarities p q1))
                       (polarities p q2))
| (q1 <==> q2) => (map lambda (_) 'pn
                       (join (polarities p q1) (polarities p q2)))
| ((some-sent-con _) (some-list args)) =>
    (flatten (map lambda (q) (polarities p q)
                  args))
| _ => []
}
```

The body of the procedure is a **match** expression that starts by checking whether p and q are identical. If so, then `['p]` is returned, as p has exactly one occurrence in q, and it is a positive one. Otherwise we check to see if q is a negation of some q_1. In that case we recursively obtain the polarities of p in q_1 and flip them (positives become negatives and vice versa, whereas positive-negative occurrences remain unchanged) to account for the outer negation. If q is a conditional, we recursively obtain the polarities of p in the antecedent and flip them (since we view the antecedent as implicitly negated), and then we concatenate those to the polarities of p in the consequent. If q is a biconditional (q_1 `<==>` q_2), we recursively obtain the polarities of p in q_1 and in q_2, and we turn all of them into positive-negative polarities. Next, if q is a conjunction or disjunction of the form

$$(\oplus\ q_1 \cdots q_n),$$

for $\oplus \in \{\mathsf{and}, \mathsf{or}\}$, we recursively compute the polarities of p in each q_i and then join them together.[28] Finally, if none of these conditions obtains, then q must be an atom distinct from p, in which case p has no occurrences in q and the empty list is returned. Here are some examples of this procedure in action:

```
> (polarities A (A & (B | C)))

List: ['p]

> (polarities A (A ==> A | B))

List: ['n 'p]

> (polarities (~ A) (A ==> ~ B))

List: []

> (polarities (B & C)
              (~ C ==> (B & C <==> D | E)))

List: ['pn]
```

28 Recall the definition of `(flatten L)` as `(foldl join [] L)`; see Section 2.6.

```
> (polarities B
              (~ B ==> (B & C <==> ~ B | E)))

List: ['p 'pn 'pn]
```

(a) Note that (polarities p q) also gives us, indirectly, the *number* of occurrences of p in q; this is simply the length of the returned list. Modify the implementation so that every item in the output list consists of a *pair* (a two-element list)

[*position polarity*]

comprising the *position* of the relevant occurrence as well as its *polarity*.[29] For instance,

`(polarities A (A & B | A & C))`

should return `[[[1 1] 'p] [[2 1] 'p]]`, indicating that there is one positive occurrence of `A` at position `[1 1]` and another one at `[2 1]`.

(b) Implement a binary procedure `polarities*` that takes a sentence p and a list of sentences L and returns a list of the form

$$[[q_1\ L_1]\ \cdots\ [q_k\ L_k]],$$

$k \geq 0$, where $q_1, \ldots, q_k$ are those members of L that contain at least one occurrence of p,[30] and where each L_i is the result of applying the preceding version of `polarities` to p and q_i. That is, L_i is a list of all the positions and polarities of p inside q_i.

(c) Implement two ternary methods `M+` and `M-` that take as inputs (i) a premise of the form (p ==> p'); (ii) a sentence q; and (iii) a position u (a list of positive integers) such that q contains an occurrence of p at u that is strictly positive or strictly negative.[31] The methods should behave as follows:

- If the occurrence of p in q at u is positive, then `M+` should derive (q ==> q'), where q' is the sentence obtained from q by replacing the said occurrence of p by p'.
- If the occurrence of p in q at u is negative, then `M-` should derive (q' ==> q), where q' is again the sentence obtained from q by replacing that occurrence of p by p'.

`M+` should fail if the occurrence of p at u is not strictly positive, and `M-` should fail if the occurrence of p at u is not strictly negative. Both methods should fail if p does not occur

---

29 Recall from Section 3.3 that the position of an occurrence of p in q is a list of integers $[i_1 \cdots i_n]$ representing the path that must be traversed to get from the root of q to the said occurrence.

30 The order in which the various items $[q_i\ L_i]$ are listed is immaterial, but there should be no duplicates in the list.

31 Meaning that the occurrence is not `'pn`.

in q at u, or if the inputs are not of the specified forms. (Note: You will want to read Section 4.10 before tackling this part of the problem, as these two methods have to be mutually recursive.) □

4.14 Proof heuristics for sentential logic

If we analyze the process that we went through in constructing the proof of Section 4.8 and try to depict it graphically, the result might look like the tree shown in Figure 4.2,[32] which we call a *goal tree*. Every node in such a tree contains a goal, which in the present context is a specification for a proof, or a *proof spec* for short. A proof spec is of the form "Derive p from β" where β is some assumption base, that is, a finite set of available assumptions, and p is the desired conclusion that must be derived from β. We use the letter τ as a variable ranging over proof specs (goals). By convention, we enclose every proof spec inside a box. Occasionally we also refer to the target conclusion p as a "goal." It will always be clear from the context whether we are using "goal" in reference to a target sentence within a proof spec or to a proof spec itself, as a whole.

The children of a proof spec τ, call them $\tau_1, \ldots, \tau_n$, $n \geq 0$, are obtained by applying a *tactic* to τ. In particular, applying a tactic to τ produces two things:

1. A partial deduction D—partial in that it may contain $n \geq 0$ applications of **force**. This partial deduction D is capable of deriving the target sentence p from the corresponding assumption base β, provided that the various gaps in it have been filled, that is, provided that the n applications of **force** have been successfully replaced by proper deductions.
2. A list of proof specs $\tau_1, \ldots, \tau_n$ that capture the proof obligations corresponding to the n applications of **force** inside D. We refer to these new proof specs as the *subgoals* generated by the tactic.

We then need to tackle these new goals $\tau_1, \ldots, \tau_n$ by applying new tactics to each τ_i, thereby expanding the goal tree to deeper levels. The process ends when the deductions at the leaves of the tree contain no occurrences of **force**, which typically happens when the deductions are single applications of `claim`. We visually mark such leaves by placing the sign $\surd$ right next to the corresponding deductions. When every branch of the goal tree has terminated at such a leaf, the process is complete. We can then scan the tree from the bottom up and put together complete definitions for the various partial deductions at every internal node, by incrementally eliminating applications of **force**, culminating with the top-level deduction at the root of the tree. Don't worry if not all of these details make full sense at this point; the ideas will start to sink in as we go along.

32 Tree structure here is indicated by indentation, which is one of several visual devices for drawing trees; see Knuth's discussion on p. 309 of his classic first volume [58].

```
Derive (~ B ==> A ==> ~ E) from β = {(A & B | A ==> C), (C <==> ~ E)}

assume (~ B)
  (!force (A ==> ~ E))

  Derive (A ==> ~ E) from {(~ B), (A & B | A ==> C), (C <==> ~ E)}

    assume A
      (!force (~ E))

    Derive (~ E) from {A, (~ B), (A & B | A ==> C), (C <==> ~ E)}

      (!cases (A & B | A ==> C))
              assume (A & B)
                (!force (~ E))
              assume (A ==> C)
                (!force (~ E)))

      Derive (~ E) from {(A & B), A, (~ B), (A & B | A ==> C), (C <==> ~ E)}

        let {_ := (!force B)}
          (!from-complements (~ E) B (~ B))

        Derive B from {(A & B), A, (~ B), (A & B | A ==> C), (C <==> ~ E)}

          let {conj := (!force (A & B))}
            (!right-and conj)

          Derive (A & B) from {(A & B), A, (~ B), (A & B | A ==> C), (C <==> ~ E)}

            (!claim (A & B)) √

      Derive (~ E) from {(A ==> C), A, (~ B), (A & B | A ==> C), (C <==> ~ E)}

        let {imp := (!left-iff (C <==> ~ E));
             _   := (!force C)}
          (!mp imp C)

        Derive C from {(A ==> C), A, (~ B), (A & B | A ==> C), (C <==> ~ E)}

          let {imp := (!force (A ==> C));
               _   := (!force A)}
            (!mp imp A)

          Derive (A ==> C) from {(A & B), A, (~ B), (A ==> C), (C <==> ~ E)}

            (!claim (A ==> C)) √

          Derive A from {(A & B), A, (~ B), (A ==> C), (C <==> ~ E)}

            (!claim A) √
```

Figure 4.2

Goal tree for the example of section 4.8.

In general, every time we are faced with a proof spec, there are two ways to proceed, one of which is *goal-driven* while the other is *information-driven*. The information-driven approach is also called *forward* or *synthetic*, because, as we will see, it moves from the premises toward the target conclusion—it tries to "synthesize" the conclusion by utilizing the premises. By contrast, the goal-driven approach is also known as *backward* or *analytic*, because it tries to move from the target conclusion toward the premises—it "analyzes" or decomposes the target sentence, usually breaking it up into smaller components. There are a number of tactics that can be used for each of these two approaches. We will first discuss backward or analytic tactics and then we will move on to forward or synthetic tactics.

Typically, discovering a proof will require a mixture of backward and forward moves. Later, after we present our collection of tactics, we will go on to formulate a *strategy* for how to apply these tactics to a given problem.[33]

4.14.1 Backward tactics

In the goal-driven approach, we ignore the set of available assumptions β (i.e., we ignore the available *information*), focusing instead on the syntactic form of the target sentence. In particular, we look at the main logical connective of that sentence, and we let that dictate the overall form that the desired deduction will take, by using the corresponding introduction mechanism or some closely related method. As we pointed out, this will only outline the skeleton of the required deduction, meaning that this deduction may contain various applications of **force**, which will in turn give rise to new (children) proof specs. Our new task will then be to tackle the proof specs created by applying the tactic.

Since there are five logical connectives in sentential logic, there are five corresponding backward tactics, depending on the logical structure of the target conclusion. They are shown below. The symbols on the right side of the tactics (enclosed within square brackets) are mnemonic names given to the tactics for future reference. In the case of `[and<-]`, for example, "`and`" indicates that the tactic is applicable to conjunctions, while the left-pointing arrow represents the direction of the tactic (backward in this case).

- *Conjunctive goals*:

Derive $(p \ \& \ q)$ from β

```
let {left := (!force p);
     right := (!force q)}                    [and<-]
  (!both left right)
```

33 A tactic is usually understood as a low-level maneuver designed to accomplish a specific and fairly narrow goal. By contrast, a strategy is a more general scheme for arranging and deploying tactics in order to arrive at the ultimate objective.

The choice of deriving the left conjunct (p) first is arbitrary. We could just as well choose to derive q first and p afterward.

Note that the number and contents of the children proof specs are uniquely determined by the applications of **force** in the tactic deduction, in combination with the evaluation semantics of Athena proofs. Specifically, there are exactly as many children proof specs as there are applications of **force**; the target sentence of each child proof spec is the argument of the corresponding **force** application; and the assumption base of each child proof spec is the parent assumption base, possibly augmented with new hypotheses or intermediate conclusions. Accordingly, because the children proof specs are obvious from the tactic's deduction, we will generally omit them from our formulation of these tactics, unless we feel that showing them explicitly adds something to the formulation.

- *Disjunctive goals*:

Derive (p | q) from β

```
let {_ := (!force p)}                    [lor<-]
  (!left-either p q)
```

There is a similar pattern, `[ror<-]`, that derives q first and then introduces (p | q) via `right-either`. As we mentioned at the end of Section 4.4 and in our discussion of `cond-def` in Section 4.9, an alternative—and often more effective—tactic for deriving a disjunction (p | q) is to treat it as the conditional (~ p ==> q) and attempt to derive that instead:

Derive (p | q) from β

```
let {cond := (!force (~ p ==> q))}                [or-if<-]
  (!cond-def cond)
```

Alternatively, we can treat (p | q) as the conditional (~ q ==> p) and attempt to derive that. As we will see later, both of these are instances of a more general *replacement* tactic.

- *Conditional goals*:

Derive (p ==> q) from β

```
assume p
  (!force q)                 [if<-]
```

Derive q from $\beta \cup \{p\}$

Note the augmented assumption base in the child proof spec.

- *Negated goals*:

  Derive (~ p) from β

```
(!by-contradiction (~ p)
  assume p                          [not<-]
    (!force false))
```

  As was mentioned before, `false` is usually derived by `absurd`, so the task of the new proof spec will usually consist in the derivation of two contradictory sentences q and (~ q), followed by an application of `absurd`. Later we will give a more specific tactic for deriving `false`.

- *Biconditional goals*:

  Derive (p <==> q) from β

```
let {left := (!force (p ==> q));
     right := (!force (q ==> p))}          [iff<-]
  (!equiv left right)
```

  As with conjunctions, the ordering of the two subgoals is immaterial.

4.14.2 Forward tactics

In the information-driven approach, we focus on the set of available assumptions β and try to find one or more sentences in β that we can use to derive the goal. Typically, the resulting proof will consist of a sequence of inference steps assembled together in a single proof block (or a **let** if we choose to use naming). For added visual clarity, when expanding a proof spec node with a forward tactic we underline the relevant assumptions in β, that is, those sentences that are used in the (partial) deduction prescribed by the tactic. We then proceed to expand the tree deeper by applying some appropriate tactic to each of the new proof specs, and so on.

The ideal situation in forward proof search occurs when the assumption base contains the target sentence, in which case we can simply `claim` it:

Derive p from $\{\ldots,\underline{p},\ldots\}$

```
(!claim p)   √                    [claim->]
```

Note that this tactic does not generate any subgoals, as there are no applications of **force** in the given deduction. Goals of this form (that generate no subgoals) will be called *primitive*. Thus, primitive goals become leaf nodes of goal trees.

Failing that, the next best-case scenario occurs when the assumption base contains a sentence from which the goal can be detached in one step, by a single application of an elimination method. For instance, if our goal is q and the assumption base contains a conjunction (p & q), then clearly we can derive q in one step by applying `right-and` to (p & q). We can depict this forward proof tactic as follows:

Derive q from $\{\ldots, \underline{(p \ \&\ q)}, \ldots\}$

`(!right-and` $(p$ `&` $q)$`)` ✓

However, this way of formulating the tactic would be too narrow. For instance, suppose that the assumption base did not contain $(p$ `&` $q)$, but it did contain, say,

$$((p \ \&\ q) \ \&\ r). \tag{4.22}$$

Here we cannot derive the goal q in one step, but we can do it in two steps: We first detach $(p$ `&` $q)$ by applying `left-and` to (4.22), and then we obtain q by applying `right-and` to $(p$ `&` $q)$.

In general, our forward proof tactics will be applicable not only when the assumption base contains a *parent* of the goal, from which the goal can be extracted in one step, but also when it contains a grandparent of the goal, or a great-grandparent, and so on, that is, when the goal is an arbitrarily deep subsentence of some member of the assumption base.[34] In such cases the application of a forward proof tactic leads to a subgoal (a new proof spec): We first need to derive the parent from the assumption base. Once that subgoal is achieved, we can derive the original goal from the parent through the appropriate elimination method.

It turns out, however, that we need not be concerned with *all* occurrences of goal parents in the assumption base, but only with *positive* occurrences.[35] That is, if a parent p of the goal occurs in some element of the assumption base β, then the subgoal of deriving p from β can only be pursued if the said occurrence of p is positive.[36] Negative occurrences can be ignored. Oftentimes that can restrict the search space considerably. We indicate that an occurrence of a sentence q inside a sentence p is positive by attaching a $^+$ superscript to it. For example,

$$p = (\cdots \texttt{(A ==> B)}^{+} \cdots)$$

signifies that the superscripted occurrence of the conditional `(A ==> B)` in the surrounding sentence p is positive. Accordingly, we reformulate the preceding proof tactic for conjunctions as follows:

34 A *child* of a sentence p is an *immediate* subsentence of p. Thus, for instance, the conjunction $p =$ `((A | B) & C)` has two children, the disjunction `(A | B)` and the atom `C`. The atoms `A` and `B` are grandchildren of p; they are subsentences of p but not *immediate* subsentences. If q is a child of p, we say that p is a *parent* of q.

35 Roughly, an occurrence of a sentence q inside a sentence p is positive (negative) iff it is within an even (odd) number of negations, but see Exercise 4.1 for a precise definition. It is legitimate to restrict attention to positive occurrences because it can be shown (although we will not do so here) that if there is any derivation of a goal from an assumption base β, then there is some derivation which only applies elimination methods to positive subsentences of β.

36 More generally, we say that a sentence p has a positive (negative) occurrence in a set of sentences β iff p has a positive (negative) occurrence in some element $q \in \beta$.

Derive q from $\beta = \{\ldots, (\cdots \underline{(p \;\&\; q)}^{+} \cdots), \ldots\}$

```
let {conj := (!force (p & q))}                           [rand->]
  (!right-and conj)
```

There is an obvious analogue for left conjuncts:

Derive p from $\beta = \{\ldots, (\cdots \underline{(p \;\&\; q)}^{+} \cdots), \ldots\}$

```
let {conj := (!force (p & q))}                           [land->]
  (!left-and conj)
```

We continue with a forward tactic involving conditionals. Suppose that the goal is to find a derivation of q from some assumption base β. If the assumption base contains a positive occurrence of a conditional of the form (p ==> q), then we can try to derive that conditional from β, then derive the antecedent p, and finally obtain q via modus ponens:

Derive q from $\beta = \{\ldots, (\cdots \underline{(p \texttt{ ==> } q)}^{+} \cdots), \ldots\}$

```
let {cond := (!force (p ==> q));
     ant  := (!force p)}                         [if->]
  (!mp cond ant)
```

Next, suppose that the parent of the goal q is a disjunction, say (q | p). Then once we derive that disjunction from the assumption base, we can perform a case analysis, one branch of which will be particularly easy. Specifically, we have to show that q follows no matter which disjunct obtains, so we have to show that (a) q follows from the first disjunct, q; and (b) q follows from the second disjunct, p. Part (a) is trivial, so (b) will usually be more challenging:

Derive q from $\{\ldots, (\cdots \underline{(q \mid p)}^{+} \cdots), \ldots\}$

```
let {disj := (!force (q | p));
     cnd1 := assume q (!claim q);                        [lor->]
     cnd2 := (!force (p ==> q))}
  (!cases disj cnd1 cnd2)
```

There is an analogous [ror->] tactic, which we do not depict here explicitly.

Finally, we have similar forward tactics involving biconditionals:

Derive q from $\beta = \{\ldots, (\cdots \underline{(p \texttt{ <==> } q)}^{+} \cdots), \ldots\}$

```
let {bcond := (!force (p <==> q));
     cond  := (!left-iff bcond);                         [liff->]
```

```
      _ := (!force p)}
  (!mp cond p)
```

As should be expected, there is a corresponding `[riff->]` tactic.

These seven forward tactics will be collectively referred to as the *extraction tactics*. Given the goal of deriving some q from some β, these tactics do three things:

1. Find some element of β with a positive occurrence of a parent of the goal q, specifically, a positive occurrence of a sentence of the following forms:
   - `(p & q)` or `(q & p)`;
   - `(p ==> q)`;
   - `(p | q)` or `(q | p)`;
   - `(p <==> q)` or `(q <==> p)`.
2. Derive that parent, and possibly other additional subgoals.
3. Detach the goal q with the proper elimination method.

The following is an example that uses several of the forward proof tactics introduced so far. Suppose we want to find a deduction D that derives the goal `C` from $\beta = \{$`(A ==> B & C)`, `(A & E)`$\}$. The conjunction `(B & C)` is the only subsentence of β that is a parent of the goal `C`, and it has a positive occurrence in β, so β is actually of the form

$$\beta = \{\texttt{(A ==> }\underline{\texttt{(B \& C)}}^{+}\texttt{)}, \texttt{(A \& E)}\},$$

and therefore the tactic `[rand->]` is applicable, so we can proceed to the subgoal of finding a deduction D' that derives the conjunction `(B & C)` from β; once that is done, we can extract the goal `C` by `right-and`.

Does β contain any parents of the new goal `(B & C)` in positive positions? The answer is affirmative—the conditional `(A ==> B & C)` is a parent of `(B & C)` and it has a positive occurrence in β (since it is in fact an element of β). So β is of the form

$$\beta = \{\underline{\texttt{(A ==> B \& C)}}^{+}, \texttt{(A \& E)}\},$$

and hence the tactic `[if->]` is applicable. We thus generate two new subgoals:

1. Derive the conditional `(A ==> B & C)` from β.
2. Derive the conditional's antecedent, `A`, from $\beta \cup \{$`(A ==> B & C)`$\}$.

Once we derive these subgoals, we can obtain `(B & C)` by modus ponens.

Now the first subgoal can be derived directly by `claim` (see tactic `[claim->]`). What about the second subgoal, `A`? The assumption base β contains only one parent of `A` in a positive position, and the only tactic that can be applied to it is `[land->]`, so we apply that tactic and generate the new subgoal `(A & E)`, which follows immediately via `claim`.

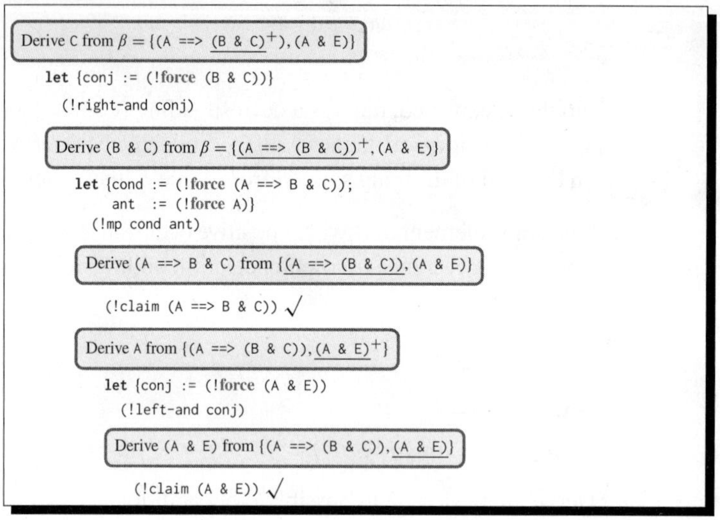

Derive C from β = {(A ==> (B & C)$^+$), (A & E)}

```
let {conj := (!force (B & C))}
  (!right-and conj)
```

Derive (B & C) from β = {(A ==> (B & C))$^+$, (A & E)}

```
let {cond := (!force (A ==> B & C));
     ant  := (!force A)}
  (!mp cond ant)
```

Derive (A ==> B & C) from {(A ==> (B & C)), (A & E)}

```
(!claim (A ==> B & C)) √
```

Derive A from {(A ==> (B & C)), (A & E)$^+$}

```
let {conj := (!force (A & E))
  (!left-and conj)
```

Derive (A & E) from {(A ==> (B & C)), (A & E)}

```
(!claim (A & E)) √
```

Figure 4.3
A goal tree generated exclusively by forward tactics.

Using the conventions we have laid down, the entire process is depicted in Figure 4.3. The following is the final assembled proof:

```
let {conj := let {cond := (!claim (A ==> B & C));
                  ant  := let {conj := (!claim (A & E))}
                            (!left-and conj)}
               (!mp cond ant)}
  (!right-and conj)
```

or, if we remove the extraneous claims:

```
let {conj := (!mp (A ==> B & C)
                  (!left-and (A & E)))}
  (!right-and conj)
```

These extraneous claims show up because, until now, the only primitive goals that we have allowed are proof specs where the assumption base contains the target sentence, which we can close off immediately by applying `claim`. For every other type of goal we would need to keep applying tactics that generate new subgoals. While that would be sensible for a computer implementation (of the kind we ask you to give in Exercise 4.40), it would be too tedious and restrictive for humans to always descend all the way down to `claim` applications. So in practice we will allow for *any* proof spec to count as primitive (to be a leaf node) as long as we produce a complete deduction for it, that is, as long as the given deduction contains no occurrences of **force**. By terminating the branches of our goal trees a little higher in this manner, we avoid proliferating `claim` applications in the final overall deduction.

A more general forward tactic involving disjunctions is the following:

Derive q from $\{\ldots,(\cdots\underline{(p_1 \mid p_2)}^{+}\cdots),\ldots\}$

```
let {disj := (!force (p1 | p2));
     cnd1 := (!force (p1 ==> q));                     [or->]
     cnd2 := (!force (p2 ==> q))}
  (!cases disj cnd1 cnd2)
```

This tactic subsumes `[lor->]` and `[ror->]`, both of which can be viewed as instances of `[or->]`. It can be used with arbitrary disjunctions in (positive positions of) the assumption base.

We close with a few tactics pertaining to contradictions. First, indirect proof, or proof by contradiction (we write $\overline{p}$ for the complement of p):

Derive p from β

```
(!by-contradiction p
  assume p̄                               [ift]
    (!force false))
```

When the goal p is a negation, this tactic reduces to the backward tactic `[not<-]`. But, unlike `[not<-]`, this tactic can be used even when the goal is not a negation.

Both `[ift]` and `[not<-]` seek to derive `false` from an assumption base. Are there any heuristics offering guidance on how to derive `false`? Clearly, if the assumption base contains two complementary sentences, then a single application of `absurd` will suffice, but what should be done in other cases? A simple heuristic is this: Look for a negation `(~` p`)` in a positive position in the assumption base, derive it, and then try to derive p:

Derive `false` from $\beta = \{\ldots,(\cdots\underline{(\sim p)}^{+}\cdots),\ldots\}$

```
let {neg := (!force (~ p));
     pos := (!force p);                      [false->]
  (!absurd pos neg)
```

The following tactic is similar and can be regarded as an instance of `[false->]`, but we list it separately for exposition purposes. The sentences p and $\overline{p}$ are complements of each other. If we already have $\overline{p}$ in the assumption base and we can also derive the positively embedded p, then clearly the assumption base is inconsistent and any goal whatsoever can be derived from it:

Derive q from $\beta = \{\ldots,\overline{p},\ldots,(\cdots p^{+}\cdots),\ldots\}$

```
let {_ := (!force p)}                          [cft]
  (!from-complements q p p̄)
```

This tactic is particularly apt when the assumption base directly contains p as well as its complement $\overline{p}$, in which case the tactic takes the following form:

Derive q from $\beta = \{\ldots,p,\ldots,(\sim p),\ldots\}$

```
(!from-complements q p (~ p))
```

4.14.3 Replacement tactics

Recall from Section 4.10 that we are allowed to replace any part of a sentence by an equivalent part. We can take advantage of such replacements both in a forward and in a backward direction. First, in a backward direction, we can transform the goal q to an equivalent goal q' that may be easier to tackle. If and when we succeed in deriving q', and assuming that q' and q are indeed equivalent in that each can be derived from the other by certain methods $M_1,\ldots,M_n$, we can then `transform` q' to the desired q:

Derive q from β

```
let {_ := (!force q')}                         [replace<-]
  (!transform q' q [M1 ··· Mn])
```

Likewise, in a forward direction, we can convert any element of the assumption base into an equivalent form and then proceed to tackle the goal anew:

Derive q from $\beta = \{\ldots,\underline{p},\ldots\}$

```
let {_ := (!transform p p' [M1 ··· Mn])}                 [replace->]
  (!force q)
```

4.14.4 Strategies for deploying the tactics

The tactics we have presented can be divided into three groups:

1. *Backward tactics*: These decompose the goal into simpler goals.
2. *Forward tactics*: These can be partitioned into the following subgroups:

   - *Reiteration tactic*: If the goal is already in the assumption base, just claim it.
   - *Extraction tactics*: Find a positive subsentence p of the assumption base that is a parent of the goal q, derive p (and possibly other subgoals), and then derive q from p (and possibly other sentences).

- *Generalized disjunction tactic*: Find some disjunction $(p_1 \mid p_2)$ that is positively embedded in the assumption base (preferably directly contained in it), derive the disjunction, and then try to show that each alternative p_i entails the goal.
- *Indirect tactic*: Assume the complement of the goal and try to derive `false`.
- *Contradiction tactics*: One of these, `[false->]`, aims to derive `false` from the assumption base, and is to be used in tandem with the indirect tactic. A related tactic, `[cft]`, derives an arbitrary goal by exploiting an inconsistent assumption base.

3. *Replacement tactics*: Transform the goal into an equivalent form that is hopefully easier to derive (`[replace<-]`), or transform the assumption base (`[replace->]`).

Note that extraction tactics do not seem able to handle two situations that, intuitively, they should handle. First, suppose that the goal is p and that a premise in the assumption base is of the form (~ $\underline{p}$ ==> q). The underlined occurrence of p in this premise is positive, since the entire conditional is equivalent to (p | q), or, alternatively, equivalent to (~ q ==> p); yet there is no extraction heuristic applicable in such a situation. Second, suppose that the goal is again p and that a premise is of the form (~ ~ p). Then there is no extraction tactic that can be applied to the premise, even though the occurrence of p here is positive and the premise is equivalent to p.

However, such cases can be handled by the forward replacement tactic. In the first example, we can `transform` the premise (~ p ==> q) to (p | q) or to (~ q ==> p), whichever we deem more promising, and then we can apply the corresponding extraction tactic to the result. For example:

Derive p from $\beta = \{\ldots, (\cdots \underline{(\sim p \texttt{ ==> } q)}^{+} \cdots), \ldots\}$

```
let {_ := (!transform (··· (~p ==> q)···)
                      (··· (~q ==> p)···) [contra-pos])}
  (!force p)
```

Likewise for double negations.

Keep in mind that when applying these tactics it is possible to start going in circles; one must be careful to avoid that. Specifically, a cycle occurs if, as you are working on a possible branch of the goal tree, you come upon a proof spec that you have already examined before on that same branch.[37] For example, consider the following proof spec:

Derive A from β = {(A ==> B), (B ==> A),...} (4.23)

37 Two proof specs are considered identical if they have identical goals and identical assumption bases.

Noting the positive occurrence of the parent `(B ==> A)`, we might choose to apply `[if->]`, thus generating the subgoal:

Derive `B` from `{(A ==> B), (B ==> A),...}`

If we now focus on the conditional `(A ==> B)` and choose to apply `[if->]` again, we will generate the subgoal

Derive `A` from `{(A ==> B), (B ==> A),...}`

which is, of course, identical to spec (4.23). If you find yourself in such a cycle you need to backtrack up to the first occurrence of the duplicate proof spec and try another avenue from that point forward (apply some other tactic to that spec). If that also goes nowhere, then we need to backtrack even further.

Cycles, of course, occur only in graphs, not in trees, so talk of cycles might appear peculiar given that we have spoken from the start of goal *trees*. In order to clarify this point, it is important to keep in mind the distinction between the process of searching for a goal tree and the tree itself. The process of searching for such a tree is most naturally depicted as a search over a graph G, which may indeed contain cycles. But if and when we obtain a solution, that solution will be a tree as we have defined it, embedded inside G. So a goal tree is essentially the desired outcome of searching the graph G.

In particular, given a proof spec τ, the objective is to construct a goal tree that has τ at the root and instances of `[claim->]` at all its leaves.[38] If we achieve this objective we say that we have *solved* the given goal. So how do we go about constructing such a tree? By searching the graph G_τ defined by τ. This graph, whose nodes are proof specs, is constructively defined by recursion as follows: If the goal is primitive, in that the desired conclusion is either a member of the given assumption base, or else `true` or `(~ false)`, then we apply the appropriate elementary tactic to it and mark the node as a leaf.[39] Otherwise, for *each* tactic t that is applicable to τ and generates n subgoals $\tau_1^t, \ldots, \tau_{n_t}^t$, there are n directed edges from τ to $\tau_1^t, \ldots, \tau_{n_t}^t$, and these edges are joined together by an arc to indicate that *all* subgoals $\tau_1^t, \ldots, \tau_{n_t}^t$ must be solved in order to solve τ. We then continue recursively,

38 For completeness, we should mention two additional backward tactics, for deriving `true` and `(~ false)`. The tactic for `true` is as follows:

Derive `true` from β

`(!true-intro)` $\checkmark$ `[true<-]`

The false-elimination tactic, `[false<-]`, is just as trivial and we omit its formulation. Goals to which these two tactics are applicable are therefore primitive (generate no new subgoals) and serve as leaves. Strictly speaking then, every leaf of a goal tree should correspond to an instance of `[claim->]` or to one of these two tactics, although, as we remarked earlier, in practice we are free to turn any proof spec into a leaf node (a primitive goal) as long as we give a complete deduction for it (one that does not contain any occurrences of **force**).

39 Thus, even if there are other tactics applicable to such goals, we do not apply them.

by constructing the directed graphs corresponding to the various τ_i^t.[40] For instance, suppose that there are exactly three tactics applicable to τ: t_1, t_2, and t_3. And suppose that t_1 creates three subgoals τ_1, τ_2 and τ_3; t_2 creates only one subgoal, τ_4; and t_3 creates two subgoals, τ_5 and τ_6. Then τ would have three groups of children, depicted as follows:

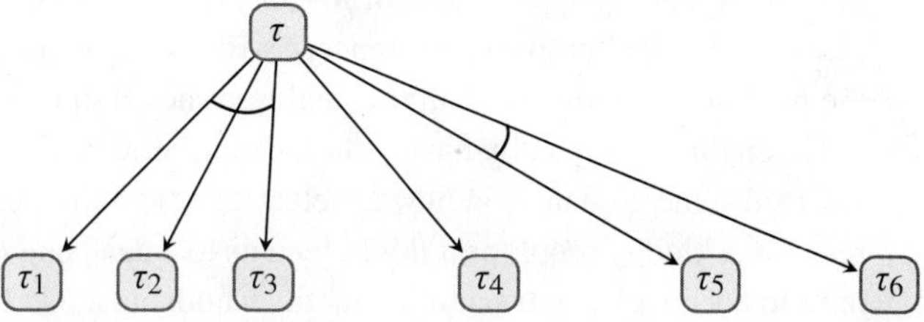

The arc joining the leftmost three edges signifies that *all* of the subgoals to which these edges point must be solved in order to solve τ. Likewise for the arc joining the rightmost two edges. The construction of G_τ would then proceed recursively with the generation of the graphs corresponding to τ_1—τ_6. Graph nodes without any children would correspond precisely to trivial goals: proof specs that can be solved by one of the few tactics that do not introduce any new subgoals.

Observe that we only need to solve *one* of these three groups of subgoals in order to solve τ. In other words, to solve τ we can:

- solve τ_1 *and* τ_2 *and* τ_3; OR
- solve τ_4; OR
- solve τ_5 *and* τ_6.

Moreover, every nonleaf node of the graph adheres to this structure: To solve the goal at such a node, we only need to solve (all of) the subgoals in *one* group of subgoals generated by applying some tactic to that node. Graphs with this sort of structure are called *AND/OR graphs* in artificial intelligence (see the chapter notes).

Formally, we say that a node τ in such a graph defines a goal tree iff one of two conditions is satisfied: Either the node has no children, or else there exists a group of k children of τ, call them $\tau_1, \ldots, \tau_k$, where the edges from τ to $\tau_1, \ldots, \tau_k$ are joined by an arc, meaning that $\tau_1, \ldots, \tau_k$ are all and only the subgoals generated by applying some tactic to τ, and, recursively, every τ_i defines a goal tree, $i = 1, \ldots, k$.

To recap then, given a proof spec τ, we need to choose strategically a tactic that is applicable to it, apply the tactic, thereby generating a number $n \geq 0$ of new proof specs (subgoals), and then repeat the process with every one of those. If at any point we feel like

40 This graph definition is *implicit*. The graph is not explicitly defined ahead of time, stored in a matrix or adjacency list (which would be impossible anyway for infinite state spaces), and then searched. Rather, the graph is incrementally constructed as needed, starting from the initial node and depending on the particular search strategy used.

we are not getting anywhere then we need to backtrack to some earlier goal and pursue an alternative tactic at that point. If and when we manage to construct a goal tree for τ as described above, we can then put together the desired top-level proof just by reading off, from bottom to top, the various deductions attached to the nodes, and substituting these for the corresponding applications of **force**. Typically we will perform a few simplifications to clean up the final proof, for instance, by removing extraneous claims. For an example, see the final proof produced from the goal tree shown in Figure 4.3.[41]

Given that we typically have a choice of several tactics that we can apply to a given goal, we need some guidance on how to select a tactic judiciously. An inauspicious choice might result in a kludgy proof, or it might lead us to a dead end or get us stuck in a cycle. Then we have to backtrack and try another tactic. Choosing cleverly which tactics to apply can avoid such pitfalls and lead to success quickly and elegantly. In what follows we list heuristics that provide some guidance on this selection task. These are not guaranteed to always work (which is why they are only heuristics after all), but they are generally helpful.

Heuristic 4.1: Tactic Ranking

Tactics should be considered for application in the following order:

- reiteration (`[claim->]`) and constant tactics (`[true<-]` and `[false<-]`);
- the complement tactic (`[cft]`);
- extraction tactics;
- replacement tactics;
- backward tactics, with the exception of `[not<-]`;
- generalized disjunction tactic;
- indirect tactic and `[not<-]`.

Replacement tactics should be used so as to make other tactics applicable, as discussed in Section 4.14.3. A particularly useful backward replacement is the transformation of a disjunctive goal into a conditional, via `cond-def`. Multiple extraction tactics might be applicable, so a selection algorithm is necessary for such cases. Specifically, when the assumption base has two or more parents of the goal in positive positions, choose the parent that is embedded in the least complex (smallest) sentence, unless there is a parent embedded in a pure conjunction, in which case that should be the selection. If two or more parents are positively embedded in sentences of the same complexity, break ties by choosing conjunctions over conditionals, conditionals over biconditionals, and any of these over disjunctions. If a

41 Athena proofs can be simplified mechanically; for the details of such a simplification algorithm refer to *Simplifying proofs in Fitch-style natural-deduction systems* [4].

cycle is encountered, backtrack to the first occurrence of the duplicate subgoal and try another tactic.

As another extended example, consider the problem of finding a proof *D* that derives the following goal from the given premises:

```
assert premise-1 := (A | (B ==> C))
assert premise-2 := (C <==> D & E)
assert premise-3 := (B & ~ A)

define goal := (B & D)
```

Let us use the strategy specified by Heuristic 4.1 in an almost-blind, algorithmic manner. The reiteration tactic is not applicable, and neither is `[cft]`. No extraction tactic can be applied either, since the goal `(B & D)` is not a subsentence of any premise. Moreover, applying the usual transformations (conditional definition, double negation, contrapositive, etc.), either in a backward or in a forward manner, would not make any extraction heuristic applicable. So, by the ranking heuristic, we move to consider backward tactics. Applying the only candidate, `[and<-]`, generates the partial deduction

```
let {left  := (!force B);
     right := (!force D)}
  (!both B D)
```

and the subgoals of deriving `B` from the three premises and `D` from the three premises along with `B`. Following the same strategy for the first subgoal leads us to use the extraction tactic `[land->]` on the third premise—we simply need to derive `(B & ~ A)` and then apply `left-and` to it. Since the derivation of `(B & ~ A)` is an immediate claim, we can express the entire deduction required to derive `B` as

```
(!left-and (B & ~ A)).
```

Note that this is the only extraction tactic applicable to the first subgoal, since the only other occurrence of `B` in the assumption base (in the first premise) is in a negative position.

We continue with the second subgoal of finding a proof that derives `D` from the three premises along with `B`, the result of the proof we found for the first subgoal. Following the same strategy, we rule out the first two entries in the ranking list of Heuristic 4.1 and go on to the extraction tactics. Now the goal `D` has only one positive occurrence in the assumption base, in the second premise, so we apply the corresponding extraction tactic, `[land->]`:

```
let {conj := (!force (D & E))}
  (!left-and conj)
```

We now have a new subgoal: Derive (D & E) from the three premises along with B. Again there is only one positive occurrence of this goal in the assumption base, again in the second premise, so we apply the corresponding extraction tactic, [liff->]:

```
let {bcond := (!force (C <==> D & E));
     cond := (!left-iff bcond);
     _ := (!force C)}
  (!mp cond C)
```

We recursively pursue the same strategy on the two new subgoals. For the subgoal of deriving the biconditional (C <==> D & E), the reiteration tactic is applicable and terminates this branch in success:

```
(!claim (C <==> D & E)).
```

Things get more interesting with the other subgoal, for deriving C. Once again we promptly rule out the first two entries of the ranking list and move to consider extraction tactics. Now, however, there are two positive occurrences of the goal C in the assumption base, one in the first premise and one in the second premise, the biconditional. However, we are currently in the midst of using that very biconditional in the reverse direction, trying to derive (D & E) via C, so attempting to derive C via (D & E) would get us in a cycle. Accordingly, we choose to pursue the first positive occurrence, applying [if->]:

```
let {cond := (!force (B ==> C));
     ant  := (!force B)}
  (!mp cond ant)
```

Now the B subgoal can be achieved immediately by reiteration, since B is already in the assumption base. For the (B ==> C) subgoal, the recommended strategy leads us to consider extraction tactics, and in this case there is only one positive occurrence of the subgoal (B ==> C) in the assumption base, in the first premise, so we use the corresponding tactic, [ror->]:

```
let {disj := (!force (A | (B ==> C)));
     cnd1 := (!force (A ==> B ==> C));
     cnd2 := assume (B | C) . (!claim (B | C))}
  (!cases disj cnd1 cnd2)
```

The first subgoal is immediate since the disjunction (A | (B ==> C)) is a premise, so we can just replace the first **force** application by a claim. So the only open subgoal left at this point is deriving (A ==> B ==> C) from the premises and previous intermediate conclusions. Again following the same strategy we end up resorting to the backward method for conditional introduction:

```
assume A
  (!force (B ==> C))
```

and we now need to derive `(B ==> C)` from the three premises and intermediate conclusions, along with the hypothesis `A`. While reiteration is not applicable here, the second-best tactic is, namely, `[cft]`, by noting that `A` is in the assumption base while its complement `(~ A)` is positively embedded in the third premise:

```
let {_ := (!force (~ A))}
  (!from-complements (B ==> C) A (~ A))
```

Finally, finding a proof that derives `(~ A)` is straightforward by applying the `[rand->]` tactic and detaching it from the third premise.[42] At this point we have successfully completed a goal tree and we can put together the top-level proof by plugging in the various discovered subproofs:

```
let {left := (!left-and (B & ~ A));
     right :=
       let {conj :=
              let {bcond := (!claim (C <==> D & E));
                   _ := let {disj := (!claim (A | (B ==> C)));
                             cnd1 := assume A
                                       (!from-complements
                                          (B ==> C)
                                          A
                                          (!right-and premise-3));
                             cnd2 := assume (B ==> C)
                                       (!claim (B ==> C));
                             cond := (!cases disj cnd1 cnd2)}
                        (!mp cond B)}
                (!mp (!left-iff bcond) C)}
         (!left-and conj)}
  (!both B D)
```

or, after a few simplifications and some general cleaning up:

```
let {left := (!left-and (B & ~ A));
     right := let {B=>C := (!cases (A | (B ==> C))
                                   assume A
                                     (!from-complements
                                        (B ==> C)
                                        A
                                        (!right-and premise-3))
                                   assume h := (B ==> C)
                                     (!claim h));
                   _ := (!mp B=>C B);
                 D&E := (!mp conclude (C ==> D & E)
                                   (!left-iff premise-2)
```

42 Note that if we tried to apply `[cft]` again we would get in a cycle. Indeed, if we were following the advice of Heuristic 4.1 completely blindly, we would enter the cycle and then we would be forced to backtrack and apply the `[rand->]` tactic.

```
                                C)}
                 (!left-and D&E)}
   (!both B D)
```

Exercise 4.2: Find a proof D that derives the following goal from the given premises:

```
define premise-1 := (A | B ==> C)
define premise-2 := (D | A | B)
define premise-3 := (~ D)
define premise-4 := (~ E ==> D)

define goal := (C <==> E)
```

Try to follow Heuristic 4.1 as closely as possible. □

Another useful heuristic is this:

Heuristic 4.2: Hypothesis Matching

When a hypothesis p is first made in a conditional proof of the form

assume p ···

scan the assumption base for premises of the form (p ==> q). If you find any, it will often be helpful to derive q by modus ponens before continuing.

After gaining some experience, you will notice that the derivation of negations is one of the most difficult proof tasks. If a goal of the form (~ p) cannot be reiterated; and there is no obvious contradiction from which it can be obtained (via [cft]); and there is no extraction heuristic that we can apply to it; then the only other two options we have, outside of a generalized case analysis, are (a) transforming the goal by a replacement tactic; or (b) reasoning by contradiction. The first of these two options can help if the body p is a compound sentence, in which case the negation sign can be pushed inward, yielding a new goal that is no longer a negation. For instance, if the goal is a negation of the form (~ (p_1 | p_2)), we can apply De Morgan's to obtain the new goal (~ p_1 & ~ p_2); if the goal is a negation of the form (~ (p_1 ==> p_2)), we can apply neg-cond to it, transforming it into (p_1 & ~ p_2); and so on. However, eventually we may be confronted with negated goals in which the negation sign is in front of an atom and cannot be moved any further, and which are not amenable to direct tactics either. In such cases we have to proceed by contradiction. This is generally a challenge because we only have one tactic for guidance here, [false->], and it is not very precisely targeted. In what follows we formulate another heuristic for handling negations that can often be a useful alternative to proof by contradiction. The basic idea is to try to massage the assumption base so as to make an extraction tactic applicable to the goal (~ p).

We have already assumed that (~ p) does not have any positive occurrences in the assumption base. However, the assumption base might contain negative occurrences of p, and that can be just as good. To take a simple example, suppose we want to derive `(~ B)` from the two premises `(B ==> A)` and `(~ A)`. There is no occurrence of `(~ B)` in the premises, so our only option is to proceed by contradiction. Now, in this simple case our heuristics would easily discover a proof,[43] but let us disregard that for the sake of the example. How can we proceed without reasoning by contradiction? We can see that the body of the desired negation, the atom `B`, has a *negative* occurrence in the assumption base, in the antecedent of the conditional `(B ==> A)`. The idea now is to transform that conditional so as to make the negation of `B` explicit, that is, so as to make a negation sign appear in front of `B`. Such a transformation will force `(~ B)` to appear in a *positive* position, since the overall polarities must remain unchanged, at which point the usual extraction tactics can take over. In this particular case, the proper transformation tactic to apply is the contrapositive, which will replace `(B ==> A)` by `(~ A ==> ~ B)`. Then `[if->]` will finish the job.

More specifically, when we are trying to derive (~ p) from an assumption base β, we find the smallest positive subsentence of β in which the body p occurs negatively; call that subsentence occurrence q. We then transform q into some sentence q' according to the rules given below. We may need to delve inside q' afterward and repeat this process until p is directly negated, or until the process is no longer applicable. The rules that dictate how q is to be transformed are shown in Figure 4.4. Inspection will show that q could not possibly be of any other form, and hence that these rules cover all possible cases. For instance, q could not be an atom, because we have assumed that q contains a negative occurrence of p, and that is impossible if q is an atom. Further, q could not be a conjunction (p_1 & p_2) or disjunction (p_1 | p_2) because we have assumed that it is the *smallest* positive subsentence of the assumption base in which p occurs negatively; this would be false if either p_1 or p_2 contained an occurrence of p. For similar reasons, q cannot be a negation of an atom or the negation of a negation. All remaining cases are handled by the given rules.

As another simple example, suppose we want to derive `(~ B)` from

```
(~ (A ==> B | C)).                                    (4.24)
```

Again, none of the first six tactics listed by Heuristic 4.1 are applicable, so we would have to proceed by contradiction. But now we can apply the heuristic we just described. The smallest positive (occurrence of a) subsentence of

```
{(~ (A ==> B | C))}
```

43 Once the assumption `B` was made, hypothesis matching would immediately derive `A`, and then `[cft]` would derive `false`.

(~ (p_1 & p_2))	⟶	(~ p_1 | ~ p_2)	(dm)
(~ (p_1 | p_2))	⟶	(~ p_1 & ~ p_2)	(dm)
(~ (p_1 ==> p_2))	⟶	(p_1 & ~ p_2)	(neg-cond)
(~ (p_1 <==> p_2))	⟶	((p_1 & ~ p_2) | (p_2 & ~ p_1))	(negated-bicond)
(p_1 ==> p_2)	⟶	(~ p_2 ==> ~ p_1)	(contra-pos)
(p_1 <==> p_2)	⟶	(~ p_1 <==> ~ p_2)	(neg-iff)

Figure 4.4
Heuristic negation transformations.

that contains a negative occurrence of B (the body of the desired negation) is the entire premise itself, (4.24). Transforming the premise by neg-cond yields

(A & ~ (B | C)). (4.25)

This still does not contain the desired negation, so we apply the heuristic again. This time, the smallest positive subsentence of the assumption base that contains a negative occurrence of B is (~ (B | C)). Applying the appropriate rule (dm) will transform (4.25) to

(A & ~ B & ~ C), (4.26)

and at this point we have a positive occurrence of the goal (~ B) and we can apply the usual extraction tactics. In this case, two successive conjunction eliminations is all we need.

We are now ready to formulate our negation heuristic more precisely as follows:

Heuristic 4.3: Proving Negations Directly

When the goal is the derivation of a negation (~ p) from β and the only recourse left by Heuristic 4.1 is proof by contradiction, consider trying the following: Find the smallest positive subsentence of β that contains a negative occurrence of p; call it q. Transform the premise that contains q by replacing q with whatever is prescribed by the rules in Figure 4.4. Repeat this process until the result contains a positive occurrence of (~ p) or until the process is no longer applicable. If the process is not applicable, then, if the body of the negation is compound, push the negation sign inward by applying the appropriate replacement tactic; otherwise proceed by way of contradiction.

Let us demonstrate with a larger example:

```
assert premise-1 := (B & D)
assert premise-2 := (B & ~ A ==> ~ C)
assert premise-3 := (B ==> ~ A)
assert premise-4 := (D & E ==> A | C)
```

```
define goal := (~ E)
```

None of the direct tactics apply here, so we are faced with the situation described by Heuristic 4.3. The smallest positive subsentence of a premise that contains a negative occurrence of the body `E` is the fourth premise itself, so we apply the corresponding transformation rule to it, the contrapositive, obtaining

`(~ (A | C) ==> ~ (D & E)).` (4.27)

We again don't have a positive occurrence of the goal `(~ E)`, so we repeat the process. This time the smallest positive subsentence of the assumption base that has a negative occurrence of `E` is the consequent of (4.27), and applying the corresponding transformation to it we obtain the following from (4.27):

`(~ (A | C) ==> ~ D | ~ E).` (4.28)

At this point we have a positive occurrence of the goal `(~ E)`, and the extraction tactic `[ror->]` is applicable. Accordingly, we attempt to infer the disjunction

`(~ D | ~ E)` (4.29)

and then we try to derive `(~ E)` from it by a case analysis. The derivation of `(~ E)` from (4.29) via case analysis is straightforward, so let us focus on the derivation of (4.29) itself. That can be obtained from (4.27) by the `[if->]` tactic, provided we can derive the antecedent `(~ (A | C))`. Now this subgoal does not have any positive occurrences in the assumption base, or any negative occurrences in positive positions, so the only alternative left is to decompose the negated goal by applying De Morgan's, which yields the conjunctive goal

`(~ A & ~ C).` (4.30)

Now applying the backward tactic `[and<-]` will yield the two subgoals `(~ A)` and `(~ C)`, both of which are readily handled by direct tactics.

Exercise 4.3: Derive the following goal from the given premises:

```
assert premises := [(A ==> B)
                    (C ==> E)
                    (~ E & A)
                    (~ C ==> B & D)]

define goal := (B & D)
```

Use the heuristics to guide your proof search. □

Exercise 4.4: Give proofs deriving the following sentences from the respective assumption bases:

(a) `(A ==> B ==> C)` from `{(A & B ==> C)}`.

(b) `(A ==> ~ E ==> ~ C)` from `{(A | B ==> C | D ==> E)}`. □

Exercise 4.5: Revisit your solution to the previous exercise. In each case, reconstruct your proof-discovery process in terms of backward and forward tactics, and analyze that process in the graph-search framework described in this section. □

4.15 ⋆ A theorem prover for sentential logic

In this section we implement a unary method, `prop-taut`, that can derive an arbitrary sentential tautology. Specifically, for any given zero-order sentence p,

`(!prop-taut` p`)`

will derive p if p is a tautology and will fail otherwise. This specification should be understood modulo computing resources such as time and memory. Given sufficient quantities of these, `prop-taut` will satisfy its specification. But no performance or efficiency claims are made, and indeed this particular method is not suitable for very large or complicated tautologies. But it is easily defined and understood, and perfectly adequate for the uses that we will make of it, mainly as a justifier for small steps of implication chains, steps that would be tedious to justify otherwise.

Note that the related problem of determining whether a sentence p follows from (is a logical consequence of) sentences $p_1, \ldots, p_n$ can be reduced to the problem of determining whether a single sentence is a tautology, because the entailment in question holds iff the conditional

$$(p_1 \ \& \ p_2 \ \& \ \cdots \ \& \ p_n \ \text{==>} \ \text{p})$$

is a tautology (Theorem 4.4). Readers who are not interested in the implementation of `prop-taut` may skip the remainder of this section.

Our implementation is based on the technique of *semantic tableaux*, which is used to determine whether a finite set of sentences is satisfiable.[44] Here we will be representing sets as lists, so the technique will be operating on lists of sentences. Because p is a tautology iff `(~` p`)` is unsatisfiable, we can use this technique to determine whether a given p is a tautology by determining whether the singleton list `[(~` p`)]` is satisfiable. If it is, then p is not a tautology, because its negation is coherent: There is a state of affairs—an interpretation—that falsifies p, that is, one satisfying the negation of p. On the other hand, if `[(~` p`)]` is *not* satisfiable then the negation of p is essentially incoherent; that is, there is no way we can interpret it that would make it true, which means that p itself is a tautology.

44 Recall that a set of sentences is satisfiable iff there is an interpretation that satisfies every element of the set.

More than that, if a list of sentences L *is* satisfiable, then this technique can actually give us an interpretation that satisfies each element of L. This interpretation will be represented as a collection of literals, where positive literals are understood to be mapped to "true" and the bodies of negative literals to "false." So the technique can be used as a satisfiability solver in the sense of Section 4.13, although for efficiency reasons, when tackling large or structurally complex sentences we recommend using Athena's primitive `sat-solve` procedure, which is implemented via external SAT solvers.

The basic operation of semantic tableaux can be understood as the incremental construction and exploration of a tree, each node of which contains a finite list of sentences. Internal nodes contain at least one sentence that is not a literal, whereas leaf nodes contain only literals. The root node contains the starting—input—list of sentences L_0 whose satisfiability we are trying to determine. The computation starts at that root node. At each node, we proceed as follows. If the node is a leaf (i.e., if it contains only literals), then we simply determine whether or not it contains two contradictory literals, and if so, we say that that particular branch (path from the root to the leaf node in question) is *closed*. This is often indicated by placing a symbol like ✗ right underneath the leaf. If, on the other hand, the leaf node does not contain any contradictory literals, then our search is over. In that case the initial list of sentences L_0 at the root is satisfiable, and indeed the set of literals appearing at the leaf node constitutes a satisfying interpretation of L_0. We then say that the branch from the root to the leaf in question is *open*, and this is often indicated by placing the mark ✓ right under the leaf. If we want additional interpretations of L_0, we can continue expanding the tree to try to find another open branch, but typically we are interested only in whether an open branch exists at all, so usually we stop the search if and when we find such a branch.

If, on the other hand, the current node u is not a leaf, that is, if the list L at u contains at least one nonliteral sentence, then we select any such sentence p and we expand the tree as follows. If p is a conjunctive sentence of the form `(`p_1 `&` p_2`)`, then we simply add to u a single child node u', whose sentences are those of L but with p replaced by p_1 and p_2 (added to the list of u' as two distinct elements). Whereas if p is a disjunction `(`p_1 `|` p_2`)`, then we attach *two* children u_1 and u_2 to u, where the sentences of u_1 are those of L but with p replaced by p_1, while the sentences of u_2 are those of L but with p replaced by p_2. We then continue the procedure recursively by expanding remaining nonleaf nodes and marking leaf nodes as open or closed in the manner we just described.

Of course there are other possibilities for what the selected nonliteral sentence p can be, other than a conjunction or disjunction. But these other possibilities ultimately lead to the same conjunctive or disjunctive analysis. For instance, if p is a conditional `(`p_1 `==>` p_2`)` then we treat it as the disjunction `(~` p_1 `|` p_2`)`. More precisely, we add to u a single new child u' with the same sentences as those of L but with p replaced by `(~` p_1 `|` p_2`)`. And if it is a biconditional `(`p_1 `<==>` p_2`)`, we treat it as the conjunction `(`p_1 `==>` p_2 `&` p_2 `==>` p_1`)`. Finally, if p is a negation, then it must be a negation of a nonatomic sentence q (for if q were an atom then p would be a literal, and we have assumed otherwise). So we now proceed by a case analysis of q, which again boils down to the same conjunctive or disjunctive treatment. If q is a disjunction (q_1 | q_2), then we treat p as the conjunction (~ q_1 & ~ q_2), so we add to u a single new child u' with the same sentences as those of L but with p replaced by

$$(\sim q_1 \ \& \ \sim q_2), \tag{4.31}$$

since we know by De Morgan's law that the negation of (q_1 | q_2) is equivalent to (4.31). If q is a conjunction (q_1 & q_2), then we treat p as the disjunction (~ q_1 | ~ q_2), again on the grounds of De Morgan's law. If q is itself a negation, so that p is of the form (~ ~ r), then we add to u a single new child with the same sentences but with p replaced by r.

It is easy to generalize this procedure, as we do in our implementation, to handle polyadic rather than just binary conjunctions and disjunctions. If p is a polyadic conjunction with n immediate subsentences, then we replace p in the list of the new child node with all n of these immediate subsentences. And if p is a polyadic disjunction with n immediate subsentences $p_1, \ldots, p_n$, then we add n new children nodes $u_1, \ldots, u_n$ to the current node u, each u_i having the same sentences as those of L but with p replaced by p_i.

It is not too difficult to show that every time this algorithm expands an internal node by adding new children to it, satisfiability is preserved. For any node u, let us write L_u for the list of sentences appearing at that node. Then, more precisely, if we expand a node u by selecting a conjunctive sentence in L_u and add a single new child u' to u in accordance with the algorithm, then $L_{u'}$ is satisfiable iff L_u is. And if we expand u by selecting a disjunctive sentence in L_u and add n new children $u_1, \ldots, u_n$ to u in accordance with the algorithm, then L_u is satisfiable iff at least one L_{u_i} is satisfiable. Thus, it follows by induction on the depth of the tree that L_0 is satisfiable iff at least one leaf list is satisfiable. If every leaf list is unsatisfiable (i.e., every branch is closed), then L_0 is unsatisfiable. Essentially, this inductive argument will be constructively embedded in the recursive definition of `prop-taut`.

As an example, suppose we want to find out whether the conclusion `(~ A)` is logically entailed by (follows from) the two sentences

```
(A | B ==> C & D)
```

and `(~ C)`. The entailment holds iff appending the negation of the conclusion to the list of the two given sentences results in an unsatisfiable list of sentences. In other words, the entailment holds iff the list

```
[(A | B ==> C & D), (~ C), (~ ~ A)]
```

is jointly unsatisfiable, that is, iff there is no interpretation that satisfies every element of this list. And that condition can be decided by the semantic tableaux procedure we just discussed. The resulting tree is shown in Figure 4.5. To facilitate reference, we have annotated each node of the tree with its Dewey path, that is, a list of integers indicating the position

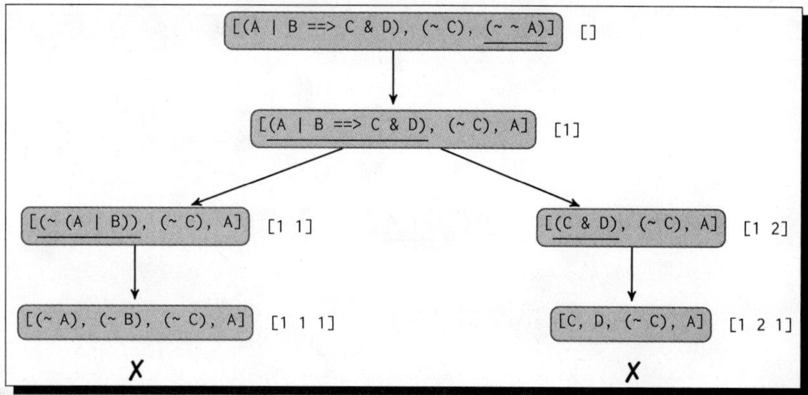

Figure 4.5
A sample semantic tableau with no open branches, indicating unsatisfiability.

of the node in the tree (see page 118 for more details on Dewey paths). We have also underlined the complex sentence we have picked at each node to expand the tree. Note that in the passage from node [1 1] to node [1 1 1] we directly replaced `(~ (A | B))` by the two sentences `(~ A)` and `(~ B)`, instead of first replacing it with the conjunction `(~ A & ~ B)` and then breaking the conjunction apart in a separate step. This is just a convenient shortcut, and our implementation below takes similar shortcuts when appropriate. Likewise, we branched directly from `[1]` to `[1 1]` and `[1 2]` instead of first replacing the conditional

```
(A | B ==> C & D)
```

by the explicit disjunction `(~ (A | B) | (C & D))` and then branching on that.

Now consider the same example except that the first premise is

```
(A & B ==> C & D)
```

rather than `(A | B ==> C & D)`. If we apply the same technique to the initial list of sentences

```
[(A & B ==> C & D), (~ C), (~ ~ A)],
```

we will manage to find a satisfying interpretation, namely {`A`, `(~ B)`, `(~ C)`}. The details, including the open branch that ends in this interpretation, are shown in Figure 4.6. (We have omitted the Dewey paths this time.)

Our implementation is packaged inside a module and is expressed as a procedure (we will present the method version shortly). The main procedure is `sat`, which takes a list of sentences *L* and determines whether they are satisfiable by continually expanding the tree rooted at *L* until it finds an open branch or until it has determined that all branches are

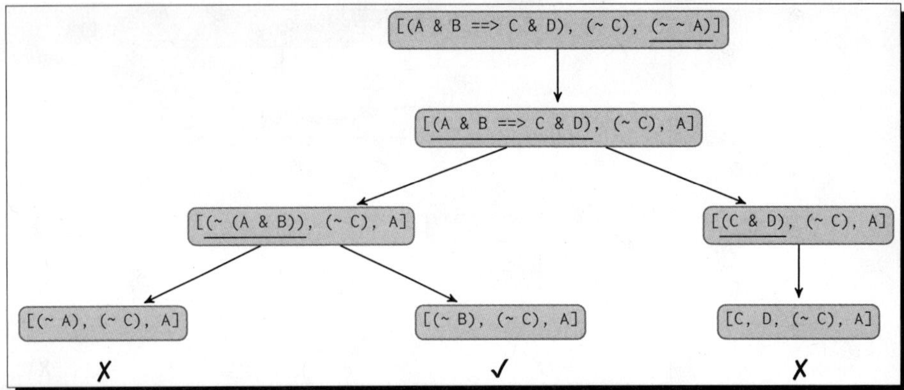

Figure 4.6
A sample semantic tableau with an open branch, indicating satisfiability.

closed. The result of sat is either false, signifying that L is not satisfiable (because all branches were closed), or else a list of literals representing an interpretation that satisfies every element of L.

```
module Sentential-Semantic-Tableaux {

define (conjunctive? p) :=
  match p {
    (and (some-list L)) => L
  | (~ (~ q)) => [q]
  | (~ (p ==> q)) => [p (~ q)]
  | (~ (or (some-list L))) => (map ~ L)
  | (p <==> q) => [(p ==> q) (q ==> p)]
  | _ => ()
  }

define (disjunctive? p) :=
  match p {
    (or (some-list L)) => L
  | (p ==> q) => [(~ p) q]
  | (~ (p <==> q)) => [(p & ~ q) (q & ~ p)]
  | (~ (and (some-list L))) => (map ~ L)
  | _ => ()
  }

define (inconsistent? L) :=
  (for-some L lambda (p) (|| (member? p [false (not true)])
                            (member? (~ p) L)))
```

```
define (literal? p) :=
  match p {
    (|| (some-atom _) (~ (some-atom _))) => true
  | _  => false
  }

define (non-literal? p) := (negate literal? p)

define (sat L) :=
  match L {
    (split L1 [(p where (non-literal? p))] L2) =>
      match (conjunctive? p) {
        (some-list args) => (sat (join L1 args L2))
      | _ => match (disjunctive? p) {
               (some-list args) =>
                 (find-some-element args
                                    (unequal-to false)
                                    lambda (d) (sat (join L1 [d] L2))
                                    lambda (x) x
                                    lambda () false)
             }
      }
  | _ => check { (inconsistent? L) => false
               | else => L}
  }

define (tautology? p) :=
  match (sat [(~ p)]) {
    false => true         # (~ p) is unsatisfiable, so p is a tautology
  | _ => false            # (~ p) is satisfiable , so p is not a tautology
  }

define (follows? p L) :=
  (tautology? (if (& L) p))

} # close module Sentential-Semantic-Tableaux
```

The module also contains a few helper procedures such as `tautology?`, for determining whether a sentence is a tautology, and `follows?`, for determining whether a sentence follows from a list of sentences. Both of these are defined in terms of `sat`. The procedure `find-some-element` is a generic list procedure (defined in Athena's `list` library file)[45] that takes a list *L*; a predicate *P* (by "predicate" here we mean any unary procedure that always returns `true` or `false`); a unary procedure *f*; and two continuations *S* and *F* (for "success"

45 The similar-sounding `find-first-element` that we defined earlier (page 234) is a binary *method*, whereas `find-some-element` is a *procedure*, which also takes a different number of arguments. Note that, while similar, `find-some-element` is also different from the `find-element` procedure on page 247, as it has a more flexible interface. (In particular, `find-element` could easily be implemented in terms of `find-some-element`.)

and "failure" respectively), the first unary and the second nullary. The procedure finds the first element x of L such that $(P\ (f\ x))$ holds (i.e., returns `true`), *if* such an element x exists in L at all. If so, then $(S\ (f\ x))$ is returned. If no such x exists, then (F) is returned. In this case, the first disjunct d is found such that `(sat (join L1 [`d`] L2))` is `unequal-to` `false`, in which case it must be a list (i.e., a satisfying interpretation). If such a disjunct d is found, then the interpretation `(sat (join L1 [`d`] L2))` is returned. If there is no such d, then `false` is returned. The `unequal-to` procedure is a simple curried utility procedure that takes a value x and returns a unary procedure that takes any y and returns `true` if y is not `equal?` to x and false otherwise. It is also defined in Athena's standard `util` library as follows:

```
define (unequal-to x) := lambda (y) (unequal? x y)

> let {f := (unequal-to 1)}
    (f 5)

Term: true

> ((unequal-to 1) 1)

Term: false
```

We illustrate with a few examples of these procedures in action:

```
declare A,B,C,D,E: Boolean

define L := [(~ (A ==> B)) (C & ~ D) (~ D ==> B)]

> (sat L)

Term: false

> (sat [(A ==> B) (C & ~ D) (~ D ==> B)])

List: [(not A) C (not D) B]

> (tautology? (A & B ==> B & A))

Term: true

> (tautology? (A & B ==> B & C))

Term: false

> (follows? B [(~ B ==> ~ C) (~ E | C) E])

Term: true
```

There are some sources of inefficiency in this implementation. Inconsistency for a list of literals L is checked by examining each element of L to see whether its negation is also in L. This algorithm is of quadratic worst-case asymptotic complexity in the length of L. We can achieve linear-time performance on average by using hash tables. Another source of inefficiency is the use of **split** patterns to find a complex (nonliteral) sentence to decompose on each iteration. This search requires traversing the current list, which can be expensive if the list is long. It is better to make `sat` take *two* lists as arguments, a list of complex sentences and a list of literals. That way we don't need to search for a complex sentence—we simply take the first element of the list of complex sentences, assuming there is one. If not, that means we have run out of complex sentences and therefore we are at a leaf, at which point we examine the second list (the one that only contains literals) for inconsistency. An additional improvement would be to further divide the list of literals into two lists, one containing positive and the other negative literals. That would speed up inconsistency checks even further. With these changes, the performance of the method can be decent as long as the starting list does not contain a lot of redundant information. If the starting list is unsatisfiable but contains hundreds or thousands of sentences, only a small number of which are responsible for the unsatisfiability, then performance will suffer greatly because the procedure may end up needlessly decomposing a whole lot of sentences, creating many choice points along the way—a deep and bushy tree—that can take up a lot of memory to store (and time too, of course, to examine all the branches). There are techniques for dealing with that issue as well. For one thing, it is not necessary to wait until we have exhausted all complex sentences in order to check for unsatisfiability, as we may well get a contradiction at a much earlier point. Using slightly smarter data structures at each node can allow for quick and frequent inconsistency checks. Second, we do not need to be blindly picking complex sentences to decompose, especially if we are given a specific goal and a list of premises. We could then pick premises that are somehow related to the goal. There are various notions of relatedness that could be used for that purpose. Such improvements, however, are beyond the scope of this discussion.

We now proceed to develop the method `prop-taut` that we mentioned in the beginning of this section. Our implementation so far has consisted only of *procedures*, which represent computations and therefore do not need to justify their operation, unlike methods. The procedures we have defined so far, and `sat` in particular, essentially perform a top-down *analysis* of the given list of sentences, an analysis that stops when we reach the leaves of the search tree—nodes that contain literals only. The method we are about to implement will perform the same top-down analysis but will follow that up with a bottom-up *synthesis* of a deductive argument that will justify the analysis and produce an actual theorem. This pattern of recursion-driven top-down analysis followed by bottom-up proof synthesis is typical of many complex Athena methods.

The theorem in this case will be the constant `false`. The method is called `refute`, and its only input is a list of premises L, meaning that each element of L must be in the assumption base when `refute` is applied. The method attempts to show that the premises in L are inconsistent by deriving a contradiction (`false`) from them. If L is satisfiable, then `refute` will fail.

The method's control flow is fairly simple. We first try to find a complex sentence in L. If we are not able to do that because L contains only literals, then the method `inconsistent-literals` is applied to it, which will derive `false` if its input list of literals is inconsistent and will fail otherwise, which will make `refute` itself fail. (Recall that a consistent set of literals at a leaf indicates satisfiability of the initial list of premises at the root, and `refute` is expected to fail when its given premises are satisfiable.) If we do find a complex sentence p in L, then p is either conjunctive or disjunctive and is respectively handled by `conjunctive-case` and `disjunctive-case`. Both of these are written in *proof-continuation-passing style*, meaning that both of them take as a trailing argument a proof continuation, namely, a method K that receives intermediate results derived by `conjunctive-case` (or `disjunctive-case`) and then performs the rest of the required reasoning in an assumption base that contains those results. This is necessary to maintain the invariant that every time `refute` is called, *all the sentences in its input list are in the assumption base*.

More precisely,

(`!conjunctive-case` p K)

will derive all conjunctive components of p, call them $p_1, \ldots, p_k$, and will then tail-recursively call the unary continuation K with the list `[`$p_1 \cdots p_k$`]` as its argument. By the "conjunctive components" of p we mean those sentences that would appear in the output list of the call (`conjunctive?` p) in accordance with the earlier definition of `conjunctive?`. Thus, if p is a conjunction, then its conjunctive components are its immediate subsentences; if p is a double negation (`~` (`~` q)), then its only conjunctive component is q; if p is of the form (`~` (p_1 `==>` p_2)) then its conjunctive components are p_1 and (`~` p_2); and so on. In the procedural implementation all we had to do was list those components, but in the method implementation we actually have to *derive* these first and then pass them on to the continuation. Note that if p does not match any of the five patterns in the body of `conjunctive-case`, then the method will fail, at which point the `try` in the body of `refute` will pass control to `disjunctive-case`.

Now the call

(`!disjunctive-case` p K)

will derive a (polyadic) disjunction from p, let's say (`or` $p_1 \cdots p_k$), and will then make a tail-recursive call to the binary continuation K with the derived disjunction (`or` $p_1 \cdots p_k$) as the first argument and the list of disjuncts `[`$p_1 \cdots p_k$`]` as the second argument:

(`!K` (`or` $p_1 \cdots p_k$) `[`$p_1 \cdots p_k$`]`).

This second argument is convenient but not necessary. We could make K receive the derived disjunction only, from which the disjuncts could then be easily extracted downstream. But the present convention makes this further processing unnecessary. The polyadic disjunction derived from p depends on the form of p, as shown in the definition of `disjunctive-case`: If p is itself a disjunction then we do not need to derive anything, we simply pass p and its disjuncts to the continuation; if p is a conditional `(`q `==>` r`)`, we derive from it the disjunction `(~` q `|` r`)` and then we pass that disjunction to the continuation, along with the list of its two disjuncts; and so on.

```
define (conjunctive-case p K) :=
  match p {
    (and (some-list _)) => (!decompose p K)

  | (q <==> r) => let {p1 := (!left-iff p);
                       p2 := (!right-iff p)}
                    (!K [p1 p2])

  | (~ (~ _)) => let {q := (!dn p)}
                   (!K [q])

  | (~ (or (some-list _))) => (!decompose (!dm p) K)

  | (~ (q ==> r)) => let {p1 := conclude q
                                  (!neg-cond1 p);
                          p2 := conclude (~ r)
                                  (!neg-cond2 p)}
                       (!K [p1 p2])
  }

define (disjunctive-case p K) :=
  match p {
    (or (some-list L)) => (!K p L)
  | (q ==> r) => let {q'|r := (!cond-def p)}
                   (!K q'|r [(complement q) r])
  | (~ (and (some-list L))) => (!K (!dm p) (map complement L))
  | (~ (q <==> r)) => (!K (!negated-bicond p) [(q & (complement r))
                                                ((complement q) & r)])
  }


define (inconsistent-literals L) :=
  match L {
    ((split _ [p] _ [q] _) where (complements? p q)) =>
      (!from-complements false p q)
  }

define (refute L) :=
  match L {
    (split L1 [(p where (non-literal? p))] L2) =>
```

```
    try { (!conjunctive-case p
               method (components)
                  (!refute (join L1 components L2)))
        | (!disjunctive-case p
               method (disjunction disjuncts)
                 (!map-method method (d)
                                assume d
                                  (!refute (join L1 [d] L2))
                             disjuncts
                             method (conds)
                               (!cases disjunction conds)))}
  | _ => (!inconsistent-literals L)
  }
```

With refute under our belt, prop-taut is easily defined:

```
define (prop-taut p) :=
  (!by-contradiction p
    assume -p := (~ p)
      (!refute [-p]))
```

4.16 Additional exercises

Your solutions to these problems may use any of the library methods in Section 4.9.

Exercise 4.6: Consider the following equivalences:

(a) (*p* & false <==> false)
(b) (*p* & true <==> *p*)
(c) (*p* | false <==> *p*)
(d) (*p* | true <==> true)
(e) ((true ==> *p*) <==> *p*)
(f) ((false ==> *p*) <==> true)
(g) ((*p* ==> true) <==> true)
(h) ((*p* ==> false) <==> ~ *p*)

For each of these, write a unary method that takes an arbitrary sentence *p* as input and proves the corresponding biconditional. □

Exercise 4.7: Prove: (B ==> ~ (A ==> B) ==> C). □

Exercise 4.8: For each of the following:

(a) {(B | (A ==> B)), A} ⊢ B
(b) {(A ==> B), (A ==> C)} ⊢ (A ==> (B & C))

(c) `{(A ==> B), (C ==> D)}` ⊢ `((A | C) ==> (B | D))`
(d) `{(A ==> (B ==> C)), (A ==> (C ==> D))}` ⊢ `(A ==> (B ==> D))`

clear the assumption base, **`assert`** the sentences that appear inside the curly braces, and then give a proof that derives the sentence on the right-hand side of the turnstile. □

Exercise 4.9: Implement a ternary method `find-some` that is similar to the binary method `find-first-element` (defined on page 234), except that (a) it takes the list argument first and the method argument second; and (b) it also takes a third argument, a nullary proof continuation K that is invoked if the application of the given method fails on every element of the given list. □

Exercise 4.10: Implement `from-false` via `from-complements` and vice versa. □

Exercise 4.11: Consider these two biconditionals:

(a) `((p & p) <==> p)`.
(b) `((p | p) <==> p)`.

Write two methods, `conj-idemp` and `disj-idemp`, that take an arbitrary sentence p as input and derive the corresponding biconditionals. □

Exercise 4.12: As in Exercise 4.8, for each of the following parts:

(a) `{(A ==> B), (~ A ==> C), (C ==> D)}` ⊢ `(B | D)`
(b) `{(A ==> B), (B ==> C), (~ C)}` ⊢ `(~ (A | B))`
(c) `{(A & B ==> C), (C | ~ B ==> D)}` ⊢ `(A ==> D)`

clear the assumption base, **`assert`** the sentences that appear inside the curly braces, and then derive the sentence on the right-hand side of the turnstile. □

Exercise 4.13: As in Exercises 4.8 and 4.12, for each of the following parts:

(a) `{}` ⊢ `(((A ==> B) ==> A) ==> A)`
(b) `{((A ==> B) ==> B), (B ==> A)}` ⊢ `A`
(c) `{}` ⊢ `(((A ==> B) | (B ==> C)) <==> (A & B ==> C))`

give a deduction that derives the conclusion on the right-hand side of the turnstile from an assumption base that includes all and only the sentences on the left-hand side. □

Exercise 4.14: Consider:

```
declare A,B,C,D,E,F,G,H,I,J: Boolean

#--- Part (a):

conclude ((A ==> B) ==> (A & C ==> B & C))
  D1
```

```
#--- Part (b):

clear-assumption-base
assert p1 := (~ ~ A ==> ~ (B <==> C))
assert p2 := (A & B)

conclude (~ C)
  D2

#--- Part (c):

clear-assumption-base
assert p1 := (~ A | (B ==> E) & (C ==> E))
assert p2 := (A & (B | C))

conclude E
  D3

#--- Part (d):

clear-assumption-base
assert p1 := ((A ==> B) &  (C ==> D))
assert p2 := (A | C)
assert p3 := ((A ==> ~ D) & (C ==> ~ B))
assert p4 := (B & ~ D ==> E)
assert p5 := (D ==> B | F)

conclude (E | F)
  D4

#--- Part (e):

clear-assumption-base
assert p1 := (A | B ==> C & D)
assert p2 := (C | E ==> ~ F & G)
assert p3 := (F | H ==> A & I)

conclude (~ F)
  D5

#--- Part (f):

clear-assumption-base
assert p1 := (A ==> B & C)
assert p2 := (B | D ==> E)
assert p3 := (~ E)

conclude  (~ D & ~ A)
  D6
```

Replace D_1–D_6 with appropriate deductions (so that the preceding stretch of code runs successfully and produces 6 theorems). □

Exercise 4.15: Implement the excluded-middle method. The application of `ex-middle` to any sentence p should derive `(`p `|` `~` p`)`. □

Exercise 4.16: Determine whether

$$(\texttt{!dm (!dm } p\texttt{))} = p$$

holds in every assumption base and for all sentences p (including those that do not match any of the expected patterns). □

Exercise 4.17: The definition of `import*` given on page 229 produces a conjunction that grows on the left. Write `import*` in such a way that it derives `((and` p_1 p_2 p_3`) ==>` p_4`)` from

$$(p_1 \texttt{ ==> } p_2 \texttt{ ==> } p_3 \texttt{ ==> } p_4);$$

more generally, so that it produces a single conjunction (a single application of `and` to a single list of sentences). □

Exercise 4.18: Define `either` as specified on page 201, in terms of `left-either` and `right-either`. □

Exercise 4.19: Implement `dm-1`, the first form of De Morgan's law (see page 217). □

Exercise 4.20: Implement a more general version of modus tollens, `mt`, that does not require the second premise to be an explicit negation of the consequent of the first conditional premise. Also, the conclusion should be the complement (rather than the direct negation) of the antecedent of the first premise. Accordingly, as before, we have:

```
(!mt (A ==> B) (~ B)) = (~ A).
```

But we also have:

```
(!mt (A ==> ~ B) B)        =    (~ A);
(!mt (~ A ==> B) (~ B))    =       A;
(!mt (~ A ==> ~ B) B)      =       A.
```

More precisely, we should have

$$(\texttt{!mt (}p \texttt{ ==> } q\texttt{)}\ \overline{q}) = \overline{p}.$$

(We are assuming, of course, that both arguments, `(`p `==>` q`)` and $\overline{q}$, are in the assumption base.) □

Exercise 4.21: This is similar to Exercise 4.20: Implement a more general version of `contra-pos` (page 213) that works with complements and can be applied in either direction, deriving, say,

```
(~ B ==> ~ A)
```

from `(A ==> B)`, as well as `(A ==> B)` from `(~ B ==> ~ A)`. □

Exercise 4.22: Derive the disjunction `((A ==> B) | (B ==> C))` from the empty assumption base. □

Exercise 4.23: Implement the method `negated-bicond`, as specified on page 226. □

Exercise 4.24: Define a method `dm-2` implementing the binary version of De Morgan outlined on page 231, where the first argument is the premise and the second argument is the goal. A more precise specification of this method is given by the following two rules:

$$\frac{\texttt{(\~{} (and } p_1 \cdots p_n\texttt{))}}{\texttt{(or } \overline{p_1} \cdots \overline{p_n}\texttt{)}}\ [R_1] \qquad \frac{\texttt{(\~{} (or } p_1 \cdots p_n\texttt{))}}{\texttt{(and } \overline{p_1} \cdots \overline{p_n}\texttt{)}}\ [R_2]$$

where $n > 1$ and for each $i = 1, \ldots, n$, either $\overline{p_i}$ is the complement of p_i or vice versa. The method should implement these rules in both directions. Specifically, if we let p be the premise at the top of rule $[R_1]$ and q the conclusion at the bottom, then both of these deductions should work:

assume p **conclude** q `(!dm-2` p q`)`

and

assume q **conclude** p `(!dm-2` q p`)`;

and likewise for $[R_2]$. The method should fail if its two arguments don't match one of these combination patterns (premise-conclusion or conclusion-premise of either $[R_1]$ or $[R_2]$). □

Exercise 4.25: Likewise, develop binary versions of `cond-def` and `bicond-def`, call them `cond-def-2` and `bicond-def-2`. For the first, if *premise* is a member of the assumption base that is either of the form `(`p `==>` q`)` or `(`p `|` q`)`; and *goal* is of the form `(`$\overline{p}$ `|` q`)` or `(`$\overline{p}$ `==>` q`)`, respectively; then

`(!cond-def-2` *premise goal*`)` (4.32)

should derive *goal*. For the second, if *premise* is a sentence in the assumption base that is of the form `(`p `<==>` q`)` and *goal* is of the form `((`p `&` q`) | (`$\overline{p}$ `&` $\overline{q}$`))` where p and $\overline{p}$ are complements of each other, as are q and $\overline{q}$; then (4.32) should derive *goal*. While if *premise* is of the form `((`p `&` q`) | (`$\overline{p}$ `&` $\overline{q}$`))` and *goal* is of the form `(`p `<==>` q`)`, where again $p, \overline{p}$ and $q, \overline{q}$ are complementary; then (4.32) should likewise derive *goal*. □

Exercise 4.26: In addition to being commutative (page 213), conjunction and disjunction are also associative, and it can be helpful to be able to go from a premise of the form `(`p` & (`q` & `r`))` to `((`p` & `q`) & `r`)` and vice versa; likewise for disjunctions. Write two methods that can carry out these two bidirectional transformations. □

Exercise 4.27: Define a recursive method `export*` that is analogous to `import*` but works in the converse direction, so that an input premise such as

$$(((p_1 \;\texttt{\&}\; p_2) \;\texttt{\&}\; p_3) \;\texttt{==>}\; p_4)$$

results in the conditional

$$(p_1 \;\texttt{==>}\; (p_2 \;\texttt{==>}\; (p_3 \;\texttt{==>}\; p_4))).$$

(Hint: It might be helpful if you first define a nonrecursive method `export`, analogous to `import`, and then implement `export*` in terms of `export`.) □

Exercise 4.28: Define the congruence methods for the remaining two sentential constructors (see page 233): `or-cong` and `iff-cong`. □

Exercise 4.29 (Random sentences): Define a binary procedure `make-random-sentence` that generates random sentences. The first argument will be the desired size of the sentence, and the second argument should be the number of atoms over which the sentence is to be built. The atoms will be Boolean variables such as `?X2:Boolean`. So an application like

```
(make-random-sentence 10 2)
```

should result in a zero-order sentence of size (roughly) 10, over the two atoms `?X1:Boolean` and `?X2:Boolean`. Such a procedure can be useful for testing purposes. □

⋆ ***Exercise 4.30:*** Write a method that takes a conjunction p in the assumption base and derives a conjunction that consists of the exact same conjuncts that appear in p,[46] but arranged in a right-chained way, so that, for instance, the input

```
((A & B) & (C & D))
```

results in `(A & (B & (C & D)))`. □

⋆⋆ ***Exercise 4.31 (Negation normal form):*** A sentence is in *negation normal form* (NNF) iff every negation sign in it appears immediately in front of an atom.

(a) Define a unary procedure `nnf?` that takes a sentence p and outputs `true` if p is in NNF and `false` otherwise.

(b) Define a unary procedure `nnf` that takes a sentence p and converts it to NNF; that is, it outputs a sentence q such that q is in NNF and is equivalent to p.

---

46 By the "conjuncts" of p we mean those subsentences of p returned by `get-conjuncts`; see Exercise 2.7.

(c) What is the (worst-case) computational complexity of your implementation of `nnf`?

(d) Define a unary method `derive-nnf` that takes an arbitrary sentence p and produces the conclusion `(`p `<==>` q`)`, where q is in negation normal form. (Hint: You will need to use double negation and the methods we developed in Section 4.9 for De Morgan's, biconditional definition, and negated conditionals.)

(e) How does the worst-case computational complexity of `derive-nnf` compare to that of the procedure `nnf`? □

** ***Exercise 4.32 (Conjunctive normal form):*** A *literal* is either an atom or the negation of an atom. A sentence is in *conjunctive normal form* (CNF) iff it is a conjunction of *clauses*, where each clause is a disjunction of literals.

(a) Define a unary procedure `cnf?` that takes a sentence p and outputs `true` if p is in CNF and `false` otherwise.

(b) Define a unary procedure `cnf` that takes a sentence p and converts it to CNF; that is, it outputs a sentence q such that q is in CNF and is equivalent to p.

(c) What is the computational complexity of your implementation of `cnf?`

(d) Define a unary method `derive-cnf` that takes an arbitrary sentence p and produces a conclusion of the form `(`p `<==>` q`)`, where q is in CNF. (Hint: You will need to use several of the methods developed in Section 4.9, including that for distributivity.)

(e) How does the computational complexity of `derive-cnf` compare to that of the procedure `cnf?` □

** ***Exercise 4.33 (Disjunctive normal form):*** A sentence is in *disjunctive normal form* (DNF) iff it is a disjunction of clauses, where each clause is a conjunction of literals.

(a) Define a unary procedure `dnf?` that takes a sentence p and outputs `true` if p is in DNF and `false` otherwise.

(b) Define a unary procedure `dnf` that takes a sentence p and converts it to DNF; that is, it outputs a sentence q such that q is in DNF and is equivalent to p.

(c) What is the computational complexity of your implementation of `dnf?`

(d) Define a unary method `derive-dnf` that takes an arbitrary sentence p and produces a conclusion of the form `(`p `<==>` q`)`, where q is in DNF.

(e) How does the computational complexity of `derive-dnf` compare to that of the procedure `dnf?` □

Exercise 4.34 (Boolean constraint propagation): Implement the pure-literal BCP algorithm of the DPLL procedure. □

** ***Exercise 4.35 (Resolution):*** Implement resolution: Define a binary method `resolve` that takes two premises of the form

$$(\texttt{or}\ p_1 \cdots p_i \cdots p_m)$$

and

$$(\texttt{or } q_1 \cdots q_j \cdots q_n),$$

with $m, n \geq 1$, and provided that either $p_i = (\sim\ q_j)$ or $q_j = (\sim\ p_i)$, it derives the conclusion

$$(\texttt{or } p_1 \cdots p_{i-1}\ p_{i+1} \cdots p_m\ q_1 \cdots q_{j-1}\ q_{j+1} \cdots q_n).$$

When $m = n = 1$, `resolve` should derive the conclusion `false`. The method should fail if one of the premises is not a disjunction or if no pair of complementary disjuncts exists. □

** ***Exercise 4.36 (De Bruijn arguments):*** Let $A_1, A_2, A_3, \ldots$ be atomic sentences. For any $n > 0$, let C_n be the conjunction $A_1 \wedge \cdots \wedge A_n$. The nth *De Bruijn argument*, for any $n > 0$, is defined as follows:

$$\frac{\begin{array}{c}(A_1 \Leftrightarrow A_2) \Rightarrow C_n \\ (A_2 \Leftrightarrow A_3) \Rightarrow C_n \\ \cdots \\ (A_n \Leftrightarrow A_1) \Rightarrow C_n\end{array}}{C_n}$$

For instance, the De Bruijn argument for $n = 3$ is:

$$\frac{\begin{array}{c}(A_1 \Leftrightarrow A_2) \Rightarrow A_1 \wedge A_2 \wedge A_3 \\ (A_2 \Leftrightarrow A_3) \Rightarrow A_1 \wedge A_2 \wedge A_3 \\ (A_3 \Leftrightarrow A_1) \Rightarrow A_1 \wedge A_2 \wedge A_3\end{array}}{A_1 \wedge A_2 \wedge A_3}$$

The nth *De Bruijn sentence* is defined as the conditional $(p_1 \wedge \cdots \wedge p_n) \Rightarrow C_n$, where $p_1, \ldots, p_n$ are the premises and C_n is the conclusion of the nth De Bruijn argument. Thus, for example, the third De Bruijn sentence is

$$\begin{array}{l}[[(A_1 \Leftrightarrow A_2) \Rightarrow A_1 \wedge A_2 \wedge A_3] \wedge \\ \ [(A_2 \Leftrightarrow A_3) \Rightarrow A_1 \wedge A_2 \wedge A_3] \wedge \\ \ [(A_3 \Leftrightarrow A_1) \Rightarrow A_1 \wedge A_2 \wedge A_3]] \Rightarrow \quad A_1 \wedge A_2 \wedge A_3.\end{array}$$

(a) Define a procedure `make-db` that takes a positive integer `n` and produces the `n`th De Bruijn sentence. (Use the variables `?A1:Boolean`, `?A2:Boolean`, ... as the underlying atoms, and the procedure `string->var` to create them.[47])

(b) Produce an informal argument showing that the nth De Bruijn argument is valid for every odd number n and invalid for every even n.

---

47 When applied to only one string argument s, `string->var` creates a polymorphic variable named s. For instance, `(string->var "x")` might produce the variable `?x:'T`. But `string->var` can be given an optional second string argument s' representing a sort, and in that case the result will be a variable named s of the sort named s'. For example, `(string->var "i" "Int")` will produce the variable `?i:Int`.

(c) Use the idea of the above argument to define a method prove-odd-db that takes any odd positive integer n and derives (proves) the nth De Bruijn sentence. □

Exercise 4.37: Change the definition of coloring-constraints so that redundant constraints of the kind discussed on page 265 are no longer generated. □

** ***Exercise 4.38 (Hilbert systems):*** In this exercise we define a Hilbert axiom system for sentential logic. The system has a small number of axiom schemas (but each with infinitely many instances) and only one inference rule, modus ponens. Three axiom schemas suffice for dealing with conditionals and negation:

$$p \Rightarrow (q \Rightarrow p),$$
$$((p \Rightarrow (q \Rightarrow r)) \Rightarrow ((p \Rightarrow q) \Rightarrow (p \Rightarrow r))),$$
$$((\neg p \Rightarrow \neg q) \Rightarrow ((\neg p \Rightarrow q) \Rightarrow p)).$$

We will set things up so that we can leverage Athena's existing proof infrastructure to write and check proofs in this axiom system. First, for the axioms: Each axiom can be represented as a *primitive method* (refer to Section 2.16, page 103).

```
primitive-method (axiom-1 sentence) :=
  match sentence {
    (p ==> (q ==> p)) => sentence
  }

primitive-method (axiom-2 sentence) :=
  match sentence {
    ((p ==> (q ==> r)) ==> ((p ==> q) ==> (p ==> r))) => sentence
  }

primitive-method (axiom-3 sentence) :=
  match sentence {
    (((~ p) ==> (~ q)) ==> (((~ p) ==> q) ==> p)) => sentence
  }

declare A, B, C, D, E: Boolean

> (!axiom-1 (B & C ==> D ==> B & C))

Theorem: (if (and B C)
             (if D
                 (and B C)))
```

Now a proof in this system is just a regular Athena proof that *only* uses mp (modus ponens), one of the three primitive methods, sequencing (either via **let** or a plain proof block), and possibly **conclude** annotations. As an example, here is a proof of (A ==> A):

```
let {step1 := (!axiom-2 ((A ==> ((A ==> A) ==> A)) ==>
                         ((A ==> A ==> A) ==> A ==> A)));
     step2 := (!axiom-1 (A ==> ((A ==> A) ==> A)));
```

```
    step3 := (!mp step1 step2);
    step4 := (!axiom-1 (A ==> A ==> A))}
  (!mp step3 step4)

Theorem: (if A A)
```

It is better to express these proofs as methods that are abstracted over input sentences. Thus, the following is a method that will take any sentence p and prove (p ==> p):

```
define (p=>p p) :=
  let {step1 := (!axiom-2 ((p ==> ((p ==> p) ==> p)) ==>
                           ((p ==> p ==> p) ==> p ==> p)));
       step2 := (!axiom-1 (p ==> ((p ==> p) ==> p)));
       step3 := conclude ((p ==> p ==> p) ==> (p ==> p))
                  (!mp step1 step2);
       step4 := (!axiom-1 (p ==> p ==> p))}
    (!mp step3 step4)

> (!p=>p (A & B))

Theorem: (if (and A B)
             (and A B))
```

This system is sound and complete if one restricts attention to negations and conditionals. (The rest of the logical connectives can either be desugared into combinations of negations and conditionals, or else additional axioms can be introduced to deal with them directly, essentially simulating introduction and elimination rules for conjunctions, disjunctions, and biconditionals via conditional axioms.)

If you think that the above is an overly complicated proof for an exceedingly simple sentence, you are right. In fact, it gets much worse. Consider the following sentence schema:

$$((p \text{ ==> } q \text{ ==> } r) \text{ ==> } (q \text{ ==> } p \text{ ==> } r)).$$

Write a proof method in this system that will take any sentences p, q, and r as inputs and will derive the above conditional. □

Exercise 4.39 (Pigeonhole principle): The pigeonhole principle states that it is impossible to place $n+1$ pigeons into n pigeonholes without at least one pigeonhole containing more than one pigeon. Or, more abstractly, that it is impossible to assign $n+1$ objects to n slots without at least one slot receiving more than one object. Formally, any function

$$f\colon \{1,\ldots,n,n+1\} \rightarrow \{1,\ldots,n\}$$

must be noninjective (not 1-1), that is, there must be two distinct numbers

$$i,j \in \{1,\ldots,n,n+1\}$$

such that $f(i) = f(j)$.

We can represent (and indeed prove) the pigeonhole principle for any given n in sentential logic with sentences of two kinds. The first states that each object is placed into *some* box; and the second that there are no two objects in any given box. The principle then simply states that the conjunction of these two sentences is inconsistent, namely, that it entails `false`. This is a refutational formulation. We could also give a direct formulation (one stating that if every object is placed into some box, then at least one box receives more than one object). But the refutational formulation is equivalent and more customary.

The following signature will be useful:

```
datatype Object := (object Int)
datatype Box := (box Int)
declare into: [Object Box] -> Boolean
```

To make things concrete, suppose that $n = 2$ (we will generalize this shortly), so we have 3 objects and 2 boxes. Then the sentences we just discussed are:

```
define all-somewhere :=
  ((object 1 into box 1 | object 1 into box 2) &
   (object 2 into box 1 | object 2 into box 2) &
   (object 3 into box 1 | object 3 into box 2))

define none-together :=
  ((object 1 into box 1 ==> ~ object 2 into box 1 & ~ object 3 into box 1) &
   (object 1 into box 2 ==> ~ object 2 into box 2 & ~ object 3 into box 2) &
   (object 2 into box 1 ==> ~ object 1 into box 1 & ~ object 3 into box 1) &
   (object 2 into box 2 ==> ~ object 1 into box 2 & ~ object 3 into box 2) &
   (object 3 into box 1 ==> ~ object 1 into box 1 & ~ object 2 into box 1) &
   (object 3 into box 2 ==> ~ object 1 into box 2 & ~ object 2 into box 2))
```

So the pigeonhole principle in this case becomes:

```
define pp-2 := (all-somewhere & none-together ==> false)
```

You could now prove that `pp-2` is a tautology. But instead of giving such a proof specifically for $n = 2$, you are asked to write a method that takes an arbitrary $n > 0$ as its argument and derives the corresponding instance of the pigeonhole principle from no assumptions. What is the complexity of your method? □

⋆⋆ ***Exercise 4.40:*** Implement the strategy we have described for applying proof tactics (or some other strategy if you prefer) as an Athena procedure or method. □

4.17 Chapter notes

Sentential logic as we understand it today is essentially Boolean algebra, and thus largely the brainchild of George Boole [11] (whence the term "Boolean" derives). At the time, Boole believed that he was formulating laws that describe how humans actually think, in the same way that the laws of mechanics might describe motion—a doctrine known as logical *psychologism* that has long since fallen into disfavor.[48] Nevertheless, Boole's work was crucial in marrying logic and algebra, and a huge leap toward making logic a branch of mathematics. Work along those lines had started much earlier, at the time of Leibniz, but it was Boole and his contemporaries (most notably De Morgan, Mill, Peirce, and Venn) who made the most seminal contributions in that direction.

While Boole and his contemporaries were the first to mathematize the semantics of sentential logic, symbolic *proofs* in that logic were introduced slightly later, in formal systems devised first by Frege [39] and then by Whitehead and Russell [112], Hilbert, and others. These systems represented *axiomatic* approaches to proof: With a number of axiom schemas and inference rules laid down, a proof becomes a finite *list* of sentences, each of which is either an instance of an axiom or else follows from previous elements in the list by an inference rule (see Exercise 4.38 for a flavor of that approach). This gets to the bare essence of a deductive argument, harking back to Euclid's axiomatization of geometry, but it is not an accurate reflection of the structure of actual mathematical proofs. Since Euclid's time, a hallmark of mathematical proofs has been the provisional making of hypotheses, possibly nested inside one another, that remain in effect only for certain stretches of discourse—in other words, the notion of assumption scope. Axiomatic systems are not able to capture that notion, nor are they good models of other staples of deductive reasoning, such as case analysis or proof by contradiction.

These key features of mathematical proofs were not properly captured until the advent of *natural deduction*, so called precisely because it does justice to deductive argumentation as actually practiced. The father of natural deduction is Jáskowski [51], although some of the same ideas were independently articulated around the same time by Gentzen in his *sequent* systems [42]. Jáskowski was the one who introduced the graphical device of enclosing the scope of an assumption in a box, something that later became almost synonymous with

---

48 Very roughly, the major arguments against it are: (1) Unlike physical laws, the principles of logic are known *a priori*, not through empirical investigation, which can neither confirm nor overthrow these principles. (2) The entities to which these laws refer are abstract (propositions), rather than physical concepts with observational import (such as mass or velocity). And (3), people do not think in accordance with the laws of sentential logic. The latter is not to say that they never do (indeed, as we mentioned earlier, *some* patterns of reasoning that conform to natural deduction rules—such as modus ponens and case analysis—are quite common, a claim that has been confirmed by research in the psychology of reasoning [84],[13]), but rather that they do so relatively rarely. And there are certain types of logical fallacies that have been found to be so systematic and robust that they strongly undermine the thesis that people reason in accordance with normatively correct models such as sentential logic [110], [43], [53].

natural deduction in logic textbooks, after becoming popularized by Fitch in the early 1950s [36], to the point where that style of proof has by now come to be known as "Fitch-style" natural deduction. Fitch-style natural deduction has been the predominant choice in logic education, adopted by many classic logic textbooks [20, 36, 55, 64]. Much of the material presented in the first third of this chapter can also be found in such textbooks, though obviously in our approach proofs are written in a computer language, which also checks them automatically for correctness. Athena's inferential core can be seen as a modern Fitch-style natural deduction system, with assumption scope represented by syntactic constructs (most notably `assume`) and corresponding formal semantics, rather than graphical devices such as boxes.

As pointed out in the text, the SAT (satisfiability) problem is of central importance in computer science, both theoretically (it was the first problem to be proven NP-complete [19]), and in a wide range of practical applications that include bioinformatics, cryptography, hardware and software verification, and operations research. There have been many improvements in SAT solving over the last fifteen years, but the core algorithm is still the Davis-Putnam-Logemann-Loveland procedure formulated in the late 1950s. The *Handbook of Satisfiability* [9] provides a good overview of the state of the art as of 2010. For more recent results, look at the proceedings of the main annual SAT conferences, such as the International Conference on Theory and Applications of Satisfiability Testing. For a broader and more theoretical introduction to the subject, see Marek's *Introduction to Mathematics of Satisfiability* [68].

Heuristics for how to construct proofs in natural deduction systems have been given by many authors, with differing degrees of completeness and rigor. Most of the tactics we have presented here (with the exception of the replacement tactic and the tactic of Heuristic 4.3 for deriving negations directly) have appeared in print elsewhere or have been informally known as folklore for a while. Similar heuristics, for instance, have been discussed by Sieg and Scheines [90], Sieg and Byrnes [89], and Portoraro [83]. Here we have formulated them so that they could be directly implemented through a heuristically informed AND/OR graph-search strategy.

AND/OR graphs are discussed in most AI textbooks; see, for example, Nilsson's *Artificial Intelligence: A New Synthesis* [79]. They are a particularly apt representation for problem solving whenever a given problem (or solution "goal") G is either primitive and hence trivial to solve, or else can be decomposed into a number $m > 0$ of subproblem groups:

$$\{G_1^1, \ldots, G_{n_1}^1\}, \ldots, \{G_1^m, \ldots, G_{n_m}^m\},$$

such that solving *every* subproblem (the "and" part) in *any* of these groups (the "or" part) is sufficient to solve G. The subproblems within each group must be independently solvable.[49] Many problems, especially in theorem proving (and reasoning in general), fit this

49 This requirement can be relaxed somewhat; some minimal dependencies between the subproblems may exist.

paradigm naturally. If numeric costs are attached to the edges, then the search can be informed by heuristics that can be used to find an optimal solution. Our ranking of tactics can be seen as a simple measure of cost, but much more sophisticated measures are possible. Logic programming, discussed in Appendix B, is a natural vehicle for searching AND/OR graphs because the underlying execution mechanism is itself an AND/OR graph search, so one can essentially offload the search to the operational semantics of logic programming. Naively implemented, however, this approach has some disadvantages: It cannot take into account numeric costs, and if one is using Prolog, its depth-first search strategy might fail to find any solution at all (it might get stuck in cycles); see Bratko's discussion of these issues [14].

Semantic tableaux were invented by Beth in the 1950s [8]. They were simplified and popularized by Smullyan in the late 1960s [91]. For more technical details, consult the relevant chapter in the *Handbook of Automated Reasoning* [85].

Some of the methods we defined in this chapter were written in *continuation-passing style*, or CPS for short. A method M is written in CPS if one of its arguments (typically the last one) is itself a method K, called a *proof continuation*. Instead of returning a theorem directly, M works by passing a number of values to K. Thus, the ultimate result of a call to M is produced by applying K to whatever values M gives to it. Continuation-passing style is a powerful tool that is frequently used in functional programming [104]. It can also be used in that same conventional sense in Athena's regular programming language, using procedures. However, CPS is particularly convenient—and indeed often necessary—in connection with Athena's assumption-base proof semantics, mostly because it presents a handy way to circumvent the requirement that every method call must result in (at most) one sentence.

5 First-Order Logic

IN this chapter we turn our attention to predicate (or *first-order*) logic. Our main objective is to describe the various Athena mechanisms for constructing proofs in predicate logic. In addition, we present heuristics for developing such proofs, we build a library of generically useful proof methods, and we also discuss the semantics of (polymorphic, many-sorted) predicate logic.

Recall from Section 2.4 that the syntax of predicate logic extends the syntax of sentential logic. So, just as before, we have atomic sentences (atoms), negations, conjunctions, disjunctions, conditionals, and biconditionals, where an atom can be an arbitrary term of sort `Boolean`. Thus, assuming that `x`, `y`, and `z` denote term variables, all of the following are legal first-order sentences:

Sentence	Type
`(x < x + 1)`	*Atom*
`(~ even 3)`	*Negation*
`(x < x + 1 & 7 =/= 8)`	*Conjunction*
`(x < y | x = y | y < x)`	*Disjunction*
`(x < y & y < z ==> x < z)`	*Conditional*
`(x + 1 < y + 1 <==> x < y)`	*Biconditional*

We assume for the remainder of this chapter that the following domains have been introduced:

```
domains Object, Element, Set
```

along with the following symbols:

```
declare even, odd, prime: [Int] -> Boolean
declare in: [Element Set] -> Boolean
declare subset: [Set Set] -> Boolean

declare P, Q, S: [Object] -> Boolean
declare R, T: [Object Object] -> Boolean
declare a, b, c, d: Object
```

We also assume that `x`, `y`, `z`, and `w` have been defined as the variables `?x`, `?y`, `?z`, and `?w`:

```
define [x y z w] := [?x ?y ?z ?w]
```

In addition to the kinds of sentences we are already familiar with from sentential logic, we now have *quantified sentences* (or simply *quantifications*): For any variable v and sentence p,

$$\texttt{(forall } v \texttt{ . } p\texttt{)} \tag{5.1}$$

and

$$\texttt{(exists } v \texttt{ . } p\texttt{)} \tag{5.2}$$

are also legal first-order sentences. We refer to p as the *body* of (5.1) and (5.2). One can also quantify over multiple variables with only one quantifier, for example, writing

```
(forall x y . x + y = y + x)
```

as a shorthand for

```
(forall x . forall y . x + y = y + x).
```

Intuitively, sentences of the form (5.1) state that *every* object of a certain sort has such-and-such property; while sentences of the form (5.2) state that *some* object (of a certain sort) has such-and-such property. For instance, the statement that every prime number greater than two is odd can be expressed as follows:

```
(forall x . prime x & x > 2 ==> odd x),
```

while the statement that there is some even prime number can be expressed as:

```
(exists x . prime x & even x).
```

Quantifiers can be combined to form more complex sentences. For instance:

```
(forall x . exists y . x subset y)
```

states that every set has some superset, while

```
(exists x . forall y . x subset y)
```

says that there is a set that is a subset of every set. Note that it is not necessary to explicitly annotate occurrences of quantified variables with their respective sorts. For instance, in the above examples it is not necessary to tell Athena that x and y range over sets. That is inferred automatically:

```
> (exists x . forall y . x subset y)

Sentence: (exists ?x:Set
            (forall ?y:Set
              (subset ?x:Set ?y:Set)))
```

When we wish to make the sort S of a quantified variable explicit, we will write quantifications as

(forall $v:S$. p)

and

(exists $v:S$. p).

In the next two sections we will present Athena's mechanisms for introducing and eliminating each type of quantified sentence. But first we briefly turn our attention to an exercise that develops more fully the idea that terms and sentences are trees, and gives computationally precise definitions of concepts such as *variable occurrences* in a sentence (both *free* and *bound* variable occurrences), free variable replacements, and alpha-equivalence. These concepts have already been introduced, so those who feel sufficiently comfortable with them can move on to the next section. But for those who don't mind some coding, this exercise is a good opportunity to consolidate these ideas in greater algorithmic detail.

Exercise 5.1: As we explained in Section 3.3 by way of examples, terms and sentences are essentially trees. We will now make this and other related ideas more precise.

(a) Define two unary procedures `term->tree` and `sent->tree` that transform a given term or sentence into an explicit tree representation. Each node in the tree will be a map m of the form

|{ 'data := x, 'pos := L, 'children := [$m_1 \cdots m_n$] }|, (5.3)

where the `'data` value x is the function symbol, variable, quantifier, or sentential constructor that appears at that node; the `'pos` value L is the Dewey path of the node in the tree, as a list of positive integers; and $m_1, \ldots, m_n$ are the maps that (recursively) correspond to the children of the node (hence leaf nodes will have $n = 0$), where each m_i is itself of the form (5.3). Thus, for instance, we should get:

```
> define t := (1 + x)

Term t defined.

> (term->tree t)

Map:
|{
'data := +
'pos := []
'children := [ |{'data := 1, 'pos := [1], 'children := []}|
               |{'data := ?x:Int, 'pos := [2], 'children := []}|
             ]
}|

> (term->tree true)
```

```
Map: |{'data := true, 'pos := [], 'children := []}|

> (sent->tree (~ x < 1))

Map:
|{
'data := not,
'pos := [],
'children := [|{'data := <,
                'pos := [1],
                'children := [|{'data := ?x:Int,
                                'pos := [1 1],
                                'children := []}|
                              |{'data := 1,
                                'pos := [1 2],
                                'children := []}|]
               }|]
}|
```

(b) Define a procedure tree-leaves that takes a tree T produced by (either) one of the above procedures and returns a list of all the leaves of T, where each leaf is represented as a record (map) of the form `|{'data :=` x`, 'pos :=` L`}|`. For instance:

```
> (tree-leaves term->tree t)

List: [|{'data := 1, 'pos := [1]}|
       |{'data := ?x:Int, 'pos := [2]}|]
```

(c) Define a procedure var-occs that takes a term t or sentence p and produces a map m whose keys are all and only the variables that occur in t (or p). The value that m assigns to such a variable v is a list of the positions at which v occurs in t (or p). For example:

```
> (var-occs forall x . exists y . x = y)

Map: |{?x:'T3097 := [[1] [2 2 1]], ?y:'T3097 := [[2 1] [2 2 2]]}|
```

(d) Define a procedure bound-var-occs that takes a sentence p and produces a map of the form described above, except that the keys now consist of all and only the variables that have bound occurrences in p. Also define a procedure free-var-occs that does the same for the variables that have free occurrences in the input p.

(e) Define a procedure alpha-equiv? that determines whether two sentences are alpha-equivalent.[1]

---

1 Of course Athena already provides an implementation of alpha-equivalence—one can simply apply equal? to two sentences. The objective here, as in the following two parts of this exercise, is to implement this functionality from scratch for instructive purposes.

(f) Define a procedure `sent-rename` that alpha-renames a given sentence.

(g) Define a procedure `(replace-var-by-term` x t p`)` that replaces every free occurrence of variable x in a sentence p by term t, performing renaming as needed to avoid variable capture. Assume that the sort of t is an instance of the most general sort of x in p. □

5.1 Working with universal quantifications

5.1.1 Using universal quantifications

A universal quantification makes a general statement, about *every* object of some sort. For instance,

```
(forall x . x < x + 1)
```
(5.4)

says that every integer is less than its successor. Hence, if we know that (5.4) holds, we should be able to conclude that any *particular* integer is less than its successor, say, that 5 is less than its successor:

```
(5 < 5 + 1)
```
(5.5)

or that 78 is less than its successor:

```
(78 < 78 + 1).
```
(5.6)

We say that the conclusions (5.5) and (5.6) are obtained from (5.4) by *universal specialization*, or *universal instantiation.*

In Athena, universal specialization is performed by the binary method `uspec`. The first argument to `uspec` is the universal quantification p that we want to instantiate; p must be in the assumption base, otherwise the application will fail. The second argument is the term with which we want to specialize p:

```
assert p := (forall x . x < x + 1)

> (!uspec p 5)

Theorem: (< 5
            (+ 5 1))

> (!uspec p 78)

Theorem: (< 78
            (+ 78 1))

> (!uspec p (2 * x))

Theorem: (< (* 2 ?x:Int)
            (+ (* 2 ?x:Int)
               1))
```

The last example shows that the instantiating term does not have to be ground.

More precisely, if p is a universal quantification `(forall v . q)` in the assumption base and t is a term, then the method call

`(!uspec p t)`

will produce the conclusion $\{v \mapsto t\}(q)$, where $\{v \mapsto t\}(q)$ is the sentence obtained from q by replacing every free occurrence of v by t. An error will occur if the result of this replacement is ill-sorted, as would happen, for instance, if we tried to specialize (5.4) with the Boolean term `true`. Note also that the sort S of the quantified variable v may be polymorphic. The application of `uspec` will work fine as long as the sort S_t of the instantiating term t is unifiable with S (recall the definition of unifiable sorts from page 51). For example, the quantified sentence might state that the double reversal of any list L of any sort is identical to L, and the instantiating term might be a list of Boolean terms, or a list of integers, or a polymorphic variable:

```
declare reverse: (T) [(List T)] -> (List T)

> assert p := (forall x . reverse reverse x = x)

The sentence
(forall ?x:(List 'S)
  (= (reverse (reverse ?x:(List 'S)))
     ?x:(List 'S)))
has been added to the assumption base.

> (!uspec p (true::false::nil))

Theorem: (= (reverse (reverse (:: true
                                  (:: false
                                      nil:(List Boolean)))))
            (:: true
                (:: false
                    nil:(List Boolean))))

> (!uspec p (78::nil))

Theorem: (= (reverse (reverse (:: 78
                                  nil:(List Int))))
            (:: 78
                                  nil:(List Int)))

> (!uspec p ?L)

Theorem: (= (reverse (reverse ?L:(List 'S)))
            ?L:(List 'S))
```

In proof systems in which alpha-convertible sentences are not viewed as identical,[2] care must be taken when performing universal specialization to avoid variable capture. In particular, the instantiating term should not contain any variables that might become accidentally bound as a result of the substitution. Violating this proviso generates an error in such systems. For instance, consider the sentence

$$\forall x \,.\, \exists y \,.\, x \neq y, \tag{5.7}$$

which essentially says that there are at least two distinct individuals. If, in such a system, we were allowed to specialize (5.7) with the variable y, we would obtain the nonsensical conclusion

$$\exists y \,.\, y \neq y.$$

The problem here is variable capture: The instantiating term (the variable y) becomes improperly bound by the existential quantifier $\exists\, y$.

One way of dealing with this issue is to disallow such universal specializations. Systems like Athena, in which the choice of a quantified variable name is immaterial, offer a more flexible alternative: They automatically avoid variable capture simply by alpha-renaming the sentence being specialized, using fresh variables. Doing so is safe because, for deductive purposes, the renamed sentence is identical to the original. For instance, here is what happens if we try to reproduce the above example in Athena:

```
> assert p := (forall ?x . exists ?y . ?x =/= ?y)

The sentence
(forall ?x:'S
  (exists ?y:'S
    (not (= ?x:'S ?y:'S))))
has been added to the assumption base.

> (!uspec p ?y)

Theorem: (exists ?v1915:'S
           (not (= ?y:'S ?v1915:'S)))
```

Athena accepted the attempt to instantiate

$$p = \texttt{(forall ?x . exists ?y . x =/= ?y)} \tag{5.8}$$

with ?y, but averted variable capture by first renaming p to something like

$$q = \texttt{(forall ?v1914 . exists ?v1915 . ?v1914 =/= ?v1915).} \tag{5.9}$$

Substituting ?y for the free occurrence of ?x in the body of p now becomes the same as substituting ?y for the free occurrence of ?v1914 in the body of (5.9), a replacement that is

2 So that, for example, $\forall x \,.\, P(x)$ and $\forall y \,.\, P(y)$ are treated as two distinct sentences.

harmless and results in the theorem displayed above. The upshot is that there is no need to be concerned about variable capture when performing universal instantiation.

Finally, it is noteworthy that in some respects a universal quantification can be understood as a large—potentially infinite—conjunction. For instance, if `B` is a unary predicate on `Boolean`, then the sentence

```
(forall ?x:Boolean . B ?x:Boolean)
```

means the same thing as

```
(B true & B false),
```

given that `true` and `false` are the only `Boolean` values. Likewise, the sentence

```
(forall ?x:Int . ?x:Int < ?x:Int + 1)
```

can be viewed as the infinite conjunction of all atoms of the form (t < t + 1), for every possible integer numeral t (positive, negative, and zero). With this analogy in mind, universal instantiation can be thought of as a generalized version of conjunction elimination.

5.1.2 Deriving universal quantifications

How do we go about proving that every object of some sort S has a property P? That is, how do we derive a goal of the form $(\forall\, v : S\,.\, P(v))$? Typically, mathematicians prove such statements by reasoning as follows:

$$\text{Consider any } I \text{ of sort } S. \text{ Then } \cdots D \cdots$$

where the name (identifier) I occurs free inside the proof D. We can regard this construction as a unary method with parameter I and body D that will take any term t of sort S and will attempt to prove $P(t)$. Clearly, if we have a method that can prove that *any* given object in the relevant domain has the property P, then we are entitled to conclude the desired $(\forall\, v{:}S\,.\, P(v))$. But how can we make sure that this "method" is indeed capable of doing that?

Well, we can try applying the method to some random term and see if it succeeds, that is, we can try evaluating the body D with some random term t of sort S in place of the parameter I, and see if we succeed in deriving $P(t)$. But what if the success is a fluke? What if D exploits some special assumptions about t that wouldn't be valid for some other term t'? Suppose, for instance, that the domain in question is that of the natural numbers, and when we apply the method to, say, `zero`, we do get the theorem P(`zero`) as output. But what if D relied on some special assumptions about `zero` in the assumption base? Even if it did not, how can we be sure that the method truly generalizes, meaning that applying the same reasoning to *any* other term t of sort `N` will also derive $P(t)$?

The trick is to choose our test term t judiciously. It must be a term with no baggage, so to speak. That is, a term about which there are definitely no special assumptions in effect.

A *fresh variable* x of sort S is just such a term; its freshness ensures that no special assumptions about it can possibly be in effect, since the assumption base will not contain any occurrences of it. Accordingly, we may regard x as representing a truly arbitrary element of the domain at hand. Hence, if the evaluation of D in an environment in which I refers to a fresh variable x succeeds in deriving the theorem $P(x)$, we may conclude that P holds for *every* object of the relevant sort. That is, we may safely conclude $(\forall\, v\!:\!S\,.\,P(v))$.

Reasoning of this kind is expressed in Athena with deductions of the form

$$\texttt{pick-any}\ I\ D. \tag{5.10}$$

We refer to D as the *body* of (5.10). To evaluate a deduction of this form in an assumption base β, we first generate a fresh variable x of sort S, where S is itself a fresh sort variable (representing a completely unconstrained sort), say, `?v135:'S47`. This ensures that x is a variable that has never been used before in the current Athena session. We then evaluate the body D in β and, importantly, in an environment in which the name I refers to the fresh variable x. We say that D represents the *scope* of that variable. If and when that evaluation results in a conclusion p, we return the quantification (forall x . p) as the final result of (5.10).

To make things concrete, consider as an example the deduction

```
pick-any x (!reflex x).                                    (5.11)
```

Recall that `reflex` is a unary primitive method that takes any term t and produces the equality (t = t). To evaluate (5.11) in some assumption base β, Athena will first generate a fresh variable of a completely general and fresh sort, say `?x18:'S35`. It will then evaluate the body of the **pick-any** deduction, namely `(!reflex x)`, in an environment in which `x` refers to `?x18:'S35`. Thus, Athena will essentially be evaluating the deduction `(!reflex ?x18:'S35)`. According to the semantics of `reflex`, that will produce the equality `(?x18:'S35 = ?x18:'S35)`. Finally, Athena will generalize over the fresh variable and return the conclusion

```
(forall ?x18:'S35 . ?x18 = ?x18)                           (5.12)
```

as the result of the entire **pick-any**. For readability purposes, however, Athena will present this conclusion as

```
(forall ?x:'S . ?x = ?x)                                   (5.13)
```

instead of (5.12). That is harmless because the two sentences are alpha-equivalent, and Athena treats alpha-equivalent sentences as identical for deductive purposes. Hence, if all goes well, the user will never actually see or have to deal with the particular fresh variable that Athena generated, ?x18:'S35.[3] The user simply enters the deduction (5.11) and the theorem (5.13) is produced:

```
> pick-any x (!reflex x)

Theorem: (forall ?x:'S
           (= ?x:'S ?x:'S))
```

This example demonstrates that **pick-any** can universally generalize not just over all the elements of a specific (ground) sort, such as the natural numbers or the Booleans, but over all the elements of infinitely many sorts. Such a polymorphic generalization can then be specialized with any terms of appropriate sorts, say, with natural numbers or with Booleans:

```
> define p := pick-any x (!reflex x)

Theorem: (forall ?x:'S
           (= ?x:'S ?x:'S))

Sentence p defined.

> (!uspec p true)

Theorem: (= true true)

> (!uspec p (2 * y))

Theorem: (= (* 2 ?y:Int)
            (* 2 ?y:Int))
```

As another example, here is a proof that the equality relation is symmetric. Recall that sym is a unary primitive method that takes an equality $(s = t)$ and returns $(t = s)$, provided that $(s = t)$ is in the assumption base:

```
1  > pick-any a
2      pick-any b
3        assume h := (a = b)
4          (!sym h)
5
6  Theorem: (forall ?a:'S
7             (forall ?b:'S
8               (if (= ?a:'S ?b:'S)
9                   (= ?b:'S ?a:'S))))
```

3 But Athena does choose a fresh variable of the form $?I\_n$, where I is the original **pick-any** name and n is a number, so that if the proof fails with an error message, it is easier to see where that particular fresh variable originated.

How was this proof evaluated? It first generated a fresh variable of a brand new and completely unconstrained sort, say `?a74:'S95`, and then proceeded to line 2 to evaluate the body of the outer **pick-any** in an environment in which the name `a` refers to `?a74:'S95`. Now the body of that **pick-any** happens to be another **pick-any**, so another fresh variable (of unconstrained sort again) is generated, say, `?b75:'S96`, and Athena proceeds to evaluate the body of the inner **pick-any** in an environment in which `b` denotes `?b75:'S96` (and `a` denotes `?a74:'S95`). The body of the inner **pick-any** is an **assume**, so Athena adds the hypothesis `h := (?a74:'S95 = ?b75:'S95)` to the assumption base (note that the two sorts `'S95` and `'S96` have been unified at this point), and goes on to evaluate the body of the **assume**, namely, the application of `sym` to the identity `h`. Since `h` is in the assumption base, `sym` succeeds and returns the conclusion `(?b75:'S95 = ?a74:'S95)`, so, backing up one level, the **assume** returns the conclusion

```
(?a74 = ?b75 ==> ?b75 = ?a74)
```

(we have omitted the variable sorts for brevity). Hence, backing up one more level, the inner **pick-any** returns the generalization

```
(forall ?b75 . ?a74 = ?b75 ==> ?b75 = ?a74),
```

or, equivalently,

```
(forall ?b . ?a74 = ?b ==> ?b = ?a74).
```

Finally, backing up one more level, the outer **pick-any** returns the generalization

```
(forall ?a74 . forall ?b . ?a74 = ?b ==> ?b = ?a74),
```

or, equivalently,

```
(forall ?a:'S . forall ?b:'S . ?a = ?b ==> ?b = ?a).
```

This analysis included a lot of low-level details. It is rarely necessary to descend to such a level when working with Athena proofs. After a little practice one can understand what a proof does quite well at the higher and more intuitive level of its source text.

Observe that both quantified variables `?a` and `?b` in the conclusion of the above proof have the same sort `'S`. Again, `'S` is a sort variable, ranging over the collection of all available sorts. The only constraint in the above sentence is that `?a` and `?b` range over the *same* sort. That sort could be anything, but it has to be the same for both variables. This constraint was discovered by sort inference in the course of evaluating the deduction.

Although not usually necessary, users may, if they wish, provide explicit sort annotations for **pick-any** identifiers, and indeed this practice sometimes makes the proof more readable, and can also simplify sort inference, especially with nested **pick-any** deductions. For instance, one may write:

```
pick-any e:Element
  pick-any s:Set
    assume h := (e in s)
      ...
```

In that case the generated fresh variable will be of the specified sort.

Nested **pick-any** deductions, incidentally, of the form

$$\texttt{pick-any}\ I_1\ \texttt{pick-any}\ I_2\ \texttt{pick-any}\ I_3\ \cdots$$

may be abbreviated as **pick-any** $I_1\ I_2\ I_3 \cdots$. Thus, for example, the previous proof could also be written as follows:

```
pick-any a b
  assume h := (a = b)
    (!sym h)
```

As another example, let us prove the tautology

```
((forall x . P x) & (forall x . Q x) ==> forall y . P y & Q y).
```

(For example, if everything is green and everything is large, then everything is both green and large.)

```
define [all-P all-Q] := [(forall x . P x) (forall x . Q x)]

> assume hyp := (all-P & all-Q)
    pick-any y:Object
      let {P-y := conclude (P y)
                     (!uspec all-P y);
           Q-y := conclude (Q y)
                     (!uspec all-Q y)}
        (!both P-y Q-y)

Theorem: (if (and (forall ?x:Object
                     (P ?x:Object))
                  (forall ?y:Object
                     (Q ?y:Object)))
             (forall ?y:Object
               (and (P ?y:Object)
                    (Q ?y:Object))))
```

Many more examples will appear later in the text and in the exercises.

We close this section by mentioning an alternative mechanism for deriving universal quantifications, the syntax form

$$\texttt{generalize-over}\ E\ D. \tag{5.14}$$

Here E is an arbitrary Athena expression whose evaluation must produce a variable v. The idea is that v appears in D, and we will generalize over v whatever conclusion we obtain from D. That is, we evaluate D in the given assumption base β, and whatever conclusion p we get from that evaluation, we generalize it over v, thereby arriving at (forall v . p) as the overall result of (5.14). There is a proviso here, namely that the variable v should not have any free occurrences in the assumption base β. This ensures that D cannot exploit any special assumptions about v that might happen to be in effect when D is evaluated, and which could lead to an invalid generalization. As an example, evaluating the deduction

```
generalize-over ?foo:Int (!reflex ?foo:Int)
```

in some assumption base β will result in (forall ?foo:Int . ?foo = ?foo), provided that ?foo:Int does not occur free in β. If it does occur free, an error will be reported.

The two forms **pick-any** and **generalize-over** are closely related. The latter is more primitive. The former can actually be defined as syntax sugar on top of **generalize-over**. Specifically, we can define **pick-any** I D as

```
let {I := (fresh-var)}
  generalize-over I D
```

Thus, **pick-any** ensures that the aforementioned proviso is satisfied by generalizing over a *fresh* variable. Either form can be used to derive universal quantifications, but **pick-any** is almost always more convenient. The **generalize-over** construct is useful primarily when writing methods for discovering proofs automatically.

5.2 Working with existential quantifications

5.2.1 Deriving existential quantifications

If we know that 2 is an even number, then clearly we may conclude that *there exists* an even number. Likewise, if we know—or have assumed—that box b is red, we may conclude that there exists a red box. In general, if we have $\{x \mapsto t\}(p)$, we may conclude (exists x . p).[4] This type of reasoning is known as *existential generalization.* In Athena, existential generalization is performed by the binary method egen. The first argument to egen is the existential quantification that we want to derive, say

(exists x . p).

4 Recall that $\{x \mapsto t\}(p)$ is the sentence obtained from p by replacing every free occurrence of x by t (renaming as necessary to avoid variable capture).

The second argument is a term t on the basis of which we are to infer `(exists` x `.` p`)`. Specifically, if $\{x \mapsto t\}(p)$ is in the assumption base, then the call

(!egen (exists x . p) t)

will derive the theorem `(exists` x `.` p`)`. The idea here is that because p holds for the object named by t, that is, because $\{x \mapsto t\}(p)$ is in the assumption base, we are entitled to conclude that there is *some* object for which p holds. We might view the term t as evidence for the existential claim.

For instance, suppose that `(even 2)` is in the assumption base. Since we know that `2` is even, we are entitled to conclude that there exists an even integer:

```
assert (even 2)

> (!egen (exists x . even x) 2)

Theorem: (exists ?x:Int
           (even ?x:Int))
```

It is an error if the required evidence is not in the assumption base:

```
> clear-assumption-base

Assumption base cleared.

> (!egen (exists x . even x) 2)

standard input:1:1: Error: Failed existential
generalization---the required witness sentence
(even 2)
is not in the assumption base.
```

5.2.2 Using existential quantifications

Suppose that we know—or that we have assumed—that an existential quantification is true, so that some sentence of the form `(exists` x `.` p`)` is in the assumption base. How can we put such a sentence to use, that is, how can we derive further conclusions with the help of such a premise?

The answer is the technique of *existential instantiation*.[5] It is very commonly used in mathematics, in the following general form:

> We have it as a given that $\exists\, x \,.\, p$, so that p holds for *some* object. Let v be a name for such an object, that is, let v be a "witness" for the existential sentence $\exists\, x \,.\, p$, so that $\{x \mapsto v\}(p)$ can be assumed to hold. Then $\cdots D \cdots$ (5.15)

5 Common alternative names for it are *existential specialization* and *existential elimination* .

where D is a deduction that proceeds to derive some conclusion q with the aid of the assumption $\{x \mapsto v\}(p)$, along with whatever other assumptions are currently operative. We refer to

$$\exists\, x \,.\, p$$

as the *existential premise*; v is called the *witness* variable; and the sentence $\{x \mapsto v\}(p)$ is called the *witness hypothesis*. We call D the *body* of the existential instantiation. It represents the *scope* of the witness hypothesis, as well as the scope of v. The conclusion q derived by the body D becomes the result of the entire proof (5.15).

The following proof is a simple and fairly typical example of how existential instantiation is used in practice. (We use conventional notation for this example, since this is a generic, language-agnostic illustration of the technique.) The proposition of interest here is that for all integers n, *if* $even(n)$ *then* $even(n+2)$ (i.e., if n is even, then so is $n+2$). Let us say that the unary predicate *even* is defined as follows:

$$(\forall\, i \,.\, even(i) \Leftrightarrow \exists\, j \,.\, i = 2 \cdot j). \tag{5.16}$$

The proof relies on the lemma

$$(\forall\, x\, y \,.\, x \cdot (y+1) = x \cdot y + x) \tag{5.17}$$

and proceeds as follows:

> Pick any n and assume $even(n)$. Then, by (5.16), we infer $(\exists\, j \,.\, n = 2 \cdot j)$, that is, there is some number, which, when multiplied by 2, yields n. *Let k stand for such a number, so that $n = 2 \cdot k$. Then, by congruence, $n+2 = (2 \cdot k) + 2$. But, by* (5.17)*, $(2 \cdot k) + 2 = 2 \cdot (k+1)$, hence, by the transitivity of equality, $n+2 = 2 \cdot (k+1)$. Therefore, by existential generalization, we obtain $(\exists\, m \,.\, n+2 = 2 \cdot m)$, and so, from* (5.16)*, we conclude $even(n+2)$.*

We have italicized the existential elimination argument in the above proof. The existential premise here is $(\exists\, j \,.\, n = 2 \cdot j)$; the witness is the variable k; and the witness premise is the equality $n = 2 \cdot k$. The conclusion of the existential instantiation is $even(n+2)$.

There are some important caveats that must be observed to ensure that this type of reasoning, as outlined in (5.15), will not lead us astray. The witness v must serve as a dummy placeholder—no special assumptions about it should be used. In particular, the body D must not rely on any previous working assumptions that might happen to contain free occurrences of v. Things can easily go wrong otherwise. Suppose, for example, that the current assumption base contains the atom $even(k)$, for some variable k, along with the sentence $(\exists\, n \,.\, odd(n))$. Now if we unwisely choose k as a witness for this existential premise, we will obtain the witness hypothesis $odd(k)$, and hence, in tandem with $even(k)$, we will be able to conclude that there is a number that is both even and odd. We will shortly see how

Athena manages to enforce this proviso automatically, so that users do not need to be concerned with it. Further, to ensure that the witness v is only used as a temporary dummy, the final conclusion q should not depend on v in any essential way. Specifically, q should not contain any free occurrences of v.

Existential instantiations in Athena are performed by deductions of the form

$$\texttt{pick-witness}\ I\ \texttt{for}\ F\ D \tag{5.18}$$

where I is a name that will be bound to the witness variable, F is a phrase that evaluates to an existential premise (exists $x : S$. p), and D is the body.[6] To evaluate (5.18) in an assumption base β, we first make sure that the existential premise (exists $x : S$. p) is indeed in β; if not, evaluation halts and a relevant error is reported. Assuming that the existential premise is in β, we begin by generating a fresh variable $v : S$, which will serve as the actual witness variable. We then construct the witness hypothesis, call it p', obtained from p by replacing every free occurrence of $x : S$ by the witness $v : S$. Finally, we evaluate the body D in the augmented assumption base $\beta \cup \{p'\}$ and in an environment in which the name I is bound to the witness variable $v : S$. If and when that evaluation produces a conclusion q, we return q as the result of the entire proof (5.18), provided that q does not contain any free occurrences of $v : S$ (it is an error if it does). The fact that the witness variable $v : S$ is freshly generated is what guarantees that the body D will not be able to rely on any special assumptions about it. The freshness of $v : S$ along with the explicit proviso that it must not occur in the conclusion q ensures that the witness is used only as a temporary placeholder. Bear in mind that I itself in (5.18) is not an Athena term variable; it is a name—an identifier—that will come to *denote* a fresh variable (the witness variable $v : S$) in the course of evaluating the body D.

As an example, let us use existential instantiation to derive the tautology

```
((exists x . ~ prime x) ==> ~ forall x . prime x)
```

```
> assume hyp := (exists x . ~ prime x)
    pick-witness w for hyp  # We now have (~ prime w)
      (!by-contradiction (~ forall x . prime x)
         assume all-prime := (forall x . prime x)
           let {prime-w := (!uspec all-prime w)}
             (!absurd prime-w (~ prime w)))

Theorem: (if (exists ?x:Int
                (not (prime ?x:Int)))
             (not (forall ?x:Int
                    (prime ?x:Int))))
```

As an example of an incorrect use of **pick-witness**, the following violates the proviso that the witness variable must not appear in the conclusion:

6 It is an error if the evaluation of F produces any value other than an existentially quantified sentence.

```
> assume hyp := (exists x . prime x)
    pick-witness w for hyp
      (!claim (prime w))

input prompt:2:5: Error: Failed existential instantiation---the
witness variable occurs free in the resulting sentence.
```

Another potential error is that the existential premise is not in the assumption base:

```
> pick-witness w for (exists x . x =/= x)
    (!true-intro)

input prompt:1:22: Error: Existential sentence to be instantiated
is not in the assumption base:
(exists ?x:'S
  (not (= ?x:'S ?x:'S))).
```

When the existential premise has multiple consecutive existentially quantified variables, we can abbreviate nested **pick-witness** deductions with the form

$$\texttt{pick-witnesses}\ I_1 \cdots I_n\ \texttt{for}\ F\ D$$

where F is a phrase that evaluates to an existential premise (exists $x_1 \cdots x_m$. p) with $m \geq n$. The semantics of **pick-witnesses** is given by desugaring to **pick-witness**. For instance,

$$\texttt{pick-witnesses}\ I_1\ I_2\ \texttt{for}\ (\texttt{exists}\ x_1\ x_2\ .\ p)\ D$$

is an abbreviation for

```
pick-witness I1 for (exists x1 x2 . p)
  pick-witness I2 for (exists x2 . {x1 ↦ I1}(p))
    D
```

Here is a sample proof that (exists x y . x < y) implies (exists y x . x < y):

```
> assume hyp := (exists x y . x < y)
    pick-witnesses w1 w2 for hyp # This gives (w1 < w2)
      let {_ := (!egen (exists x . x < w2) w1)}
        (!egen (exists y x . x < y) w2);;

Theorem: (if (exists ?x:Real
               (exists ?y:Real
                 (< ?x:Real ?y:Real)))
             (exists ?y:Real
               (exists ?x:Real
                 (< ?x:Real ?y:Real))))
```

Sometimes it is convenient to give a name to the witness hypothesis and then refer to it by that name inside the body of the **pick-witness**. This can be done by inserting a name (an identifier) before the body *D* of the **pick-witness**. That identifier will then refer to the witness premise inside *D*. For example, the proof

pick-witness w (exists x . x = x) wp *D*

will give the name wp to the witness premise, so that every free occurrence of wp within *D* will refer to the witness premise (w = w). Thus, for instance, one of our earlier proofs could be written as follows:

```
> assume hyp := (exists x . ~ prime x)
    pick-witness w for hyp -prime-w
    # We now have -prime-w := (~ P w) in the a.b.
      (!by-contradiction (~ forall x . prime x)
        assume all-prime := (forall x . prime x)
          let {prime-w := (!uspec all-prime w)}
            (!absurd prime-w -prime-w))

Theorem: (if (exists ?x:Int
                 (not (prime ?x:Int)))
             (not (forall ?x:Int
                    (prime ?x:Int))))
```

This can also be done for **pick-witnesses** deductions; an identifier appearing right before the body will denote the witness premise. For example:

pick-witnesses w1 w2 for (exists x y . x =/= y) wp *D*.

Note that, intuitively, existential quantifications correspond to (potentially infinite) disjunctions. For instance, let B be a unary predicate on Boolean:

```
declare B: [Boolean] -> Boolean
```

To say that there is some Boolean value for which B holds:

(exists ?x:Boolean . B ?x:Boolean)) (5.19)

is simply to say that B holds for true *or* B holds for false:

(B true | B false). (5.20)

Sentences (5.19) and (5.20) have the same content. Likewise, to say that there is some integer that is prime:

(exists ?x:Int . prime ?x:Int) (5.21)

is really the same as saying that 0 is prime *or* 1 is prime, *or* -1 is prime, *or* 2 is prime, *or* -2 is prime, and so on:

```
(prime 0 | prime 1 | prime (- 1) | prime 2 | ···).
```
(5.22)

Accordingly, we should expect existential elimination and introduction to intuitively correspond to disjunction elimination and introduction, respectively. Consider, for example, an instantiation of the existential claim that `B` holds for *some* Boolean value:

```
pick-witness I for (exists x . B x)
  D
```

Such a proof could also be expressed as a disjunction elimination:

```
(!cases (B true | B false)
  assume (B true)
    D1
  assume (B false)
    D2)
```

The chief difference is that the first proof is more efficient because it does not consider every possible case separately. Rather, it abstracts what reasoning is common to D_1 and D_2 into what is essentially one single *method* parameterized over I. Also, the first proof does not use an explicit **`assume`**, as the insertion of the witness hypothesis into the assumption base is performed automatically. Thus, we can explain the evaluations of D_1 and D_2 as the evaluation of one and the same deduction, D, but under two different lexical bindings: with I denoting `true` in the first case and with I denoting `false` in the second. Indeed, if we only needed to work with *finite* domains, then we could formally define the evaluation semantics of **`pick-witness`** proofs by desugaring such proofs into `cases` applications. Such desugaring is not possible in the general case, which includes infinite domains, but nevertheless it is still instructive to keep the analogy between existential instantiation and disjunction elimination in mind.

We close by mentioning an alternative mechanism for existential instantiation, the syntax form

```
(with-witness E F D).
```

The expression E must produce a variable v, which will be used as the witness variable of the existential instantiation. Accordingly, this variable must not have any free occurrences in the assumption base. The phrase F must produce an existential quantification `(exists` x `.` p`)`, which will serve as the existential premise and must therefore be in the assumption base. Finally, D is the body of the instantiation, which, when evaluated in the given assumption base augmented with the witness hypothesis $\{x \mapsto v\}(p)$, must yield a conclusion q that does not contain any free occurrences of v. The sentence q then becomes the conclusion of the entire existential instantiation.

This is, essentially, a more primitive version of **pick-witness**. It stands to **pick-witness** in the same relationship that **generalize-over** stands to **pick-any**. Specifically, we can desugar **pick-witness** I **for** F D as

```
let {I := (fresh-var)}
  (with-witness I F D)
```

5.3 Some examples

In this section we present some simple examples of quantifier reasoning illustrating the mechanisms introduced in the previous section. We first write each derived sentence in conventional notation, and then we show how to prove that sentence in Athena.

$$(\forall x \,.\, P(x) \wedge Q(x)) \Rightarrow (\forall y \,.\, P(y)) \wedge (\forall y \,.\, Q(y))$$

```
assume hyp := (forall x . P x & Q x)
  let {all-P := pick-any y:Object
                  conclude (P y)
                    (!left-and (!uspec hyp y));
       all-Q := pick-any y:Object
                  conclude (Q y)
                    (!right-and (!uspec hyp y))}
    (!both all-P all-Q)
```

$$(\exists x \,.\, P(x) \wedge Q(x)) \Rightarrow (\exists y \,.\, P(y)) \wedge (\exists y \,.\, Q(y))$$

```
assume hyp := (exists x . P x & Q x)
  pick-witness w for hyp wp # we now have wp := (P w & Q w) in the a.b.
    let {Pw := (!left-and wp);
         Qw := (!right-and wp);
         some-P := (!egen (exists y . P y) w);
         some-Q := (!egen (exists y . Q y) w)}
      (!both some-P some-Q)
```

$$((\forall x \,.\, P(x)) \vee (\forall x \,.\, Q(x))) \Rightarrow (\forall y \,.\, P(y) \vee Q(y))$$

```
assume hyp := ((forall x . P x) | (forall x . Q x))
  pick-any y
    (!cases hyp
           assume case-1 := (forall x . P x)
             conclude (P y | Q y)
               (!either (!uspec case-1 y) (Q y))
           assume case-2 := (forall x . Q x)
             conclude (P y | Q y)
```

```
                (!either (P y) (!uspec case-2 y)))
```

$$(\exists x \,.\, P(x) \lor Q(x)) \Rightarrow (\exists x \,.\, P(x)) \lor (\exists x \,.\, Q(x))$$

```
assume hyp := (exists x . P x | Q x)
  pick-witness w for hyp wp
  # we now have wp := (P w | Q w) in the a.b.
    (!cases (P w | Q w)
            assume (P w)
              let {some-P := (!egen (exists x . P x) w)}
                (!either some-P (exists x . Q x))
            assume (Q w)
              let {some-Q := (!egen (exists x . Q x) w)}
                (!either (exists x . P x) some-Q))
```

$$(\forall x \,.\, P(x) \Rightarrow Q(x)) \Rightarrow ((\forall x \,.\, P(x)) \Rightarrow (\forall x \,.\, Q(x)))$$

```
assume hyp1 := (forall x . P x ==> Q x)
  assume hyp2 := (forall x . P x)
    pick-any a
      conclude (Q a)
        let {Pa=>Qa := (!uspec hyp1 a);
             Pa := (!uspec hyp2 a)}
          (!mp Pa=>Qa Pa)
```

Note that, strictly speaking, the theorem produced by this proof is printed as

$$(\forall x \,.\, P(x) \Rightarrow Q(x)) \Rightarrow ((\forall x \,.\, P(x)) \Rightarrow (\forall a \,.\, Q(a))),$$

but this is of course equivalent (in fact identical, for deductive purposes) to:

$$(\forall x \,.\, P(x) \Rightarrow Q(x)) \Rightarrow ((\forall x \,.\, P(x)) \Rightarrow (\forall x \,.\, Q(x))).$$

We could easily produce the latter representation if we so prefer, either by using x in place of a in the above proof, or else by wrapping the present **pick-any** proof inside a **conclude** that is quantified over x:

```
assume hyp1 := (forall x . P x ==> Q x)
  assume hyp2 := (forall x . P x)
    conclude (forall x . Q x)
      pick-any a
        conclude (Q a)
          let {Pa=>Qa := (!uspec hyp1 a);
               Pa := (!uspec hyp2 a)}
            (!mp Pa=>Qa Pa)
```

$$((\exists x \,.\, P(x)) \vee (\exists x \,.\, Q(x))) \Rightarrow (\exists x \,.\, P(x) \vee Q(x))$$

```
assume hyp := ((exists x . P x) | (exists x . Q x))
  let {goal := (exists x . P x | Q x)}
    (!cases
      hyp
      assume case-1 := (exists x . P x)
        pick-witness w for case-1 # we now have (P w) in the a.b.
          let {Pw|Qw := (!either (P w) (Q w))}
            (!egen goal w)
      assume case-2 := (exists x . Q x)
        pick-witness w for case-2 # we now have (Q w) in the a.b.
          let {Pw|Qw := (!either (P w) (Q w))}
            (!egen goal w))
```

A closer look at this proof reveals some duplication of effort: The reasoning is essentially identical in both cases, yet it is repeated almost verbatim for each case separately. We can easily factor out the commonalities in a single method that we can then reuse accordingly. The savings afforded by such abstraction are minimal in this case (as the proof is so small to begin with), but it is nevertheless instructive to carry out the factoring anyway. In larger proofs the benefits can be more substantial:

```
assume hyp := ((exists x . P x) | (exists x . Q x))
   let {goal := (exists x . P x | Q x);
        M := method (ex-premise)
               assume ex-premise
                 pick-witness w for ex-premise
                   let {Pw|Qw := (!either (P w) (Q w))}
                     (!egen goal w)}
     (!cases hyp (!M (exists x . P x))
                 (!M (exists x . Q x)))
```

The next section presents several more examples of custom-written first-order methods.

Exercise 5.2: Each of the following listings asserts a number (possibly zero) of premises and defines a goal:

Listing 5.3.1

```
assert premise-1 := (forall x y . x R y)

define goal := (forall x . x R x)
```

Listing 5.3.2

```
define goal := (forall x . exists y . x = y)
```

Listing 5.3.3

```
assert premise-1 := (forall x . P x | Q x ==> S x)
assert premise-2 := (exists y . Q y)

define goal := (exists y . S y)
```

Listing 5.3.4

```
assert premise-1 := (exists x . P x & Q x)
assert premise-2 := (forall y . P y ==> S y)

define goal := (exists x . S x & Q x)
```

Listing 5.3.5

```
assert premise-1 := (~ exists x . Q x)
assert premise-2 := (forall x . P x ==> Q x)

define goal := (~ exists x . P x)
```

Listing 5.3.6

```
assert premise-1 := (forall x . P x ==> Q x)
assert premise-2 := (exists x . S x & ~ Q x)

define goal := (exists x . S x & ~ P x)
```

Listing 5.3.7

```
assert premise-1 := (forall x . x R x ==> P x)
assert premise-2 := (exists x . P x ==> ~ exists y . Q y)

define goal := ((forall x . Q x) ==> ~ exists z . z R z)
```

Listing 5.3.8

```
define goal :=
 ((exists x . P x | Q x) <==> (exists x . P x) | (exists x . Q x))
```

Derive each goal from the corresponding premises. □

5.4 Methods for quantifier reasoning

The four introduction and elimination mechanisms for quantifiers that we have discussed so far (the methods uspec and egen and the deduction forms **pick-any** and **pick-witness**), in tandem with the introduction and elimination mechanisms for the sentential connectives presented in Chapter 4, constitute a complete proof system for first-order logic. That is, if any sentence p follows logically from an assumption base β, then there is some proof D composed from these mechanisms that can derive p from β.[7] However, as we pointed out in the previous chapter, if we had to limit ourselves to these primitive mechanisms when writing proofs, our job would be much more difficult than it needs to be.

Luckily, that is not the case. We can write our own proof tools by combining the primitive deductive mechanisms in various flexible ways. Athena's semantics guarantees that these new tools will never go wrong, meaning that they will never produce a conclusion that is not logically entailed by the assumption base. It will become evident throughout the rest of the book that the ability to write methods that automate recurrent patterns of reasoning is crucial for managing the complexity of proof development. In the previous chapter we defined a number of methods that are useful for sentential reasoning. In this section we present some methods that are useful for quantifier reasoning. Of course, many more are possible. These examples will only give a flavor of what can be accomplished.

- *Multiple universal specialization and existential generalization*:

  Suppose that we have a premise p with $k \geq 0$ universal quantifiers at the front. We often want to perform universal specialization on p in one fell swoop, with a *list* of k terms [$t_1 \cdots t_k$], instead of having to apply uspec k separate times. That functionality can easily be programmed as a recursive method:

```
define uspec* :=
  method (premise terms)
    match terms {
      [] => (!claim premise)
    | (list-of t rest) => (!uspec* (!uspec premise t) rest)
    }
```

  Recall the semantics of nested method calls from Section 2.10.2. Since the call

```
(!uspec premise t)
```

  appears in an argument position of the outer method call, its conclusion will be added to the assumption base prior to the application of the outer method (uspec*). This ensures the continued satisfaction of the method's precondition, namely, that the first argument

7 A more precise definition of what it means for a sentence p to follow logically from an assumption base β is given in Section 5.6.

of uspec* should be a *premise*, meaning that it should be in the assumption base at the time when uspec* is applied. An equivalent way of writing this method is this:

```
define uspec* :=
  method (premise terms)
    match terms {
      [] => (!claim premise)
    | (list-of t rest) => let {premise' := (!uspec premise t)}
                            (!uspec* premise' rest)
    }
```

The virtue of the first version is conciseness.

An example of uspec* in action:

```
assert premise := (forall x y . x = y ==> y = x)

> (!uspec* premise [1 2])

Theorem: (if (= 1 2)
             (= 2 1))
```

The method can accept a list of fewer than k terms, in which case the universal quantification is only partially instantiated:

```
assert <-transitivity := (forall x y z . x < y & y < z ==> x < z)

> (!uspec* <-transitivity [1.7 2.9])

Theorem: (forall ?v1805:Real
           (if (and (< 1.7 2.9)
                    (< 2.9 ?v1805:Real))
               (< 1.7 ?v1805:Real)))
```

If the second argument is the empty list of terms, then the application of uspec* has the same effect as applying claim to the first argument. Note that uspec* is also known as instance; it was first introduced on page 115. (Except that instance can also take a single term as its second argument, so it can be used in place of uspec as well.)

The ability to existentially generalize over multiple terms in one step is likewise useful, and this is possible with the method egen*. For instance, if we have (1 < 2) in the assumption base, we can derive (exists x y . x < y) in one step, simply by citing the terms 1 and 2:

```
(!egen* (exists x y . x < y) [1 2]).
```

The order of the existential quantifiers corresponds to the order in which the terms are listed, meaning that the generalization over x is to be based on 1, while the generalization over y is based on 2. In general, for $k > 0$,

(!egen* (exists $x_1 \cdots x_k$. p) [$t_1 \cdots t_k$])

derives the conclusion `(exists` $x_1 \cdots x_k$ `.` p`)`, provided that

$$\{x_k \mapsto t_k\}(\cdots\{x_1 \mapsto t_1\}(p)\cdots)$$

is in the assumption base (an error occurs otherwise).

Exercise 5.3: Implement the method `egen*`. □

- *Forward Horn clause inference*: In practice, the most useful—and common—universal quantifications are of the form

$$\texttt{(forall } x_1 \cdots x_n \texttt{ . } p \texttt{ ==> } q\texttt{)} \tag{5.23}$$

and

$$\texttt{(forall } x_1 \cdots x_n \texttt{ . } p \texttt{ <==> } q\texttt{).} \tag{5.24}$$

A few examples:

```
(forall x y . x = y ==> y = x);                    (5.25)
(forall x y . x parent y ==> x ancestor y);        (5.26)
(forall x y . x > 0 & y > 0 ==> x - y < x);        (5.27)
(forall x y . x <= y <==> x = y | x < y).          (5.28)
```

Many more examples will be encountered throughout the book. It is not an exaggeration to say that most mathematical results, as well as the large majority of logical formulas that arise in system modeling, are expressed in one of these two forms, particularly the first one, (5.23). Sentences of that form are called *Horn clauses*.[8] We will also refer to them as Horn *rules*, or, when there is no risk of confusion, simply as "rules." Note that a sentence of the second form, (5.24), can be regarded as the conjunction[9] of the following two Horn clauses:

$$\texttt{(forall } x_1 \cdots x_n \texttt{ . } p \texttt{ ==> } q\texttt{)} \tag{5.29}$$

and

$$\texttt{(forall } x_1 \cdots x_n \texttt{ . } q \texttt{ ==> } p\texttt{).} \tag{5.30}$$

One of the most common things that we want to do with a Horn rule of the form (5.23) is to apply it (or to "fire" it, in the terminology of rule systems) on some specific terms $t_1, \ldots, t_n$, that is, to derive the appropriate instance of the conclusion q, given that *the corresponding instance of the antecedent p has already been established.* For instance, suppose that we know that `peter` is a parent of `mary`, so that the atom

8 This is actually a generalized conception of Horn clauses. In the usual formulation, the antecedent of a Horn clause is a conjunction of $k > 0$ atoms (perhaps only one), known as the *body* of the clause, and the consequent is a single atom, called the *head* of the clause; see Appendix B for more details. Here, by contrast, we can have arbitrary sentences in the antecedent and consequent positions.

9 We ask you to prove the equivalence of (5.24) and the conjunction of (5.29) and (5.30) in Exercise (5.12).

```
(peter parent mary)
```

is in the assumption base. It then becomes evident that rule (5.26) is applicable, and specifically that we can use it to infer the conclusion

```
(peter ancestor mary).
```

We refer to this process as "firing" (5.26) on the terms `peter` and `mary`. This type of inference with Horn rules is also called *forward*, because we proceed from the antecedent of (an instance of) the rule to the consequent, in contrast to the backward sort of inference search that occurs in logic programming, and which proceeds from consequent to antecedent. (We discuss logic programming in Appendix B.)

Thus, "firing" a Horn rule

(forall $x_1 \cdots x_n$. p ==> q)

proceeds in two stages. First, a list of terms [$t_1 \cdots t_n$] is used to specialize the universal quantification, with each t_i replacing x_i, $i = 1, \ldots, n$. Then, we perform modus ponens on the instantiated rule and its antecedent. For instance, the effect of the above example can be achieved by the following proof:

```
domain Person

declare parent, ancestor: [Person Person] -> Boolean

declare peter, mary: Person

assert ancestor-rule := (forall x y . x parent y ==> x ancestor y)

assert fact := (peter parent mary)

let {rule-instance :=
       conclude (peter parent mary ==> peter ancestor mary)
         (!uspec* ancestor-rule [peter mary])}
  (!mp rule-instance fact)
```

This is the kind of thing we would like to do with a single method call. Specifically, we would like this type of process to be carried out with a single call of the form

(!fire R [$t_1 \cdots t_n$]),

where R is a Horn rule and $t_1 \cdots t_n$ are arbitrary terms of the proper sorts. In the above example, the call

```
(!fire ancestor-rule [peter mary])
```

should derive `(peter ancestor mary)`—provided, again, that the precondition

```
(peter parent mary)
```

is in the assumption base.

The following is a first attempt at coding `fire`:

```
define fire :=
  method (rule terms)
    match rule {
      (forall (some-list _) (_ ==> _)) =>
        let {rule-instance := (!uspec* rule terms)}
          (!mp rule-instance (antecedent rule-instance))
    }
```

The procedure `antecedent` (defined in the Athena library) simply returns the antecedent of a given conditional:

```
define antecedent :=
  lambda (cond)
    match cond {
      (p ==> _) => p
    }
```

This implementation always expects the antecedent of the instantiated rule to be in the assumption base. Oftentimes, however, the antecedent of a Horn clause is a conjunction of several sentences $p_1, \ldots, p_k$—clause (5.27) is an example—and at the time when we invoke `fire` we do not quite have the entire conjunction of the relevant instances of $p_1, \ldots, p_k$ in the assumption base; we only have the instances themselves. For instance, in the case of clause (5.27), we might have the atomic components `(x > 0)` and `(y > 0)` in the assumption base, but not their conjunction. What we need to do in such cases is assemble the instances in a conjunction before we apply modus ponens. We can accommodate such cases by slightly changing the structure of the above implementation: Instead of directly passing the antecedent of the appropriate rule instance to modus ponens, we first pass it to a method that *derives* the antecedent:

```
define fire :=
  method (rule terms)
    match rule {
      (forall (some-list _) (_ ==> _)) =>
        let {rule-instance := (!uspec* rule terms)}
          (!mp rule-instance (!prove-antecedent rule-instance))
    }
```

Now, if we simply defined `prove-antecedent` as follows:

```
define prove-antecedent :=
  method (cond)
    match cond {
      (p ==> _) => (!claim p)
    }
```

then our new implementation would have the exact same effect as the preceding one. But we can make fire more powerful by independently making prove-antecedent more resourceful. Specifically, the method will first check to see whether the antecedent is already in the assumption base. If it is, it will simply claim it. If it is not, then it will try to derive it depending on its logical structure:

```
define prove-components-of :=
  method (p)
    check {
      (holds? p) => (!claim p)
    | else => match p {
            (p1 & p2) =>
              (!both (!prove-components-of p1)
                     (!prove-components-of p2))
          | (p1 | p2) =>
              try {
                (!either (!prove-components-of p1) p2)
              | (!either p1 (!prove-components-of p2))
              }
          }
    }

define prove-antecedent :=
  method (p)
    (!prove-components-of (antecedent p))
```

So if the antecedent instance is not in the assumption base, we check to see if it is a conjunction or a disjunction. If it is a conjunction, we recursively derive each conjunct and then apply both. If it is a disjunction, we try to (recursively) derive one of the disjuncts, and then we apply either. This is a very simple theorem-proving strategy that nevertheless works fairly well in practice. A more sophisticated version of the same idea is used to implement the conditional-rewriting reasoning performed by the chain method that will be introduced in the next chapter.

It is convenient to extend fire so that it works with universally quantified biconditionals as well, that is, with rules of the form (5.24). Since the method is only given the rule and a list of terms, it cannot know ahead of time which side of the instantiated biconditional will be in the assumption base, so we use backtracking (via **try**) to find out:

```
1 define fire :=
2   method (rule terms)
3     match rule {
4       (forall (some-list _) (_ ==> _)) =>
5         let {rule-instance := (!uspec* rule terms)}
6           (!mp rule-instance (!prove-antecedent rule-instance))
7     | (forall (some-list _) (_ <==> _)) =>
8         let {rule-instance := (!uspec* rule terms)}
```

```
 9          try {
10            let {cond-1 := (!left-iff rule-instance)}
11              (!mp cond-1 (!prove-antecedent cond-1))
12          | let {cond-2 := (!right-iff rule-instance)}
13              (!mp cond-2 (!prove-antecedent cond-2))
14          }
15      | (forall _ _) => (!uspec* rule terms)
16      }
```

For example, supposing that we have `(a <= b)` in the assumption base, applying `fire` to rule (5.28) and the list `[a b]` will derive the conclusion

```
(a < b | a = b).                                                    (5.31)
```

Conversely, if we had (5.31) in the assumption base but not `(a <= b)`, then the same application would derive the theorem `(a <= b)`.[10] Also, note the addition of line 15 in the code listing; for extra flexibility, when the body of the given rule p is neither a conditional nor a biconditional, the method simply instantiates p on the given terms.

- *Quantifier negation*: A negation sign in front of a quantifier Q can always be pushed inward provided we change Q from universal to existential or vice versa. For instance, from

  ```
  (~ forall x . prime x)                                            (5.32)
  ```

  we may infer

  ```
  (exists x . ~ prime x).                                           (5.33)
  ```

  Such an inference is valid: If not every integer is prime, then there must be some integer that is not prime. Likewise, if `immortal` is a unary predicate on `Person` (with signature `[Person] -> Boolean`), then from

  ```
  (~ exists x . immortal x)                                         (5.34)
  ```

  we should be able to infer

  ```
  (forall x . ~ immortal x).                                        (5.35)
  ```

  The rules that license such inferences are collectively referred to as the *quantifier negation* rules. In customary notation they can be depicted as follows:

  $$\frac{\neg \forall x \,.\, p}{\exists x \,.\, \neg p} \qquad\qquad \frac{\neg \exists x \,.\, p}{\forall x \,.\, \neg p}$$

  The rules are bidirectional: Each can also license passage from a sentence of the form shown below the line to the sentence above the line. Thus, for instance, from (5.33) we

10 If both `(a <= b)` and (5.31) are in the assumption base then there would be little point in applying the method, but nevertheless, *if* the method were applied, which theorem would be derived?

can infer (5.32), and from (5.35) we can derive (5.34). If we recall the analogy between universal quantification and conjunction, and between existential quantification and disjunction, then the quantifier negation rules can be viewed as generalized versions of De Morgan's rules from sentential logic.

In the vein of sentential methods such as `cond-def` (page 222), we will implement versions of the above rules that use complementation instead of negation, as these tend to be more convenient in practice:

$$\frac{\neg \forall x \,.\, p}{\exists x \,.\, \overline{p}} \qquad\qquad \frac{\neg \exists x \,.\, p}{\forall x \,.\, \overline{p}}$$

We first define a separate unary method for each of the four different directions, which will do the heavy lifting. Once we have these, we can define a single quantifier-negation method that takes the appropriate action based on the form of the input. We implement three of these methods here and leave the fourth as an exercise.

```
define (qn-1 premise) :=
  match premise {
    (~ (forall x p)) => (!force (exists x . complement p))
  }

define (qn-2 premise) :=
  match premise {
    (exists x p) =>
      (!by-contradiction (~ forall x . complement p)
        assume hyp := (forall x . complement p)
          pick-witness w for premise p-w
            let {p'-w := (!uspec hyp w)}
              (!from-complements
                false
                p'-w
                p-w))
  }

define (qn-3 premise) :=
  match premise {
    (~ (exists x p)) =>
      conclude (forall x . complement p)
        pick-any y
          let {p-y := (replace-var x y p)}
            (!by-contradiction (complement p-y)
              assume p-y
                (!absurd (!egen (exists y p-y) y)
                         premise))
  }

define (qn-4 premise) :=
  match premise {
```

```
    (forall x p) =>
      (!by-contradiction (~ exists x . complement p)
        assume hyp := (exists x . complement p)
          pick-witness w for hyp -p-w
            (!from-complements false -p-w (!uspec premise w)))
  }
```

Exercise 5.4: Implement qn-1. □

Our top-level method for quantifier negation now does a simple dispatch:

```
define qn :=
  method (premise)
    match premise {
      (~ (forall _ _)) => (!qn-1 premise)
    | (exists _ _)     => (!qn-2 premise)
    | (~ (exists _ _)) => (!qn-3 premise)
    | (forall _ _)     => (!qn-4 premise)
    }
```

- *Quantifier swapping*: Two leading quantifiers of the same kind can always be swapped. For instance,

  `(forall x y . x < y)`

  is equivalent to

  `(forall y x . x < y).`

  In addition, a universal quantifier immediately following an existential quantifier can always be pulled to the front of the existential quantifier. For example, from

  `(exists x . forall y . x <= y)`

  we can derive

  `(forall y . exists x . x <= y).`

  It is convenient to have a method that can perform such transformations automatically:

```
define (quant-swap premise) :=
  match premise {
    (forall _ (forall _ _)) =>
      pick-any y x
        (!uspec* premise [x y])
  | (exists x (exists y body)) =>
      pick-witnesses w_x w_y for premise
        (!egen* (exists y x . body) [w_y w_x])
  | (exists x (forall _ _)) =>
      pick-any y
```

```
      pick-witness w for premise wp
        let {q := (!uspec wp y)}
          (!egen (exists x (replace-var w x q)) w)
}
```

The last case, when the input premise is of the form (exists x . forall y . p), is somewhat subtle. The reader should trace the reasoning of the method on some specific examples, such as (exists x . forall y . x <= y).

- *Quantifier distribution*: Universal quantifiers distribute over conjunction and existential quantifiers distribute over disjunction. In other words, each of the two following inference rules is bidirectional and can be applied either on the premise above the line to derive the conclusion below the line or conversely:

$$\frac{\forall x \,.\, p_1 \wedge p_2}{(\forall x \,.\, p_1) \wedge (\forall x \,.\, p_1)} \qquad \frac{\exists x \,.\, p_1 \vee p_2}{(\exists x \,.\, p_1) \vee (\exists x \,.\, p_1)}$$

  We have already seen proofs implementing limited versions of these rules in Section 5.3. A more general unary method that can work in any one of the four directions, quant-dist, is defined in Exercise 5.14. That method can also accept a premise of the form $\exists x \,.\, p_1 \wedge p_2$ and derive $(\exists x \,.\, p_1) \wedge (\exists x \,.\, p_2)$; and, finally, from a premise of the form $(\forall x \,.\, p_1) \vee (\forall x \,.\, p_1)$ it can derive the conclusion $(\forall x \,.\, p_1 \vee p2)$.

- *Replacement rule*: It is useful to extend the replacement method transform of Section 4.10 so that it can handle quantifiers. Consider, for instance, the following two sentences:

```
define p :=  (forall x . exists y . x R y ==> ~ (P x & Q y))

define q := (forall z . exists w . ~ ~ z R w ==> ~ P z | ~ Q w)
```

  We should be able to infer q from p, and vice versa, simply by citing the two bidirectional methods bdn and dm. For instance, assuming that p is in the assumption base,

```
(!transform p q [bdn dm])
```

  should produce q. Conversely, if q is in the assumption base,

```
(!transform q p [bdn dm])
```

  should produce p.

  After reviewing the implementation of transform, it might appear that we need to implement congruence methods for the two quantifiers, call them ugen-cong and egen-cong, analogous to the sentential methods or-cong, iff-cong, and so on. That is correct, but the presence of variables complicates the situation somewhat. For instance, in the case of p and q above, we can hardly expect to prove the equivalence of the quantified bodies

```
(?x R ?y ==> ~ (P ?x & Q ?y))
```

and

```
(~ ~ ?z R ?w ==> ~ P ?z | ~ Q ?w)
```

and then apply the quantifier congruence methods, because the two bodies are actually *not* equivalent, since, for instance, (?x R ?y) is not derivable from (~ ~ ?z R ?w).

We can get around this issue by making the interface of the quantifier congruence methods, ugen-cong and egen-cong, a bit more flexible. Say we are interested in proving the equivalence of

$$(\forall x \, . \, p) \tag{5.36}$$

and

$$(\forall y \, . \, q) \tag{5.37}$$

on the basis of n methods $M_1, \ldots, M_n$. Instead of expecting the bodies p and q to have already been shown equivalent prior to the invocation of ugen-cong, we will pass a binary method M to ugen-cong, which can then be used dynamically to prove the equivalence of appropriately renamed versions of p and q. That equivalence will then enable the derivation of the outer equivalence of (5.36) and (5.37):

```
define (ugen-cong p q M) :=
  match [p q] {
    [(forall v1 p-body) (forall v2 q-body)] =>
      conclude (p <==> q)
        (!equiv assume p
                  pick-any x
                    let {p-body' := (!uspec p x);
                         q-body' := (replace-var v2 x q-body);
                         bicond := conclude (p-body' <==> q-body')
                                     (!M p-body' q-body')}
                      (!mp (!left-iff bicond) p-body')
                assume q
                  pick-any x
                    let {q-body' := (!uspec q x);
                         p-body' := (replace-var v1 x p-body);
                         bicond := conclude (q-body' <==> p-body')
                                     (!M q-body' p-body')}
                      (!mp (!left-iff bicond) q-body'))
  }
```

The implementation of a similar ternary method egen-cong for existential quantifiers is left as an exercise.

Exercise 5.5: Define egen-cong. □

Finally, we extend prove-equiv as follows:

```
define prove-equiv :=
  method (p q methods)
    try {
      (!find-first-element method (M)
                          (!equiv assume p (!M p)
                                  assume q (!M q))
                          methods)
    | check {
        (equal? p q) => (!ref-equiv p)
      | else => let {M := method (p1 p2)
                            (!prove-equiv p1 p2 methods)}
                match [p q] {
                  [(~ p1) (~ q1)] =>
                    (!not-cong (!prove-equiv p1 q1 methods))
                | [((some-sent-con pc) p1 p2) (pc q1 q2)] =>
                    (!(choose-cong-method pc)
                      (!prove-equiv p1 q1 methods)
                      (!prove-equiv p2 q2 methods))
                | [(forall _ _) (forall _ _)] =>
                    (!ugen-cong p q M)
                | [(exists _ _) (exists _ _)] =>
                    (!egen-cong p q M)
                }
      }
    }
```

5.5 Proof heuristics for first-order logic

We will now expand the collection of tactics described in Section 4.14 with some new entries: two backward tactics for introducing quantifiers and some forward tactics for eliminating them. We will also slightly extend the notion of a proof spec. Instead of "Derive p from β," the new format for a proof spec will be

$$\text{Derive } p \text{ from } (\beta, \boldsymbol{T})$$

where $\boldsymbol{T}$ is a finite set of terms, which we call the *proof terms*. We explain the role of this set below.

Writing first-order proofs often requires choosing appropriate terms with which to specialize universal quantifications (via uspec), or from which to existentially generalize (via egen). In theory, there are infinitely many choices for such terms, since there are infinitely many terms of any given sort.[11] However, in practice these terms come from a finite pool of candidates, namely, the set of all terms t such that:

11 Even if there are only finitely many *ground* terms of a certain sort, there are still infinitely many variables available for it.

1. t occurs in the assumption base or in the goal;[12]
2. t is a fresh variable introduced by a universal generalization (a **generalize-over** or **pick-any**), or a witness variable introduced by an existential instantiation (a **with-witness** or **pick-witness**).

The first of these two collections of terms is fixed once we are given an assumption base and a goal. But the second one grows dynamically as we attempt more universal generalizations and existential instantiations. The tactics described below keep track of these additions, ensuring that we have access to an appropriate set of proof terms at any given point in the proof search. These terms are then used by other tactics for existential generalization and universal specialization.

5.5.1 Backward tactics for quantifiers

The following is the standard backward tactic for universal quantifications:

Derive (forall v . p) from (β, T)

```
pick-any I
  (!force2 {v ↦ I}(p) T ∪ {I})                    [forall<-]
```

where I is a name of our choosing (typically different from all other names we have previously used for similar purposes in the course of a proof attempt). Informally, this tactic can be read as follows: To derive a universal quantification (forall v . p), pick a fresh variable named I and attempt to find a proof that derives $\{v \mapsto I\}(p)$. Note that the variable denoted by I becomes a member of the set of available proof terms in the new subgoal.

We use **force2** here as a binary version of **force** simply in order to keep track of the set of proof terms in the formulation of the subgoals. The exact representation of that set in Athena is immaterial as long as it supports the usual operations, so we will not specify it in detail here. We'll simply assume there is a procedure set that takes a list of values and makes a set out of them. We can define **force2** as follows:

```
define (force2 p terms) := (!force p)
```

The following is the backward tactic for existential quantifications:

Derive (exists v . p) from $(\beta, T = \{\ldots, t, \ldots\})$

```
let {_ := (!force2 {v ↦ t}(p) T)}                  [exists<-]
  (!egen (exists v . p) t)
```

12 Unless the said occurrence of t also contains bound variable occurrences.

It can be understood as follows: To derive an existentially quantified sentence

$$\texttt{(exists } v \texttt{ . } p\texttt{)}, \tag{5.38}$$

choose a proof term t and try to show that p holds for t, or more precisely, try to derive $\{v \mapsto t\}(p)$. If we manage that, we can infer the original goal (5.38) by existential generalization. There may be several proof terms available, in which case the tactic calls for a nondeterministic choice. In practice, we use whatever term seems most promising (i.e., whatever term produces a subgoal that can be tackled by a high-priority tactic).

5.5.2 Forward tactics for quantifiers

Before we describe the forward quantifier tactics, we need to introduce two new concepts. First, let us say that an occurrence of a sentence q inside a sentence p is *universal* iff it is not inside the scope of any positively embedded[13] existential quantifiers or negatively embedded universal quantifiers. We will often use a starred subscript to indicate that an occurrence of q inside p is universal: $p = (\cdots q_* \cdots)$.

Example 5.1: The following highlighted subsentence occurrences are universal in their respective supersentences p_1–p_6:

p_1 = `(P a & Q b)`
p_2 = `(forall x . forall y . x R y ==> P x)`
p_3 = `(forall x . P x | forall y . Q y)`
p_4 = `(forall x . P x | forall y . Q y)`
p_5 = `(P a & forall x . P x ==> ~ exists y . ~ x R y)`
p_6 = `(~ exists x . P x ==> ~ Q x)`

The reason is that they are not within the scope of any positive existential or negative universal quantifier occurrences. However, the following highlighted subsentence occurrence is not universal in p:

$p =$ `(P a | exists x . P x ==> x R y)`.

That is because `(x R y)` lies within the scope of an existential quantifier that is positively embedded in p. Likewise, the occurrence of `(P x)` is not universal in the following sentence:

$p =$ `(~ forall x . P x)`,

because it is within the scope of a universal quantifier that is negatively embedded in p. □

13 All of the concepts and algorithms involving polarities that were introduced in the sentential setting can be straightforwardly extended to the first-order case; see Exercise 5.6.

Second, if q is a universal sentence occurrence inside p, we write $UFV(q,p)$ for the set of all and only those free variables of q that fall within the scope of some quantifier in p.[14] We refer to the elements of $UFV(q,p)$ as the *universal free variables* of q in p.

Example 5.2: Consider the sentences $p_1, \ldots, p_6$ above, and let q_i refer to the highlighted sentence that occurs in p_i. We then have:

$$\begin{array}{lcl} UFV(q_1,p_1) & = & \emptyset \\ UFV(q_2,p_2) & = & \{\mathtt{x}\} \\ UFV(q_3,p_3) & = & \{\mathtt{y}\} \\ UFV(q_4,p_4) & = & \emptyset \\ UFV(q_5,p_5) & = & \{\mathtt{x},\mathtt{y}\} \\ UFV(q_6,p_6) & = & \{\mathtt{x}\} \end{array}$$

Observe the difference between $UFV(q_3,p_3)$ and $UFV(q_4,p_4)$. □

We are now ready to present the new tactics, starting with one for eliminating existential quantifiers:

Derive r from $(\beta = \{\ldots,(\cdots \texttt{(exists } v \texttt{ . } q\texttt{)}^{+}_{*} \cdots),\ldots\}, \boldsymbol{T})$

```
let {_ := (!force2 (exists v . q) T)}                    [exists->]
  pick-witness I for (exists v . q)
    (!force2 r T ∪ {I})
```

where I is a (typically fresh) name of our choosing. This is one of the highest-priority tactics, as we will see later. It advises us to eliminate existential quantifiers: If we see an existential quantification positively embedded in a universal position in the assumption base, derive it and then "unpack" it—eliminate the existential quantifier (via **pick-witness**). Thus, the first subgoal is to derive the existential quantification itself. This may well be trivial, as the existential quantification is often directly contained in the assumption base, in which case we do not have to derive it at all. The second subgoal is to derive the original r, but only after introducing a fresh witness variable for the existential sentence. Therefore, this second subgoal has access to an additional premise (the witness premise $\{v \mapsto w\}(q)$, where w is the variable denoted by I),[15] and an additional proof term (the variable denoted by I).

Note that the requirement that the existential quantification be universally embedded in p ensures that the tactic is always applied in an outside-to-inside order. For instance, if p is

14 That is, these variables occur free in q when q is considered in isolation, but they are bound when q is considered as a subsentence of p. Also note that since the occurrence of q in p is universal, the quantifiers in question have to be essentially universal, meaning that they must be either universal quantifiers that are positively embedded in p or existential quantifiers that are negatively embedded in it.

15 If the existential quantification `(exists` v `.` q`)` was not initially in β, then the second subgoal also has access to that sentence. However, the witness premise $\{v \mapsto I\}(q)$ is much more likely to play an important role inside D_2 than the sentence `(exists` v `.` q`)`.

the sentence (exists x . exists y . x R y), then the tactic is only applicable to the outer existential quantification, as the inner one is not in a universal position.

Universal quantifications that are negatively embedded in the assumption base are essentially existential quantifications, and should also be unpacked as soon as possible. Thus, in a sense, the following tactic is the dual of [exists->] (recall that we write $\overline{q}$ for the complement of q):

Derive r from ($\beta = \{\ldots, (\cdots$ (forall v . $q)^{-}_{*} \cdots), \ldots, T$)

```
let {p1 := (!force2 (~ forall v . q) T);                    [exists2->]
     p2 := (!qn p1)}
  pick-witness I for p2
    (!force2 r T ∪ {I}))
```

Finally, we introduce a forward tactic involving universal quantifiers. It recommends specializing a universal quantification in the assumption base with some available proof term:

Derive r from ($\beta = \{\ldots, (\cdots$ (forall v . $q)^{+}_{*} \cdots), \ldots\}, T = \{\ldots, t, \ldots\}$)

```
let {p := (!force2 (forall v . q) T);                    [forall->]
     _ := (!uspec p t)}
  (!force2 r T)
```

In this case, too, we have a dual tactic, which is applicable when we have an existential quantification negatively embedded in the assumption base. We then try to derive the negation of the existential sentence, transform it into a universal quantification by moving the negation sign inward, and specialize it:

Derive r from ($\{\ldots, (\cdots$ (exists v . $q)^{-}_{*} \cdots), \ldots\}, T = \{\ldots, t, \ldots\}$)

```
let {p1 := (!force2 (~ exists v . q) T);                    [forall2->]
     p2 := (!qn p1);
     _ := (!uspec p2 t)}
  (!force2 r T)
```

[forall->] and its dual are relatively seldom used because their application cannot be easily controlled and can render the search process inefficient. Later we will introduce a more precisely targeted forward tactic for universal quantifiers.

That is, by and large, all we need in the way of new tactics. However, before we can start using these we need to do a little more work. First, we need to make some small modifications to the sentential tactics to ensure that they can be properly used in the new setting of first-order logic. One obvious change is that every proof spec now contains a set of proof terms $\boldsymbol{T}$ in addition to the assumption base β, so the tactics need to be reformulated to reflect that. This is a perfectly straightforward exercise that we will carry out for illustration purposes only for the backward conjunction tactic, [and<-]:

Derive (p & q) from (β, T)

```
let {left := (!force2 p T);                                    [and<-]
     right := (!force2 q T)}
  (!both left right)
```

This is the only change that is necessary for the backward tactics, and also for the following forward tactics: reiteration; the indirect and complement tactics; and the replacement tactics. The generalized disjunction tactic ([or->]), the tactic for deriving false ([false->]), and the extraction tactics need more extensive renovations. Let us start with the extraction tactics.

Recall that the idea behind the extraction tactics was to search the assumption base for a positive occurrence of an appropriate *parent* of the goal. We then attempt to (a) derive the parent, and (b) detach the goal from the parent by the proper elimination method. Parenthood was defined syntactically; for instance, a conjunctive parent of p was understood as any conjunction (p_1 & p_2) such that at least one p_i is *syntactically identical* to p. We will need to enrich this notion of parenthood by relaxing the requirement of syntactic identity, replacing it with the more flexible *matching* relation.[16]

To motivate this change, suppose that we need to derive the goal (P zero) from the premise[17]

```
(forall x . P x & Q x).
```

If we used syntactic identity as our criterion for parenthood, then no extraction tactic would be applicable here, since, under that understanding of parenthood, the premise does not contain any positive occurrences of any parents of the goal. However, intuitively, we can see that the premise does contain a positive occurrence of a generalized kind of parent of the goal, namely, the sentence

```
(P x & Q x).                                                    (5.39)
```

This is a "generalized" sort of parent in that an *instance* of (5.39) is a regular parent of the goal (P zero). In particular, the instance of (5.39) that is obtainable through the substitution {x ↦ zero} is a regular parent of the goal, in the old sense. The instance in question, of course, is the conjunction (P zero & Q zero).

---

16 Recall from Section 2.14.8 that a sentence q *matches* a sentence p iff there is a substitution θ such that $\theta(p) = q$.

17 In this case we could discover a proof even without modifying the extraction tactics as we are about to do, but disregard that for now.

So from now on we will be searching the assumption base for positive occurrences of such generalized goal parents. To do this, we will need to determine if the goal *matches* various positive subsentences of the assumption base. As we saw in Section 2.14.8, there is an efficient procedure for sentence matching, which we can use to make such determinations. However, there are a couple of additional caveats that need to be observed in the present context. First, we only consider as potential parents sentences that are *universally* positive in some member of the assumption base. For instance, assuming that sentences (5.40) and (5.41) below are in the assumption base, neither of the highlighted subsentences will be considered as a possible parent of the goal `(P zero)`, even though the goal matches a child of each, because neither highlighted subsentence occurs in a universally positive position:

```
(exists x . P x & Q x)                                    (5.40)

(~ forall y . ~ (x R y ==> P y))                          (5.41)
```

Second, if a goal r does successfully match an appropriate subsentence q of some element p of the assumption base, the domain of the resulting substitution θ must only contain universal free variables. More precisely, we must have $Dom(\theta) \subseteq UFV(q,p)$. This condition is intended to ensure that free variables in the assumption base are not treated as universally quantified variables. For instance, suppose that the goal is `(zero R zero)` and that the assumption base contains the premise

$p =$ `(forall x . P x ==> x R y)`,

which has a free occurrence of `y`. Even though the goal matches the consequent `(x R y)` under the substitution

$\{\texttt{x} \mapsto \texttt{zero}, \texttt{y} \mapsto \texttt{zero}\}$,

we will *not* regard the conditional `(P x ==> x R y)` as a generalized parent of the goal, even though the first proviso above is satisfied (i.e., the conditional occurs in a universally positive position in p). The reason is that the occurrence of the variable `y` in the consequent is not universally quantified. It is, rather, a free occurrence, and hence for our present purposes it should be treated essentially as a constant—not as a universally quantified variable. Under that understanding, the goal does not match the consequent.

To sum up, suppose that q is a specific occurrence of a subsentence of p, and let r be any sentence. We say that *r properly matches q in p under θ*, written $PM(r,q,p,\theta)$, iff r matches q under θ in the usual sense and, in addition, the two foregoing conditions hold: (a) q occurs in a universally positive position in p; and (b) $Dom(\theta) \subseteq UFV(q,p)$. It is easy to determine whether $PM(r,q,p,\theta)$ holds for any given p, q, r, and θ; and, more importantly, given any p and r, it is easy to determine whether there exists some subsentence q of p and some θ such that $PM(r,q,p,\theta)$.[18]

18 Strictly speaking, q is a subsentence *occurrence*, and would therefore be represented by something like a pair consisting of a sentence and a position.

Accordingly, faced with the task of deriving a goal r from some β (and set of proof terms $\boldsymbol{T}$), our general algorithm for applying extraction tactics now becomes:

1. Find some p in β with a universally positive occurrence of a generalized parent of the goal r, specifically, a universally positive occurrence of a sentence q of the following form:
   - $(p_1$ `&` $p_2)$, where r properly matches some p_i in p under some θ;
   - $(p_1$ `==>` $p_2)$, where r properly matches p_2 in p under some θ;
   - $(p_1$ `|` $p_2)$, where r properly matches some p_i in p under some θ;
   - $(p_1$ `<==>` $p_2)$, where r properly matches some p_i in p under some θ.
2. Derive the corresponding substitution instance of that parent, and possibly other additional subgoals.
3. Detach the goal r from that substitution instance of the parent by the proper elimination method.

Graphical formulations of the new versions of these tactics can help to clarify their structure. Here is the case of `[rand->]`:

Derive r from $(\beta = \{\ldots, p = (\cdots (p_1$ `&` $p_2)^{+}_{*} \cdots), \ldots\}, \boldsymbol{T})$

```
let {conj := (!force2 θ((p1 & p2)) T)}  # where r properly matches p2 in p under θ
  (!right-and conj)                                            [rand->]
```

The rest of the extraction tactics can be modified likewise (with one small lacuna that will be discussed soon). We only need to add one new extraction tactic dealing with universal quantifiers, and we are done:

Derive r from $(\beta = \{\ldots, p = (\cdots$ `(forall` v `.` $q)^{+}_{*} \cdots), \ldots\}, \boldsymbol{T})$

```
let {lemma := (!force2 θ((forall v . q)) T)}  # where r properly matches q in p under θ
  (!uspec lemma θ(v))                                          [forall3->]
```

This new tactic treats any universal quantification `(forall` v `.` q`)` as a generalized parent of all substitution instances of its body q. The dual of this tactic is as follows:

Derive r from $(\beta = \{\ldots, (\cdots$ `(exists` v `.` $q)^{-}_{*} \cdots), \ldots\}, \boldsymbol{T})$

```
let {p1 := (!force2 θ((~ exists v . q)) T);                    [forall4->]
     p2 := (!qn p1)}  # where r properly matches q̄ in p under θ
  (!uspec θ((forall v . q̄)) θ(v))
```

To apply this tactic, we look for a premise p with an existential quantification (`exists` v . q) embedded in a universally negative position and such that the goal r properly matches $\overline{q}$ in p.

These two are the forward universal-quantifier tactics that are more commonly used; as we remarked earlier, `[forall->]` and its dual are applied more sparingly. Let us look at some examples.

Example 5.3: Suppose that we want to derive the goal `(R a b)` from the premise

$p =$ `(P c & forall x . forall y . x R y)`.

Thus, letting $\beta = \{p\}$ and $\boldsymbol{T} = \{\mathsf{a}, \mathsf{b}, \mathsf{c}\}$, the initial proof spec is:

Derive `(a R b)` from $(\beta, \boldsymbol{T})$

Let q be the following subsentence of p: `(x R y)`. We observe that the goal `(a R b)` properly matches q in p under the substitution

$$\theta_1 = \{\mathsf{x} \mapsto \mathsf{a}, \mathsf{y} \mapsto \mathsf{b}\},$$

and that the immediately surrounding universal quantification, `(forall y . x R y)`, is in a universally positive position in p:

`(P c & forall x . forall y . x R y`$^{+}_{*}$`).` (5.42)

Therefore, the `[forall3->]` tactic is applicable and generates the following proof spec:

Derive θ_1(`(forall y . x R y)`) = `(forall y . a R y)` from $(\beta, \boldsymbol{T})$

(5.43)

Once the subgoal `(forall y . a R y)` has been derived, we will perform a `uspec` on it with $\theta_1(\mathsf{y}) = \mathsf{b}$, as the tactic prescribes, to obtain our top goal, `(a R b)`. We continue with spec (5.43). Now let q be the following subsentence of premise p: `(forall y . x R y)`. We observe that the current goal `(forall y . a R y)` properly matches q in p under $\theta_2 = \{\mathsf{x} \mapsto \mathsf{a}\}$, and that the immediately surrounding universal quantification has universally positive polarity:

`(P c & forall x . forall y . x R y`$^{+}_{*}$`).` (5.44)

Hence, `[forall3->]` is again applicable and leads to the following new proof spec:

Derive θ_2(`(forall x . forall y . x R y)`) = `(forall x . forall y . x R y)` from $(\beta, \boldsymbol{T})$

(5.45)

As `[forall3->]` states, once this new goal has been derived we can obtain the subgoal

`(forall y . a R y)`

by performing `uspec` on it with $\theta_2(\mathtt{x}) = \mathtt{a}$. Finally, (5.45) is immediately handled by `[rand->]`, since this latest goal properly matches the right conjunct of the premise p under the empty substitution, and is thus obtainable by a single application of `right-and`. The complete final proof is:

```
let {lemma := let {lemma := (!right-and (P c & forall x y . x R y))}
                (!uspec lemma a)}
  (!uspec lemma b)
```

or, equivalently:

```
let {p1 := (!right-and (P c & forall x y . x R y));
     p2 := (!uspec p1 a)}
  (!uspec p2 b)
```

(after unfolding the nested **`let`** deduction). □

Example 5.4: Suppose that we want to derive the goal `(P b)` from the singleton

$$\beta = \{p = \texttt{(\~{} (Q a | exists x . \~{} P x))}\},$$

so that $\boldsymbol{T} = \{\mathtt{a}, \mathtt{b}\}$ initially. We observe that the existential quantification

$$\texttt{(exists x . \~{} P x)} \tag{5.46}$$

is negatively embedded in a universal position in p, and that the goal properly matches the complement of the body of (5.46), `(P x)`, in p, under the substitution $\theta = \{\mathtt{x} \mapsto \mathtt{b}\}$. Accordingly, following tactic `[forall4->]`, we generate the subgoal

$$\boxed{\text{Derive } \texttt{(\~{} exists x . \~{}P x)} \text{ from } (\beta, \boldsymbol{T})} \tag{5.47}$$

Once we have `(~ exists x . ~ P x)`, we transform it to `(forall x . P x)` via `qn` and then perform `uspec` on it with $\theta(\mathtt{x}) = \mathtt{b}$, which will derive the desired `(P b)`. Simple sentential reasoning will easily take care of (5.47). □

Example 5.5: Suppose that we want to derive the goal

$$r = \texttt{(exists y . Q y)}$$

from the following pair of premises:

$$\beta = \{p_1 = \texttt{(forall x . P x ==> Q x)}, p_2 = \texttt{(P a)}\}.$$

The initial set of available proof terms is $\boldsymbol{T} = \{\mathtt{a}\}$ and the starting proof spec is:

$$\boxed{\text{Derive } r \text{ from } (\beta, \boldsymbol{T})} \tag{5.48}$$

The tactic `[exists<-]` is applicable here, which says that to prove a sentence of the form `(exists` v `.` q`)`, take the body q, replace every free occurrence of v in it by an available proof term t, and try to derive the result. If and when we succeed, we can use existential generalization from t to infer `(exists` v `.` q`)`. Here there is only one available proof term, a, so the tactic gives rise to the subgoal:

$$\boxed{\text{Derive } \texttt{(Q a)} \text{ from } (\beta, \boldsymbol{T})} \tag{5.49}$$

We observe that the goal `(Q a)` properly matches the consequent of p_1 in p_1 under $\theta = \{\texttt{x} \mapsto \texttt{a}\}$, and that the parent of that consequent is positively embedded in p_1. Accordingly, `[if->]` generates the subgoals

$$\boxed{\text{Derive } \theta(\texttt{(P x ==> Q x)}) = \texttt{(P a ==> Q a)} \text{ from } (\beta, \boldsymbol{T})} \tag{5.50}$$

and

$$\boxed{\text{Derive } \theta(\texttt{(P x)}) = \texttt{(P a)} \text{ from } (\beta \cup \{\texttt{(P a ==> Q a)}\}, \boldsymbol{T})}. \tag{5.51}$$

The first subgoal is readily handled by `[forall3->]` owing to premise p_1, while the second subgoal is handled by reiteration. The final proof is:

```
let {lemma := (!uspec (forall x . P x ==> Q x) a);
     _ := (!mp (P a ==> Q a) (P a))}
  (!egen (exists y . Q y) a)
```

(after some routine simplifications). □

Example 5.6: Let us use the tactics to derive the goal

$$r = \texttt{((exists x . P x) ==> (exists y . Q y))}$$

from the premise

$$p = \texttt{(forall x . P x ==> Q x)}.$$

The initial set of proof terms is $\boldsymbol{T} = \emptyset$, so the top proof spec is:

$$\boxed{\text{Derive } r \text{ from } (\{p\}, \emptyset)} \tag{5.52}$$

None of the high-ranking tactics are applicable, so we proceed by the backward tactic `[if<-]`. Thus, we assume the antecedent of the goal and try to prove the consequent:

$$\boxed{\text{Derive } \texttt{(exists y . Q y)} \text{ from } (\{p\} \cup \{\texttt{(exists x . P x)}\}, \emptyset)} \tag{5.53}$$

We now have an existential quantification as an available assumption, so we proceed to unpack it via `[exists->]`. The top-level proof sketch now looks as follows:

```
assume (exists x . P x)
  pick-witness w for (exists x . P x)  # We now have (P w)
    (!force2 (exists y . Q y) (set [w]))
```

with the new proof spec:

$$\boxed{\text{Derive } \texttt{(exists y . Q y)} \text{ from } (\{p\} \cup \{\texttt{(exists x . P x)}, \texttt{(P w)}\}, \{\texttt{w}\})} \tag{5.54}$$

The set of available proof terms now contains `w`. Hence, the backward tactic for existential quantifications is applicable, `[exists<-]`, whereby we try to derive the body of the existential goal with a proof term (in this case `w`) in place of the quantified variable:

$$\boxed{\text{Derive } \texttt{(Q w)} \text{ from } (\{p\} \cup \{\texttt{(exists x . P x)}, \texttt{(P w)}\}, \{\texttt{w}\})} \tag{5.55}$$

The new goal `(Q w)` properly matches the consequent of p's body in p under the substitution $\theta = \{\texttt{x} \mapsto \texttt{w}\}$, so applying the tactic `[if->]` yields the new specs:

$$\boxed{\text{Derive } \theta(\texttt{(P x ==> Q x)}) = \texttt{(P w ==> Q w)} \text{ from } (\beta', \{\texttt{w}\})} \tag{5.56}$$

and

$$\boxed{\text{Derive } \theta(\texttt{(P x)}) = \texttt{(P w)} \text{ from } (\beta' \cup \{\texttt{(P w ==> Q w)}\}, \{\texttt{w}\})} \tag{5.57}$$

where $\beta' = \{p\} \cup \{\texttt{(exists x . P x)}, \texttt{(P w)}\}$. Finally, (5.56) and (5.57) are respectively handled by `[forall3->]` and `[claim->]`. □

Example 5.7: We want to derive the goal $r =$ `(exists z . z R z | P z)` from the two premises

$$\beta = \{p_1 = \texttt{(forall x . ~ P x ==> Q x)}, p_2 = \texttt{(exists y . ~ Q y)}\}.$$

The initial set of available proof terms is thus $\emptyset$. We proceed to instantiate the existential quantification p_2, following tactic `[exists->]`:

```
pick-witness w for (exists y . ~ Q y)      # We now have (~ Q w)
  (!force2 (exists z . z R z | P z) (set [w]))
```

generating the subgoal:

$$\boxed{\text{Derive } r \text{ from } (\beta \cup \{\texttt{(~ Q w)}\}, \{\texttt{w}\})} \tag{5.58}$$

Now we can use the backward tactic for existential quantifiers. We only have one proof term available, `w`, so we will try to derive `(w R w | P w)`:

```
pick-witness w for (exists y . ~ Q y)
  let {lemma := (!force2 (w R w | P w) (set [w]))}
    (!egen (exists z . z R z | P z) w)
```

Hence, our new proof spec is:

$$\boxed{\text{Derive } \texttt{(w R w | P w)} \text{ from } (\beta \cup \{\texttt{(~ Q w)}\}, \{\texttt{w}\})} \tag{5.59}$$

No direct tactic is applicable here, so we transform the goal to a conditional and apply the backward tactic for conditionals:

```
pick-witness w for (exists y . ~ Q y)
  let {lemma := let {lemma' := assume (~ w R w)
                                   (!force2 (P w) (set [w]))}
                 (!cond-def lemma')}
    (!egen (exists z . z R z | P z) w)
```

where the new proof spec is:

$$\boxed{\text{Derive } \texttt{(P w)} \text{ from } (\beta \cup \{\texttt{(~ Q w)}, \texttt{(~ w R w)}\}, \{\texttt{w}\})} \tag{5.60}$$

The goal `(P w)` properly matches `(P x)` in the first premise under $\{\texttt{x} \mapsto \texttt{w}\}$. Following the discussion on page 283, we `transform` the first premise to

`(forall x . ~ Q x ==> P x)`

via `contra-pos` and then apply the `[if->]` tactic. This will yield the subgoals

`(~ Q w ==> P w)`

and `(~ Q w)`, which are directly derivable by `[forall3->]` and `[claim->]`. The complete definition of the final proof is left as an exercise. □

Example 5.8: We want to derive the goal

$r =$ `(forall x . (exists y . x R y) ==> exists y . y R x)`

from

$\beta = \{p =$ `(forall x . forall y . x R y ==> y R x)`$\}$.

The comments in the following proof sketch indicate the tactics that were applied at the respective points, along with other relevant information:

```
1  pick-any a                                  # [forall<-], T = {a}
2    assume hyp := (exists y . a R y)          # [if<-]
3      pick-witness w for hyp                  # [exists->]: we now have (a R w)
4        conclude goal := (exists y . y R a)   # and we now have T = {w, a}
5          let {_ := (!force2 (w R a)          # [exists<-]
6                              (set [w a]))}
7            (!egen goal w)
```

We now have to find a proof that derives `(w R a)` from

$\beta \cup \{$`(exists y . a R y)`, `(a R w)`$\}$

with $\boldsymbol{T} = \{\texttt{a}, \texttt{w}\}$. An extraction tactic is applicable here, as the goal properly matches the consequent of premise p in p under the substitution

$$\theta = \{\texttt{y} \mapsto \texttt{w}, \texttt{x} \mapsto \texttt{a}\}.$$

Hence, following [if->], we will attempt to derive:

1. the corresponding substitution instance of the conditional parent,

$$\theta((\texttt{x R y ==> y R x})) = (\texttt{a R w ==> w R a});$$

2. the corresponding substitution instance of the antecedent, (a R w).

The first subgoal is an instance of premise p and is readily handled by two successive applications of [forall3->]. The second subgoal is in the assumption base, so the reiteration tactic prescribes a claim. In summary, and with a bit of proof simplification, the desired proof becomes:

```
let {lemma := (!uspec p [a w])}
  (!mp lemma (a R w))
```

Observe that this is the only forward tactic used in the proof search. Until this point, only backward tactics were applied: [forall<-], [if<-], and [exists<-]. The latter tactic presented us with a choice point. Since $\boldsymbol{T} = \{\texttt{w}, \texttt{a}\}$ at that particular juncture (line 5 of the first listing), we could choose to apply [exists<-] with w (which is what we did), *or* with a, which would have produced the subgoal (a R a). What would have happened had we made that choice first? If we applied tactics in a completely mechanical fashion, the proof search would have failed as follows. First, we would have tried the extraction tactic [if->], since the goal (R a a) properly matches the consequent of p in p under

$$\theta = \{\texttt{y} \mapsto \texttt{a}, \texttt{x} \mapsto \texttt{a}\}.$$

We would then have had to derive the corresponding antecedent, which would again have been (a R a). That would get us in a cycle, so we would have had to backtrack and try another tactic. The only other tactic would be indirect proof, and that would also get us nowhere. So eventually we would have to backtrack and try [exists<-] with w, which would succeed as shown. □

As we mentioned earlier, our current formulation of the extraction tactics has a small lacuna that needs to be addressed. Here is the issue: When we (properly) match the goal r against some positive subsentence q of the assumption base, thereby obtaining some matching substitution θ, the parent of q, call it q', might contain variables *that do not appear in the domain of* θ. Hence, when we come to apply θ to the parent q', the resulting subgoal $\theta(q')$ will not be fully instantiated.

As an example, let β comprise the following two premises (both of which are true in the domain of natural numbers, where (pos n) means that n is nonzero):

```
(forall x y . x < y ==> pos y)

(forall x . zero < S x)
```

Now consider the goal $r =$ `(pos S a)`, for some constant a. This goal properly matches the consequent of the first premise under $\theta = \{$`y` $\mapsto$ `(S a)`$\}$. Hence, by the `[if->]` tactic, we would need to prove the following two subgoals:

$$\theta(\texttt{(x < y ==> pos y)}) \tag{5.61}$$

and

$$\theta(\texttt{(x < y)}). \tag{5.62}$$

But now note that these subgoals contain the variable x, and that θ does not assign any term to x, since x $\notin Dom(\theta)$. Accordingly, neither of these subgoals would be fully instantiated and the tactic would fail. This situation occurs whenever the parent q' contains strictly more (universal free) variables than its child q (against which we have matched the goal r).

To guard against such a scenario, we need to force θ to be total by choosing arbitrary terms from $\boldsymbol{T}$ to assign to the extra variables. (If there are none, fresh variables of appropriate sorts can be used.) In this particular example, $\boldsymbol{T} = \{\texttt{zero}, \texttt{a}\}$, so we can assign either `zero` or `a` to `x`. If we make the first choice, so that $\theta = \{\texttt{x} \mapsto \texttt{zero}, \texttt{y} \mapsto \texttt{(S a)}\}$, we would generate the subgoals

```
(zero < S a ==> pos S a)
```

and

```
(zero < S a),
```

both of which are readily derivable through our tactics. In general, if the parent has k extra variables, there will be $k \cdot |\boldsymbol{T}|$ possible choices for θ, which means that there will be $k \cdot |\boldsymbol{T}|$ different ways in which the tactic can be applied to those particular sentence occurrences.[19]

Example 5.9: Let β contain these two premises:

```
define premise-1 := (exists x . P x & forall y . Q y ==> x R y)

define premise-2 := (forall x . P x ==> forall y . S y ==> ~ x R y)
```

From β we want to derive the following:

```
define goal := (forall x . P x & Q x ==> ~ S x)
```

We begin by eliminating the existential quantifier from the first premise, after which we apply two backward tactics, thus arriving at the following proof sketch:

```
pick-witness w for premise-1
  pick-any a
    assume (P a & Q a)
      (!force2 (~ S a) (set [a w]))
```

19 If we were implementing these tactics as a computer program, we would be able—in some cases—to avoid blind guessing by computing θ through Prolog-style backward chaining, a technique that we discuss in Appendix B. A similar approach could be used to avoid random guessing in applications of `[exists<-]`.

We now have to derive `(~ S a)` from

$$\beta \cup \{\texttt{(P w \& forall y . Q y ==> w R y)}, \texttt{(P a \& Q a)}\}$$

and $\boldsymbol{T} = \{\texttt{w}, \texttt{a}\}$. The only options here are indirect proof by contradiction, or an application of the negation heuristic 4.3. To apply that heuristic in the first-order setting, we look for the smallest positive subsentence of β *in a universal position* that contains a negative occurrence of a sentence that is a generalization of the sentence we are trying to negate. We then apply the same transformations to that sentence as before, until an extraction tactic becomes applicable or we can no longer perform any transformations. In this case, for the goal `(~ S a)`, we see that the second premise contains a negative occurrence of `(S y)`, of which `(S a)` is a proper instance under $\theta = \{\texttt{y} \mapsto \texttt{a}\}$. Applying the contrapositive rule (see page 291), we transform the second premise to

```
(forall x . P x ==> forall y . x R y ==> ~ S y)      (5.63)
```

and now the extraction tactic `[if->]` becomes applicable, with the goal `(~ S a)` properly matching the consequent of the inner conditional under $\theta = \{\texttt{y} \mapsto \texttt{a}\}$. However, we note that the parent of that consequent has an extra universal free variable, `x`, so to get a total matching substitution we must assign a term to `x` as well. Here we have a choice point. Since $\boldsymbol{T} = \{\texttt{w}, \texttt{a}\}$, we can try `w` or `a`. Let's go with the first, which gives us

$$\theta = \{\texttt{x} \mapsto \texttt{w}, \texttt{y} \mapsto \texttt{a}\}.$$

Thus, `[if->]` generates the two subgoals

```
(w R a ==> ~ S a)      (5.64)
```

and

```
(w R a).      (5.65)
```

The first of these is obtainable from (5.63) by three more extraction tactic applications: `[forall3->]`, `[if->]`, and `[forall3->]` again. The second will give rise to the subgoal `(P w)`, which follows directly from the witness premise. Finally, subgoal (5.65) can be extracted from the first premise by the `[if->]` tactic, which will require the two subgoals:

```
(Q a ==> w R a)      (5.66)
```

and

```
(Q a).      (5.67)
```

The first can be inferred by a universal specialization of the second conjunct of the witness premise (tactics `[forall3->]` and `[rand->]`); while (5.67) is derivable from the assumption `(P a & Q a)`, also by `[rand->]`. □

We close this section by reformulating the `[false->]` and `[or->]` tactics, starting with the former:

Derive false from $(\{\ldots, p = (\cdots (\sim q)_*^+ \cdots), \ldots\}, T = \{\ldots, t_1, \ldots, t_n, \ldots\})$

```
let {p1 := (!force2 θ((~ q)) T);   # where θ = {x1 ↦ t1,...,xn ↦ tn}, {x1,...,xn} = UFV(q,p)
     p2 := (!force2 θ(q) T)}                       [false->]
  (!absurd p1 p2)
```

A particular application of this tactic is determined by two choices: First we select a negation embedded in a universally positive position of the assumption base; and then we choose a substitution θ. The second choice is made nondeterministically. Clearly, the more proof terms there are available and the larger the set $UFV(q,p)$, the more choices there will be. In a mechanical implementation this could lead to a lot of backtracking, and some mechanism would be necessary for eliminating certain choices.

Example 5.10: Let β consist of the following two premises, which state that R is irreflexive and transitive:

```
define premise-1 := (forall x . ~ x R x)

define premise-2 := (forall x y z . x R y & y R z ==> x R z)
```

From these two premises we want to derive the conclusion that R is asymmetric:

```
define goal := (forall x y . x R y ==> ~ y R x)
```

The tactics that give rise to the following proof sketch are shown in the comments:

```
pick-any a b                           # Two successive applications of [forall<-]
  assume (a R b)                       # [if<-]: we now have T = [a b]
    (!by-contradiction (~ b R a) # [not<-]
      assume (b R a)                   # [if<-]
        (!force2 false (set [a b]))
```

Thus, we now have the following proof spec:

Derive false from $(\beta \cup \{$(a R b), (b R a)$\}, \{$a,b$\})$

Let us try [false->]. The only negation in a universally positive position in the assumption base is that of the first premise, and there are two choices for the corresponding substitution: $\theta_1 = \{\text{x} \mapsto \text{a}\}$ and $\theta_2 = \{\text{x} \mapsto \text{b}\}$. We try θ_1 first. Thus, we have the two subgoals:

Derive (~ a R a) from $(\beta \cup \{$(a R b), (b R a)$\}, \{$a,b$\})$

and

Derive (a R a) from $(\beta \cup \{$(a R b), (b R a)$\}, \{$a,b$\})$

The first of these is an immediate instance of the first premise, and thus readily handled by `[forall3->]`. The extraction tactic `[if->]` is applicable on the second subgoal, as `(a R a)` properly matches the consequent `(x R z)` in the second premise under $\theta = \{\mathsf{x} \mapsto \mathsf{a}, \mathsf{z} \mapsto \mathsf{a}\}$. However, the parent of that consequent, the conditional

```
(x R y & y R z ==> x R z),
```

has an extra universal free variable, `y`, so we need to assign a term to it. It is easy to see that assigning `b` to `y` will succeed, so that

$$\theta = \{\mathsf{x} \mapsto \mathsf{a}, \mathsf{y} \mapsto \mathsf{b}, \mathsf{z} \mapsto \mathsf{a}\}.$$

Applying `[if->]` with this θ will eventually produce the subgoals `(a R b)` and `(b R a)`, both of which are in the assumption base. Had we chosen to assign `a` to `y` instead, the tactic would have failed (by leading to a cycle, as can be verified). □

Finally, the first-order version of the generalized case analysis tactic, `[or->]`, is this:

Derive r from $(\beta = \{\ldots, p = (\cdots (p_1 \mid p_2)^{+}_{*} \cdots), \ldots\}, \boldsymbol{T} = \{\ldots, t_1, \ldots, t_n, \ldots\})$

```
let {d := (!force2 θ((p1 | p2)) T);
     c1 := assume θ(p1)   # where θ = {x1 ↦ t1,...,xn ↦ tn}, {x1,...,xn} = UFV((p1 | p2),p)
             (!force2 r T);
     c2 := assume θ(p2)
             (!force2 r T)}           [or->]
  (!cases d c1 c2)
```

A typical use of it is illustrated in the following example.

Example 5.11: Let β consist of these three premises:

```
define premise-1 := (forall x . P x | Q x)
define premise-2 := (forall x . P x ==> S x)
define premise-3 := (forall x . Q x ==> S x)
```

We want to derive `(S a)` from β. It should be clear that no extraction, backward, or standard replacement tactics are applicable, but `[or->]` is. The only universal free variable of the disjunction in the first premise is `x`, so we need to choose an available proof term t to substitute for `x`. Since we have $\boldsymbol{T} = \{\mathsf{a}\}$, the constant `a` is our only choice, so the relevant substitution is $\theta = \{\mathsf{x} \mapsto \mathsf{a}\}$. Following the tactic, we generate the three subgoals:

Derive `(P a | Q a)` from $(\beta, \{\mathsf{a}\})$

Derive `(S a)` from $(\beta \cup \{$`(P a)`, `(P a | Q a)`$\}, \{\mathsf{a}\})$

Derive `(S a)` from ($\beta \cup \{$`(Q a)`, `(P a | Q a)`$\}$, $\{$`a`$\}$)

The first subgoal is directly handled by `[forall3->]`, while the extraction tactic `[if->]` on the second and third premises will dispose of the second and third subgoals, respectively.

In this example the choice of θ was obvious, as there was only one proof term available. If there are many, then the goal r can provide some guidance: Those elements of $\boldsymbol{T}$ that occur in r are more likely to figure in θ than the rest. □

5.5.3 Proof strategy for first-order logic

The overall strategy for deploying tactics carries over unchanged from sentential logic (Section 4.14.4), with two minor additions. First, we now try to eliminate existential quantifiers if no extraction tactics are applicable; and second, we try the unrestricted universal instantiation tactic `[forall->]` and its dual `[forall2->]` before we try proof by contradiction. Thus, the general ranking is now as follows:

1. reiteration (`[claim->]`) and constant tactics (`[true<-]` and `[false<-]`);
2. the complement tactic (`[cft]`);
3. extraction tactics, including `[forall3->]` and `[forall4->]` ;
4. existential instantiation tactics (`[exists->]` and `[exists2->]`);
5. replacement tactics;
6. backward tactics (including `[forall<-]`), with the exception of `[not<-]`;
7. generalized disjunction tactic;
8. indirect tactic, `[not<-]`, or the negation heuristic.

Note that a new type of looping is now possible: We can perform existential instantiation on the same sentence *ad infinitum*. Because a fresh witness variable will be used each time, the new proof spec generated on each iteration will be, strictly speaking, different from all previous proof specs; it will contain a new witness premise, and there will be a new proof term too (the fresh witness variable). Hence, our old definition of cycle does not cover this kind of repetition. To avoid this type of looping, we need to make sure that an existential quantification is never instantiated more than once on the same subproof level. It might need to be repeatedly instantiated in disjoint subproofs, say, in the branches of a case analysis; but it should not be instantiated in any D_1 and D_2 such that D_2 is a subproof of D_1 or vice versa.

Rarely, if we are getting nowhere with any of these tactics, we might need to augment the set of proof terms $\boldsymbol{T}$ with a new term, typically a fresh variable. Consider, for instance, proving `(exists x . P x)` from `(forall x . P x)`. Here we have $\boldsymbol{T} = \emptyset$ and none of the above tactics will manage to derive the goal. Populating the set of proof terms with any term t, such as a fresh variable, will allow the tactics to gain traction.

5.6 First-order logic semantics

In this section we present the formal semantics of Athena's first-order logic, giving precise definitions of first-order counterparts of notions such as interpretation, logical consequence (entailment), tautology, and so on, which were first defined for sentential logic in Section 4.12. We start by introducing the semantics of *monomorphic* first-order logic, where all terms have ground sorts. The semantics of polymorphic logic will then be given by reduction to the monomorphic case. (Note: Some of the material in this section requires some familiarity with elementary universal algebra.)

Fix a symbol set γ consisting of a possibly infinite collection of monomorphic sorts Σ and a possibly infinite collection F of function symbols over Σ, each with a unique signature.[20] For brevity, and also to adhere to more customary terminology, in what follows we will refer to this pair of Σ and F as an Athena *vocabulary*. We assume that Σ always contains the datatypes `Boolean` and the domains `Int`, `Real`, and `Ide`, and F always contains the numeric operators `+`, `-`, `*`, and `/`. In addition to these primitive sorts, Σ may also contain complex monomorphic sorts, such as `(List (List Boolean))`, produced by applying sort constructors to other monomorphic sorts. For the remainder of this section, and unless we explicitly indicate otherwise, when we speak of an Athena term t (or sentence p), t (or p) should be understood in the context of this vocabulary. More precisely, the function symbols that occur in t (or p) should be assumed to be elements of F, and hence any sorts that occur in t (or p) should be assumed to be in Σ.

An *interpretation* I of such a vocabulary consists of:

1. a unique nonempty set S^I for each sort $S \in \Sigma$; we refer to S^I as the *carrier* (or the interpretation) of the sort S. And:
2. a unique function
$$f^I : S_1^I \times \cdots \times S_n^I \rightarrow S^I$$
for each function symbol $f \in F$ with signature `[`$S_1 \cdots S_n$`] -> `S; we refer to f^I as "the interpretation" of f under I.[21] When $n = 0$ (i.e., when f is a constant symbol), f^I will simply be an element of S^I.

The primitive sorts and symbols mentioned above always receive the expected interpretations, which will be elaborated shortly. The carriers assigned to other sorts can be arbitrary, although there are certain restrictions on the carriers of datatypes and structures, which will

20 Thus, every sort that appears in the domain or range of a function symbol in F is an element of Σ.

21 Formally, I can be defined as an ordered pair of functions i_1 and i_2, where i_1 maps every sort S in Σ to some nonempty set (the carrier of S), and i_2 maps F to a set of functions (the interpretations of the various function symbols). We write S^I for $i_1(S)$ and f^I for $i_2(f)$.

be discussed later. The functions assigned to nonprimitive function symbols can likewise be arbitrary, again with the exception of datatype and structure constructors. A constructor c with signature $[S_1 \cdots S_n]$ -> S is always interpreted as a particularly simple function, one that takes n terms $t_1, \ldots, t_n$ of sorts $S_1, \ldots, S_n$ as arguments and returns the new term $(c\ t_1 \cdots t_n)$ as output. This, of course, presupposes that the carriers assigned to $S_1, \ldots, S_n$ consist of terms. We will make sure of that.

In the context of a given interpretation I, a *variable assignment* χ is a total function from the set of all variables of the form $v{:}S$, with $S \in \Sigma$, to elements of the corresponding carriers. When $\boldsymbol{a}$ is an element of S^I, we write

$$\chi[v : S \mapsto \boldsymbol{a}]$$

for the assignment that maps the variable $v{:}S$ to $\boldsymbol{a}$ and agrees with χ everywhere else. Now, let t be any Athena term over the given vocabulary. We define *the value of t under I and χ*, denoted by $I_\chi(t)$, through the following structural recursion:

$$I_\chi(v : S) = \chi(v : S);$$
$$I_\chi((f\ t_1 \cdots t_n)) = f^I(I_\chi(t_1), \ldots, I_\chi(t_n)).$$

Note that we use prefix syntax on the left-hand side of the second equation to emphasize that $(f\ t_1 \cdots t_n)$, the argument to I_χ, is an Athena term. On the right-hand side, we are applying an abstract mathematical function f^I to n values $a_1, \ldots, a_n$, so for that application we use conventional mathematical notation: $f^I(a_1, \ldots, a_n)$. A straightforward induction will show that if t is a term of sort S, then $I_\chi(t)$ is an element of the carrier S^I.

Let us say that two variable assignments χ_1 and χ_2 *agree* on a set of monomorphic variables V iff $\chi_1(v) = \chi_2(v)$ for all $v \in V$. And let us write $V(t)$ for the set of variables that occur in a term t, and $FV(p)$ for the set of variables that have free occurrences in a sentence p. The following result is proved by structural induction on terms.

Lemma 5.1

If χ_1 and χ_2 agree on $V(t)$, then $I_{\chi_1}(t) = I_{\chi_1}(t)$.

We now define the satisfaction relation for monomorphic first-order logic. Specifically, we say that *an interpretation I satisfies a sentence p with respect to a variable assignment χ* iff $I \models_\chi p$, where the latter is defined by structural recursion on p as shown in Figure 5.1. We write $I \not\models_\chi p$ to mean that it is not the case that $I \models_\chi p$. For a (possibly infinite) set of sentences Φ, we define $I \models_\chi \Phi$ iff $I \models_\chi p$ for every $p \in \Phi$. We say that p is *true* in (or "under") I iff $I \models_\chi p$ for every χ. A set of sentences Φ_1 logically *entails* (or *implies*) a set of sentences Φ_2, written $\Phi_1 \models \Phi_2$, iff for every I and χ we have $I \models_\chi \Phi_2$ whenever $I \models_\chi \Phi_1$; we then also say that Φ_2 is a logical *consequence* of Φ_1. When Φ_2 is a singleton $\{p\}$, we write $\Phi_1 \models p$ interchangeably with $\Phi_1 \models \{p\}$, and we say that Φ_1 logically entails (or implies) p, or that p logically *follows from* Φ_1. Further, Φ is *satisfiable* iff there is some

$I \models_\chi$ (=$_S$ t_1 t_2)	iff	$I_\chi(t_1) = I_\chi(t_2)$
$I \models_\chi (R\ t_1 \cdots t_n)$	iff	$R^I(I_\chi(t_1), \ldots, I_\chi(t_n)) = \texttt{true}^I$
$I \models_\chi$ (not p)	iff	$I \not\models_\chi p$
$I \models_\chi$ (and $p_1 \cdots p_n$)	iff	$I \models_\chi p_i$ for every $i = 1, \ldots, n$
$I \models_\chi$ (or $p_1 \cdots p_n$)	iff	$I \models_\chi p_i$ for some $i = 1, \ldots, n$
$I \models_\chi$ (if p_1 p_2)	iff	$I \not\models_\chi p_1$ or $I \models_\chi p_2$
$I \models_\chi$ (iff p_1 p_2)	iff	$I \models_\chi p_1$ and $I \models_\chi p_2$ or $I \not\models_\chi p_1$ and $I \not\models_\chi p_2$
$I \models_\chi$ (forall $v{:}S$ p)	iff	$I \models_{\chi[v:S \mapsto \mathbf{a}]} p$ for every $\mathbf{a} \in S^I$
$I \models_\chi$ (exists $v{:}S$ p)	iff	$I \models_{\chi[v:S \mapsto \mathbf{a}]} p$ for some $\mathbf{a} \in S^I$

Figure 5.1
Definition of the satisfaction relation for monomorphic many-sorted first-order logic.

I and χ such that $I \models_\chi \Phi$; otherwise Φ is *unsatisfiable*. Again, when Φ is a singleton $\{p\}$ we speak directly of p as satisfiable or unsatisfiable. Finally, a *tautology* is a sentence p such that $I \models_\chi p$ *for every* I and χ. The following result generalizes the previous lemma from terms to sentences. It is likewise proved by structural induction, this time on sentences.

Theorem 5.1

If χ_1 and χ_2 agree on $FV(p)$, then $I \models_{\chi_1} p$ iff $I \models_{\chi_2} p$.

Since any two assignments agree on the empty set of variables, we conclude:

Corollary 5.1

If $FV(p) = \emptyset$ then for all χ_1 and χ_2, $I \models_{\chi_1} p$ iff $I \models_{\chi_2} p$.

Note that equality receives special treatment in these semantics. We assume that for every $S \in \Sigma$, F contains a binary identity symbol $=_S$ with the signature:

```
[S S] -> Boolean
```

and we require that I interprets $=_S$ as the identity relation on S, so that for any terms t_1 and t_2 of sort S we have:

$$I_\chi \models (t_1 =_S t_2) \Leftrightarrow I_\chi(t_1) = I_\chi(t_2).$$

To build up some computational intuitions about these ideas, let us implement the definition of the satisfaction relation as a procedure that will take a sentence p, an interpretation I, and an assignment χ, and, modulo some limitations to be discussed shortly, will produce true or false depending on whether or not $I \models_\chi p$. First we need to decide how to encode an interpretation I. We will represent I as an Athena map that:

- maps sorts (or more precisely, their names, represented as strings) to lists of objects, which, viewed as sets, will serve as the carriers of the respective sorts;
- maps function symbols to procedures that take carrier elements as inputs and produce carrier elements as outputs, in a way that respects the sort constraints imposed by the symbols' signatures.

A variable assignment will also be represented as an Athena map from variables to carrier elements. Of course an Athena map has a finite domain, whereas a variable assignment was defined as an infinite function whose domain is the set of *all* Athena variables (of the relevant monomorphic sorts). But for the purpose of determining whether $I_\chi \models p$ for a given p, we are justified in restricting our attention to a finite number of variables owing to Theorem 5.1, which assures us that the only variables that matter for that purpose are the free variables of p.

With these conventions in place, we need to implement two ternary procedures: tv and sat. The former will take as input a term t over Σ and F; an interpretation I of Σ and F, represented as just described; and a finite map χ assigning carrier elements to the variables of t; and will compute the value of t under I and χ, that is, $I_\chi(t)$.[22] The sat procedure will take a sentence p over Σ and F; an interpretation I of Σ and F; and a finite map χ from variables to carrier elements; and will output true if $I \models_\chi p$ and false otherwise:

```
define (tv t I asgn) :=
  match t {
    (some-var _) => (asgn t)
  | (= s t) => ((tv s I asgn) equals? (tv t I asgn))
  | ((some-symbol f) (some-list args)) =>
      (app-proc (I f)
                (map lambda (t) (tv t I asgn)
                     args))
  }

define (quant-content q) :=
 match q {
   forall => for-each | exists => for-some
 }

define (sat p I asgn) :=
    match p {
      (some-atom A) => (tv A I asgn)
    | (not q) => (negate (sat q I asgn))
    | (and (some-list args)) => (for-each args lambda (q) (sat q I asgn))
```

22 More precisely, it will compute the value of t under I and *all* variable assignments that agree with χ on the variables of t; Lemma 5.1 entails that the value of t under I is invariant over all such assignments.

```
  | (or (some-list args)) => (for-some args lambda (q) (sat q I asgn))
  | (if p1 p2) => (sat (~ p1 | p2) I asgn)
  | (iff p1 p2) => (sat ((p1 ==> p2) & (p2 ==> p1)) I asgn)
  | ((some-quant q) x body) => ((quant-content q)
                                  (I (sort-of x))
                                  lambda (a)
                                    (sat body I (Map.add asgn [[x a]])))
  }
```

Observe that we fix the interpretation of equality. The primitive `app-proc` is a higher-order binary procedure that takes any procedure f of n arguments and any list of n values `[`$V_1 \cdots V_n$`]` and produces whatever outcome is obtained by applying f to $V_1 \cdots V_n$:

```
> (app-proc plus [1 2])

Term: 3

> (app-proc root [(+ 1  2)])

Symbol: +

> (app-proc join [[1] [2 3] [4]])

List: [1 2 3 4]
```

As an example, consider the vocabulary introduced by the following module:[23]

```
module T {
  domain D
  declare +: [D D] -> D
  declare <: [D D] -> Boolean
}
```

Here is an interpretation of that vocabulary. We omit the standard components that are part of every interpretation (Booleans, integers, reals, and elementary operations on them):

```
define I := |{"T.D" := [[] [1] [2] [1 1] [1 2] [2 1] [2 2]],
              T.+ := lambda (l1 l2) (rd (l1 join l2)),
              T.< := prefix? }|
```

Thus, the carrier of `T.D` contains exactly 7 objects, the integer lists `[]`, `[1]`, $\cdots$, `[2 2]`. `I` interprets `T.+` as a binary operation that takes any two lists l_1 and l_2 from this carrier and

23 Modules are discussed in Chapter 7. For now you can think of a module M as a separate namespace with its own sorts and function symbols.

outputs the list obtained by concatenating l_1 with l_2 and removing any duplicate occurrences from left to right.[24] Thus, for instance:

```
 ([1] (I T.+) [2])    →    [1 2]
  ([] (I T.+) [1])    →    [1]
([1 1] (I T.+) [1])   →    [1]
([1 2] (I T.+) [2])   →    [1 2]
```

and so on. It can be verified that this operation is closed over this 7-element set, meaning that the result of performing this operation on any two elements from this set is also an element of the same set. Finally, `T.<` is interpreted as the prefix relation on lists (restricted on the above set of 7 lists). The primitive binary procedure `prefix?` returns `true` or `false` depending on whether or not its first input list is a prefix of the second:

```
> ([5] prefix? [5 7 9])

Term: true

> ([] prefix? [])

Term: true

> ([5] prefix? [7 5 9])

Term: false
```

Now let us go ahead and define some sentences in `T`:

```
extend-module T {

  define p1 := (forall x y . x + y = y + x)
  define p2 := (forall x y z . x + (y + z) = (x + y) + z)
  define p3 := (forall x y . x < y ==> ~ y < x)
  define p4 := (forall x y . x < y & x =/= y ==> ~ y < x)
  define p5 := (forall x y z . x < y & y < z ==> x < z)

}
```

We can now ask whether *I satisfies* sentences `T.p1` through `T.p5` with respect to some variable assignment. Because all five of these sentences contain no free variables, it follows from Corollary 5.1 that it makes no difference which variable assignment we give as an argument to `sat`. We thus use the empty map:

---

24 The unary procedure `rd` does just that: It removes duplicate element occurrences from any given list in a stable manner (from left to right, leaving the original list order intact).

```
define (sat? I) := lambda (p) (sat p I |{}|)

> (map (sat? I) [T.p1 T.p2 T.p3 T.p4 T.p5])

List: [false true false true true]
```

So we see that `T.p1` and `T.p3` are false under this interpretation, but the rest of the sentences are true. `T.p1` is false because `T.+` is not commutative according to `I`: Applying this operation to `[1]` and `[2]` gives `[1 2]`, but applying it to `[2]` and `[1]` gives `[2 1]`, which is a different result according to our treatment of equality. However, it can be verified that this operation is associative, hence `T.p2` is true. `T.p3` is false because the prefix relation is not asymmetric. In fact it is reflexive (every list is a prefix of itself), so it could not possibly be asymmetric. It is, however, anti-symmetric, meaning that if two lists are prefixes of each other then they must be identical, which is the proposition expressed by `T.p4`. The prefix relation is also transitive, so `T.p5` is also true.

A limitation of this implementation is that it can only represent interpretations with finite carriers, as every sort is mapped to—interpreted by the elements of—an Athena list, which can only have finitely many members. We could remove this limitation by using lazily evaluated infinite streams instead of lists, implemented by higher-order procedures, which could represent countably infinite carriers, such as the set of all integers. But in that case an application of `sat` might never terminate, as it might need to examine all elements of an infinite stream. More importantly, however, even if we used streams we would still be limited to countably infinite sets, which are insufficient for general mathematical purposes, as the intended interpretations of many standard mathematical theories require uncountably many objects. Indeed, the primitive domain `Real` in Athena is assigned the set of all real numbers as its carrier, a paradigmatically uncountable set. And interpreting a theory of sets such as ZF (Zermelo-Fraenkel set theory [38]) would require even larger—mind-bogglingly larger—collections of mathematical entities.

We now come to the interpretation of monomorphic datatypes. Unlike the plain domains of a vocabulary, the carrier assigned to a datatype is not arbitrary. Once we fix the carriers of all other (nondatatype) sorts in a vocabulary, the carrier of a datatype becomes uniquely determined. A monomorphic datatype T may involve $n \geq 0$ other sorts $S_1, \ldots, S_n$ introduced prior to T, as dictated by the signatures of T's constructors, let us call these $c_1, \ldots, c_k$ (where we must have $k > 0$). Now, the definition of T may also refer to T itself, if T happens to be recursive, but let us denote by $S_1, \ldots, S_n$ all and only the sorts *other than* T that appear in the signatures of the constructors $c_1, \ldots, c_k$ of T. Some of these sorts $S_1, \ldots, S_n$ may be domains, others may be primitive sorts or datatypes (such as `Int` or `Boolean`), and others may be previous user-defined datatypes. For instance, the monomorphic datatype `(List Real)`, which we might write out as

$$\mathtt{List_{Real}} := \mathtt{nil_{Real}} \mid (\mathtt{::_{Real}}\ \mathtt{Real}\ \mathtt{List_{Real}}),$$

involves the primitive domain `Real`. Assuming that each such S_j has received an interpretation, meaning that the carrier S_j^I has been specified, our task now is to specify the carrier for T, T^I. That set is defined as the free algebra built by the constructors $c_1, \ldots, c_k$ and with

$$S_1^I, \ldots, S_n^I$$

as generator sets. Equivalently, we may regard the elements of $S_1^I, \ldots, S_n^I$ as either typed "variables" or as typed constant symbols annotated with their respective sorts,[25] and we can then understand T^I as the many-sorted term algebra constructed by $c_1, \ldots, c_k$ over the set of typed variables (or constant symbols)

$$\bigcup_{j=1}^{n} S_j^I,$$

and specifically as the set of all terms from this algebra that are of sort T. To ensure that the algebra in question exists and is nonempty, we require that there must be at least one (finite) term of sort T constructed by some c_i.

No special complications arise when T happens to be one of a number of mutually recursive datatypes $T_1, \ldots, T_m$. In that case, too, we define T^I to be the free algebra of its constructors generated over the carriers of all sorts that occur in the definition other than $T_1, \ldots, T_m$. For example, consider two mutually recursive datatypes for multi-way integer trees:

```
datatypes Tree_Int := null_Int
                    | (node_Int Int List_Tree_Int) &&

          List_Tree_Int := nil_Tree_Int
                    | (::_Tree_Int Tree_Int List_Tree_Int)
```

We then have the following many-sorted term signature (written in conventional notation):

```
     null_Int:   Tree_Int
     node_Int:   Int × List_Tree_Int → Tree_Int
 nil_Tree_Int:   List_Tree_Int
  ::_Tree_Int:   Tree_Int × List_Tree_Int → List_Tree_Int
```

and, taking the set of all integers $\mathbb{Z}$ as the set of our "variables" of sort `Int`, we consider the many-sorted term algebra over the above signature that is generated by $\mathbb{Z}$, which contains "terms" such as:

25 The quotes were inserted because this notion of a variable has little in common with the conventional syntactic notion of a variable, or with that of a constant symbol for that matter. For instance, all real numbers may be thought of as variables (or constant symbols) in this context.

```
                    null_Int
              (node_Int 5 nil_Tree_Int)
(node_Int 78 (::_Tree_Int (node_Int 5 nil_Tree_Int) nil_Tree_Int))
                        ⋮
```

Although we write down these terms as strings, keep in mind that these are tree structures—essentially partial orders—over arbitrary sets, namely, the carriers whose elements are serving as generators. In fact, these carriers can be uncountable (e.g., the real numbers), and in that case the corresponding set of "terms" will also be uncountably infinite. So these are really quite abstract mathematical objects. At any rate, the carriers of `Tree_Int` and `List_Tree_Int` consist of precisely the "terms" of this algebra. Specifically, the carrier of `Tree_Int` consists of all and only those terms of this algebra whose top symbol is one of the two constructors of `Tree_Int` (or equivalently, all and only those terms of the algebra that have sort `Tree_Int`); and the carrier of `List_Tree_Int` consists of all and only those terms whose top symbol is one of the other two constructors. Here too, the rejection of cyclic dependencies and the requirement that each datatype contains at least one constructor ensures that the term algebra corresponding to each datatype is nonempty.

Finally, a monomorphic structure T is interpreted as the quotient of the free algebra induced by some fixed congruence relation. Thus, assuming a congruence relation $\equiv$ on the underlying free algebra, the elements of the structure's carrier (T^I) will be all and only the equivalence classes of $\equiv$. It is typical practice to redefine the identity relation for a structure precisely as the intended congruence relation. Examples are given for the structures of finite sets and maps in Chapter 10. If no such relation is specified for a given structure, then the standard identity relation is taken as the default congruence, which means that every equivalence class will contain exactly one term and the semantics will revert to the usual free-algebra semantics of datatypes.

For polymorphic sentences, semantic notions such as interpretations and satisfaction are defined by reduction to the monomorphic case. Specifically, suppose now that the vocabulary is polymorphic, so that the collection of sorts Σ may contain polymorphic sorts (as well as monomorphic ones), and F may also contain polymorphic function symbols. As discussed in Chapter 2, every polymorphic sort S in Σ can be viewed as a (possibly infinite) *set of monomorphic sorts*, called *instances* of S, obtainable by consistently replacing sort variables in S with monomorphic sorts from Σ. For a given polymorphic sort $S \in \Sigma$, let us write $\widehat{S}$ to denote the set of all of its monomorphic instances (relative to Σ). If $S \in \Sigma$ is monomorphic, we define $\widehat{S}$ as the singleton $\{S\}$. We then define

$$\widehat{\Sigma} = \bigcup_{S \in \Sigma} \widehat{S}.$$

We may refer to $\widehat{\Sigma}$ as the *grounding* of Σ. In words, $\widehat{\Sigma}$ is the collection of *all* (and only) the monomorphic sorts obtainable from Σ.

Likewise, every polymorphic function symbol f can be viewed as a possibly infinite set of monomorphic function symbols, also said to be instances of f, and whose signatures are obtainable from the signature of f by consistently replacing sort variables with monomorphic sorts from Σ. A "symbol instance" of this form is distinguished from f by being indexed (subscripted) by its monomorphic signature. For example, suppose we have introduced a polymorphic function symbol for list reversal with the following declaration:

```
declare rev: (S) [(List S)] -> (List S).
```

Then `rev` will have the following monomorphic symbols among its instances:

$$\texttt{rev}_{\texttt{[[(List Int)] (List Int)]}}, \tag{5.68}$$

$$\texttt{rev}_{\texttt{[[(List (List Ide))] (List (List Ide))]}}, \tag{5.69}$$

$$\texttt{rev}_{\texttt{[[(List Boolean)] (List Boolean)]}}, \tag{5.70}$$

where for typesetting simplicity we write a signature $[S_1 \cdots S_n]$ -> S as the two-element list $[[S_1 \cdots S_n]\ S]$ whenever we are using the signature as a subscript. Written more conventionally, the signatures of these three symbol instances are as follows:

```
[(List Int)] -> (List Int),
[(List (List Ide))] -> (List (List Ide)),
[(List Boolean)] -> (List Boolean).
```

For a polymorphic function symbol f and a set of monomorphic sorts Σ, we define $G_\Sigma(f)$ as the set of all its monomorphic instances with respect to Σ, that is, the set of all monomorphic function symbols that can be derived from f as just described, by replacing sort variables in the signature of f with ground sorts from Σ. If f is monomorphic, we set

$$G_\Sigma(f) = \{f\}.$$

Finally, for a set of polymorphic function symbols F and a set of polymorphic sorts Σ, we define

$$G_\Sigma(F) = \bigcup_{f \in F} G_{\widehat{\Sigma}}(f).$$

We can now define an interpretation of a polymorphic vocabulary consisting of Σ and F as an interpretation of the monomorphic vocabulary consisting of $\widehat{\Sigma}$ and $G_\Sigma(F)$. Accordingly, to give an interpretation of Σ and F is just to give an interpretation of $\widehat{\Sigma}$ and $G_\Sigma(F)$. Of course, the interpretation of symbol instances such as (5.68)–(5.70) will typically be identical save for the sort differences.

Recall that a sentence p is polymorphic iff it contains at least one sort variable, or more precisely, iff at least one variable occurrence in p or one constant symbol occurrence in p has a nonground sort. Informally, p is polymorphic iff you see at least one sort variable when Athena prints out p. Otherwise p is monomorphic. Note that p might contain occurrences of polymorphic function symbols but it might not itself be polymorphic. For example, the sentence `('a = ?x)` contains an occurrence of the polymorphic equality symbol =, but it is not itself a polymorphic sentence, because no leaves of it (viewing the sentence as a tree) have polymorphic sorts. This can be confirmed when Athena prints out the sentence; we then observe that there are no sort variables anywhere. Likewise for the sentence `(?l = 1::nil)`, which has not one but two polymorphic symbol occurrences, one occurrence of the identity symbol and one of the polymorphic list constructor `::`. That sentence, too, is monomorphic:

```
> ('a = ?x)

Term: (= 'a ?x:Ide)

> (?l = 1 :: nil)

Term: (= ?l:(List Int)
         (:: 1
             nil:(List Int)))
```

By contrast, the following sentence is polymorphic because at least one of its leaves (occurrence of a constant symbol or variable) has a polymorphic sort:

```
> (not (?x = ?y))

Sentence: (not (= ?x:'T47 ?y:'T47))
```

A polymorphic sentence p like this one can be grounded by consistently replacing all sort variables at its leaves with ground sorts. We refer to the monomorphic sentence thus obtained as a ground *instance* of p. For example, two ground instances of the above sentence are:

```
> (not (?x:Boolean = ?y:Boolean))

Sentence: (not (= ?x:Boolean ?y:Boolean))

> (not (?x:(List (List Real)) = ?y:(List (List Real))))

Sentence: (not (= ?x:(List (List Real))
                  ?y:(List (List Real))))
```

If Σ is a set of ground sorts, then $G_\Sigma(p)$ is the set of all ground instances of p with respect to Σ. The set $G_\Sigma(p)$ will be infinite iff Σ is infinite.

Now let p be a monomorphic sentence over Σ and F that might nevertheless contain occurrences of polymorphic symbols, like the examples that were just given. Then p can be uniquely rewritten into a sentence p' over $\widehat{\Sigma}$ and $G_\Sigma(F)$, called *the elaboration of* p, and denoted by $E(p)$, by replacing every occurrence of a polymorphic function symbol f in p with a unique instance f' of f in $G_\Sigma(F)$, as follows. Let t be the term rooted at the said occurrence of f in p, and let $t_1, \ldots, t_n$ be the immediate subterms of t (we will have $n = 0$ iff t is a polymorphic constant symbol). We then replace the said occurrence of f by that unique instance of f determined by (a) the (monomorphic) sorts of $t_1, \ldots, t_n$ and (b) the monomorphic sort of the said occurrence of t. For example, if p is the sentence

```
(not (= ?x:Int ?y:Int)),
```

then the elaboration of p, $E(p)$, is the sentence

$$\texttt{(not (}=_{\texttt{[[Int Int] Boolean]}}\texttt{ ?x:Int ?y:Int)),}$$

which is obtained from p by replacing the sole occurrence of the polymorphic identity symbol = by its instance

$$=_{\texttt{[[Int Int] Boolean]}}.$$

Essentially, $E(p)$ is produced from p by tagging every occurrence of a polymorphic function symbol in p with the unique monomorphic signature determined by the local context of that occurrence. Note that $E(p)$ is not a sentence over the original signature F. Rather, it is a sentence over $G_\Sigma(F)$, the set of fully elaborated symbols we obtain from the symbols in F by grounding their signatures with respect to $\widehat{\Sigma}$.

Finally, for any polymorphic sentence p over Σ and F, we define $GE(p)$, the set of all *elaborated ground instances* of p, as follows:

$$GE(p) = \{E(q) \mid q \in G_\Sigma(p)\}.$$

Moreover, given:

- an interpretation I of a polymorphic vocabulary Σ and F, that is, an interpretation of the monomorphic vocabulary $\widehat{\Sigma}$ and $G_\Sigma(F)$;
- a variable assignment χ, namely, a total function mapping every variable $v : S$ with $S \in \widehat{\Sigma}$ to an element of S^I; and
- a polymorphic sentence p over Σ and F;

we define

$$I_\chi \models p \text{ iff } I_\chi \models GE(p). \tag{5.71}$$

In other words, I satisfies a polymorphic sentence p with respect to a variable assignment χ iff I satisfies every elaborated ground instance of p with respect to χ. The rest of the monomorphic definitions of semantic notions given earlier carry over unchanged to the polymorphic case, ultimately grounded in (5.71). Thus, for a (possibly infinite) set of polymorphic sentences Φ, we write $I_\chi \models \Phi$ iff $I_\chi \models p$ for every $p \in \Phi$. We say that a polymorphic sentence p is *true* in a given interpretation I iff $I_\chi \models p$ for every χ. (Hence, for example, the sentence

```
(forall ?x:'S . ?x:'S = ?x:'S)
```

is true in every interpretation.) A set of polymorphic sentences Φ_1 logically entails a set of polymorphic sentences Φ_2 iff for every I and χ we have $I \models_\chi \Phi_2$ whenever $I \models_\chi \Phi_1$; a set of polymorphic sentences Φ is satisfiable iff there exist I and χ such that $I \models_\chi \Phi$; a polymorphic sentence p is a tautology iff $I \models_\chi \{p\}$ for every I and χ; and so on.

Based on these definitions, Athena proofs can be shown to be sound and complete for polymorphic many-sorted logic in the following sense. Fixing a vocabulary, a lexical environment, and a store, let us write $\beta \vdash D \hookrightarrow p$ to mean that evaluating an Athena deduction D in assumption base β produces the sentence p. We then have:

Theorem 5.2: Soundness

If $\beta \vdash D \hookrightarrow p$ then $\beta \models p$.

Theorem 5.3: Completeness

If $\beta \models p$ then there exists a deduction D such that $\beta \vdash D \hookrightarrow p$.

The definitions in this section have assumed flat namespaces for sorts and function symbols but can easily accommodate modules, as for semantic purposes the sorts and symbols introduced in a module can always be flattened (they can be viewed as sorts and symbols introduced at the top level—outside of any modules—but with fully qualified names as dictated by the given nesting of modules).

We close with some brief remarks on the interpretation of Athena's primitive domains and datatypes. The interpretation of `Boolean` is the standard two-element domain of truth values, which we may identify with the two term constructors of `Boolean`, namely, `true` and `false`. The interpretation of `Ide` is the set of all strings (finite sequences of characters) that can serve as Athena meta-identifiers, and every constant symbol of sort `Ide` is interpreted as the corresponding string (so that `'x` is literally interpreted as the string `x`). The interpretation of `Real` is the set of all real numbers, and that of `Int` is the set of all integers (a subset of the reals). All constant numerals receive their obvious interpretations; for example, `3.14` is interpreted as the corresponding rational number. The four binary symbols (`+`, `-`, `*`, and `/`) take reals and produce reals, but these symbols are also overloaded in the standard sense of operator overloading, and thus represent the only instances of ad hoc polymorphism in Athena. Accordingly, there are essentially two different sets of function symbols involved here, one set with integer-based signatures and the other with real-based signatures. The interpretations of the symbols in each of these two families are obvious. Resolution is performed dynamically, based on the sorts of the arguments. The minus symbol can be used as a unary symbol as well, in which case it is interpreted either as integer or as real negation, depending again on the sort of the argument:

```
> (?x:Int + 1)

Term: (+ ?x:Int 1)

> (?x:Real + 1)

Term: (+ ?x:Real 1)

> (?x + 1.0)

Term: (+ ?x:Real 1.0)

> (?x - 1)

Term: (- ?x:Int 1)

> (- ?x:Real)

Term: (- ?x:Real)

> (- ?x:Int)

Term: (- ?x:Int)
```

5.7 Additional exercises

Exercise 5.6: Extend the procedure `polarities` (defined on page 269) to the first-order case. Then:

(a) Modify that implementation as specified in the first part of Exercise 4.1, so that every item in the output list consists of a two-element list

[*position polarity*]

comprising the *position* of the relevant occurrence as well as its *polarity*.

(b) Implement the first-order version of the binary procedure `polarities*`, as specified in the second part of Exercise 4.1.

(c) Extend the methods `M+` and `M-` of the third part of that same exercise to handle quantified sentences. □

Exercise 5.7: Similarly to Exercise 5.2, each of the following listings asserts some premises (possibly none) and defines a goal:

Listing 5.7.1

```
define goal :=
  ((forall x . P x <==> Q x) ==> (forall x . P x) <==> (forall x . Q x))
```

Listing 5.7.2

```
declare A: Boolean

assert premise-1 := (exists x . c R x & A)
assert premise-2 := (exists x . Q x & x T x)
assert premise-3 := (forall x . A & Q x ==> ~ S x)

define goal := (exists y . ~ S y & y T y)
```

Listing 5.7.3

```
assert premise-1 := (exists x . P x & forall y . Q y ==> x R y)
assert premise-2 := (forall x . P x ==> forall y . S y ==> ~ x R y)

define goal := (forall  x .  Q x ==> ~ S x)
```

Listing 5.7.4

```
declare f: [Object] -> Object

assert premise-1 := (forall x . x R x)

assert premise-2 := (forall x . f x = f f x)

define goal := (exists y . y R f y)
```

For each of these, clear the assumption base and then derive the `goal` from the corresponding premises. □

Exercise 5.8: Write a proof that derives the sentence

```
(exists x . P x | ~ P x)
```

from the empty assumption base. □

Exercise 5.9: One of the transformations implemented by the quantifier-swapping method of Section 5.4 is this:

$$(\exists x \,.\, \forall y \,.\, p) \longrightarrow (\forall y \,.\, \exists x \,.\, p).$$

Show that the converse transformation is not valid. □

Exercise 5.10: Write a unary method `M` that takes an arbitrary term t (of some sort `Object`) and derives the following theorem: `(P t <==> exists x . x = t & P x)`. □

Exercise 5.11: Define a binary method `pick-all-witnesses` whose first argument is an existentially quantified premise with an arbitrary number $n \geq 0$ of existential quantifiers at the front, followed by a sentence q that is not existentially quantified. The idea here is to use **`pick-witness`** n times on the given premise in order to extract n witnesses (fresh variables) $w_1, \ldots, w_n$ for the n existential quantifiers, along with a corresponding instance q' of the body q. The second argument of `pick-all-witnesses` is a binary method M that will receive (a) the list of witnesses $w_1, \ldots, w_n$, in that order; and (b) the sentence q', which must be in the assumption base when M is called. The result of `pick-all-witnesses` will be whatever result is obtained by applying M to these two arguments.

Note that if n were fixed (statically known), we could implement this method simply by nesting n **`pick-witness`** deductions, followed by a call to M. But the point here is that the number n of leading existential quantifiers is unknown, hence some form of programmatic iteration is needed. □

* ***Exercise 5.12:*** Consider again sentence (5.24):

```
(forall x1 ··· xn . p <==> q)
```

and the conjunction of (5.29) and (5.30):

```
((forall x1 ··· xn . p ==> q) & (forall x1 ··· xn . q ==> p)).
```

Prove that the two are equivalent. More specifically, write a unary method that can accept either of these two sentences as a premise and derive the other. (Hint: Because n is not statically fixed, you will need some form of iteration akin to what was used in Exercise 5.11, involving a similar proof continuation as an argument.) □

Exercise 5.13: Define a unary method `move-quant` implementing the following inference rules:

(a)
```
  (forall x . p & q)
  ---------------------
  ((forall x . p) & q)
```
provided that x does not occur free in q

(b)

$$\frac{\texttt{(forall } x \texttt{ . } p \texttt{ | } q\texttt{)}}{\texttt{((forall } x \texttt{ . } p\texttt{) | } q\texttt{)}}$$

provided that x does not occur free in q

(c)

$$\frac{\texttt{(exists } x \texttt{ . } p \texttt{ \& } q\texttt{)}}{\texttt{((exists } x \texttt{ . } p\texttt{) \& } q\texttt{)}}$$

provided that x does not occur free in q

(d)

$$\frac{\texttt{(exists } x \texttt{ . } p \texttt{ | } q\texttt{)}}{\texttt{((exists } x \texttt{ . } p\texttt{) | } q\texttt{)}}$$

provided that x does not occur free in q

The method should be bidirectional, capable of applying the inference rules in both directions. □

Exercise 5.14 (Quantifier Distribution): Define a method `quant-dist` implementing the rules for quantifier distribution presented in Section 5.4.

** ***Exercise 5.15 (The halting problem):*** In this exercise we formalize the halting problem [26] and prove its undecidability.[26] In doing so, we will need to talk about and quantify over *algorithms*, so we begin by introducing a corresponding domain:

```
domain Algorithm
```

Algorithms take inputs of various sorts (e.g., integers, strings, Booleans, pairs, lists, and, quite importantly, even algorithms), so we introduce a polymorphic datatype representing algorithm inputs:

```
datatype (Input S) := (input S)
```

Thus, for instance, `(input 4)` represents a single integer input (specifically, the number 4), while `(input (pair 4 true))` represents an ordered-pair input, specifically, an ordered pair consisting of the integer 4 and the Boolean `true` (see Section 10.1 for more details on ordered pairs). Likewise, algorithms can generate results of various sorts, so we introduce a polymorphic datatype for outputs:

```
datatype (Output S) := (answer S)
```

Next, we introduce a ternary predicate `outputs` that holds between an algorithm A, an input x, and an output y, iff A, when applied to input x, eventually produces output y. Because inputs and outputs are independently polymorphic, the predicate is parameterized over two sort variables:

---

26 This exercise was inspired by Burkholder [15], although our formalization here is different from—and, we believe, simpler than—the one given there.

```
declare outputs: (S, T) [Algorithm (Input S) (Output T)] -> Boolean
```

We also introduce a binary predicate halts-on that is intended to obtain between an algorithm A and an input x iff A halts on input x:

```
declare halts-on: (S) [Algorithm (Input S)] -> Boolean
```

We say that an algorithm A *decides* whether an algorithm B halts on an input x iff:

1. If B does halt on x, then when the pair (B,x) is given to A as input, A outputs true.
2. If, on the other hand, B does not halt on x, then when (B,x) is given to A as input, A outputs false.

In short, A takes (B,x) as input and outputs true iff B halts on x. We formalize this relation as follows:

```
declare decides-halting: (S) [Algorithm Algorithm S] -> Boolean

define [A B A' B'] := [?A ?B ?A' ?B']

assert* decides-halting-def :=
 ((decides-halting A B x) <==>
     (B halts-on input x ==>   outputs A (input B @ x) answer true)
   & (~ B halts-on input x ==> outputs A (input B @ x) answer false))
```

To solve the halting problem mechanically, we must have an algorithm A that is capable of deciding whether *any* algorithm B halts on *any* input x. Let us say that such an algorithm A is a *halting decider*:

```
declare halting-decider: (S) [(Algorithm S)] -> Boolean

assert halting-decider-def :=
  (forall A . halting-decider A <==> forall B x .
                                       (decides-halting A B x))
```

Note that so far we have not made any substantive assertions. Both of the above are conservative definitions, which essentially hold by stipulation. They are guaranteed to be consistent. In fact, the only real premise we will need to assert for the proof is the following:

```
1 assert premise :=
2   (forall A .
3     exists B .
4       forall x .
5        ((outputs A (input x @ x) (answer true))  ==> ~ B halts-on input x)
6       & ((outputs A (input x @ x) (answer false)) ==> B halts-on input x))
```

What does this sentence say, and why is it true? It says that for any algorithm A, there is an algorithm B that takes an input x and behaves as follows. It applies A to the input pair (x, x); if and when that application outputs the answer `true`, B goes into an infinite loop, and thus fails to halt on x (the outcome described on line 5); but if and when the application of A to (x, x) outputs the answer `false`, B halts (with some unspecified output), as specified on line 6. The premise does not specify the behavior of B for any other possibility.

The premise is true simply because for any algorithm A, the existence of B can be ensured *by construction*. Given A, we can specify an algorithm B that behaves as described: B accepts an input x and runs A on the input pair (x, x). If the execution of A never terminates for that input pair, then of course B will also go on forever. But if A does halt, we examine its output and act as follows. If the output is `true`, we deliberately get into an infinite loop; if the output is `false`, we halt. In fact, if we regard algorithms as unary procedures, we can easily implement this in Athena as a higher-order procedure that takes A as input first and then produces the required algorithm B:

```
define loop := lambda () (loop)

define make-B :=
  lambda (A)
    lambda (x)
      match (A (x @ x)) {
        true   => (loop)
      | false => 'ok
      }
```

The problem asks you to prove that there exists no algorithm for deciding the halting problem, or in our notation:

```
define goal := (~ exists A . halting-decider A)
```

That is, give a proof D that derives `goal` from the three asserted sentences. □

Exercise 5.16 (Russell's paradox): Consider:

```
domain D
declare M: [D D] -> Boolean
```

and prove:

```
(~ exists x . forall y . y M x <==> ~ y M y).        (5.72)
```

When we interpret `M` as the membership relation on sets, so that (x `M` y) iff set x is a member of set y, then (5.72) states that there is no set whose members are all and only those sets that do not contain themselves. When we interpret `D` as a set of men and (x `M` y) as the statement that x is shaved by y, then the sentence says that there is no man who shaves all and only those men who do not shave themselves. (That is the popularized "barber" version of Russell's paradox.) It is a remarkable fact that many truths of set theory, such as the nonexistence of a set that contains all and only those sets that do not contain themselves, can be formulated and proved as tautologies of pure logic. For additional examples refer to the text by Kalish and Montague [55]. □

Exercise 5.17 (Drinker's principle): Prove the following:

$$(\exists x . Dx \Rightarrow \forall y . Dy).$$

This is colloquially referred to as *the drinker's principle*: There is always some individual x such that, if x is drinking, then everyone is drinking. □

Exercise 5.18 (Theory of Measurement): This exercise is based on a paper by Patrick Suppes [98] on the theory of empirical measurement of extensive quantities, and in the present context, specifically on the measurement of mass. The vocabulary of the theory contains a binary function symbol $\star$ and a binary predicate Q. Intuitively, the domain of the intended interpretation consists of a large set of physical objects whose masses are to be measured by a scale. We interpret

$$(x \ \mathtt{Q} \ y)$$

as the proposition that when we place x on the left pan of the scale and y on the right, we observe one of two outcomes: Either the scale is perfectly balanced, which means that x and y have identical mass, or else y outweighs x:

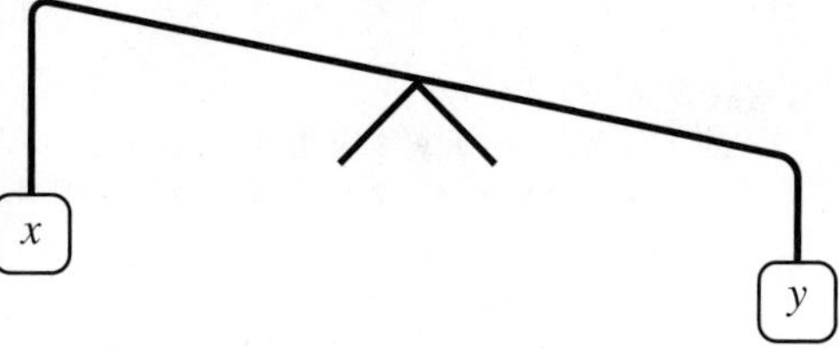

In other words, $(x \ \mathtt{Q} \ y)$ holds iff the observed mass of x is less than or equal to that of y.[27] The operator $\star$ can be viewed as a way to produce a new object by combining two existing objects. Intuitively, we can think of $(x \star y)$ as the operation of placing x followed by y on a pan of the scale. The theory is formalized through the following five axioms, written here in conventional notation:

27 Of course, we could interpret Q directly as the usual numeric less-than relation, provided we also introduce a unary function symbol m from objects to real numbers, intended to represent their mass measurements. However, that would be putting the cart before the horse for the purpose of theories such as that of Suppes, whose task is a conceptual investigation of the fundamental conditions that a physical system must satisfy *in order* for numeric measurement to be possible (and meaningful). For more background on measurement theory, refer to the historical overview of the field by Díez [32].

[A1]	$(\forall\, x,y,z\,.\, x \; \mathtt{Q} \; y \wedge y \; \mathtt{Q} \; z \Rightarrow x \; \mathtt{Q} \; z)$
[A2]	$(\forall\, x,y,z\,.\, (x * y) * z \; \mathtt{Q} \; x * (y * z))$
[A3]	$(\forall\, x,y,z\,.\, x \; \mathtt{Q} \; y \Rightarrow x * z \; \mathtt{Q} \; z * y)$
[A4]	$(\forall\, x,y\,.\, \neg\, x \; \mathtt{Q} \; y \Rightarrow \exists\, z\,.\, x \; \mathtt{Q} \; y * z \wedge y * z \; \mathtt{Q} \; x)$
[A5]	$(\forall\, x,y\,.\, \neg\, x * y \; \mathtt{Q} \; x)$

You are asked to derive the following theorems from these axioms:

[T1]	$(\forall\, x\,.\, x \; \mathtt{Q} \; x)$
[T2]	$(\forall\, x,y\,.\, x * y \; \mathtt{Q} \; y * x)$
[T3]	$(\forall\, x_1,x_2,x_3,x_4\,.\, x_1 \; \mathtt{Q} \; x_2 \wedge x_3 \; \mathtt{Q} \; x_4 \Rightarrow x_1 * x_3 \; \mathtt{Q} \; x_2 * x_4)$
[T4]	$(\forall\, x,y,z\,.\, x * (y * z) \; \mathtt{Q} \; (x * y) * z)$
[T5]	$(\forall\, x,y\,.\, x \; \mathtt{Q} \; y \vee y \; \mathtt{Q} \; x)$
[T6]	$(\forall\, x,y,z\,.\, x * z \; \mathtt{Q} \; y * z \Rightarrow x \; \mathtt{Q} \; y)$
[T7]	$(\forall\, x,y,z\,.\, y * z \; \mathtt{Q} \; u \wedge x \; \mathtt{Q} \; y \Rightarrow x * z \; \mathtt{Q} \; u)$
[T8]	$(\forall\, x,y,z\,.\, u \; \mathtt{Q} \; x * z \wedge x \; \mathtt{Q} \; y \Rightarrow u \; \mathtt{Q} \; y * z)$

In Athena notation, derive `theorem-1` through `theorem-8` of module `MT`:

```
module MT {
  domain D
  declare Q: [D D] -> Boolean
  declare *: [D D] -> D

  define [x y z z' u v x1 x2 x3 x4 x5] := [?x:D ?y:D ··· ?x5:D]

  define [A1 A2 A3 A4 A5] :=
    (map close [(x Q y & y Q z ==> x Q z)
                ((x * y) * z Q x * (y * z))
                (x Q y ==> x * z Q z * y)
                (~ x Q y ==> exists z . x Q y * z & y * z Q x)
                (~ x * y Q x)])

  assert [A1 A2 A3 A4 A5]

  define [theorem-1 theorem-2 theorem-3 theorem-4
          theorem-5 theorem-6 theorem-7 theorem-8] :=
    (map close [(x Q x)
                (x * y Q y * x)
                (x1 Q x2 & x3 Q x4 ==> x1 * x3 Q x2 * x4)
                (x * (y * z) Q (x * y) * z)
                (x Q y | y Q x)
                (x * z Q y * z ==> x Q y)
                (y * z Q u & x Q y ==> x * z Q u)
                (u Q x * z & x Q y ==> u Q y * z)])
}
```

You are encouraged to write your own auxiliary methods to facilitate reasoning in this theory. □

5.8 Chapter notes

Important patterns of inference with quantifiers had already been introduced and studied by Aristotle in his logic of syllogisms [3], but they were limited to sentences with single quantifier occurrences, and therefore to unary predicates.[28] Quantified logic as we understand it today did not emerge until the work of Frege, Peano, and Russell in the late 19th and early 20th centuries. Frege [39], in particular, was the first to explicitly introduce a mathematically precise concept of a logical quantifier.

Multiple quantifier nesting allowed for a huge leap in expressiveness from Aristotle's unary predicates to arbitrary relations. Complicated propositions with intricate logical structure became readily expressible in predicate logic, which has turned out to be capable of capturing all of mathematics and even much of natural language. Frege's work also enabled a much clearer distinction between syntax and semantics, a distinction that was soon sharpened and investigated in depth by Hilbert, Tarski, and others, giving rise to the classic metatheoretic results for first-order logic, such as soundness, completeness, and compactness, as well as the negative results of Gödel. Well-known mathematical logic textbooks with rigorous discussions of such results include Kleene's *Mathematical Logic* [56] and Mendelson's *Introduction to Mathematical Logic* [73]. For a more contemporary exposition, see the textbook by Ebbinghaus et al. [34]. Recursive definitions of satisfaction relations, like the one shown in Figure 5.1, are specifically due to Tarski's seminal work [101]. Exercise 5.18 also appears in a classic logic textbook by Suppes [99]. Another text that has aged eminently well is Tarski's *Introduction to Logic and to the Methodology of the Deductive Sciences* [102], which first appeared (in Polish) in 1936. It is highly recommended as a general—and very accessible—introduction to logic.

In its usual formulations, first-order logic is *single-sorted*; there are no sorts (or types).[29] It is assumed that there is only one universe of discourse U, so every n-ary function symbol takes all its arguments from U and produces an element of U. Yet types were integral to logic from the beginning, and figured prominently in the work of Russell and even of Frege (though in the latter case not in the conventional contemporary sense). First-order logic enriched with a sort system results in so-called *many-sorted* (or "multi-sorted") first-order logic, which is at the heart of Athena. The first formulation of many-sorted logic as we understand it today appears to have been given by Wang in 1952 [109]. Manzano [67] argues

28 Although Aristotle viewed quantified statements as binary relations obtaining between unary predicates ("terms").

29 In this text we use types to classify Athena's denotable values, and sorts to classify Athena terms. Terms are just one type of value in Athena.

that many-sorted first-order logic is distinguished by its naturalness and favorable theoretical properties, and that most reasonable logical systems (including higher-order logic) have natural translations into many-sorted first-order logic. Meinke and Tucker also single out many-sorted first-order logic as a unique logic that "stands beside category theory and type theory as a very general mathematical framework which can: (i) support foundational studies; (ii) provide tools for solving specific problems; and (iii) be processed by machines. It is simpler to use than category theory and type theory and closer to applications" [72] (p. xvi). Higher-order logic has many of the same advantages as many-sorted first-order logic, but at the expense of certain algorithmic and theoretical complications, such as an undecidable unification problem and deductive incompleteness.

Athena further enhances many-sorted first-order logic with Hindley-Milner-style *polymorphism* [74], essentially the same type system found in programming languages like ML and Haskell. Athena discovers the most general sort for a given term using the same type-reconstruction inference algorithm invented by Milner. The advantages of parametric polymorphism for logic and proofs are largely the same as they are for programming: We only need to prove a theorem once, using the most general possible sorts for its quantified variables; after that we can directly apply the theorem to derive conclusions about any objects whose sorts are compatible with the most general sorts. If we know that reversing a list l twice produces l, where l is a list of *any* sort whatsoever, then we can directly derive the same conclusion when l is a list of integers, or a list of trees, or a list of lists of integers, and so on. With few exceptions (such as the natural numbers), most of the theories and proofs that appear in subsequent chapters are polymorphic. Athena's reduction of the semantics of polymorphic multi-sorted first-order logic to the monomorphic case is similar to that carried out in the semantics of the Gödel programming language [46]. For more background on universal algebra, consult the excellent introduction by Meinke and Tucker [71] and the references given there.

Implication Chaining

IN this chapter we extend the chaining style of proof from equational logic to full first-order logic. Proofs expressed in this style tend to be particularly readable and will be used widely throughout the rest of the book.

6.1 Implication chains

Recall the general form of a proof by equational chaining:

```
(!chain [t1 = t2      J1
            = t3      J2
            ⋮
            = tn+1    Jn]).
```

The goal here is to "connect" the starting term t_1 with the final term t_{n+1} through the identity relation, that is, to derive

$$t_1 = t_{n+1}. \tag{6.1}$$

This is done in a number of steps $n > 0$, where each step derives an intermediate identity $t_i = t_{i+1}$, $i = 1, \ldots, n$, by citing J_i as its justification (we do not need to be concerned here with the exact nature of that justification). Provided that each step in the chain goes through, the desired result (6.1) finally follows from the transitivity of the identity relation. Ultimately, it is this transitivity that makes equational chaining work, and it is this transitivity, along with the inherently tabular format of `chain`, that makes these proofs so perspicuous.

But identity is not the only important logical relation that is transitive. Implication and equivalence are also transitive. Capitalizing on that observation, we can extend the chaining style of proof from identities to implications and equivalences, reaping similar benefits in notational clarity.

Let us consider implications first. The general format here is very similar to equational chaining, except that the chain links are now sentences p_i rather than terms t_i, and the symbol ==> takes the place of =:

```
(!chain [p1 ==> p2      J1
            ==> p3      J2
            ⋮
            ==> pn+1    Jn]).
```

The larger idea remains the same: The goal is to "connect" the starting point p_1 with the end point p_{n+1}, this time through the implication relation, that is, to derive

$$p_1 \Rightarrow p_{n+1}. \tag{6.2}$$

This is done in a number of steps $n > 0$, where each step derives an intermediate implication $p_i \Rightarrow p_{i+1}$, $i = 1, \ldots, n$, by citing J_i as its justification (we will discuss shortly exactly what kinds of justification are appropriate here; for now we only need to keep in mind that each step in the chain must include some justification item, whatever that might be, just like equational chaining). Provided that each step in the chain goes through, the desired conditional (6.2) ultimately follows from the transitivity of implication.

The justification J_i for a step p_i ==> p_{i+1} typically consists of a unary method M that can be applied to p_i in order to produce p_{i+1}. That is, M is a method that can derive the right-hand side of the step from the left-hand side. More precisely, M is such that *if* p_i is in the assumption base, then (!M p_i) will produce p_{i+1}. Here is an example:

```
> (!chain [(A & B) ==> A left-and])

Theorem: (if (and A B)
             A)
```

As with equational chaining, we prefer to enclose justification items inside lists, so we would typically write the above as follows:

```
> (!chain [(A & B) ==> A [left-and]])

Theorem: (if (and A B)
             A)
```

The reason for the list notation is that a justification item may consist of more than one method (and it may also include sentences, as we will soon see), so in the general case we need to group all of the given entries—methods and/or sentences—in a list.

Here is a more interesting example:

```
> (!chain [(A & ~~ B) ==> (~~ B) [right-and]
                      ==> B       [dn]])

Theorem: (if (and A
                  (not (not B)))
             B)
```

This chain has two steps. On the first step we use `right-and` to derive `(~ ~ B)` from

`(A & ~ ~ B),`

on the assumption that the latter holds; and on the second step we derive `B` from `(~ ~ B)` by `dn`.

Note that even though an implication chain produces a conditional conclusion, there is no explicit use of **assume** anywhere. It is the implementation of chain that uses **assume**, not its clients. Specifically, the implementation starts by assuming the first sentence of the chain, p_1, and proceeds to derive the second element of the chain, p_2, by using the given method(s). If successful, this produces the conclusion $p_1 \Rightarrow p_2$. Then the same process is repeated for the next pair of the chain: The conclusion $p_2 \Rightarrow p_3$ is derived, by assuming p_2 and inferring p_3 through the given methods. We continue in that fashion, until finally transitivity is used to produce the output $p_1 \Rightarrow p_{n+1}$.

Anonymous methods can appear inline in the justification list of a given step:

```
> (!chain [(A & ~~ B) ==> B [method (p) (!dn (!right-and p))]])

Theorem: (if (and A
                  (not (not B)))
             B)

> (!chain [(forall ?x . ?x = ?x) ==> (1 = 1) [method (p) (!uspec p 1)]])

Theorem: (if (forall ?x:'S
                (= ?x:'S ?x:'S))
             (= 1 1))
```

In the first chain, **method** (p) (!dn (!right-and p)) was applied to the hypothesis (A & ~ ~ B) to produce the conclusion B in one step. Inlining very small anonymous methods inside a chain step like that may be acceptable, but readability is compromised if the method is more complicated. Even the first of the two examples above would arguably be cleaner if the definition of the method were pulled out of the chain:

```
> let {M := method (p) (!dn (!right-and p))}
    (!chain [(A & ~~ B) ==> B [M]])

Theorem: (if (and A
                  (not (not B)))
             B)
```

The justification list of a given step p_i ==> p_{i+1} may contain extraneous methods that fail to derive p_{i+1} from p_i. The implementation of chain will try every given method until one succeeds, disregarding those that fail. For instance, if we add superfluous information to each step of the previous example, the proof will still work:

```
> (!chain [(A & ~~ B) ==> (~~ B) [right-and dn mp]
                      ==> B       [left-and dn]])

Theorem: (if (and A
                  (not (not B)))
             B)
```

This is a useful feature because oftentimes we know that a certain *set* of resources (methods and/or premises) justify a given step, but it would be onerous to specify *exactly which* elements of that set are needed for the step; we prefer to have Athena make that determination for us.[1]

A justifying method for a step p_i ==> p_{i+1} need not be unary, taking p_i as its only argument and deriving p_{i+1}. Occasionally it makes sense to feed both the left-hand side premise p_i *and* the goal p_{i+1} as two distinct arguments to a justifying method. Methods of either type (unary or binary) are acceptable as justifications for an implication step. When a binary method is used, `chain` passes it the premise p_i as its first argument and the goal p_{i+1} as its second argument. For instance:

```
> let {M := method (premise goal) (!right-and premise)}
    (!chain [(A & ~~ B) ==> (~~ B) [M]
                        ==> B       [dn]])

Theorem: (if (and A
                  (not (not B)))
             B)
```

In this simple example the binary method `M` ignores its second argument, the goal, because we already (statically) know what that goal is—it's the second conjunct of the premise! But in more complex scenarios we may not know the goal ahead of time, and in such cases it can be helpful to make the justifying method binary, giving it access to the premise as well as to the goal.

For example, we often want to take steps p_i ==> p_{i+1} that pass from p_i to an arbitrarily complicated conjunction p_{i+1} that contains p_i as one of its conjuncts, while all the other conjuncts of p_{i+1} are already known or assumed to hold. We want a general-purpose method, call it `augment`, that can justify such steps, so that we can write, for instance:

```
assert A, B

> (!chain [C ==> (A & B & C) [augment]])

Theorem: (if C
             (and A
                  (and B C)))

> (!chain [C ==> (C & B) [augment]])

Theorem: (if C
             (and C B))
```

1 In fact this feature is not just useful but necessary, because, as we will see shortly, sometimes one justifier is necessary for one part of the implication step and another justifier is needed for another part of the same step (such cases arise often in structural implication steps; see Section 6.3). So the implementation of `chain` must be able to accept an arbitrarily large list of justifiers and automatically determine which of these may be used where.

```
> (!chain [C ==> (and A A B C A B) [augment]])

Theorem: (if C
             (and A A B C A B))
```

The natural way to define augment is as a binary method that takes p_i as its first argument and the conjunction p_{i+1} as its second argument and then derives p_{i+1} by conj-intro:[2]

```
define augment :=
  method (premise conjunctive-goal)
    (!conj-intro conjunctive-goal)
```

Another example is the method existence, which allows us to pass from any sentence p to an existential generalization of p, that is, to a sentence of the form (exists $v_1 \cdots v_n$. q) such that p can be obtained from q by substituting appropriate terms for the free occurrences of the variables $v_1 \cdots v_n$. For instance:

```
domain Person
declare likes: [Person Person] -> Boolean
declare Mary, Peter: Person

> (!chain [(Mary likes Peter) ==>
           (exists x . x likes Peter) [existence]])

Theorem: (if (likes Mary Peter)
             (exists ?x:Person
               (likes ?x:Person Peter)))

> (!chain [(zero < S zero) ==> (exists x y . x < y) [existence]])

Theorem: (if (< zero
                (S zero))
             (exists ?x:N
               (exists ?y:N
                 (< ?x:N ?y:N))))
```

The most straightforward way to implement existence is as a binary method that takes the premise p as its first argument and the desired existential generalization as its second argument:

```
define existence :=
  method (premise eg-goal)
    match (match-sentences premise (quant-body eg-goal)) {
```

2 Recall (footnote 16) that conj-intro is a unary method that takes as input an arbitrarily complicated conjunction p and derives p, provided that all of p's conjuncts are in the assumption base. It can be regarded as a generalization of both.

```
    (some-sub sub) => (!egen* eg-goal (sub (qvars-of eg-goal)))
  }
```

(The procedure quant-body returns the body of a quantified sentence, while qvars-of takes a quantified sentence of the form (Q $v_1 \cdots v_n$. p), where p does not start with the quantifier Q, and returns the list of variables [$v_1 \cdots v_n$]. Both are defined in Athena's library. Finally, recall Exercise 5.3 for the definition of egen*.) Users can implement their own methods for implication steps, with either unary or binary interfaces.

Sometimes an implication step p_i ==> p_{i+1} can only go through if we combine the premise p_i with additional, earlier information, namely, with sentences that were previously asserted, assumed, or derived.[3] Suppose, for example, that we know (A ==> B) to be the case, say, because we have asserted it. Then, intuitively, we should be able to write something like the following:

```
assert (A ==> B)

(!chain [(~ B) ==> (~ A) M])
```

The question now is what sort of justification method M could work here. It is clear that modus tollens (mt) is involved, but the issue seems to be that M only has access, via its arguments, to local information, namely, to the left-hand side of the step (or, if we express M as a binary method, to the right-hand side of the step as well). So how can we make M take (A ==> B) into account?

The answer lies in a sort of partially evaluated application of modus tollens, whereby we define M so that (A ==> B) already appears inside its body:

```
1  assert A=>B := (A ==> B)
2
3  > (!chain [(~ B) ==> (~ A) [method (p) (!mt A=>B p)]])
4
5  Theorem: (if (not B)
6               (not A))
```

It should be clear that

```
method (p) (!mt A=>B p)
```

is a unary method which, when applied to the left-hand side of the implication step on line 3, namely (~ B), will successfully derive the right-hand side, (~ A), as required by the specification of chain.

3 Uses of augment are actually of this sort, but the key difference of a step p_i ==> p_{i+1} [augment] from the cases we are about to discuss is that the "earlier information" is all embedded in and retrievable from the right-hand side, p_{i+1}, and is therefore local to the step, unlike the forthcoming examples.

The following example uses the same technique to combine the left-hand side of a chain step with *two* previous pieces of information via the ternary method cases:

```
assert A=>C := (A ==> C)
assert B=>C := (B ==> C)

> (!chain [(A | B) ==> C [method (p) (!cases p A=>C B=>C)]])

Theorem: (if (or A B)
             C)
```

However, it is tedious to write justifying methods in this long form every time we want to combine a left-hand side with previous information through a method M of $k > 1$ arguments (such as mt, cases, etc.). It is better to have a single generic mechanism that lets us specify M and the $k-1$ nonlocal arguments, and constructs the appropriate method automatically. The binary procedure with is such a mechanism. It takes the k-ary method M as its first argument and a list of the $k-1$ nonlocal arguments as its second argument, and produces the appropriate method required by the implementation of chain. Using with in infix notation,[4] the preceding example involving mt can be written as follows:

```
> (!chain [(~ B) ==> (~ A) [(mt with [A=>B])]])

Theorem: (if (not B)
             (not A))
```

Informally, this step says: "We derive the goal (~ A) by applying mt to the left-hand premise (~ B) *and* to the nonlocal premise (A ==> B), in some appropriate order." Or, somewhat more precisely, "we derive the right-hand side (~ A) from the left-hand side, (~ B), through a method that is obtained from mt by fixing its other argument to be (A ==> B)."

The cases example above can be expressed as follows:

```
assert A=>C := (A ==> C)
assert B=>C := (B ==> C)

> (!chain [(A | B) ==> C [(cases with [A=>C B=>C])]])

Theorem: (if (or A B)
             C)
```

This chaining step can likewise be understood as follows: Derive C by applying cases to the left-hand side (A | B) along with the two (nonlocal) sentences A=>C and B=>C.[5]

Another advantage of this approach is that the nonlocal arguments can be listed in an arbitrary order. The implementation of with will try different permutations to discover one

4 Recall that binary procedures can be used in infix by default.

5 Again, by "nonlocal" we mean a sentence that appears on neither side of the step.

that succeeds. For instance, in the above chain, we could list the two nonlocal arguments in reverse order and the proof would still go through:

```
> (!chain [(A | B) ==> C [(cases with [B=>C A=>C])]])

Theorem: (if (or A B)
             C)
```

By contrast, earlier, when we expressed the justifying method of this example in long form, the arguments to cases had to be given in the exact right order:

```
(!chain [(A | B) ==> C [method (p) (!cases (A | B)
                                           A=>C
                                           B=>C)]])
```

Also, the data values in the list argument of `with` can be of arbitrary type, not just sentences. For instance:

```
> (!chain [(forall x . x = x) ==> (1 = 1) [(uspec with [1])]])

Theorem: (if (forall ?x:'S
               (= ?x:'S ?x:'S))
             (= 1 1))
```

When the list has only one element, we can drop the square brackets altogether and simply write the element by itself:

```
assert A=>B := (A ==> B)

> (!chain [(~ B) ==> (~ A) [(mt with A=>B)]])

Theorem: (if (not B)
             (not A))

> (!chain [(forall x . x = x) ==> (1 = 1) [(uspec with 1)]])

Theorem: (if (forall ?x:'S
               (= ?x:'S ?x:'S))
             (= 1 1))
```

6.2 Using sentences as justifiers

It is natural to allow sentences to appear as justifications of implication steps, particularly sentences that we will call *rules*, namely, sentences of the following form:

$$(\texttt{forall}\ v_1 \cdots v_k\ .\ p_1\ \texttt{\&}\ \cdots\ \texttt{\&}\ p_n \Rightarrow q_1\ \texttt{\&}\ \cdots\ \texttt{\&}\ q_m) \tag{6.3}$$

where $k, n \geq 0$, $m > 0$.[6] Consider, for instance, a universally quantified premise that expresses the symmetry of marriage:

```
declare married-to: [Person Person] -> Boolean

assert* marriage-symmetry := (x married-to y ==> y married-to x)
```

We should be able to proceed from a sentence of the form (*s* `married-to` *t*) to the conclusion (*t* `married-to` *s*) simply by citing `marriage-symmetry`. Inference steps of this form are exceedingly common. The implementation of `chain` allows for such steps, as the following example demonstrates:

```
> (!chain [(Ann married-to Tom)
        ==> (Tom married-to Ann) [marriage-symmetry]])

Theorem: (if (married-to Ann Tom)
             (married-to Tom Ann))
```

The implementation realizes that the starting premise `(Ann married-to Tom)` *matches* the antecedent of the cited rule, `marriage-symmetry`, under the substitution

`?x:Person --> Ann, ?y:Person --> Tom,`

and proceeds to instantiate the rule with these bindings and perform modus ponens on the result of the instantiation and the starting premise. This sequence of actions, wherein a starting premise is matched against the antecedent of a rule, resulting in a substitution, and then the rule is instantiated under that substitution and "fired" via modus ponens, is a fundamental mode of reasoning,[7] and steps of that form are very common in implication chains.[8] Here is another example:

```
assert* <-tran := (x < y & y < z ==> x < z)

> (!chain [(x < 3.14 & 3.14 < 5.2) ==> (x < 5.2) [<-tran]])

Theorem: (if (and (< ?x:Real 3.14)
                  (< 3.14 5.2))
             (< ?x:Real 5.2))
```

---

6 The form of (6.3) suggests that rule conjunctions must be binary (with longer conjunctions expressed as right-associative compositions of binary conjunctions), but in fact polyadic conjunctions are also allowed.

7 Recall the relevant discussion of forward Horn clause inference from Section 5.4.

8 Strictly speaking, `chain` will match both the left-hand and right-hand sides of the step against the antecedent and consequent of the cited rule, respectively. This is necessary because in some cases the antecedent might not have enough information to give us the proper substitution. The forthcoming example on page 407, with the step `true ==> (is-male father Ann)`, illustrates that situation.

When the antecedent of the rule is a binary conjunction, the instantiated conjuncts can be listed in any order on the left-hand side of the step. Thus, for example, the following works just as well:

```
> (!chain  [(3.14 < 5.2 & x < 3.14) ==> (x < 5.2) [<-tran]])

Theorem: (if (and (< 3.14 5.2)
                  (< ?x:Real 3.14))
             (< ?x:Real 5.2))
```

The implementation of chain will notice that the left-hand side matches the antecedent of the rule modulo the commutativity of conjunction, and will rearrange the conjuncts as needed.

Biconditionals can also be used as rules. Athena will automatically extract the appropriate conditional for the step at hand:

```
declare empty: [Set] -> Boolean

define [s s1 s2] := [?s:Set ?s1:Set ?s2:Set]

assert* empty-def := (empty s <==> forall x . ~ x in s)

> (!chain [(empty s) ==> (forall x . ~ x in s) [empty-def]])

Theorem: (if (empty ?s:Set)
             (forall ?x:Element
               (not (in ?x:Element ?s:Set))))

> (!chain [(forall x . ~ x in s) ==> (empty s) [empty-def]])

Theorem: (if (forall ?x:Element
               (not (in ?x:Element ?s:Set)))
             (empty ?s:Set))
```

In addition, chain allows for rules of the following form:

$$(\texttt{forall}\ v_1 \cdots v_k\ .\ p_1\ |\ \cdots\ |\ p_n \Rightarrow q_1\ \&\ \cdots\ \&\ q_m), \tag{6.4}$$

where $k, n \geq 0$ and $m > 0$.[9] For example:

```
assert* R := (s1 = null | s2 = null ==> s1 intersection s2 = null)

> (!chain [(x = null | y = null) ==> (x intersection y = null) [R]])

Theorem: (if (or (= ?x:Set null)
                 (= ?y:Set null))
```

9 Again, more flexibility is actually allowed in that polyadic disjunctions may also be used.

```
            (= (intersection ?x:Set ?y:Set)
               null))
```

A step of this form will go through as long as the antecedent matches *some* disjunct in the antecedent:

```
> (!chain [(x = null) ==> (x intersection y = null) [R]])

Theorem: (if (= ?x:Set null)
             (= (intersection ?x:Set ?y:Set)
                null))
```

The implementation of chain actually allows for some flexibility in using a rule of the form (6.3) as a justifier for a step p ==> q. Specifically, if the left-hand side p matches the antecedent of the rule and the right-hand side q matches *some* conjunct of the consequent, the step will go through. For example, consider a rule whose consequent has three conjuncts:

```
declare child, parent: [Person Person] -> Boolean

assert* R := (father x = y ==> male y & y parent x & x child y)
```

Then all of the following chain steps are successful:

```
> (!chain [(father Ann = Tom) ==> (Tom parent Ann) [R]])

Theorem: (if (= (father Ann)
                Tom)
             (is-parent-of Tom Ann))

> (!chain [(father Ann = Tom) ==> (male Tom) [R]])

Theorem: (if (= (father Ann)
                Tom)
             (male Tom))

> (!chain [(father Ann = Tom) ==> (Ann child Tom) [R]])

Theorem: (if (= (father Ann)
                Tom)
             (child Ann Tom))
```

This is just a convenience that can save us the effort of detaching the conjunct later in a separate step. More interestingly, chain also allows us to use the rule in the contrapositive direction: If p matches the *complement* of *some* consequent conjunct, and q matches either the complement of the antecedent or else

$$(\overline{p_1} \mid \cdots \mid \overline{p_n})$$

where $\overline{p_i}$ is the complement of p_i, then the step goes through. For instance:

```
declare A, B, C, D, E, F: Boolean

assert R := (A & B & C ==> D & E & F)

> (!chain [(~ (D & E & F)) ==> (~ (A & B & C)) [R]])

Theorem: (if (not (and D
                       (and E F)))
             (not (and A
                       (and B C))))

> (!chain [(~ F) ==> (~ (A & B & C)) [R]])

Theorem: (if (not F)
             (not (and A
                       (and B C))))

> (!chain [(~ E) ==> (~A | ~B | ~C) [R]])

Theorem: (if (not E)
             (or (not A)
                 (or (not B)
                     (not C))))
```

The first application of chain, on line 5, is a typical use of the contrapositive. The second application, on line 12, saves some effort in that it first tacitly infers the negated consequent, (~ (D & E & F)), from the given (~ F). The third application, on line 18, goes further in that it follows up with an application of De Morgan's law.

Similar remarks apply to rules of the form (6.4). First, the right-hand side of the step may only match one of the consequent's conjuncts, not all of them. Again, this allows for tacit conjunction simplification:

```
assert* R := (A | B | C ==> D & E & F)

> (!chain [A ==> E [R]])

Theorem: (if A E)
```

Second, in the contrapositive direction, it is possible for the left-hand side to match the complement of *some* consequent conjunct and for the right-hand side to match either the complement of the antecedent or the conjunction of the complements of some of the antecedent's disjuncts:

```
assert* R := (A | B | C ==> D & E & F)

> (!chain [(~ E) ==> (~ B) [R]])
```

```
Theorem: (if (not E)
             (not B))

> (!chain [(~ D) ==> (~B & ~C) [R]])

Theorem: (if (not D)
             (and (not B)
                  (not C)))
```

It is possible to use "rules" without an explicit antecedent, so that $n = 0$ in the general form (6.3). The implementation of chain will treat such a degenerate rule as a conditional whose antecedent is true:

```
assert R := (forall x . male father x)

> (!chain [true ==> (male father Ann) [R]])

Theorem: (if true
             (male (father Ann)))
```

This example works because Athena transforms the sentence

```
(forall x . male father x)
```

into the logically equivalent

```
(forall x . true ==> male father x).
```

We will see that implication chains starting with true are particularly handy with a variant of chain called chain-> that returns the last element of the chain as its conclusion.

We have said that chain accepts sentences ("rules") as justifiers for implication steps, in addition to methods, but internally it is all methods; chain will actually "compile" a given rule such as (6.3) into a method that can take any instance of the left-hand side of the rule as input and will derive the appropriately instantiated right-hand side of the rule as its conclusion.

6.2.1 Nested rules

Sometimes rules of the form (6.3) are buried inside larger, enclosing rules. A typical example is provided by definitions of relation symbols, which are often of the form

$$\texttt{(forall } x_1 \cdots x_n \texttt{ . (R } x_1 \cdots x_n\texttt{) <==> } C\texttt{)} \qquad (6.5)$$

where C is itself a rule of the form (6.3). For instance:

```
assert* subset-definition :=
  (s1 subset s2 <==> forall x . x in s1 ==> x in s2)
```

Here the body of the definition is

```
(forall x . x in s1 ==> x in s2).
```

When `s1` and `s2` take on particular values, this becomes a rule that can be used in chaining steps. For example, suppose we know that a certain set `A` is a subset of some set `B`. Then we can derive the following specialized, nested rule from `subset-definition`:

```
(forall x . x in A ==> x in B).
```

We would like the ability to use this specialized rule in an implication step without having to explicitly derive it first, simply by citing the *enclosing rule* as our justifier—in this case, `subset-definition`. Thus, if the assumption base contains `(A subset B)`, we would like the following chain to produce the conditional stating that if `e` is in `A` then `e` is in `B` (where `e` is some term of sort `Element`):

```
(!chain [(e in A) ==> (e in B) [subset-definition]]).        (6.6)
```

If implication steps could *only* use rules of the form (6.3) as justifiers, then steps such as the above could not work. We would first have to derive the nested rule separately, by specializing `subset-definition` with `A` and `B`, and then use that rule in the above step:

```
let {nested-rule :=
       (!chain-> [(A subset B)
             ==>  (forall x . x in A ==> x in B) [subset-definition]])}
  (!chain [(e in A) ==> (e in B) [nested-rule]])
```

(See Section 6.4 for a description of `chain->`.) Alternatively, using nested chains (Section 6.8), we could write:

```
(!chain [(e in A) ==> (e in B)
                          [(forall x . x in A ==> x in B) <==
                             (A subset B) [subset-definition]]])
```

Either way, we would have to explicitly derive the inner, specialized rule first, and *then* use that as a justifier for the desired implication step.

This should not be necessary, however. We should be able to take chaining steps such as (6.6) directly, without having to explicitly derive the specialized inner rule first, and Athena allows this, for example:

```
declare A,B: Set

assert (A subset B)

pick-any element
 (!chain [(element in A) ==> (element in B) [subset-definition]])
```

```
Theorem: (forall ?element:Element
           (if (in ?element:Element A)
               (in ?element:Element B)))
```

In general, Athena will accept as a justifier for an implication step any nested rule of the form

$$\texttt{(forall } x_1 \cdots x_n \texttt{ . } p \texttt{ <==> (forall } y_1 \cdots y_m \texttt{ . } p_1 \texttt{ ==> } p_2\texttt{))}, \tag{6.7}$$

as well as

$$\texttt{(forall } x_1 \cdots x_n \texttt{ . } p \texttt{ ==> (forall } y_1 \cdots y_m \texttt{ . } p_1 \texttt{ ==> } p_2\texttt{))}, \tag{6.8}$$

and will automatically derive and use a properly specialized instance of the inner rule,

$$\texttt{(forall } y_1 \cdots y_m \texttt{ . } p_1 \texttt{ ==> } p_2\texttt{)},$$

provided that the corresponding instance of the outer antecedent, p, is in the assumption base.[10]

6.3 Implication chaining through sentential structure

There is another noteworthy way of enabling an implication-chain step from p to q: By using the structure of p and q. The simplest case occurs when p and q are identical, in which case the step will succeed even without any justifiers. The next simplest case occurs when both p and q are atomic sentences of the form $(R\ s_1 \cdots s_n)$ and $(R\ t_1 \cdots t_n)$, respectively. Then, if J is a justifier (e.g., a list of identities and/or conditional identities) licensing the conclusions $(s_i = t_i)$ for $i = 1, \ldots, n$, then the step

$$(R\ s_1 \cdots s_n) \texttt{ ==> } (R\ t_1 \cdots t_n)\ J \tag{6.9}$$

will succeed (in some appropriate assumption base β).[11] For example:

```
define [\/ /\] := [union intersection]
assert* R1 := (x \/ y = y \/ x)
assert* R2 := (x /\ null = null)

> (!chain [(s1 /\ null subset s2 \/ s3)
       ==> (null subset s3 \/ s2)          [R1 R2]])

Theorem: (if (subset (intersection ?s1:Set null)
                     (union ?s2:Set ?s3:Set))
```

10 Of course the outer rule—(6.7) or (6.8)—must itself be in the assumption base.

11 More precisely, what we mean by J "licensing the conclusions $s_i = t_i$ for $i = 1, \ldots, n$" is that, for each such i, the equational step `(!chain [`s_i `=` t_i J`])` should go through in β (where, again, β is the assumption base in which the step (6.9) is taking place).

```
          (subset null
                  (union ?s3:Set ?s2:Set)))
```

By using the given identities, R1 and R2, the system was able to equate corresponding terms on each side:

1. the first subterm of the left-hand side, `(s1 /\ null)`, was equated with the first subterm of the right-hand side, `null`, via R2; and
2. the second subterm of the left-hand side, `(s2 \/ s3)`, was equated with the second subterm of the right-hand side, `(s3 \/ s2)`, via R1.

This is just a form of relational congruence. The same result could be obtained through `rcong`, but we would first need to establish the identities of the respective terms explicitly. This shorthand is more convenient.[12]

Two other structural cases occur when p and q are of the form

$$(\odot\ p_1 \cdots p_n) \text{ and } (\odot\ q_1 \cdots q_n)$$

respectively, where $\odot$ is either the conjunction or disjunction constructor. Such cases are handled recursively: If the justifier J can enable each step (p_i ==> q_i) for $i = 1, \ldots, n$, then the step (p ==> q) goes through. For instance:

```
assert* marriage-symmetry := (x married-to y ==> y married-to x)
assert* union-comm := (x \/ y = y \/ x)

> (!chain [(Tom married-to Ann | s1 \/ s2 = s3)
          ==> (Ann married-to Tom | s2 \/ s1 = s3)   [marriage-symmetry
                                                      union-comm]])

Theorem: (if (or (married-to Tom Ann)
                 (= (union ?s1:Set ?s2:Set)
                    ?s3:Set))
             (or (married-to Ann Tom)
                 (= (union ?s2:Set ?s1:Set)
                    ?s3:Set)))
```

Here the implementation of `chain` will realize that the two sides of the implication step are disjunctions, and will attempt to use the given justifiers to recursively relate the corresponding disjuncts: `(Tom married-to Ann)` with `(Ann married-to Tom)`, and

```
(s1 \/ s2 = s3)
```

with `(s2 \/ s1 = s3)`. Both of these will succeed with the given information. The example would work just as well if it involved conjunctions instead. Ultimately, steps of this form succeed owing to the validity of the following inference rule, where $\odot \in \{\wedge, \vee\}$:

12 The implementation of `chain` does use `rcong` for steps of this form.

$$\frac{(p_1 \texttt{ ==> } q_1) \quad \cdots \quad (p_n \texttt{ ==> } q_n)}{((\odot\ p_1 \cdots p_n) \texttt{ ==> } (\odot\ q_1 \cdots q_n))} \quad [SC]$$

Note that this rule is valid only for $\odot \in$ {and, or}. It is *not* valid for $\odot \in$ {not, if, iff}.

The two remaining structural cases occur when p and q are both quantified sentences, respectively of the form

$$(Q\ x\ p') \text{ and } (Q\ y\ q')$$

for $Q \in$ {forall, exists}, in which case chain will continue its structural work recursively, with the given justifier, on appropriately renamed variants of p' and q'. For example:

```
> (!chain [(forall s1 . s1 \/ null = null)
       ==> (forall s2 . null \/ s2 = null)  [union-comm]])

Theorem: (if (forall ?s1:Set
                (= (union ?s1:Set null)
                   null))
              (forall ?s2:Set
                (= (union null ?s2:Set)
                   null)))
```

6.4 Using chains with chain-last

Sometimes when we put together an implication chain of the form

$$
\begin{array}{lll}
\texttt{(!chain [}p_1 \texttt{ ==> } p_2 & J_1 & \\
\qquad\qquad\quad \texttt{==> } p_3 & J_2 & \\
\qquad\qquad\qquad \vdots & & \qquad (6.10)\\
\qquad\qquad\quad \texttt{==> } p_{n+1} & J_n\texttt{])} &
\end{array}
$$

we are operating in an assumption base that contains the initial sentence, p_1. In such cases we are usually not content with merely showing that p_1 implies p_{n+1}, which is the result that would be returned by (6.10). Instead, we want to derive the final element of the chain, p_{n+1}. That can be done just as above, but with a method named chain-last rather than chain. An alternative (and shorter) name that is often used for chain-last is chain->. This method works just as chain does, except that after the implication $p_1 \Rightarrow p_{n+1}$ is produced, modus ponens is used on it and p_1 to derive p_{n+1}.[13] For instance, the chain-last call below will produce the conclusion B:

13 If you view the chain of implications as a line of dominoes, you can think of the effect of chain-last as tapping the first domino, p_1, to topple (derive) the last piece, p_{n+1}.

```
> assume hyp := (A & ~~ B)
    (!both (!chain-last [hyp ==> (~~ B) [right-and]
                             ==> B      [dn]])
           (!left-and hyp))

Theorem: (if (and A
                  (not (not B)))
             (and B A))
```

Thus, a call of the form

$$\texttt{(!chain-last [}p_1 \texttt{ ==> } p_2\ J_1 \quad \cdots \quad p_n \texttt{ ==> } p_{n+1}\ J_n\texttt{])}$$

is equivalent to:

$$\texttt{(!mp (!chain [}p_1 \texttt{ ==> } p_2\ J_1 \quad \cdots \quad p_n \texttt{ ==> } p_{n+1}\ J_n\texttt{]) } p_1\texttt{)}.$$

And conversely,

$$\texttt{(!chain [}p_1 \texttt{ ==> } p_2\ J_1 \quad \cdots \quad p_n \texttt{ ==> } p_{n+1}\ J_n\texttt{])}$$

is equivalent to

$$\textbf{assume}\ p_1\ \texttt{(!chain-last [}p_1 \texttt{ ==> } p_2\ J_1 \quad \cdots \quad p_n \texttt{ ==> } p_{n+1}\ J_n\texttt{])}.$$

Note that, unless it is exceptionally long, a `chain-last` application rarely needs a conclusion annotation, since its conclusion is immediately apparent: It is always the last link of the chain.

The rest of the book contains many other examples similar to the preceding proof of

`(A & ~ ~ B ==> B & A),`

where applications of `chain-last` (and other `chain` variants) are mixed in with other Athena constructs: appearing as arguments to enclosing method calls, inside **assume** bodies, and so on. Of course, in this particular example the proof could be expressed more directly as a single chain:

```
> (!chain [(A & ~~ B) ==> (~~ B & A) [comm]
                      ==> (B & A)    [dn]])

Theorem: (if (and A
                  (not (not B)))
             (and B A))
```

By default, `chain-last` can be used on an implication chain that starts with `true` even if the assumption base does not contain `true` explicitly. This is useful in tandem with the aforementioned convention, whereby, for chaining purposes, a nonconditional rule is treated as a conditional with `true` as its antecedent.

```
assert* R := (male father x)

> (!chain-> [true ==> (male father Ann) [R]])

Theorem: (male (father Ann))
```

6.5 Backward chains and chain-first

Implication chains can be written in reverse, by using the symbol <== instead of ==>. This can be useful in showing how a goal decomposes into something different (and hopefully simpler). As an example, suppose we have the following properties in the assumption base, where `Mod` denotes the remainder function on natural numbers:

```
declare pos: [N] -> Boolean
declare less: [N N] -> Boolean [<]
declare Mod: [N N] -> N [%]
define [x y z] := [?x:N ?y:N ?z:N]

assert* mod-< := (pos y ==> x % y < y)
assert* less-asymmetric := (x < y ==> ~ y < x)
assert* <-S-2 := (x < y ==> x < S y)
```

Suppose now we want to prove that for any natural numbers `a` and `b`, the successor of `b` is not less than `(a % b)`: `(~ S b < a % b)`. The following chain demonstrates how this somewhat complex goal reduces to the simple atom `(pos b)`:

```
pick-any a:N b:N
  (!chain [(~ S b < a % b) <== (a % b < S b)     [less-asymmetric]
                           <== (a % b < b)       [<-S-2]
                           <== (pos b)           [mod-<]])
```

This is operationally equivalent to reversing the links of the chain and using forward rather than backward implications:

```
(!chain [(pos b) ==> (a % b < b)       [mod-<]
                 ==> (a % b < S b)     [<-S-2]
                 ==> (~ S b < a % b) [less-asymmetric]])
```

For the purposes of proof development, however, the two are not the same. To write the second (forward) chain, we have to know the key starting point ahead of time: `(pos b)`. Essentially, we must have already figured out the reasoning in its entirety and then simply written it down. That rarely happens in practice. In this case the backward version may be a closer reflection of how the proof would actually be built, because we can *start with the goal*, which we do know from the beginning, and then incrementally transform it by inspecting the assumption base for appropriate information. In this case, presumably, we would notice that the goal matches the consequent of `less-asymmetric`, and that would lead us to take the backward step of transforming the goal to the appropriate instance of the corresponding antecedent: `(a % b < S b)`. Then we might notice that this new goal matches the consequent of `<-S-2`, and that would lead us to transform that goal to the relevant instance of the corresponding antecedent: `(a % b < b)`. Finally, we would observe that this goal matches the consequent of `mod-<`, and that would lead us, with one last backward step, to the goal `(pos b)`, which might already be known to hold. This process is known as *backward chaining*. We have already seen it in connection with the backward tactics we discussed when we studied proof heuristics, and we will explore it again in the context of logic programming (in Appendix B).

In symmetry with `chain-last`, there is a `chain-first` method that can be used to derive the first element of a backward chain, provided that the last one is in the assumption base. An alternative name for `chain-first` is `chain<-`. And also as before, when the last sentence is `true`, `chain-first` will succeed even if `true` is not in the assumption base. For example:

```
1   define [x y z] := [?x ?y ?z]
2
3   assert* p-def := (x parent y <==> x = father y | x = mother y)
4   assert* gp-def := (x grandparent z <== x parent y & y parent z)
5
6   assert* fact1 := (Mary = mother Bob)
7   assert* fact2 := (Peter = father Mary)
8
9   > (!chain-first
10       [(Peter grandparent Bob)
11    <== (Peter parent Mary & Mary parent Bob)           [gp-def]
12    <== (Peter = father Mary & Mary = mother Bob)       [p-def]
13    <== (true & true)                                    [fact1 fact2]
14    <== true                                             [augment]])
15
16  Theorem: (grandparent Peter Bob)
```

(The second and third steps of this example, on lines 12 and 11, demonstrate the structural implication chaining discussed in Section 6.3.) Essentially the same proof could also be written in a somewhat shorter form by avoiding any reference to `true` and starting from the given facts:

```
 (!chain-first
   [(Peter grandparent Bob)
<== (Peter parent Mary & Mary parent Bob) [gp-def]
<== (Peter = father Mary & Mary = mother Bob)])
```

Athena will realize that the first element of the chain is a conjunction of facts and will derive that conjunction automatically.

6.6 Equivalence chains

We can use chain to put together equivalence chains just as well, by using the symbol <==> instead of ==>. A chain call of the form

```
(!chain [p1 <==> p2      J1
            <==> p3      J2
              ⋮
            <==> pn+1    Jn])
```

(6.11)

will derive the biconditional (p_1 <==> p_{n+1}), provided that each step p_i <==> p_{i+1} J_i goes through, $i = 1, \ldots, n$. Everything that we have said so far about implication steps applies here as well, with one additional caveat: The relevant justifying methods in J_i must be bidirectional, that is, they must not only be able to derive the right-hand side from the left-hand side, but conversely as well. Here is an example:

```
> (!chain [(~~ A & (B ==> C)) <==> ((B ==> C) & ~~ A)  [comm]
                               <==> ((~ C ==> ~B) & A)  [contra-pos bdn]])

Theorem: (iff (and (not (not A))
                   (if B C))
              (and (if (not C)
                       (not B))
                   A))
```

Both of these steps went through because the justifying methods are bidirectional. An error would occur if, say, we replaced the bidirectional version of double negation, bdn, with the regular unidirectional version, dn, since we would then be unable to derive (~ ~ A) from A.

As suggested by the second step of the previous example, everything that we have said about structural implication steps carries over to equivalence steps. Another example:

```
assert* R1 := (x \/ y = y \/ x)
assert* R2 := (x /\ null = null)

> (!chain [(s /\ null subset s1 \/ s2)
       <==> (null subset s2 \/ s1)          [R1 R2]])

Theorem: (iff (subset (intersection ?S:Set null)
                      (union ?S1:Set ?S2:Set))
              (subset null
                      (union ?S2:Set ?S1:Set)))
```

Structural equivalence chaining is actually more flexible, as the analogue of rule [*SC*] is more widely applicable:

$$\frac{(p_1 \texttt{<==>} q_1) \quad \cdots \quad (p_n \texttt{<==>} q_n)}{((\odot\ p_1 \cdots p_n) \texttt{ <==> } (\odot\ q_1 \cdots q_n))} \tag{6.12}$$

Rule (6.12) holds for $\odot \in$ {not, and, or, if, iff}, not just for $\odot \in$ {and, or}, as was the case with [*SC*]. Thus, for example:

```
> (!chain [(~ (A & B)) <==> (~ (B & A)) [comm]])

Theorem: (iff (not (and A B))
              (not (and B A)))
```

The quantified analogues of the rule hold as well:

$$\frac{(p \texttt{<==>} q)}{(Q\ v\,.\,p \texttt{ <==> } Q\ v\,.\,q)} \tag{6.13}$$

for $Q \in$ {forall, exists}. For example:

```
assert* R := (s1 \/ s2 = s2 \/ s1)

> (!chain [(forall x y z . x subset y \/ z)
      <==> (forall x y z . x subset z \/ y)  [R]])

Theorem: (iff (forall ?x:Set
                (forall ?y:Set
                  (forall ?z:Set
                    (subset ?x:Set
                            (union ?y:Set ?z:Set)))))
              (forall ?x:Set
                (forall ?y:Set
                  (forall ?z:Set
                    (subset ?x:Set
                            (union ?z:Set ?y:Set))))))
```

In fact, chain implements a somewhat stronger formulation of (6.13) that allows for alpha-equivalence. For example, the following variant of the preceding example works just as well:

```
> (!chain [(forall x y z . x subset y \/ z)
      <==> (forall u v w . u subset w \/ v)  [R]])

Theorem: (iff (forall ?x:Set
                (forall ?y:Set
                  (forall ?z:Set
                    (subset ?x:Set
                            (union ?y:Set ?z:Set)))))
```

```
(forall ?u:Set
  (forall ?v:Set
    (forall ?w:Set
      (subset ?u:Set
              (union ?w:Set ?v:Set)))))
```

6.7 Mixing equational, implication, and equivalence steps

Equivalence and implication steps can be mixed in a chain. In particular, it is possible to switch from an equivalence step to an implication step:

$$
\begin{array}{lll}
(\texttt{!chain [}p_1 & \texttt{<==>}\ p_2 & J_1 \\
 & \vdots & \\
 & \texttt{<==>}\ p_{i+1} & J_i \\
 & \texttt{==>}\ p_{i+2} & J_{i+1} \\
 & \vdots & \\
 & \texttt{==>}\ p_{n+1} & J_n\texttt{]}).
\end{array}
$$

For example:

```
> (!chain [(A & B | A & C) <==> (A & (B | C)) [dist]
                           ==> A             [left-and]])

Theorem: (if (or (and A B)
                 (and A C))
             A)
```

We can also switch from implication steps to equivalence steps:

```
> (!chain [(A & B | A & C) <==> (A & (B | C)) [dist]
                           ==> A             [left-and]
                          <==> (~~ A)        [bdn]])

Theorem: (if (or (and A B)
                 (and A C))
             (not (not A)))
```

If there is even one implication step, however, the chain of equivalences is broken, and the conclusion of the chain cannot be a biconditional; at best, it will be a conditional. Hence, strictly speaking, there is little reason to mix equivalence and implication steps. The above proof, for instance, could just as well be written with implication steps only:

```
(!chain [(A & B | A & C) ==> (A & (B | C)) [dist]
                         ==> A             [left-and]
                         ==> (~~ A)        [bdn]])
```

Nevertheless, the former version conveys more information to the reader: It makes it clear that the first two elements of the chain are equivalent (owing to `dist`), as are the last two elements of the chain (owing to `bdn`).

Backward implication steps can also be mixed with equivalence steps. Athena will adjust the "direction" of the result accordingly. In the presence of backward implication steps, the antecedent and consequent will be the last and first elements of the chain, respectively:

```
> (!chain [(~~ A) <==> A               [bdn]
                  <==  (A & (B | C))   [left-and]
                  <==> (A & B | A & C)  [dist]])

Theorem: (if (or (and A B)
                 (and A C))
             (not (not A)))
```

More interestingly, equational steps can be mixed with implication and/or equivalence steps. In one direction, we can switch from equational steps to implication steps:

$$
\begin{array}{rll}
\texttt{(!chain-> [}t_1 = t_2 & & J_1 \\
\vdots & & \\
= t_{n+1} & & J_n \\
\texttt{==>}\, p_1 & & J_{n+1} \\
\vdots & & \\
\texttt{==>}\, p_{m+1} & & J_{n+m+1}\texttt{]).}
\end{array}
\qquad (6.14)
$$

This can be useful when we first establish an identity $t_1 = t_{n+1}$ by rewriting and then that identity becomes transformed to some conclusion p_1, either via some appropriate method(s) or because it matches the antecedent of some rule and can therefore be used to derive the rule's consequent. After that point, implication chaining proceeds normally. The final result of (6.14) will be the last element of the chain, p_{m+1}. For example, suppose that we have defined the `divides` relation as shown below; and suppose we know that multiplication distributes over addition. We can then prove that a divides $(a \cdot b) + (a \cdot c)$, for all natural numbers a, b, and c, with the following mixed chain:

```
declare Plus:    [N N] -> N [+]
declare Times:   [N N] -> N [*]
declare divides: [N N] -> Boolean
define [x y z k] := [?x:N ?y:N ?z:N ?k:N]
```

```
assert* divides-def := (x divides y <==> exists z . x * z = y)

assert* times-dist := (x * (y + z) = x * y + x * z)

pick-any a b c
  (!chain-> [(a * (b + c)) = (a * b + a * c)          [times-dist]
         ==> (exists k . a * k = a * b + a * c)   [existence]
         ==> (a divides a * b + a * c)            [divides-def]])
```

In the reverse direction, we can switch from implication/equivalence steps to equational steps. For example, a call of the following form will derive the conclusion $s = t_m$, provided that the initial sentence p_1 is in the assumption base.

$$
\begin{array}{llll}
\texttt{(!chain-> [}p_1 \texttt{ ==> } p_2 & & J_1 & \\
\quad\vdots & & & \\
\quad\texttt{==> (}s \texttt{ = } t_1\texttt{)} & & J_{n+1} & \\
 & \texttt{= } t_2 & J_{n+2} & \quad (6.15)\\
 & \quad\vdots & & \\
 & \texttt{= } t_m & J_{n+m}\texttt{]).} &
\end{array}
$$

If chain is used instead of chain-> in (6.15), then the conditional $p_1 \Rightarrow s = t_m$ is returned.[14] For example:

```
declare Minus:   [N N] -> N [-]
declare pos:   [N] -> Boolean

assert* R1  := (x * zero = zero)
assert* R2  := (x - x = zero)

pick-any a b c d
 (!chain [(pos a & b = c * (d - d)) ==> (b = c * (d - d)) [right-and]
                                      = (c * zero)    [R2]
                                      = zero          [R1]])

Theorem: (forall ?a:N
           (forall ?b:N
             (forall ?c:N
               (forall ?d:N
                 (if (and (pos ?a:N)
                          (= ?b:N
                             (Times ?c:N
```

14 If chain were used instead of chain-> in the case of (6.14), the conditional $t_1 = t_{n+1} \Rightarrow p_{m+1}$ would be returned. While this is fine as far as it goes, it means that all the work of the equational part of the chain would be in vain, since the same result could be produced by skipping the equational steps and simply starting the implication chain with the hypothesis $t_1 = t_{n+1}$. In general, chain applications that start with equational steps and then switch to implication steps are unnecessary.

```
                                  (Minus ?d:N ?d:N))))
                   (= ?b:N zero))))))
```

Multiple switches can occur in the same chain, from implication steps to equational steps, back to implication (and/or equivalence) steps, and so on, arbitrarily many times:

```
> pick-any a b c d
    (!chain [(pos a & b = c * (d - d)) ==> (b = c * (d - d)) [right-and]
                                        = (c * zero)    [R2]
                                        = zero          [R1]
                                      ==> (zero = b)    [sym]])

Theorem: (forall ?a:N
           (forall ?b:N
             (forall ?c:N
               (forall ?d:N
                 (if (and (pos ?a:N)
                          (= ?b:N
                             (Times ?c:N
                                    (Minus ?d:N ?d:N))))
                     (= zero ?b:N))))))
```

Athena allows for switching from implication to identity steps even when the final conclusion is an atomic sentence other than an identity. For instance:

```
declare less: [N N] -> Boolean [<]

pick-any a b c d
  (!chain [(pos a & b < c * (d - d)) ==> (b < c * (d - d)) [right-and]
                                      = (c * zero)    [R2]
                                      = zero          [R1]])

Theorem: (forall ?a:N
           (forall ?b:N
             (forall ?c:N
               (forall ?d:N
                 (if (and (pos ?a:N)
                          (less ?b:N
                                (Times ?c:N
                                       (Minus ?d:N ?d:N))))
                     (less ?b:N zero))))))
```

In this example, rewriting starts on line 4 with the atom `(b < c * (d - d))`. The rightmost child of this atom, `c * (d - d)`, is rewritten in two steps into the term `zero`, and the conclusion `(b < zero)` is thereby generated by congruence. Instead of `(b < c * (d - d))`, we could have started with an arbitrary atom $R(t_1, \ldots, t_n)$, and we could have then proceeded in the same fashion to rewrite the last term, t_n, into some other term $t_n{}'$ in a number of equational steps, eventually producing the result $R(t_1, \ldots, t_n{}')$.

6.8 Chain nesting

It is possible to insert a chain inside the justification list of another chain, to an arbitrary depth. The rationale for such nesting arises as follows. Sometimes, a step deep inside a long chain must rely for its justification on a result p that is not in the assumption base in which the chain itself is evaluated:

```
(!chain [t1 = t2      J1
             ⋮
           = ti+1     [p]
             ⋮
           = tn+1     Jn]).
```
(6.16)

(Here the chain is equational, but similar remarks apply to implication chains or mixed chains.) In this case, the passage from t_i to t_{i+1} relies on p, but p is not in the assumption base in which (6.16) is to be evaluated. One way to get around this difficulty is to establish p *prior* to the chain, by some proof D, so that it can then become available as a lemma to (6.16), say, by putting (6.16) inside a **let**:

```
let {_ := conclude p
            D}
  (6.16)
```

This approach works, but its drawback is that it might obscure the structure of the argument: p is established first, but it does not get used until considerably later in the proof, in justifying the step from t_i to t_{i+1} inside (6.16). It would be preferable if we could somehow inline the derivation of p right where it is needed, in the justification of the said step.[15] We would then have no need for an enclosing **let**; we could instead express the chain directly in the following form:

```
(!chain [t1 = t2      J1
             ⋮
           = ti+1     [Derive p here]
             ⋮
           = tn+1     Jn])
```

While the implementation of chain does not allow for arbitrary deductions inside justification lists, it does allow for abbreviated subchains. So if p can be obtained by a chain, there is probably a way to inline that chain above, placing it right where it says "*Derive p here.*"

15 Assuming that p is only needed at that one point. If p is used at multiple locations inside the chain, then it would be preferable to derive it first and make it available as a lemma inside the chain.

By "abbreviated" we mean that the inlined chain will not be a complete method call of the form (!chain ···) or (!chain-> ···), etc.; only the steps of the chain will be inlined.

As a concrete example from a real proof, consider the following goal:

```
declare leq: [N N] -> Boolean [<=]

define plus-minus-cancel :=
  (forall x y . y <= x ==> x = (x - y) + y)
```

This is a useful result, proven by induction on x. For the basis, we need to prove

```
(forall y . y <= zero ==> zero = (zero - y) + y)
```

The available premises from which we are to derive this result are as follows:

```
assert* R1 := (zero - x = zero)
assert* R2 := (x + zero = x)
assert* R3 := (x <= zero ==> x = zero)
```

Now here is one possible proof of the basis step:

```
conclude (forall y . y <= zero ==> zero = (zero - y) + y)
  pick-any y
    assume hyp := (y <= zero)
      let {lemma := (!chain-> [hyp ==> (y = zero) [R3]])}
        (!chain-> [((zero - y) + y)
                 = ((zero - y) + zero)      [lemma]
                 = (zero + zero)            [R1]
                 = zero                     [R2]
               ==> (zero = (zero - y) + y) [sym]])
```

An alternative is to inline the chain derivation of lemma right where it is needed:

```
1 conclude (forall y . y <= zero ==> zero = (zero - y) + y)
2   pick-any y
3     assume hyp := (y <= zero)
4       (!chain-last [((zero - y) + y)
5                   = ((zero - y) + zero)      [(y = zero) <== hyp [R3]]
6                   = (zero + zero)            [R1]
7                   = zero                     [R2]
8                 ==> (zero = (zero - y) + y) [sym]])
```

In this version there was no need for the **let**, since the lemma (y = zero) was derived (from hyp) at the step where it was needed, on line 5. A backward step <== was used for the derivation, for readability, though a forward step would work as well:

```
conclude (forall y . y <= zero ==> zero = (zero - y) + y)
  pick-any y
    assume hyp := (y <= zero)
      (!chain-> [((zero - y) + y)
               = ((zero - y) + zero)      [hyp ==> (y = zero) [R3]]
               = (zero + zero)            [R1]
```

```
                 = zero                          [R2]
               ==> (zero = (zero - y) + y) [sym]])
```

The justification list `[hyp ==> (y = zero) [R3]]` is essentially an abbreviated application of `chain->` (more precisely, it has the exact form of an argument to `chain->`). Likewise, the list `[(y = zero) <== hyp [R3]]` in the previous proof can be viewed as an abbreviated application of `chain-<`. In general, the direction of the arrows determines whether Athena understands a nested chain as a tacit application of `chain->` or `chain-<`.[16]

6.9 Exercises

Exercise 6.1: Consider the following problems:

(a)

```
assert* premise-1 := (B | (A ==> B))
assert* premise-2 := A

conclude B
  D1
```

(b)

```
assert* premise-1 := (~ B ==> ~ C)
assert* premise-2 := (A & B | ~ ~ C)

conclude B
  D2
```

(c)

```
assert* premise-1 := (~ exists x . Q x)
assert* premise-2 := (forall x . P x ==> Q x)

conclude (~ exists x . P x)
  D3
```

(d)

```
assert* prem-1 := (forall x . x R x ==> P x)
assert* prem-2 := ((exists x . P x) ==> (~ exists y . Q y))

conclude ((forall x . Q x) ==> ~ exists z . z R z)
  D4
```

16 These are the only two options for nested implication or equivalence chains. Such chains are always treated as applications of `chain-last` or `chain-first`, never as applications of `chain`.

(e)

```
assert* prem-1 := (exists x . P x)
assert* prem-2 := (exists x . Q x)
assert* prem-3 := (forall x . P x ==> forall y . Q y ==> x R y)

conclude (exists x y . x R y)
  D5
```

Find appropriate deductions $D_1, \ldots, D_5$, in chaining style, for each of the above. □

Exercise 6.2 (ite: If-Then-Else operator): Athena has `ite` predeclared as a polymorphic ternary function symbol with the following signature:

```
declare ite: (S) [Boolean S S] -> S
```

The meaning of `ite` is captured by the following two axioms, which are also predefined and contained in the initial assumption base:

```
assert* ite-axioms := [((ite true x _) = x)
                       ((ite false _ x) = x)]
```

Thus, this operator can be viewed as implementing an "if-then-else" mechanism for conditional branching: If b is `true` then (`ite` b t_1 t_2) "returns" t_1, while if b is `false` then (`ite` b t_1 t_2) returns t_2. We refer to b as the *guard* of the `ite` term.

In this book we do not make extensive use of `ite` because all of the functions we study are definable just as well with more generic constructs. However, `ite` is a very useful operator and quite pervasive in software and hardware modeling. In this exercise we provide some brief information on how `ite` works in Athena and illustrate its use in a simple problem.

The `ite` operator is natively supported both by `eval` and by `chain`.[17] To illustrate, introduce a `min` operator on natural numbers as follows:[18]

```
load "nat-less"

declare min: [N N] -> N [[int->nat int->nat]]

overload < N.<

define [x y] := [?x:N ?y:N]

assert* min-def := [(x min y = (ite (x < y) x y))]
```

17 It is also specially handled in translating Athena theories to external theorem provers such as first-order ATPs and SMT solvers.

18 We load the file `lib/main/nat-less.ath` (we omit the `lib/main/` part because Athena looks there by default), as that is where a less-than relation `N.<` on the natural numbers is introduced and defined, inside a module named `N`. We will have more to say about the contents of that module in subsequent chapters.

```
> (eval 0 min 1)

Term: zero

> (eval 2 min 1)

Term: S zero
```

We can nest `ite` occurrences to an arbitrary depth. Constructions of the form

(ite b_1 s_1 (ite b_2 t_2 $(\cdots)$))

are common. However, such explicit nesting can be cumbersome to read and write. Athena provides an `ite*` procedure that takes a list of guard-term pairs, each such pair separated by `-->` and with the final guard possibly being the wildcard, and automatically produces the correctly nested `ite` term. For instance, suppose we want to define binary search on search trees of natural numbers:

```
datatype BT := null | (node N BT BT)

declare search: [N BT] -> BT
```

Then the interesting recursive clause of the definition could be written as follows:

```
((search x (node y L R)) = ite* [(x = y) --> (node y L R)
                                | (x < y) --> (search x L)
                                | _       --> (search x R)])
```

(In Chapter 11 and later chapters we define similar functions using another construct, `fun`.)

In equational chaining, `ite` is handled essentially by desugaring into conditional equations. As long as the corresponding guard holds (is in the assumption base), an `ite` term can be rewritten into the corresponding term. As an example, we ask you to prove the following: `(forall x y . x min y = x | x min y = y)`. □

III PROOFS ABOUT FUNDAMENTAL DATATYPES

In this part, we introduce and prove many properties of natural numbers (Chapter 8), integers (Chapter 9), and other fundamental discrete structures (Chapter 10). Not only will these developments serve well for practicing the proof methods introduced in Parts I and II, but they will also provide infrastructure supporting the algorithm and abstraction studies in Parts IV and V, and the programming language studies in Part VI.

In Chapter 8 we continue the study of proof methods for the natural numbers that we began in Chapter 3, concentrating here on ordering properties. In Chapter 9 we turn to the integers and demonstrate how proofs can be simplified by creating mappings back and forth between different data representations. The final chapter of this part develops the theories of some fundamental data types, such as sets and maps. Much of this material is often covered in discrete mathematics courses. The presentation here is entirely computer-based, yet with little departure from traditional textbook formulations.

The theorems presented in these chapters will be numerous and diverse enough that it is worth first giving some attention to Athena module features, in Chapter 7. Modules provide a convenient mechanism for managing namespaces and for packaging large numbers of axioms, theorems, proofs, and other data and code into properly organized components. In examples, we recast some of the development from Chapter 3 involving addition on natural numbers by declaring the operator `+` inside a module named `N`. When we introduce the integers in Chapter 9, we will be able to reuse the operator `+` for integer addition by declaring it inside another module named `Z`. In contexts where we need to use only addition on natural numbers or only addition on integers, we can arrange to write simply `+`, but when we need to use both, we can distinguish them as `N.+` and `Z.+`.[1] Again in Chapter 10, in studying fundamental discrete structures, we will achieve greater notational flexibility by using modules than would be possible without them.

---

1 Overloading (Section 2.13) can permit the use of `+` for both operations within the same term, as is often done in conventional mathematical notation and in programming languages, but one must be careful not to overdo it and make reading such terms more difficult than it would be with the longer notations.

7 Organizing Theory Development with Athena Modules

MODULES are used to partition Athena code into separate namespaces. We begin by describing how to introduce a module and how to refer to names introduced inside a module. Some examples are drawn from a recasting of some of the natural number development in Chapter 3 using modules. We further introduce nested modules and illustrate their use as a convenient way of encapsulating axioms and theorems for function symbols. Brief examples are then given to illustrate a few additional module features, and the chapter closes with a note on indentation conventions for modules.

7.1 Introducing a module

A module with name *M* is introduced with the following syntax:

module *M* {· · ·}

The region between the matching curly braces is called the *scope* of the module. The contents of that scope can consist of any nonempty sequence of any of the Athena directives we have seen so far: sort declarations, datatype or value definitions, and so on, possibly including other module definitions (called *nested* modules). Here is an example:

```
> module A {

    define x := 'foo

    define L := [1 2 3]

  }

Term A.x defined.

List A.L defined.

Module A defined.
```

Module `A` does two simple things: It defines `x` as the term `'foo` and `L` as the list `[1 2 3]`. To retrieve one of these values after the module has been defined, we use its *qualified name*, expressed using the dot notation: We write the name of the module, followed by a dot, followed by the name that was given to the value inside the module.

```
> A.x

Term: 'foo

> A.L
```

```
List: [1 2 3]

> A.z

input prompt:1:1: Error: Could not find a value for A.z.
```

An error was reported when we tried to evaluate `A.z`, since the name `z` was not defined inside `A`. The definitions given inside a module M can be inspected by the procedure `module->string`, which takes as input the name of a module and returns a properly formatted and indented printed representation of the module's bindings:

```
> (print (module->string "A"))

A --> module { ### 2 bindings
        x --> 'foo
        L --> [
                1
                2
                3
               ]
      }
Unit: ()
```

As this example illustrates, all names introduced inside a module are by default visible to the outside, once the module has been successfully defined. Accordingly, it is not necessary to specify any "export" directives; everything is automatically exported. However, there is a way to override this default behavior by prefixing a definition with the **private** keyword. A private definition will not be visible outside of the module in which it appears.

```
> module A {

    private define a := 99

    define b := (a plus 1)
  }

Term A.b defined.

Module A defined.

> A.b

Term: 100

> A.a

input prompt:1:1: Error: Could not find a value for A.a.
```

When a module is defined, its contents are processed sequentially, from top to bottom, so forward references are not meaningful and are flagged as errors:

```
> module A {

    define x := foo

    define foo := 3
  }

input prompt:3:15: Error: Could not find a value for foo.
Module A was not successfully defined.
```

This is similar to the top-down way in which **load** processes the contents of a file. Of course, mutually recursive definitions are still possible using **letrec**. Also, as this last example demonstrates, when an error occurs during the processing of a module's definition, the definition is considered to have failed, and therefore no module will be created as a result of the (attempted) definition. If the module whose definition failed is embedded inside other module definitions, then all those outer module definitions will fail as well.[1]

7.2 Natural numbers using modules

To show some of these module features in actual use, and to help introduce other features, let us redo some of the development of the natural number datatype and functions:

```
datatype N := zero | (S N)

assert (datatype-axioms "N")

module N {
  declare one, two: N

  define [m n n' x y z k] := [?m:N ?n:N ?n':N ?x:N ?y:N ?z:N ?k:N]

  assert one-definition := (one = S zero)
  assert two-definition := (two = S one)

  define S-not-zero     := (forall n . S n =/= zero)
  define one-not-zero   := (one =/= zero)
  define injective      := (forall m n . S m = S n <==> m = n)

} # close module N
```

1 When a module definition fails, the global lexical environment, the global assumption base, and the global symbol set are reverted to the values they had prior to the definition of the outermost module containing the failing definition. That is, Athena will transitively roll back any intermediate effects to these three semantic parameters that might have come about as a result of processing the module definition(s) up to the point of failure. However, side effects to the store (and obviously, any output side effects) will not be undone.

Here the module name is N, the same identifier we used to name the natural number datatype. We could have used a different identifier to name the module; using the same name is just a matter of convenience. If we write a proof outside the scope of module N, we use qualified names; for example:

```
> conclude N.S-not-zero
    pick-any n
      (!chain->
        [true ==> (zero =/= S n) [(datatype-axioms "N")]
              ==> (S n =/= zero) [sym]])

Theorem: (forall ?n:N
           (not (= (S ?n)
                   zero)))
```

We can in turn use this sentence, since it is now a theorem, in the proof of N.one-not-zero.

```
> (!by-contradiction N.one-not-zero
    assume (N.one = zero)
      let {is := conclude (S zero = zero)
                   (!chain [(S zero)
                            = N.one                    [N.one-definition]
                            = zero                     [(N.one = zero)]]);
           is-not := (!chain-> [true
                                ==> (S zero =/= zero) [N.S-not-zero]])}
        (!absurd is is-not))

Theorem: (not (= N.one zero))
```

If we had included the proofs within the scope of the module, we could have use unqualified names. Another way to write the proofs using unqualified names is to use an **open** directive:

```
open N

conclude S-not-zero
  pick-any n
    (!chain->
      [true ==> (zero =/= S n) [(datatype-axioms "N")]
            ==> (S n =/= zero) [sym]])

(!by-contradiction one-not-zero
   assume (one = zero)
     let {is := conclude (S zero = zero)
                  (!chain
                    [(S zero)
                    = one           [one-definition]
                    = zero          [(one = zero)]]);
          is-not := (!chain->
                       [true
```

```
          ==> (S zero =/= zero) [S-not-zero]])}
  (!absurd is is-not))
```

open M brings all the names introduced inside module M into the current scope, so that they may be used without qualification. Multiple **open** directives may be issued, and several modules $M_1, \ldots, M_n$ can be opened with a single **open** directive:

$$\textbf{open}\ M_1, \ldots, M_n$$

However, every time we **open** a module we increase the odds of naming confusion and conflicts, so excessive use of this directive is discouraged.

7.3 Extending a module

Modules are dynamic, in the sense that one can reopen a module's scope and enter additional content into it at any time. This is done with **extend-module**, for example:

```
extend-module N {

  define nonzero-S :=
    (forall n . n =/= zero ==> exists m . n = S m)

  define S-not-same := (forall n . S n =/= n)

  by-induction nonzero-S {
    zero => assume (zero =/= zero)
              (!from-complements (exists m . zero = S m)
                                 (!reflex zero)
                                 (zero =/= zero))
  | (S k) => assume (S k =/= zero)
                let {_ := (!reflex (S k))}
                  (!egen (exists m . S k = S m) k)
  }

  by-induction S-not-same {
    zero => conclude (S zero =/= zero)
              (!instance S-not-zero zero)
  | (S m) => let {ihyp := (S m =/= m)}
               (!chain-> [ihyp
                     ==> (S S m =/= S m) [injective]])
  }

} # close module N
```

In this case we have included the proofs within the module and are thus able to use unqualified names.

7.4 Modules for function symbols

Although we could continue to put natural number properties directly into module N, another alternative is to create a *nested module* within module N for properties of each function symbol that we specify. For example, we can begin to do so for addition of natural numbers, as follows:

```
extend-module N {
  declare +: [N N] -> N
  module Plus {

    assert right-zero    := (forall n . n + zero = n)
    assert right-nonzero := (forall m n . n + S m = S (n + m))

  } # close module Plus
} # close module N
```

Within module Plus, we can use the unqualified names right-zero and right-nonzero. Outside of module Plus, but within module N, we must use names qualified with Plus: Plus.right-zero and Plus.right-nonzero. And outside of N, we must use the fully qualified names N.Plus.right-zero and N.Plus.right-nonzero.

We now extend Plus with a couple of sentences that we intend as theorems:

```
extend-module N {
  extend-module Plus {
    define left-zero    := (forall n . zero + n = n)
    define left-nonzero := (forall n m . (S m) + n = S (m + n))
  } # close module Plus
} # close module N
```

Exercise 7.1: Prove

N.Plus.left-zero

using N.Plus.right-zero and N.Plus.right-nonzero. □

Exercise 7.2: Derive N.Plus.left-nonzero from the same two axioms. □

We could continue with additional extensions and corresponding proofs. See the file lib/main/nat-plus.ath, where such a development is carried out. For multiplication, see the development of a Times module in lib/main/nat-times.ath.

7.5 Additional module features

Static scoping applies as usual, so new definitions inside a module overshadow earlier definitions in everything that follows, including any subsequent modules:

```
module M1 {
  define x := 2

  module M2 {

    define y := (x plus 3)
    define x := 99

    module M3 {

      define y := (x plus 1)

    } # close module M3

  } # close module M2

} # close module M1

> M1.M2.M3.y

Term: 100
```

Note that **open** is transitive. If a module M_1 is opened inside M_2, and then M_2 is opened inside M_3, M_1 will also be opened inside M_3:

```
module M1 {
  define a1 := 1
}

module M2 {
  open M1
  define a2 := (a1 plus 2)
}

module M3 {
  open M2
  define a3 := (a2 plus a1)
}

> M3.a3

Term: 4
```

The value of a name *I* defined at the top level can always be retrieved by writing Top.*I*. For instance:

```
define foo := 2

module M {
  define foo := 3;
  define y := (foo plus Top.foo)
}

> M.y

Term: 5
```

Top is not quite a proper module (it would have to contain itself otherwise!), but it can be treated as such for most practical purposes. For instance,

```
(print (module->string "Top"))
```

produces the expected results. To ensure that Top is always treated distinctly and consistently, a user-defined module cannot be named Top.[2]

7.6 Additional module procedures

We close by describing a few more primitive procedures that may be useful in working with modules.

1. The unary procedure module-size takes the name of a module, as a string, and returns the number of bindings in the corresponding module.
2. The unary procedure module-domain takes the name of a module, as a string, and returns a list of all and only those identifiers, represented as strings, that are defined inside the module. For instance:

```
module M {
  define [x y] := [1 2]
}

> (map-proc println (module-domain "M"))

x

y

Unit: ()
```

2 Of course Top can still be used to name a value, as module names and value-denoting identifiers comprise different namespaces.

3. The binary procedure `apply-module` is a programmatic version of the dot notation. It takes the name of a module M and any identifier I defined inside M and produces the corresponding value, $M.I$. The module name can itself be nested, written in dot notation:

```
module M1 { module M2 { define L := [] }}

> (apply-module "M1.M2" "L")

List: []
```

All of the above procedures generate an error if their first string argument does not represent a module.

7.7 A note on indentation

In the brief examples of modules in this chapter, we have indented the contents, which is consistent with a general policy of indenting code inside curly braces. In practice, however, strictly adhering to this policy for modules can become inconvenient, as the contents of a module often extend over hundreds or thousands of lines. Consequently, in the rest of the book and in the Athena libraries, we indent the contents of modules in short examples, but otherwise we just start at the first character of the line.

8 Natural Number Orderings

IN this chapter, we return to the study of proof methods motivated by properties of the natural numbers. In Chapter 3 we concentrated on equality properties, but here we turn to *ordering* properties. In the process, we will be putting into further practice some of the techniques we studied in the previous chapters. These ordering properties also prepare us for dealing with integer addition and subtraction in Chapter 9.

We start by constructively defining a "strictly-less-than" operator, $<$, on natural numbers. From that operator and equality, we easily obtain definitions of other commonly used ordering operators ($\leq$, $>$, $\geq$). Although only a few axioms are required to define $<$, there are many results that can be derived as theorems about it and the other ordering operators.

Subtraction of natural numbers is closely related to their ordering, so we include a constructive definition of this operator and a few theorems about it. These results will be used in defining integer addition and subtraction in Chapter 9, and for defining quotient and remainder operators in Chapter 13.

We also introduce *ordered lists* of natural numbers, a subset of the list datatype we studied in Chapter 3. These are defined with the aid of a predicate on lists, a technique also used in the final section of the chapter to define *binary search trees* as a subset of a *binary tree* datatype. With binary trees that have natural numbers in their nodes, we can take advantage of the natural-number ordering properties to achieve more efficient searching than with the linear-list datatype we studied in Chapter 3, as we will see in detail in Chapter 11.[1] But first we have to define binary trees and introduce a corresponding induction principle—distinguished from the induction principles for natural numbers and lists in having *two induction hypotheses*, one for each of a node's subtrees—and use it in proving several important properties, including one that involves the concept of an "inorder tree walk." The chapter concludes with the definition of binary search trees and the proof of an elegant theorem connecting binary search trees, inorder tree walks, and ordered lists.

8.1 Properties of natural number ordering functions

We begin with properties of a strict ordering operator $<$. Since the definition of this operator does not involve equality, there is much less of a role for equality chaining in the relevant proofs. But we will get plenty of practice with implication chaining, case analysis, and proof by contradiction.

To define strict inequality on natural numbers, we introduce the following symbol declaration and module nested within module `N`. Within module `N.Less` we define three axioms and a number of sentences we intend to prove as theorems:

1 In Chapters 14 and 15 we will generalize to binary search trees over any domain with an ordering relation satisfying the axioms of a strict weak order relation (Section 14.9.6).

```
1  extend-module N {
2    open Plus
3
4    declare <: [N N] -> Boolean [[int->nat int->nat]]
5
6    define [m n n' x y z k] := [?m:N ?n:N ?n':N ?x:N ?y:N ?z:N ?k:N]
7
8    module Less {
9
10     assert* def := [(zero < S n)
11                     (~ _ < zero)
12                     (S m < S n <==> m < n)]
13
14     define [zero<S not-zero injective] := def
15
16     define irreflexive  := (forall n . ~ n < n)
17     define <S           := (forall n . n < S n)
18     define =zero        := (forall n . ~ zero < n ==> n = zero)
19     define zero<        := (forall n . n =/= zero <==> zero < n)
20     define S1           := (forall x y . S x < y ==> x < y)
21     define S2           := (forall x y . x < y ==> x < S y)
22     define S4           := (forall m n . S m < n ==> exists n' . n = S n')
23     define S-step       := (forall x y . x < S y & x =/= y ==> x < y)
24     define discrete     := (forall n . ~ exists x . n < x & x < S n)
25
26   } # close module Less
```

Each of the defined properties (lines 16–24) can be proved by induction. The first two are quite easy (their proofs and those that follow are expressed within the scope of the module `N` extension, but outside the scope of module `N.Less`):

```
by-induction Less.irreflexive {
  zero =>   (!chain-> [true ==> (~ zero < zero)      [Less.not-zero]])
| (S n) => (!chain-> [(~ n < n) ==> (~ S n < S n) [Less.injective]])
}

by-induction Less.<S {
  zero =>   (!chain-> [true ==> (zero < S zero)     [Less.zero<S]])
| (S n) => (!chain-> [(n < S n) ==> (S n < S S n) [Less.injective]])
}
```

More typically, proofs about < require case splitting in one or both of the basis case and inductive step, and proof by contradiction is also frequently useful, as in the following proof of `Less.S1`.

```
by-induction Less.S1 {
  zero =>
  conclude (forall y . S zero < y ==> zero < y)
```

```
      pick-any y
        assume Szero<y := (S zero < y)
          (!two-cases
            assume y=zero := (y = zero)
              (!by-contradiction (zero < y)
                assume (~ zero < y)
                  let {-Szero<y :=
                         conclude (~ S zero < y)
                           (!chain->
                             [true ==> (~ S zero < zero) [Less.not-zero]
                                   ==> (~ S zero < y)    [y=zero]])}
                    (!absurd Szero<y -Szero<y))
            assume y!=zero := (y =/= zero)
              let {has-predecessor :=
                     (!chain-> [y!=zero
                               ==> (exists m . y = S m)    [nonzero-S]])}
                pick-witness m for has-predecessor
                  (!chain-> [true ==> (zero < S m)         [Less.zero<S]
                                  ==> (zero < y)           [(y = S m)]]))
| (S n) =>
    let {ind-hyp := (forall y . S n < y ==> n < y)}
      conclude (forall y . S S n < y ==> S n < y)
        pick-any y
          assume Less := (S S n < y)
            (!two-cases
              assume (y = zero)
                (!by-contradiction (S n < y)
                  assume (~ S n < y)
                    let {not-Less :=
                           (!chain-> [true
                                     ==> (~ S S n < zero)   [Less.not-zero]
                                     ==> (~ S S n < y)      [(y = zero)]])}
                      (!absurd Less not-Less))
              assume nonzero := (y =/= zero)
                let {has-predecessor :=
                       (!chain->
                         [nonzero
                        ==> (exists m . y = S m)             [nonzero-S]])}
                  pick-witness m for has-predecessor
                    # we now have (y = S m)
                    (!chain->
                      [(S S n < y) ==> (S S n < S m)    [(y = S m)]
                                   ==> (S n < m)        [Less.injective]
                                   ==> (n < m)          [ind-hyp]
                                   ==> (S n < S m)      [Less.injective]
                                   ==> (S n < y)        [(y = S m)]]))
}
```

Note that in both the basis case and the inductive step we prove the existentially quantified sentence (exists m . y = S m) and pick a witness for it. We will return to this example shortly to see a simpler way of writing these parts of the proof.

The proof of Less.S2 is much simpler:

```
by-induction Less.S2 {
  zero =>
    conclude (forall y . zero < y ==> zero < S y)
      pick-any y
        assume (zero < y)
          (!chain-> [true ==> (zero < S y)    [Less.zero<S]])
| (x as (S m)) =>
    conclude (forall y . x < y ==> x < S y)
      pick-any y
        (!chain [(S m < y)
              ==> (m < y)                      [Less.S1]
              ==> (S m < S y)                  [Less.injective]])
}
```

The key point to note about this proof is that the induction hypothesis wasn't needed. We didn't really need to set up the proof with **by-induction** at all; we could have written it instead with a case analysis and in the nonzero case proved (exists m . y = S m) and picked a witness for it, just as we did in the proof of Less.S1. But, by comparison with the above proof, the result would be a bit longer and more complex. There is a third alternative that retains the simplicity of this proof but makes it clearer that there is no induction involved: using Athena's **datatype-cases** form instead of **by-induction**. There is no other change needed in the proof:

```
datatype-cases Less.S2 {
  zero =>
    conclude (forall y . zero < y ==> zero < S y)
      pick-any y
        assume (zero < y)
          (!chain-> [true ==> (zero < S y) [Less.zero<S]])
| (S m) =>
    conclude (forall y . S m < y ==> S m < S y)
      pick-any y
        (!chain [(S m < y)
              ==> (m < y)                    [Less.S1]
              ==> (S m < S y)                [Less.injective]])
}
```

Returning to Less.S1, we can use **datatype-cases** for the case splitting within both the basis case and the inductive step, thus significantly simplifying the proof:

```
by-induction Less.S1 {
  zero =>
    datatype-cases (forall y . S zero < y ==> zero < y) {
```

```
      zero => assume less := (S zero < zero)
                let {-less := (!chain-> [true
                                        ==> (~ S zero < zero) [Less.not-zero]])}
                  (!from-complements (zero < zero)
                                        less
                                        -less)
    | (S m) => assume (S zero < S m)
                 (!chain-> [true ==> (zero < S m)          [Less.zero<S]])
    }
| (S n) =>
    let {ind-hyp := (forall y . S n < y ==> n < y)}
      datatype-cases (forall y . S S n < y ==> S n < y) {
        zero =>
          assume less := (S S n < zero)
            let {-less := (!chain-> [true
                                    ==> (~ S S n < zero)      [Less.not-zero]])}
              (!from-complements (S n < zero)
                                    less
                                    -less)
      | (S m) =>
          (!chain [(S S n < S m) ==> (S n < m)             [Less.injective]
                                 ==> (n < m)               [ind-hyp]
                                 ==> (S n < S m)           [Less.injective]])
      }
} # by-induction Less.S1
```

Exercise 8.1: Prove `Less.S-step`, *without* using **`datatype-cases`** for case analysis. □

Exercise 8.2: Prove `Less.S-step`, with case splitting done using **`datatype-cases`** in both the basis case and the inductive step. □

⋆ ***Exercise 8.3:*** Prove `Less.discrete`. *Hint:* Start immediately with a proof by contradiction in both the basis and the inductive step. This yields an existential premise; pick a witness `x` and split into zero and nonzero cases of `x`. □

8.1.1 Trichotomy properties

Strict inequality defines a *total ordering* on the natural numbers. The property expressing this is called *trichotomy*, reflecting the fact that one of three possibilities always holds. While it could be expressed as a disjunction of the three possibilities, other forms that are often more convenient in proofs are stated here:

```
extend-module Less {
  define trichotomy  := (forall m n . ~ m < n & m =/= n ==> n < m)
  define trichotomy1 := (forall m n . ~ m < n & ~ n < m ==> m = n)
  define trichotomy2 := (forall m n . m = n <==> ~ m < n & ~ n < m)
}
```

We give the proof of only the first and leave the other two as exercises.

```
by-induction Less.trichotomy {
  zero =>
    pick-any n
      assume (~ zero < n & zero =/= n)
        conclude (n < zero)
          let {has-pred := (!chain->
                             [(zero =/= n)
                          ==> (n =/= zero)                  [sym]
                          ==> (exists k . n = S k)          [nonzero-S]])}
            pick-witness k for has-pred
              let {less := (!chain->
                             [true
                          ==> (zero < S k)                  [Less.zero<S]
                          ==> (zero < n)                    [(n = S k)]]);
                   -less := (~ zero < n)}
                (!from-complements
                   (n < zero) less -less)
| (m as (S k)) =>
    let {ind-hyp := (forall n . ~ k < n & k =/= n ==> n < k)}
      datatype-cases (forall n . ~ m < n & m =/= n ==> n < m) {
        zero => assume (~ m < zero & m =/= zero)
                  (!chain-> [true ==> (zero < m)            [Less.zero<S]])
      | (n as (S n')) =>
          assume hyp := (~ m < n & m =/= n)
            (!chain->
              [hyp ==> (~ k < n' & k =/= n')                [Less.injective
                                                             S-injective]
                   ==> (n' < k)                             [ind-hyp]
                   ==> (S n' < S k)                         [Less.injective]])
      }
}
```

Exercise 8.4: Prove `Less.trichotomy1` (using `Less.trichotomy` as a lemma). □

Exercise 8.5: Prove `Less.trichotomy2` (using `Less.trichotomy1` as a lemma). □

8.1.2 Transitive and asymmetric properties

The next collection of properties of < concerns transitivity:

```
extend-module Less {
  define transitive   := (forall x y z . x < y & y < z ==> x < z)
  define transitive1  := (forall x y z . x < y & ~ z < y ==> x < z)
  define transitive2  := (forall x y z . x < y & ~ x < z ==> z < y)
  define transitive3  := (forall x y z . ~ y < x & y < z ==> x < z)
}
```

If we attempt to prove `transitive` by straightforwardly inducting on `x`, it turns out that within both the basis case and inductive step, case splitting on zero and nonzero values is required for both `y` and `z`. While it is possible to obtain the proof in this way (see Exercise 8.31), such proofs can be tedious both to write and to read. In such circumstances, it is sometimes helpful to try inducting on another variable (rather than the first), to see if that allows a simpler proof. Athena's **`by-induction`** form does not provide a direct way to specify the variable to induct on, but we can get around this simply by restating the property with a different order of the quantifiers, proving the restated property by induction on its first variable, then deriving the originally stated property by substitution. We follow this approach in the following proof:

```
# A version with the easiest-to-induct-on variable first:
conclude transitive0 :=
  (forall z x y . x < y & y < z ==> x < z)
by-induction transitive0 {
  zero =>
    pick-any x y
      assume (x < y & y < zero)
        let {-y<0 := (!chain->
                       [true ==> (~ y < zero)        [Less.not-zero]])}
          (!from-complements
            (x < zero) (y < zero) -y<0)
| (z as (S n)) =>
    let {ind-hyp := (forall x y .
                       x < y & y < n ==> x < n)}
      pick-any x y
        assume (x < y & y < z)
          conclude (x < z)
            let {_ := conclude (x < n)
                        (!two-cases
                          assume (y = n)
                            (!chain->
                              [(x < y) ==> (x < n)  [(y = n)]])
                          assume (y =/= n)
                            (!chain->
                              [(y =/= n)
                           ==> (y < S n & y =/= n)  [augment]
                           ==> (y < n)              [Less.S-step]
                           ==> (x < y & y < n)      [augment]
                           ==> (x < n)              [ind-hyp]]))}
              (!chain-> [(x < n) ==> (x < S n)      [Less.S2]])
}

conclude Less.transitive
  pick-any x y z
    (!chain [(x < y & y < z) ==> (x < z)            [transitive0]])
```

How can we discover which variable is easiest to induct on, other than just by trying them one by one? In Section 13.2 we will discuss a heuristic that is often useful in making the choice.

Another key property of < is *asymmetry*:

```
extend-module Less {
  define asymmetric   := (forall m n . m < n ==> ~ n < m)
}
```

With the transitive and irreflexive properties at hand to use as lemmas, `Less.asymmetric` doesn't even require induction:

```
conclude Less.asymmetric
  pick-any x y
    assume (x < y)
      (!by-contradiction (~ y < x)
         assume (y < x)
           let {x<x  := (!chain-> [(x < y & y < x)
                                  ==> (x < x)          [Less.transitive]]);
                -x<x := (!chain-> [true
                                  ==> (~ x < x)        [Less.irreflexive]])}
             (!absurd x<x -x<x))
```

That asymmetry can be proved from transitivity and irreflexivity without induction suggests that the relation between these properties has greater generality than just for the natural numbers. In fact, transitivity and irreflexivity are the defining axioms of what mathematicians call a *strict partial order*, which has a myriad of useful instances. We will encounter the general case of strict partial orders in Chapter 14 as part of our study of proofs at an abstract level.

The following property is an easy consequence of `Less.S1` and irreflexivity:

```
extend-module Less {
  define S-not-< := (forall n . ~ S n < n)
}

conclude Less.S-not-<
  pick-any n
    (!by-contradiction (~ S n < n)
       assume (S n < n)
         (!absurd
           (!chain-> [(S n < n) ==> (n < n)  [Less.S1]])
           (!chain-> [true ==> (~ n < n)     [Less.irreflexive]])))
```

Exercise 8.6: Prove `Less.transitive1`. □

Exercise 8.7: Prove `Less.transitive2`. □

Exercise 8.8: Prove `Less.transitive3`. □

8.1.3 Less-equal properties

Other ordering operators—≤, >, and ≥—can be defined in terms of < and =. We restrict our attention here to ≤ and at once introduce a substantial collection of its properties (note that this code continues within the scope of module `N`):

```
declare <=: [N N] -> Boolean [[int->nat int->nat]]

module Less= {

  assert definition := (forall x y . x <= y <==> x < y | x = y)

  define Implied-by-<       := (forall m n . m < n ==> m <= n)
  define Implied-by-equal := (forall m n . m = n ==> m <= n)
  define reflexive      := (forall n . n <= n)
  define zero<=         := (forall n . zero <= n)
  define S-zero-S-n     := (forall n . S zero <= S n)
  define injective      := (forall n m . S n <= S m <==> n <= m)
  define not-S          := (forall n . ~ S n <= n)
  define S-not-equal    := (forall k n . S k <= n ==> k =/= n)
  define discrete       := (forall m n . m < n ==> S m <= n)
  define transitive     := (forall x y z . x <= y & y <= z ==> x <= z)
  define transitive1    := (forall x y z . x < y & y <= z ==> x < z)
  define transitive2    := (forall x y z . x <= y & y < z ==> x < z)
  define S1             := (forall n m . n <= m ==> n < S m)
  define S2             := (forall n m . n <= m ==> n <= S m)
  define S3             := (forall n . n <= S n)
  define trichotomy1    := (forall m n . ~ n <= m ==> m < n)
  define trichotomy2    := (forall m n . ~ n < m ==>  m <= n)
  define trichotomy3    := (forall m n . n < m ==> ~ m <= n)
  define trichotomy4    := (forall m n . n <= m ==> ~ m < n)
  define trichotomy5    := (forall m n . m <= n & n <= m ==> m = n)
  define Plus-cancellation := (forall k m n . m + k <= n + k ==> m <= n)
  define Plus-k         := (forall k m n . m <= n ==> m + k <= n + k)
  define Plus-k1        := (forall k m n . m <= n ==> m <= n + k)
  define k-Less=        := (forall k m n . n = m + k ==> m <= n)
  define zero2          := (forall n . n <= zero ==> n = zero)
  define not-S-zero     := (forall n . ~ S n <= zero)
  define S4             := (forall m n . S m <= n ==> exists n' . n = S n')
  define S5             := (forall n m . n <= S m & n =/= S m ==> n <= m)
  define =zero          := (forall m . m < one ==> m = zero)
  define zero<=one      := (forall m . m = zero ==> m <= one)
} # close module Less=
```

Only `Less=.definition` needs to be asserted as an axiom; the rest are theorems provable from `Less=.definition` and properties of < and equality. It should be noted that induction is not necessary for any of these proofs. Instead, they provide good practice working with disjunctions and, in many cases, proof by contradiction.

We consider here only a few of the proofs and leave the rest as exercises. The first two properties follow simply from the definition, but offer examples of using `alternate` in implication chains.

```
conclude Less=.Implied-by-<
  pick-any m n
    (!chain [(m < n) ==> (m < n | m = n)    [alternate]
                     ==> (m <= n)           [Less=.definition]])

conclude Less=.Implied-by-equal
  pick-any m:N n:N
    (!chain [(m = n) ==> (m < n | m = n)    [alternate]
                     ==> (m <= n)           [Less=.definition]])
```

We continue with `Less=.reflexive`, taking advantage of the fact that an identity of the form $(s = s)$ can serve as the first link of any implication chain:

```
conclude Less=.reflexive
  pick-any n
    (!chain-> [(n = n)
               ==> (n < n | n = n)    [alternate]
               ==> (n <= n)           [Less=.definition]])
```

To prove `Less=.zero<=`, we split into zero and nonzero cases, for which the best tool at hand is **datatype-cases**:

```
datatype-cases Less=.zero<= {
  zero =>
   (!chain-> [(zero = zero) ==> (zero <= zero) [Less=.Implied-by-equal]])
| (S n) =>
   (!chain-> [true ==> (zero < S n)            [Less.zero<S]
                   ==> (zero <= S n)           [Less=.Implied-by-<]])
}
```

Exercise 8.9: Prove `Less=.S-zero-S-n`. *Hint:* Split into zero and nonzero cases using **datatype-cases**. □

The proof of the next property, `Less=.injective`, provides a nice illustration of the power of equivalence chaining:

```
conclude Less=.injective
  pick-any n m
    (!chain [(S n <= S m)
        <==> (S n < S m | S n = S m)  [Less=.definition]
        <==> (n < m | n = m)          [Less.injective S-injective]
```

```
    <==> (n <= m)                     [Less=.definition]])
```

Since Less=.not-S is a (quantified) negation, we use proof by contradiction:

```
conclude Less=.not-S
  pick-any n
    (!by-contradiction (~ S n <= n)
      assume Sn<=n := (S n <= n)
        let {disjunction :=
               (!chain-> [Sn<=n ==> (S n < n | S n = n) [Less=.definition]])}
          (!cases disjunction
            assume Sn<n := (S n < n)
              let {-Sn<n := (!chain-> [true ==> (~ Sn<n) [Less.S-not-<]])}
                (!absurd Sn<n -Sn<n)
            assume Sn=n := (S n = n)
              let {-Sn=n := (!chain-> [true ==> (~ Sn=n) [S-not-same]])}
                (!absurd Sn=n -Sn=n)))
```

The following proof of Less=.discrete provides an interesting example of generating an existential property in order to contradict the negated one in Less.discrete:

```
conclude Less=.discrete
  pick-any m n
    assume (m < n)
      (!by-contradiction (S m <= n)
        assume -Sm<=n := (~ S m <= n)
          let {in-between := (exists k . m < k & k < S m)}
            (!absurd
              (!chain-> [-Sm<=n
                     ==> (~ (S m < n | S m = n))    [Less=.definition]
                     ==> (~ S m < n & S m =/= n)    [dm]
                     ==> (n < S m)                  [Less.trichotomy]
                     ==> (m < n & n < S m)          [augment]
                     ==> in-between                 [existence]])
              (!chain-> [true ==> (~ in-between)    [Less.discrete]])))
```

A short proof of Less=.zero2 is possible using dsyl (disjunctive syllogism; see page 215):

```
conclude Less=.zero2
  pick-any n
    assume hyp := (n <= zero)
      (!dsyl (!chain-> [hyp  ==> (n < zero | n = zero) [Less=.definition]])
             (!chain-> [true ==> (~ n < zero)          [Less.not-zero]]))
```

The proof of Less=.not-S-zero is another simple example of proof by contradiction:

```
conclude Less=.not-S-zero
  pick-any n
    (!by-contradiction (~ S n <= zero)
      assume hyp := (S n <= zero)
        (!absurd
          (!chain-> [hyp  ==> (S n = zero)    [Less=.zero2]])
          (!chain-> [true ==> (S n =/= zero) [S-not-zero]])))
```

Exercise 8.10: Prove `N.Less=.trichotomy1` and `N.Less=.trichotomy2`. □

Exercise 8.11: Prove `N.Less=.trichotomy3` and `N.Less=.trichotomy4`. □

8.1.4 Combining ordering and arithmetic

We conclude this section by considering a simple case of interaction between ordering operators and natural number addition.

```
extend-module Less {
  define Plus-cancellation := (forall k m n . m + k < n + k ==> m < n)
  define Plus-k := (forall k m n . m < n ==> m + k < n + k)
}

extend-module Less= {
 define Plus-cancellation := (forall k m n . m + k <= n + k ==> m <= n)
 define Plus-k := (forall k m n . m <= n ==> m + k <= n + k)
}
```

We introduce these properties together, as it is useful in proving them to consider both cancellation properties first, then the other two.

```
by-induction Less.Plus-cancellation {
  zero =>
    pick-any m n
      (!chain [(m + zero < n + zero)
           ==> (m < n)                   [Plus.right-zero]])
| (S j) =>
    let {ind-hyp := (forall m n . m + j < n + j ==> m < n)}
      pick-any m n
        (!chain [(m + S j < n + S j)
             ==> (S (m + j) < S (n + j)) [Plus.right-nonzero]
             ==> (m + j < n + j)         [Less.injective]
             ==> (m < n)                 [ind-hyp]])
}
```

Exercise 8.12: Prove the following cancellation property of `N.+`:

```
define =-cancellation := (forall k m n . m + k = n + k ==> m = n)
```

Put the definition and proof within (an extension of) the `N.Plus` module. □

Exercise 8.13: Prove `Less=.Plus-cancellation`. *Hint:* Both `Less.Plus-cancellation` and `Plus.=-cancellation` (Exercise 8.12) are likely to be of use. □

For the proof of `Less.Plus-k` we use `Less=.Plus-cancellation` as well as a couple of the trichotomy properties.

```
conclude Less.Plus-k
  pick-any k m n
    assume hyp1 := (m < n)
      let {goal := (m + k < n + k)}
        (!by-contradiction goal
          (!chain [(~ goal)
               ==> (n + k <= m + k)    [Less=.trichotomy2]
               ==> (n <= m)            [Less=.Plus-cancellation]
               ==> (~ m < n)           [Less=.trichotomy4]
               ==> (hyp1 & ~ m < n)    [augment]
               ==> false               [prop-taut]]))
```

Exercise 8.14: Prove `Less=.Plus-k`. □

Similar properties hold between ordering relations and natural number multiplication. See the Athena library file `lib/main/nat-less.ath`.

Some of the remaining properties stated in the `Less=` module are given as exercises (Section 8.6). Proofs of the others may be found in `lib/main/nat-less.ath`.

8.2 Natural number subtraction

In this section we define subtraction on the natural numbers and prove some useful results about it.

Customary subtraction is not closed on the natural numbers: when the second argument is larger than the first, the result is not a natural number but a negative integer. For instance, subtracting 5 from 3 yields -2. In Chapter 9 we will study proof methods for integers, but here we continue to deal only with natural numbers and stipulate that when the second argument is larger than the first, the result is 0. For all other arguments n and m, the result will be, as expected, the difference $n - m$.[2] The definition can be given by three axioms, the first two handling the cases when one of the two arguments is 0, and the third for the case when both arguments are greater than zero.

2 The function just described is sometimes called the *monus* operation.

```
declare -: [N N] -> N [200 [int->nat int->nat]]

module Minus {

  assert* axioms := [(zero - x = zero)
                     (x - zero = x)
                     (S x - S y = x - y)]

  define [zero-left zero-right both-nonzero] := axioms
```

These axioms can be used to deduce the value of (`n - m`) for any canonical terms n and m. For instance, here is a proof that subtracting 3 from 4 yields 1:

```
> (!chain [(S S S S zero - S S S zero)
         = (S S S zero - S S zero)         [both-nonzero]
         = (S S zero - S zero)             [both-nonzero]
         = (S zero - zero)                 [both-nonzero]
         = (S zero)                        [zero-right]])

Theorem: (= (N.- (S (S (S (S zero))))
                 (S (S (S zero))))
            (S zero))
```

Thus, when both arguments are nonzero, of the form (`S n`) and (`S m`), then (`S m - S n`) is identical to the result of recursively computing (`n - m`). If the second argument is larger than the first, successive recursions will eventually bring down the first argument to `zero` while the second is still nonzero, and at that point `Minus.zero-left` will produce `zero` as the result. (We will prove this formally by induction shortly.) We can use the evaluator to test the axioms on various other inputs. Note that we can apply `-` directly to integer numerals (because we have introduced the symbol `-` so that it automatically converts integer numeral arguments into their equivalent natural number representations):

```
> (eval S S zero - zero)

Term: (S (S zero))

> (eval 5 - 2)

Term: (S (S (S zero)))

> (eval 5 - 5)

Term: zero
```

An important property of subtraction is the following form of *cancellation*: When the second argument y is no larger than the first, x, then

$$x = (x - y) + y.$$

That is, if we first subtract y from x and then add y back to the result, we get x again.

```
define Plus-Cancel := (forall y x . y <= x ==> x = (x - y) + y)
```

This simple property is quite useful in arithmetic proofs. To prove it, we use structural induction on the outer universally quantified variable. We begin with a proof sketch, setting up the induction cases but using **force** for the inductive step so that we can go ahead and test the proof of the basis case:

```
by-induction Plus-Cancel {
  zero =>
    conclude (forall x . zero <= x ==> x = (x - zero) + zero)
      pick-any x
        assume (zero <= x)
          (!chain-> [((x - zero) + zero)
                  = (x + zero)              [zero-right]
                  = x                       [Plus.right-zero]
              ==> (x = (x - zero) + zero)   [sym]])
| (y as (S y')) =>
    let {ind-hyp := (forall x . y' <= x ==> x = (x - y') + y')}
      (!force (forall x . S y' <= x ==> x = (x - S y') + S y'))
}
```

In the basis case,

```
(forall x . zero <= x ==> x = (x - zero) + zero),
```

the assumption `(zero <= x)` is not actually needed (and it is true for all `x` anyway), but in the proof we assume it in order to match the form of the sentence we are proving. To prove the conclusion using `chain`, we start with the more complex right-hand side and reduce it to the left-hand side, then reverse the resulting identity using symmetry.

We've seen that in many cases we can prove an identity, if not directly by equality chaining then by an induction in which both the basis case and the inductive step are proved by equality chaining. Here we have succeeded with the basis case equality, but when we examine the equality to be proved in the inductive step, namely `(x = (x - S y') + S y')`, we again cannot find any axiom that would apply. Of course, we should also examine our stock of previously derived results about these functions (i.e., about `N.+` and `N.-`) to see whether any of them can help us to prove our goal. But none of the theorems we have proved about `N.+` helps, and we have not yet proved any results involving `N.-`.

In this situation a useful heuristic is to attempt another case analysis, further breaking into cases in which equality chaining or some other method can finally be employed. But where can we look for guidance in choosing how to break the problem down into cases? Let us look closely at the defining axioms for `N.-`, to try to see how to introduce cases in which they could be applied.

In order to rewrite the underlined subterm in

```
((x - S y') + S y')
```

using `Minus.zero-left`, we need `x` to be `zero`. To rewrite it using `Minus.both- nonzero`, we need `x` to be of the form `(S x')` for some `x'`. We can most easily break the proof into these two cases using **`datatype-cases`**. We initially write a proof sketch using **`force`** in each of the two cases.

```
by-induction Plus-Cancel {
  zero =>
    conclude (forall x . zero <= x ==> x = (x - zero) + zero)
      pick-any x
        assume (zero <= x)
          (!chain-> [((x - zero) + zero)
                   = (x + zero)              [zero-right]
                   = x                       [Plus.right-zero]
               ==> (x = (x - zero) + zero)   [sym]])
| (y as (S y')) =>
    let {ind-hyp := (forall x . y' <= x ==> x = (x - y') + y')}
      datatype-cases (forall x . S y' <= x ==> x = (x - S y') + S y') {
        zero =>     (!force (S y' <= zero ==> zero = (zero - S y') + S y'))
      | (S x') => (!force (S y' <= S x' ==> S x' = (S x' - S y') + S y'))
      }
}
```

Now the proof can get more traction because some of the defining axioms of `Minus` are directly applicable—we made sure they were by splitting the proof into these two cases!—and hence we can now start using chaining or other proof tools and see how far we get.

It turns out we can get quite far. In the `zero` case we can use proof by contradiction, since the assumption `(S y' <= zero)` contradicts the natural number property `Less=.not-S-zero`. We can thus derive the desired conclusion (or, in fact, any conclusion) using `from-complements`:

```
assume hyp := (S y' <= zero)
  (!from-complements
     (zero = (zero - S y') + S y')
     hyp
     (!chain-> [true ==> (~ hyp)  [Less=.not-S-zero]]))
```

In the `(S x')` case, `((S x' - S y') + S y')` can be reduced to `((x' - y') + S y')` using `Minus.both-nonzero`, and then to `(S ((x' - y') + y'))` using `Plus.right-nonzero`. At this point the key is to realize that the inductive hypothesis might be applicable to the subterm `((x' - y') + y')`, because the term we are subtracting and then adding is `y'`, which is smaller than the term `y` ≡ `(S y')` on which we are currently inducting. Indeed, all we have to do in order to apply the inductive hypothesis is make sure that the corresponding antecedent holds, namely, that we have `(y' <= x')`. But this follows easily from the assumption `(S y' <= S x')`, using `Less=.injective`. (Again, see page 447.) Thus, the following succeeds in proving the `(S y')` subcase of the inductive step:

```
assume hyp := (S y' <= S x')
  let {y'<=x' := (!chain->
                    [hyp ==> (y' <= x')          [Less=.injective]])
    (!chain [((S x' - S y') + S y')
           = ((x' - y') + S y')                  [both-nonzero]
           = (S ((x' - y') + y'))                [Plus.right-nonzero]
           = (S x')                              [y'<=x' ind-hyp]
       ==> (S x' = (S x' - S y') + S y')         [sym]])
```

Substituting these proofs for the occurrences of `force` in our proof sketch, and making more use of some pattern variables, we obtain a complete proof:

```
by-induction Plus-Cancel {
  zero =>
    conclude (forall x . zero <= x ==> x = (x - zero) + zero)
      pick-any x
        assume (zero <= x)
          (!chain-> [((x - zero) + zero)
                   = (x + zero)                       [zero-right]
                   = x                                [Plus.right-zero]
                ==> (x = (x - zero) + zero)           [sym]])
| (y as (S y')) =>
    let {IH := (forall x . y' <= x ==>
                         x = (x - y') + y')}
     datatype-cases (forall x . y <= x ==>
                              x = (x - y) + y) {
       zero => assume hyp := (y <= zero)
                 (!from-complements
                   (zero = (zero - S y') + S y')
                   hyp
                   (!chain->
                     [true ==> (~ hyp)                [Less=.not-S-zero]]))
     | (x as (S x')) =>
         conclude (y <= x ==> x = (x - y) + y)
           assume hyp := (y <= x)
             let {y'<=x' := (!chain->
                               [hyp ==> (y' <= x') [Less=.injective]])}
               (!chain->
                 [((S x' - S y') + S y')
                = ((x' - y') + S y')                  [both-nonzero]
                = (S ((x' - y') + y'))                [Plus.right-nonzero]
                = (S x')                              [y'<=x' IH]
              ==> (S x' = (S x' - S y') + S y')       [sym]])
     }
}
} # close module Minus
```

Generalizing, we arrive at the following useful principle:

Heuristic 8.1: Case Analysis Based on Conditional Definitions

When the proof goal involves a function f that is conditionally defined according to a number of mutually exclusive conditions $C_1, \ldots, C_n$, try to structure the proof as a case analysis that distinguishes n appropriate instances $C_1', \ldots, C_n'$ of the conditions $C_1, \ldots, C_n$.

In this case the function was `N.-` and the relevant conditions were two, depending on whether the second argument to `N.-` is `zero` or not. We will soon encounter more occasions illustrating this powerful heuristic.

Exercise 8.15: Prove `N.Less=.S1`. □

Exercise 8.16: Prove `N.Less=.S2` and `N.Less=.S3`. □

Exercise 8.17: Prove `Less=.transitive`. □

Exercise 8.18: Prove `N.Less=.k-Less=`. □

We now extend module `N.Minus` in order to state and prove additional properties of `N.-`.

```
extend-module Minus {
```

Exercise 8.19: Prove the following alternative characterization of <=:

```
define alt-<=-characterization :=
  (forall x y . x <= y <==> exists z . y = x + z)
```

using `N.Minus.Plus-Cancel` and properties of `N.<=`. *Hint:* Review properties of `N.<=` (see page 447) for possible application in this proof. □

Another fundamental property of subtraction is that if the second argument is greater than zero but not greater than the first, then the result of the subtraction will be strictly less than the first argument:

$$0 < y \leq x \Rightarrow (x - y) < x.$$

We can state and prove this result as follows:

```
define <-left := (forall x y . zero < y  & y <= x ==> x - y < x)

conclude <-left
  pick-any x y
    assume (zero < y & y <= x)
      let {goal := (x - y < x)}
        (!by-contradiction goal
```

```
     assume (~ goal)
       (!absurd
         (!chain-> [(zero < y)
                ==> (zero + x < y + x)       [Less.Plus-k]
                ==> (x < y + x)              [Plus.left-zero]])
         (!chain-> [(~ goal)
                ==> (x <= x - y)             [Less=.trichotomy1]
                ==> (x + y <= (x - y) + y)   [Less=.Plus-k]
                ==> (x + y <= x)             [(y <= x) Plus-Cancel]
                ==> (~ x < x + y)            [Less=.trichotomy4]
                ==> (~ x < y + x)            [Plus.commutative]])))
```

The next two results pertain to the cases when the second argument is equal to or greater than the first argument. In both cases, the result of the subtraction is `zero`. We start with the case when both arguments are identical.

```
define second-equal := (forall x . x - x = zero)

by-induction second-equal {
  zero => (!chain [(zero - zero) = zero    [zero-left]])
| (S k) =>
    let {ind-hyp := (k - k = zero)}
      (!chain [(S k - S k) = (k - k)      [both-nonzero]
                           = zero          [ind-hyp]])
}
```

It is possible, and an interesting exercise, to prove `second-equal` without using induction:

Exercise 8.20: Prove `second-equal` without using induction. *Hint:* Start with `(x <= x)` and use `Plus-Cancel` and some properties of <=. □

In general, many results whose proofs would require induction if given directly from first principles (defining axioms), can be derived by noninductive proofs if they make use of previous results that were proved inductively. In the following case, however, when the second argument is greater than the first, we are unable to use `Plus-Cancel` because its precondition is not satisfied. Accordingly, the result is proved by induction, where, in the inductive step case we proceed as we did in the proof of `Plus-Cancel` with a further case analysis based on whether the remaining variable is zero or nonzero:

```
define second-greater := (forall x y . x < y ==> x - y = zero)

by-induction second-greater {
  zero =>
    conclude (forall y . zero < y ==> zero - y = zero)
      pick-any y
        assume (zero < y)
          (!chain [(zero - y) = zero                [zero-left]])
```

```
| (x as (S x')) =>
    let {ind-hyp := (forall y . x' < y ==> x' - y = zero)}
      datatype-cases (forall y . x < y ==> x - y = zero) {
         zero => assume hyp := (S x' < zero)
                   (!from-complements
                     (S x' - zero = zero)
                     hyp
                     (!chain-> [true ==> (~ hyp)  [Less.not-zero]]))
      | (S y') => assume hyp := (S x' < S y')
                    let {x'<y' := (!chain->
                                    [hyp
                                 ==> (x' < y')     [Less.injective]])}
                      (!chain [(S x' - S y')
                             = (x' - y')           [both-nonzero]
                             = zero                [x'<y' ind-hyp]])
      }
}
```

It is convenient to formulate a more general result that applies whenever the second argument is greater than *or* equal to the first:

```
define second-greater-or-equal :=
  (forall x y . x <= y ==> x - y = zero)

conclude second-greater-or-equal
  pick-any x:N y
    assume h := (x <= y)
      (!cases (!chain<- [(x < y | x = y) <== h [Less=.definition]])
         (!chain [(x < y) ==> (x - y = zero)    [second-greater]])
         assume (x = y)
           (!chain [(x - y) = (x - x)           [(x = y)]
                            = zero              [second-equal]]))
```

We will also need the following simple property:

$$\forall x, y, z \,.\, x = y + z \Rightarrow x - y = z.$$

```
define Plus-Minus-property := (forall x y z . x = y + z ==> x - y = z)

conclude Plus-Minus-property
  pick-any x y z
    assume h := (x = y + z)
      let {p1 := (!chain-> [h ==> (y <= x)          [Less=.k-Less=]
                              ==> (x = (x - y) + y)  [Plus-Cancel]]);
           p2 := (!chain-> [h ==> (x = z + y)       [Plus.commutative]])}

        (!chain-> [((x - y) + y) = x               [p1]
                                 = (z + y)         [p2]
                  ==> (x - y = z)                  [Plus.=-cancellation]])
```

From Plus-Minus-property we easily obtain another cancellation property:

```
define cancellation :=  (forall x y . (x + y) - x = y)

conclude cancellation
  pick-any x y
    (!chain->
      [(x + y = x + y)  ==> ((x + y) - x = y) [Plus-Minus-property]])

} # close module Minus
```

Another result of interest is the property of distributivity of multiplication over subtraction:

$$x \cdot y - x \cdot z = x \cdot (y - z).$$

⋆ ***Exercise 8.21:*** Apply properties of N.* (see the extensions of the N.Times module in the Athena library files lib/main/nat-times.ath and lib/main/nat-less.ath) and results of this section to prove:

```
define Times-Distributivity := (forall x y z . x * y - x * z = x * (y - z))
```

While induction is not necessary, case analysis (using two-cases) is. □

We now close module N. In the next section we will be working inside the List module.

```
} # close module N
```

8.3 Ordered lists

In Section 3.9, we introduced a datatype List-N for representing lists of natural numbers, and dealt with equations between List-N terms. Later in that section we generalized the datatype to (List S), for an arbitrary sort S. Actually, in asserting and proving equalities between such lists, S cannot be completely arbitrary since we use properties of the equality operator, =, on elements of S. But since the equality properties are built into Athena's logic and proof methods, we sometimes overlook the important role that they play. With ordering properties of $<$, on the other hand, there are no such built-in assumptions—and there shouldn't be, since there are important domains (such as complex numbers) that lack a natural ordering (but may be ordered artificially in several different ways). In this section, we consider lists of natural numbers and use the order operators < and <= on natural numbers, which we have already axiomatized and developed proofs about in the preceding sections. Later, in Chapter 14, we will see how to deal with (List S), where S can be any sort that satisfies certain minimal ordering axioms (specifically, those of a *strict weak order*) that are just sufficient to allow algorithms for such operations as merging and sorting to work correctly.

In preparation for defining ordered lists, we first introduce a couple of predicates that will be used in their definition, and derive some lemmas about them. First, we define a *membership* predicate on lists as follows.

```
extend-module List {

  declare in: (S) [S (List S)] -> Boolean [[id (alist->clist id)]]

  module in {
    assert* def := [(~ x in nil)
                    (x in h::t <==> x = h | x in t)]

    define [empty nonempty] := def
  }
```

Exercise 8.22: Prove the following lemmas:

```
define head := (forall x L . x in x::L)
define tail := (forall x y L . x in L ==> x in y::L)
```

Put the definitions and proofs within module `List.in`. □

Exercise 8.23: Prove the following theorem:

```
define of-join := (forall L M x . x in L ++ M <==> x in L | x in M)
```

(Recall that `List.++` is defined as an alias for `List.join`.) Put the definition and proof within module `List.in`. □

We introduce ordered lists of natural numbers not as a new datatype but as a subset of `(List N)` values *defined by a predicate*, `List.ordered`. We define this predicate constructively, by first introducing the following binary predicate, which compares an `N` value with a `(List N)` value (continuing within an extension of the `List` module):

```
declare <=L: [N (List N)] -> Boolean [[int->nat (alist->clist int->nat)]]
```

We define (x `<=L` L) to hold iff x is less than or equal to the first element of L or if L is empty:

```
module <=L {

  assert* def := [(x <=L nil)
                  (x <=L h::_ <==> x <= h)]

  define [empty nonempty] := def
```

Each of the following simple lemmas about <=L was suggested by a subgoal in one of the proofs about ordered lists that come later in this section. Although their proofs make good exercises (and are given as such in the additional exercises section), you may want to skip even reading them at this point and attempt to work in a more top-down manner (i.e., wait until you are working on those later proofs to see if you recognize the need for such lemmas, then state and prove them at that point).

Continuing within module List.<=L:

```
  define left-transitive :=
    (forall L x y . x <= y & y <=L L ==> x <=L L)

  define before-all-implies-before-first :=
    (forall L x . (forall y . y in L ==> x <= y) ==> x <=L L)

  define append :=
    (forall L M x . x <=L L & x <=L M ==> x <=L L ++ M)

} # close module <=L
```

Now we can define the ordered predicate as follows:

```
declare ordered: [(List N)] -> Boolean [[(alist->clist int->nat)]]

module ordered {

  assert* def := [(ordered nil)
                  (ordered h::t <==> h <=L t & ordered t)]

  define [empty nonempty] := def

} # close module ordered
```

Exercise 8.24: Prove the following lemmas (the proofs are short and easy):

```
define head := (forall L x . ordered x::L ==> x <=L L)
define tail := (forall L x . ordered x::L ==> ordered L)
```

Put these lemmas and their proofs within module List.ordered. □

Here are some useful theorems about ordered lists:

```
  define first-to-rest-relation :=
    (forall L x y .  ordered x::L & y in L ==> x <= y)

  define cons :=
    (forall L x . ordered L & (forall y . y in L ==> x <= y)
                ==> ordered x::L)
```

```
define append :=
  (forall L M . ordered L &
                ordered M &
                (forall x y . x in L & y in M ==> x <= y)
            ==> ordered L ++ M)
```

And here is a proof of the first:

```
by-induction first-to-rest-relation {
  nil =>
    pick-any x y
      assume (ordered x::nil & y in nil)
        let {not-in := (!chain-> [true ==> (~ y in nil) [in.empty]])}
          (!from-complements (x <= y) (y in nil) not-in)
| (L as (z :: M)) =>
  let {ind-hyp := (forall x y . ordered x::M & y in M ==> x <= y)}
    conclude (forall x y . ordered x::L & y in L ==> x <= y)
      pick-any x:N y:N
        assume (ordered x::z::M & y in z::M)
          let {p0 := (!chain->
                        [(ordered x::z::M)
                     ==> (x <=L z::M &
                          ordered z::M)                 [nonempty]
                     ==> (x <=L z::M & z <=L M &
                          ordered M)                    [nonempty]
                     ==> (x <= z  &
                          z <=L M &
                          ordered M)                    [<=L.nonempty]]);

               p1 := (!chain-> [p0 ==> (ordered M)      [prop-taut]]);

               p2 := (!chain-> [p0
                             ==> (x <= z & z <=L M)     [prop-taut]
                             ==> (x <=L M)              [<=L.left-transitive]
                             ==> (x <=L M & ordered M)  [augment]
                             ==> (ordered x::M)         [nonempty]])}
        (!cases (!chain<- [(y = z | y in M) <==
                           (y in z::M)                  [in.nonempty]])
           assume (y = z)
             (!chain-> [p0 ==> (x <= z)                 [left-and]
                           ==> (x <= y)                 [(y = z)]])
             (!chain [(y in M)
                  ==> (p2 & y in M)                     [augment]
                  ==> (x <= y)                          [ind-hyp]]))
} # by-induction
```

Exercise 8.25: Prove `List.ordered.cons`. □

⋆ ***Exercise 8.26:*** Prove `List.ordered.append`. *Hint:* You will need to use `<=L.append`, among other properties of `<=L`. □

Before turning to the next topic, we close the `List.ordered` and `List` modules.

```
  } # close module ordered
} # close module List
```

8.4 Binary search trees

Another important datatype related to ordering properties is that of *binary search trees*, a subset of binary trees satisfying a condition relating elements of the tree by their order. First, let's define binary trees:

```
datatype (BinTree S) :=  null
                       | (node (BinTree S) S (BinTree S))

assert (datatype-axioms "BinTree")
```

Here, as we did with the `List` datatype in Chapter 3, we have defined `BinTree` with a sort parameter, `S`. The declaration says that a value of sort `(BinTree S)` is either the constant `null` or a term with constructor `node` and three arguments: two of sort `(BinTree S)`, recursively, and one of sort `S`. We wish to interpret `null` as an empty tree (no elements), and `(node` L x R`)` as a tree with element x at its root, L as its "left subtree," and R as its "right subtree." See Figure 8.1 for an example.

The first function we define on binary trees is a predicate for testing whether a value of sort `S` is present in a `(BinTree S)`. (The corresponding membership predicate for `(List S)` was introduced in Exercises 8.22 and 8.23.)

```
module BinTree {

  define ++ := List.++

  define [x x' y T T' L R] :=
         [?x ?x' ?y ?T:(BinTree 'S1) ?T':(BinTree 'S2)
          ?L:(BinTree 'S3) ?R:(BinTree 'S4)]

  declare in: (S) [S (BinTree S)] -> Boolean

  module in {

    assert* def :=
      [(~ x in null)
       (x in (node L y R) <==> x = y | x in L | x in R)]
```

```
define tree1 :=
  (node (node null
              2
              (node (node null
                          3
                          null)
                    5
                    null))
        7
        (node (node null
                    11
                    null)
              13
              null))
```

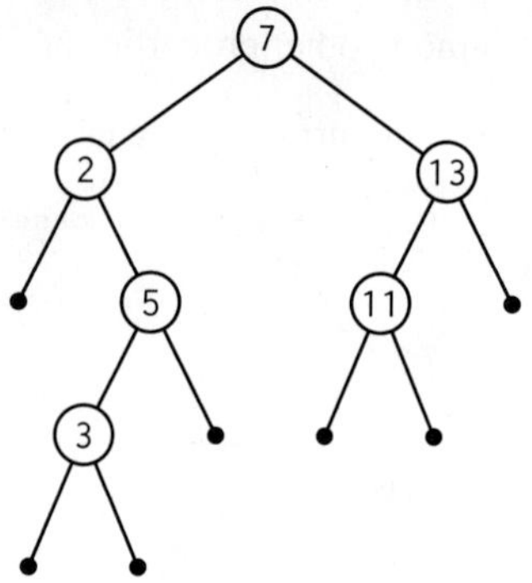

Figure 8.1
A binary tree with Int elements, together with a corresponding graphical depiction.

```
  define [empty nonempty] := def

} # close module in
} # close module BinTree
```

There is, of course, an induction principle for binary trees that we will come to momentarily, but in the following exercise it is not required; implication chaining suffices for the proofs.

Exercise 8.27: Prove the following simple lemmas:

```
define root  := (forall x L y R . x = y  ==> x in (node L y R))
define left  := (forall x L y R . x in L ==> x in (node L y R))
define right := (forall x L y R . x in R ==> x in (node L y R))
```

Put the definitions and proofs inside module BinTree.in. □

Next we look at one of the "tree walk" operations for traversing through all the nodes of a tree, called an "inorder" walk (the other main tree walks are called "preorder" and "postorder"). We define a function inorder from (BinTree S) to (List S) that produces a list of the tree elements ordered so that the root element appears between the elements of the left subtree and those of the right subtree, and recursively the elements are in this order within each subtree.

```
extend-module BinTree {

  declare inorder: (S) [(BinTree S)] -> (List S)

  module inorder {
```

```
  assert* def :=
      [(inorder null = nil)
       (inorder (node L x R) = (inorder L) ++ x::inorder R)]

  define [empty nonempty] := def

}
```

As an example, we can apply `inorder` to the binary tree in Figure 8.1.

```
> (eval inorder tree1)

Term: (:: 2
          (:: 3
              (:: 5
                  (:: 7
                      (:: 11
                          (:: 13
                              nil:(List Int)))))))
```

We first relate `BinTree.inorder` to `BinTree.in`:

```
overload in List.in

extend-module inorder {

  define in-correctness := (forall T x . x in inorder T <==> x in T)
```

Note the overloading, which allows `in` to denote both `List.in` and `BinTree.in`, disambiguated by the sort of its second argument. In the theorem `in-correctness`, therefore, the first occurrence of `in` stands for `List.in` while the second stands for `BinTree.in`.

To prove `BinTree.in-correctness`, we can separately prove each of the implications, then combine the results with `equiv`.

```
define in-correctness-1 := (forall T x . x in inorder T ==> x in T)
define in-correctness-2 := (forall T x . x in T ==> x in inorder T)
```

To prove each of these implications, we will need induction. For binary trees we have a mathematical induction principle whose clauses correspond to the two tree constructors: There is a basis case corresponding to `null` and an induction step with *two induction hypotheses* corresponding to the "left" subtree L and "right" subtree R in (`node` L x R):

Principle 8.1: Mathematical Induction for Binary Trees

To prove $\forall\, T \,.\, P(T)$ where T ranges over binary trees, it suffices to prove:

1. *Basis Case*: $P(null)$.
2. *Induction Step*: $\forall\, L\, R \,.\, P(L) \wedge P(R) \Rightarrow \forall\, x \,.\, P(node(L, x, R))$.

In the induction step, the assumptions $P(L)$ and $P(R)$ are called the *induction hypotheses*.

As we have seen with natural numbers and lists, Athena is able to derive from the definition of the BinTree datatype the cases that must be handled by an inductive proof over binary trees. During evaluation, it also automatically enters both induction hypotheses into the assumption base so that they are available without having to explicitly assume them in the inductive step case. Here is a proof of in-correctness-1:

```
by-induction in-correctness-1 {
  null =>
  pick-any x
    assume (x in inorder null)
    let {in-nil := (!chain->
                     [(x in inorder null)
                  ==> (x in nil)                  [empty]]);
         -in-nil := (!chain-> [true
                           ==> (~ x in nil)       [List.in.empty]])}
      (!from-complements (x in null) in-nil -in-nil)
| (T as (node L y R)) =>
  let {ind-hyp1 := (forall x . x in inorder L ==> x in L);
       ind-hyp2 := (forall x . x in inorder R ==> x in R)}
  pick-any x
    assume hyp := (x in inorder T)
      let {D := (!chain-> [hyp
                       ==> (x in (inorder L) ++
                                 y::inorder R)   [nonempty]
                       ==> (x in inorder L |
                             x in y::inorder R)   [List.in.of-join]
                       ==> (x in inorder L |
                             x = y |
                             x in inorder R)      [List.in.nonempty]])}

      (!cases D
        (!chain [(x in inorder L)
             ==> (x in L)                         [ind-hyp1]
             ==> (x in T)                         [in.left]])
        (!chain [(x = y) ==> (x in (node L y R))  [in.root]])
        (!chain [(x in inorder R)
             ==> (x in R)                         [ind-hyp2]
             ==> (x in T)                         [in.right]]))
```

```
}
} # close module inorder
} # close module BinTree
```

* ***Exercise 8.28:*** Prove `in-correctness-2` and combine the result with `in-correctness-1` to conclude `in-correctness`. □

Exercise 8.29: (a) Define a function `List.count`, for counting the number of occurrences of a given value in a given list, as follows:

```
extend-module List {

  declare count: (S) [S (List S)] -> N

  module count {

    assert* count-def :=
    [((count x nil) = zero)
     ((count x h::t) = S (count x t) <== x = h)
     ((count x h::t) = (count x t) <== x =/= h)]

    define [empty more same] := count-def
```

Then prove the following theorem:

```
    define of-join :=
      (forall L M x . (count x L ++ M) = (count x L) + (count x M))
  } # close module count
} # close module List
```

(b) Similarly, declare `BinTree.count` with the signature:

```
extend-module BinTree {

  declare count: (S) [S (BinTree S)] -> N

  overload count List.count

} # close module BinTree
```

The overloading allows `count` to denote both `List.count` and `BinTree.count`, disambiguated by the sort of its second argument. Give an axiomatization of `BinTree.count` that defines it as the number of occurrences of a given value in a binary tree.

(c) Then prove the following theorem:

```
extend-module BinTree {
  extend-module inorder {

    define count-correctness :=
      (forall T x .  (count x (inorder T)) =  (count x T))

  } # close module inorder
} # close module BinTree
```

□

Finally, we focus on binary search trees, a special type of binary trees: those satisfying a predicate, `BST`. A general definition for elements of any sort for which an appropriate comparison function is defined will be given in Section 15.1, but here we define `BST` for natural numbers. As we often do elsewhere, we first give a constructive definition of `BST`, and then we derive from that definition a more abstract characterization theorem that we use in subsequent proofs. The constructive definition is given in terms of two binary predicates: `no-smaller` and `no-larger`. The first takes a binary tree of natural numbers T and a natural number n and returns `true` iff every element of T is not smaller than (i.e., larger than or equal to) n. The `no-larger` predicate is similar except that it holds iff every element of T is smaller than or equal to n. The characterization theorem has two parts, which are here directly asserted as axioms. Their derivations from `BinTree.BST.definition` are left as an exercise.

```
extend-module BinTree {

  define [x y z T L R] := [?x:N ?y:N ?z:N ?T:(BinTree N) ?L:(BinTree N)
                           ?R:(BinTree N)]

  define [< <=] := [N.< N.<=]

  declare BST: [(BinTree N)] -> Boolean

  declare no-smaller, no-larger: [(BinTree N) N] -> Boolean

  assert* no-smaller-def :=
    [(null no-smaller _)
     ((node L y R) no-smaller x <==> x <= y &
                                      L no-smaller x &
                                      R no-smaller x)]

  assert* no-larger-def :=
    [(null no-larger _)
     ((node L y R) no-larger x <==> y <= x &
                                     L no-larger x &
                                     R no-larger x)]

  module BST {
```

```
    assert* definition :=
      [(BST null)
       (BST (node L x R) <==> BST L & L no-larger x &
                                BST R & R no-smaller x)]

    # The two parts of the characterization result, asserted here as axioms:

    assert empty := (BST null)

    assert nonempty :=
      (forall L y R .
        BST (node L y R) <==>
          BST L &
          (forall x . x in L ==> x <= y) &
          BST R &
          (forall z . z in R ==> y <= z))

  } # close module BST
} # close module BinTree
```

The binary tree in Figure 8.1 is an example of a natural number binary search tree, if you replace its `Int` values by corresponding `N` values.

In Chapter 11, we will introduce a *binary search function* whose domain is binary search trees. Aside from its importance as a more efficient search algorithm than is possible with lists, that function will also provide a nice first example of significant issues in specifying and proving algorithm correctness.

⋆ ***Exercise 8.30:*** Derive `empty` and `nonempty` from the constructive definition of `BST`. □

8.5 Summary and a connecting theorem

In this chapter, we have focused on:

- Defining inequality relations on natural numbers with (constructive) axioms and using proofs of selected theorems about these relations for further practice with induction, case analysis, implication chaining, and proof by contradiction. Looking ahead, we will see in Chapter 14 how we can raise some of these theorems to an abstract level, from which they have many instantiations to other concrete domains.
- Defining subtraction of natural numbers and proving theorems that will be useful in defining integer addition and subtraction in Chapter 9, and division operators in Chapter 13.
- Extending the discussion in Chapter 3 of the `List` datatype to properties of ordered lists over the natural numbers.

- Introducing binary trees over the natural numbers and a corresponding induction principle.
- Defining binary search trees over natural numbers, in preparation for our first example of algorithm specification and correctness issues in Chapter 11.

We conclude this chapter with the statement and proof of a theorem connecting three of the main ordering concepts we have studied—namely, that the `inorder` function applied to a binary search tree produces an ordered list (proved here only for natural numbers).

```
extend-module BinTree {
extend-module BST {

define is-ordered := (forall T . BST T ==> List.ordered inorder T)

define [x1 x2] := [?x1:N ?x2:N]

define ordered := List.ordered

by-induction is-ordered {
  (T as null:(BinTree N)) =>
    assume (BST null)
      (!chain->
        [true
     ==> (ordered nil:(List N))                          [empty]
     ==> (ordered inorder T)                             [inorder.empty]])
| (node L:(BinTree N) y:N R:(BinTree N)) =>
    let {ind-hyp1 := (BST L ==> ordered inorder L);
         ind-hyp2 := (BST R ==> ordered inorder R);
         inorder-in-correct := inorder.in-correctness;
         <=-transitivity := N.Less=.transitive;
         smaller-in-left := (forall x . x in L ==> x <= y);
         larger-in-right := (forall z . z in R ==> y <= z)}
     assume h := (BST (node L y R))
       conclude goal := (ordered inorder (node L y R))
         let {p1 := (!chain->
                       [h
                    ==> (BST L &
                         smaller-in-left &
                         BST R &
                         larger-in-right)                [nonempty]]);
              p2 := (!chain-> [p1
                          ==> (BST L)                    [left-and]
                          ==> (ordered inorder L)        [ind-hyp1]]);

              p3 := (!chain-> [p1
                           ==> (BST R)                   [prop-taut]
                           ==> (ordered inorder R)       [ind-hyp2]]);
              p4 := (!chain-> [p1
                           ==> smaller-in-left           [prop-taut]]);
              p5 := (!chain-> [p1
```

```
                    ==> larger-in-right              [prop-taut]]);
      p6 := conclude (forall x1 x2 .
                        x1 in inorder L &
                        x2 in y :: inorder R
                        ==> x1 <= x2)
                pick-any u v
                  assume h1 := (u in inorder L &
                                v in y :: inorder R)
                    let {disjunction :=
                          (!chain->
                            [h1
                        ==> (u in inorder L &
                             (v = y |
                              v in inorder R))  [List.in.nonempty]
                        ==> (u in L &
                             (v = y | v in R))  [inorder-in-correct]
                        ==> ((u in L & v = y) |
                             (u in L & v in R)) [prop-taut]])}
                      (!cases disjunction
                        assume (u in L & v = y)
                          (!chain->
                            [(u in L)
                        ==> (u <= y)                  [smaller-in-left]
                        ==> (u <= v)                  [(v = y)]])
                        (!chain [(u in L &
                                  v in R)
                            ==> (u <= y &
                                  y <= v)             [smaller-in-left
                                                       larger-in-right]
                            ==> (u <= v)              [<=-transitivity]]));
      p7 := conclude (forall z .
                       z in inorder R
                       ==> y <= z)
                pick-any z
                  (!chain [(z in inorder R)
                      ==> (z in R)                    [inorder-in-correct]
                      ==> (y <= z)                    [larger-in-right]])}
    (!chain->
      [(p3 & p7)
    ==> (ordered (y :: inorder R))                    [List.ordered.cons]
    ==> (p2 & (ordered (y :: inorder R)))             [augment]
    ==> (p2 & (ordered (y :: inorder R)) & p6)        [augment]
    ==> (ordered ((inorder L) ++
                  (y::inorder R)))                    [List.ordered.append]
    ==> goal                                          [inorder.nonempty]])
} # by-induction
} # close module BST
} # close module BinTree
```

8.6 Additional exercises

⋆ ***Exercise 8.31:*** In Section 8.1.2 we proved `Less.transitive` by reordering its quantifiers so that `z` was first, inducting on `z`, and then getting the property with the original quantifier ordering by substitution. Instead, prove it directly by inducting on its first quantified variable, `x`. *Hint:* Case splitting may be necessary on zero and nonzero cases of both `y` and `z`. It is easiest to do all such inner case splits with **`datatype-cases`**. □

Exercise 8.32: Prove `N.Less=.not-S`. □

Exercise 8.33: Prove `N.Less=.S-not-equal`. □

Exercise 8.34: Prove `N.Less=.trichotomy5`. □

Exercise 8.35: We can define a function returning the *maximum* of a pair of natural numbers as follows:

```
extend-module N {
  declare max: [N N] -> N [[int->nat int->nat]]

  module Max {

    assert* def := [(y < x ==> x max y = x)
                    (~ y < x ==> x max y = y)]

    define [less2 not-less2] := def

  } # close module Max
} # close module N
```

The names `Max.less2` and `Max.not-less2` are short for "the second argument of `max` is (is not) less than the first."

(a) Prove that `max` is *idempotent*; that is, `(forall x . x max x = x)`.

(b) Prove that `max` is commutative.

⋆ (c) Prove that `max` is associative. □

Exercise 8.36: Prove `List.<=L.left-transitive`. □

Exercise 8.37: Prove `List.<=L.before-all-implies-before-first`. □

Exercise 8.38: Prove `List.<=L.append`. □

8.7 Chapter notes

In this chapter, we have defined ordering relations on natural numbers and proved several results about these relations. These proofs provide further practice with high-level proof methods, and also serve as preparation for later generalization to an abstract level (in Chapters 14 and 15).

For these purposes we have not been concerned with automated proofs, but in some contexts more automation would be desirable, such as software verification or test case generation, where there may be many complex conjectures involving inequality relations that need to be proved or falsified. Robert Shostak's 1977 paper [87] discusses some of the earliest work on a class of formulas called Presburger Arithmetic and improves an approach called the Sup-Inf method into an efficient decision procedure for an important subclass of Presburger formulas. An even more restricted but still useful subset of inequalities called separation formulas is decided by Shostak's 1981 method based on computing loop residues [88]. More recently, researchers such as Strichman et al. [97] have turned away from procedures specifically developed for such formulas in favor of transforming them into propositional satisfiability (SAT) problems, taking advantage of advances in the efficiency of SAT solvers. SMT solvers have further streamlined reasoning about (linear) inequalities [28]. For a discussion of SMT solving in Athena refer to the "Automated Theorem Proving" appendix on the book's website.

9 Integer Representations and Proof Mappings

It may seem trivial to advance from natural numbers to integers: Just attach a sign, positive or negative, to a natural number and define addition, subtraction (this time, the customary definition instead of the special one introduced for natural numbers in Section 8.2), and multiplication for these signed numbers.[1] As we will see, however, there are complications. The main problem we encounter in some proofs is the large number of cases that must be considered when the signs of several integer variables are taken into account. The best way around this difficulty seems to be to finesse it by first doing some proofs using an entirely different representation of integers, as *ordered pairs of natural numbers*, and then completing the proofs with the aid of mappings back and forth between the signed and pair representations. The key technique illustrated—multiple representations and mappings between them—is well worth study, since it serves as a valuable approach to proof in many other contexts.[2]

The chapter concludes with a brief treatment of formal power series with integer coefficients. Addition of such power series has some of the same properties as integer addition, such as commutativity and associativity, an observation we will use in Chapter 14 as a starting point for discussion of abstract structures.

9.1 Declarations and axioms

We begin by defining the integers as a structure named `Z` and introduce a module `Z` to hold further declarations, axioms, and theorems. Note that we use a structure rather than a datatype because the corresponding term algebra is not free: Some distinct constructor terms will in fact represent identical values (namely, `(pos zero)` and `(neg zero)`). We do not fully characterize the identity conditions on this structure here, as these are not needed for the purposes of this chapter, but see Exercise 9.6 for such a development.

```
structure Z := (pos N) | (neg N)

expand-input pos, neg [int->nat]
```

1 If we were only interested in building a theory of integers, a more direct alternative would be to bypass the natural numbers altogether and instead provide an axiomatization of the built-in domain `Int`, including a primitive method for mathematical induction. However, one of the main objectives of this chapter is to introduce and illustrate the uses of homomorphic mappings, which is why we explicitly represent integers in terms of natural numbers.

2 And not just to proof—multiple representations of a data type and transformations between them are a recurring theme in computer science. Consider, for example, polynomials represented in coefficient form and in point-value form. Each has its own pros and cons for solving problems like polynomial evaluation and arithmetic, but by being able to switch back and forth efficiently via the DFT (Discrete Fourier Transform) algorithm, we are often able to get the best of both worlds.

```
set-precedence (pos neg) (plus 1 (get-precedence *))

assert* [(pos x = pos y <==> x = y)
         (neg x = neg y <==> x = y)
         ((exists y . x = pos y) | (exists y . x = neg y))]

module Z {
  open N
  declare zero, one: Z
  assert zero-definition := (zero = pos N.zero)
  assert zero-property := (zero = neg N.zero)
  assert one-definition := (one = pos N.one)
  define [a b c] := [?a:Z ?b:Z ?c:Z]
```

Next, we introduce integer addition, with axioms in terms of natural number addition and ordering (+ from Chapter 3 and < from Chapter 8), and subtraction (-, from Section 8.2):

```
declare +: [Z Z] -> Z

module Plus {

  overload + N.+
  define [x y] := [?x:N ?y:N]

  assert* axioms :=
    [(pos x + pos y = pos (x + y))                    #pos-pos
     (  x < y ==> pos x + neg y = neg (y - x))  #pos-neg-case1
     (~ x < y ==> pos x + neg y = pos (x - y))  #pos-neg-case2
     (  x < y ==> neg x + pos y = pos (y - x))  #neg-pos-case1
     (~ x < y ==> neg x + pos y = neg (x - y))  #neg-pos-case2
     (neg x + neg y = neg (x + y))]                   #neg-neg

  define [pos-pos pos-neg-case1 pos-neg-case2
          neg-pos-case1 neg-pos-case2 neg-neg] := axioms

  (eval pos 1 + neg 3)
  # This produces the result (neg (S (S zero))), i.e. -2.

} # close module Plus
```

Next come negation and subtraction:

```
declare negate: [Z] -> Z [120]

module Negate {

  assert* definition := [(negate pos x = neg x)
                         (negate neg x = pos x)]

  define [positive negative] := definition
```

```
} # close module Negate

declare -: [Z Z] -> Z

module Minus {

  assert* definition :=  (a - b =  a + negate b)

} # close module Minus
```

9.2 First proofs of integer properties

This brings us to our first proof involving integer operations:

```
extend-module Plus {
open Minus
overload - N.-

define Right-Inverse := (forall a . a + negate a = zero)

datatype-cases Right-Inverse {
  (pos x) =>
    conclude (pos x + negate pos x = zero)
      let {_ := (!chain-> [true ==> (~ x < x)     [Less.irreflexive]])}
        (!chain [(pos x + negate pos x)
              --> (pos x + neg x)                 [Negate.positive]
              --> (pos (x - x))                   [pos-neg-case2]
              --> (pos N.zero)                    [N.Minus.second-equal]
              <-- zero                            [zero-definition]])
| (neg x) =>
    conclude (neg x + negate neg x = zero)
      let {_ := (!chain-> [true ==> (~ x < x)     [Less.irreflexive]])}
        (!chain [(neg x + negate neg x)
              --> (neg x + pos x)                 [Negate.negative]
              --> (neg (x - x))                   [neg-pos-case2]
              --> (neg N.zero)                    [N.Minus.second-equal]
              <-- zero                            [zero-property]])
}
```

Proofs that `Z.zero` is a right and left identity for `Z.Plus` proceed in a similar way:

```
define Right-Identity := (forall a . a + zero = a)
define Left-Identity :=  (forall a . zero + a = a)

datatype-cases Right-Identity {
  (pos x) =>
    conclude (pos x + zero = pos x)
      (!chain [(pos x + zero)
```

```
       --> (pos x + pos N.zero)          [zero-definition]
       --> (pos (x + N.zero))            [pos-pos]
       --> (pos x)                       [N.Plus.right-zero]])
| (neg x) =>
    conclude (neg x + zero = neg x)
      let {_ := (!chain-> [true
                   ==> (~ x < N.zero)    [Less.not-zero]])}
        (!chain [(neg x + zero)
           --> (neg x + pos N.zero)      [zero-definition]
           --> (neg (x - N.zero))        [neg-pos-case2]
           --> (neg x)                   [N.Minus.zero-right]])
}

datatype-cases Left-Identity {
  (pos x) =>
    conclude (zero + pos x = pos x)
      (!chain [(zero + pos x)
         --> (pos N.zero + pos x)        [zero-definition]
         --> (pos (N.zero + x))          [pos-pos]
         --> (pos x)                     [N.Plus.left-zero]])
| (neg x) =>
    conclude (zero + neg x = neg x)
      (!chain [(zero + neg x)
         --> (neg N.zero + neg x)        [zero-property]
         --> (neg (N.zero + x))          [neg-neg]
         --> (neg x)                     [N.Plus.left-zero]])
}
} # close module Plus
```

9.3 Another integer representation

When we consider associativity of addition,

```
extend-module Plus {
  define associative := (forall a b c . (a + b) + c = a + (b + c))
} # close module Plus
```

the presence of the third variable increases the number of sign combinations to consider from four to eight, and most of those require subcasing based on the ordering relation between variables. For proofs of those subcases, we would need several lemmas about reassociating subtraction terms, making the overall proof exercise very long and tedious. Instead, we can simplify the task considerably by turning to a different integer representation, as an ordered pair of natural numbers, or actually as an *equivalence class of ordered pairs of natural numbers*. For example, 3 can be represented as $\{(3,0),(4,1),(5,2),\ldots\}$, the class of all natural-number pairs in which the first is 3 greater than the second, while -5 can be represented as $\{(0,5),(1,6),(2,7),\ldots\}$, the class of all natural number pairs in which the second is 5 greater than the first. The advantage of this representation is that addition can be performed without checking signs; we just add corresponding pair components.

We use the structure NN to represent integers as pairs of natural numbers:

```
structure NN := (nn N N)
```

Thus, the constructor nn essentially serves as an ordered-pair constructor. We use the symbol @ as an infix alias for nn, and we force @ to convert integer numeral arguments into their corresponding natural-number representations. We remain within the module Z in what follows. We use the letters a, b, and c to range over NN, and n1, n2, m1, and m2 to range over N.

```
overload @ lambda (x y) (nn (int->nat x) (int->nat y))

module NN {

  overload + N.+

  define [a b c] := [?a:NN ?b:NN ?c:NN]

  declare +': [NN NN] -> NN [110]

  overload + +'

  module Plus {

    define [n m n1 n2 m1 m2] := [?n:N ?m:N ?n1:N ?n2:N ?m1:N ?m2:N]

    assert* definition := (n1 @ n2 + m1 @ m2 = (n1 + m1) @ (n2 + m2))
```

So, for example, adding 3 and -5, represented as (say) $(3,0)$ and $(1,6)$, yields $(4,6)$, one of the pairs representing -2. Observe that while we introduce the symbol +' specifically to designate addition on NN, we overload + so that it can represent +' *or* N.+ *or* Z.+, depending on the sorts of its argument terms. Keep in mind also that @ has higher precedence than +, so a term like (n1 @ n2 + m1 @ m2) denotes the application of Z.NN.+' to (nn n1 n2) and (nn m1 m2).

Proving associativity of addition in this representation is easy; it is just a matter of applying the definition and invoking associativity of natural number addition:

```
    define associative := (forall a b c . (a + b) + c = a + (b + c))
    define nat-plus-assoc := N.Plus.associative

    datatype-cases associative {
      (nn n1 n2) =>
        datatype-cases (forall b c . (n1 @ n2 + b) + c =
                                     n1 @ n2 + (b + c)) {
          (nn m1 m2) =>
```

```
          datatype-cases (forall c . (n1 @ n2 + m1 @ m2) + c =
                                      n1 @ n2 + (m1 @ m2 + c)) {
            (nn k1 k2) =>
              (!chain
                [((n1 @ n2 + m1 @ m2) + k1 @ k2)
             --> (((n1 + m1) @ (n2 + m2)) + k1 @ k2)     [definition]
             --> (((n1 + m1) + k1) @ ((n2 + m2) + k2))  [definition]
             --> ((n1 + (m1 + k1)) @ (n2 + (m2 + k2)))  [nat-plus-assoc]
             <-- ((n1 @ n2) + ((m1 + k1) @ (m2 + k2)))  [definition]
             <-- ((n1 @ n2) + (m1 @ m2 + k1 @ k2))      [definition]])
          }
        }
      }
  } # close module Plus
} # close module NN
```

Of course, this does not automatically grant us associativity of addition as defined in the signed representation. To get that, we must first prove that we can faithfully map back and forth between the two representations, and that proof itself will turn out to be somewhat involved, in that it still requires an extensive case analysis. However, once proved correct, the mappings back and forth will be useful for proving many other integer properties besides associativity, and the ideas involved are applicable in many other situations.

9.4 Mappings between the signed and pair representations

We declare and axiomatize the mappings as follows (continuing inside module Z):

```
declare Z->NN: [Z] -> NN
declare NN->Z: [NN] -> Z

module Z-NN {
  overload (+ N.+) (- N.-)
  define [x y] := [?x:N ?y:N]

  assert* to-pos   := (Z->NN pos x = x @ N.zero)
  assert* to-neg   := (Z->NN neg x = N.zero @ x)
  assert* from-pos := (~ x < y ==> NN->Z x @ y = pos (x - y))
  assert* from-neg := (x < y ==> NN->Z  x @ y = neg (y - x))
```

Next we show an inverse relationship between the two mappings:

```
  define inverse := (forall a . NN->Z Z->NN a = a)

  datatype-cases inverse {
    (pos x) => {
      (!chain-> [true ==> (~ x < N.zero)           [N.Less.not-zero]]);
      (!chain [(NN->Z Z->NN pos x)
```

```
          --> (NN->Z x @ N.zero)                          [to-pos]
          --> (pos (x - N.zero))                          [from-pos]
          --> (pos x)                                     [N.Minus.zero-right]])
   }
 | (neg x) =>
   (!two-cases
     assume (x = N.zero)
       let {_ := (!chain-> [true
                           ==> (~ N.zero < N.zero) [N.Less.irreflexive]])}
         (!chain [(NN->Z Z->NN neg x)
                --> (NN->Z N.zero @ x)                    [to-neg]
                --> (NN->Z N.zero @ N.zero)               [(x = N.zero)]
                --> (pos (N.zero - N.zero))               [from-pos]
                --> (pos N.zero)                          [N.Minus.zero-right]
                <-- zero                                  [zero-definition]
                --> (neg N.zero)                          [zero-property]
                <-- (neg x)                               [(x = N.zero)]])
     assume (x =/= N.zero)
       let {D := (!chain-> [true
                          ==> (N.zero <= x)               [N.Less=.zero<=]
                          ==> (N.zero < x |
                               N.zero = x)                [N.Less=.definition]])}
         (!cases D
            assume (N.zero < x)
              (!chain [(NN->Z Z->NN neg x)
                     --> (NN->Z N.zero @ x)               [to-neg]
                     --> (neg (x - N.zero))               [from-neg]
                     --> (neg x)                          [N.Minus.zero-right]])
            assume (N.zero = x)
              (!from-complements
                (NN->Z Z->NN neg x = neg x)
                (!sym (N.zero = x))
                (x =/= N.zero))))
 }
 # datatype-cases
} # close module Z-NN
```

9.5 Additive homomorphism property

The next step is to show that `Z->NN` is what mathematicians call an *additive homomorphism*, namely, a mapping from one domain to another that preserves the meaning of addition:

```
(forall a b . Z->NN (a + b) = (Z->NN a) + (Z->NN b)).
```

Before we can prove this property, we need to formally state and assert the equivalence of arbitrary `nn` values to ones we consider to be canonical representatives. It is natural to choose as canonical those pairs in which at least one of the members is 0.

```
module NN-equivalence {
  overload - N.-

  define [n1 n2] := [?n1:N ?n2:N]

  assert* case1 := (n1 < n2 ==> n1 @ n2 = N.zero @ (n2 - n1))

  assert* case2 := (~ n1 < n2 ==>  n1 @ n2 = (n1 - n2) @ N.zero)

} # close module NN-equivalence
```

Now we can prove the additive-homomorphism property. The proof is quite long, due to the extensive case analysis required, but each case is straightforward. The following is just a proof sketch, since it uses **force** in all cases except the first; in Exercise 9.8 you are asked to complete the proof by replacing all uses of **force** with proper deductions.

```
extend-module Z-NN {

define [+ +'] := [NN.+ NN.+']

define additive-homomorphism :=
   (forall a b . Z->NN (a + b) = (Z->NN a) + (Z->NN b))

datatype-cases additive-homomorphism {
  (pos x) =>
    datatype-cases (forall b . Z->NN (pos x + b) =
                               (Z->NN pos x) + (Z->NN b)) {
      (pos y) =>
        (!combine-equations
         (!chain [(Z->NN (pos x + pos y))
               --> (Z->NN (pos (x + y)))                  [Plus.pos-pos]
               --> ((x + y) @ N.zero)                     [to-pos]])
         (!chain [((Z->NN pos x) + (Z->NN pos y))
               --> ((x @ N.zero) +  (y @ N.zero))         [to-pos]
               --> ((x + y) @ (N.zero + N.zero))          [NN.Plus.definition]
               --> ((x + y) @ N.zero)                     [N.Plus.right-zero]]))
    | (neg y) =>
        (!two-cases
          assume (x < y)
            (!force (Z->NN (pos x + neg y) =
                    (Z->NN pos x) + (Z->NN neg y)))
          assume (~ x < y)
            (!force (Z->NN (pos x + neg y) =
                    (Z->NN pos x) + (Z->NN neg y))))
    }
| (neg x) =>
    datatype-cases (forall b . Z->NN (neg x + b) =
                               (Z->NN neg x) + (Z->NN b)) {
      (pos y) =>
```

```
      (!two-cases
        assume (x < y)
         let {_ := (!chain-> [(x < y)
                              ==> (~ y < x)          [N.Less.asymmetric]])}
         (!force (Z->NN (neg x + pos y) =
                  (Z->NN neg x) + (Z->NN pos y)))
        assume (~ x < y)
         let {D := (!chain->
                    [(~ x < y)
                 ==> (y <= x)                        [N.Less=.trichotomy2]
                 ==> (y < x | y = x)                 [N.Less=.definition]])}
         (!cases D
           assume (y < x)
             (!force (Z->NN (neg x + pos y) =
                      (Z->NN neg x)  + (Z->NN pos y)))
           assume (y = x)
             let {_ := (!chain-> [true
                                  ==> (~ x < x)      [N.Less.irreflexive]])}
             (!force (Z->NN (neg x + pos y) =
                      (Z->NN neg x) + (Z->NN pos y)))))
    | (neg y) =>
      (!force (Z->NN (neg x + neg y) =
               (Z->NN neg x) + (Z->NN neg y)))
    }
} # datatype-cases additive-homomorphism
} # close module Z-NN
```

9.6 Associativity and commutativity of integer addition

Finally, we are ready to prove the associativity of addition in the signed representation. The key ingredients are the inverse and additive-homomorphism properties of our mappings, and the reader will no doubt be happy to see that the proof is strictly by equational chaining; no further case analysis is required!

```
extend-module Plus {
  define [+ +'] := [NN.+ NN.+']

  conclude associative
    pick-any a:Z b:Z c:Z
      let {f:(OP 1) := Z->NN;
           g:(OP 1) := NN->Z;
           f-application :=
            conclude (f ((a + b) + c) =  f (a + (b + c)))
              (!chain
                [(f ((a + b) + c))
              --> ((f (a + b)) + (f c))       [Z-NN.additive-homomorphism]
```

```
          --> ((f a + f b) + f c)            [Z-NN.additive-homomorphism]
          --> (f a + (f b + f c))            [NN.Plus.associative]
          <-- (f a + (f (b + c)))            [Z-NN.additive-homomorphism]
          <-- (f (a + (b + c)))              [Z-NN.additive-homomorphism]])}
    conclude ((a + b) + c = a + (b + c))
      (!chain [((a + b) + c)
            <-- (g f ((a + b) + c))          [Z-NN.inverse]
            --> (g f (a + (b + c)))          [f-application]
            --> (a + (b + c))                [Z-NN.inverse]])

} # close module Plus
```

Now consider commutativity of integer addition:

```
extend-module Plus {
  define commutative := (forall a b . a + b = b + a)
}
} # close module Z
```

Exercise 9.1: Prove `Z.Plus.commutative`, following the same strategy as with associativity: Define and prove commutativity in the `NN` representation, then map the result to the `Z` representation using the additive homomorphism and inverse properties. □

Exercise 9.2: Prove `Z.Plus.commutative` by working directly in the signed representation (which is more feasible than it would be for associativity, since there are not so many cases to consider). □

9.7 Power series

A formal power series P is an infinite sum of the form

$$P(x) \;=\; \sum_{i\geq 0} p_i x^i = p_0 + p_1 x^1 + p_2 x^2 + p_3 x^3 + \cdots$$

Addition of two such power series P and Q is defined in terms of addition of their corresponding coefficients: If

$$Q(x) \;=\; \sum_{i\geq 0} q_i x^i = q_0 + q_1 x^1 + q_2 x^2 + q_3 x^3 + \cdots$$

then the sum $R = P + Q$ is the power series

$$R(x) \;=\; \sum_{i\geq 0} r_i x^i = r_0 + r_1 x^1 + r_2 x^2 + r_3 x^3 + \cdots$$

where $r_i = p_i + q_i$.

We can regard a power series P as being defined by its coefficients $p_0, p_1, p_2, p_3, \ldots$, so that the monomials x^i are, in effect, just placeholders. By treating such an infinite sequence of coefficients as a function on natural numbers, writing $p(i)$ instead of p_i, we have a representation that we can work with entirely in terms of functions. Thus, to add two power series represented by coefficient functions p and q, we simply form the function r such that $r(i) = p(i) + q(i)$ for all $i \in N$.

Because of the "for all $i \in N$," this formulation doesn't actually give us an algorithm for computing the sum of two power series. On the other hand, since we *can* formally reason about such universal quantifications, we can use this means to precisely define the meaning of sums and other arithmetic operations on power series. Such a nonalgorithmic specification can be quite useful as a standard against which to judge correctness when we consider special cases of power series where we *can* do arithmetic algorithmically—such as polynomials, which can be regarded as power series in which there are only a finite number of nonzero coefficients.

So let us proceed to define a bit of power series arithmetic in terms of this representation. As a technical matter, we introduce an `apply` function, so that we can quantify over functions p while remaining in (many-sorted) first-order logic: Instead of $p(i)$ we write `(apply` p i`)`. We introduce `at` as an infix-friendly shorthand for `apply`, writing `(`p `at` i`)` interchangeably with `(apply` p i`)`. For concreteness, we assume the coefficients of our power series are integers.

```
module ZPS {

domain (Fun N Z)
declare zero: (Fun N Z)
declare apply: [(Fun N Z) N] -> Z
define at := apply

define +' := Z.+
define zero' := Z.zero

define [p q r i] := [?p:(Fun N Z) ?q:(Fun N Z) ?r:(Fun N Z) ?i:N]

assert* equality := (p = q <==> forall i . p at i =  q at i)

assert* zero-definition := (zero at i = zero')
```

Next, we define addition:

```
declare +: [(Fun N Z) (Fun N Z)] -> (Fun N Z)

module Plus {
assert* definition := ((p + q) at i = (p at i) +' (q at i))
```

For our first power series theorems, we have

```
define right-identity := (forall p . p + zero = p)
define left-identity :=  (forall p . zero + p = p)

conclude right-identity
  pick-any p
    let {lemma := pick-any i
                    (!chain
                      [((p + zero) at i)
                     = ((p at i) +' (zero at i))  [definition]
                     = ((p at i) +' zero')        [zero-definition]
                     = (p at i)                   [Z.Plus.Right-Identity]])}
      (!chain-> [lemma ==> (p + zero = p)         [equality]])
} # close module Plus
} # close module ZPS
```

Exercise 9.3: Prove `ZPS.Plus.left-identity`. □

We also leave as exercises the proofs of the following theorems:

```
extend-module ZPS {
  define commutative := (forall p q . p + q = q + p)
  define associative := (forall p q r . (p + q) + r = p + (q + r))
}
```

Exercise 9.4: Prove `ZPS.Plus.commutative`. □

Exercise 9.5: Prove `ZPS.Plus.associative` □

We now define negation and subtraction:

```
extend-module ZPS {

declare negate: [(Fun N Z)] -> (Fun N Z)

module Negate {

  assert* definition := ((negate p) at i = Z.negate (p at i))

} # close module Negate

declare -: [(Fun N Z) (Fun N Z)] -> (Fun N Z)

module Minus {

  assert* definition := (p - q = p + negate q)

} # close module Minus
} # close module ZPS
```

The next theorems show that we have additive inverses:

```
extend-module ZPS {
extend-module Plus {

  define Plus-definition := definition
  open Negate
  open Minus

  define right-inverse := (forall p . p + (negate p) = zero)
  define left-inverse  := (forall p . (negate p) + p = zero)

  conclude right-inverse
    pick-any p
      let {lemma :=
            pick-any i
              (!chain
                [((p + negate p) at i)
               = ((p at i) +' ((negate p) at i))    [Plus-definition]
               = ((p at i) +' Z.negate (p at i))    [Negate.definition]
               = zero'                              [Z.Plus.Right-Inverse]
               = (zero at i)                        [zero-definition]])}
          (!chain-> [lemma ==> (p + negate p = zero) [equality]])
  } # close module Plus
} # close module ZPS
```

9.8 Summary and looking ahead

In this chapter we began to develop the integers as a datatype, or actually as two datatypes, one using a signed representation and the other using a pair of natural numbers. With the signed representation, proofs about the addition operation would be long and tedious to carry out, but they are much simpler with the pair representation. To complete the picture formed by working in both representations, we needed to define mappings back and forth between them and show that the algebraic properties of the arithmetic operators involved are preserved by the mappings. Specifically, we showed that we had defined *additive homomorphisms*—mappings from one domain to another that preserve the meaning of addition. Homomorphic mappings are a fundamental tool for proving algebraic properties, whose value extends far beyond the integer addition example.

A few more properties of integer addition and subtraction are proved in the additional exercises. Of course, there are many other basic properties to consider, especially when multiplication is introduced, but we defer such further development until Chapter 14. There we will bring into play axioms, theorems, and proofs at the more abstract level of algebraic structures, such as monoid, group, ring, and integral domain. Then, among many other potential applications, we will be able to derive integer properties as instances of the more abstract properties.

The topic of the last section, power series with integer coefficients, and its special case of polynomials with integer coefficients, have widespread applications in mathematics and computer science, but the main purpose of our brief treatment is to provide another concrete example of the algebraic structures to be discussed in Chapter 14.

9.9 Additional exercises

Exercise 9.6: Provide a constructive definition of equality on `Z` via an auxiliary relation `eq`:

```
extend-module Z {
declare eq: [Z Z] -> Boolean
```

and then explicitly **`assert`** that two integers are equal iff they are related by `eq`. □

Exercise 9.7: Prove `ZPS.Plus.left-inverse`. □

* ***Exercise 9.8:*** Complete the proof of `Z-ZNN.additive-homomorphism` (Section 9.5) by replacing all uses of **`force`** with actual proofs. □

Exercise 9.9: Prove the following property:

```
extend-module Z {
  extend-module Plus {
    define Left-Inverse := (forall a . (negate a) + a = zero)
  } # close module Plus
} # close module Z
```

In the proof, use the commutativity of addition (introduced as `Z.Plus.commutative` in Exercise 9.1); **`assert`** it if you haven't done the proof. □

Exercise 9.10: Give another proof of `Z.Plus.Left-Inverse`, this time *without* using commutativity of addition. □

Exercise 9.11: Prove the following property:

```
define double-negation := (forall a . negate negate a = a)
```

Put the definition and proof in an extension of module `Z.Plus`. □

Exercise 9.12: Prove the following property:

```
define unique-negation := (forall a b . a + b = zero ==> negate a = b)
```

Put the definition and proof in an extension of module `Z.Plus`. □

Exercise 9.13: Prove the following property:

```
define neg-plus := (forall a b . negate (a + b) = negate a + negate b)
```

Put the definition and proof in an extension of module `Z.Plus`. □

10 Fundamental Discrete Structures

IN this chapter we introduce some fundamental data types and prove a number of useful theorems about them. All of these data types are of central importance in computer science, particularly for the modeling and verification of digital systems. They resurface in key roles in many formal specification and verification languages; they may do so in somewhat different guises, but their fundamental properties remain largely invariant.

A common thread throughout this chapter is the constructive definition of data type operations and the subsequent derivation of more abstract characterization theorems for those operations. Having constructive definitions available, typically formulated with conditional rewrite rules, means that one can *execute* operations on arbitrary inputs and observe the results, which is exceedingly useful for building up intuitions. It also means that various nontrivial conjectures can be mechanically tested with the `falsify` procedure, which is put to use repeatedly in this chapter.

Much of the material we develop here, especially in the section on sets, is standard fare for discrete mathematics courses. The difference is that the present development is entirely machine-readable. Not only are the specifications executable, but more importantly, all of the proofs are machine-checkable as well.

This chapter demonstrates that structured and high-level computerized proofs are often viable even without the benefit of external ATPs. Moreover, the proofs given here are a testament to the pervasiveness and clarity of the tabular proof notation of the `chain` method. Many of the proofs are isomorphic to their English textbook counterparts. For instance, we prove a large number of set-theoretic results that are also derived in the book *Axiomatic Set Theory* by Patrick Suppes [100]. In many cases the Athena proofs given here and those given in the book are virtually identical. As an example, consider the following theorem (where \ denotes the set-difference operation):

$$A \setminus (A \cap B) = A \setminus B.$$

Here is the textbook proof (p. 29, Section 2.3), verbatim:

Theorem 1: $A \setminus (A \cap B) = A \setminus B$

PROOF: Let x be an arbitrary element. Then

$x \in A \setminus (A \cap B)$	$\Leftrightarrow$	$x \in A \ \& \sim (x \in A \cap B)$	by Theorem 31
	$\Leftrightarrow$	$x \in A \ \& \sim (x \in A \ \& \ x \in B)$	by Theorem 12
	$\Leftrightarrow$	$x \in A \ \& \ (x \notin A \vee x \notin B)$	by sentential logic
	$\Leftrightarrow$	$(x \in A \ \& \ x \notin A) \vee (x \in A \ \& \ x \notin B)$	by sentential logic
	$\Leftrightarrow$	$x \in A \ \& \ x \notin B$	by sentential logic
	$\Leftrightarrow$	$x \in A \setminus B$	by Theorem 31 ■

Theorems 31 and 12 are previously derived results in the textbook, respectively:

$$\forall x, A, B \,.\, x \in A \setminus B \Leftrightarrow x \in A \;\&\; x \notin B \tag{10.1}$$

and

$$\forall x, A, B \,.\, x \in A \cap B \Leftrightarrow x \in A \;\&\; x \in B. \tag{10.2}$$

These are essentially characterization theorems for set difference and intersection, respectively. And here is the corresponding Athena proof:

```
pick-any A B
  (!set-identity-intro-direct
    pick-any x
      (!chain [(x in A \ (A /\ B))
          <==> (x in A & ~ x in A /\ B)                  [DC]
          <==> (x in A & ~ (x in A & x in B))            [IC]
          <==> (x in A & (~ x in A | ~ x in B))          [prop-taut]
          <==> ((x in A & ~ x in A) | (x in A & ~ x in B)) [prop-taut]
          <==> (x in A & ~ x in B)                       [prop-taut]
          <==> (x in A \ B)                              [DC]]))
```

Here, DC and IC are Athena formulations of the characterization theorems for difference and intersection, namely (10.1) and (10.2) respectively, also previously derived, which are written as:

```
conclude DC := (forall A B x . x in A \ B <==> x in A & ~ x in B)
  ...

conclude IC := (forall A B x . x in A /\ B <==> x in A & x in B)
  ...
```

Modulo trivial notational differences, the two proofs are virtually identical in structure and level of detail.

Of course, not all proofs can be written largely as chains, and even when that is possible, the results do not always attain this level of clarity and simplicity. Nevertheless, both equational and implicational chains are widely applicable, and they can greatly clarify the reasoning behind a proof.

All sections in this chapter follow the same general pattern: They start with some brief remarks on the general ideas behind the data type at hand and with a subsection on representation and notation, which defines the data type as an inductive structure and lays down notational conventions such as precedence levels for constructors, input expansions and output transformations, and the like. This is followed by additional subsections that introduce the most important operations of the data type and derive various results about them. Exercises are sprinkled throughout. Most of them have complete solutions included in the solutions appendix available on the book's web site.

All of these theories are developed inside their own modules, but the modules for pairs and options are open at the top level by default, as their contents are both widely used and relatively meager (and thus do not pollute the global namespace to any significant degree).

10.1 Ordered pairs

10.1.1 Representation and notation

Intuitively, an ordered pair of two objects a and b, often written in conventional notation as (a,b), is a pair of a and b in which order matters: a is the *first* element of the pair and b is the second. It follows that, if $a \neq b$, then (a,b) and (b,a) are two distinct ordered pairs, even though they contain the same elements. Indeed, the fundamental property of ordered pairs is that (a_1,b_1) is the same pair as (a_2,b_2) iff $a_1 = a_2$ and $b_1 = b_2$. We will see that when we formalize ordered pairs as an algebraic datatype, this property follows directly from the datatype's axioms.

We introduce ordered pairs with a polymorphic datatype `Pair`, which has only one constructor, the binary symbol `pair`. We also declare two selectors,[1] one for retrieving the left and another for retrieving the right element of an ordered pair:

```
datatype (Pair S T) := (pair pair-left:S pair-right:T)
```

All of the code of this section is introduced inside a predefined module `Pair` that is loaded when Athena first starts:

```
module Pair {

      ⋮

}
```

Because ordered pairs are so widely used, this module is by default opened at the top level.

We introduce the symbol `@` as an alias for the constructor `pair`, so we can write pairs (a,b) in infix notation as `(`a `@` b`)`. We want the pair constructor to bind quite tightly, so we give it a high precedence level:

```
set-precedence pair 310

define @ := pair
```

Occasionally we also want to write ordered pairs using square-bracket list notation, writing `[1 2]` for `(pair 1 2)`, for example. Conversely, we will want to transform `(pair` s t`)`

1 See Section A.5 for a general discussion of selectors.

terms into two-element Athena lists of the form [s t]. The procedures lst->pair and pair->lst below let us do both. They are defined in terms of two more general ternary procedures that also allow us to specify two preprocessing procedures to be respectively applied to the two list (or pair) elements before the pair (respectively, list) is produced. Both lst->pair and pair->lst use the identity procedure for both positions, but we will see examples later (e.g., in Chapter 18) where less trivial procedures are given as values for these arguments.

```
define (lst->pair-general x pre-left pre-right) :=
  match x {
    [v1 v2] => (pair (pre-left v1) (pre-right v2))
  | _ => x
  }

define (lst->pair x) :=  (lst->pair-general x id id)

define (pair->lst-general x pre-left pre-right) :=
  match x {
    (pair v1 v2) => [(pre-left v1) (pre-right v2)]
  | _ => x
  }

define (pair->lst x) := (pair->lst-general x id id)
```

Finally, we define some variable abbreviations and assert the relevant datatype and selector axioms:

```
define [x y z a b w p p1 p2] :=
       [?x ?y ?z ?a ?b ?w ?p:(Pair 'S 'T) ?p1:(Pair 'S1 'T1)
        ?p2:(Pair 'S2 'T2)]

assert pair-axioms :=
  (datatype-axioms "Pair" joined-with selector-axioms "Pair")
```

10.1.2 Results and methods

We first check to make sure that the aforementioned fundamental property of ordered pairs is an element of pair-axioms:

```
define fp := (forall x y z w . x @ y = z @ w <==> x = z & y = w)

> (member? fp pair-axioms)

Term: true
```

Let us now prove a few theorems:

```
conclude pair-theorem-1 :=
     (forall p . p = (pair-left p) @ (pair-right p))
  datatype-cases pair-theorem-1 {
    (p as (pair x y)) =>
      (!chain [p = ((pair-left p) @ (pair-right p))  [pair-axioms]])
  }

conclude pair-theorem-2 :=
    (forall x y z w . x @ y = z @ w <==> y @ x = w @ z)
  pick-any x y z w
    (!chain [(x @ y = z @ w) <==> (x = z & y = w)     [pair-axioms]
                             <==> (y = w & x = z)     [comm]
                             <==> (y @ x = w @ z)     [pair-axioms]])
```

We introduce a swap function that swaps the elements of an ordered pair:

```
declare swap:  (S, T) [(Pair S T)] -> (Pair T S)

assert* swap-def := [(swap x @ y = y @ x)]

> (eval swap swap 1 @ 'a)

Term: (pair 1 'a)

> (eval swap swap swap 1 @ 'a)

Term: (pair 'a 1)

conclude swap-theorem-1 :=
  (forall x y . swap swap x @ y = x @ y)
pick-any x y
  (!chain [(swap swap x @ y) = (swap y @ x) [swap-def]
                             = (x @ y)      [swap-def]])
```

What if we wanted to state the theorem by quantifying directly over ordered pairs? In other words, consider the following formulation of this result:

```
(forall p . swap swap p = p).
```

How would we go about proving this version? We could proceed by considering an arbitrary pair p, but then we would need to somehow get a hold of the left and right elements of p and work with them directly, as required by the defining axioms of swap:

```
conclude swap-theorem-1b := (forall p . swap swap p = p)
 pick-any p
  let {E := (!chain-> [true ==> (exists x y . p = x @ y) [pair-axioms]])}
    pick-witnesses x y for E  # we now have (p = x @ y)
      (!chain-> [(swap swap x @ y)
                = (swap y @ x)        [swap-def]
```

```
                 = (x @ y)              [swap-def]
             ==> (swap swap p = p)  [(p = x @ y)]])
```

This idiom is so frequent that it is well worth abstracting into a method that hides the details involving existential quantifiers and their manipulation. In particular, oftentimes we need to prove a result of the form

$$\forall p : (\texttt{Pair}\ S\ T)\ .\ \cdots p \cdots \tag{10.3}$$

that is universally quantified over all pairs p, where p ranges over (Pair S T) for some S and T. While we could proceed as above, the steps involving the existential quantification and its unpacking would be tedious. It is better—and more customary—to pick an arbitrary x of sort S and an arbitrary y of sort T and prove instead the result for the term (x @ y), deriving the following variant of (10.3):

$$\forall x{:}S\ y{:}T\ .\ \cdots (x\ @\ y) \cdots \tag{10.4}$$

Once we have (10.4), we can readily obtain (10.3) with a standard bit of reasoning similar to that used in the proof of swap-theorem-1b.

That reasoning is encapsulated in a unary method pair-converter, defined inside the Pair module, that takes a premise of the form (10.4) and derives a conclusion of the form (10.3), where the body $\cdots p \cdots$ of (10.3) is obtained from the body of the premise, $\cdots (x\ @\ y) \cdots$, by replacing every occurrence of the term (x @ y) by p.[2] With this method at hand we could have proven swap-theorem-1b in a more natural style as follows:

```
conclude swap-theorem-1b := (forall p . swap swap p = p)
  (!pair-converter
     pick-any x y
       (!chain [(swap swap x @ y) = (swap y @ x) [swap-def]
                                  = (x @ y)      [swap-def]]))
```

Indeed, if swap-theorem-1 is already in place, we can get swap-theorem-1b in one line:

```
> conclude swap-theorem-1b := (forall p . swap swap p = p)
    (!pair-converter swap-theorem-1)

Theorem: (forall ?p:(Pair 'S 'T)
           (= (swap (swap ?p:(Pair 'S 'T)))
              ?p:(Pair 'S 'T)))
```

Exercise 10.1: Prove the sentence:

```
(forall p . swap p = (pair-right p) @ (pair-left p))
```

by using pair-converter. □

2 Some caution is needed: If the body of (10.4) contains free occurrences of x or y that are *not* inside a term (x @ y), then these occurrences will not be replaced by anything, and the result will most likely be meaningless, if harmless.

The conversion method works in the opposite direction as well. For example, given a premise of the form (10.3), it will derive a result of the form (10.4). However, the other direction is more common in practice. The method also handles existential quantifiers, so it will go from a premise of the form

$$\exists p : (\texttt{Pair}\ S\ T)\ .\ \cdots p \cdots$$

to a conclusion of the form

$$\exists x{:}S\ y{:}T\ .\ \cdots (x\ \texttt{@}\ y) \cdots$$

and conversely.

10.2 Options

10.2.1 Representation and notation

Option values are generated by the following polymorphic datatype:

```
datatype (Option S) := NONE | (SOME option-val:S)
```

Intuitively, this definition says that a value of sort (Option *S*) is either NONE or of the form (SOME *v*) for some value *v* of sort *S*. In the latter case, applying the selector function option-val to (SOME *v*) retrieves the value *v*. The symbol SOME is given low precedence (110), since we want its scope to extend as far to the right as possible, as illustrated below.

```
> (SOME 'foo :: nil)

Term: (SOME (:: 'foo
                nil:(List Ide)))

> (?x = SOME 3.14)

Term: (= ?x:(Option Real)
         (SOME 3.14))

> NONE

Term: NONE:(Option 'T169)

> (eval option-val SOME 99 - 1)

Term: 98
```

Options are useful in a number of contexts, most notably for error handling. The intuition here is this: If a function *f* that is normally supposed to produce a value of sort *S* generates an error, it can instead return NONE as its output; that will explicitly indicate the occurrence of an error. If, on the other hand, all goes well and f successfully computes a result v of sort S, it can return the output (SOME v), indicating that the computation succeeded and resulted in v. Hence, to make f handle errors explicitly, we modify its output sort by lifting it from S to (Option S).

It is important to keep in mind the datatype axioms for Option:

1. NONE is distinct from any value of the form (SOME v);
2. (SOME v_1) is identical to (SOME v_2) iff v_1 is identical to v_2; and
3. every option value is either equal to NONE or to a value of the form (SOME v).

For ease of reference, we give the name opt-axioms to the list of these axioms:

```
assert opt-axioms := (datatype-axioms "Option")
```

10.2.2 Some useful results

We conclude this section with some useful option lemmas. These are derived inside a module named Options when Athena first starts, but that module too is subsequently opened at the top level. We prove a few of these and leave the rest as exercises.

```
define [x y z] := [?x ?y ?z]

conclude option-lemma-1 := (forall x y . x = SOME y ==> x =/= NONE)
  pick-any x y
    assume hyp := (x = SOME y)
      (!chain-> [true ==> (NONE =/= SOME y)  [opt-axioms]
                      ==> (NONE =/= x)       [hyp]
                      ==> (x =/= NONE)       [sym]])

conclude option-lemma-2 :=
          (forall x . x =/= NONE ==> exists y . x = SOME y)
  pick-any x
    assume hyp := (x =/= NONE)
      (!chain->
        [true ==> (x = NONE | exists y . x = SOME y) [opt-axioms]
              ==> (exists y . x = SOME y)            [(dsyl with hyp)]])

define option-lemma-2-conv :=
  (forall x . (forall y . x =/= SOME y) ==> x = NONE)

define option-lemma-3 :=
  (forall x . x = NONE ==> ~ exists y . x = SOME y)

define option-lemma-4 :=
   (forall x y . x = NONE ==> x =/= SOME y)
```

```
conclude option-lemma-5 :=
  (forall x y z . x = SOME y & y =/= z ==> x =/= SOME z)
 pick-any x y z
   assume h := (x = SOME y & y =/= z)
     (!chain-> [h ==> (y =/= z)              [right-and]
                  ==> (SOME y =/= SOME z) [opt-axioms]
                  ==> (x =/= SOME z)      [h]])

define opt-lemmas :=
  [option-lemma-1 option-lemma-2 option-lemma-2-conv
   option-lemma-3 option-lemma-4 option-lemma-5]

define option-results := (join opt-axioms opt-lemmas)
```

Exercise 10.2: Supply the missing proofs. □

10.3 Sets, relations, and functions

10.3.1 Representation and notation

Finite sets are introduced through a polymorphic structure (Set S) that is inductively generated from two constructors: a nullary constructor null, representing the empty set of sort S, and a binary constructor insert, which adds an element of sort S into a set of sort (Set S).

```
structure (Set S) :=  null | (insert S (Set S))
```

This structure definition, and all subsequent code and proofs in this section, can be found in an Athena module named Set, which is part of the Athena library. Unlike Options and Pair, this module is not open by default. The rest of this section is written as if we were developing the contents of Set from scratch at the top level.

Observe that this definition appears to be structurally identical to that of finite *lists*:

```
datatype (List S) :=  nil | (:: S (List S))
```

with the roles of nil and :: played by null and insert, respectively. There is one crucial difference, however: Lists are freely generated (hence the **datatype** keyword), whereas sets are only inductively generated (hence the **structure** keyword). Free generation means that two lists are equal iff they are identical as terms. Thus, (:: 1 (:: 2 nil)) and

(:: 2 (:: 1 nil))

represent different lists. This is reflected in the datatype-axioms of List, one of which is:

```
(forall h1 t1 h2 t2 . h1 :: t1 = h2 :: t2 ==> h1 = h2 & t1 = t2)
```

and whose contrapositive is:

```
(forall h1 t1 h2 t2 . h1 =/= h2 | t1 =/= t2 ==> h1 :: t1 =/= h2 :: t2)
```

By contrast, the analogous axiom for sets does not hold. Indeed, we will have to ensure that `(insert 1 (insert 2 null))` and

`(insert 2 (insert 1 null))`

are identical in order to respect the *extensionality principle*, which states that two sets are equal iff they contain the same elements.

It will be convenient to be able to transform square-bracket-enclosed Athena lists into sets and vice versa, so we can write, for instance, `[1 2 3]` as a shorthand for the set

`(1 insert 2 insert 3 insert null).`

The `Set` module introduces two such procedures, `alist->set` and `set->alist`:

```
1  define (alist->set L) :=
2    match L {
3      [] => null
4    | (list-of x rest) => (insert (alist->set x) (alist->set rest))
5    | _ => L
6    }
```

Recursing on both the head and the tail of the list allows us to convert nested lists into nested sets:

```
> (?x = alist->set [[1 2]])

Term: (= ?x:(Set (Set Int))
         (insert (insert 1
                         (insert 2
                                 null:(Set Int)))
                 null:(Set (Set Int))))
```

We define a converse procedure as well, to go from finite sets to Athena lists, which also removes duplicates from the final output list:

```
define (set->alist-aux s) :=
  match s {
    null => []
  | (insert x rest) => (add (set->alist-aux x) (set->alist-aux rest))
  | _ => s
  }

define (set->alist s) :=
  match (set->alist-aux s) {
```

```
    (some-list L) => (dedup L)
  | _ => s
  }

> (set->alist 1 insert 2 insert 1 insert null)

List: [1 2]
```

We expand the input space of insert so that it can accept Athena lists as second arguments:

```
expand-input insert [id alist->set]
```

And in the interest of brevity and readability, we define ++ as an alias for insert:

```
define ++ := insert
```

We can now write sets in a succinct and fairly natural notation:

```
> (1 ++ [2 3])

Term: (insert 1
              (insert 2
                      (insert 3
                              null:(Set Int))))
```

We also set the precedence of ++ to be relatively high, and define some abbreviations for term variables: x, y, z and h, h' as polymorphic variables of completely general sorts, and the rest specifically as (polymorphic) set variables:

```
set-precedence ++ 210

define [x y z h h' s s' t t' s1 s2 s3 A B C D E] :=
       [?x ?y ?z ?h ?h' ?s:(Set 'T1) ··· ?t:(Set 'T3) ··· ?A:(Set 'T8) ···]
```

10.3.2 Set membership, the subset relation, and set identity

The first relation we introduce is the most fundamental one: the set membership relation. We expand its input space so that it can take square-bracket-enclosed lists as second arguments (which will be converted to set terms via alist->set). And we define the relation recursively, much in the same way that a similar relation for list membership would be defined:

```
declare in: (S) [S (Set S)] -> Boolean [[id alist->set]]

assert* in-def :=
  [(~ _ in null)
   (x in h ++ t <==> x = h | x in t)]
```

The first axiom simply says that nothing is a member of the empty set (null), or more precisely, that everything is not a member of null. The second axiom says that an element x is a member of a set of the form (h ++ t) iff either x is equal to h or else x is a member of t. Let us use eval to confirm that these axioms have the intended effect:

```
> (eval 23 in [1 5 23 98])

Term: true

> (eval 23 in [1 5 98])

Term: false
```

The following two lemmas are simple but handy:

```
conclude null-characterization := (forall x . x in [] <==> false)
  pick-any x
    (!equiv
      assume hyp := (x in [])
        (!absurd hyp (!chain-> [true ==> (~ x in []) [in-def]]))
      assume false
        (!from-false (x in [])))

conclude in-lemma-1 := (forall x A . x in x ++ A)
  pick-any x A
    (!chain-> [(x = x) ==> (x in x ++ A) [in-def]])
```

We often use upper-case shorthands for characterization theorems such as the above:

```
define NC := null-characterization
```

We call this a "characterization" result for null, by the way, because it provides necessary and sufficient membership conditions for null. We will derive several other characterization results in what follows.

We continue with another fundamental relation, built on top of the membership relation: the *subset* relation, customarily represented by the symbol $\subseteq$. A set A is a subset of a set B iff every element of A is also an element of B. In symbols:

$$\forall A, B \,.\, A \subseteq B \Leftrightarrow (\forall x \,.\, x \in A \Rightarrow x \in B). \tag{10.5}$$

(As in other parts of the book, we occasionally use conventional mathematical notation, typeset in LaTeX, for increased readability. The correspondence with Athena sentences is immediate if one knows how to interpret the primitive relation symbols; in this case the symbol $\in$ corresponds to in and $\subseteq$ to subset.)

We could define subset directly by asserting (10.5). However, such a definition would have a drawback: It would not be constructive, due to the unrestricted universal quantifier that occurs on the right-hand side of the biconditional. And without a constructive definition, typically given by recursion, evaluation will not work; for example, `eval` will not be able to reduce a term like `([1] subset [2 1 3])` to `true`. We want `eval` to be able to handle such terms, not only in order to test our axioms, but also in order to test conjectures. For that reason, we will instead give a constructive definition of `subset`. Once we have such a definition, we will then *derive* (10.5) from it.

This is a general strategy that we will pursue with subsequent function and relation symbols in this chapter (and beyond), as we already have done in previous chapters. For instance, instead of defining set-theoretic union by directly asserting that an element x belongs to the union of A and B iff it belongs to A or it belongs to B, we instead give a recursive definition of union and then inductively derive the said property from it.[3] In general, when there is a choice between a constructive definition D_1 and a nonconstructive definition D_2, we will typically prefer D_1 and then proceed to derive D_2 from D_1, usually by induction. In this case the required recursive definition can be given as follows:

```
declare subset: (S) [(Set S) (Set S)] -> Boolean [[alist->set alist->set]]

assert* subset-def :=
  [([] subset _)
   (h ++ t subset A <==> h in A & t subset A)]
```

We can use `eval` to confirm that the definition works as expected:

```
> (eval [1 2] subset [3 2 4 1 5])

Term: true

> (eval [1 2] subset [3 2])

Term: false
```

We now go on to derive the nonconstructive characterization of the subset relation from this definition. This is done in two steps, beginning with the left-to-right direction stating that if A is a subset of B (according to `subset-def`) then for all x, if x belongs to A then x also belongs to B. Like most other characterization results in this section, it is proved by structural induction:

```
1  define subset-characterization-1 :=
2    by-induction (forall A B . A subset B ==> forall x . x in A ==> x in B) {
3      null => pick-any B
4                assume (null subset B)
```

3 Of course, this is possible only because we are dealing exclusively with finite sets here. No algorithm (at least in the conventional Church-Turing sense of the term) could compute the union of two arbitrary—possibly uncountably—infinite sets.

```
5          pick-any x
6            (!chain [(x in null) ==> false               [NC]
7                                 ==> (x in B)            [prop-taut]])
8   | (A as (insert h t)) =>
9       pick-any B
10        assume hyp := (A subset B)
11          let {IH := (forall B . t subset B ==>
12                           forall x . x in t ==> x in B);
13               _ := (!chain-> [hyp ==> (t subset B) [subset-def]])}
14            pick-any x
15              assume (x in A)
16                (!cases (!chain<- [(x = h | x in t)
17                                   <== (x in A)          [in-def]])
18                  assume (x = h)
19                    (!chain-> [hyp ==>  (h in B)         [subset-def]
20                                   ==>  (x in B)         [(x = h)]])
21                  (!chain [(x in t) ==> (x in B)         [IH]]))
22  }
```

Note that the inference on line 14 is necessary for the last `chain` application (on line 21). The converse direction is likewise proved by structural induction. With both directions available, the classical (nonconstructive) subset characterization is easy to derive.

```
define subset-characterization-2 :=
  (forall A B . (forall x . x in A ==> x in B) ==> A subset B)

define subset-characterization :=
  (forall s1 s2 . s1 subset s2 <==> forall x . x in s1 ==> x in s2)
```

Exercise 10.3: Derive the two preceding sentences. □

We define `SC` as a handy abbreviation for `subset-characterization`:

```
define SC := subset-characterization
```

It will be very useful to have an introduction method for goals of the form (*s* `subset` *t*), based on `subset-characterization`. The method will take a premise of the form

(`forall x . x in` *s* `==> x in` *t*)

and will use `SC` to derive the desired goal (*s* `subset` *t*):

```
define subset-intro :=
  method (p)
    match p {
      (forall x ((x in A) ==> (x in B))) =>
```

```
      (!chain-> [p ==> (A subset B) [SC]])
  }
```

Recall that when an inductively generated structure is not free, we need to define the identity relation on that structure; that is, we need to state exactly when two terms of that structure are to count as identical. In this case, the identity relation on sets is defined as should be expected: Two sets are equal iff each is a subset of the other.

```
assert* set-identity := (A = B <==> A subset B & B subset A)
```

Recall also that when a structural identity axiom of this kind is asserted, Athena automatically adjusts `eval` to ensure that it adheres to the new definition. Thus, we now have:

```
> (eval 1 ++ 2 ++ [] = 2 ++ 1 ++ [])

Term: true
```

If we did not have a constructive definition of `subset`, we would also not have a constructive set identity, and the `eval` call above would fail.

An alternative and nonconstructive characterization of set identity is the following:

```
define set-identity-characterization :=
  (forall A B . A = B <==> forall x . x in A <==> x in B)

define SIC := set-identity-characterization
```

In our case we can readily derive `SIC` from `set-identity` and `SC`; the proof is left as an exercise.

Exercise 10.4: Prove `set-identity-characterization`. □

Similar to `subset-intro`, we define a method `set-identity-intro` that takes two premises of the form (s `subset` t) and (t `subset` s) and derives (s = t):

```
define set-identity-intro :=
  method (p1 p2)
    match [p1 p2] {
      [(A subset B) (B subset A)] =>
        (!chain-> [p1 ==> (p1 & p2) [augment]
                      ==> (A = B)   [set-identity]])
    }
```

A more direct introduction method, based on `SIC`, is also useful. This one takes a premise of the form

$$\forall x \,.\, x \in A \Leftrightarrow x \in B$$

and derives the conclusion $A = B$:

```
define set-identity-intro-direct :=
  method (premise)
    match premise {
      (forall x ((x in A) <==> (x in B))) =>
        (!chain-> [premise ==> (A = B) [SIC]])
    }
```

Using SIC we can also derive a couple of handy alternative characterizations of the empty set:

```
conclude NC-2 := (forall A . A = null <==> forall x . ~ x in A)
pick-any A
  (!chain [(A = null)
      <==> (forall x . x in A <==> x in null)   [SIC]
      <==> (forall x . x in A <==> false)       [NC]
      <==> (forall x . ~ x in A)                [prop-taut]])

define NC-3 := (forall A . A =/= null <==> exists x . x in A)
```

The proof of NC-3 is left as an exercise. The following will be a convenient abbreviation, defined as a procedure:

```
define (non-empty s) := (s =/= null)
```

Some standard results about the subset relation can be readily derived now: reflexivity, antisymmetry, and transitivity. We prove transitivity here and leave the other two as exercises.

```
define subset-reflexivity := (forall A . A subset A)

define subset-antisymmetry :=
  (forall A B . A subset B & B subset A ==> A = B)

conclude subset-transitivity :=
  (forall A B C . A subset B & B subset C ==> A subset C)
    pick-any A B C
      assume (A subset B & B subset C)
        (!subset-intro
          pick-any x
            (!chain [(x in A)
                 ==> (x in B)   [SC]
                 ==> (x in C)   [SC]]))
```

Exercise 10.5: Supply the missing proofs. □

The *proper subset* relation is defined as follows:

```
declare proper-subset: (S) [(Set S) (Set S)] -> Boolean
                           [[alist->set alist->set]]

assert* proper-subset-def :=
  [(s1 proper-subset s2 <==> s1 subset s2 & s1 =/= s2)]

> (eval [1 2] proper-subset [2 3 1])

Term: true
```

A characterization theorem and some useful lemmas follow; their proofs are left as exercises.

```
define proper-subset-characterization :=
  (forall s1 s2 . s1 proper-subset s2 <==>
                  s1 subset s2 & exists x . x in s2 & ~ x in s1)

define PSC := proper-subset-characterization

define proper-subset-lemma :=
  (forall A B x . A subset B & x in B & ~ x in A ==> A proper-subset B)

define in-lemma-2 := (forall h t . h in t ==> h ++ t = t)

define in-lemma-3 := (forall x h t . x in t ==> x in h ++ t)

define in-lemma-4 := (forall A x y . x in A ==> y in A <==> y = x | y in A)
```

Exercise 10.6: Supply the missing proofs. □

We end this section by introducing an operation for forming singletons:

```
declare singleton: (S) [S] -> (Set S)

assert* singleton-def := [(singleton x = x ++ null)]

conclude singleton-characterization :=
  (forall x y . x in singleton y <==> x = y)
 pick-any x y
  (!chain [(x in singleton y)
       <==> (x in y ++ null)       [singleton-def]
       <==> (x = y | x in null)    [in-def]
       <==> (x = y | false)        [NC]
       <==> (x = y)                [prop-taut]])
```

10.3.3 Set operations

Next, we introduce three important binary set operations: union, intersection, and difference:

```
declare union, intersection, diff:
   (S) [(Set S) (Set S)] -> (Set S) [120 [alist->set alist->set]]

define [\/ /\ \] := [union intersection diff]
```

As we did with subset, we will define these operations constructively and then deductively derive their classical nonconstructive characterizations. We begin with union:

```
assert* union-def :=
 [([] \/ s = s)
  (h ++ t \/ s = h ++ (t \/ s))]
```

We transform the output of eval so that sets are printed out in square-bracket notation:

```
transform-output eval [set->alist]
```

We use eval to test the definition:

```
> (eval [1 2 3] \/ [4 5 6 3])

List: [1 2 3 4 5 6]
```

Note that the strong typing of these operations means that we cannot, for instance, form the union of a set of integers and a set of identifiers. So while something like

```
(exists x . x in [1 2 3] \/ [4 5])
```

is perfectly sensible, the apparently similar

```
(exists x . x in [1 2 3] \/ ['a 'b])
```

leads to an error, because the term `([1 2 3] \/ ['a 'b])` is ill-sorted. For those familiar with typed functional languages, this is similar to the strong typing of lists in languages such as ML and Haskell, which prohibits, among other things, the concatenation of a list of integers and a list of characters. The workaround behind such restrictions is also similar: If we are interested in sets containing two different sorts of things S and T, say numbers and meta-identifiers, we can introduce a new algebraic datatype U representing the (explicitly tagged) union of S and T and proceed to work directly with sets of sort U instead. Using this approach, for instance, we could represent the above set as follows:

```
datatype U := (int Int) | (ide Ide)

(exists x . x = [(int 1) (int 2) (int 3)] \/ [(ide 'a) (ide 'b)])
```

With some input expansions automating the injection of the sort tags, we could write out terms of this augmented sort quite naturally.

The classical characterization of union is:

```
(forall A B x . x in A \/ B <==> x in A | x in B).
```
(10.6)

We derive (10.6) in two steps, one for each direction of the biconditional:

```
conclude union-characterization-1 :=
  (forall A B x . x in A \/ B ==> x in A | x in B)
 by-induction union-characterization-1 {
   null => pick-any B x
            (!chain [(x in null \/ B)
                 ==> (x in B)                [union-def]
                 ==> (x in null | x in B)    [alternate]])
 | (A as (h insert t)) =>
     let {IH := (forall B x . x in t \/ B ==> x in t | x in B)}
       pick-any B x
         (!chain [(x in A \/ B)
               ==> (x in h ++ (t \/ B))        [union-def]
               ==> (x = h | x in t \/ B)       [in-def]
               ==> (x = h | x in t | x in B)   [IH]
               ==> ((x = h | x in t) | x in B) [prop-taut]
               ==> (x in A | x in B)           [in-def]])
  }
```

The converse, and the ultimate derivation of the nonconstructive union characterization, are left as exercises.

```
define union-characterization-2 :=
 (forall A B x . x in A | x in B ==> x in A \/ B)

define union-characterization  :=
  (forall A B x . x in A \/ B <==> x in A | x in B)

define UC := union-characterization
```

Exercise 10.7: Derive the preceding two sentences. □

We continue with intersection:

```
assert* intersection-def :=
 [(null /\ s = null)
  (h ++ t /\ A = h ++ (t /\ A) <== h in A)
  (h ++ t /\ A = t /\ A <== ~ h in A)]

> (eval [1 2 1] /\ [5 1 3])

List: [1]
```

Once again, we derive the classical characterization of the operation,

```
(forall A B x . x in A /\ B <==> x in A & x in B),
```
(10.7)

in two steps:

```
conclude intersection-characterization-1 :=
  (forall A B x . x in A /\ B ==> x in A & x in B)
by-induction intersection-characterization-1 {
  null =>
    pick-any B x
      (!chain [(x in null /\ B)
           ==> (x in null)                              [intersection-def]
           ==> false                                    [NC]
           ==> (x in null & x in B)                     [prop-taut]])
| (A as (insert h t)) =>
  let {IH := (forall B x . x in t /\ B ==> x in t & x in B)}
    pick-any B x
      (!two-cases
        assume (h in B)
          (!chain
            [(x in (h ++ t) /\ B)
         ==> (x in h ++ (t /\ B))                       [intersection-def]
         ==> (x = h | x in t /\ B)                      [in-def]
         ==> (x = h | x in t & x in B)                  [IH]
         ==> ((x = h | x in t) & (x = h | x in B)) [prop-taut]
         ==> (x in A & (x in B | x in B))               [in-def (h in B)]
         ==> (x in A & x in B)                          [prop-taut]])
        assume (~ h in B)
          (!chain [(x in A /\ B)
               ==> (x in t /\ B)                        [intersection-def]
               ==> (x in t & x in B)                    [IH]
               ==> (x in A & x in B)                    [in-def]]))
    }
```

As usual, the two-cases analysis in the inductive step is dictated by the corresponding case split in intersection-def. The same goes for the converse, which is left as an exercise. The classical characterization of intersection, which we will abbreviate as IC, is then readily derived from the two directions.

```
define intersection-characterization-2 :=
 (forall A B x . x in A & x in B ==> x in A /\ B)

define intersection-characterization :=
 (forall A B x . x in A /\ B <==> x in A & x in B)

define IC := intersection-characterization
```

Exercise 10.8: Derive intersection-characterization-2 and IC. □

The following is an immediate corollary of `IC`:

```
conclude intersection-subset-theorem :=
  (forall A B . A /\ B subset A)
pick-any A B
  (!subset-intro
    pick-any x
      (!chain [(x in A /\ B)
           ==> (x in A)        [IC]]))
```

Finally, we go through the same process for the set-theoretic difference operation:

```
assert* diff-def :=
 [(null \ _ = null)
  (h ++ t \ A = t \ A <== h in A)
  (h ++ t \ A = h ++ (t \ A) <== ~ h in A)]

> (eval [1 2 3] \ [3 1])

List: [2]
```

We derive the following:

```
define diff-characterization-1 :=
  (forall A B x . x in A \ B ==> x in A & ~ x in B)

define diff-characterization-2 :=
  (forall A B x . x in A & ~ x in B ==> x in A \ B)

conclude diff-characterization :=
  (forall A B x . x in A \ B <==> x in A & ~ x in B)
  pick-any A B x
    (!equiv
      (!chain [(x in A \ B)
           ==> (x in A & ~ x in B) [diff-characterization-1]])
      (!chain [(x in A & ~ x in B)
           ==> (x in A \ B)          [diff-characterization-2]]))

define DC := diff-characterization
```

As we did with union and intersection, we break the proof into two independent lemmas, one for each direction of the biconditional.

Exercise 10.9: Prove `diff-characterization-1` and `diff-characterization-2`. □

We also introduce an operation that removes an element from a set, using - as an alias for it:

```
declare remove: (S) [(Set S) S] -> (Set S) [- [alist->set id]]

assert* remove-def :=
  [(null - _ = null)
   (h ++ t - x = t - x <== x = h)
   (h ++ t - x = h ++ (t - x) <== x =/= h)]

(eval [1 2 3 2 5] - 2)

> List: [1 3 5]
```

Deriving the corresponding characterization theorem for this operation is left as an exercise:

```
define remove-characterization :=
  (forall A x y . y in A - x <==> y in A & y =/= x)

define RC := remove-characterization
```

Some simple but useful corollaries of `RC` follow; the proofs are left as exercises.

```
define remove-corollary :=   (forall A x   . ~ x in A - x)
define remove-corollary-2 := (forall A x   . ~ x in A ==> A - x = A)
define remove-corollary-3 := (forall A x y . x in A & y =/= x ==> x in A - y)
define remove-corollary-4 := (forall A x y  . ~ x in A ==> ~ x in A - y)
define remove-corollary-5 :=
  (forall A B x . A subset B & ~ x in A ==> A subset B - x)
```

Exercise 10.10: Supply the missing proofs. □

We have now derived characterization theorems for most of the fundamental set-theoretic functions and relations: union, intersection, set difference, the subset and identity relations, and removal. We can proceed to prove some important results, starting with the commutativity and associativity of union and intersection:

```
conclude intersection-commutes := (forall A B . A /\ B = B /\ A)
 pick-any A B
   (!set-identity-intro-direct
     pick-any x
       (!chain [(x in A /\ B) <==> (x in A & x in B)  [IC]
                              <==> (x in B & x in A)  [prop-taut]
                              <==> (x in B /\ A)      [IC]]))
```

The proof of union commutativity is similar.

```
define union-commutes := (forall A B . A \/ B = B \/ A)
```

Using the direct introduction method for set identity, as above, tends to result in readable proofs because the method requires a biconditional, and these can often be derived neatly with equivalence chains. By contrast, if we were to use the binary introduction method that requires two `subset` premises, `set-identity-intro`, the proof would need to derive the two `subset` directions separately:

```
conclude intersection-commutes := (forall A B . A /\ B = B /\ A)
 pick-any A B
   let {A-subset-of-B :=
          (!subset-intro
             pick-any x
               (!chain [(x in A /\ B)
                    ==> (x in A & x in B) [IC]
                    ==> (x in B & x in A) [prop-taut]
                    ==> (x in B /\ A)     [IC]]));
        B-subset-of-A :=
         (!subset-intro
            pick-any x
              (!chain [(x in B /\ A)
                   ==> (x in B & x in A) [IC]
                   ==> (x in A & x in B) [prop-taut]
                   ==> (x in A /\ B)     [IC]]))
        }
      (!set-identity-intro A-subset-of-B B-subset-of-A)
```

Even with this approach, however, we could use method abstraction to shorten the proofs considerably. In particular, observe that the reasoning behind the second lemma in the above proof is the same as that of the first lemma. We could abstract this reasoning into a common method which would then be used to derive both lemmas as follows:

```
conclude intersection-commutes := (forall A B . A /\ B = B /\ A)
  let {M := method (A B) # derive (A /\ B subset B /\ A)
              (!subset-intro
                 pick-any x
                   (!chain [(x in A /\ B)
                        ==> (x in A & x in B) [IC]
                        ==> (x in B & x in A) [prop-taut]
                        ==> (x in B /\ A)     [IC]]))}
    pick-any A B
      (!set-identity-intro (!M A B) (!M B A))
```

This is a shorter and more elegant proof. In general, when two or more stretches of reasoning can be seen to follow the same general pattern, that pattern can often be factored out into a single reusable method. This can be seen as a formalization of the colloquial mathematical idiom that derives a result "by parity of reasoning," or by citing an earlier proof and stating that the derivation of the present result is "symmetric" or similar to the one given earlier. That said, in what follows we will generally prefer to use the direct method of set-identity introduction.

We proceed with the associativity of intersection and union; the proof of the latter is similar to that of the former and we omit it.

```
conclude intersection-associativity :=
  (forall A B C . A /\ (B /\ C) = (A /\ B) /\ C)
 pick-any A B C
   (!set-identity-intro-direct
     pick-any x
       (!chain [(x in A /\ B /\ C)
           <==> (x in A & x in B /\ C)         [IC]
           <==> (x in A & x in B & x in C)     [IC]
           <==> ((x in A & x in B) & x in C) [prop-taut]
           <==> ((x in A /\ B) & x in C)       [IC]
           <==> (x in (A /\ B) /\ C)           [IC]]))

define union-associativity := (forall A B C . A \/ B \/ C = (A \/ B) \/ C)
```

Idempotence is also useful to derive for both operations:

```
conclude /\-idempotence := (forall A . A /\ A = A)
  pick-any A
    (!set-identity-intro-direct
      pick-any x
        (!chain [(x in A /\ A)
            <==> (x in A & x in A)  [IC]
            <==> (x in A)           [prop-taut]]))
```

Union idempotence is likewise established, and defined as `\/-idempotence`.

The following is a useful result specifying necessary and sufficient conditions for the union of two sets to be empty:

```
conclude union-null-theorem :=
  (forall A B . A \/ B = null <==> A = null & B = null)
pick-any A B
  (!chain [(A \/ B = null)
      <==> (forall x . x in A \/ B <==> x in null)       [SIC]
      <==> (forall x . x in A \/ B <==> false)           [NC]
      <==> (forall x . x in A | x in B <==> false)       [UC]
      <==> (forall x . ~ x in A & ~ x in B)              [prop-taut]
      <==> ((forall x . ~ x in A) & (forall x ~ x in B)) [quant-dist]
      <==> (A = null & B = null)                         [NC-2]])
```

We continue with two distributivity theorems. The first one is this:

$$\forall A, B, C \,.\, A \cup (B \cap C) = (A \cup B) \cap (A \cup C).$$

```
conclude distributivity-1 :=
  (forall A B C . A \/ (B /\ C) = (A \/ B) /\ (A \/ C))
  pick-any A B C
    (!set-identity-intro-direct
      pick-any x
```

```
      (!chain [(x in A \/ (B /\ C))
           <==> (x in A | x in B /\ C)                   [UC]
           <==> (x in A | x in B & x in C)               [IC]
           <==> ((x in A | x in B) & (x in A | x in C)) [prop-taut]
           <==> (x in A \/ B & x in A \/ C)              [UC]
           <==> (x in (A \/ B) /\ (A \/ C))              [IC]]))
```

The distributivity result above shows that union distributes over intersection. Our next distributivity result shows that intersection distributes over union. Its proof is similar and left as an exercise.

```
define distributivity-2 :=
  (forall A B C . A /\ (B \/ C) = (A /\ B) \/ (A /\ C))
```

Exercise 10.11: Derive `distributivity-2`. □

We continue with some theorems about set-theoretic difference:

```
conclude diff-theorem-1 := (forall A . A \ A = null)
  pick-any A
    (!set-identity-intro-direct
      pick-any x
        (!chain [(x in A \ A)
             <==> (x in A & ~ x in A) [DC]
             <==> false               [prop-taut]
             <==> (x in null)         [NC]]))

conclude diff-theorem-2 :=
  (forall A B C . B subset C ==> A \ C subset A \ B)
pick-any A B C
  assume (B subset C)
    (!subset-intro
      pick-any x
        (!chain [(x in A \ C)
              ==> (x in A & ~ x in C) [DC]
              ==> (x in A & ~ x in B) [SC]
              ==> (x in A \ B)        [DC]]))
```

Note that if we transpose the order of the two terms in the consequent of the last theorem, the result does not hold:

```
define p := (forall A B C . B subset C ==> A \ B subset A \ C)

> (falsify p 20)

List: ['success
|{
?A:(Set Int) := (insert 1 null:(Set Int))
```

```
?B:(Set Int) := null:(Set Int)

?C:(Set Int) := (insert 1 null:(Set Int))
}|]
```

We go on to derive a number of additional results relating the three fundamental set operations of union, intersection, and difference.

```
define diff-theorem-3 := (forall A B . A \ (A /\ B) = A \ B)

conclude diff-theorem-4 :=
  (forall A B . A /\ (A \ B) = A \ B)
    pick-any A B
      (!set-identity-intro-direct
        pick-any x
          (!chain [(x in A /\ (A \ B))
                <==> (x in A & x in A \ B)          [IC]
                <==> (x in A & ~ x in B)            [DC]
                <==> (x in A & ~ x in B)            [prop-taut]
                <==> (x in A \ B)                   [DC]]))

define diff-theorem-5 := (forall A B . (A \ B) \/ B = A \/ B)
define diff-theorem-6 := (forall A B . (A \/ B) \ B = A \ B)
define diff-theorem-7 := (forall A B . (A /\ B) \ B = null)

conclude diff-theorem-8 :=
  (forall A B . (A \ B) /\ B = null)
    pick-any A B
      (!set-identity-intro-direct
        pick-any x
          (!chain [(x in (A \ B) /\ B)
                <==> (x in A \ B & x in B)             [IC]
                <==> ((x in A & ~ x in B) & x in B)    [DC]
                <==> false                             [prop-taut]
                <==> (x in null)                       [NC]]))

define diff-theorem-9 := (forall A B C . A \ (B \/ C) = (A \ B) /\ (A \ C))

conclude diff-theorem-10 :=
  (forall A B C . A \ (B /\ C) = (A \ B) \/ (A \ C))
    pick-any A B C
      (!set-identity-intro-direct
        pick-any x
          (!chain [(x in A \ (B /\ C))
                <==> (x in A & ~ x in B /\ C)                        [DC]
                <==> (x in A & ~ (x in B & x in C))                  [IC]
                <==> (x in A & (~ x in B | ~ x in C))                [prop-taut]
                <==> ((x in A & ~ x in B) | (x in A & ~ x in C))     [prop-taut]
                <==> (x in A \ B | x in A \ C)                       [DC]
                <==> (x in (A \ B) \/ (A \ C))                       [UC]]))
```

```
define diff-theorem-11 := (forall A B . A \ (A \ B) = A /\ B)
define diff-theorem-12 := (forall A B . A subset B ==> A \/ (B \ A) = B)

conclude subset-theorem-1 :=
  (forall A B . A subset B ==> A \/ B = B)
pick-any A B
  assume (A subset B)
    (!set-identity-intro-direct
      pick-any x
        (!chain [(x in A \/ B)
            <==> (x in A | x in B) [UC]
            <==> (x in B | x in B) [prop-taut SC]
            <==> (x in B)          [prop-taut]]))

define subset-theorem-2 := (forall A B . A subset B ==> A /\ B = A)

define intersection-subset-theorem' :=
  (forall A B C . A subset B /\ C <==> A subset B & A subset C)

conclude union-lemma-2 :=
  (forall A B x . x ++ (A \/ B) = A \/ x ++ B)
pick-any A B x
  (!chain [(x ++ (A \/ B))
         = (x ++ (B \/ A)) [union-commutes]
         = ((x ++ B) \/ A) [union-def]
         = (A \/ (x ++ B)) [union-commutes]])

conclude absorption-1 :=
  (forall x A . x in A <==> x ++ A = A)
  pick-any x A
    (!equiv
      assume hyp := (x in A)
        (!set-identity-intro-direct
          pick-any y
            (!chain [(y in x ++ A)
                <==> (y = x | y in A)      [in-def]
                <==> (y in A | y in A)     [hyp prop-taut]
                <==> (y in A)              [prop-taut]]))
      assume (x ++ A = A)
        (!chain-> [true ==> (x in x ++ A) [in-lemma-1]
                        ==> (x in A)      [SIC]]))
```

Note that on the second step of the equivalence chain in the last of these proofs, we specify both `hyp` and `prop-taut` as justifiers. Neither would be sufficient by itself.

Exercise 10.12: Supply the missing proofs. □

The union analogue of `intersection-subset-theorem'` does not hold:

```
> (falsify (forall A B C . A subset B \/ C <==> A subset B | A subset C) 10)

List: ['success
        |{?A:(Set Int) :=
           insert 1
                  (insert 2
                         null:(Set Int)))
          ?B:(Set Int) := (insert 1
                               null:(Set Int))
          ?C:(Set Int) := (insert 2
                               null:(Set Int))
      }|]
```

However, a weaker version does hold:

```
define union-subset-theorem :=
  (forall A B C . A subset B | A subset C ==> A subset B \/ C)
```

Exercise 10.13: Prove `union-subset-theorem`. □

10.3.4 Cartesian products

In this section we develop the theory of cartesian products, which will become the foundation for a theory of relations and functions. The core cartesian-product operation will be a polymorphic binary function `X` that accepts as inputs two sets of sorts S and T and produces an output set of sort `(Pair` S T`)`, that is, a set of ordered pairs of elements from S and T. We give a constructive definition of this operation in terms of a simpler operation, `paired-with`, that takes a single element x of sort S and a set A of sort T and produces a set of sort `(Pair` S T`)`, by prepending x in front of every element of A. Thus, for instance, applying `paired-with` to 3 and the set $\{2, 8\}$ will produce the set of pairs $\{(3, 2), (3, 8)\}$ (using conventional notation for sets and ordered pairs).

```
declare paired-with: (S, T) [S (Set T)] -> (Set (Pair S T))
                            [130 [id alist->set]]

assert* paired-with-def :=
  [(_ paired-with null = null)
   (x paired-with h ++ t = x @ h ++ (x paired-with t))]

> (eval 3 paired-with [2 8])

List: [(pair 3 8) (pair 3 2)]
```

The following proof derives the set-theoretic characterization of `paired-with` from its constructive definition:

```
conclude paired-with-characterization :=
  (forall B x y a . x @ y in a paired-with B <==> x = a & y in B)
 by-induction paired-with-characterization {
   null => pick-any x y a
                (!chain [(x @ y in a paired-with null)
                    <==> (x @ y in null)                    [paired-with-def]
                    <==> false                              [NC]
                    <==> (x = a & false)                    [prop-taut]
                    <==> (x = a & y in null)                [NC]])
 | (B as (insert h t)) =>
     pick-any x y a
       let {IH := (forall x y a .
                     x @ y in a paired-with t <==> x = a & y in t)}
        (!chain
          [(x @ y in a paired-with h ++ t)
       <==> (x @ y in a @ h ++ (a paired-with t))          [paired-with-def]
       <==> (x @ y = a @ h | x @ y in a paired-with t) [in-def]
       <==> (x = a & y = h | x @ y in a paired-with t) [pair-axioms]
       <==> (x = a & y = h | x = a & y in t)                [IH]
       <==> (x = a & (y = h | y in t))                      [prop-taut]
       <==> (x = a & y in B)                                [in-def]])
  }

define PWC := paired-with-characterization
```

Exercise 10.14: The following is a useful result:

```
define paired-with-lemma-1 :=
  (forall A  x . x paired-with A = null ==> A = null)
```

Prove this lemma. □

We can now define cartesian products constructively in terms of `paired-with`:

```
declare product: (S, T) [(Set S) (Set T)] -> (Set (Pair S T))
                                        [150 [alist->set alist->set]]

define X := product

assert* product-def :=
  [(null X _ = null)
   (h ++ t X A = h paired-with A \/ t X A)]

> (eval [1 2] X ['a 'b])

List: [(pair 1 'a) (pair 2 'a)  (pair 1 'b) (pair 2 'b)]
```

The classical set-theoretic characterization of this operation can be derived from the preceding definition by induction; the proof is left as an exercise.

```
define cartesian-product-characterization :=
  (forall A B a b . a @ b in A X B <==> a in A & b in B)

define CPC := cartesian-product-characterization
```

An alternative characterization is occasionally useful:

```
define cartesian-product-characterization-2 :=
  (forall x A B . x in A X B <==> exists a b . x = a @ b & a in A & b in B)

define CPC-2 := cartesian-product-characterization-2
```

Exercise 10.15: Derive CPC and CPC-2. □

We derive a number of theorems:

```
conclude product-theorem-1 :=
  (forall A B . A X B = null ==> A = null | B = null)
datatype-cases product-theorem-1 {
  null =>
    pick-any B
      (!chain [(null X B = null)
           ==> (null = null)                      [product-def]
           ==> (null = null | B = null)           [prop-taut]])
 | (A as (insert h t)) =>
     pick-any B
       (!chain
         [(h ++ t X B = null)
      ==> (h paired-with B \/ t X B = null)       [product-def]
      ==> (h paired-with B = null & t X B = null) [union-null-theorem]
      ==> (B = null)                              [paired-with-lemma-1]
      ==> (h ++ t = null | B = null)              [prop-taut]])
}

define product-theorem-2 :=
  (forall A B . non-empty A & non-empty B ==> A X B = B X A <==> A = B)

define product-theorem-3 :=
  (forall A  B C . non-empty A &  A X B subset A X C ==> B subset C)

define product-theorem-4 :=
  (forall A B C . B subset C ==> A X B subset A X C)

conclude product-theorem-5 :=
 (forall A B C . A X (B /\ C) = A X B /\ A X C)
pick-any A B C
  (!set-identity-intro-direct
    (!pair-converter
      pick-any x y
        (!chain [(x @ y in A X (B /\ C))
```

```
          <==> (x in A & y in B /\ C)                    [CPC]
          <==> (x in A & y in B & y in C)                [IC]
          <==> ((x in A & y in B) & (x in A & y in C))   [prop-taut]
          <==> (x @ y in A X B & x @ y in A X C)         [CPC]
          <==> (x @ y in A X B /\ A X C)                 [IC]])))

define product-theorem-6 := (forall A B C . A X (B \/ C) = A X B \/ A X C)

define product-theorem-7 := (forall A B C . A X (B \ C) = A X B \ A X C)
```

Exercise 10.16: Supply the missing proofs. □

10.3.5 Relations

From now on we will understand *a binary relation from A to B* precisely as *a set of ordered pairs* (a,b) *with* $a \in A$ *and* $b \in B$. This is the usual mathematical definition of relations. Thus, in what follows, instead of "binary relations" we will be talking about sets of ordered pairs. We introduce a few convenient variables for binary relations:

```
define [R R1 R2 R3 R4] :=
       [?R:(Set (Pair 'T14 'T15)) ?R1:(Set (Pair 'T16 'T17)) ···]
```

Let us now begin by defining the *domain* and *range* of a given relation. The domain of a relation R is the set of left elements of all the ordered pairs in R; the range of R is the set of all right elements of those pairs. In conventional mathematical notation:

$$dom(R) = \{a \mid \exists\, b \,.\, (a,b) \in R\} \tag{10.8}$$

and

$$range(R) = \{b \mid \exists\, a \,.\, (a,b) \in R\}. \tag{10.9}$$

We give constructive definitions of both as follows:

```
declare dom: (S, T) [(Set (Pair S T))] -> (Set S) [150 [alist->set]]

assert* dom-def :=
  [(dom null = null)
   (dom x @ _ ++ t = x ++ dom t)]

> (eval dom [('a @ 1) ('b @ 2) ('c @ 98)])

List: ['a 'b 'c]

declare range: (S, T) [(Set (Pair S T))] -> (Set T) [150 [alist->set]]

assert* range-def :=
  [(range null = null)
```

```
  (range _ @ y ++ t = y ++ range t)]

> (eval range [('a @ 1) ('b @ 2) ('c @ 98)])

List: [1 2 98]
```

From these definitions, the conventional characterizations (10.8) and (10.9) can be derived by induction. These characterizations, whose proofs are left as exercises, are defined below along with a few simple but useful lemmas relating the membership relation to `dom` and `range`:

```
conclude in-dom-lemma-1 :=
  (forall R a x y . a = x ==> a in dom x @ y ++ R)
pick-any R a x y
  (!chain [(a = x) ==> (a in x ++ dom R)        [in-def]
                   ==> (a in dom x @ y ++ R)  [dom-def]])

define in-range-lemma-1 :=
  (forall R a x y . a = y ==> a in range x @ y ++ R)

conclude in-dom-lemma-2 :=
  (forall R x a b . x in dom R ==> x in dom a @ b ++ R)
pick-any R x a b
  (!chain    [(x in dom a @ b ++ R)
         <== (x in a ++ dom R)        [dom-def]
         <== (x in dom R)             [in-def]])

define in-range-lemma-2 :=
  (forall R y a b . y in range R ==> y in range a @ b ++ R)

define dom-characterization :=
  (forall R x . x in dom R <==> exists y . x @ y in R)

define range-characterization :=
  (forall R y . y in range R <==> exists x . x @ y in R)

define [DOMC RANC] := [dom-characterization range-characterization]
```

Exercise 10.17: Derive `in-range-lemma-1` and `in-range-lemma-2`. □

Exercise 10.18: Derive `DOMC` and `RANC`. □

We now derive a number of results about domains and ranges. The proofs of some of these are left as exercises. We first note that both operations distribute over unions:

```
conclude dom-theorem-1 :=
  (forall R1 R2 . dom (R1 \/ R2) = dom R1 \/ dom R2)
pick-any R1 R2
```

```
  (!set-identity-intro-direct
    pick-any x
      (!chain
        [(x in dom (R1 \/ R2))
    <==> (exists y . x @ y in R1 \/ R2)                          [DOMC]
    <==> (exists y . x @ y in R1 | x @ y in R2)                  [UC]
    <==> ((exists y . x @ y in R1) | (exists y . x @ y in R2)) [quant-dist]
    <==> (x in dom R1 | x in dom R2)                             [DOMC]
    <==> (x in dom R1 \/ dom R2)                                 [UC]]))

define range-theorem-1 :=
  (forall R1 R2 . range (R1 \/ R2) = range R1 \/ range R2)
```

Exercise 10.19: Prove range-theorem-1. □

These results might lead us to suspect that the two operations distribute over intersections as well, for example, that the following holds:

```
define dom-intersection-conjecture :=
  (forall R1 R2 . dom (R1 /\ R2) = dom R1 /\ dom R2)
```

However, when we put the conjecture to the test we see that it fails:

```
> (falsify dom-intersection-conjecture 10)

List: ['success
|{
?R1:(Set (Pair Int Int)) := (insert (pair 1 1)
                                    null:(Set (Pair Int Int)))
?R2:(Set (Pair Int Int)) := (insert (pair 1 2)
                                    null:(Set (Pair Int Int)))
}|]
```

The produced counterexample shows that the conjecture does not hold when $R_1 = \{(1,1)\}$ and $R_2 = \{(1,2)\}$. In that case we have $dom(R_1 \cap R_2) = dom(\emptyset) = \emptyset$, while

$$dom(R_1) \cap dom(R_2) = \{1\} \cap \{1\} = \{1\},$$

so the two sides are unequal. But weaker versions of the conjecture do hold for both operations:

```
conclude dom-theorem-2 :=
  (forall R1 R2 . dom (R1 /\ R2) subset dom R1 /\ dom R2)
pick-any R1 R2
  (!subset-intro
    pick-any x
      (!chain
        [(x in dom (R1 /\ R2))
     ==> (exists y . x @ y in R1 /\ R2)                          [DOMC]
```

```
    ==> (exists y . x @ y in R1 & x @ y in R2)                    [IC]
    ==> ((exists y . x @ y in R1) & (exists y . x @ y in R2)) [quant-dist]
    ==> (x in dom R1 & x in dom R2)                               [DOMC]
    ==> (x in dom R1 /\ dom R2)                                   [IC]]))

define range-theorem-2 :=
  (forall R1 R2 . range (R1 /\ R2) subset range R1 /\ range R2)

define dom-theorem-3 :=
  (forall R1 R2 . dom R1 \ dom R2 subset dom (R1 \ R2))

define range-theorem-3 :=
  (forall R1 R2 . range R1 \ range R2 subset range (R1 \ R2))
```

Exercise 10.20: Supply the missing proofs. □

Next, we introduce the *converse* of a binary relation R. This is simply the relation that holds between a and b iff R holds between b and a. We introduce the double-minus sign `--` as an alias for this operation:

```
declare conv: (S, T) [(Set (Pair S T))] -> (Set (Pair T S))
                                           [210 [alist->set]]
define -- := conv

assert* conv-def :=
  [(-- null = null)
   (-- x @ y ++ t = y @ x ++ -- t)]

> (eval -- [(1 @ 'a) (2 @ 'b) (3 @ 'c)])

List: [(pair 'a 1) (pair 'b 2) (pair 'c 3)]
```

As usual, the characterization theorem for this operation is derived by induction:

```
conclude converse-characterization :=
  (forall R x y . x @ y in -- R <==> y @ x in R)
by-induction converse-characterization {
  null => pick-any x y
            (!chain [(x @ y in -- null)
                <==> (x @ y in null)     [conv-def]
                <==> false               [NC]
                <==> (y @ x in null)     [NC]])
| (R as (insert (pair a b) t)) =>
    let {IH := (forall x y . x @ y in -- t <==> y @ x in t)}
      pick-any x y
        (!chain [(x @ y in -- R)
            <==> (x @ y in b @ a ++ -- t)          [conv-def]
            <==> (x @ y = b @ a | x @ y in -- t)   [in-def]
            <==> (y @ x = a @ b | x @ y in -- t)   [pair-theorem-2]
```

```
              <==> (y @ x = a @ b | y @ x in t)        [IH]
              <==> (y @ x in R)                        [in-def]])
}
define CC := converse-characterization
```

A number of useful results can be derived:

```
define converse-theorem-1 := (forall R . -- -- R = R)

conclude converse-theorem-2 :=
  (forall R1 R2 . -- (R1 /\ R2) = -- R1 /\ -- R2)
 pick-any R1 R2
   (!set-identity-intro-direct
     (!pair-converter
        pick-any x y
          (!chain [(x @ y in -- (R1 /\ R2))
              <==> (y @ x in R1 /\ R2)              [CC]
              <==> (y @ x in R1 & y @ x in R2)      [IC]
              <==> (x @ y in -- R1 & x @ y in -- R2) [CC]
              <==> (x @ y in -- R1 /\ -- R2)        [IC]])))

define converse-theorem-3 := (forall R1 R2 . -- (R1 \/ R2) = -- R1 \/ -- R2)
define converse-theorem-4 := (forall R1 R2 . -- (R1 \ R2) = -- R1 \ -- R2)
```

We continue with the important notion of *relation composition*. The composition of a relation R_1 from S_1 to S_2 with a relation R_2 from S_2 to S_3, which we will denote by (R_1 o R_2), is the set of all and only those ordered pairs (x @ z) for which there is a y such that (x @ y) is in R_1 and (y @ z) is in R_2. Roughly speaking, the composition relates two elements x and z iff there is some "intermediate" node y between x and z. For example, after we define composition we will have:

```
> let {R1 := [('nyc @ 'boston) ('austin @ 'dc)];
       R2 := [('boston @ 'montreal) ('dc @ 'chicago) ('chicago @ 'seattle)]}
    (eval R1 o R2)

List: [(pair 'nyc 'montreal) (pair 'austin 'chicago)]
```

We give a constructive definition of composition in two stages. First we define a simpler operation, composed-with, that can only compose a given ordered pair (a @ b) with a binary relation R. The result is another relation that contains all and only those pairs (a @ c) such that (b @ c) is in R:

```
declare composed-with: (S1, S2, S3)
   [(Pair S1 S2) (Set (Pair S2 S3))] -> (Set (Pair S1 S3))
   [200 [id alist->set]]

assert* composed-with-def :=
```

```
  [(_ composed-with null = null)
   (x @ y composed-with z @ w ++ t =
    x @ w ++ (x @ y composed-with t) <== y = z)

   (x @ y composed-with z @ w ++ t =
    x @ y composed-with t <== y =/= z)]

> (eval 1 @ 2 composed-with [(2 @ 5) (7 @ 8) (2 @ 3)])

List: [(pair 1 5) (pair 1 3)]

define composed-with-characterization :=
  (forall R x y z w . w @ z in x @ y composed-with R <==> w = x & y @ z in R)
```

The inductive derivation of the characterization result is left as an exercise.

We can now define the main composition operator as follows:

```
declare o: (S1, S2, S3) [(Set (Pair S1 S2)) (Set (Pair S2 S3))]
                      -> (Set (Pair S1 S3)) [200 [alist->set alist->set]]

assert* o-def :=
 [(null o _ = null)
  (x @ y ++ t o R = x @ y composed-with R \/ t o R)]

> (eval [('nyc @ 'boston) ('houston @ 'dallas) ('austin @ 'dc)] o
        [('boston @ 'montreal) ('dallas @ 'chicago) ('dc @ 'nyc)] o
        [('chicago @ 'seattle) ('montreal @ 'london)])

List: [(pair 'nyc 'london) (pair 'houston 'seattle)]
```

A characterization theorem and various useful results about composition can now be derived; most proofs are left as exercises:

```
define o-characterization :=
  (forall R1 R2 x z . x @ z in R1 o R2 <==>
                      exists y . x @ y in R1 & y @ z in R2)

define OC := o-characterization

define compose-theorem-1 := (forall R1 R2 . dom R1 o R2 subset dom R1)

define compose-theorem-2 :=
  (forall R1 R2 R3 R4 . R1 subset R2 & R3 subset R4
                        ==> R1 o R3 subset R2 o R4)

define compose-theorem-3 :=
  (forall R1 R2 R3  . R1 o (R2 \/ R3) = R1 o R2 \/ R1 o R3)

define compose-theorem-4 :=
  (forall R1 R2 R3  . R1 o (R2 /\ R3) subset R1 o R2 /\ R1 o R3)
```

```
define compose-theorem-5 :=
  (forall R1 R2 R3 . R1 o R2 \ R1 o R3 subset R1 o (R2 \ R3))

define composition-associativity :=
   (forall R1 R2 R3 . R1 o R2 o R3 = (R1 o R2) o R3)

conclude compose-theorem-6 :=
  (forall R1 R2 . -- (R1 o R2) = -- R2 o -- R1)
pick-any R1 R2
  (!set-identity-intro-direct
    (!pair-converter
      pick-any x y
        (!chain [(x @ y in -- (R1 o R2))
            <==> (y @ x in R1 o R2)                          [CC]
            <==> (exists z . y @ z in R1 & z @ x in R2)      [OC]
            <==> (exists z . z @ y in -- R1 & x @ z in -- R2) [CC]
            <==> (exists z . x @ z in -- R2 & z @ y in -- R1) [prop-taut]
            <==> (x @ y in -- R2 o -- R1)                     [OC]])))
```

We continue with the notion of relation *restriction*. Restricting a relation R to a set of elements A yields a relation consisting of all and only those pairs $(x$ @ $y)$ in R such that $x \in A$. We again define this relation incrementally, starting with an operation that restricts a relation to a single element:

```
declare restrict1: (S, T) [(Set (Pair S T)) S] -> (Set (Pair S T))
                                                   [200 [alist->set id]]

define ^1 := restrict1

assert* restrict1-def :=
  [(null ^1 _ = null)
   (x @ y ++ t ^1 z = x @ y ++ (t ^1 z) <== x = z)
   (x @ y ++ t ^1 z = t ^1 z            <== x =/= z)]

> (eval [(1 @ 'a) (2 @ 'b) (1 @ 'c)] ^1 1)

List: [(pair 1 'a) (pair 1 'c)]
```

A characterization theorem is derived by structural induction; the proof is omitted.

```
define restrict1-characterization :=
 (forall R x y a . x @ y in R ^1 a <==> x @ y in R & x = a)
```

We now define the regular restriction operator as follows:

```
declare restrict: (S, T) [(Set (Pair S T)) (Set S)] ->
                         (Set (Pair S T)) [200 [alist->set alist->set]]
```

```
define ^ := restrict

assert* restrict-def :=
[(R ^ null = null)
 (R ^ h ++ t = R ^1 h \/ R ^ t)]

> (eval [(1 @ 'a) (2 @ 'b) (3 @ 'c) (1 @ 'f)] ^ [1 2])

List: [(pair 1 'a) (pair 1 'f) (pair 2 'b)]
```

A characterization theorem and some useful results follow:

```
define restrict-characterization :=
 (forall A R x y . x @ y in R restrict A <==> x @ y in R & x in A)

define RSC := restrict-characterization

define restriction-theorem-1 :=
 (forall R A B . A subset B ==> R ^ A subset R ^ B)

define restriction-theorem-2 :=
 (forall R A B . R ^ (A /\ B) = R ^ A /\ R ^ B)

conclude restriction-theorem-3 :=
 (forall R A B . R ^ (A \/ B) = R ^ A \/ R ^ B)
pick-any R A B
  (!set-identity-intro-direct
    (!pair-converter
      pick-any x y
        (!chain [(x @ y in R ^ (A \/ B))
             <==> (x @ y in R & x in A \/ B)                        [RSC]
             <==> (x @ y in R & (x in A | x in B))                  [UC]
             <==> ((x @ y in R & x in A) | (x @ y in R & x in B)) [prop-taut]
             <==> (x @ y in R ^ A | x @ y in R ^ B)                 [RSC]
             <==> (x @ y in R ^ A \/ R ^ B)                         [UC]])))

define restriction-theorem-4 := (forall R A B . R ^ (A \ B) = R ^ A \ R ^ B)

define restriction-theorem-5 :=
  (forall R1 R2 A . (R1 o R2) ^ A = (R1 ^ A) o R2)
```

The *image* of a relation R on a set A is the set of all y such that $(x$ @ $y) \in R$ for some $x \in A$. This is the same as the range of the restriction of R on A:

```
declare image: (S, T) [(Set (Pair S T)) (Set S)] -> (Set T)
                      [200 [alist->set alist->set]]

assert* image-def := [(R ** A = range R ^ A)]

> (eval [(1 @ 'a) (2 @ 'b) (3 @ 'c)] ** [1 3])
```

```
List: ['a 'c]
```

We derive a characterization theorem and some useful results:

```
define image-characterization :=
  (forall R A y . y in R ** A <==> exists x . x @ y in R & x in A)

define IMGC := image-characterization

define image-theorem-1 :=
  (forall R A B . R ** (A \/ B) = R ** A \/ R ** B)

define image-theorem-2 :=
  (forall R A B . R ** (A /\ B) subset R ** A /\ R ** B)

define image-theorem-3 :=
  (forall R A B . R ** A \ R ** B subset R ** (A \ B))

define image-theorem-4 :=
  (forall R A B . A subset B ==> R ** A subset R ** B)

define image-theorem-5 :=
  (forall R A . R ** A = null <==> dom R /\ A = null)

define image-theorem-6 :=
  (forall R A . dom R /\ A subset -- R ** R ** A)

define image-theorem-7 :=
  (forall R A B . (R ** A) /\ B subset R ** (A /\ -- R ** B))
```

10.3.6 Set cardinality

We define the size or *cardinality* of a set as follows:

```
declare card: (S) [(Set S)] -> N [[alist->set]]

define S := N.S  # Natural-number successor

assert* card-def :=
  [(card null = zero)
   (card h ++ t = card t    <== h in t)
   (card h ++ t = S card t <== ~ h in t)]

transform-output eval [nat->int]

> (eval card [1 2 3] \/ [4 7 8])

Term: 6
```

A number of useful results involving cardinality are proven in the `Set` module. We note a few of them below; most of the proofs are omitted.

```
define card-theorem-1 := (forall x . card singleton x = S zero)

conclude card-theorem-2 :=
  (forall A x . ~ x in A ==> card A < card x ++ A)
pick-any A x
  assume hyp := (~ x in A)
     (!chain-> [true ==> (card A < S card A)      [N.Less.<S]
                     ==> (card A < card x ++ A) [card-def]])

define minus-card-theorem := (forall A x . x in A ==> card A = S card A - x)

define subset-card-theorem := (forall A B . A subset B ==> card A <= card B)

define proper-subset-card-theorem :=
  (forall A B . A proper-subset B ==> card A < card B)

conclude intersection-card-theorem-1 :=
  (forall A B . card A /\ B <= card A)
pick-any A B
  (!chain-> [true ==> (A /\ B subset A)          [intersection-subset-theorem]
                  ==> (card A /\ B <= card A) [subset-card-theorem]])

define intersection-card-theorem-2 := (forall A B . card A /\ B <= card B)

define intersection-card-theorem-3 :=
  (forall A B x . ~ x in A & x in B ==> card (x ++ A) /\ B = S card A /\ B)

define union-card :=
  (forall A B . card A \/ B = ((card A) + (card B)) - card A /\ B)

define diff-card-lemma :=
  (forall A B . card A = (card A \ B) + (card A /\ B))

conclude diff-card-theorem :=
  (forall A B . card A \ B =  (card A) - card A /\ B)
pick-any A B
  (!chain->
    [true
 ==> (card A = (card A \ B) + card A /\ B) [diff-card-lemma]
 ==> ((card A \ B) + card A /\ B = card A) [sym]
 ==> (card A \ B = (card A) - card A /\ B) [N.Minus.Plus-Minus-properties]])
```

10.3.7 Powersets

The powerset of a set A is the set of all subsets of A. We introduce this operation as a function symbol with the following signature:

```
declare powerset: (S) [(Set S)] -> (Set (Set S)) [[alist->set]]
```

We will define this operation constructively in two stages. First we introduce a binary operation `in-all` that takes an arbitrary element x of sort S and an arbitrary set A of sets of sort S, and produces the set of sets obtained from A by inserting x in every set in A:

```
declare in-all: (S) [S (Set (Set S))] -> (Set (Set S)) [[id alist->set]]

assert* in-all-def :=
  [(_ in-all null = null)
   (x in-all A ++ t = (x ++ A) ++ (x in-all t))]

> (eval 7 in-all [[1 2] [] [4 5 6]])

List: [[7 1 2] [7] [7 4 5 6]]
```

The following is a characterization theorem for `in-all`:

```
define in-all-characterization :=
  (forall U s x . s in x in-all U <==> exists B . B in U & s = x ++ B)
```

We can now define `powerset` recursively in terms of this operation:

```
assert* powerset-def :=
  [(powerset null = singleton null)
   (powerset x ++ t = (powerset t) \/ (x in-all powerset t))]

> (eval powerset [1 2 3])

List: [[] [3] [2] [2 3] [1] [1 3] [1 2] [1 2 3]]
```

The corresponding characterization theorem is this:

```
define powerset-characterization :=
  (forall A B . B in powerset A <==> B subset A)

define POSC := powerset-characterization
```

A number of useful theorems involving the powerset operator are derived:

```
conclude ps-theorem-1 := (forall A . null in powerset A)
  pick-any A
   (!chain-> [true ==> (null subset A)       [subset-def]
                   ==> (null in powerset A) [POSC]])

conclude ps-theorem-2 := (forall A . A in powerset A)
  pick-any A
    (!chain-> [true ==> (A subset A)       [subset-reflexivity]
                    ==> (A in powerset A) [POSC]])
```

```
conclude ps-theorem-3 :=
  (forall A B . A subset B <==> powerset A subset powerset B)
pick-any A B
  (!equiv assume (A subset B)
          (!subset-intro
            pick-any C
              (!chain [(C in powerset A)
                   ==> (C subset A)              [POSC]
                   ==> (C subset B)              [subset-transitivity]
                   ==> (C in powerset B)         [POSC]]))
          assume (powerset A subset powerset B)
            (!chain-> [true ==> (A in powerset A)  [ps-theorem-2]
                            ==> (A in powerset B)  [SC]
                            ==> (A subset B)       [POSC]]))

conclude ps-theorem-4 :=
  (forall A B . powerset A /\ B = (powerset A) /\ (powerset B))
pick-any A B
  (!set-identity-intro-direct
    pick-any C
      (!chain
        [(C in powerset A /\ B)
    <==> (C subset A /\ B)                      [POSC]
    <==> (C subset A & C subset B)              [intersection-subset-theorem']
    <==> (C in powerset A & C in powerset B) [POSC]
    <==> (C in (powerset A) /\ (powerset B)) [IC]]))
```

Remarkably, the union analogue of ps-theorem-4 does not hold:

```
define p := (forall A B . powerset A \/ B = (powerset A) \/ (powerset B))

> (falsify p 10)

List: ['success
|{
?A := (insert 1 null)

?B := (insert 2 null)

}|]

> (eval powerset [1] \/ [2])

List: [[] [2] [1] [1 2]]

> (eval (powerset [1]) \/ (powerset [2]))

List: [[] [1] [2]]
```

But a weaker version of the result does hold:

```
define ps-theorem-5 :=
  (forall A B . (powerset A) \/ (powerset B) subset powerset A \/ B)
```

Exercise 10.21: Prove `ps-theorem-5`. □

10.4 Maps

10.4.1 Representation and notation

Finite maps are another pervasive data type and an indispensable tool in our mathematical arsenal for the modeling and analysis of discrete systems. While we could develop the theory of such maps as part of our theory of sets, along the lines of Section 10.3.5, the direct approach we take here will be more expedient. However, sets will be widely used in this theory, as they often are elsewhere. For instance, the domain of a map will be defined to return a *set* of keys, while the range of a map will be a set of values. We will use upper-case letters from the beginning of the alphabet as variables ranging over finite sets of arbitrary sorts:

```
define [A B C] := [?A:(Set.Set 'S1) ?B:(Set.Set 'S2) ?C:(Set.Set 'S3)]
```

(All of the code in this section appears inside a `FMap` module that is part of the Athena library, but as was the case with the material on sets, this section too is written as if the contents of `FMap` were developed interactively from scratch and at the top level.)

Maps are intended to represent finite functions from arbitrary keys to arbitrary values. Conventional algorithmic treatments of the subject focus on different implementations of maps and on the computational complexity of the fundamental map operations in each of these implementations (red-black trees vs. hash tables, etc.). Here we are primarily interested in maps as an abstract data type (the so-called *dictionary* data type), so our focus will be on the specification and deductive analysis of some of the essential properties and relationships of the fundamental map operations. As usual, we will define these operations constructively, so we will be able to evaluate map insertions, deletions, and so on. But, as elsewhere, the emphasis will be on formal analysis rather than computational complexity.

We introduce maps as a polymorphic inductive structure, abstracted over both the sort of the keys and that of the values:

```
structure (Map Key Value) := empty-map
                           | (update (Pair Key Value) (Map Key Value))
```

Note that the general shape of this structure is also similar to that of lists, with `empty-map` playing the role of `nil` and `update` the role of `::` (the "cons" operation). The intuition here is that `update` is used to incrementally add new key-value bindings to a map, with the bindings represented as ordered pairs. These key-value pairs are added at the front of the map, so, like lists, maps grow to the left. The construction process always starts with the empty map. If a new key-value binding is added to a map *m* that shares a key with an already existing binding in *m*, then the new (leftmost) binding takes precedence. Thus, even though the older pair is not removed from the new map, it is shadowed by the more recent binding. This will become more precise soon, when we come to define how a map is applied to a given key. Also, because this is a structure rather than a datatype, we will have to precisely define what it means for two maps to be identical. As you might guess, two maps will be the same iff they produce the same values when applied to the same keys.

As with other sorts that are structurally similar to lists, it will be notationally convenient to have procedures that can convert Athena lists to maps and conversely. We start with a procedure for converting lists to `Map` terms. Key-value bindings are expressed as either two-element key-value lists `[k v]`, or, for enhanced notational effect, as three-element lists of the form `[k --> v]`:[4]

```
define (alist->map L) :=
  match L {
    [] => empty-map
  | (list-of (|| [k --> v] [k v]) rest) => (update (pair k v)
                                                   (alist->map rest))
  | _ => L
  }
```

The converse procedure will also come handy:

```
define (map->alist m) :=
  match m {
    empty-map => []
  | (update (pair k v) rest) => (add [k --> v] map->alist rest)
  | _ => m
  }
```

We also introduce a procedure for converting bindings given in such list notation to ordered pairs:

```
define (alist->pair inner-1 inner-2) :=
  lambda (L)
    match L  {
      [a b] =>       ((inner-1 a) @ (inner-2 b))
    | [a --> b] => ((inner-1 a) @ (inner-2 b))
    | _ => L
    }
```

4 Recall that --> is a built-in polymorphic nullary symbol; it is used as a simple notational placeholder in appropriate contexts.

We now expand the input range of update so that it can receive bindings as lists of the above form in the first argument position; and lists representing maps in the above notation in the second position:

```
expand-input update [(alist->pair id id) alist->map]
```

Since we will be dealing with sets and set operations a good deal, we introduce a few abbreviations for their fully qualified names:

```
define [null ++ in subset \/ /\ \ -] :=
  [Set.null Set.++ Set.in Set.subset Set.\/ Set./\ Set.\ Set.-]
```

More importantly, we overload ++ so that it can be used as an alias for the Map constructor update in appropriate contexts:

```
overload ++ update
```

Thus:

```
> (['a --> 1] ++ [])

Term: (update (pair 'a 1)
              empty-map:(Map Ide Int))

> (75 ++ [])

Term: (Set.insert 75
                  Set.null:(Set.Set Int))
```

We also transform the output of eval so that it prints out sets and maps as Athena lists:

```
transform-output eval [Set.set->alist map->alist]
```

Finally, we introduce a few convenient variable names:

```
define [key key1 key2 k k' k1 k2] := [?key ?key1 ?key2 ?k ?k' ?k1 ?k2]

define [val val1 val2 v v' v1 v2 x x1 x2] :=
       [?val ?val1 ?val2 ?v ?v' ?v1 ?v2 ?x ?x1 ?x2]

define [m m1 m2 m3 rest rest1] :=
       [?m:(Map 'S1 'S2) ?m1:(Map 'S3 'S4) ...]

define [s1 s2 s3 A B C] := [Set.s1 Set.s2 Set.s3 Set.A Set.B Set.C]
```

10.4.2 Map operations and theorems

With that notational preamble behind us, we are ready to introduce the first operation on maps, which is also the most fundamental one: that of *applying* a map to a given key. This is also known as *looking up* the value associated with a key. And here we must face the question of what value to specify as the result when the given key does not exist in the map. Barring subsorting and explicit error elements, there are three main alternatives. One is to return an arbitrary (but fixed) value. Another is to say nothing on the matter—to leave the value of the lookup operation unspecified when the key is not in the map. And another is to return an `Option` value of the form `(SOME` v`)` when the given key *is* bound to a value v in the map, and `NONE` otherwise. Here we take this latter approach; we examine the first alternative in Section 10.4.3.

```
declare apply: (Key, Value) [(Map Key Value) Key] -> (Option Value)
                            [110 [alist->map id]]

define at := apply
```

We have specified an input converter for the first argument position that allows us to write the maps in the list notation we described earlier:

```
> ([['a --> 1] ['b --> 2] ['c --> 3]] at 'a)

Term: (apply (update (pair 'a 1)
                     (update (pair 'b 2)
                             (update (pair 'c 3)
                                     empty-map:(Map Ide Int))))
             'a)
```

We now define map application as follows:

```
assert* apply-def :=
  [([] at _ = NONE)
   ([key val] ++ _ at x = SOME val     <== key = x)
   ([key _] ++ rest at x = rest at x <== key =/= x)]
```

We test the definition to make sure it behaves sensibly:

```
define ide-map := [['a --> 1] ['b --> 2] ['c --> 3] ['a --> 99]]

> (eval ide-map at 'a)

Term: (SOME 1)

> (eval ide-map at 'b)

Term: (SOME 2)

> (eval ide-map at 'foo)

Term: NONE: (Option Int)
```

The following is a simple but useful lemma:

```
conclude apply-lemma-1 :=
  (forall key val rest x .
     [key val] ++ rest at x = NONE ==> rest at x = NONE)
   pick-any key val rest x
    let {m := ([key val] ++ rest);
         hyp := (m at x = NONE);
         goal := (rest at x = NONE)}
      assume hyp
        (!two-cases
          (!chain [(key = x)
               ==> (m at x = SOME val)  [apply-def]
               ==> (m at x =/= NONE)    [option-results]
               ==> (hyp & ~hyp)         [augment]
               ==> goal                 [prop-taut]])
          (!chain [(key =/= x)
               ==> (m at x = rest at x) [apply-def]
               ==> (NONE = rest at x)   [hyp]
               ==> goal                 [sym]]))
```

As usual, the conditions of the defining equations of the operation give rise to a corresponding case analysis inside the proof. Two more simple but useful lemmas concerning application are given next; their proofs are left as exercises.

Exercise 10.22: Consider:

```
define apply-lemma-2 :=
  (forall k v rest x .
     [k v] ++ rest at x =/= NONE <==> k = x | rest at x =/= NONE)

define apply-lemma-3 :=
  (forall m k v1 v2 . m at k = SOME v1 & m at k = SOME v2 ==> v1 = v2)
```

Prove both lemmas. □

We continue with an operation that removes a key from a map:

```
1  declare remove: (Key, Value) [(Map Key Value) Key] -> (Map Key Value)
2                               [- 120 [alist->map id]]
3
4  left-assoc -
5
6  assert* remove-def :=
7    [(empty-map - _ = empty-map)
8     ([key _] ++ rest - key = rest - key)
9     (key =/= x ==> [key val] ++ rest - x = [key val] ++ (rest - x))]
```

Observe the second equation, on line 8: A recursive call is needed even when we find the key to be removed at the front of the map, because there may be additional occurrences of the key further inside the map. Also note that we have declared the subtraction operator here, `-`, to associate to the left, so that we can write something like `(m - k1 - k2)` and have it parsed as `((m - k1) - k2)`.

```
> (eval ide-map - 'a)

List: [['b --> 2] ['c --> 3]]

> (eval ide-map - 'a - 'c)

List: [['b --> 2]]

> (eval ide-map - 'a - 'b at 'c)

Term: (SOME 3)
```

The following proves that the removal operation achieves the desired effect:

```
conclude remove-correctness :=
  (forall m x . m - x at x = NONE)
by-induction remove-correctness {
  empty-map =>
    pick-any x
      (!chain [([] - x at x)
               = ([] at x)                               [remove-def]
               = NONE                                    [apply-def]])
| (m as (update (pair key val) rest)) =>
    let {IH := (forall x . rest - x at x = NONE)}
      pick-any x
        (!two-cases
          assume case1 := (key = x)
            (!chain [(m - x at x)
                     = (m - key at key)                  [case1]
                     = (rest - x at x)                   [case1 remove-def]
                     = NONE                              [IH]])
          assume case2 := (key =/= x)
            (!chain [(m - x at x)
                     = ([key val] ++ (rest - x) at x) [remove-def]
                     = (rest - x at x)                   [apply-def]
                     = NONE                              [IH]]))
}
```

This is only one half, the important one, of `remove`'s correctness. We also need to prove that removing a key from a map does not affect any of the other keys:

```
define remove-correctness-2 :=
  (forall m k k' . k =/= k' ==> m - k at k' = m at k')
```

That proof is given in the exercises.

Exercise 10.23: Prove `remove-correctness-2`. □

Next, we define the *domain* of a map m as the set of all keys in m:

```
1  declare dom: (Key, Value) [(Map Key Value)] -> (Set.Set Key) [[alist->map]]
2
3  assert* dom-def :=
4    [(dom empty-map = null)
5     (dom [k _] ++ rest = k ++ dom rest)]
6
7  > (eval dom ide-map)
8
9  List: ['a 'b 'c]
```

Observe the overloaded use of ++ in the defining equation of line 5: On the left-hand side, ++ is used as an alias for `update`, a map operation; whereas on the right-hand side it is used as an alias for `Set.insert`, a set operation.

We define the *size* of a map m as the cardinality of its domain, namely, the number of keys in m:

```
declare size: (S, T) [(Map S T)] -> N [[alist->map]]

assert* size-def := [(size m = Set.card dom m)]

transform-output eval [nat->int]

> (eval size ide-map)

Term: 3

> (eval size ide-map - 'a - 'b)

Term: 1
```

A few results about domains follow. The first two are simple lemmas, but the third is a more fundamental characterization theorem; the fourth one is a useful property of the domain of a map obtained from a key removal. The proofs of three of these results are left as exercises.

```
define dom-lemma-1 := (forall k v rest . k in dom [k v] ++ rest)

define dom-lemma-2 := (forall m k v . dom m subset dom [k v] ++ m)

conclude dom-characterization :=
  (forall m k . k in dom m <==> m at k =/= NONE)
by-induction dom-characterization {
  (m as empty-map) =>
    pick-any k
```

```
      (!equiv
        (!chain [(k in dom m)
               ==> (k in null)                    [dom-def]
               ==> false                          [Set.NC]
               ==> (m at k =/= NONE)              [prop-taut]])
         assume hyp := (m at k =/= NONE)
           (!chain-> [true
                  ==> (m at k = NONE)             [apply-def]
                  ==> (m at k = NONE & hyp)       [augment]
                  ==> false                       [prop-taut]
                  ==> (k in dom m)                [prop-taut]]))
| (m as (update (pair x y) rest)) =>
     let {IH := (forall k . k in dom rest <==> rest at k =/= NONE)}
       pick-any k
         (!chain [(k in dom m)
             <==> (k in x ++ dom rest)            [dom-def]
             <==> (k = x | k in dom rest)         [Set.in-def]
             <==> (k = x | rest at k =/= NONE)    [IH]
             <==> (x = k | rest at k =/= NONE)    [sym]
             <==> (m at k =/= NONE)               [apply-lemma-2]])
}

define dom-lemma-3 := (forall m k . dom (m - k) subset dom m)
```

Exercise 10.24: Prove the three lemmas above. □

We now introduce an operation that converts a map into a set of ordered pairs. This operation will be used to define identity conditions for maps:

```
declare map->set: (K, V) [(Map K V)] -> (Set.Set (Pair K V)) [[alist->map]]

assert* map->set-def :=
  [(map->set empty-map = null)
   (map->set [k v] ++ rest = [k v] ++ map->set (rest - k))]

> (eval map->set ide-map)

List: [(pair 'a 1) (pair 'b 2) (pair 'c 3)]
```

It is noteworthy that the definition of this function is not primitive recursive, meaning that the argument to the recursive call (in the second equation) is not a subterm of one of the constructor arguments on the left-hand side, such as `rest`, but rather the result of applying *another function* to such subterms, in this case the result of applying the `remove` operation (`-`) to `rest` and `k`.[5] The upshot is that sentences of the form (`forall` m `.` $\cdots$ `map->set` m $\cdots$) cannot be derived by ordinary structural induction on

5 For a discussion of primitive recursion in the context of inductive data types and recursively defined functions on them, see *Walther Recursion* by McAllester and Arkoudas [69].

m. Rather, we have to use another form of induction, *measure induction*, which requires that some quantity associated with the arguments of the function decreases strictly with every recursive call. In this case it is the *size* of the argument that decreases with every recursive call to map->set. The measure-induction method, a convenient variant of strong induction on the natural numbers, defined in the module strong-induction and discussed in Section 12.9, can be used to derive such goals, and indeed that method is used in the present setting to derive a key characterization result for map->set (see Exercise 10.26).

We now define two maps to be identical iff they are identical as sets:

```
assert* map-identity := (m1 = m2 <==> map->set m1 = map->set m2)
```

We can immediately put the identity conditions to the test:

```
> let {map1 := (alist->map [['b --> 2] ['a --> 1]]);
       map2 := (alist->map [['a --> 1] ['b --> 2] ['b --> 3]])}
    (eval map1 = map2)

Term: true

> let {map1 := (alist->map [['b --> 2] ['a --> 1]]);
       map2 := (alist->map [['b --> 2]])}
    (eval map1 = map2)

Term: false
```

Note that we could have introduced map->set prior to our definition of dom, and then defined the domain of a map as the set-theoretic domain of the corresponding binary relation:

```
(dom m = Set.dom map->set m).
```

That would have further underlined the close connection between the two theories, and would have also had the advantage of reuse. However, the more direct approach taken here has its own advantages, most notably the simplification of certain proof steps. We will illustrate the other approach shortly when we come to define the *range* of a map as the set-theoretic range of the binary relation obtained by map->set.

Exercise 10.25: Consider an inverse operation transforming sets of ordered pairs into maps:

```
declare set->map: (S, T) [(Set.Set (Pair S T))] -> (Map S T)
                         [[Set.alist->set]]

assert* set->map-def :=
  [(set->map null = empty-map)
   (set->map (k @ v) ++ A = [k v] ++ set->map A)]
```

What, if anything, might be wrong with this definition? □

A major characterization theorem needs to be proved at this point, namely, that two maps are identical in the sense we just described iff they are identical in the intuitive sense, that is, iff they produce identical outputs when applied to identical inputs:

```
define identity-characterization :=
  (forall m1 m2 . m1 = m2 <==> forall k . m1 at k = m2 at k)
```

This is fairly straightforward to derive once we have derived a closely related characterization of `map->set`:

```
define map->set-characterization :=
  (forall m k v . k @ v in map->set m <==> m at k = SOME v)
```

* ***Exercise 10.26:*** Prove `map->set-characterization` and `identity-characterization`. □

Let us now define the *range* of a map as follows:

```
declare range: (K, V) [(Map K V)] -> (Set.Set V) [[alist->map]]

assert* range-def := [(range m = Set.range map->set m)]

> (eval range ide-map)

List: [1 2 3]
```

Most of the results we derived about domains have natural analogues for ranges, albeit with some important differences.

```
conclude range-lemma-1 :=
  (forall m v . v in range m <==> exists k . k @ v in map->set m)
pick-any m v
  (!chain [(v in range m)
      <==> (v in Set.range map->set m)      [range-def]
      <==> (exists k . k @ v in map->set m) [Set.range-characterization]])

conclude range-characterization :=
  (forall m v . v in range m <==> exists k . m at k = SOME v)
pick-any m v
  (!chain [(v in range m)
      <==> (exists k . k @ v in map->set m) [range-lemma-1]
      <==> (exists k . m at k = SOME v)      [map->set-characterization]])

conclude range-lemma-2 :=
   (forall k v rest . v in range [k v] ++ rest)
pick-any k v rest
  (!chain<- [(v in range [k v] ++ rest)
         <== (v in Set.range map->set [k v] ++ rest)     [range-def]
         <== (v in Set.range k @ v ++ map->set rest - k) [map->set-def]
```

```
        <== (v in v ++ Set.range map->set rest - k)     [Set.range-def]
        <== (v = v | v in Set.range map->set rest - k)  [Set.in-def]
        <== (v = v)                                      [alternate]])
```

A subtle difference is that the analogue of `dom-lemma-2` does *not* hold:

```
define range-lemma-conjecture :=
  (forall m k v . range m subset range [k v] ++ m)

> (falsify range-lemma-conjecture 10)

List: ['success
|{
?k:Int := 1
?m:(Map Int Int) := (update (pair 1 1)
                            empty-map:(Map Int Int))
?v:Int := 0
}|]
```

This should not be surprising. A key can always be rebound to a different value in a map, thereby possibly removing the previous value from the range of the map (assuming that no other key maps to that same value). But this does not hold for domains: If a key is in a map, then no further additions will ever remove that key from the map's domain.

Exercise 10.27: Prove: `(forall m k . range m - k subset range m)`. □

We continue with an operation that *restricts* a map to a given set of keys, using the symbol `|^` as an infix alias for it:

```
declare restricted-to: (S, T) [(Map S T) (Set.Set S)] -> (Map S T)
                              [|^ 120 [alist->map Set.alist->set]]

assert* restrict-def :=
   [(empty-map |^ _ = empty-map)
    (k in A ==> [k v] ++ rest |^ A = [k v] ++ (rest |^ A))
    (~ k in A ==> [k v] ++ rest |^ A = rest |^ A)]

> (eval ide-map |^ ['b 'c])

List: [['b --> 2] ['c --> 3]]
```

We prove that the domain of any restriction of a map *m* is always a subset of the domain of *m*:

```
conclude restriction-theorem-1 := (forall m A . dom m |^ A subset A)
by-induction restriction-theorem-1 {
  empty-map =>
    pick-any A
```

```
      (!Set.subset-intro
        pick-any x
          (!chain [(x in dom empty-map |^ A)
               ==> (x in dom empty-map)       [restrict-def]
               ==> (x in null)                [dom-def]
               ==> false                      [Set.NC]
               ==> (x in A)                   [prop-taut]]))
 | (m as (update (pair k v) rest)) =>
     pick-any A
       let {IH := (forall A . dom rest |^ A subset A);
            lemma := (!chain-> [true ==> (dom rest |^ A subset A) [IH]])}
         (!two-cases
           assume (k in A)
             (!Set.subset-intro
               pick-any x
                 (!chain [(x in dom m |^ A)
                      ==> (x in dom [k v] ++ (rest |^ A)) [restrict-def]
                      ==> (x in k ++ dom rest |^ A)       [dom-def]
                      ==> (x = k | x in dom rest |^ A)    [Set.in-def]
                      ==> (x in A | x in dom rest |^ A)   [(k in A)]
                      ==> (x in A | x in A)               [Set.SC]
                      ==> (x in A)                        [prop-taut]]))
           assume (~ k in A)
             (!Set.subset-intro
               pick-any x
                 (!chain [(x in dom m |^ A)
                      ==> (x in dom rest |^ A)          [restrict-def]
                      ==> (x in A)                      [Set.SC]])))
 }
```

Exercise 10.28: Consider:

```
define restriction-theorem-2 :=
  (forall m A . dom m subset A ==> m |^ A = m)
```

Prove this result. □

The above operation restricts a map on a subset of its domain. There is an analogous operation that restricts a map on a subset of its range. Let us call this *range restriction* and denote it by the symbol ^|:

```
declare range-restriction: (S, T) [(Map S T) (Set.Set T)] -> (Map S T)
                                  [150 ^| [alist->map Set.alist->set]]

assert* range-restriction-def :=
  [(empty-map ^| _ = empty-map)
   ([k v] ++ rest ^| A = [k v] ++ (rest ^| A) <== v in A)
   ([k v] ++ rest ^| A = rest ^| A <== ~ v in A)]
```

For example:

```
define eye-color := [['bob --> 'brown]  ['tom --> 'blue] ['lisa --> 'green]
                      ['peter --> 'blue] ['ann --> 'brown]]

> (eval eye-color ^| ['blue])

List: [['tom --> 'blue] ['peter --> 'blue]]
```

We introduce a useful ternary predicate, agree-on, which holds between two maps m_1 and m_2 and a set of keys A iff m_1 and m_2 produce the same value for every key in A. There is no recursion involved in the definition of this agreement predicate. The definition is essentially an abbreviation: m_1 and m_2 agree on A iff their restrictions on A are identical:

```
declare agree-on: (S, T) [(Map S T) (Map S T) (Set.Set S)] -> Boolean
                  [[alist->map alist->map Set.alist->set]]

assert* agree-on-def := [((agree-on m1 m2 A) <==> m1 |^ A = m2 |^ A)]

> (eval (agree-on ide-map ide-map ['a 'b]))

Term: true

> (eval (agree-on [['a --> 1] ['b --> 2]]
                  [['b --> 3] ['a --> 1]]
                  ['b]))

Term: false
```

We have:

```
conclude agree-on-sym :=
  (forall m1 m2 A . (agree-on m1 m2 A) <==> (agree-on m2 m1 A))
pick-any m1 m2 A
  (!chain [(agree-on m1 m2 A)
      <==> (m1 |^ A = m2 |^ A)  [agree-on-def]
      <==> (m2 |^ A = m1 |^ A)  [sym]
      <==> (agree-on m2 m1 A)   [agree-on-def]])
```

We close this section with two more useful binary operations on maps: *overriding* and *composition*. Overriding is like union except that preference is given to the values of the rightmost map. In other words, applying the operation of overriding to two maps m_1 and m_2 results in a map that contains all and only the bindings of both maps, except that for bindings with a shared key k, preference is given to m_2 (the rightmost operand), meaning that the value associated with k in the resulting map is the same value that m_2 associates with k. We use the symbol ** as an alias for this operation:

```
declare override: (S, T) [(Map S T) (Map S T)] -> (Map S T)
                          [** [alist->map alist->map]]

assert* override-def :=
  [(m ** [] = m)
   (m ** [k v] ++ rest = [k v] ++ (m ** rest))]

> (eval [[1 --> 'a] [2 --> 'b]] ** [[1 --> 'foo] [3 --> 'c]])

List: [[1 --> 'foo] [3 --> 'c] [1 --> 'a] [2 --> 'b]]
```

The recursive definition of overriding stipulates that the empty map is a right identity element for the operation, but we can readily prove that it is also a left identity element:

```
by-induction (forall m . [] ** m = m) {
  (m as empty-map) =>
    (!chain [(empty-map ** m) = empty-map [override-def]])
| (m as (update (pair k v) rest)) =>
    (!chain [(empty-map ** m)
            = ([k v] ++ (empty-map ** rest))  [override-def]
            = ([k v] ++ rest)                 [([] ** rest = rest)]])
}
```

Exercise 10.29: Consider the following three conjectures (where `k`, `m1`, and `m2` are universally quantified in all three sentences):

```
(k in dom m2 ==> (m1 ** m2) at k = m2 at k);
   (dom m2 ** m1 = (dom m2) \/ (dom m1));
(range m2 ** m1 = (range m2) \/ (range m1)).
```

Prove or disprove each of these. □

One way to define the composition of a map m_2 with a map m_1 is as follows (there are alternative notions of composition, see below): For every key k in the domain of m_1, we retrieve the corresponding value v and then apply m_2 to it, provided that v is in the domain of m_2; if not, we ignore the binding of k to v. For instance, if m_1 and m_2 are, respectively, the following maps:

```
[['paris --> 'france] ['tokyo --> 'japan] ['cairo --> 'egypt]]
```

and

```
[['france --> 'europe] ['algeria --> 'africa] ['japan --> 'asia]],
```

then composing m_2 with m_1 would yield:

```
[['paris --> 'europe] ['tokyo --> 'asia]].
```

Note that the binding ['cairo --> 'egypt] is not included in the result, nor are any bindings from m_2. The following is a constructive definition of this operation; we use o as an infix alias for it.

```
declare compose: (S1, S2, S3) [(Map S2 S3) (Map S1 S2)] -> (Map S1 S3)
                               [o [alist->map alist->map]]

assert* compose-def :=
  [(_ o empty-map = empty-map)
   (m o [k v] ++ rest = [k v'] ++ (m o rest) <== m at v = SOME v')
   (m o [k v] ++ rest = m o rest <== m at v = NONE)]
```

It is not necessary for both maps to share the sorts of their keys and values. All that is needed is for the sort of the values of the first map to be the same as the sort of the keys of the second map. Thus:

```
> let {m1 := [[1 --> 'a] [2 --> 'b] [1 --> 'c]];
       m2 := [['a --> true] ['b --> false] ['foo --> true]]}
    (eval m2 o m1)

List: [[1 --> true] [2 --> false]]
```

And the foregoing example:

```
define capitals :=
  [['paris --> 'france] ['tokyo --> 'japan] ['cairo --> 'egypt]]

define countries :=
  [['france --> 'europe] ['algeria --> 'africa] ['japan --> 'asia]]

> (eval countries o capitals)

List: [['paris --> 'europe] ['tokyo --> 'asia]]
```

Composition is not commutative, and, as defined here, it is not associative either:

```
> (falsify (forall m1 m2 . m1 o m2 = m2 o m1) 10)

List: ['success
|{
?m1:(Map Int Int) := (update (pair 1 1)
                             empty-map:(Map Int Int))
?m2:(Map Int Int) := (update (pair 1 2)
                             empty-map:(Map Int Int))
}|]


> (falsify (forall m1 m2 m3 . m1 o (m2 o m3) = (m1 o m2) o m3) 10)

List: ['success
|{
```

```
?m1:(Map Int Int) := (update (pair 1 1)
                             empty-map:(Map Int Int))
?m2:(Map Int Int) := (update (pair 1 2)
                             (update (pair 1 1)
                                     empty-map:(Map Int Int)))
?m3:(Map Int Int) := (update (pair 1 1)
                             empty-map:(Map Int Int))
}|]
```

When keys and values are of the same sort, composition can be iterated indefinitely:

```
declare iterate: (S, S) [(Map S S) N] -> (Map S S)
                        [^^ [alist->map int->nat]]

define n := ?n:N

assert* iterate-def :=
        [(m ^^ zero = m)
         (m ^^ N.S n =  m o (m ^^ n))]

> let {m := (alist->map [[1 --> 2] [2 --> 3] [3 --> 1]]);
       _ := (print "\nm iterated once:    " (eval m ^^ 1));
       _ := (print "\nm iterated twice:   " (eval m ^^ 2));
       _ := (print "\nm iterated thrice: " (eval m ^^ 3))}
    (print "\nAre m and m^^3 identical?: " (eval m = m ^^ 3))

m iterated once:    [[1 --> 3] [2 --> 1] [3 --> 2]]
m iterated twice:   [[1 --> 1] [2 --> 2] [3 --> 3]]
m iterated thrice:  [[1 --> 2] [2 --> 3] [3 --> 1]]
Are m and m^^3 identical?:  true

Unit: ()
```

Exercise 10.30: The following states that domain restriction distributes over `**`:

```
(forall m2 m1 A . (m1 ** m2) |^ A = m1 |^ A ** m2 |^ A).
```

Prove this sentence. □

Exercise 10.31: Define a binary compatibility relation on maps, `<->`, so that m_1 `<->` m_2 iff m_1 and m_2 agree on the intersection of their domains:

```
declare compatible: (K, V) [(Map K V) (Map K V)] -> Boolean
                           [<-> [alist->map alist->map]]

assert* compatible-def :=
  [(m1 <-> m2 <==> agree-on m1 m2 (dom m1) /\ (dom m2))]
```

Prove or disprove the claim that <-> is an equivalence relation (namely, reflexive, symmetric, and transitive). □

Another notion of composition, applicable when keys and values are of the same sort, can be defined as follows:

```
declare o': (S) [(Map S S) (Map S S)] -> (Map S S)
                                          [[alist->map alist->map]]

assert* o'-def :=
  [(m o' [] = m)
   (m o' [k v] ++ rest = [k v'] ++ (m o' rest) <== m at v = SOME v')
   (m o' [k v] ++ rest = [k v] ++ (m o' rest)  <== m at v = NONE)]
```

This composition operator is not commutative either, as can be readily verified.

Exercise 10.32: Prove or disprove the claim that this alternative composition operator is associative. □

10.4.3 Default maps

In this subsection we outline an alternative approach to formalizing maps, which is similar to—but in certain ways more flexible than—arrays that are everywhere initialized with a default value. One advantage of this approach is that it maps keys to values directly, rather than to value options: If a key is not in the map's domain, the lookup operation simply returns the map's default value (say `zero`, if the values are natural numbers). In Athena, these maps are the only ones that may appear inside constraints to be solved by SMT (satisfiability-modulo-theories) algorithms; see the online appendix on automated theorem proving.These maps are also used in Chapter 18 to represent semantic states. They are defined in Athena's library in the top-level module `DMap`.

10.4.3.1 Representation and notation

Default maps are defined as the following polymorphic inductive structure, abstracted over the sort variables `K` and `V` (the sorts of keys and values, respectively):

```
structure (DMap K V) := (empty-map V) | (update (Pair K V) (DMap K V))

set-precedence empty-map 250
```

Note that this is virtually identical to the `Map` structure (page 533), with one important difference: The irreflexive constructor `empty-map`, which gets a relatively high precedence, is now unary rather than nullary (a constant). The sole argument of `empty-map` is a value—an element of sort `V`—that represents the default value that will be produced when a key is not present in the map. Other than that, these maps work as before: `update` incrementally adds new bindings to a map, a binding being simply an ordered pair of a key and a value. These maps also grow to the left. A new binding added by `update` might shadow an older binding with the same key.

As before, we also provide procedures that can serve as input expanders and output transformers, allowing us to input and output default maps in bracket-list notation. The input expander is `alist->dmap` ("Athena list to default map") and the output transformer is `dmap->alist` ("default map to Athena list"). In Athena (bracket) list notation, a default map is represented as a two-element list

$$[d\ pairs] \tag{10.10}$$

where d is the default value of the map and *pairs* is a list of pairs representing the map's bindings (with bindings on the left taking precedence). Each binding is written as either [*k* *v*] or as [*k* --> *v*]. The unary procedure `alist->dmap` is defined in terms of an auxiliary procedure `alist->dmap-general`, which, in addition to taking an Athena list of the form (10.10) as input, also takes a "preprocessor" procedure as a second argument and applies it to all keys and values, giving us a chance to, say, convert natural numbers to integers on the fly as we are converting the list to a default map term. The default value of this argument is the identity, meaning that no preprocessing of keys and values is done by default. The converse procedure, `dmap->alist`, is likewise defined in terms of the binary `dmap->alist-general`:

```
define (alist->dmap-general L preprocessor) :=
  match L {
    [d (some-list pairs)] =>
       letrec {loop := lambda (L)
                         match L {
                           [] => (empty-map d)
                         | (list-of (|| [k --> v] [k v]) rest) =>
                             (update (pair (preprocessor k)
                                           (preprocessor v))
                                     (loop rest))}}
         (loop pairs)
  | _ => L
  }

define (alist->dmap L) := (alist->dmap-general L id)

define (dmap->alist-general m preprocessor) :=
  letrec {loop := lambda (m pairs)
                    match m {
                      (empty-map d) => [d (rev pairs)]
                    | (update (pair k v) rest) =>
                        (loop rest (add [(preprocessor k) -->
                                         (preprocessor v)]
                                        pairs))
                    | _ => m}}
```

```
    (loop m [])

define (dmap->alist m) := (dmap->alist-general m id)
```

Thus, we have:

```
define dm := (alist->dmap [0 [['a --> 1] ['b --> 2] ['a --> 99]]])

> dm

Term: (DMap.update (pair 'a 1)
                   (DMap.update (pair 'b 2)
                                (DMap.update (pair 'a 99)
                                             (DMap.empty-map 0))))

> (dmap->alist dm)

List: [0 [['a --> 1] ['b --> 2] ['a --> 99]]]
```

It is convenient to also have a canonical output transformer that converts a default map term into Athena list notation as above, except that only one binding per key is shown, the leftmost one, which is the one that matters given that bindings with the same key further to the right are obsolete (shadowed):

```
define (remove-from m k) :=
  match m {
    (empty-map _) => m
  | (update (binding as (pair key val)) rest) =>
      check {(k equal? key) => (remove-from rest k)
            | else => (update binding (remove-from rest k))}
  }

define (dmap->alist-canonical-general m preprocessor) :=
  letrec {loop := lambda (m pairs)
                    match m {
                      (empty-map d) => [d (rev pairs)]
                    | (update (pair k v) rest) =>
                        (loop (remove-from rest k)
                              (add [(preprocessor k) -->
                                    (preprocessor v)]
                                   pairs))
                    | _ => m}}
    (loop m [])

> (dmap->alist-canonical-general dm id)

List: [0 [['a --> 1] ['b --> 2]]]
```

We issue some directives to make `update` more notationally flexible and define a number of variable abbreviations:

```
expand-input update [(alist->pair id id) alist->dmap]
overload ++ update
set-precedence ++ 210

define [key k k' k1 k2 d d' val v v' v1 v2] := [?key ···]
define [h t A B] := [Set.h Set.t Set.A Set.B]
define [m m' m1 m2 rest] := [?m:(DMap 'S1 'S2) ···]
```

10.4.3.2 Main operations, identity conditions, and results

The chief operation is of course key lookup:

```
declare apply: (K, V) [(DMap K V) K] -> V [110 [alist->dmap id]]

define at := apply

assert* apply-def :=
  [(empty-map d at _ = d)
   ([k v] ++ rest at x = v <== k = x)
   ([k v] ++ rest at x = rest at x <== k =/= x)]
```

Let's test the definition:

```
define dm := (alist->dmap [0 [['a --> 1] ['b --> 2] ['c --> 3]]])

> (eval dm at 'a)

Term: 1

> [(eval dm at 'b) (eval dm at 'c)]

List: [2 3]

> (eval dm at 'foo)

Term: 0
```

Let's define a function to get the default value of any map:

```
declare default: (K, V) [(DMap K V)] -> V [200 [alist->dmap]]

assert* default-def :=
  [(default empty-map d = d)
   (default _ ++ rest = default rest)]

> (eval default dm)

Term: 0
```

Key removal is defined as follows:

```
declare remove: (S, T) [(DMap S T) S] -> (DMap S T)
                        [- 120 [alist->dmap id]]

left-assoc -

assert* remove-def :=
  [(empty-map d - _ = empty-map d)
   ([key _] ++ rest - key = rest - key)
   (key =/= x ==> [key val] ++ rest - x = [key val] ++ (rest - x))]

> (eval dm - 'b)

Term: (DMap.update (pair 'a 1)
                   (DMap.update (pair 'c 3)
                                (DMap.empty-map 0)))
```

The domain of a default map is a bit more subtle. We count only those keys that are bound to a value other than the map's default value. Importantly, we must then make sure that subsequent bindings of that key are removed:

```
declare dom: (K, V) [(DMap K V)] -> (Set.Set K) [[alist->dmap]]

assert* dom-def :=
  [(dom empty-map _ = null)
   (dom [k v] ++ rest = dom (rest - k) <== v = default rest)
   (dom [k v] ++ rest = k ++ dom rest <== v =/= default rest)]

define dm :=
   (alist->dmap [0 [['a --> 1] ['b --> 0] ['c --> 3] ['b --> 99]]])

> (eval dom dm)

Term: (Set.insert 'a
                  (Set.insert 'c
                              Set.null:(Set.Set Ide)))
```

The size of a default map is defined as the cardinality of its domain:

```
declare size: (S, T) [(DMap S T)] -> N [[alist->dmap]]

assert* size-def := [(size m = Set.card dom m)]
```

We list some pertinent results below, leaving most proofs as exercises:

```
define remove-correctness-0 := (forall m x . ~ x in dom m - x)

conclude remove-correctness-1 := (forall m x . m - x at x = default m)
by-induction remove-correctness-1 {
  (m as (empty-map d)) =>
     pick-any x
```

```
      (!chain [(m - x at x)
            = (m at x)                                [remove-def]
            = d                                       [apply-def]
            = (default m)                             [default-def]])
| (m as (update (pair k v) rest)) =>
    let {IH := (forall x . rest - x at x = default rest)}
      pick-any x
        (!two-cases
           assume (k = x)
            (!chain [(m - x at x)
                   = (m - k at k)                     [(k = x)]
                   = (rest - k at k)                  [remove-def]
                   = (default rest)                   [IH]
                   = (default m)                      [default-def]
                    ])
           assume (k =/= x)
             (!chain [(m - x at x)
                    = ((k @ v) ++ (rest - x) at x) [remove-def]
                    = (rest - x at x)                 [apply-def]
                    = (default rest)                  [IH]
                    = (default m)                     [default-def]]))
}

define remove-correctness-2 :=
  (forall m k x . k =/= x ==> m - k at x = m at x)

define remove-correctness-3 :=
  (forall m k . default m = default m - k)

define dom-lemma-1 :=
  (forall k v rest . v =/= default rest ==> k in dom [k v] ++ rest)

define dom-lemma-2 :=
  (forall m k v . v =/= default m ==> dom m subset dom [k v] ++ m)

define dom-lemma-2b :=
  (forall m x k v . v =/= default m & x in dom m ==> x in dom [k v] ++ m)
```

In addition to size, a very simplistic notion of map length is also useful for strong inductive proofs:

```
define [< <=] := [N.< N.<=]

declare len: (K, V) [(DMap K V)] -> N [[alist->dmap]]

assert* len-def :=
  [(len empty-map _ = zero)
   (len [_  _] ++ rest = S len rest)]

define len-lemma-1 := (forall m k v . len m < len (k @ v) ++ m)
```

```
define len-lemma-2 := (forall m k . len m - k <= len m)
define len-lemma-3 := (forall key val k rest .
                         len rest - k < len key @ val ++ rest)

define lemma-D := (forall m k . k in dom m <==> m at k =/= default m)
```

```
define dom-lemma-3 := (forall m k . dom m - k subset dom m)

conclude dom-lemma-3 := (forall m k . dom (m - k) subset dom m)
pick-any m:(DMap 'K 'V) k:'K
  (!Set.subset-intro
    pick-any x:'K
      let {[rc1 rc2 rc3] := [remove-correctness-1
                             remove-correctness-2
                             remove-correctness-3]}
        assume hyp := (x in dom m - k)
          (!two-cases
            assume (x = k)
              let {L := (!chain-> [true
                                 ==> (m - k at k = default m) [rc1]])}
                (!chain-> [hyp
                       ==> (k in dom m - k)                  [(x = k)]
                       ==> (m - k at k =/= default m - k)    [lemma-D]
                       ==> (m - k at k =/= default m)        [rc3]
                       ==> (L & m - k at k =/= default m)    [augment]
                       ==> false                             [prop-taut]
                       ==> (x in dom m)                      [prop-taut]])
            assume (x =/= k)
              (!chain-> [hyp
                     ==> (m - k at x =/= default m - k)      [lemma-D]
                     ==> (m at x =/= default m - k)          [rc2]
                     ==> (m at x =/= default m)              [rc3]
                     ==> (x in dom m)                        [lemma-D]])))

define dom-corollary-1 :=
  (forall key val k rest .
    k in dom rest - key ==> k in dom [key val] ++ rest)
```

The following will extract a set of key-value pairs from a map. Note that, consistent with the definition of dom, we only include those bindings whose values are different from the map's default value:

```
declare dmap->set: (K, V) [(DMap K V)] -> (Set.Set (Pair K V))
                          [[alist->dmap]]

assert* dmap->set-def :=
  [(dmap->set empty-map _ = null)
```

```
  (dmap->set k @ v ++ rest = dmap->set rest - k
         <== v = default rest)

  (dmap->set k @ v ++ rest = (k @ v) ++ dmap->set rest - k
         <== v =/= default rest)]
```

The following are useful results:

```
define dms-lemma-1 := (forall k v . k @ v in dmap->set m ==> k in dom m)

define dms-lemma-1b :=
   (forall m k . ~ k in dom m ==> forall v . ~ k @ v in dmap->set m)

define dms-lemma-1b' :=
  (forall m k . ~ k in dom m ==> ~ exists v . k @ v in dmap->set m)

define dms-theorem-1 :=
  (forall m k v . k @ v in dmap->set m ==> m at k = v)

define dms-theorem-2 :=
  (forall m k . ~ k in dom m ==> m at k = default m)

define dms-lemma-2 :=
  (forall m k k' . k in dom m & k =/= k' ==> k in dom m - k')

define dms-lemma-3 :=
  (forall m key val .
    val =/= default m ==> dom key @ val ++ m = key ++ dom m - key)

define dms-theorem-3 :=
  (forall m k . k in dom m ==> exists v . k @ v in dmap->set m)

define at-characterization :=
  (forall m k v . m at k = v <==> k @ v in dmap->set m |
                                  ~ k in dom m & v = default m)
```

We now assert the following identity conditions for default maps:

```
assert* dmap-identity :=
 (forall m1 m2 . m1 = m2 <==> default m1 = default m2 &
                              dmap->set m1 = dmap->set m2)
```

In other words, two maps are identical iff (a) their default values are the same, and (b) they have the exact same sets of bindings, as these are computed by dmap->set.

A notion of agreement on a set of keys is introduced as follows:

```
declare agree-on: (S, T) [(DMap S T) (DMap S T) (Set.Set S)] -> Boolean
                         [[alist->dmap alist->dmap Set.alist->set]]
```

```
assert* agree-on-def :=
  [(agree-on m1 m2 null)
   ((agree-on m1 m2 h Set.++ t) <==> m1 at h = m2 at h &
                                     (agree-on m1 m2 t))]

> let {dm1 := (alist->dmap [0 [['a --> 1] ['b --> 2] ['c --> 3]]]);
       dm2 := (alist->dmap [9 [['a --> 1] ['b --> 8] ['c --> 3]]])}
   [(eval agree-on dm1 dm2 ['a 'c])
    (eval agree-on dm1 dm2 ['a 'b])]

List: [true false]

define agreement-characterization :=
  (forall A m1 m2 . (agree-on m1 m2 A) <==>
                    forall k . k in A ==> m1 at k = m2 at k)

define downward-agreement-lemma :=
  (forall B A m1 m2 . (agree-on m1 m2 A) & B subset A ==>
                      (agree-on m1 m2 B))
```

And a restriction operation:

```
declare restricted-to: (S, T) [(DMap S T) (Set.Set S)] -> (DMap S T)
                               [150 |^ [alist->dmap Set.alist->set]]

assert* restrict-def :=
   [(empty-map d |^ _ = empty-map d)
    (k in A ==> [k v] ++ rest |^ A = [k v] ++ (rest |^ A))
    (~ k in A ==> [k v] ++ rest |^ A = rest |^ A)]
```

As a nonconstructive characterization of identity, we prove that two maps are identical iff their default values are the same and they map *all* keys to the same values:

```
define identity-characterization :=
  (forall m1 m2 . m1 = m2 <==> default m1 = default m2 &
                              forall k . m1 at k = m2 at k)
```

The explicit presence of the condition

```
default m1 = default m2                                    (10.11)
```

might seem odd. After all, one might expect that condition to be redundant in view of the seemingly stronger condition

```
forall k . m1 at k = m2 at k.                              (10.12)
```

But that is actually not the case when there are only finitely many possible keys and the domain of the maps includes all of them. For example, consider:

```
define m1 :=  (alist->dmap [0 [[true --> 1] [false --> 0]]])
define m2 :=  (alist->dmap [9 [[true --> 1] [false --> 0]]])
```

For these two maps, with `Boolean` keys, condition (10.12) holds but (10.11) does not, and hence the two are not identical (according to `dmap-identity`). However, when there are infinitely many possible keys, then (10.12) does entail (10.11).

10.5 Chapter notes

Ordered pairs are an extremely fundamental concept in mathematics, as well as in programming. They are discussed—at least in passing—in most mathematics textbooks, especially discrete-mathematics textbooks, and in virtually all set-theory books (see below for some references). Their importance is only matched by their modesty: There is not much to say about them. The main property that we want is for two putative ordered pairs to be identical iff their first and second components are respectively identical. As long as that holds, the corresponding objects can serve as ordered pairs. In our case, as we saw, this property follows immediately from structural axioms applicable to all algebraic datatypes. In programming, ordered pairs are the workhorse of Lisp-like languages, where lists are formed by chaining together ordered pairs, but they are widely used in most languages, including popular contemporary languages like C++ and Python.

Options have been used primarily in type theory and functional programming languages like ML and Haskell, but they are also useful in specification and proofs. Values of "nullable types" in languages such as Java (essentially pointers to heap-allocated objects) can be viewed as a less rigorous version of the same basic idea.

Sets are another extremely fundamental concept.[6] All branches of modern mathematics are ultimately based on set theory, in that all standard mathematical objects (functions, sequences, relations, etc.) can be viewed as sets. For expositions of classical set theory see *Foundations of Set Theory* by A. A. Fraenkel et al. [38]; *Axiomatic Set Theory* by Patrick Suppes [100], which contains many of the theorems derived in this chapter; and *Set Theory* by Thomas Jech [52]. These books cover classical set theory, mostly as developed in the early 20th century by Zermelo and Fraenkel, known as *Zermelo-Fraenkel set theory*, or simply as ZF. ZF is untyped, and "naive" untyped set theories are susceptible to certain inconsistencies, such as Russell's paradox [49]. ZF avoids such paradoxes by postulating

6 In some respects they are more fundamental even than ordered pairs, as ordered pairs can be defined as sets. One of the earliest purely set-theoretic definitions of an ordered pair (a,b) was famously given by Kuratowski [62] in 1921:

$$(a,b) = \{\{a,b\},\{a\}\}.$$

However, instead of analyzing it in terms of sets, it is possible to take the notion of an ordered pair as primitive, along with the notion of function, and develop mathematics on that basis.

additional axioms. In typed systems, such as that of Athena, this is not necessary. Indeed, Russell originally developed type theory precisely in order to avert the various set-theoretic paradoxes that emerged in the beginning of the last century. Outside of pure mathematics, sets play a fundamental role in specification. The specification language Z [33] [50], for instance, is based on (typed) set theory.

Here we have developed a theory of finite sets, but most of the proofs would carry over unchanged to a slight modification of the theory that could accommodate infinite sets. Of course, in that case we couldn't have constructive definitions of set operations, so our axioms would have to start with what are now derived characterization theorems. For example, a defining axiom for the intersection operation would assert that an element belongs to $(A \cap B)$ iff it belongs both to A and to B.

Maps are a standard *abstract data type* (ADT); they are also known as *dictionaries* or *associative arrays*. ADTs are usually formulated and studied in an imperative, stateful style that allows for destructive modification. For instance, in the classical imperative style, an operation that removes a key from a map does its work as a "side effect" by modifying the *state* of the map on which it is applied. The type of the operation's result is something like `void` or `unit`, indicating that there is no proper result returned by the operation. In stateful ADTs, operations are typically divided into three classes:[7]

1. *creators* and *producers*;
2. *observers*; and
3. *mutators*.

The creators of an ADT T create new values of T from values of other types, while its producers create new values of T from other values of T (possibly along with non-T values). Observers take T values as arguments and return values of other types as results. Finally, mutators statefully *modify* values of type T, without returning new values.

In this chapter we have given a *functional* or *algebraic* formulation of the dictionary ADT (and of the set ADT as well). In functional ADT formulations, there are no mutators.[8] Operations such as key removal, which would be modeled as mutators in a stateful formulation of the ADT, are formulated here as *pure mathematical functions* instead, which take the old version of the map as input and return the new ("modified") map as output:

```
declare remove: (K, V) [(Map K V) K] -> (Map K V)
```

In the algebraic setting, they can be thought of as producers rather than mutators.

---

7 Four if one bothers to distinguish creators from producers.

8 Creators and producers are usually represented as *constructors*, such as `update` and `empty-map`. Note that constructors representing producers are always reflexive. Not all producers are constructors, however. Key removal, for instance, is modeled as a producer. Observers are operations such as `size`.

Algebraic formulations of ADTs tend to be more mathematically tractable and easier to reason about. Perhaps the first algebraic formulation of a dictionary-like ADT[9] was given in a seminal 1962 paper by John McCarthy [70], consisting of

1. three sorts: `Value`, `Index`, and `Array`;
2. two functions, *read* and *write*, with the following types:

$$write : \texttt{Array} \times \texttt{Index} \times \texttt{Value} \longrightarrow \texttt{Array}$$
$$read : \texttt{Array} \times \texttt{Index} \longrightarrow \texttt{Value}$$

3. and two axioms, in the form of conditional equations:

$$i = j \Rightarrow read(write(a,i,v),j) = v$$
$$i \neq j \Rightarrow read(write(a,i,v),j) = read(a,j)$$

This humble array theory is automated by SMT solvers and plays a key role in modern software verification and program analysis [28]. We will return to this theory, using different notation, in Section 16.1.

9 Arrays and maps are not quite the same thing, and our theory of maps is considerably different from McCarthy's theory, but abstracted indices can be seen as keys of sorts.

IV PROOFS ABOUT ALGORITHMS

One of the most important applications of logic in computer science is proving the correctness of an algorithm. In this part, we explore this topic in terms of three examples, a binary search algorithm, an exponentiation algorithm, and a greatest common divisor algorithm. We already have at hand most of the tools of logic that we need, and indeed, upon reflection, we can see that we even have done some proofs of algorithm correctness already. Even our axiomatic definitions of functions such as `+`, `*`, and `**` on natural numbers can be considered to embody algorithms, but as such they are rather trivial. These "algorithms" have little practical value but are useful mainly as simple starting points for understanding the somewhat more complex issues in other settings. But a potential obstacle to moving on to algorithms of greater complexity and practicality is the way we have expressed function definitions as a set of axioms, which might seem unfamiliar to most readers with programming experience in mainstream languages (or even in research languages). To narrow this gap, at least to a modest extent, we introduce the `fun` procedure, which permits one to express algorithms in a style that is a step closer to one of the traditional programming paradigms, functional programming. At the same time, we remain firmly based in logic, since a definition using `fun` translates straightforwardly into a set of first-order axioms.

Our first example is the correctness of a binary search algorithm on binary search trees. The `binary-search` function we define is very naturally expressed with conditional control and recursion, and thus can be elegantly expressed in a functional style (like that of `fun`).

In Chapter 12, our second example of algorithm correctness, we study an exponentiation algorithm that requires substantially fewer multiplications than the straightforward one we studied in Chapter 3. Such algorithms are more commonly expressed with iterative (looping) control structures and assignment statements. Here we stick with a recursive definition easily expressible with `fun`, but this formulation could serve as a guide to programming the algorithm in an iterative, memory-updating form if that is preferred. We introduce one additional proof tool in Chapter 12, strong induction, and a variant of it, measure induction.

For our third example of algorithm correctness, in Chapter 13, we consider Euclid's algorithm for computing greatest common divisors. In preparation, we define division and remainder operations on natural numbers and prove a few of their key properties. These operations and the Euclidean algorithm itself are quite naturally expressed with recursion. Strong induction plays an important role here too.

All three algorithms are also interesting for historical reasons. Which is the oldest? Which was the first to be accompanied by a correctness proof? You might be surprised by the answers, which we discuss in the chapter notes at the end of Chapter 13.

11 A Binary Search Algorithm

11.1 Defining the algorithm

Binary search is an algorithm capable of searching a binary tree containing n items with only $O(\log n)$ comparisons, assuming the tree is balanced (roughly speaking, this means that each path from the root to a leaf node is about $\log_2 n$ in length). To define the binary search function on binary trees of natural numbers,[1] we start as usual with its declaration:

```
extend-module BinTree {

  declare binary-search: [N (BinTree N)] -> (BinTree N)

  module binary-search {
# ...
```

Recall that in module BinTree we have already defined the following variables:

```
define [x y z T L R] :=
  [?x:N ?y:N ?z:N ?T:(BinTree N) ?L:(BinTree N) ?R:(BinTree N)]
```

To define how binary-search is computed, we call upon a library procedure named fun to translate a simpler input into defining axioms:

```
assert axioms :=
  (fun
   [(binary-search x (node L y R)) =
        [(node L y R)           when (x = y)
         (binary-search x L)    when (x < y)
         (binary-search x R)    when (x =/= y & ~ x < y)]
    (binary-search x null) = null])

define [at-root go-left go-right empty] := axioms
```

The effect of the **assert** of the fun together with the **define** is exactly the same as if we had written

```
assert* at-root :=
  (x = y ==> (binary-search x (node L y R)) = (node L y R))

assert* go-left :=
  (x < y ==> (binary-search x (node L y R)) = (binary-search x L))

assert* go-right :=
  (x =/= y & ~ x < y
```

1 In Chapter 15 we give a more general treatment, with binary trees over any sort with a strict weak ordering relation (an abstraction discussed in Chapter 14).

```
  ==> (binary-search x (node L y R)) = (binary-search x R))

assert* empty := ((binary-search x null) = null)
```

For the details of how fun transforms its argument into defining axioms, see Exercise 11.11. From the binary-search example, we see that using fun affords important advantages over writing the axioms directly. An obvious advantage is that fun produces the universal closure of each clause of the definition, which saves writing the quantifiers explicitly. But the more important advantage comes from the fact that when there are several different axioms for a single argument pattern, corresponding to different conditions, one can write the left-hand side only once and put the right-hand sides and their enabling conditions in a list. This syntax is similar to multipart definitions in conventional mathematical notation, where one might write

$$binary\text{-}search(x, node(L,y,R)) = \begin{cases} node(L,y,R) & \text{where } x = y; \\ binary\text{-}search(x,L) & \text{where } x < y; \\ binary\text{-}search(x,R) & \text{where } x \neq y \text{ and } \sim x < y. \end{cases}$$

Actually, one would probably write the third condition simply as the equivalent $y < x$, but writing it as the conjunction of the negations of the preceding conditions, as we have, makes it syntactically explicit that the condition is disjoint from all of the previous ones. Another reason this is helpful is that when we structure proofs using case analysis based on the same predicates that appear in the axioms, all of the required conditions will be in the assumption base when needed. For the binary-search axioms, the skeletal structure can be:

```
(!two-cases
  assume (x = y)
    # here at-root can be applied
      ...
  assume (x =/= y)
    (!two-cases
      assume (x < y)
        # here go-left can be applied, since (x < y) is
        # in the assumption base
        ...
      assume (~ x < y)
        # here go-right can be applied, since both (x =/= y)
        # and (~ x < y) are now in the assumption base
        ...
    ))
```

This is the structure we use in the proofs of correctness theorems about binary-search in the next section.

In conventional mathematics, one often simply writes "otherwise" for the last condition, but the `fun` procedure requires writing the last condition in full.[2] It is worth the extra typing to have the condition expressed fully, for readability and for purposes of reference when writing proofs. The lack of an "otherwise" option also means that the clauses could be reordered without changing the meaning, just as is the case with the generated axioms or with axiom lists or individually asserted axioms.

11.1.1 Efficiency considerations

While the order of presentation makes no difference for the use of axioms in proofs, it can have a significant effect on efficiency when the axioms are used for computation. In Athena, axioms defining a function *f* are compiled into code, which is then used by the `eval` procedure to power the evaluation of ground terms. Let's look at the code that is compiled from the axioms generated by our `fun` definition of `binary-search`:[3]

```
define (binary-search' t1 t2) :=
  match [t1 t2] {
      [x3 (node x1 x4 x2)] =>
        check {(=' x3 x4) => (node x1 x4 x2)
              | (N.Less.<'' x3 x4) => (binary-search' x3 x1)
              | else => (binary-search' x3 x2)}
    | [x1 null:(BinTree N)] => null:(BinTree N)
    }
```

The axioms have been compiled into a **match** expression with two clauses, one for each of the two constructors of the `BinTree` datatype, which constitute a case analysis for the structure of the second argument to `binary-search`. The clause corresponding to the `node` constructor includes a further case analysis implemented by a **check**, which first tests whether the key is equal to the root value, and then whether it is less than it. But note that the last clause of the **check** expression does not include a proper condition, having a catch-all **else** guard instead, whereas a completely straightforward translation of the corresponding clause of the `fun` definition, namely

```
(binary-search x R) when (x =/= y & ~ x < y)
```

would result in:

```
check {(=' x3 x4) => (node x1 x4 x2)
      | (N.Less.<'' x3 x4) => (binary-search' x3 x1)
      | (&& (negate (=' x3 x4)) (negate (N.Less.<'' x3 x4)))  =>
          (binary-search' x3 x2)}
```

---

2 But note that other similar mechanisms in Athena do not require this. For example, the `ite*` procedure described in Exercise 6.2 allows for a trailing "catch-all" guard written as a wildcard.

3 Athena usually compiles the definition of a function symbol *f* into a procedure named *f*' or *f*''.

But because each conjunct in the last guard negates a condition already tested in previous clauses, the guard simplifies to `(&& true true)`, and thus it can be replaced by **`else`**.

We see that such simplifications depend crucially on the order of the given conditions. There is actually a different order for this example that results in more efficient computation (in terms of the number of comparisons done on average), as we will see in Section 11.4. One might conclude from such considerations that we need to be aware of all the details of how axioms are compiled into Athena code, and beyond that, the details of all the transformations it may undergo on the way to machine code (as, for example, if the Athena code is compiled into ML and thence to C and finally to machine code). But for most purposes, the following guarantees about the compiled Athena code suffice: (1) the order in which **`check`** clauses (or potentially **`match`** clauses with **`where`** guards) are listed is the same as the order in which the axioms with the corresponding conditions are presented; and (2) each guard in a clause is simplified by removing the negation of any guard of a preceding clause. These guarantees ensure that lines in a `fun` serve, for purposes of computation, essentially like an "if-then-else" or "cond" construct in conventional high-level languages.

11.1.2 Correspondence to definitions in other languages

With that understanding, and in a programming language that allows datatype constructor patterns as formal parameters of function definitions, such as ML or Haskell, one can write the binary search definition in a form very similar to the way we have written it using `fun`. In ML, for example, we could write

```
datatype BinTree = null | node of BinTree * int * BinTree

fun binarySearch(x,node(L,y,R)) =
      if x = y then
         node(L,y,R)
      else if x < y then
         binarySearch(x,L)
      else
         binarySearch(x,R)
  | binarySearch(x,null) = null
```

Most languages, however, lack such pattern matching, and we would thus be restricted to using only variables as formal parameters. Then, in the body of the function, we would need to replace occurrences of pattern variables with the corresponding accessor function call or other component access notation. In C++, for example, we would write something like

```
struct node;
typedef node* BinTree;

struct node {
```

```
  BinTree left; int key; BinTree right;
  node(BinTree L, int y, BinTree R) : left(L), key(y), right(R) {}
};

BinTree binarySearch(int x, BinTree T)
{
  if (T == NULL)
    return NULL;
  if (x == T->key)
    return T;
  if (x < T->key)
    return binarySearch(x, T->left);
  return binarySearch(x, T->right);
}
```

(Note that in this case the test for the empty tree must be done first, since otherwise the expression `T->key`, etc., would cause an error when `T` is a null pointer.)

11.1.3 Interface design

Before going on to proofs, let us note one additional factor in algorithm design that is often just as important as computational efficiency, namely, the interface. In the case of `binary-search`, it would be simpler to return a Boolean value, `true` if the target value is found, `false` otherwise. (Most significantly, that interface would make some of the issues we have to deal with in the upcoming proofs disappear.) By returning a node or `null`, however, we still have an easy test for success—`((binary-search` x T`) =/= null)`—and we make it possible to continue with other computations at the position in the tree where the search value is found, without recomputing that position. For example, if we wanted to output all elements in the tree within a range from x to y, where $x < y$, we could use `binary-search` to find x and from that node do an inorder tree walk until y is encountered. (A tree walk starting from an interior node needs the tree to have parent links as well as left and right subtree links, so this could not be done with the simple form of tree considered here, but the binary search algorithm would remain the same if parent links were included.)

11.1.4 Testing with evaluation

As in earlier cases, it is a good idea to test our definition of binary search using evaluation. To make it easy to define trees for test cases, we can arrange to use decimal integer notation instead of canonical natural numbers, using the techniques introduced in Section 3.12. We first define tree input and output conversion procedures as follows (to do this we get back to the `BinTree` level, since these procedures are useful for working with other functions besides `binary-search`):

```
} # close module binary-search

define nat-tree->int-tree :=
  lambda (T)
    match T {
      (node L x R) => (node (nat-tree->int-tree L)
                            (nat->int x)
                            (nat-tree->int-tree R))
    | null:(BinTree N) => null:(BinTree Int)
    | _ => T
    }

define int-tree->nat-tree :=
  lambda (T)
    match T {
      (node L x R) => (node (int-tree->nat-tree L)
                            (int->nat x)
                            (int-tree->nat-tree R))
    | null:(BinTree Int) => null:(BinTree N)
    | _ => T
    }
```

We can then use input expansion and output transformation to work with trees such as the one defined in Figure 8.1:

```
expand-input binary-search [int->nat int-tree->nat-tree]
transform-output eval [nat-tree->int-tree]

> (eval (binary-search 5 tree1))

Term: (node (node null:(BinTree Int)
                  3
                  null:(BinTree Int))
            5
            null:(BinTree Int))

> (eval (binary-search 11 tree1))

Term: (node null:(BinTree Int)
            11
            null:(BinTree Int))

> (eval (binary-search 12 tree1))

Term: null:(BinTree Int)
```

11.2 First correctness properties

The first correctness properties we consider concern what the result returned by (binary-search x T) implies about the presence or absence of x in T.

```
extend-module binary-search {

  define found :=
    (forall T . BST T ==>
      forall x L y R . (binary-search x T) = (node L y R) ==> x = y & x in T)

  define not-found :=
    (forall T . BST T ==> forall x . (binary-search x T) = null ==> ~ x in T)
```

The proof of found will be by induction on binary trees. Let's first write the proof of the basis case, encapsulating it in the following method:

```
define tree-axioms := (datatype-axioms "BinTree")

define (binary-search-found-base) :=
 conclude (BST null ==>
             forall x L y R .
               (binary-search x null) = (node L y R)
               ==> x = y & x in null)
    assume (BST null)
      pick-any x:N L:(BinTree N) y:N R:(BinTree N)
        assume i := ((binary-search x null) = (node L y R))
          let {p  := (!chain [null:(BinTree N)
                             = (binary-search x null)    [empty]
                             = (node L y R)              [i]]);
               -p := (!chain-> [true
                            ==> (null =/= (node L y R))  [tree-axioms]])}
            (!from-complements (x = y & x in null) p -p)
```

Now we can check the basis case proof before going on to the induction step:

```
> (!binary-search-found-base)

Theorem: (if (BinTree.BST null:(BinTree N))
             (forall ?x:N
               (forall ?L:(BinTree N)
                 (forall ?y:N
                   (forall ?R:(BinTree N)
                     (if (= (BinTree.binary-search ?x:N
                                                   null:(BinTree N))
                            (node ?L:(BinTree N)
                                  ?y:N
```

```
                              ?R:(BinTree N)))
                    (and (= ?x:N ?y:N)
                         (BinTree.in ?x:N
                                     null:(BinTree N)))))))))
```

We therefore continue with the induction step. To simplify the notation somewhat, we introduce a property procedure, found-property, which allows us to express the desired found result as follows:

```
(forall T . BST T ==> found-property T).
```

After applying the definition of BST to (node L y R) and combining the result with the induction hypotheses, the proof proceeds by case analysis, setting up cases according to whether x is found at the root, in the left subtree, or in the right subtree, and applying the corresponding binary-search axiom in each case.

```
define [x1 y1 L1 R1] := [?x1:N ?y1:N ?L1:(BinTree N) ?R1:(BinTree N)]

define (found-property T) :=
  (forall x L1 y1 R1 .
     (binary-search x T) = (node L1 y1 R1) ==> x = y1 & x in T)

define binary-search-found-step :=
 method (T)
   match T {
     (node L:(BinTree N) y:N R:(BinTree N)) =>
       let {[ind-hyp1 ind-hyp2] := [(BST L ==> found-property L)
                                    (BST R ==> found-property R)]}
       assume hyp := (BST T)
         conclude (found-property T)
           let {p0 := (BST L &
                       (forall x . x in L ==> x <= y) &
                       BST R &
                       (forall z . z in R ==> y <= z));
                _ := (!chain-> [hyp ==> p0                  [BST.nonempty]]);
                fpl := (!chain-> [p0 ==> (BST L)            [prop-taut]
                                     ==> (found-property L) [ind-hyp1]]);
                fpr := (!chain-> [p0 ==> (BST R)            [prop-taut]
                                     ==> (found-property R) [ind-hyp2]])}
          pick-any x:N L1 y1:N R1
            let {subtree := (node L1 y1 R1)}
             assume hyp' := ((binary-search x T) = subtree)
              conclude (x = y1 & x in T)
                (!two-cases
                   assume (x = y)
                     (!both conclude (x = y1)
                              (!chain->
                             [T = (binary-search x T)    [at-root]
                                = subtree                [hyp']
```

```
                         ==> (y = y1)                  [tree-axioms]
                         ==> (x = y1)                  [(x = y)]])
                     conclude (x in T)
                          (!chain-> [(x = y)
                                 ==> (x in T)          [in.root]]))
            assume (x =/= y)
               (!two-cases
                 assume (x < y)
                   (!chain-> [(binary-search x L)
                           = (binary-search x T) [go-left]
                           = subtree             [hyp']
                         ==> (x = y1 & x in L)   [fpl]
                         ==> (x = y1 & x in T)   [in.left]])
                 assume (~ x < y)
                   (!chain-> [(binary-search x R)
                           = (binary-search x T) [go-right]
                           = subtree             [hyp']
                         ==> (x = y1 & x in R)   [fpr]
                         ==> (x = y1 & x in T)   [in.right]])))
  }
```

The complete proof is then:

```
> by-induction found {
    null                => (!binary-search-found-base)
  | (T as (node _ _ _)) => (!binary-search-found-step T)
}

Theorem: (forall ?T:(BinTree N)
           (if (BinTree.BST ?T:(BinTree N))
               (forall ?x:N
                 (forall ?L:(BinTree N)
                   (forall ?y:N
                     (forall ?R:(BinTree N)
                       (if (= (BinTree.binary-search ?x:N
                                                     ?T:(BinTree N))
                              (node ?L:(BinTree N)
                                    ?y:N
                                    ?R:(BinTree N)))
                           (and (= ?x:N ?y:N)
                                (BinTree.in ?x:N
                                            ?T:(BinTree N))))))))))
```

Turning to the proof of `not-found`, we will write the basis case and induction step proofs inline. The basis case is an immediate consequence of `BinTree.in.empty`. In the induction step, we again start by applying the definition of `BST` to `T` and combining the result with the induction hypotheses. The conclusion we seek being a negative, we naturally use proof by contradiction, but there are several different cases to be considered, and we must find a contradiction within each. We define a property procedure (not-found-prop) for this proof, too.

```
define (not-found-prop T) :=
  (forall x . (binary-search x T) = null ==> ~ x in T)

by-induction not-found {
  null =>
    assume (BST null)
      conclude (not-found-prop null)
        pick-any x:N
          assume ((binary-search x null) = null)
            (!chain-> [true ==> (~ x in null)    [in.empty]])
| (T as (node L y:N R)) =>
    let {[p1 p2]              := [(not-found-prop L) (not-found-prop R)];
         Less=-tricho-4       := N.Less=.trichotomy4;
         [ind-hyp1 ind-hyp2] := [(BST L ==> p1) (BST R ==> p2)]}
    assume hyp := (BST T)
      conclude (not-found-prop T)
        let {smaller-in-left := (forall x . x in L ==> x <= y);
             larger-in-right := (forall z . z in R ==> y <= z);
             p0 := (BST L &
                    smaller-in-left &
                    BST R &
                    larger-in-right);
             _ := (!chain-> [hyp ==> p0                 [BST.nonempty]]);
             _ := (!chain-> [p0 ==> smaller-in-left    [prop-taut]]);
             _ := (!chain-> [p0 ==> larger-in-right    [prop-taut]]);
             _ := (!chain-> [p0
                        ==> (BST L)                    [prop-taut]
                        ==> (not-found-prop L)         [ind-hyp1]]);
             _ := (!chain-> [p0
                        ==> (BST R)                    [prop-taut]
                        ==> (not-found-prop R)         [ind-hyp2]])}
        pick-any x
          assume hyp' := ((binary-search x T) = null)
            (!by-contradiction (~ x in T)
              assume (x in T)
                let {disj := (!chain-> [(x in T)
                                    ==> (x = y  |
                                         x in L |
                                         x in R)      [in.nonempty]])}
                (!two-cases
                  assume (x = y)
                   (!absurd
                     (!chain [null:(BinTree N)
                           = (binary-search x T)      [hyp']
                           = T                        [at-root]])
                     (!chain-> [true
                           ==> (null =/= T)           [tree-axioms]]))
                 assume (x =/= y)
```

```
        (!two-cases
         assume (x < y)
           (!cases disj
             assume (x = y)
               (!absurd (x = y) (x =/= y))
             assume (x in L)
              (!absurd
                (x in L)
                (!chain->
                  [(binary-search x L)
                 = (binary-search x T)           [go-left]
                 = null                          [hyp']
              ==> (~ x in L)                     [p1]]))
             assume (x in R)
              (!absurd
                (x < y)
                (!chain->
                  [(x in R)
               ==> (y <= x)                      [larger-in-right]
               ==> (~ x < y)                     [Less=-tricho-4]])))
         assume (~ x < y)
           (!cases disj
            assume (x = y)
              (!absurd (x = y) (x =/= y))
            assume (x in L)
              (!absurd
                (x =/= y)
                (!chain-> [(x in L)
                       ==> (x <= y)              [smaller-in-left]
                       ==> (x < y |
                            x = y)               [N.Less=.definition]
                       ==> (~ x < y &
                            (x < y |
                             x = y))             [augment]
                       ==> (x = y)               [prop-taut]]))
            assume (x in R)
              (!absurd
                (x in R)
                (!chain->
                 [(binary-search x R)
                = (binary-search x T)            [go-right]
                = null                           [hyp']
              ==> (~ x in R)                     [p2]])))))
} # by-induction
```

Exercise 11.1: In the proof given for `not-found`, we structured the case analysis along the lines discussed on page 564 (and used also in the proof of `found`). Within two of the cases there are nested case analyses based on the disjunction (x = y | x in L | x in R). Modify the proof to do a top-level case analysis based on those cases. □

11.3 Specifying requirements on a function to be defined

We started the last section with a constructive definition of binary-search and proved two theorems about it that give some evidence that our definition is correct. Suppose, though, that we start with axioms found and not-found and consider them to be a *specification of requirements* that a binary search function, yet to be written, must satisfy. We could then give a different definition for binary-search (e.g., see Section 11.4) and try to use it to prove found and not-found. The question then arises whether these two requirements are sufficient. That is, could they be satisfied by a definition of binary-search that is not satisfactory in some other respect besides what is explicitly mandated by the two requirements? If so, what additional requirements, if any, do we need in order to "tighten up" the function's specification? The following exercises explore this question.

Exercise 11.2: The two requirements found and not-found only pertain to the presence or absence of the search value in the tree, based on the function's result. Consider not-found, for example. Perhaps we should add as another requirement the implication in the other direction:

```
define not-in-implies-null-result :=
  (forall T . BST T ==> forall x . ~ x in T ==> (binary-search x T) = null)
```

Show that separately requiring this implication is unnecessary, by proving it follows from found and one of the axioms in (datatype-axioms "BinTree"). In the proof, do not use any of the defining axioms for binary-search (so that you establish that *any* definition of binary-search that satisfies found also satisfies not-in-implies-null-result). □

Exercise 11.3: Prove the following theorem:

```
define not-found-iff-not-in :=
 (forall T . BST T ==> forall x . (binary-search x T) = null <==> ~ x in T)
```

This theorem combines the implications in not-found and not-in-implies-null-result into an equivalence. □

Exercise 11.4: Prove the following:

```
define in-implies-node-result :=
  (forall T .
     BST T ==>
       forall x .
         x in T ==> exists L R . (binary-search x T) = (node L x R))
```

You shouldn't need to directly use any of the axioms for `binary-search`, just `(datatype-axioms "BinTree")` and the two theorems `found` and `not-found`. □

Exercise 11.5: Exercise 11.4 shows that if we consider `found` and `not-found` to be requirements that a `binary-search` function must satisfy, then `(binary-search` x T`)` must return a node result and the value in the node must be x. But if the node returned is `(node` L x R`)`, what about the subtrees L and R? Could they be different from the left and right subtrees of the node in T where x is found? □

11.4 Correctness of an optimized binary search algorithm

The binary search algorithm studied in the previous sections is the one traditionally presented in textbooks, perhaps because it seems natural to first compare for equality of the search value with the current node's key. Notice, however, that until the equality comparison succeeds, the algorithm expends two comparisons per tree level. And in a balanced n-element tree, most searches will require descending through almost $\log_2 n$ levels (since about $n/2$ elements reside in the bottom level, $n/4$ at the next lowest level, etc.) Thus, on average, a total of almost $2 \log_2 n$ comparisons will be required. Now consider the following alternative definition:

```
assert axioms :=
  (fun
   [(binary-search x (node L y R)) =
        [(binary-search x L)  when (x < y)
         (binary-search x R)  when (y < x)
         (node L y R)         when (~ x < y & ~ y < x)]
    (binary-search x null) = null])

define [go-left go-right at-root empty] := axioms
```

With this control structure, if the search value is in the left subtree, only one comparison is required to descend a level; if it is in the right subtree, two are required. Assuming it is equally likely to take one path or the other at each level, a total of about $1.5 \log_2 n$ comparisons are required, on average—a 25 percent reduction from the traditional version of the algorithm.

This is a significant optimization, but before adopting it as the preferred version of binary search, we should be sure that it is correct. This is a small example, but it is illustrative of one of the most important applications of logic and proof in computer science, correctness of algorithm optimizations, so we examine it closely here.

Exercise 11.6: Prove `found` using the optimized version of `binary-search`. *Hint:* Much of the top-level structure of the proof given in the text can be retained, but you should restructure the case analysis as suggested by the new `when` conditions. □

Exercise 11.7: Prove `not-found` using the optimized version of `binary-search`. *Hint:* Same as in Exercise 11.6. □

Exercise 11.8: Another way to establish `found` and `not-found` using the optimized version of `binary-search` is to start with the new axioms of the optimized version and prove the axioms of Section 11.1 as lemmas. Then the original proofs given in that section will still work. That is, from the new axioms first prove

```
define at-root' :=
  (forall x L y R .
    x = y ==> (binary-search x (node L y R)) = (node L y R))

define go-right' :=
  (forall x L y R .
    x =/= y & ~ x < y ==>
    (binary-search x (node L y R)) = (binary-search x R))
```

(`go-left` is unchanged). Then you will be able to reprove `not-found` with

```
let {[at-root go-right] := [at-root' go-right']}
  conclude not-found
    # ...   same proof as in Section 11.2.
```

and similarly for `found`. □

At this point we can close both modules:

```
  } # close module binary-search
} # close module BinTree
```

11.5 Summary and looking ahead

In this chapter we have begun the study of algorithm specification and correctness issues, in terms of a classic binary search algorithm that operates on binary search trees. We expressed the algorithm using the `fun` procedure, which provides a format for expressing function definitions that resembles traditional syntax forms for such definitions in functional languages, while remaining firmly based in logic.

The treatment of binary search in this chapter is not the end of the story. We will return to the topic twice more, first in Chapter 15 with an algorithm that works with more general value orderings than just for natural numbers, as a first illustration of proof issues with abstract (or generic) algorithms. Then in Chapter 16 we will specify and prove correct an abstract binary search algorithm that works on memory ranges rather than binary trees. In these treatments we continue to highlight specification issues—how to state correctness properties precisely enough for proof, without restricting applicability.

11.6 Additional exercises

Exercise 11.9: Suppose we wish to define a `replace` function on lists such that `(replace` L x y`)` returns a copy of L except that all occurrences of x are replaced by y. Declare `replace` and define it using `fun`. □

Exercise 11.10: Based on the solution to the previous exercise, prove the following theorems:

```
extend-module List {
  define sanity-check1 :=
    (forall L x y . x =/= y  ==> (count x (replace L x y)) = zero)
  define sanity-check2 :=
    (forall L x y .
      x =/= y  ==> (count y (replace L x y)) = (count x L) + (count y L))
} # close module List
```

⋆ ***Exercise 11.11:*** The procedure `fun` is used to define either a predicate or a nonpredicate. In either case, it takes one argument, a list containing *defining clauses*, and translates it into a list `[`s_1 ... s_k`]` where each s_i is a sentence. A defining clause for a nonpredicate function f is a triplet (three consecutive list elements): a term l of the form $(f\ t_1\ \ldots\ t_n)$, called a *left-hand-side* term; the symbol =; and a *right-hand-side term* or *list*. In the left-hand-side term, the number of arguments n must be the same as f's arity, and each argument t_i must be either a variable or a term of the sort required by f's i^{th} argument position (so that the whole term is legal). In the case of a right-hand-side term r, r must be a term whose sort is the return sort of f and the corresponding sentence is the closure of $(l = r)$. In the case of a right-hand-side list, it must consist of a sequence of triplets, each triplet consisting of a right-hand-side term r, followed by the meta-identifiers `'when`, followed by a guard p. For each such triplet, the corresponding sentence is the closure of $(p \texttt{ ==> } l = r)$. (We can allow writing `when` instead of `'when` with `define when := 'when`.)

(a) Implement `fun` according to the above specification. (You do not need to implement it for predicates.)

(b) Extend the specification of `fun` to handle predicates, and extend your implementation to satisfy it. □

12 A Fast Exponentiation Algorithm

In this chapter we study an algorithm for efficiently computing x^n, given x from a suitable domain and a natural number n. For the present we assume, for simplicity, that the domain of x is also the natural numbers, a restriction we will relax when we return to this example in Chapter 15.

12.1 Mathematical background

We first defined the function ** in Chapter 3. We restate that definition here, this time inside a module:

```
transform-output eval [nat->int]

extend-module N {

  define [x y m n r] := [?x:N ?y:N ?m:N ?n:N ?r:N]

  open Times

  declare **: [N N] -> N [400 [int->nat int->nat]]

  module Power {

    assert* def := [(x ** zero = one)
                    (x ** S n = x * x ** n)]

    define [if-zero if-nonzero] := def

    (print "\n2 raised to the 3rd: " (eval 2 ** 3))
  } # close module Power

} # close module N
```

Such a definition corresponds directly to what we would write in conventional notation as an inductive definition of exponentiation:

$$x^0 = 1;$$
$$x^{n+1} = x \cdot x^n.$$

Of course, such an inductive definition can itself be regarded as a recursive algorithm for computing the function. While such an algorithm taken directly from the function definition can be used in simple cases (such as when multiplication is a cheap computation and n is a small number), there are other exponentiation algorithms that are much more efficient. One important case is an algorithm that requires, instead of n multiplications, only approximately $\log_2 n$ multiplications, based on the following identities:

$$n = 2\lfloor n/2 \rfloor, \text{ if } n \text{ is even};$$
$$n = 2\lfloor n/2 \rfloor + 1, \text{ if } n \text{ is odd}.$$

(In Section 12.4 we take these identities as the definitions of even and odd.)

Thus, if n is even,

$$\begin{aligned} x^n &= x^{2\lfloor n/2 \rfloor} \\ &= (x^{\lfloor n/2 \rfloor})^2, \end{aligned}$$

and if n is odd,

$$\begin{aligned} x^n &= x^{2\lfloor n/2 \rfloor + 1} \\ &= x^{2\lfloor n/2 \rfloor} \cdot x \\ &= (x^{\lfloor n/2 \rfloor})^2 \cdot x. \end{aligned}$$

Letting (half n) $= \lfloor n/2 \rfloor$ and (square x) $= x \cdot x$, fast-power can be defined as follows. (This definition will be properly given in Section 12.6, as we first need to introduce and derive some useful results for half and even; the definition we give here is only intended as a preview, to facilitate the discussion in the next section.)

```
extend-module N {

 declare fast-power: [N N] -> N [[int->nat int->nat]]

 module fast-power {

  assert def :=
    (fun
      [(fast-power x n) =
        [one                                    when (n = zero)
         (square (fast-power x half n))         when (n =/= zero & even n)
         ((square (fast-power x half n)) * x)  when (n =/= zero & ~ even n)]])

  define [if-zero nonzero-even nonzero-odd] := def
 } # close module fast-power
} # close module N
```

Using this recursive algorithm, the exponent is approximately halved at each step, which means that only about $\log_2 n$ steps are required. In this formulation, one multiplication is used for the squaring and, when n is odd, one more for the factor x. Thus, the total number of multiplications is bounded by about $2\log_2 n$. (More precisely, it is $\lfloor \log_2 n \rfloor + \nu(n) + 1$, where $\nu(n)$ is the number of one bits in the binary representation of n.)

12.2 Strong induction

Thus, `fast-power` has some justification for the "fast" part of its name, but what about the "power" part—does it really compute x^n? We can, in fact, rigorously prove that `(fast-power` x n`)` $= x^n$ for all x and n, using mathematical induction, but the proof is most readily carried out using a different formulation of induction than the one discussed in Section 3.8 on page 136. In contrast to that principle, which is often referred to as "ordinary induction," we have the following "strong induction" principle:

Principle 12.1: Strong Mathematical Induction for Natural Numbers

To prove $\forall\, n\,.\, P(n)$ where n ranges over the natural numbers, it suffices to prove:

$$\forall\, n\,.\, [\forall\, k\,.\, k < n \Rightarrow P(k)] \;\Rightarrow\; P(n).$$

The assumption $[\forall\, k\,.\, k < n \Rightarrow P(k)]$ is called the *strong induction hypothesis*.

Thus, whereas in ordinary induction the induction hypothesis applies to the single value n immediately preceding $n+1$ (or, in a slight variant, the single value $n-1$ immediately preceding n), the strong induction hypothesis assumes $P(k)$ for *all* preceding values $k = 0, \ldots, n-1$. This stronger assumption is just what is needed for proofs about natural number recurrence relations that recur back to one or more values that are not necessarily immediately preceding the current value n—such as $\text{half}(n)$ in the case of our recurrence for fast exponentiation.

Writing f for `fast-power`, we may therefore argue as follows to prove $f(x,n) = x^n$ for all x and n. Letting x and n be arbitrary natural numbers, there are two cases:

1. If $n = 0$ then $f(x,n) = f(x,0) = 1 = x^n$.
2. If $n \neq 0$ there are two subcases:

   a. If n is even, then

$$\begin{aligned}
f(x,n) &= \text{square}(f(x,\text{half}(n))) && \text{defn. of } f \\
&= \text{square}(f(x,n/2)) && \text{defns. of even and half} \\
&= \text{square}(x^{n/2}) && \text{strong induction hypothesis} \\
&= (x^{n/2})^2 && \text{defn. of square} \\
&= x^{(n/2)\cdot 2} && \text{property of exponentiation} \\
&= x^n && \text{property of multiplication}
\end{aligned}$$

b. If n is odd, then

$$
\begin{aligned}
f(x,n) &= \text{square}(f(x,\text{half}(n))) \cdot x && \text{defn. of } f \\
&= \text{square}(f(x,(n-1)/2)) \cdot x && \text{defns. of odd and half} \\
&= \text{square}(x^{(n-1)/2}) \cdot x && \text{strong induction hypothesis} \\
&= (x^{(n-1)/2})^2 \cdot x && \text{defn. of square} \\
&= x^{((n-1)/2) \cdot 2} \cdot x && \text{property of exponentiation} \\
&= x^{n-1} \cdot x && \text{property of multiplication} \\
&= x^n && \text{property of exponentiation}
\end{aligned}
$$

Using this example we can clarify several points about strong induction that sometimes cause confusion:

1. **Where is the basis case?** First, it may appear that the formulation of the strong induction principle is missing the "basis case," $P(0)$, that is required in ordinary induction. But in fact, it is necessary to prove $P(0)$ in order to prove the induction step in the principle:

$$\forall\, n \,.\, [\forall\, k \,.\, k < n \Rightarrow P(k)] \;\Rightarrow\; P(n).$$

To see why, let's consider what it says for $n = 0$:

$$[\forall\, k \,.\, k < 0 \Rightarrow P(k)] \;\Rightarrow\; P(0).$$

Since there are no natural numbers less than 0, the condition

$$[\forall\, k \,.\, k < 0 \Rightarrow P(k)]$$

is vacuously true, so the induction step in this case reads:

$$\text{true} \Rightarrow P(0),$$

which is logically equivalent to $P(0)$. In other words, the strong induction hypothesis is no help when $n = 0$, since there are no natural numbers less than 0. So a proof by strong induction must always be broken into separate cases: $n = 0$, which must be proved without the strong induction hypothesis since it is useless in that case; and $n \neq 0$, where we can use it. This is just what we did in the above proof. (In that proof we had separate even and odd subcases when $n \neq 0$, but such further subcasing is specific to this problem and not always necessary.)

2. **Does it have to be that "strong"?** That is, in the induction step, do we really need to assume $P(k)$ for *all* $k < n$? In our proof about f, we needed the assumption for only one value, half(n), and it is fairly common for recurrences to recur on only one or two values. It may seem that strong induction is more of a "sledgehammer" than we need. But sometimes we do have recurrence relations that use previous values in less predictable ways, as we'll see with Euclid's greatest common divisor algorithm. See also Exercise 13.9. This leads, however, to the final point.

3. **Is it really a stronger principle than ordinary induction?** That is, are there properties of the natural numbers that can be proved with strong induction that cannot be proved with ordinary induction? The answer, perhaps surprisingly, is *no*. We will show this in Section 12.8 by defining a proof method that can transform any strong induction proof into an ordinary one formulated in terms of Athena's `by-induction` construct. It is left as an exercise to express this transformation in traditional mathematical notation.

Although we have given a proof of correctness of our fast exponentiation algorithm in traditional mathematical terms, there are good reasons to carry out the exercise in a formal system like Athena. The first reason is to check the details more rigorously (a task that should be taken more seriously than it often is). Although the example as formulated so far is pretty close to the elementary mathematics on which it is based, there are still a large number of details involved that require care in stating and confirming with proof. Second, there are further optimizations that can be applied to improve the computational efficiency of the algorithm, and some of this "fine tuning" introduces complexities that require subtle reasoning, as we shall see when we consider a more refined version of the algorithm in Chapter 15. Such complexities are often easy to overlook or misinterpret, thus introducing errors, unless a machine-checkable correctness proof is carried out. Finally, this example is a good setting in which to explore applications of many of the logic principles we have been studying in the previous chapters, plus the new "strong" variety of mathematical induction.

Without further ado, therefore, let us proceed to formulate our fast exponentiation algorithm example in Athena. We first need to introduce declarations, defining axioms, and some lemmas for functions that are used in defining the exponentiation algorithm, namely `half`, `even`, and `odd`. We will also need some additional lemmas about `**` beyond those introduced in Chapter 3.

12.3 Properties of half

In the previous section we defined half$(n) = \lfloor n/2 \rfloor$, where the operator `/` is defined over the real numbers. But for present purposes that is not necessary. We could instead directly define a division operator on integers, or just on natural numbers. In fact we will do that in Section 13.1, but for now we define half directly by axioms, as follows:

```
extend-module N {

  declare half: [N] -> N [[int->nat]]

  module half {

    assert* def :=
      [(half zero = zero)
```

```
    (half S zero = zero)
    (half S S n = S half n)]

  define [if-zero if-one nonzero-nonone] := def

  (print "\nHalf of 20: " (eval half 20)
         "\nand half of 21: " (eval half 21) "\n")
```

Here are a couple of simple properties of `half` that we will need:

```
define double    := (forall n . half (n + n) = n)
define times-two := (forall n . half (two * n) = n)
```

The proof of the first is by induction, but it requires a slight variant on the ordinary induction principle, with *two* basis cases instead of one:

Principle 12.2: Mathematical Induction for Natural Numbers—Variant

To prove $\forall\, n\, .\, P(n)$ where n ranges over the natural numbers, it suffices to prove:

1. *First Basis Case*: $P(0)$.
2. *Second Basis Case*: $P(1)$.
3. *Induction Step*: $\forall\, n\, .\, P(n) \Rightarrow P(n+2)$.

With the original formulation of induction on the natural numbers, we can visualize how, starting from 0 and repeatedly using the induction step to advance by one successor operation, we exhaust all the natural numbers, as in the following diagram:

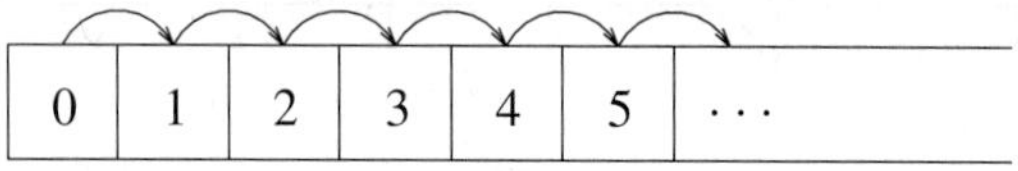

0	1	2	3	4	5	...

Our variant with two basis cases and advancing by two in the induction step can correspondingly be visualized as exhausting all the natural numbers as follows:

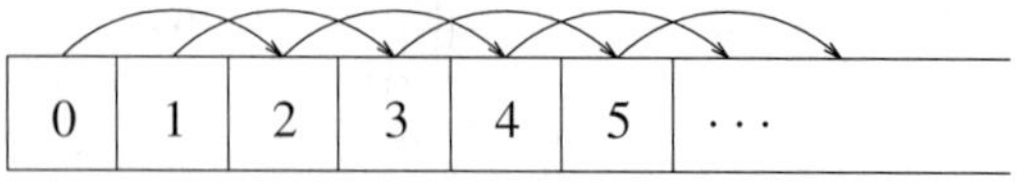

0	1	2	3	4	5	...

In Athena, this "two-step" variant requires no new machinery; the built-in **`by-induction`** form is sufficient, since it allows subcasing of the proof in any way that exhausts the entire set of natural numbers. Here is the proof of `double`:

```
by-induction double {
  zero => (!chain [(half (zero + zero))
                   --> (half zero)          [Plus.right-zero]
                   --> zero                 [if-zero]])
| (S zero) =>
    (!chain [(half (S zero + S zero))
             --> (half S (S zero + zero))   [Plus.right-nonzero]
             --> (half S S (zero + zero))   [Plus.left-nonzero]
             --> (half S S zero)            [Plus.right-zero]
             --> (S half zero)              [nonzero-nonone]
             --> (S zero)                   [if-zero]])
| (S (S m)) =>
    let {IH := (half (m + m) = m)}
      (!chain
        [(half (S S m + S S m))
       --> (half S (S S m + S m))           [Plus.right-nonzero]
       --> (half S S (S S m + m))           [Plus.right-nonzero]
       --> (S half (S S m + m))             [nonzero-nonone]
       --> (S half S (S m + m))             [Plus.left-nonzero]
       --> (S half S S (m + m))             [Plus.left-nonzero]
       --> (S S half (m + m))               [nonzero-nonone]
       --> (S S m)                          [IH]])
}
```

Now the proof of `half.times-two` is a simple equality chain:

```
conclude times-two
  pick-any x
    (!chain [(half (two * x))
         --> (half (x + x))    [Times.two-times]
         --> x                 [double]])
```

Exercise 12.1: Prove the following:

```
define twice := (forall x . two * half S S x = S S (two * half x))
```

Note: you do not need induction. □

* ***Exercise 12.2:*** Prove the following property of `half`:

```
define two-plus := (forall x y . half (two * x + y) = x + half y)
```

□

We also need ordering properties relating `(half n)` to `n`:

```
define less-S := (forall n . half n < S n)
define less :=   (forall n . n =/= zero ==> half n < n)
```

The first is used as a lemma in the proof of the second, which in turn is used in the proof of correctness of our fast exponentiation algorithm.

```
by-induction less-S {
  zero => (!chain-> [true
                 ==> (zero < S zero)                  [Less.<S]
                 ==> (half zero < S zero)             [if-zero]])
| (S zero) =>
    let {1/2<1 := (!chain-> [true
                          ==> (zero < S zero)         [Less.<S]
                          ==> (half S zero < S zero) [if-one]])}
      (!chain-> [true
              ==> (S zero < S S zero)                 [Less.<S]
              ==> (S zero < S S zero & 1/2<1)         [augment]
              ==> (half S zero < S S zero)            [Less.transitive]])
| (S (S m)) =>
    let {IH := (half m < S m);
         m+2<m+3 := (!chain-> [true
                            ==> (S S m < S S S m)     [Less.<S]])}
      (!chain-> [IH
              ==> (S half m < S S m)                  [Less.injective]
              ==> (half S S m < S S m)                [nonzero-nonone]
              ==> (half S S m < S S m & m+2<m+3)      [augment]
              ==> (half S S m < S S S m)              [Less.transitive]])
}

datatype-cases less {
  zero => assume (zero =/= zero)
            (!from-complements (half zero < zero)
                               (!reflex zero)
                               (zero =/= zero))
| (S zero) =>
    assume (S zero =/= zero)
      (!chain-> [true
              ==> (zero < S zero)                   [Less.<S]
              ==> (half S zero < S zero)            [if-one]])
| (S (S m)) =>
    assume (S S m =/= zero)
      (!chain-> [true
              ==> (half m < S m)                    [less-S]
              ==> (S half m < S S m)                [Less.injective]
              ==> (half S S m < S S m)              [nonzero-nonone]])
}
} # close module half
} # close module N
```

The chapter exercises in Section 12.11 include proofs of other properties of `half`.

12.4 Properties of odd and even

Predicates for testing the parity of a number—whether it is odd or even—can be defined in a variety of ways, but here we take advantage of half:

```
extend-module N {
  declare even, odd: [N] -> Boolean [[int->nat]]

  module EO {

    assert* even-definition := [(even x <==> two * half x = x)]

    assert* odd-definition :=  [(odd  x <==> two * (half x) + one = x)]


    (print "\nIs 20 even?: " (eval even 20))
    (print "\nIs 20 odd?: "  (eval odd 20))
```

Here are some useful properties of these predicates:

```
define even-zero := (even zero)
define odd-one := (odd S zero)
define even-S-S := (forall n . even S S n <==> even n)
define odd-S-S := (forall n . odd S S n <==> odd n)
define odd-if-not-even  := (forall n . ~ even n ==> odd n)
define not-odd-if-even := (forall n . even n ==> ~ odd n)
define even-iff-not-odd := (forall n . even n <==> ~ odd n)
define not-even-if-odd := (forall n . odd n ==> ~ even n)
define half-nonzero-if-nonzero-even :=
  (forall n . n =/= zero & even n ==> half n =/= zero)
define half-nonzero-if-nonone-odd :=
  (forall n . n =/= one & odd n ==> half n =/= zero)
define even-twice := (forall n . even (two * n))
define even-square := (forall n . even n <==> even square n)
} # close module EO
```

Of these properties, only EO.odd-if-not-even is directly used in the correctness proof of our fast exponentiation algorithm. Its proof follows, preceded by proofs of three properties needed as lemmas.

```
extend-module EO {

conclude even-zero
  (!chain-> [(two * half zero)
         --> ((half zero) + (half zero))        [Times.two-times]
         --> (zero + zero)                      [half.if-zero]
```

```
      --> zero                                   [Plus.right-zero]
      ==> (even zero)                            [even-definition]])

conclude even-S-S
  pick-any n
    (!chain [(even S S n)
        <==> (two * half S S n = S S n)          [even-definition]
        <==> (S S (two * half n) = S S n)        [half.twice]
        <==> (S (two * half n) = S n)            [S-injective]
        <==> (two * half n = n)                  [S-injective]
        <==> (even n)                            [even-definition]])

conclude odd-S-S
  pick-any n
    (!chain [(odd S S n)
        <==> (two * (half S S n) + one = S S n)  [odd-definition]
        <==> (S S (two * half n) + one = S S n)  [half.twice]
        <==> (S S S (two * half n) = S S n)      [Plus.right-one]
        <==> (S S (two * half n) = S n)          [S-injective]
        <==> (S (two * half n) = n)              [S-injective]
        <==> ((two * half n) + one = n)          [Plus.right-one]
        <==> (odd n)                             [odd-definition]])

by-induction odd-if-not-even {
  zero => (!chain [(~ even zero)
              ==> (even zero & ~ even zero)      [augment]
              ==> (odd zero)                     [prop-taut]])
| (S zero) =>
    assume (~ even S zero)
      (!chain<-
        [(odd S zero)
     <== (two * (half S zero) + one = S zero)    [odd-definition]
     <== (S (two * half S zero) = S zero)        [Plus.right-one]
     <== (S (two * zero)         = S zero)       [half.if-one]
     <== (S zero                 = S zero)       [Times.right-zero]])
| (S (S m)) =>
    let {IH := (~ even m ==> odd m)}
      (!chain [(~ even S S m)
          ==> (~ even m)                         [even-S-S]
          ==> (odd m)                            [IH]
          ==> (odd S S m)                        [odd-S-S]])
}
} # close module EO
```

Exercise 12.3: Prove `EO.odd-one`. □

Exercise 12.4: Prove `EO.even-twice`. □

Exercise 12.5: Prove `EO.not-odd-if-even`. □

⋆ ***Exercise 12.6:*** We define the *square* function as follows:

```
# Still in module N
declare square: [N] -> N [[nat->int]]
module square {
  assert* def := [(square x = x * x)]
}
```

Prove the following property of even: □

```
extend-module EO {
  define even-square := (forall x . even x <==> even square x)
}
```

Exercises proving the other stated odd and even properties are in Section 12.11.

12.5 Properties of power

There are a few additional properties of ** that we will need for the correctness proof of the more efficient power algorithm.

```
extend-module Power {
  define Plus-case   := (forall m n x . x ** (m + n) = x ** m * x ** n)
  define left-one    := (forall n . one ** n = one)
  define right-one   := (forall n . n ** one = n)
  define right-two   := (forall n . n ** two = n * n)
  define left-times  := (forall n x y . (x * y) ** n = x ** n * y ** n)
  define right-times := (forall m n x . x ** (m * n) = (x ** m) ** n)
  define two-case    := (forall n . square n = n ** two)
} # close module Power
```

Some of these appeared in exercises in Chapter 3; we have restated them here as part of the Power module. The previous proofs also need to be redone using the module notation in referring to axioms and lemmas. For example,

```
extend-module Power {
  by-induction Plus-case {
    zero =>
      conclude (forall n x . x ** (zero + n) = x ** zero * x ** n)
        pick-any n x
          (!chain [(x ** (zero + n))
                 --> (x ** n)                    [Plus.left-zero]
                 <-- (one * x ** n)              [Times.left-one]
                 <-- (x ** zero * x ** n)        [if-zero]])
  | (m as (S k)) =>
      let {IH := (forall n x . x ** (k + n) = x ** k * x ** n)}
        conclude (forall n x . x ** (m + n) = x ** m * x ** n)
```

```
    pick-any n x
      (!combine-equations
        (!chain [(x ** ((S k) + n))
             --> (x ** (S (k + n)))              [Plus.left-nonzero]
             --> (x * x ** (k + n))              [if-nonzero]
             --> (x * (x ** k * x ** n))         [IH]])
        (!chain [(x ** (S k) * x ** n)
             --> ((x * (x ** k)) * x ** n)  [if-nonzero]
             --> (x * (x ** k * x ** n))    [Times.associative]]))
  }
} # close module Power
```

Exercise 12.7: Prove `Power.left-one`. □

Exercise 12.8: Prove `Power.right-one`. □

Section 12.11 contains exercises to prove `Power.left-times` and `Power.right-times`.

12.6 Properties of fast-power

Now recall the definition we gave at the beginning of the chapter for computing x^n in only $k = \log_2 n$ steps:

```
declare fast-power: [N N] -> N [[int->nat int->nat]]

module fast-power {

assert axioms :=
  (fun
    [(fast-power x n) =
      [one                                   when (n = zero)
       (square (fast-power x half n))        when (n =/= zero & even n)
       ((square (fast-power x half n)) * x)  when (n =/= zero & ~ even n)]])

define [if-zero nonzero-even nonzero-odd] := axioms

# prints out 8:

(print "\n2 raised to the 3rd with fast-power: " (eval (fast-power 2 3)))
```

The property we wish to prove is that `fast-power` computes the same results as `**`.

```
define correctness := (forall n x . (fast-power x n) = x ** n)
```

As with the proof given earlier in conventional mathematical terms, we use strong induction,which is available in Athena as a binary method `strong-induction.principle` that takes the following arguments:

1. The sentence p that we are seeking to derive.
2. A unary method M that derives the strong induction step of the proof.

Given these two arguments, an application of `strong-induction.principle` will derive p.

To prove `correctness` we thus apply `strong-induction.principle` to a suitably defined step method. To save space, we locally define `^`, `sq`, and `hf` as aliases for `fast-power`, `square`, and `half`.

```
1  define [^ sq hf] := [fast-power square half]
2
3  define step :=
4   method (n)
5     assume ind-hyp := (forall m . m < n ==> forall x . x ^ m = x ** m)
6       conclude (forall x . x ^ n = x ** n)
7         pick-any x
8           (!two-cases
9             assume (n = zero)
10              (!chain [(x ^ n)
11                   --> one                                  [if-zero]
12                   <-- (x ** zero)                          [Power.if-zero]
13                   <-- (x ** n)                             [(n = zero)]])
14            assume (n =/= zero)
15              let {p1 :=
16                     conclude
17                       p := (forall x . x ^ hf n =
18                                          x ** hf n)
19                         (!chain-> [(n =/= zero)
20                                    ==> (hf n < n)          [half.less]
21                                    ==> p                   [ind-hyp]]);
22                   p2 :=
23                     conclude (sq (x ^ hf n) =
24                                 x ** (two * hf n))
25                       (!chain
26                         [(sq (x ^ hf n))
27                      --> (sq (x ** hf n))                  [p1]
28                      --> ((x ** hf n) *
29                           (x ** hf n))                     [square.def]
30                      <-- (x ** ((hf n) + hf n))            [Power.Plus-case]
31                      <-- (x ** (two * hf n))               [Times.two-times]])}
32                (!two-cases
33                  assume (even n)
34                    (!chain
35                      [(x ^ n)
36                  --> (sq (x ^ hf n))                       [nonzero-even]
37                  --> (x ** (two * hf n))                   [p2]
38                  --> (x ** n)                              [EO.even-definition]])
39                  assume (~ even n)
40                    let {_ := (!chain-> [(~ even n)
```

```
41                                ==> (odd n)       [EO.odd-if-not-even]])}
42              (!chain
43                [(x ^ n)
44               --> ((sq (x ^ hf n)) * x)          [nonzero-odd]
45               --> ((x ** (two * hf n)) * x)      [p2]
46               <-- ((x ** (two * hf n)) *
47                   (x ** one))                    [Power.right-one]
48               <-- (x ** ((two * hf n) + one))    [Power.Plus-case]
49               --> (x ** n)                       [EO.odd-definition]])))
50
51  (!strong-induction.principle correctness step)
```

Rather than defining the step method separately, we could write the whole proof as an application of `strong-induction.principle` with the step method defined inline:

```
(!strong-induction.principle correctness
  method (n)
    # same as lines 3-46 of the preceding listing
)
```

Before continuing, we close modules `N.fast-power` and `N`:

```
} # close module fast-power
} # close module N
```

12.7 Tail recursion, a potential optimization

The recursive calls in the definition of `fast-power` (page 590) are examples of *embedded recursion*. In general, a call to a procedure f is said to be *embedded* with respect to the body of its innermost enclosing procedure g iff some computation is performed by g *after* the call to f is complete. In this case we have $f = g$, so the call is a recursive one, and there is indeed further work done after the result of either recursive call is obtained: The result is squared in both cases, and in the odd case we also have the multiplication by `x`. By contrast, if the innermost enclosing procedure g performs no further computation after a call to f (other than simply returning the computed result), then that call is said to be a *tail call*. A recursive function definition could have both tail and embedded recursive calls; if it has no embedded recursive calls, it is said to be a *tail-recursive definition*.[1] For example,

1 Note that tail-recursiveness is not a property of a mathematical function—viewed as a set of tuples—but a property of a particular definition or algorithm. One and the same function can be computed by many different algorithms, having different definitions, some of which may be tail-recursive and others not. In higher-order functional languages *every* function can be defined tail-recursively, because, with the explicit use of continuations, every algorithm can be transformed into a tail-recursive form that computes the same function.

the definition of binary-search in Section 11.1 is tail-recursive, because neither of its two recursive calls is embedded:

```
assert axioms :=
 (fun
  [(binary-search x null) = null
   (binary-search x (node L y R)) =
     [(node L y R)          when (x = y)
      (binary-search x L)   when (x < y)
      (binary-search x R)   when (x =/= y & ~ x < y)]])

define [empty at-root go-left go-right] := axioms
```

In the context of pure equational definitions like these, determining whether a recursive call is embedded is a straightforward syntactic check: If a recursive function call has a proper superterm in a defining axiom, or if it occurs in a guard, then the call is embedded, otherwise it is a tail call.

Such distinctions are important because there is an easy and effective optimization of tail-recursive calls: A compiler or interpreter can replace such a call by a sequence of assignments to its formal parameters, followed by a jump to the beginning of the function. Consider, for example, the C++ code that we wrote as the direct translation of the definition of binary-search.

```
BinTree binarySearch(int x, BinTree T) {
  if (T == NULL)
    return NULL;
  if (x == T->key)
    return T;
  if (x < T->key)
    return binarySearch(x, T->left);
   // (T->key < x)
   return binarySearch(x, T->right);
}
```

Tail-recursion removal would compile this code as though it had been written as follows:

```
BinTree binarySearch(int x, BinTree T) {
  start:
  if (T == NULL)
    return NULL;
  if (x == T->key)
    return T;
  if (x < T->key) {
    T = T->left;
    goto start;
  }
  // (T->key < x)
  T = T->right;
  goto start;
}
```

or, in a more commendable style using a structured control statement:

```
BinTree binarySearch(int x, BinTree T) {
  while (T != NULL && x != T->key) {
    if (x < T->key)
      T = T->left;
    else
      // (T->key < x)
      T = T->right;
  }
  return T;
}
```

since at the level of machine code this loop would also be implemented with a jump to its first instruction.

If programmers could rely on tail-recursion removal—and in the case of some languages, like Scheme or ML, they can—then they could write functions with tail-recursive calls and be assured that they would be no less efficient than if written with the equivalent loops.[2]

With embedded recursion, on the other hand, it is not possible to optimize away the usual function call bookkeeping (involving pushing arguments and return address onto the runtime stack and popping arguments off the stack when control is transferred back to the caller). However, in many cases it is possible to replace embedded recursion with tail recursion, by means of the technique of using an extra argument, called an *accumulator*. This argument is so called because it is used to accumulate the result that is ultimately returned. If we move the computation that occurs after an embedded call into computation of the value passed as the accumulator argument, the call becomes tail-recursive. In the case of fast-power, we can apply this technique as follows:

```
extend-module N {

declare fast-power-accumulate: [N N N] -> N [[int->nat int->nat int->nat]]

module fast-power-accumulate {

define fpa := fast-power-accumulate

assert axioms :=
  (fun
    [(fpa r x n) =
      [r                               when (n = zero)
       (fpa r (x * x) (half n))        when (n =/= zero & even n)
       (fpa (r * x) (x * x) (half n))  when (n =/= zero & ~ even n)]])

define [if-zero nonzero-even nonzero-odd] := axioms
```

2 This is true for Scheme and ML because the language standards require all implementations to perform the optimization. Unfortunately that is not the case for most languages, C++ included, so for those languages there is strong incentive to program in terms of loops rather than tail recursion.

With this definition we have the following theorem:

```
define correctness :=
  (forall n r x . (fpa r x n) = r * x ** n)
} # close module fast-power-accumulate
```

The proof is left as an exercise. If we still want an exponentiation function with the same two-argument interface as before, we can define it as

```
extend-module fast-power {

  define fpa := fast-power-accumulate
  assert* definition := [((fast-power x n) = (fpa one x n))]

} # close module fast-power
} # close module N
```

and reprove `fast-power.correctness` using this new definition.

The definition of `fast-power-accumulate` is tail-recursive, and therefore is equivalent to a loop. But does that make it necessarily more efficient than the original embedded-recursion version? Perhaps surprisingly, and unfortunately, the answer is no. We may have even made the computation *much less* efficient! The reason is that there is an unnecessary squaring at the last step, when $n = 1$. For example, (fpa 1 x 13) results in calls (fpa x x^2 6), (fpa x x^4 3), (fpa x^5 x^8 1), and (fpa x^{13} x^{16} 0) before returning x^{13}, but at the last step $x^8 \cdot x^8$ is computed but not used. There is another unnecessary multiplication, by 1. If multiplication is a fixed time computation, as it may be with fixed precision or floating point or modular arithmetic, this may be no worse than the original embedded recursion version, which also has two unneeded multiplications, $1 \cdot 1$ and $1 \cdot x$. But if the time for multiplication can grow with the size of its arguments, as is the case with unlimited precision integers, or, even more extremely, with polynomials or matrices with unlimited precision coefficients, it can be far more costly to perform that last squaring in the above version of the algorithm versus the multiplications with small values in the original algorithm. In fact, the time for that single multiplication might dominate the time for all the other multiplications combined.

So in this example we do not yet have an optimization. But we can, after all, achieve one if we perform other transformations in addition to replacing embedded recursion with tail recursion. We shall postpone that discussion, however, until we have seen how to write the various algorithms at an abstract level, where the domain of x, rather than just being natural numbers, is any domain satisfying the axioms of the algebraic theory known as a monoid. Then, the wider applicability of the resulting algorithms can better justify the substantial effort required to work out the details of the transformations and prove them correct. This development appears in Chapter 15.

12.8 Transforming strong induction into ordinary induction

The `strong-induction.principle` method works by transforming a strong induction proof into an ordinary induction proof expressed in terms of **`by-induction`**. In the following code this transformation is carried out within the body of another method, `strong-induction.principle-lemma-proof`, which proves a lemma, that *p* follows from the strong induction step for *p*. Then all the `strong-induction.principle` method itself has to do is: (1) confirm that the step method passed to it establishes the strong induction step, and (2) apply the lemma to obtain *p*.

```
module strong-induction {

define < := N.<

define (conclusion p n) := (urep (rename p) [n])

define (hypothesis p n) :=
  (forall ?m' (if (< ?m' n) (conclusion p ?m')))

define (step p) :=
  (forall ?n . (hypothesis p ?n) ==> (conclusion p ?n))

define (principle-lemma p) := ((step p) ==> p)

define principle-lemma-proof :=
  method (p)
    conclude (principle-lemma p)
      assume (step p)
        let {sublemma :=
               by-induction (forall x . (hypothesis p x)) {
                 zero =>
                   conclude (hypothesis p zero)
                     pick-any y:N
                       assume (y < zero)
                         (!from-complements (conclusion p y)
                           (y < zero) (!instance N.Less.not-zero [y]))
               | (S x') =>
                   let {ind-hyp := (hypothesis p x')}
                     conclude (hypothesis p (S x'))
                       pick-any y:N
                         assume (y < S x')
                           (!two-cases
                             assume (y = x')
                               (!chain->
                                 [ind-hyp
                             ==> (hypothesis p y)     [(y = x')]
                             ==> (conclusion p y)     [(step p)]])
                             (!chain
```

```
                              [(y =/= x')
                          ==> (y < S x' & y =/= x') [augment]
                          ==> (y < x')              [N.Less.S-step]
                          ==> (conclusion p y)      [ind-hyp]]))
          }}
        conclude p
          pick-any x:N
            (!chain->
              [sublemma
           ==> (hypothesis p x) [(uspec with x)]
           ==> (conclusion p x) [(step p)]])

define principle :=
  method (p step-method)
    let {lemma := (!principle-lemma-proof p);
         sp := conclude (step p)
                 pick-any n:N
                   (!step-method n)}
      (!chain-> [sp ==> p [lemma]])

} # close module strong-induction
```

The helper procedure urep is defined in Athena's library as follows:

```
define (urep p terms) :=
  match [p terms] {
    [(forall x q) (list-of t more)]  => (urep (replace-var x t q) more)
  | _ => p
  }
```

12.9 Measure induction

As we have formulated it here, strong induction can only be used to derive universal generalizations over the natural numbers. Sometimes, however, we need to derive a universal generalization

$$goal = \forall\, x:S \,.\, P(x) \tag{12.1}$$

over some sort S other than the natural numbers, by strong induction *over some natural-number quantity associated with $x:S$*. (You can think of P here as a procedure that takes a term of sort S and produces a sentence.) For instance, x might range over lists and we might want to prove the goal via strong induction on the length of the list; or x might range over finite sets and we might want to proceed by strong induction on set cardinality; and so on. Indeed, the quantity in question need not involve just x, but could well involve additional terms as well, such as further quantified variables. We give such an example in Section 10.4.

Nevertheless, we can always transform *goal* into another sentence *goal′* that is explicitly generalized over natural numbers; use regular strong induction over the natural numbers to derive *goal′*; and then finally derive the original *goal* from *goal′*. For example, if x ranges over lists and we want to proceed by strong induction on the length of the list, we can transform (12.1) into:

$$goal' = \forall\, n\texttt{:N} \,.\, \forall\, x\texttt{:(List 'S)} \,.\, \texttt{length}(x) = n \Rightarrow P(x). \tag{12.2}$$

Once *goal′* is derived by conventional strong induction, *goal* can be obtained as follows:

```
conclude goal
  pick-any x:(List 'S)
    let {body := (!uspec* goal' [(length x) x])}
      (!mp body (!reflex (length x)))
```

However, having to perform this sort of transformation manually every time would be unduly cumbersome. Moreover, the proof of *goal′* would itself be somewhat complicated because of the presence of the new quantified variable n:N, which does not even appear in the original goal. It would thus be preferable to hide all this plumbing and allow the proof to proceed directly by reasoning on the quantity of interest.

Method abstraction proves to be the right tool here as well. The module `strong-induction` defines a ternary method called `measure-induction` that hides this complexity. The three arguments of this method are as follows:

1. The first argument is the universally quantified goal (12.1) that we wish to derive, whose variable $x{:}S$ ranges over an arbitrary sort S.
2. The second argument is the quantity of interest, Q, represented as a unary procedure that takes any term of sort S and produces a natural-number term (a term of sort N). Typically this will be a function symbol from S to N.
3. Finally, the third argument must be, essentially, the theorem of interest, namely, a sentence in the assumption base of the following form:

$$\forall\, y{:}S \,.\, IH(y) \Rightarrow P(y), \tag{12.3}$$

where IH is the following unary procedure, written here in conventional notation:

$$IH(y) \equiv \forall\, x{:}S \,.\, Q(x) < Q(y) \Rightarrow P(x). \tag{12.4}$$

The IH procedure constructs the strong induction hypothesis for a given $y{:}S$, namely, a sentence to the effect that the property P holds for all elements $x{:}S$ whose quantity Q is strictly less than that of y. The expected theorem then, (12.3), must state that for *all* $y{:}S$, *if* the said inductive hypothesis holds for y, *then* P holds for y. If that sentence is in the assumption base, then the first argument—the desired goal—is derived and returned as a theorem. The derivation proceeds via the three-step technique outlined above. By expressing the inductive hypothesis directly in terms of the quantity Q, without introducing any extraneous variables ranging over the natural numbers, the proof of the required sentence (12.3) can be expressed more naturally.

12.10 Summary and looking ahead

In this chapter we have continued the study of algorithm specification and correctness issues, using an efficient algorithm for exponentiation as the primary example. Leading up to the correctness proof, we dealt with a number of simple but important natural number concepts such as parity and halving.

We have again used the `fun` procedure rather than sets of axioms to express algorithms. Although we are restricted in using `fun` (or sets of axioms) to conditional and recursive control structures rather than loops and assignments, we discussed how tail recursion can achieve the same efficiency as those iterative constructs.

The fast exponentiation algorithm has also provided the setting for introducing another important proof tool, strong mathematical induction on the natural numbers. A variant of this technique, measure induction, allows one to apply strong mathematical induction to other datatypes just by supplying an appropriate mapping to natural numbers.

In Chapter 15 we will reconsider fast exponentiation, proving correctness of an abstract algorithm that is not only more widely applicable but also more highly optimized.

12.11 Additional exercises

Exercise 12.9: Prove `EO.even-iff-not-odd` (defined in Section 12.4). ☐

Exercise 12.10: Prove `Power.left-times`. ☐

Exercise 12.11: Prove `Power.right-times`. ☐

⋆ ***Exercise 12.12:*** Prove `fast-power-accumulate.correctness`. ☐

Exercise 12.13: With `fast-power` redefined in terms of `N.fast-power-accumulate`, as in Section 12.7, prove `N.fast-power.correctness`. ☐

Exercise 12.14: It was mentioned in Section 12.7 that the embedded recursion version of `faster-power` does two unnecessary multiplications. Another possible optimization strategy is to keep the embedded recursion but try to eliminate the unnecessary multiplications. We can in fact do that by inserting a test for $n = 1$, as in the following redefinition:

```
extend-module N {
extend-module fast-power {
assert axioms :=
  (fun
```

```
  [(fast-power x n) =
     [one                 when (n = zero)
      x                   when (n = one)
      (square (fast-power x (half n)))
                          when (n =/= zero & n =/= one & even n)
      ((square (fast-power x (half n))) * x)
                          when (n =/= zero & n =/= one & ~ even n)]])
define [if-zero if-one nonzero-nonone-even nonzero-nonone-odd] := axioms
} # close module fast-power
```

But we must ask, as we did regarding the tail-recursive version, is this really an optimization? Not if multiplication is a fixed-cost operation, since the extra test doubles the number of test instructions executed. A better solution is discussed in Chapter 15. Nevertheless, for practice in using the strategy adopted for correctness of a `binary-search` variant in Exercise 11.8, of proving the axioms of the original version as lemmas, prove `fast-power.correctness` again using that strategy with the definition given here. □

⋆ ***Exercise 12.15:*** Study the transformation done in the`strong-induction.principle` proof method (Section 12.8). Then express this transformation and its proof in conventional mathematical notation. □

⋆ ***Exercise 12.16:*** There is a variant of the strong induction principle in which the case $P(0)$ is considered separately:

Principle 12.3: Strong Mathematical Induction for Natural Numbers—Variant

To prove $\forall\, n\, .\, P(n)$ where n ranges over the natural numbers, it suffices to prove:

1. *Basis Case*: $P(0)$.
2. *Strong Induction Step*:

$$\forall\, n\, .\, [\forall\, k\, .\, k < n+1 \Rightarrow P(k)] \Rightarrow P(n+1).$$

The assumption $[\forall\, k\, .\, k < n+1 \Rightarrow P(k)]$ is called the *strong induction hypothesis*.

One reason for making this separation is, as discussed earlier in the chapter, that strong induction requires one to consider $P(0)$ as a special case when proving (the original formulation of) the strong induction step anyway, and making that requirement explicit in the formulation of the principle gives additional guidance for the case analysis and allows greater modularity in expressing the proof. (On the other hand, it is an elegant attribute of `strong-induction.principle` that the entire proof can be packaged in a single method.)

Implement this variant as a ternary method `strong-induction.principle2` that takes:

1. The sentence $p = \forall n\, .\, P(n)$ that we are seeking to derive;

2. a nullary method `basis` that derives $P(0)$, the basis case of the proof; and
3. a unary method `step-method` that derives the strong induction step of the proof.

Given these three arguments, an application of `strong-induction.principle2` should derive the desired sentence p. (*Hint:* The lemma proved by the `lemma-proof` method can be applied in `strong-induction.principle2`, just as it was in `strong-induction.principle`. Before you apply it, you just have to establish `(step` p`)` using `basis` and `step-method`. Consider using the **`datatype-cases`** construct to set up the zero and nonzero cases.)

Test your method by suitably defining methods `basis` and `step2` inside the `fast-power` module for proving `fast-power.correctness`. □

13 Euclid's Algorithm for Greatest Common Divisors

We continue in this chapter with proofs of properties of division, leading up to another significant algorithm correctness example, for Euclid's algorithm for computing greatest common divisors.

13.1 Quotient and remainder

In this section we define the division and remainder functions on the natural numbers, and prove several of their properties that will be needed later.

```
extend-module N {

  declare /, %: [N N] -> N [300 [int->nat int->nat]]

  module Div {
    assert* def :=  [(x < y ==> x / y = zero)
                     (~ x < y & zero < y ==> x / y = S ((x - y) / y))]

    define [basis reduction] := def
  }

  module Mod {
    assert* def := [(x < y ==> x % y = x)
                    (~ x < y & zero < y ==> x % y = (x - y) % y)]

    define [basis reduction] := def
  }

  transform-output eval [nat->int]
```

Proposition `Div.basis` says that if the first argument (the *dividend*) is less than the second argument (the *divisor*), then the result of the division is 0. Thus, for instance, dividing 3 by 5 yields 0:

```
> (eval S S S zero / S S S S S zero)

Term: 0
```

(Note that `S` has precedence 350, thus binding tighter than both of these new function symbols.) Since we have declared these two functions so that they can take integer numerals as arguments and convert them to natural numbers, we can also write:

```
> (eval 3 / 5)

Term: 0
```

The second axiom says that if the dividend x is not less than the divisor y *and* if the divisor is positive, then the result of the division is one more than the result of dividing $x-y$ by y. This axiom highlights the fact that division is really repeated subtraction: to divide a number x by a positive number $y \leq x$, we repeatedly subtract y from x, keeping track of the tally, until what is left of the dividend is less than y. We can use `eval` to test that our definition produces the expected results:

```
> (eval 2 / 2)

Term: 1

> (eval 50 / 2)

Term: 25

> (eval 6 / 2)

Term: 3

> (eval 10 % 3)

Term: 1
```

Note that the axioms do not specify an output value when the divisor is zero.[1]

```
> (eval 2 / 0)

Unable to reduce the term:

(N./ (S (S zero))
     zero)

to a normal form.
```

---

1 Nevertheless, a term such as `(S zero / zero)` is perfectly legitimate, and indeed, as we have pointed out before, it has to denote a natural number. It's just that our theory does not specify exactly *which* number that is. So, unless our theory is inconsistent, we cannot prove

`(S zero / zero =` *t*`)`

for any canonical *t*, but we can prove, for example,

`(S zero / zero = S zero / zero)`

by a single application of reflexivity, since every term is identical to itself. Underspecification is a convenient compromise solution to a thorny problem (that of undefined values), but its most serious conceptual drawback is that we have to admit terms like `(S zero / zero)` as denoting actual (if unspecified) values. This is rather counterintuitive and does not reflect common mathematical practice. Mathematicians would not say that $1/0$ is a natural (or a real) number, albeit an unspecified one. Rather, they say that $1/0$ does not denote a number *at all*. Later, in Section 17.2, we will examine some alternative approaches to this issue.

This is an example of deliberately underconstraining a function definition. The axioms for / are satisfied by infinitely many binary functions on the natural numbers, each producing a different result when the divisor is zero. However, the two given axioms ensure that all these functions behave identically when the second argument is nonzero (regardless of the first argument).

Likewise, in the axioms for the remainder function, %, no result is specified when the second argument is zero.

```
> (eval 1 % 0)

Unable to reduce the term:

(N.% (S zero)
     zero)

to a normal form.
```

13.2 The division algorithm

In this section we state and prove an important result that is traditionally known as *the division algorithm*, although this is somewhat of a misnomer, as the result is not an algorithm (it is a theorem), nor does it suggest an algorithm. The result is this:

```
define division-algorithm :=
  (forall x y . zero < y ==> (x / y) * y + x % y = x & x % y < y)
```

Some brief experimentation with this goal will show that a noninductive proof attempt is not likely to succeed. But which variable, x or y, should we attempt to induct on? Should we try to use ordinary induction or strong induction? The following heuristic principle can provide some guidance in answering both questions.

Heuristic 13.1: Induction Variable and Ordinary or Strong Induction Choice

When the proof goal is of the form

$$\forall\, x_1 \cdots x_n \cdots \;.\; \cdots f(x_1,\ldots,x_n)\cdots$$

and the size of the i-th argument, x_i, can be reduced by an application of an axiom for f, consider proving the goal

$$\forall\, x_i\, x_1 \cdots x_{i-1}\, x_{i+1} \cdots x_n \cdots \;.\; \cdots f(x_1,\ldots,x_n)\cdots$$

by induction on x_i. Moreover, if the size of x_i decreases by one, try using ordinary induction; if it decreases by more than one, try strong induction.

In this case, the argument whose size decreases with each recursive call of / and % is the first one, so we don't need to change the order of the universally quantified variables. (We will see a case in Section 13.4 where we do need to change the order.) The use of strong induction is indicated since the size of the first argument of / may decrease by more than one with each recursive call.

We begin with a proof sketch that merely sets up the strong induction skeleton:

```
conclude goal := division-algorithm
  (!strong-induction.principle goal
    method (x)
      assume ind-hyp := (strong-induction.hypothesis goal x)
        conclude (strong-induction.conclusion goal x)
          pick-any y
            assume (zero < y)
              conclude goal' := ((x / y) * y + x % y = x & x % y < y)
                (!force goal'))
```

Here we have called upon the conclusion procedure defined in the strong-induction module to state the desired conclusion. With a first argument of the form $\forall\, n\, .\, P(n)$ and second argument t, conclusion produces $P(t)$. In this case, it gives:

```
(forall y . zero < y ==> (x / y) * y + x % y = x & x % y < y)
```

Even more useful is the strong-induction.hypothesis procedure for stating the induction hypothesis. Recall that the strong induction hypothesis for proving $\forall\, n\, .\, P(n)$ is

$$\forall\, k\, .\, k < n \Rightarrow P(k),$$

so in this case it is

```
(forall x' .
  x' < x ==>
   forall y . zero < y ==> (x' / y) * y + x' % y = x' & x' % y < y)
```

Using these procedures both shortens the statement of the induction hypothesis and conclusion and avoids a potential source of error.[2]

To proceed with the proof, we introduce a case analysis depending on whether or not (x < y). The decision to perform this particular case analysis is made by deploying Heuristic 8.1, this time in connection with the definitions of / and %. The conjuncts of the goal contain occurrences of the terms (x / y) and (x % y), and in order to establish these conjuncts we will most likely have to apply the definitions of / and %. If we look at these

2 With ordinary induction proofs, there are no predefined counterparts of strong-induction's hypothesis and conclusion procedures (though they could be easily written), but there is generally less need for them because the induction hypothesis is simpler. Nevertheless, readers may wish to introduce a property procedure specific to the goal and express the induction hypothesis and desired conclusion in terms of it, as we do in many examples in this book.

definitions we see that they are given conditionally, depending on whether the first argument is less than the second, and that is precisely the chosen condition for our case analysis, in accordance with the heuristic.

In the (x < y) case, we do not need the induction hypothesis and the proof is straightforward:

```
conclude goal := division-algorithm
  (!strong-induction.principle goal
    method (x)
      assume ind-hyp := (strong-induction.hypothesis goal x)
        conclude (strong-induction.conclusion goal x)
          pick-any y
            assume (zero < y)
              conclude goal' := ((x / y) * y + x % y = x & x % y < y)
                (!two-cases
                  assume (x < y)
                    let {x/y=0 := (!chain->
                                    [(x < y)
                                  ==> (x / y = zero)       [Div.basis]]);
                         x%y=x := (!chain->
                                    [(x < y)
                                  ==> (x % y = x)          [Mod.basis]])}
                      (!both
                        (!chain [((x / y) * y + x % y)
                                = (zero * y + x)           [x/y=0 x%y=x]
                                = x                        [Times.left-zero
                                                            Plus.left-zero]])
                        (!chain-> [(x < y)
                                  ==> (x % y < y)          [x%y=x]]))
                  assume (~ x < y)
                    (!force goal')))
```

In the (~ x < y) case, we have to apply the reduction axioms for / and %, and show that in the resulting terms, the value in the first argument of these functions, namely (x - y), is less than x, so that the induction hypothesis can be applied:

```
let {p1 := (!chain-> [(~ x < y & zero < y)
                  ==> (x / y = S ((x - y) / y))    [Div.reduction]]);
     p2 := (!chain-> [(~ x < y & zero < y)
                  ==> (x % y = (x - y) % y)        [Mod.reduction]]);
     p3 := (!chain-> [(~ x < y) ==> (y <= x)       [Less=.trichotomy2]]);
     p4 := (!chain-> [(zero < y & y <= x)
                  ==> (x - y < x)                  [Minus.<-left]
                  ==> (forall z . zero < z ==>
                         (((x - y) / z) * z + (x - y) % z
                        = x - y & (x - y) % z < z)) [ind-hyp]]);
     ...
```

From that point on it is a matter of straightforward calculation using properties of multiplication, addition, and subtraction to complete the proof:

```
let {...
    p4 := ...
    (and p5a p5b) := (!chain->
                       [(zero < y)
                    ==> (((x - y) / y) * y +
                           (x - y) % y = x - y
                         & (x - y) % y < y)    [p4]]);
    p6 := (!chain [((x / y) * y + x % y)
                  = ((S ((x - y) / y)) * y
                         + (x - y) % y)        [p1 p2]
                  = ((y + ((x - y) / y) * y)
                       + (x - y) % y)          [Times.left-nonzero]
                  = (y + (((x - y) / y) * y
                       + (x - y) % y))         [Plus.associative]
                  = (y + (x - y))              [p5a]
                  = ((x - y) + y)              [Plus.commutative]
                  = x                          [p3 Minus.Plus-Cancel]])}
  (!chain-> [p5b ==> (x % y < y)               [p2]
                 ==> (p6 & (x % y < y))        [augment]])))
```

Note, in particular, the use of `Minus.Plus-Cancel`, which is permitted since `(y <= x)`.

It is a simple exercise to prove the following two corollaries:

```
define division-algorithm-corollary1 :=
  (forall x y . zero < y ==> (x / y) * y + x % y = x)

define division-algorithm-corollary2 := (forall x y . zero < y ==> x % y < y)
```

As these two names are rather long, we introduce two abbreviations for them:

```
define [DAC-1 DAC-2] := [division-algorithm-corollary1
                         division-algorithm-corollary2]
```

Exercise 13.1: Prove `division-algorithm-corollary1` and `division-algorithm-corollary2`. □

13.3 Divisibility

We say that a number *y* *divides* a number *x* iff the division of *x* by *y* leaves a `zero` remainder. The following definition introduces `divides` as a binary relation symbol:

```
declare divides: [N N] -> Boolean  [300 [int->nat int->nat]]

module divides {

  assert* def :=
    [(y = zero ==> y divides x <==> x = zero)
```

```
      (zero < y ==> y divides x <==> x % y = zero)]

  define [left-zero left-positive] := def
```

The relation "y divides x" is often written $y \mid x$, but we avoid that notation since the vertical bar is commonly used in Athena to represent disjunction.

Another definition frequently given is this: y `divides` x iff there is some number z such that $y \cdot z = x$. We instead prove this as a *characterization theorem* (as we did in many cases in Chapter 10 and elsewhere):

```
  define characterization :=
    (forall x y . y divides x <==> exists z . y * z = x)
} # close module divides
```

Once we have proved this theorem, we can use it to derive many other results about the divides relation, just as if we had taken it as the definition of divisibility. The reason we prefer the above definition of `divides` in terms of `%` and equality is that it is *constructive*, so it allows testing with `eval`, while defining in terms of existence would not.

```
> (eval 3 divides 6)

Symbol: true

> (eval 3 divides 7)

Symbol: false

> (eval 3 divides 8)

Symbol: false

> (eval 3 divides 9)

Symbol: true

> (eval 0 divides 3)

Symbol: false

> (eval 0 divides 0)

Symbol: true
```

Note that for (y `divides` x) we have specified a result when y is 0, in contrast to our underspecification of (x `/` y) and (x `%` y). The reason for this decision will become clear when we come to tackle the proof of the cancellation theorem.

13.3.1 A cancellation lemma

The following cancellation result for division into a product will play a role in the proof of the characterization theorem for `divides`:

```
extend-module Div {
  define cancellation := (forall x y . zero < y ==> (x * y) / y = x)
}
```

Exercise 13.2: Prove `Div.cancellation`. *Hint*: Use induction. □

13.3.2 Proof of the characterization theorem

In the proof of `characterization`, we split into a couple of cases, and within each we use `equiv` to divide the proof into the two relevant implications:

```
extend-module divides {
  conclude characterization
    pick-any x y
      (!two-cases
        assume (zero < y)
          (!equiv
            (!force (y divides x ==> exists z . y * z = x))
            (!force ((exists z . y * z = x) ==> y divides x)))
        assume (~ zero < y)
          (!equiv
            (!force (y divides x ==> exists z . y * z = x))
            (!force ((exists z . y * z = x) ==> y divides x))))
}
```

In the `(zero < y)` case, the proofs of both directions of the equivalence involve using the definition of `divides` and the first corollary of the division algorithm. We show the proof of the second direction and leave the other as an exercise.

```
assume h := (exists z . y * z = x)
  pick-witness z for h y*z=x
    (!by-contradiction (y divides x)
      assume -y|x := (~ y divides x)
        let {p := (!chain-> [(zero < y)
                         ==> (y divides x <==> x % y = zero)
                                                [left-positive]])}
         (!absurd
           (!chain->
             [-y|x
         ==> (x % y =/= zero)          [p]
         ==> (zero < x % y)            [Less.zero<]
         ==> (zero + (x / y) * y <
               x % y + (x / y) * y)    [Less.Plus-k]
         ==> ((x / y) * y <
               (x / y) * y + x % y)    [Plus.left-zero
```

```
                                  Plus.commutative]
        ==> ((x / y) * y < x)    [division-algorithm-corollary1]
        ==> (y * (x / y) < y * z) [y*z=x Times.commutative]
        ==> (x / y < z)          [Times.<-cancellation]
        ==> ((y * z) / y < z)    [y*z=x]
        ==> (z < z)              [Times.commutative
                                  Div.cancellation]])
         (!chain->
           [true
        ==> (~ z < z)            [Less.irreflexive]])))
```

Exercise 13.3: Fill in the proof of the `(y divides x ==> exists z . y * z = x)` sub-case in the `(zero < y)` case. □

Exercise 13.4: Complete the proof of the characterization theorem by filling in the proof of the second top-level case, `(~ zero < y)`. This is where the `divides.left-zero` axiom comes into play; without it, the characterization theorem would have to be conditional on a positive divisor. □

13.3.3 Additional properties of divisibility

It is easy to see that every number x divides 0, since $x \cdot 0 = 0$ for all x:

```
extend-module divides {
conclude right-zero := (forall x . x divides zero)
  pick-any x
    (!chain-> [true
            ==> (x * zero = zero)              [Times.right-zero]
            ==> (exists y . x * y = zero)      [existence]
            ==> (x divides zero)               [characterization]])
```

The `divides` relation is reflexive, antisymmetric, and transitive. We prove these properties below:

```
conclude reflexive := (forall x . x divides x)
  pick-any x
    (!chain-> [true ==> (x * one = x)          [Times.right-one]
                    ==> (exists y . x * y = x) [existence]
                    ==> (x divides x)          [characterization]])
```

It is convenient to introduce a method, `divides.elim`, for eliminating `divides` in favor of its characterizing existential sentence:

```
define elim :=
  method (x y)
    let {v := (fresh-var (sort-of x))}
      (!chain->
        [(x divides y) ==> (exists v . x * v = y) [characterization]])
```

We can then apply this method to obtain the existential and pick a witness for it in the same line, as in the following proofs:

```
define antisymmetric :=
  (forall x y . x divides y & y divides x ==> x = y)

conclude antisymmetric
  pick-any x y
    assume (x divides y & y divides x)
      pick-witness u for (!elim x y) x*u=y
        pick-witness v for (!elim y x) y*v=x
        (!two-cases
          assume x=zero := (x = zero)
            (!chain-> [x*u=y
                   ==> (zero * u = y)          [x=zero]
                   ==> (zero = y)              [Times.left-zero]
                   ==> (x = y)                 [x=zero]])
          assume x!=zero := (x =/= zero)
            let {0<x := (!chain->
                        [x!=zero
                     ==> (zero < x)            [Less.zero<]]);
                 u=1 := (!chain->
                        [x = (y * v)           [y*v=x]
                           = ((x * u) * v)     [x*u=y]
                           = (x * (u * v))     [Times.associative]
                     ==> (x * (u * v) = x)     [sym]
                     ==> (u * v = one)         [p1 Times.identity-lemma1]
                     ==> (u = one)             [Times.identity-lemma2]])}
              (!chain [x = (x * one)           [Times.right-one]
                         = (x * u)             [u=1]
                         = y                   [x*u=y]]))

define transitive :=
  (forall x y z . x divides y & y divides z ==> x divides z)

conclude transitive
  pick-any x y z
    assume (x divides y & y divides z)
      pick-witness u for (!elim x y) x*u=y
        pick-witness v for (!elim y z) y*v=z
          (!chain->
            [(x * (u * v)) = ((x * u) * v)     [Times.associative]
                           = (y * v)           [x*u=y]
                           = z                 [y*v=z]
        ==> (exists w . x * w = z)             [existence]
        ==> (x divides z)                      [characterization]])
```

The following are two other properties of interest: (a) If x divides y or x divides z, then x divides the product $y \cdot z$; and (b) x divides y and x divides z iff x divides y and x divides the sum $y+z$. A useful corollary variant of (b) is only concerned with the left-to-right direction: (b') if x divides y and x divides z then x divides $y+z$. We formulate these properties as follows:

```
define product-lemma :=
  (forall x y z . x divides y | x divides z ==> x divides y * z)

define sum :=
  (forall x y z . x divides y & x divides z  <==>
                  x divides y & x divides (y + z))

define sum-lemma1 :=
  (forall x y z . x divides y & x divides z ==> x divides (y + z))

define sum-lemma2 :=
  (forall x y z . x divides y & x divides (y + z) ==> x divides z)
```

Note that the strict converse of `sum-lemma1` does not hold; the condition `(x divides y)` is crucial:

```
define sum-lemma1' :=
 (forall x y z . x divides (y + z) ==> x divides y & x divides z)

> (falsify sum-lemma1' 50)

List: ['success |{?x:N := (S (S zero)), ?y:N := (S zero), ?z:N := (S zero)}|]
```

We prove the first lemma below, and leave the proof of the others as an exercise. (The third and fourth can be used in the proof of the second.)

Exercise 13.5: Prove `divides.sum-lemma1`, `divides.sum-lemma2`, and `divides.sum`. □

One way of tackling the proof of `divides.product-lemma` is to proceed directly: Assume that x divides y *or* that x divides z, and then go on to show that x must divide the product $y \cdot z$. Doing so would necessitate a case analysis on whether x divides y or z, and showing that x divides their product in either case. Building a single monolithic proof in this manner, however, is not the most elegant approach. We can save a good deal of effort by being a bit more clever and exploiting the symmetry inherent in the combination of the disjunctive antecedent and the commutativity of the multiplication that appears in the consequent. Specifically, suppose that we first prove a slightly weaker, nondisjunctive result:

```
define product-left-lemma :=
  (forall x y z . x divides y ==> x divides y * z)

conclude product-left-lemma
  pick-any x y z
    assume (x divides y)
      pick-witness u for (!elim x y) x*u=y
```

```
    (!chain->
      [(y * z) = ((x * u) * z)          [x*u=y]
               = (x * (u * z))          [Times.associative]
   ==> (x * (u * z) = y * z)            [sym]
   ==> (exists w . x * w = y * z)       [existence]
   ==> (x divides y * z)                [characterization]])
```

We can then easily derive divides.product-lemma by invoking divides.product-left-lemma in each of the two cases, taking care to swap the product in the second case:

```
conclude product-lemma
  pick-any x y z
    assume hyp := (x divides y | x divides z)
      conclude goal := (x divides y * z)
        (!cases hyp
           (!chain
              [(x divides y) ==> goal               [product-left-lemma]])
           (!chain
              [(x divides z) ==> (x divides z * y) [product-left-lemma]
                             ==> goal               [Times.commutative]]))
```

We will also need:

```
define first-lemma :=
 (forall x y z . zero < y & z divides y & z divides x % y ==> z divides x)
```

The following picture provides a geometric demonstration of this result:

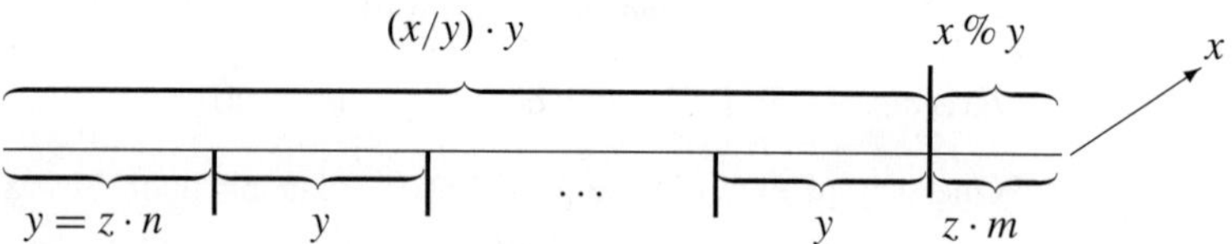

The long horizontal line represents x, which, by the division algorithm, equals the sum of $(x/y) \cdot y$ and $x \,\%\, y$. Since z divides y (i.e., there is some n such that $y = z \cdot n$), it follows from divides.product-lemma that z divides $(x/y) \cdot y$. Moreover, by assumption, z divides $x \,\%\, y$. Therefore, by divides.sum-lemma1 we can infer that z divides x. (In fact, the picture makes clear what the exact relationship is: $x = z \cdot [((x/y) \cdot n) + m]$.) A symbolic proof can be given as follows:

```
conclude first-lemma
  pick-any x y z
    assume (zero < y & z divides y & z divides x % y)
      conclude goal := (z divides x)
        pick-witness u for (!elim z y) z*u=y
          pick-witness v for (!elim z (x % y)) z*v=x%y
```

```
   (!chain->
    [x = ((x / y) * y +
          x % y)                     [(zero < y) DAC-1]
       = ((x / y) * (z * u) +
          z * v)                     [z*u=y z*v=x%y]
       = (((x / y) * u) * z +
          v * z)                     [Times.commutative
                                      Times.associative]
       = (((x / y) * u + v) * z)     [Times.right-distributive]
       = (z * ((x / y) * u + v))     [Times.commutative]
  ==> (z * ((x / y) * u + v) = x)    [sym]
  ==> (exists w . z * w = x)         [existence]
  ==> goal                           [characterization]])
```

Another useful set of properties is this: For some binary operation *f*, if *x* divides both *y* and *z*, then *x* divides (*y* *f* *z*). For instance, if *x* divides both *y* and *z*, then it also divides their difference:

```
define Minus-lemma :=
  (forall x y z . x divides y & x divides z ==> x divides (y - z))

conclude Minus-lemma
  pick-any x y z
    assume (x divides y & x divides z)
      pick-witness u for (!elim x y) x*u=y
        pick-witness v for (!elim x z) x*v=z
          (!chain->
           [(y - z) = (x * u - x * v)      [x*u=y x*v=z]
                    = (x * (u - v))        [Minus.Times-Distributivity]
        ==> (x * (u - v) = y - z)          [sym]
        ==> (exists w . x * w = y - z)     [existence]
        ==> (x divides (y - z))            [characterization]])
```

A similar property holds for the remainder operation: If *x* divides *y* and *z* then it also divides (*y* % *z*), provided that *z* is positive.

Exercise 13.6: Consider:

```
define Mod-lemma :=
  (forall x y z . x divides y & x divides z & zero < z ==> x divides y % z)
```

Prove this lemma. □

With these results in our toolbox, we close modules `N.divides` and `N`.

```
  } # close module divides
} # close module N
```

13.4 Euclid's algorithm

We are now ready to formulate and prove the correctness of Euclid's algorithm for computing the greatest common divisor of two natural numbers. We begin by introducing a new binary function, `euclid`, and defining its behavior with a recursive `fun` definition:

```
extend-module N {

declare euclid: [N N] -> N  [[int->nat int->nat]]

module Euclid {
  assert axioms :=
    (fun [(euclid x y) = [x                           when (y = zero)
                          (euclid y (x % y))  when (y =/= zero)]])

  define [base reduction] := axioms
```

It is not difficult to see that these axioms constitute a proper definition, in that they adhere to the guidelines discussed in Section 3.10. First, the two axioms are mutually exclusive and jointly exhaustive, meaning that for any given canonical terms s and t, exactly one of the two axioms can be used to derive a conclusion of the form

$$((\texttt{euclid}\ s\ t) = t').$$

Moreover, the two axioms cannot give rise to an infinite chain of the form

$$(\texttt{euclid}\ s_1\ t_1) \rightarrow (\texttt{euclid}\ s_2\ t_2) \rightarrow (\texttt{euclid}\ s_3\ t_3) \rightarrow \cdots$$

for canonical terms s_i, t_i. This follows from theorem `division-algorithm-corollary2`, which assures us that if y is positive then $(x\ \%\ y)$ is strictly less than y. Therefore, in any chain of the preceding form, if a step

$$(\texttt{euclid}\ s_i\ t_i) \rightarrow (\texttt{euclid}\ s_{i+1}\ t_{i+1})$$

is licensed by `Euclid.reduction`, then the term t_{i+1} will be strictly less than t_i. Therefore, in any such chain, the second argument will eventually become `zero`, at which point `Euclid.base` will terminate the recursion by outputting the first argument.

Let us now examine the behavior of `euclid` in more detail. We can start by "running" the definition on various canonical terms using `eval`, to see what results we get:

```
> (eval euclid 6 3)

Term: 3

> (eval euclid 6 9)

Term: 3
```

```
> (eval euclid 16 12)

Term: 4

> (eval euclid 25 5)

Term: 5

> (eval euclid 48 64)

Term: 16

> (eval euclid 1 0)

Term: 1
```

Our definition thus appears to give correct results. We will now go on to prove that in fact it always gives correct results. Specifically, we will prove that for all x and y, `(euclid x y)` is the "greatest" common divisor of x and y (the use of scare quotes will be explained later).

We will break the proof into two parts. First we will show that `(euclid x y)` is a common divisor of x and y. And second, that it is the greatest such common divisor.

We begin by defining a simple binary procedure `is-common-divisor`, which takes a natural number `z` and a list `[x y]` of two natural numbers and outputs a sentence stating that `z` divides both `x` and `y`:

```
define (is-common-divisor z terms) :=
  match terms {
    [x y] => (z divides x & z divides y)
  }
```

As we have already seen at many previous points in the book, sentence-producing procedures can be very useful. They are often a simpler alternative than having to introduce new function symbols and then axiomatize them (as we would have to do in this case if we declared `is-common-divisor` as a predicate symbol).

We can then define the first sentence that we want to derive as follows:

```
define common-divisor :=
  (forall x y . (euclid x y) is-common-divisor [x y])
```

which expands to

```
(forall x y . (euclid x y) divides x & (euclid x y) divides y)
```

In this case, the argument of euclid whose size decreases with each recursive call is the second one.[3] Therefore, in accordance with Heuristic 13.1, instead of attempting an inductive proof of common-divisor, we will try to prove

```
define common-divisor' :=
  (forall y x . (euclid x y) is-common-divisor [x y])
```

And since the second argument decreases by an arbitrary amount rather than just by one, we choose to attempt the proof by strong induction.

As we did with the division-algorithm proof, we start with a sketch that sets up the strong induction proof structure, this time also sketching the case analysis based on whether y is zero or nonzero.

```
conclude goal := common-divisor'
 (!strong-induction.principle goal
   method (y)
    assume ind-hyp := (strong-induction.hypothesis goal y)
      conclude (strong-induction.conclusion goal y)
        pick-any x
          conclude goal' := ((euclid x y) is-common-divisor [x y])
            (!two-cases
              assume y=0 := (y = zero)
                (!force C)
              assume y!=0 := (y =/= zero)
                (!force C)))
```

The case analysis was suggested by Heuristic 8.1. The first case, when y is zero, is straightforward:

```
assume y=0 := (y = zero)
  (!both (!chain->
           [true ==> (x divides x)                 [divides.reflexive]
                 ==> ((euclid x y) divides x)      [base y=0]])
         (!chain->
           [true ==> ((euclid x y) divides zero) [divides.right-zero]
                 ==> ((euclid x y) divides y)     [y=0]]))
```

In the second case we have to show that (x % y < y) so that we can apply the induction hypothesis. For this purpose we have at hand division-algorithm-corollary2 (page 608).

```
assume y!=0 := (y =/= zero)
  let {0<y := (!chain-> [(y =/= zero)
                     ==> (zero < y)                [Less.zero<]]);
       divides-both := (!chain->
                         [0<y
                      ==> (x % y < y)              [DAC-2]
                      ==> (forall x' .
                            (euclid x' (x % y))
```

3 This follows from division-algorithm-corollary2.

```
                        is-common-divisor
                          [x' (x % y)])          [ind-hyp]
                 ==> ((euclid y (x % y))
                        is-common-divisor
                          [y (x % y)])           [(uspec with y)]
                 ==> ((euclid x y)
                        is-common-divisor
                          [y (x % y)])           [reduction y!=0]]);
        divides-y := (!left-and divides-both)}
     (!chain-> [(0<y & divides-both)
           ==> ((euclid x y) divides x)          [divides.first-lemma]
           ==> ((euclid x y) divides x & divides-y) [augment]])
```

For the second part of the correctness proof we will show that `(euclid` x y`)` is the greatest common divisor of x and y, meaning that if *any* number z is a common divisor of x and y, then z is also a divisor of `(euclid` x y`)`.[4] More precisely, our goal is the following:

```
define greatest :=
  (forall x y z . z is-common-divisor [x y] ==> z divides (euclid x y))
```

As in the previous part, we follow Heuristic 13.1 and decide to use strong induction on the second argument to `euclid`. Accordingly, we will prove the following equivalent variant of the above goal:

```
define greatest' :=
  (forall y x z . z is-common-divisor [x y] ==> z divides (euclid x y))
```

The proof is shown below. To save space, we define `icd` as an abbreviation for `is-common-divisor`.

```
define icd := is-common-divisor

conclude goal := greatest'
 (!strong-induction.principle goal
   method (y)
    assume ind-hyp := (strong-induction.hypothesis goal y)
      conclude (strong-induction.conclusion goal y)
        pick-any x z
          assume (z is-common-divisor [x y])
            conclude (z divides (euclid x y))
              (!two-cases
                assume y=0 := (y = zero)
                  (!chain->
                   [(z divides x)
                ==> (z divides (euclid x y))          [y=0 base]])
```

4 We do not always have `(`z `<= (euclid` x y`))` when z divides x and y. Demonstrating that is left as an exercise.

```
              assume y!=0 := (y =/= zero)
               let {0<y := (!chain->
                             [(y =/= zero)
                          ==> (zero < y)                [Less.zero<]]);
                    z|x%y := (!chain->
                               [0<y
                            ==> (z divides x &
                                 z divides y & 0<y) [augment]
                            ==> (z divides x % y)   [divides.Mod-lemma]]);
                    p := (!chain->
                          [0<y
                       ==> (x % y < y)                  [DAC-2]
                       ==> (forall x' z .
                              z icd [x' (x % y)]
                               ==>
                             z divides
                                (euclid x'
                                        (x % y))) [ind-hyp]
                       ==> (z icd [y (x % y)]
                               ==>
                              z divides
                                (euclid y
                                  (x % y)))        [(uspec*
                                                     with [[y z]])]])}
                 (!chain-> [(z divides y & z|x%y)
                        ==> (z divides
                              (euclid y (x % y))) [p]
                        ==> (z divides
                             (euclid x y))         [y!=0 reduction]])))
```

Finally, we combine these two main results into a single statement of correctness of Euclid's algorithm:

```
define correctness :=
  (forall x y . (euclid x y) is-common-divisor [x y] &
               forall z . z is-common-divisor [x y] ==>
                          z divides (euclid x y))

conclude correctness
  pick-any x y
    (!both (!chain->
             [true
         ==> ((euclid x y) is-common-divisor    [x y])
                                                [common-divisor']])
         pick-any z
           (!chain [(z is-common-divisor [x y])
                ==> (z divides (euclid x y))    [greatest']]))
} # close module Euclid
} # close module N
```

13.5 Summary

In preparation for specifying and proving the correctness of Euclid's classic algorithm, we constructively defined quotient and remainder functions and a divisibility relation on natural numbers. Because the definitions were constructive, we were able to test them using evaluation (with the `eval` procedure). We also proved a characterization theorem for the divisibility relation in terms of the existence statement that is often used to define that relation. The characterization theorem was in turn a primary tool for deriving several other useful properties of divisibility.

Strong induction proofs, introduced in the previous chapter, came into play in this chapter as well, both for the proof of the division algorithm (which, as mentioned, is not actually an algorithm but a theorem relating the quotient and remainder of a division), and for the correctness of Euclid's algorithm.

In the additional exercises we explore a few further properties of divisibility and touch on one of the most important topics of number theory: primality.

13.6 Additional exercises

Exercise 13.7: The division algorithm implies that for any x and y, there are numbers q and r such that

$$(q * y) + r = x, \tag{13.1}$$

namely, $q = x \; / \; y$ and $r = x \; \% \; y$. Show that these numbers are *unique*, which is to say that *for all* q and r, if (13.1) holds then $q = x \; / \; y$ and $r = x \; \% \; y$. □

Exercise 13.8: Show that the result of the Euclidean algorithm is not always the *greatest* common divisor of its inputs, in terms of natural number ordering. □

⋆ ***Exercise 13.9:*** We can define prime numbers (nonconstructively) as follows:

```
open N
open List

define [m n k L] := [?m ?n ?k ?L]

declare prime: [N] -> Boolean [[int->nat]]

module prime {

  assert definition :=
    (forall n .
      prime n <==> one < n &
                   forall m . m divides n ==> m = one | m = n)

}
```

⋆ (a) Prove the following:

```
define not-prime :=
  (forall n . one < n & ~ prime n <==> exists m . one < m &
                                                  m < n &
                                                  m divides n)
```

(*Hint*: Consider using **datatype-cases** for the direction ==>.)

(b) Define the product of a nonempty list of natural numbers by

```
declare product: [(List N)] -> N [[(alist->clist int->nat)]]

module product {

  assert* def := [(product nil = one)
                  (product n::L = n * product L)]
```

Prove the following two properties of product:

```
  define of-singleton := (forall n . product n::nil = n)

  define of-join :=
    (forall L M . product L ++ M = (product L) * (product M))

} # close module product
```

⋆ (c) Consider the following nonconstructive definition of a predicate all-primes:

```
declare all-primes: [(List N)] -> Boolean  [[(alist->clist int->nat)]]

module all-primes {
  assert definition :=
    (forall L . all-primes L <==> forall k . k in L ==> prime k)
}
```

Prove, using strong induction, that every natural number greater than 1 is the product of (one or more) primes:

```
define product-of-primes :=
  (forall n . one < n ==> exists L . n = product L & all-primes L)
```

Note: This is a case, as with Euclid's algorithm, where a value to induct on in a strong-induction proof could be *any* of the values smaller than n. Also, you will need to apply the strong induction hypothesis on *two* smaller values. □

13.7 Chapter notes

Euclid's algorithm is often cited as one of the oldest known algorithms (Euclid lived around 300 years BCE, but the algorithm is believed to have been known earlier by Eudoxus of Cnidus and by Aristotle). In fact, the central idea behind the fast exponentiation algorithm of the previous chapter is even older: An Egyptian manuscript believed to have been written about 1650 BCE reveals that mathematicians of that era used such a method, not for raising a number to a power (using fewer multiplications), but for multiplying two numbers (using fewer additions).

The first proof of algorithm correctness, however, had to wait until the notion of mathematical proof was developed by Greek mathematicians. If you consider each of the geometric constructions in Euclid's *Elements* as describing an algorithm, not only are many algorithms presented, but they are all also proved correct! The greatest common divisor algorithm, phrased as it is in the *Elements* as a geometric construction, is no exception.

Binary search algorithms are much more recent. Although we started with an algorithm on tree structures in Chapter 11 and will later consider one that operates on sequential storage (among other memory ranges), historically it was the reverse. Knuth [57] attributes the first public mention of binary search on sequential storage to John Mauchly in 1946 and the earliest algorithms on tree structures to independent discovery by several people in the 1950s. Knuth credits the idea of avoiding a separate test for equality to H. Bottenbruch in 1962, in a binary search algorithm on sequential storage. We used that strategy in the optimized binary search on trees (Section 11.4), and we will use it again in the abstract memory range algorithm considered in Section 16.10.

Most of these historical notes are recounted in greater detail by Stepanov and Rose in *From Mathematics to Generic Programming* [94], a recent book with a mathematics-based approach to programming that analyzes these three algorithms and others much as we do in this book, but with less emphasis on proofs (none on writing computer-based proofs).

V PROOFS AT AN ABSTRACT LEVEL

In this part, we explore the advantages of abstraction and specialization in developing and applying theories, first for traditional, static mathematical theories in Chapter 14. We continue with theories about algorithms in Chapter 15, illustrating the approach with abstracted developments of the binary-search and exponentiation algorithms of Part IV. Chapter 16 then shows how to reason about abstractions that explicitly include memory locations and allow for updating their contents, illustrated with iterator operations and abstract (generic) algorithms modeled after those in the C++ Standard Template Library (STL).

In Athena, we can express abstract theories using a *structured theory* data structure. Proofs about the theories introduced in this way are expressed with parameterized methods that follow certain conventions, which are outlined and illustrated in Chapter 14. Structured theories suffice to express algebraic theories such as monoids, groups, and rings; relational theories such as partial and total orders; memory abstractions such as individual locations and memory ranges; and abstract algorithms on each of these kinds of theories.

14 Abstract Structures

IN this chapter, we introduce abstract axioms and theorems and show how to organize them using a *structured theory* data structure that will also be used for abstract algorithm properties. We begin with algebraic theories that generalize some of the concrete data structures we have encountered in previous chapters. These algebraic theories are the mathematical basis of many kinds of numeric and nonnumeric computation, and readers are encouraged to explore beyond the examples given in this and the next chapter. In Section 14.9, we turn to theories that generalize concrete ordering relations such as $<$ and $\leq$ on natural numbers or integers. These theories provide a sound basis for generalization of searching and sorting algorithms to an abstract level. We will see only a couple of examples, both binary-search algorithms, one in the next chapter for searching binary trees and the other in Chapter 16 for searching a range of memory locations, but Section 16.11 contains suggestions for further study.

14.1 Group properties

In Chapter 9 we studied two distinct data types, `Z` (integers) and `ZPS` (formal power series with integer coefficients). Focusing on properties of their respective addition operators, `Z.+` and `ZPS.+`, we see that they share several common properties (in what follows `a`, `b`, and `c` are defined as ?a, ?b, and ?c, respectively):

- Associativity:

```
(forall a b c . (a  Z.+ b) Z.+ c = a Z.+ (b Z.+ c))
(forall a b c . (a  ZPS.+ b) ZPS.+ c = a ZPS.+ (b ZPS.+ c))
```

- Left and Right Identity:

```
(forall a . Z.zero Z.+ a = a)
(forall a . a Z.+ Z.zero = a)

(forall a . ZPS.zero ZPS.+ a = a)
(forall a . a ZPS.+ ZPS.zero = a)
```

- Left and Right Inverses:

```
(forall a . (Z.negate a) Z.+ a = Z.zero)
(forall a . a Z.+ (Z.negate a) = Z.zero)

(forall a . (ZPS.Negate a) ZPS.+ a = ZPS.zero)
(forall a . a ZPS.+ (ZPS.Negate a) = ZPS.zero)
```

In each case, the properties are the same except for the function names. We have already seen how such similarities, called *homomorphisms*, can be exploited in proofs, when we used the additive homomorphism property between the signed and pair representations of integers in Section 9.5. Yet these two structures are obviously not the same in the sense of having all other properties related homomorphically, the most notable difference being that there is no mapping from `ZPS` onto `Z` that preserves addition.

Such observations have long held a special interest for mathematicians, as they have far-reaching consequences for understanding different mathematical structures and being able to conjecture and prove additional results about them. In mathematics, a set of axioms together with theorems that follow from the axioms is called a *theory*. In this chapter and the next we explore how we can define and use theories in computer science, beginning with theories based on traditional numeric, algebraic, and relational notions, but continuing with ones inspired by recognition of commonalities among algorithms.

A *group* is a structure with a distinguished "zero" element and a binary operation satisfying the associative, identity, and invertibility properties discussed below.[1] Both `Z` and `ZPS` are groups, and while both are numeric in nature, there are many completely nonnumeric examples; we study one such nonnumeric case in Section 14.8. We could formally introduce *group theory* in Athena as shown below (temporarily; we'll show a better way shortly). We first introduce some variable names that will be visible for the remainder of this chapter:

```
define [x y z] := [?x ?y ?z]

module Group {
  declare +: (S) [S S] -> S [200]
  declare <0>: (S) [] -> S
  declare U-: (S) [S] -> S    # Unary minus
  declare -: (S) [S S] -> S  # Binary minus

  define associative := (forall x y z . (x + y) + z = x + (y + z))
  define left-identity := (forall x . <0> + x = x)
  define right-identity := (forall x . x + <0> = x)
  define left-inverse := (forall x . (U- x) + x = <0>)
  define right-inverse := (forall x . x + (U- x) = <0>)
  define minus-definition := (forall x y . x - y = x + U- y)
} # close module Group
```

Although we regard the properties listed in the module as the axioms of group theory, we do not assert them into the assumption base. Instead, if we have *proved* that a homomorphic image of each of these properties is a theorem, then we will be able to use proof methods

---

1 In the context of a theory T, the term "structure" refers to an arbitrary set along with a number of operations and predicates on that set which may or may not satisfy T's axioms. So in this sense a structure can be viewed as an interpretation of T's vocabulary, as defined in Section 5.6. A *model* of T is a structure that satisfies T's axioms. When discussing theories, we use the term "property" essentially as a synonym for "sentence." More precisely, we say that a sentence p expresses a property of a theory T iff p is true in every structure that satisfies T's axioms.

associated with the theory to prove new theorems, rather than having to write their proofs from scratch for every structure that models the theory.

The main idea behind our approach is that of a *symbol map*, which is literally a map from function symbols to function symbols, that is, an Athena map whose keys and values are function symbols, such as

```
|{+ := *}|.
```

A reasonable symbol map m must satisfy a number of requirements: It must preserve arities (so that it cannot map a binary symbol to a unary symbol); it must be sort-consistent (so that, for example, if m maps f to g and f has signature `[Ide Ide] -> Boolean`, then g cannot have a signature like `[Boolean Ide] -> Real`); and so on. However, it will not be necessary to define or enforce these conditions explicitly.

It is convenient to extend a symbol map m into a unary procedure r_m that can accept an arbitrary term t or sentence p and replace every occurrence of a function symbol f in t (or in p) by the symbol $(m\ f)$, provided that f is in the domain of m. Occurrences of function symbols that are not in the domain of m are left unchanged. We can perform this extension with the procedure `renaming`, which takes as input a symbol map m and produces as output the procedure r_m. For instance:

```
define m := |{+ := *}|

define r_m := (renaming m)

> (r_m (1 + x))

Term: (* 1 ?x:Int)

> (r_m (1 / (2 + 3)))

Term: (/ 1
         (* 2 3))

> (r_m (forall x . x <= x + 1))

Sentence: (forall ?x:Int
            (<= ?x:Int
                (* ?x:Int 1)))
```

The procedure r_m produced by `renaming` can also be applied to a list `[`$V_1 \cdots V_n$`]` of terms and/or sentences, in which case it will produce `[`$(r_m\ V_1)\ \cdots\ (r_m\ V_n)$`]`:

```
> (r_m [(x = x + y) (forall x . x <= x + 1)])

List: [
(= ?x
   (* ?x ?y))
```

```
(forall ?x:Int
  (<= ?x
      (* ?x 1)))
]
```

We say that r_m is an "extension" of m because we can still apply it to an individual function symbol f in the domain of m and get back $(m\ f)$. If f is not in m's domain, we get back f again:

```
> (r_m +)

Symbol: *

> (r_m /)

Symbol: /
```

We use the term *adapter* to refer both to a symbol map m and to the procedure r_m we get by applying renaming to m. The context will always clarify the intended use.

For an example that will be closer in spirit to the theories with which we will be working, let us assume that modules Z and ZPS from Chapter 9 have been loaded in the current session. We can then say:

```
set-flag print-var-sorts "off"

define Z-Additive-Group :=
  (renaming |{Group.+ := Z.+, Group.<0> := Z.zero,
              Group.U- := Z.negate, Group.- := Z.-}|)

> (Z-Additive-Group Group.associative)

Sentence: (forall ?x:Z
            (forall ?y:Z
              (forall ?z:Z
                (= (Z.+ (Z.+ ?x ?y)
                        ?z)
                   (Z.+ ?x
                        (Z.+ ?y ?z))))))

> define ZPS-Additive-Group :=
  (renaming |{Group.+ := ZPS.+, Group.<0> := ZPS.zero,
              Group.U- := ZPS.Negate, Group.- := ZPS.-}|)

> (ZPS-Additive-Group Group.associative)

Sentence: (forall ?x:(ZPS.Fun N Z)
            (forall ?y:(ZPS.Fun N Z)
              (forall ?z:(ZPS.Fun N Z)
```

```
                    (= (ZPS.+ (ZPS.+ ?x
                                     ?y)
                              ?z)
                       (ZPS.+ ?x
                              (ZPS.+ ?y
                                     ?z))))))
```

The ST module, which we discuss in the next section, defines the special adapter no-renaming as the identity procedure. Note, finally, that we issued the directive **set-flag** print-var-sorts "off" at the top of this code listing in order to suppress the printing of variable sorts in Athena's output (with the exception of variables immediately following quantifier occurrences), in the interest of brevity. This directive will remain in effect for the rest of this chapter.

14.2 Theory refinement

Before we go further with group properties, we should note that there are many interesting structures that have some, but not all, of the properties group theory requires. While Z is a group under addition, it is not under multiplication (the inverse properties fail to hold); and the structure we have studied most, the natural numbers, is not a group even under addition (again because the inverse properties fail). As a nonnumeric example, consider lists under concatenation (the List.join function defined in Section 3.9): List.join is associative, and nil acts as a left and right identity, but once again there are no inverses.

So we see that having the properties of associativity and left and right identities, but not inverses, is common to several structures. There being in fact many such structures, mathematicians have invented a theory that unites them: *monoid*. As a consequence, another way of defining group theory is to say that it is monoid theory (i.e., it requires the monoid properties, associativity, and left and right identity) and also requires left and right inverses. This relationship between groups and monoids is called *refinement*: Theory B is a refinement of theory A iff B's axioms are a superset of A's axioms, and none of the new axioms of B are theorems of A. Thus, every model of B is also a model of A, but there are models of A that are not models of B. We call A a *superior* of B.

Theory refinement is an important notion:

1. If B is a refinement of A and p is a theorem derivable from the axioms of A, then it is also derivable from the axioms of B. This follows because (1) the axioms of B are a superset of those of A; and (2) classical logic is monotonic, which means (as we noted in Section 4.3) that any property derivable from some β is also derivable from every superset $\beta' \supseteq \beta$. Thus, there is no need to reprove p for B; we can rely on the proof already recorded for A.

2. We can take 1 as a basis for extending our approach to organizing a library of axioms, theorems, and proofs. The essential idea is to associate each theorem and its proof with a certain theory in a *theory refinement hierarchy*, chosen as the one that has just the axioms needed in the proof—and no more.

Consequently, rather than defining group theory monolithically as we did above, we prefer to define it hierarchically using a data structure we call a *structured theory*, which is implemented in the module `ST` (for "Structured Theory"). The following are the six most important procedures implemented in `ST`:

- `make-theory`: Viewing structured theories as an ADT (abstract data type), this is a *creator* of that ADT (refer to the terminology described in Section 10.5).
- `add-axioms`: This can be used to add axioms to an existing theory; it is a *mutator* of the theory ADT.
- `add-theorems`: This is used to add theorems (or more precisely, *proof methods* capable of deriving theorems) to an existing theory; it is also a mutator.
- `theory-name`: This returns the name of a theory; it is an observer of the theory ADT.
- `theory-superiors`: This returns the list of (all) superiors of a given theory; it is also an observer.
- `theory-axioms`: This observer returns a list of (all) axioms of a theory.
- `theory-theorems`: This observer returns a list of (all) theorems of a theory.

These procedures are visible at the top level, so we can just write, for example, `make-theory` instead of `ST.make-theory`. We illustrate some of these procedures by showing how they can be used to introduce group theory:

```
module Semigroup {
  declare +: (S) [S S] -> S [200]

  define associative := (forall x y z . (x + y) + z = x + (y + z))

  define theory := (make-theory [] [associative])
}

module Identity {
  open Semigroup
  declare <0>: (S) [] -> S

  define left-identity :=  (forall x . <0> + x = x)
  define right-identity := (forall x . x + <0> = x)

  define theory := (make-theory [] [left-identity right-identity])
```

```
}

module Monoid {
  open Identity

  define theory := (make-theory ['Semigroup 'Identity] [])
}

module Group {
  open Monoid

  declare U-: (S) [S] -> S     # Unary minus
  declare  -: (S) [S S] -> S   # Binary minus

  define right-inverse      := (forall x . x + U- x = <0>)
  define minus-definition := (forall x y . x - y = x + U- y)

  define theory := (make-theory ['Monoid] [right-inverse minus-definition])
}
```

Specifically, when *sup* is a list of existing theories (or their names, represented as meta-identifiers or strings) and *L* is a list of sentences, the call

`(make-theory` *sup L*`)`

creates a structured theory with the elements of *L* as its *axioms* and the theories specified in *sup* as its direct *superiors*, namely, the theories of which the new theory is a direct refinement. We say that the sentences in *L* are the *direct* or *top* axioms of the new theory. The entire set of axioms for the new theory includes these top axioms as well as (recursively) all of the axioms of all the superiors. Likewise, we refer to the elements of *sup* as the direct or "top" superiors of the new theory. The new theory's entire set of superiors includes the theories in *sup* as well as (recursively) all of *their* superiors. That set is what the procedure `theory-superiors` returns (as a list of theories).[2] In the case of `Semigroup`, there is a single axiom, named `associative`, and there are no superiors. `Monoid` refines superiors `Semigroup` and `Identity`, without adding any new axioms of its own, and `Group` refines `Monoid` (while adding an axiom about right inverses and one that defines the minus operation).

Theories must be constructed inside modules, and the name that is (automatically) given to a theory produced by `make-theory` is simply the fully qualified name of the module in which `make-theory` was applied:

2 There is also a procedure `ST.top-theory-superiors` that only returns the direct superiors of a theory. Likewise, `ST.top-theory-axioms` and `ST.top-theory-theorems` return only the top axioms and top theorems, respectively, of a given theory.

```
> (println (theory-name Group.theory))

Group

Unit: ()
```

All procedures in module ST that expect theories as inputs can also accept theory names instead, and those names can be represented either as strings or as meta-identifiers. That is, if a procedure in ST expects as an argument a value of the structured-theory ADT, we can instead pass it the name of that theory. For instance, instead of writing

```
(theory-axioms Monoid.theory)
```

we can instead write, more succinctly, `(theory-axioms 'Monoid)`. The two are equivalent:

```
> (theory-axioms 'Monoid)

List: [
(forall ?x:'T45752
  (forall ?y:'T45752
    (forall ?z:'T45752
      (= (Semigroup.+ (Semigroup.+ ?x ?y)
                      ?z)
         (Semigroup.+ ?x
                      (Semigroup.+ ?y ?z))))))

(forall ?x:'T45763
  (= (Semigroup.+ Identity.<0> ?x)
     ?x))

(forall ?x:'T45769
  (= (Semigroup.+ ?x Identity.<0>)
     ?x))
]

> (theory-axioms Monoid.theory equals? theory-axioms "Monoid")

Term: true
```

Before continuing, we define Z-Additive-Group and ZPS-Additive-Group as before, but picking up the new, structured definition of the Group module:

```
define Z-Additive-Group :=
  (renaming |{Group.+ := Z.+, Group.<0> := Z.zero,
              Group.U- := Z.negate, Group.- := Z.-}|)

define ZPS-Additive-Group :=
  (renaming |{Group.+ := ZPS.+, Group.<0> := ZPS.zero,
              Group.U- := ZPS.Negate, Group.- := ZPS.-}|)
```

14.3 Writing proofs at the level of a theory

It may seem that there is a mistake in defining `Group`: We left out the `left-inverse` axiom. We did that intentionally, however, because we can *prove* the `left-inverse` property from the other group theory axioms, and extend the theory with the theorem and its proof. There are, in fact, several useful theorems we can prove from the group axioms:

```
extend-module Group {
  define left-inverse      := (forall x . (U- x) + x = <0>)
  define double-negation := (forall x . U- U- x = x)
  define unique-negation := (forall x y . x + y = <0> ==> U- x = y)
  define neg-plus          := (forall x y . U- (x + y) = (U- y) + (U- x))
```

Here is a method that proves the first of these theorems:

```
  define left-inverse-proof :=
    method (theorem adapt)
      let {[_ _ chain _ _] := (proof-tools adapt theory);
           [+ U- <0>] := (adapt [+ U- <0>])}
        conclude (adapt theorem)
          pick-any x
            (!chain
             [((U- x) + x)
          <-- (((U- x) + x) + <0>)                      [right-identity]
          --> ((U- x) + (x + <0>))                      [associative]
          <-- ((U- x) + (x + ((U- x) + U- U- x)))       [right-inverse]
          <-- ((U- x) + ((x + U- x) + U- U- x))         [associative]
          --> ((U- x) + (<0> + U- U- x))                [right-inverse]
          <-- (((U- x) + <0>) + U- U- x)                [associative]
          --> ((U- x) + U- U- x)                        [right-identity]
          --> <0>                                       [right-inverse]])
```

Before discussing the conventions used in defining this method, and the auxiliary procedure `proof-tools`, we note that we can now extend `Group.theory` by adding the `left-inverse` theorem and its proof method using `add-theorems`:

```
  (add-theorems 'Group |{left-inverse := left-inverse-proof}|)

} # close module Group
```

In general, when T is a structured theory (or the name thereof) and m is a map from sentences to methods,

(`add-theorems` T m)

extends T by associating every sentence p in the domain of m with the corresponding method `(m p)`. The precise form that such methods must take is described in Section 14.4. As we will see, a single proof method may be capable of deriving several different theorems, and for that reason, the map m may have an entire list of sentences $[p_1 \cdots p_n]$ as a single key, so we can write something like:

`(add-theorems T |{[p1 ··· pn] := M}|).`

Note that `add-theorems` is only used for its side effects; it returns the unit value `()` as its result.

We can at any time retrieve an appropriately renamed instance of a `Group` axiom or theorem with the `get-property` procedure. For example, with `Z-Additive-Group` as defined above,

```
> (get-property Group.left-identity Z-Additive-Group Group.theory)

Sentence: (forall ?x:Z
            (= (Z.+ Z.zero ?x)
               ?x))
```

In general,

`(get-property` *p adapter starting-theory*`)`

searches the theory refinement hierarchy for p, using *starting-theory* as the starting point for the search. (As usual, *starting-theory* can be specified either as a value returned by `make-theory` or by its name.) The top sentences (axioms and theorems) of *starting-theory* are searched, followed by a recursive search of each superior of *starting-theory*. If `get-property` finds a sentence that is alpha-equivalent to p, it applies the given *adapter* to it and returns the result. An error will occur if the search ends without finding p, or if applying *adapter* results in an ill-sorted sentence.

Note that we would get the same result that we got from the above `get-property` call if we simply wrote

`(Z-Additive-Group Group.left-identity).`

However, `get-property` can give a different result—an error—if the given property is *not* in the given theory, as would happen for example for the call:

`(get-property Group.left-identity Z-Additive-Group 'Semigroup).`

An error would be reported in that case because neither `Semigroup.theory` nor any of its superiors (there are none anyway) contains `Group.left-identity`.

While `get-property` is just a procedure, there is a corresponding *method*, `prove-property`, which takes the same arguments in the same order and conducts the same search and adaptation, except that `prove-property` tries to *prove* the resulting sentence using the methods that have been stored in the various theories via `add-theorems`. Let us assume that the proofs shown in Chapter 9 have been done in the current session, but not Exercise 9.9 or 9.10, which were to prove the left inverse property (with or without using the commutative property). Here we can prove it just by adapting the abstract-level proof given above:

```
> (!prove-property Group.left-inverse Z-Additive-Group 'Group)

Theorem: (forall ?x:Z
           (= (Z.+ (Z.negate ?x) ?x)
              Z.zero))
```

We can also apply `prove-property` to axioms:

```
> (!prove-property Group.left-identity Z-Additive-Group 'Group)

Theorem: (forall ?x:Z
           (= (Z.+ Z.zero ?x)
              ?x))
```

We can easily check whether an instance of a theory axiom is in the assumption base:

```
> (holds? (get-property Group.associative Z-Additive-Group 'Group))

Term: true
```

For complete checking of all of a theory's axioms, properly adapted, we could write

`(hold? (Z-Additive-Group (theory-axioms Group.theory)))`.

But another predefined procedure, `print-instance-check`, also prints the (adapted) axioms as they are checked:

```
> (print-instance-check Z-Additive-Group 'Group)

Checking
(forall ?x:Z
  (= (Z.+ ?x
          (Z.negate ?x))
     Z.zero))

Checking
(forall ?x:Z
  (forall ?y:Z
    (= (Z.- ?x ?y)
       (Z.+ ?x
            (Z.negate ?y)))))

Checking
```

```
(forall ?x:Z
  (forall ?y:Z
    (forall ?z:Z
      (= (Z.+ (Z.+ ?x ?y)
              ?z)
         (Z.+ ?x
              (Z.+ ?y ?z))))))

Checking
(forall ?x:Z
  (= (Z.+ Z.zero ?x)
     ?x))

Checking
(forall ?x:Z
  (= (Z.+ ?x Z.zero)
     ?x))

Unit: ()
```

Exercise 14.1: Show that lists as defined in Chapter 3 are a monoid under concatenation, by defining an appropriate theory instance and calling `print-instance-check` with it. □

14.4 Abstract proof method conventions

There are a few conventions that need to be followed when programming a method to prove theorems at an abstract level. We illustrate these with a proof method that includes the proofs of each of the three `Group.theory` theorems other than `left-inverse`:

```
1  extend-module Group {
2    define proofs :=
3      method (goal adapt)
4        let {[get prove chain chain-> chain<-] := (proof-tools adapt 'Group);
5             [+ U- <0>] := (adapt [+ U- <0>])}
6          match goal {
7            (val-of double-negation) =>
8              conclude (adapt goal)
9                pick-any x:(sort-of <0>)
10                 (!chain [(U- U- x)
11                        <-- (<0> + U- U- x)                [left-identity]
12                        <-- ((x + U- x) + U- U- x)         [right-inverse]
13                        --> (x + ((U- x) + U- U- x))       [associative]
14                        --> (x + <0>)                      [right-inverse]
15                        --> x                              [right-identity]])
16         | (val-of unique-negation) =>
17             conclude (adapt goal)
18               pick-any x:(sort-of <0>) y:(sort-of <0>)
```

```
19             let {LI := (!prove left-inverse)}
20               assume hyp := (x + y = <0>)
21                 (!chain [(U- x)
22                      <-- ((U- x) + <0>)                       [right-identity]
23                      <-- ((U- x) + (x + y))                   [hyp]
24                      <-- (((U- x) + x) + y)                   [associative]
25                      --> (<0> + y)                            [LI]
26                      --> y                                    [left-identity]])
27       | (val-of neg-plus) =>
28           conclude (adapt goal)
29             pick-any x y
30               let {UN := (!prove unique-negation)}
31                 (!chain-> [((x + y) + ((U- y) + (U- x)))
32                      <-- (x + ((y + (U- y)) + (U- x)))  [associative]
33                      --> (x + (<0> + U- x))                   [right-inverse]
34                      --> (x + U- x)                           [left-identity]
35                      --> <0>                                  [right-inverse]
36                      ==> (U- (x + y) = (U- y) + (U- x)) [UN]])
37       }
38
39   (add-theorems 'Group
40                 |{[double-negation unique-negation neg-plus] := proofs}|)
41 } # close module Group
```

First, the method should accept exactly two arguments, *goal* and *adapt*, where *goal* is the sentence to be proved and *adapt* is an adapter procedure (what we get when we apply `renaming` to a symbol map). This adapter should be applied to each of the symbols declared in the module containing the theory (including those imported from superior theories with **open**), as in line 5. The objective of the method is to prove (*adapt goal*). The method may contain multiple subproofs, in which case the desired *goal* is used as the discriminant of a **match** deduction to select the right subproof, as in lines 6–37 in the preceding method. But note that nothing prevents us from deriving these three results with three separate proof methods if we prefer to do so. For instance, instead of defining a single monolithic `proofs` method as we did above, we could have derived the results separately as follows:

```
extend-module Group {

  define (double-negation-proof goal adapt) :=
      let {[get prove chain chain-> chain<-] := (proof-tools adapt theory);
           [+ U- <0>] := (adapt [+ U- <0>])}
        conclude (adapt double-negation)
          pick-any x:(sort-of <0>)
            (!chain [(U- U- x)
                 <-- (<0> + U- U- x)                      [left-identity]
                 <-- ((x + U- x) + U- U- x)               [right-inverse]
                 --> (x + ((U- x) + U- U- x))             [associative]
                 --> (x + <0>)                            [right-inverse]
                 --> x                                    [right-identity]])
```

```
  define (unique-negation-proof goal adapt) :=
    ...

  define (neg-plus-proof goal adapt) :=
    ...

  (add-theorems 'Group |{double-negation := double-negation-proof,
                          unique-negation := unique-negation-proof,
                          neg-plus        := neg-plus-proof}|)
}
```

Which alternative we prefer is largely a matter of style. The advantage of packing the proofs of several theorems into one method is that we only have to call `proof-tools` and adapt the theory's symbols once. The main drawback is that we end up with larger methods.

In line 4 of the earlier code listing, with the aid of the predefined procedure `proof-tools`, we defined a procedure, `get`, and four methods, `prove`, `chain`, `chain->`, and `chain<-`. The `proof-tools` procedure is defined as follows:

```
define (proof-tools adapt theory) :=
  let {get := lambda (p) (get-property p adapt theory);
       prove := method (p) (!prove-property p adapt theory);
       chain := method (L) (!chain-help get L 'none);
       chain-> := method (L) (!chain-help get L 'last);
       chain<- := method (L) (!chain-help get L 'first)}
    [get prove chain chain-> chain<-]
```

Thus, line 4 of the `proofs` method provides local, specialized versions of `get-property`, `prove-property`, and the equality and implication chaining methods. Usually not all of these are needed; in `left-inverse-proof` we only needed `chain`, and in `proofs` we only need `prove` and two of the chaining methods. But since we can define them all in one line using `proof-tools`, we will do that by convention from now on. If we want to emphasize that some of these five values are not needed for the method at hand, we will use wildcards in the corresponding places of the defining list pattern, as when we wrote

```
let {[_ _ chain _ _] := (proof-tools adapt theory)
```

in the earlier definition of `left-inverse-proof`, which only needed `chain` out of the five values returned by `proof-tools`.

The `get` procedure and the four methods are handy tools for simplifying the way that external properties can be named in chain steps that need them. The `prove` method simply calls the `prove-property` method (see page 636), passing it `p` (`prove`'s argument), `adapt`, and the theory to be used as the starting point in the theory structure for searches for `p` (which in this case is `Group.theory`). Thus, it finds the first occurrence of `p` and uses the proof method accompanying it to derive `(adapt p)`.

The `get` procedure (line 4) is similar but only retrieves `p` using `get-property` (page 636); it is used in an abstract-level proof when `p` is known either to be an axiom of the theory or to have already been proved using `prove` (or `prove-property` directly). Thus, in either case, `p` is already in the assumption base.

Usually one does not invoke `get` directly, but with the locally-specialized version of `chain`, `chain->`, or `chain<-`. As shown in the definition of `proof-tools`, these methods are defined in terms of the predefined method `chain-help`, which uses `get` to convert every occurrence it encounters of a sentence *p* to `(get` *p*`)`. This is what allows one to write chain-step justifications like `[right-identity]` instead of the more verbose `[(get right-identity)]` or

```
[(get-property right-identity adapt theory)].
```

The last argument to `chain-help` is `'none`, `'first`, or `'last`, corresponding to `chain`, `chain<-`, or `chain->`, respectively.

Of course, `proof-tools` and the other mechanisms and conventions described above are applicable only when we plan to express the desired method using chaining. Otherwise no infrastructure is needed (or available), and the only requirement is that the proof method must take *goal* and *adapt* as its two arguments and must derive (*adapt goal*) as its conclusion.

14.5 Dynamic evolution of theories

Structured theories are *dynamic*: One can *evolve* a theory at any time by adding new axioms with `add-axioms` or new theorems and their proofs with `add-theorems`. An important aspect of the design of Athena's structured theories is that *theory evolution respects theory refinement*. That is, if theory *B* is a refinement of theory *A* (directly or indirectly), and we evolve *A* with new theorems and their proofs, then these theorems and proofs are available to all instances of *B* as well as those of *A*.

As a simple example, consider the following property:

```
(forall w x y z . ((w + x) + y) + z = w + (x + (y + z))).
```

This property depends only on the associativity of +, and thus in the theory hierarchy it can be introduced and its proof can be given at the level of `Semigroup`:

```
extend-module Semigroup {
  define swing-right :=
    (forall w x y z . ((w + x) + y) + z = w + (x + (y + z)))

  define swing-right-proof :=
    method (_ adapt)
      let {[_ _ chain _ _] := (proof-tools adapt 'Semigroup);
```

```
          + := (adapt +)}
        conclude (adapt swing-right)
          pick-any w x y z
            (!chain [(((w + x) + y) + z)
                 --> ((w + (x + y)) + z)    [associative]
                 --> (w + ((x + y) + z))    [associative]
                 --> (w + (x + (y + z)))    [associative]])

  (add-theorems 'Semigroup |{swing-right := swing-right-proof}|)
} # close module Semigroup
```

We can now benefit from this theorem and proof when working with any refinement of `Semigroup`; for example:

```
> (!prove-property Semigroup.swing-right Z-Additive-Group 'Group)

Theorem: (forall ?w:Z
           (forall ?x:Z
             (forall ?y:Z
               (forall ?z:Z
                 (= (Z.+ (Z.+ (Z.+ ?w ?x)
                              ?y)
                         ?z)
                    (Z.+ ?w
                         (Z.+ ?x
                              (Z.+ ?y ?z))))))))
```

Here, the search for the property `swing-right` has gone up the hierarchy from its starting point in `Group.theory` to `Monoid.theory`, and finally to `Semigroup.theory`. The proof found there is then adapted using the `Z-Additive-Group` adapter, which includes the renaming of `Group.+` to `Z.+`. While evaluating the proof, `associative` is looked up, adapted in terms of `Z.+`, and determined to be in the assumption base (having been proved in the development of the `Z` data type). This allows all applications of associativity in the `chain` call to succeed, thereby proving the desired goal.

In short, we have the ability to *incrementally* develop a theory library without having to edit the files containing the original definitions, and we can do so with great economy by associating theorems and their proofs with the most abstract structure—the one closest to the base of the hierarchy—that provides the axioms and theorems used in its proof.

14.6 Testing abstract proofs

When we use `add-theorems` to extend a structured theory T with a new proof method M, we often want to go ahead and test M immediately, without having to wait until we first obtain a concrete instance of T. We can do that with the binary procedure `test-proofs`. Its first argument is a list of sentences $p_1, \ldots, p_n$ whose proofs we want to check; and the second argument is T, the theory to which $p_1, \ldots, p_n$ were added (via `add-theorems`). The implementation of `test-proofs` will then find the proof method M_i associated with each p_i in T; and it will apply M_i to p_i and the identity adapter `no-renaming`. If the proof goes through, it will report success; otherwise it will print a relevant error message. But before it starts evaluating the stored proofs, `test-proofs` will make sure to add all axioms of T in the assumption base, since these will usually be needed by the stored proofs. More precisely, for each p_i, `test-proofs` will evaluate the deduction

assume `(and* (theory-axioms` T`))` `(!`M_i p_i `no-renaming)`.

Earlier, for example, immediately after we issued the command

`(add-theorems 'Group |left-inverse := left-inverse-proof|)`,

we could have tested the added proof as follows:

```
> (test-proofs [Group.left-inverse] 'Group)

Testing proof of:

(forall ?x:'T45800
  (= (Semigroup.+ (Group.U- ?x)
                  ?x)
     Identity.<0>))
 ...

Proof worked.

Unit: ()
```

Likewise, we could test the proofs of the other three theorems we added to group theory as follows:

```
> (test-proofs [Group.double-negation
                Group.unique-negation Group.neg-plus] 'Group)

Testing proof of:

(forall ?x:'T45806
  (= (Group.U- (Group.U- ?x))
     ?x))
 ...

Proof worked.

Testing proof of:
```

```
(forall ?x:'T45814
  (forall ?y:'T45814
    (if (= (Semigroup.+ ?x ?y)
           Identity.<0>)
        (= (Group.U- ?x)
           ?y))))
  ...

Proof worked.

Testing proof of:

(forall ?x:'T45832
  (forall ?y:'T45832
    (= (Group.U- (Semigroup.+ ?x ?y))
       (Semigroup.+ (Group.U- ?y)
                    (Group.U- ?x)))))
  ...

Proof worked.

Unit: ()
```

If we want to test the proofs of *all* the theorems of a given theory, we can use the unary procedure `test-all-proofs`, which takes a theory T as input and runs `test-proofs` on a list comprising all of T's theorems.

14.7 Group theory refinements

Mathematicians have devised many useful refinements of group theory. Let us take a few of these as examples of extending the small refinement hierarchy we have so far.

14.7.1 Abelian group theory

First, consider commutativity:

```
(forall x y . x + y = y + x).
```

We should first ask whether the group axioms imply this property, in which case we should simply evolve `Group.theory` to include it as a theorem along with its proof. While the integer operator `Z.+` and the power-series operator `ZPS.+` both have this property (and we proved it in Chapter 9), there are structures that have all the group properties but for which commutativity fails. (One example is invertible square matrices under multiplication.) This means that commutativity is *independent* of the group axioms, so it makes sense to define a refinement of group theory that includes this axiom. And in fact, the *Abelian monoid* and *Abelian group* theories[3] add only this property and nothing more:

```
module Abelian-Monoid {
  open Monoid
  define commutative := (forall x y . x + y = y + x)
  define theory := (make-theory ['Monoid] [commutative])
}

module Abelian-Group {
  open Group
  define commutative := (forall x y . x + y = y + x)
  define theory := (make-theory ['Group] [commutative])
}
```

Commutativity allows for a shorter proof of `left-inverse` and a more natural statement of `neg-plus`:

```
1  extend-module Abelian-Group {
2    define left-inverse-proof :=
3      method (_ adapt)
4        let {[_ _ chain _ _] := (proof-tools adapt theory);
5             [+ U- <0>] := (adapt [+ U- <0>])}
6          conclude (adapt left-inverse)
7            pick-any x
8              (!chain [((U- x) + x)
9                       --> (x + (U- x))          [commutative]
10                      --> <0>                   [right-inverse]])
11
12   (add-theorems theory |{left-inverse := left-inverse-proof}|)
13
14   define neg-plus := (forall x y . U- (x + y) = (U- x) + (U- y))
15
16   define neg-plus-proof :=
17     method (goal adapt)
18       let {[_ _ chain _ _] := (proof-tools adapt theory);
19            [+ U- <0>] := (adapt [+ U- <0>])}
20         conclude (adapt neg-plus)
21           pick-any x y
22             let {group-version := (!prove-property Group.neg-plus
23                                                    adapt
24                                                    Group.theory)}
25               (!chain [(U- (x + y))
26                        --> ((U- y) + (U- x))  [group-version]
27                        --> ((U- x) + (U- y))  [commutative]])
28
29   (add-theorems theory |{neg-plus := neg-plus-proof}|)
30 } # close module Abelian-Group
```

3 Named for Norwegian mathematician Niels Henrik Abel (1802–1829).

Note that the `add-theorems` call on line 12 places a new occurrence of `left-inverse` in the refinement hierarchy, and it is this new occurrence that will be found when starting from `Abelian-Group.theory`. Thus, its associated proof, captured by this new `left-inverse-proof` method, will be evaluated. When starting at the `Group.theory` level, on the other hand, the old occurrence of `left-inverse` and its associated proof method will be found, and that proof does not depend on commutativity.

In the case of `neg-plus`, in lines 22–24 we still find and evaluate the `Group` level proof, then use commutativity to finish the proof of the restated property.

14.7.2 Multiplicative theories

The next refinement step we take is from `Group` to `Ring`, a theory that captures many of the properties of numeric and algebraic structures like `Z`, `ZPS`, polynomials in one or more variables, etc., and not only properties of their addition operator, but also of their multiplication operator. Specifically, a ring is a structure S involving an addition operator `+` and a multiplication operator `*`, as well as an additive "zero" element `<0>` and a multiplicative identity element `<1>`, such that S is an Abelian group under `+` and `<0>` and a semigroup under `*`.[4] Thus, in some sense, a structured theory of rings must include two different versions of `Semigroup` as its superiors, one pertaining to addition and the other to multiplication. For this and other similar purposes, we often need a version of a structured theory in which the original function symbols are replaced with new ones. Of course, we could always define such a version directly:

```
module Multiplicative-Semigroup {
  declare *: (S) [S S] ->  S [300]

  define associative := (forall x y z . (x * y) * z = x * (y * z))

  define theory := (make-theory [] [associative])
}
```

Thus, `Multiplicative-Semigroup.associative` expresses associativity of `*`, whereas `Semigroup.associative` expressed it for `+`. We could then go ahead and replicate the existing hierarchy with multiplicative versions of each theory; for instance, we could introduce `Multiplicative-Identity`, `Multiplicative-Monoid` (which would refine `Multiplicative-Semigroup` and `Multiplicative-Identity`), `Multiplicative-Group`, and so on. If we took that approach, however, we would need to repeat not only the axioms of each theory but also all theorems and their proofs, restating all of them in terms of `*` rather than `+`. Worse yet, if we go on to prove new theorems in one of the additive theories, we would need to replicate the results at the corresponding multiplicative theory, and vice versa.

4 And further, multiplication distributes over addition.

Instead, we can avoid such duplication of effort by allowing superiors of structured theories to be *shared* or *adapted* theories. You can think of a shared theory T as a pair whose first element is a pointer to another (already existing) structured theory T', called the *base* theory of T, and whose second element is an adapter m. The fact that we are using a "pointer" to the base theory T' means that we are free to extend T' as we see fit, and all extensions will automatically be available to the shared theory T. For instance, we can express a multiplicative semigroup as a shared theory consisting of the regular (additive) semigroup theory as its base, along with a renaming that "translates" + to *.

More specifically, we obtain a shared theory by applying the binary procedure `adapt-theory` to an existing theory and an adapter. For example, we can get a multiplicative version of `Semigroup.theory` as follows:

```
module MSG {
  declare *: (S) [S S] ->  S [300]

  define theory := (adapt-theory 'Semigroup |{Semigroup.+ := *}|)
}
```

Thus, `MSG.theory` is a copy of `Semigroup.theory`, adapted through `|Semigroup.+ := *|`.

If T is a shared theory (one produced by `adapt-theory`), we can obtain its symbol map by using the procedure `get-symbol-map`:

```
> (get-symbol-map 'MSG)

Map: |{Semigroup.+ := MSG.*}|
```

If one tries to retrieve a property from a shared theory T that is expressed in terms of T's base theory, T's adapter is automatically applied to the result, even if no adapter is explicitly specified in the `get-property` call. For example:

```
> (get-property Semigroup.associative no-renaming 'MSG)

Sentence: (forall ?x:'T67119
            (forall ?y:'T67119
              (forall ?z:'T67119
                (= (MSG.* (MSG.* ?x ?y)
                          ?z)
                   (MSG.* ?x
                          (MSG.* ?y ?z))))))
```

Likewise when the theory argument to `prove-property` is an adapted theory. In those cases the renaming that is actually used in the proof is always the composition of the renaming specified in the `prove-property` call with the adapter of the shared theory. Thus, for instance:

```
> (!prove-property Semigroup.swing-right (renaming |{MSG.* := N.*}|) 'MSG)

Theorem: (forall ?w:N
           (forall ?x:N
             (forall ?y:N
               (forall ?z:N
                 (= (N.* (N.* (N.* ?w ?x)
                              ?y)
                         ?z)
                    (N.* ?w
                         (N.* ?x
                              (N.* ?y ?z))))))))
```

Here is another example—a multiplicative monoid:

```
module MM {
  declare <1>: (S) [] -> S

  define theory :=
    (adapt-theory 'Monoid |{Semigroup.+ := MSG.*, Monoid.<0> := <1>}|)
}

> (print-theory 'MM)

MM.theory:

Axioms:

Sentence: (forall ?x:'T61241
            (forall ?y:'T61241
              (forall ?z:'T61241
                (= (MSG.* (MSG.* ?x ?y)
                          ?z)
                   (MSG.* ?x
                          (MSG.* ?y ?z))))))

Sentence: (forall ?x:'T61250
            (= (MSG.* MM.<1> ?x)
               ?x))

Sentence: (forall ?x:'T61259
            (= (MSG.* ?x MM.<1>)
               ?x))

Theorems:

Sentence: (forall ?w:'T61340
            (forall ?x:'T61340
              (forall ?y:'T61340
                (forall ?z:'T61340
                  (= (MSG.* (MSG.* (MSG.* ?w ?x)
```

```
                                    ?y)
                              ?z)
                        (MSG.* ?w
                              (MSG.* ?x
                                    (MSG.* ?y ?z)))))))
```

Although we will not need it for ring theory, we could also write:

```
module MG {
  declare Inv: (S) [S] -> S
  declare /: (S) [S S] -> S

  define theory :=
    (adapt-theory 'Group |{Semigroup.+ := MSG.*, Monoid.<0> := MM.<1>,
                           Group.U- := Inv, Group.- := /}|)
}
```

14.7.3 Ring theory

We can now formulate ring theory as follows:

```
module Ring {
  define [+ *] := [Semigroup.+ MSG.*]

  define right-distributive :=
    (forall x y z . (x + y) * z = x * z + y * z)

  define left-distributive :=
    (forall x y z . z * (x + y) = z * x + z * y)

  define theory :=
    (make-theory ['Abelian-Group 'MSG]
                 [right-distributive left-distributive])
}
```

In this case there are two superiors in the refinement hierarchy: `Abelian-Group.theory`, which brings in the Abelian group properties of the additive operators +, `<0>`, `U-`, and `-`; and `MSG.theory`, which brings in the associative property of *. The two distributive properties are added at the top level.

There are several other important refinements of `Ring`, given here with the names commonly used in the mathematical literature.

```
module Commutative-Ring {
  define * := MSG.*

  define *commutative := (forall x y . x * y = y * x)
```

```
  define theory := (make-theory ['Ring] [*commutative])
}

module Ring-With-Identity {
  define theory := (make-theory ['MM 'Ring] [])
}

module Commutative-Ring-With-Identity {
  define theory :=
    (make-theory ['Ring-With-Identity 'Commutative-Ring] [])
}
```

Thus, in the name "commutative ring" the adjective refers to commutativity of the ring's multiplication operator. Similarly, in the name "ring with identity" the identity is that of multiplication.

An example of a ring with identity that is not commutative is provided by square matrices with real coefficients under matrix addition and multiplication. (Earlier we mentioned invertible square matrices as an example of a nonabelian group, but here we are only requiring monoid properties of the multiplication operator, so it is unnecessary to restrict the domain to invertible matrices.)

The following exercises extend the development of the integer datatype begun in Chapter 9 to prove that it is a commutative ring with identity.

Exercise 14.2: Extend module `Z` with a declaration of `*` as a binary operator on `Z` and define a module `Times` holding its properties, initialized with axioms that determine its meaning as integer multiplication. □

We begin by showing that `Z.*` is commutative. Note that we could wait to prove this until after we have proved that the integer datatype is a ring, leaving it as the final step of showing it is a commutative ring, but it is useful to do it first as a "warmup" exercise before tackling associativity, which has the extra complexity of a third variable.

Exercise 14.3: Extend `Z.*` with the commutative property and prove it. *Hint:* Unlike the case with addition, the sign of the product of two signed numbers is determined independently of the sizes of the numbers, so the proof does not require considering so many different cases. Thus, it is not necessary to map to and from the pair representation, `NN`, of Section 9.4. (But we will come to another situation where the mapping approach is best, in Exercises 14.5–14.7.) □

* ***Exercise 14.4:*** Extend `Z.*` with the associative property and prove it. *Hint:* As in the previous exercise, this associative property can best be proved in the signed integer representation, `Z`. (Although there are eight sign-combination cases to consider, each is straightforward.) □

The last two properties needed in order to have a ring are the two distributive properties. Due to the presence of addition in these properties, a proof directly in terms of Z would be complicated by the number of sign combinations and natural-number inequality cases to consider. Instead, the following exercises develop a proof based on mapping each distributive property to the pair representation, proving it there, and mapping back. We can use the mappings already defined, and their inverse relation, from Section 9.4.

⋆ ***Exercise 14.5:*** Define the multiplication operator `Z.NN.*'` for integers in the pair representation `Z.NN`. As a definition, you can assert it into the assumption base, but you must then show that it is consistent with the definition of `Z.*`: Define a *multiplicative homomorphism* analogous to the additive homomorphism introduced in Section 9.5, and show that `Z.*` and your `Z.NN.*'` jointly satisfy it. □

⋆ ***Exercise 14.6:*** Extend `Z.NN.Times` with the right distributive property and prove it. □

⋆ ***Exercise 14.7:*** Extend `Z.Times` with the distributive properties and prove them. (*Hints:* Prove the right distributive property by mapping to `Z.NN` and back, as was done with the commutative and associative properties of `Z.+` in Section 9.6. The left distributive property then follows easily by commutativity.) □

Upon completion of these exercises, we are able to check that Z is a commutative ring:

```
> define Integer-Ring :=
    (renaming |{Semigroup.+ := Z.+, MSG.* := Z.*, Monoid.<0> := Z.zero,
                Group.U- := Z.negate, Group.- := Z.-}|)

Procedure Integer-Ring defined.

> (print-instance-check Integer-Ring Commutative-Ring.theory)

Checking
 (forall ?x:Z
  (forall ?y:Z
    (= (Z.* ?x ?y)
       (Z.* ?y ?x))))

Checking
 (forall ?x:Z
  (forall ?y:Z
    (forall ?z:Z
      (= (Z.* ?z
              (Z.+ ?x ?y))
         (Z.+ (Z.* ?z ?x)
              (Z.* ?z ?y))))))
...
Checking
 (forall ?x:Z
  (forall ?y:Z
```

```
    (forall ?z:Z
      (= (Z.* (Z.* ?x ?y)
              ?z)
         (Z.* ?x
              (Z.* ?y ?z))))))
```

Exercise 14.8: Extend `Z.*` with right and left identity properties and prove them, then use `print-instance-check` to check that `Z` is a commutative ring with identity. □

14.7.4 Integral domain

A set S with additive identity 0 and multiplication operator $\times$ is said to have *zero divisors* iff there are nonzero $x, y \in S$ such that $x \times y = 0$. In `Z` there are no zero divisors, and this property of the integers is the inspiration for further refinements of ring theory (one of which, `Integral-Domain`, expresses the connection in its name).

```
module No-Zero-Divisors {
  define [* <0>] := [MSG.* Monoid.<0>]

  define no-zero-divisors :=
    (forall x y . x * y = <0> ==> x = <0> | y = <0>)

  define theory := (make-theory [] [no-zero-divisors])
}

module Ring-With-No-Zero-Divisors {
  define theory :=
    (make-theory ['Ring 'No-Zero-Divisors.theory] [])
}

module Integral-Domain {
  define theory :=
    (make-theory ['Commutative-Ring-With-Identity 'No-Zero-Divisors] [])
}
```

Exercise 14.9: Extend `Z.*` with a property expressing that `Z` has no zero divisors, and prove it. Then use `print-instance-check` to check that `Z` is an integral domain. □

14.7.5 Algebraic theory diagram

Figure 14.1 depicts the theory hierarchy developed in the preceding sections. A directed arrow means the target node is a refinement of the source node. Note that these diagrams can be produced mechanically: (`ST.draw-theory` *T*) will draw the refinement hierarchy for theory *T*, while (`ST.draw-all-theories`) will draw all existing theories.[5]

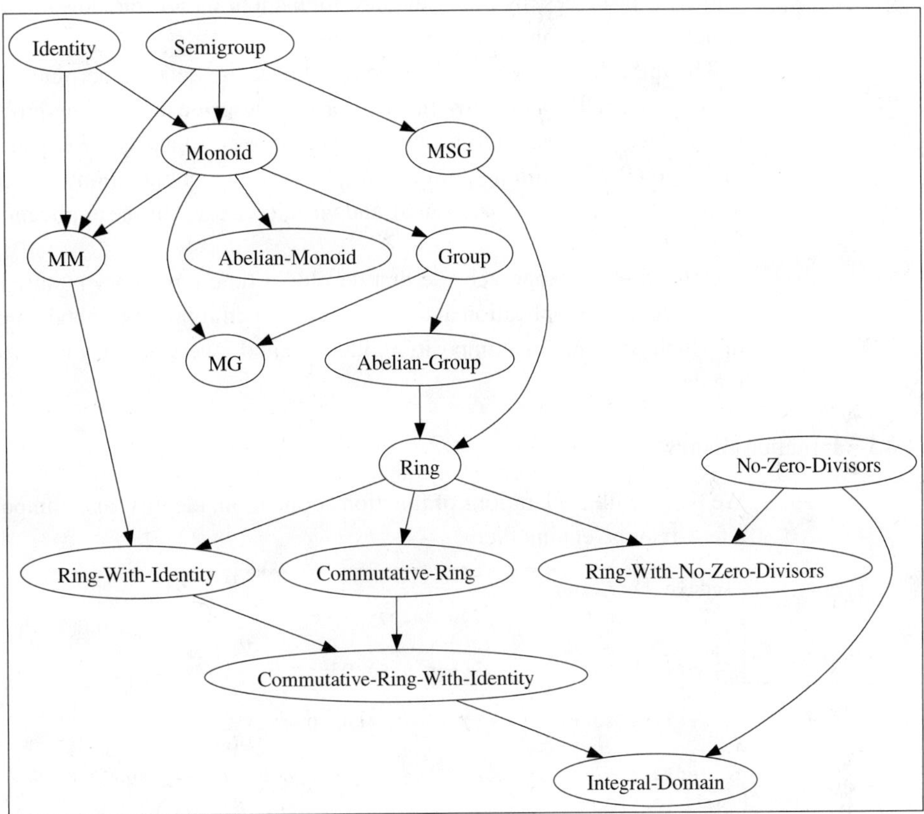

Figure 14.1
Refinement hierarchy of algebraic theories

14.8 ⋆ Permutations as a group

In this section we develop yet another group example, *permutations*. We will see three important differences from the additive integer group of the previous section. First, it is a

5 The implementation uses the Graphviz `dot` program to produce the diagrams. You must first call (`Graph-Draw.set-dot-executable` *path*), where *path* is a string representing the location where the `dot` binary resides on your machine. To display the diagrams, you must also specify a graph viewer by calling `Graph-Draw.set-viewer` with a string argument representing the path to a program capable of displaying `.gif` files. Any major browser will do, so you can write, for instance, `(Graph-Draw.set-viewer "/home/.../Mozilla Firefox/firefox")`.

nonnumeric example, further illustrating the generality of group theory. Second, we can develop permutations not just as a single new instance of group theory, but as a whole family of instances, by doing the development at an abstract level: a *permutation theory*, which itself has many nonisomorphic instances.

The third difference is a more technical one: We define a permutation as a certain kind of function, called a *bijective* function, and to describe the relevant properties in a straightforward way it is useful to be able to quantify over functions. We deal with this as we did in Section 9.7 in defining power series, by defining an operation for applying a function to a value. We name the operation `at` and write (f `at` x) for the application of a function f to x.

Following this approach, we first develop a function theory mainly in terms of properties of function application and composition, then refine it to obtain a permutation theory in which permutation composition, identity, and inversion obey the group theory requirements.

14.8.1 Function theory

We begin with declarations of function application, identity, and composition, followed by a few axioms relating them.

```
module Function {
  domain (Fun Domain Codomain)

  declare at: (C, D) [(Fun D C) D] -> C

  declare identity: (D) [] -> (Fun D D)

  declare o: (D, C, B) [(Fun C B) (Fun D C)] -> (Fun D B)

  set-precedence o (plus 10 (get-precedence at))

  define [f g h x x' y] := [?f ?g ?h ?x ?x' ?y]

  define identity-definition := (forall x . identity at x = x)

  define compose-definition := (forall f g x . (f o g) at x = f at (g at x))

  define function-equality :=
    (forall f g . f = g <==> forall x . f at x = g at x)

  define theory :=
    (make-theory []
        [identity-definition compose-definition function-equality])
```

Our eventual goal is showing that the subclass of functions called permutations form a group under composition, so we look first at associativity and identity properties.

```
define associative := (forall f g h . (f o g) o h = f o (g o h))

define right-identity :=  (forall f . f o identity = f)

define left-identity :=   (forall f . identity o f = f)

define Monoid-theorems := [associative right-identity left-identity]
```

The proof of `associative` just requires repeated use of the definition of `o` to prove the associativity property when the compositions are applied to a value `x`, then extracting the desired equality using the `function-equality` axiom.

```
define proofs :=
  method (theorem adapt)
    let {[_ _ chain chain-> _] := (proof-tools adapt theory);
         [at identity o] := (adapt [at identity o]);
         [cd id] := [compose-definition identity-definition]}
      match theorem {
        (val-of associative) =>
          pick-any f g h
            let {all-x := pick-any x
                            (!chain
                              [((f o g) o h at x)
                           --> ((f o g) at h at x)    [cd]
                           --> (f at g at h at x)     [cd]
                           <-- (f at (g o h) at x)    [cd]
                           <-- ((f o (g o h)) at x)   [cd]])}
              (!chain-> [all-x
                     ==> ((f o g) o h = f o (g o h)) [function-equality]])
      | (val-of right-identity) => (!force (adapt theorem))
      | (val-of left-identity)  => (!force (adapt theorem))}

  (add-theorems theory |{Monoid-theorems := proofs}|)
} # close module Function
```

Exercise 14.10: Fill in the proofs of `right-identity` and `left-identity` (i.e., replace the uses of `force` with actual proofs.) □

Exercise 14.11: Show that functions as defined in `Function` are a monoid under composition, by defining an appropriate theory instance and calling `print-instance-check`. □

The next step toward showing we have a group is defining inverses, but not all functions have inverses. To proceed, we must somehow restrict ourselves to the subclass of functions that do have inverses. The key conditions for defining this class are captured in the following predicates (under the same names as are commonly used in mathematics texts):

```
extend-module Function {
  declare surjective, injective, bijective: (D, C) [(Fun D C)] -> Boolean

  define surjective-definition :=
    (forall f . surjective f <==> forall y . exists x . f at x = y)

  define injective-definition :=
    (forall f . injective f <==> forall x y . f at x = f at y ==> x = y)

  define bijective-definition :=
    (forall f . bijective f <==> surjective f & injective f)

  (add-axioms theory [surjective-definition injective-definition
                      bijective-definition])
```

We will prove in the next subsection that in order for a function to have an inverse it is sufficient for it to be bijective. In preparation, we have the following theorems:

```
define identity-surjective := (surjective identity)

define identity-injective := (injective identity)

define identity-bijective := (bijective identity)

define compose-surjective-preserving :=
  (forall f g . surjective f & surjective g ==> surjective f o g)

define compose-injective-preserving :=
  (forall f g . injective f & injective g ==> injective f o g)

define compose-bijective-preserving :=
  (forall f g . bijective f & bijective g ==> bijective f o g)

define Inverse-theorems :=
  [identity-surjective identity-injective identity-bijective
   compose-surjective-preserving compose-injective-preserving
   compose-bijective-preserving]
```

The proofs of the first and fourth theorems are included in the following method:

```
define proofs-1 :=
  method (theorem adapt)
    let {[_ _ _ chain-> _] := (proof-tools adapt theory);
         [at identity o]  := (adapt [at identity o]);
         [cd id] := [compose-definition identity-definition]}
    match theorem {
      (val-of identity-surjective) =>
        let {SDI := (!instance surjective-definition [identity]);
             all-y :=
              pick-any y
                (!chain->
```

```
            [(identity at y) --> y          [id]
         ==> (exists x . identity at x = y) [existence]])}
    (!chain-> [all-y ==>
            (surjective identity)          [SDI]])
| (val-of compose-surjective-preserving) =>
  pick-any f g
    assume (surjective f & surjective g)
      let {f-case :=
            (!chain->
             [(surjective f)
          ==> (forall y .
                exists x . f at x = y)     [surjective-definition]]);
           g-case :=
            (!chain->
             [(surjective g)
            ==> (forall y .
                exists x . g at x = y)     [surjective-definition]]);
           all-y :=
             pick-any y
               let {f-case-y :=
                      (!chain->
                       [true
                    ==> (exists y' .
                           f at y' = y)     [f-case]])}
               pick-witness y' for f-case-y
                  let {g-case-y' :=
                         (!chain->
                          [true
                       ==> (exists x .
                              g at x = y')   [g-case]])}
                     pick-witness x for g-case-y'
                     (!chain->
                      [(f o g at x)
                   --> (f at g at x)        [cd]
                   --> (f at y')            [(g at x = y')]
                   --> y                    [(f at y' = y)]
                   ==> (exists x .
                          f o g at x = y)   [existence]])}

         (!chain-> [all-y
               ==> (surjective f o g)      [surjective-definition]])

   }

  (add-theorems theory |{[identity-surjective
                          compose-surjective-preserving] := proofs-1}|)
} # close module Function
```

Exercise 14.12: Prove `identity-injective` and `identity-bijective`. □

Exercise 14.13: Prove `compose-injective-preserving`. □

Exercise 14.14: Prove `compose-bijective-preserving`. □

14.8.2 Permutation theory

To continue toward our goal of showing that permutations form a group under composition, we introduce a new parameterized domain, `(Perm D)`, as the basis of a `Permutation` refinement of `Function` theory. A key axiom of the new theory says that a permutation—a `(Perm D)` value—when considered as a function—a `(Fun D D)` value—is bijective.

```
module Permutation {
  open Function
  domain (Perm D)

  declare perm->fun: (D) [(Perm D)] -> (Fun D D)
  declare fun->perm: (D) [(Fun D D)] -> (Perm D)

  set-precedence (perm->fun fun->perm) 350

  define [p q r f x y] := [?p:(Perm 'D1) ?q:(Perm 'D2) ?r:(Perm 'D3)
                            ?f:(Fun 'D4 'D5) ?x ?y]

  define is-bijective := (forall p . bijective perm->fun p)

  define fun->fun := (forall p . fun->perm perm->fun p = p)

  define perm->perm :=
    (forall f . bijective f ==> perm->fun fun->perm f = f)

  declare o: (D) [(Perm D) (Perm D)] -> (Perm D)
  declare identity: (D) [] -> (Perm D)

  define o' := Function.o
  set-precedence o' (plus 10 (get-precedence perm->fun))

  define identity' := Function.identity

  define compose-definition :=
    (forall p q . p o q = fun->perm (perm->fun p o' perm->fun q))

  define identity-definition := (identity = fun->perm identity')

  define theory :=
    (make-theory ['Function]
                 [is-bijective fun->fun perm->perm
                  compose-definition identity-definition])
```

The functions perm->fun and fun->perm can be viewed as sort conversions mapping (Perm D) values to (Fun D D) values and vice versa. The use of such conversions is a technical device for treating (Perm D) as a subsort of (Fun D D). Note that the second axiom says that we can convert any permutation to a function and back to a permutation, obtaining the same permutation. The other way, however, as stated in the third axiom, requires starting with a bijective function. This is a condition that must be discharged at any point in a proof where the term (perm->fun (fun->perm f)) appears, as we will see below.

We have Permutation versions of Function.o and Function.identity. Each of these new functions is axiomatized in terms of its Function counterpart. The first set of theorems we state and prove are the monoid properties.

```
define associative    := (forall p q r . (p o q) o r = p o (q o r))
define right-identity := (forall p . p o identity = p)
define left-identity  := (forall p . identity o p = p)

define Monoid-theorems := [associative right-identity left-identity]
```

The proof of the associative property uses an intricate combination of the function definitions from the Permutation level; the is-bijective constraint; and the compose-associative and compose-bijective- preserving properties from the Function level. These properties are woven together with perm->fun and fun->perm.

```
define proofs :=
  method (theorem adapt)
    let {[get prove chain chain-> chain<-] := (proof-tools adapt theory);
         [identity' o' at identity o fun->perm perm->fun] :=
           (adapt [identity' o' at identity o fun->perm perm->fun]);
         [id cd] := [identity-definition compose-definition]}
    match theorem {
      (val-of associative) =>
         let {CA := (!prove Function.associative);
              CBP := (!prove compose-bijective-preserving)}
         pick-any p:(Perm 'S) q:(Perm 'S) r:(Perm 'S)
           let {_ := (!chain-> [true
                          ==> (bijective perm->fun q) [is-bijective]]);
                _ := (!chain->
                       [true
                    ==> (bijective perm->fun p)          [is-bijective]
                    ==> (bijective perm->fun p &
                          bijective perm->fun q)         [augment]
                    ==> (bijective
                            (perm->fun p o' perm->fun q))  [CBP]]);
                _ := (!chain->
                       [true
                    ==> (bijective perm->fun r)          [is-bijective]
                    ==> (bijective perm->fun q &
                          bijective perm->fun r)         [augment]
                    ==> (bijective
                            (perm->fun q o' perm->fun r))  [CBP]])}
```

```
        (!combine-equations
         (!chain
          [((p o q) o r)
        --> ((fun->perm (perm->fun p o' perm->fun q)) o r)    [cd]
        --> (fun->perm
              ((perm->fun fun->perm
                (perm->fun p o' perm->fun q)) o' perm->fun r)) [cd]
        --> (fun->perm
              ((perm->fun p o' perm->fun q) o' perm->fun r))  [perm->perm]
        --> (fun->perm
              (perm->fun p o' (perm->fun q o' perm->fun r)))   [CA]])
         (!chain
          [(p o (q o r))
        --> (p o (fun->perm (perm->fun q o' perm->fun r)))   [cd]
        --> (fun->perm
              (perm->fun p o'
               (perm->fun
                fun->perm
                 (perm->fun q o' perm->fun r))))              [cd]
        --> (fun->perm
              (perm->fun p o' (perm->fun q o' perm->fun r)))  [perm->perm]
              ]))
    | (val-of right-identity) => (!force (adapt theorem))
    | (val-of left-identity) =>  (!force (adapt theorem))
    }

    (add-theorems theory |{Monoid-theorems := proofs}|)

} # close module Permutation
```

Exercise 14.15: Fill in the proofs of `right-identity` and `left-identity`. □

We are now in a position to introduce an inverse function, axiomatizing it in terms of the `at` function of the superior theory.

```
extend-module Permutation {
  declare at: (D) [(Perm D) D] -> D

  define at' := Function.at

  define at-definition := (forall p x . p at x = (perm->fun p) at' x)

  declare inverse: (D) [(Perm D)] -> (Perm D)

  define inverse-definition :=
    (forall p x y . p at x = y ==> (inverse p) at y = x)

  declare div: (D) [(Perm D) (Perm D)] -> (Perm D)
```

```
  define div-definition := (forall p q . p div q = p o inverse q)

  (add-axioms theory [at-definition inverse-definition div-definition])

  define consistent-inverse :=
    (forall p x x' y . p at x = y & p at x' = y ==> x = x')

  define right-inverse-lemma :=
    (forall p . (perm->fun p) o' (perm->fun inverse p) = identity')

  define right-inverse := (forall p . p o inverse p = identity)

  define Inverse-theorems :=
    [consistent-inverse right-inverse-lemma right-inverse]
}
```

The first two theorems prepare the way for the third, which is the actual right-inverse property for permutations. The `consistent-inverse` result justifies the definition of inverse, in the sense of showing that defining inverse as was done in the `inverse-definition` axiom does not introduce any inconsistency.

```
extend-module Permutation {
  define [bij-def inj-def] := [bijective-definition injective-definition]
  define at-def := at-definition

  define proofs :=
   method (theorem adapt)
     let {[_ prove chain chain-> _] := (proof-tools adapt theory);
          [at' identity' o' at identity o fun->perm perm->fun inverse] :=
            (adapt [at' identity' o' at identity o fun->perm perm->fun
                    inverse]);
          cd := compose-definition}
       match theorem {
         (val-of consistent-inverse) =>
           pick-any p x x' y
             let {inj := (!chain->
                           [true
                        ==> (bijective perm->fun p)           [is-bijective]
                        ==> (injective perm->fun p)           [bij-def]])}
               assume (p at x = y & p at x' = y)
                 let {p1 := (!chain->
                               [(p at x) = y                  [(p at x = y)]
                                         = (p at x')          [(p at x' = y)]
                           ==> ((perm->fun p) at' x =
                                  (perm->fun p) at' x')       [at-def]]);
                      p2 := (!chain->
                               [inj
                           ==> (forall x x' .
                                  (perm->fun p) at' x =
                                  (perm->fun p) at' x'
```

```
                              ==> x = x')                   [inj-def]])}
                  (!chain-> [p1 ==> (x = x')                [p2]])

      | (val-of right-inverse-lemma) => (!force (adapt theorem))
      | (val-of right-inverse) => (!force (adapt theorem))
      }

  (add-theorems theory |{Inverse-theorems := proofs}|)

} # close module Permutation
```

* ***Exercise 14.16:*** Fill in the proofs of `right-inverse-lemma` and `right-inverse`. □

We can now finally show that permutations form a group:

```
assert (theory-axioms Permutation.theory)

define perm-theorem :=
    method (p) (!prove-property p no-renaming Permutation.theory)

(!perm-theorem Permutation.associative)
(!perm-theorem Permutation.right-identity)
(!perm-theorem Permutation.left-identity)
(!perm-theorem Permutation.consistent-inverse)
(!perm-theorem Permutation.right-inverse-lemma)
(!perm-theorem Permutation.right-inverse)

define perm-group :=
   (renaming |{Group.+ := Permutation.o, Group.<0> := Permutation.identity,
               Group.U- := Permutation.inverse, Group.- := Permutation.div}|)

(print-instance-check perm-group Group.theory)
```

The output from the `print-instance-check` call is

```
Checking
(forall ?x:(Permutation.Perm 'T127048)
  (= (Permutation.o ?x
                    (Permutation.inverse ?x))
     Permutation.identity))

Checking
(forall ?x:(Permutation.Perm 'T127069)
  (forall ?y:(Permutation.Perm 'T127069)
    (= (Permutation.div ?x
                        ?y)
       (Permutation.o ?x
                      (Permutation.inverse ?y)))))
```

```
Checking
(forall ?x:(Permutation.Perm 'T127086)
  (forall ?y:(Permutation.Perm 'T127086)
    (forall ?z:(Permutation.Perm 'T127086)
      (= (Permutation.o (Permutation.o ?x
                                       ?y)
                        ?z)
         (Permutation.o ?x
                        (Permutation.o ?y
                                       ?z))))))

Checking
(forall ?x:(Permutation.Perm 'T127048)
  (= (Permutation.o Permutation.identity
                    ?x)
     ?x))

Checking
(forall ?x:(Permutation.Perm 'T127048)
  (= (Permutation.o ?x
                    Permutation.identity)
     ?x))
```

Finally, as we did in the integer arithmetic case, we can immediately obtain further theorems from Group theory.

```
define perm-group-prop := method (p)
  (!prove-property p perm-group Group.theory)

> (!perm-group-prop Group.left-inverse)

Theorem: (forall ?x:(Permutation.Perm 'S)
           (= (Permutation.o (Permutation.inverse ?x)
                             ?x)
              Permutation.identity))

> (!perm-group-prop Group.double-negation)

Theorem: (forall ?x:(Permutation.Perm 'S)
           (= (Permutation.inverse (Permutation.inverse ?x))
              ?x))

> (!perm-group-prop Group.unique-negation)

Theorem: (forall ?x:(Permutation.Perm 'S)
           (forall ?y:(Permutation.Perm 'S)
             (if (= (Permutation.o ?x
                                   ?y)
                    Permutation.identity)
                 (= (Permutation.inverse ?x)
```

```
                    ?y))))
> (!perm-group-prop Group.neg-plus)

Theorem: (forall ?x:(Permutation.Perm 'S)
           (forall ?y:(Permutation.Perm 'S)
             (= (Permutation.inverse (Permutation.o ?x
                                                    ?y))
                (Permutation.o (Permutation.inverse ?y)
                               (Permutation.inverse ?x)))))
```

14.9 Ordering properties at an abstract level

In previous sections we have seen how we can exploit the commonality of properties such as associativity, commutativity, existence of identity and inverse values, etc., among many different algebraic structures to allow great economy in the development of theories about them, using theory refinement and evolution. Let us now turn our attention to *ordering relations* that generalize the $<$ and $\leq$ relations on natural numbers that we studied in Chapter 8. Our goal will be to develop theories that have proved especially useful in understanding, generalizing, and optimizing ordering-related algorithms, such as sorting and searching algorithms, ensuring their correct and efficient application not only to numeric values but to many other data types that are pervasive in computer science.

14.9.1 Binary-Relation

We will want to work not only with an abstract strictly-less-than relation $<$ but also with its inverse relation $>$, and similarly for $\leq$ and $\geq$. In line with our objective of introducing axioms and theorems at the level of greatest possible generality, we start by defining the inverse of a relation in a `Binary-Relation` theory that will then be refined in several different ways.

```
module Binary-Relation {
  declare R, R': (T) [T T] -> Boolean

  define [x y z] := [?x ?y ?z]

  define inverse-def := (forall x y . x R' y <==> y R x)

  define theory := (make-theory [] [inverse-def])
}
```

14.9.2 Irreflexive

Abstracting from the irreflexive property of `N.<`, we introduce the following refinement of `Binary-Relation` theory, in which we prove as a simple lemma that the inverse relation is also irreflexive:

```
module Irreflexive {
  open Binary-Relation

  define irreflexive := (forall x . ~ x R x)

  define theory := (make-theory ['Binary-Relation] [irreflexive])

  define inverse := (forall x . ~ x R' x)

  define proof :=
    method (theorem adapt)
      let {[_ _ _ chain-> _] := (proof-tools adapt theory);
           [R R'] := (adapt [R R'])}
        match theorem {
          (val-of inverse) =>
            pick-any x
              (!chain-> [true ==> (~ x R x)    [irreflexive]
                              ==> (~ x R' x)  [inverse-def]])
        }

  (add-theorems theory |{inverse := proof}|)
}

> (test-all-proofs 'Irreflexive)

Testing proof of:

(forall ?x:'T45954
  (not (Binary-Relation.R' ?x ?x)))
 ...

Proof worked.
```

14.9.3 Transitive

Another property of `N.<` that we can generalize is transitivity.

```
module Transitive {
  open Binary-Relation

  define transitive := (forall x y z . x R y & y R z ==> x R z)

  define theory := (make-theory ['Binary-Relation] [transitive])
```

```
}
```

Exercise 14.17: Prove that the inverse relation `R'` of a transitive relation `R` is also transitive: Extend `Transitive` to include the theorem and its proof, and test the proof with `test-all-proofs`. □

14.9.4 Strict partial order

Combining the irreflexive and transitive properties yields a *strict partial order*:

```
module Strict-Partial-Order {
  open Irreflexive , Transitive

  define theory := (make-theory ['Irreflexive 'Transitive] [])
}
```

Note that the inverse relation `R'` of a strict partial order `R` is also a strict partial order, since we have shown that it has the irreflexive and transitive properties.

A strict partial order also has *asymmetry*. We prove this, along with the property that two elements in the relation cannot be equal, as follows:

```
extend-module Strict-Partial-Order {
  define asymmetric := (forall x y . x R y ==> ~ y R x)

  define implies-not-equal := (forall x y . x R y ==> x =/= y)

  define proofs :=
    method (theorem adapt)
      let {[_ _ _ chain-> _] := (proof-tools adapt theory);
           [R R'] := (adapt [R R'])}
        match theorem {
          (val-of asymmetric) =>
            pick-any x y
              assume (x R y)
                (!by-contradiction (~ y R x)
                  assume (y R x)
                    (!absurd
                      (!chain-> [(y R x)
                                 ==> (x R y & y R x)          [augment]
                                 ==> (x R x)                  [transitive]])
                      (!chain-> [true ==> (~ x R x)           [irreflexive]])))
        | (val-of implies-not-equal) =>
            pick-any x y
              assume (x R y)
                (!by-contradiction (x =/= y)
                  assume (x = y)
                    let {xRx :=  (!chain-> [(x R y)
```

```
                                        ==> (x R x)      [(x = y)]]);
                  -xRx := (!chain-> [true
                                        ==> (~ x R x)    [irreflexive]])}
                (!absurd xRx -xRx))
       }

  (add-theorems theory |{[asymmetric implies-not-equal] := proofs}|)
}
```

We can also adapt the same proofs to prove that the inverse relation has the same two properties.

```
assert (theory-axioms 'Strict-Partial-Order)

define inverse-renaming :=
  (renaming |{Binary-Relation.R := Binary-Relation.R'}|)

> (!prove-property Irreflexive.inverse no-renaming 'Irreflexive)

Theorem: (forall ?x:'S
           (not (Binary-Relation.R' ?x ?x)))

> (!prove-property Transitive.inverse no-renaming 'Transitive)

Theorem: (forall ?x:'S
           (forall ?y:'S
             (forall ?z:'S
               (if (and (Binary-Relation.R' ?x ?y)
                        (Binary-Relation.R' ?y ?z))
                   (Binary-Relation.R' ?x ?z)))))

> (!prove-property Strict-Partial-Order.asymmetric
                   inverse-renaming 'Strict-Partial-Order)

Theorem: (forall ?x:'S
           (forall ?y:'S
             (if (Binary-Relation.R' ?x ?y)
                 (not (Binary-Relation.R' ?y ?x)))))

> (!prove-property Strict-Partial-Order.implies-not-equal
                   inverse-renaming 'Strict-Partial-Order)

Theorem: (forall ?x:'S
           (forall ?y:'S
             (if (Binary-Relation.R' ?x ?y)
                 (not (= ?x ?y)))))
```

14.9.5 Nonstrict partial orders

Generalizing from N.<=, we develop a theory of nonstrict *partial orderings*. Again, we build this theory by refinement.

```
module Reflexive {
  open Binary-Relation

  define reflexive := (forall x . x R x)

  define theory := (make-theory ['Binary-Relation] [reflexive])

  define inverse := (forall x . x R' x)

  define proof :=
    method (theorem adapt)
      let {[_ _ _ chain-> _] := (proof-tools adapt theory);
           [R R'] := (adapt [R R'])}
        match theorem {
          (val-of inverse) =>
            pick-any x
              (!chain-> [true ==> (x R x)  [reflexive]
                              ==> (x R' x) [inverse-def]])
        }

  (add-theorems theory |{inverse := proof}|)
}

module Preorder {

  open Transitive, Reflexive

  define theory := (make-theory ['Transitive 'Reflexive] [])
}
```

Besides the transitive and reflexive properties of a preorder, a partial order also has the *antisymmetric* property:

```
module Antisymmetric {
  open Binary-Relation

  define antisymmetric := (forall x y . x R y & y R x ==> x = y)

  define theory := (make-theory ['Binary-Relation] [antisymmetric])
}

module Partial-Order {
  open Preorder, Antisymmetric

  define Theory := (make-theory ['Preorder 'Antisymmetric] [])
}
```

Exercise 14.18: Prove that the inverse relation `R'` of an antisymmetric relation `R` is also antisymmetric: Extend `Antisymmetric` to include the theorem and its proof, and test the proof with `test-all-proofs`. □

Note that the inverse relation `R'` of a partial order `R` is also a partial order, since we have shown that it has the transitive, reflexive, and antisymmetry properties.

`Strict-Partial-Order` and `Partial-Order` are equivalent in generality, in the sense that one can start with a strict partial order relation R_1 and based on it define a relation R_2 that satisfies the partial order properties, and vice versa. We will prove this, but first let's make sure we can express properties involving both relations without confusion. With `Ring` theory and its refinements, we introduced multiplicative versions of some theories by mapping `+` to `*`, allowing us to express properties such as distributivity that involved both, while also sharing some axioms, theorems, and proofs about them. Here we take a similar approach. We define `SPO.theory` via a shared theory based on `Strict-Partial-Order.theory`:

```
module SPO {
  declare <, >: (T) [T T] -> Boolean

  define sm := |{Binary-Relation.R := <, Binary-Relation.R' := >}|
  define renaming := (renaming sm)

  define theory := (adapt-theory 'Strict-Partial-Order sm)
}
```

We can show that `SPO.>` as well as `SPO.<` is a strict partial order:

```
assert (theory-axioms 'SPO)

> (!prove-property Irreflexive.inverse no-renaming 'SPO)

Theorem: (forall ?x:'S
           (not (SPO.> ?x ?x)))

> (!prove-property Transitive.inverse no-renaming 'SPO)

Theorem: (forall ?x:'S
           (forall ?y:'S
             (forall ?z:'S
               (if (and (SPO.> ?x ?y)
                        (SPO.> ?y ?z))
                   (SPO.> ?x ?z)))))
```

Similarly, we define `PO` as follows:

```
module PO {

  declare <=, >=: (T) [T T] -> Boolean
```

```
  define sm := |{Binary-Relation.R := <=, Binary-Relation.R' := >=}|
  define renaming := (renaming sm)

  define theory := (adapt-theory 'Partial-Order sm)
}
```

Now we can prove that we can get a partial order from a strict partial order by defining $x \leq y$ as $x < y$ or $x = y$. From this definition and the axioms of `SPO.theory`, we can prove that the axioms of `PO.theory` are satisfied:

```
module PO-from-SPO {

 define [x y z] := [?x ?y ?z]

 define [< <=] := [SPO.< PO.<=]

 define <=-definition := (forall x y . x <= y <==> x < y | x = y)

 (add-axioms 'SPO [<=-definition])

 define implied-by-less := (forall x y . x < y ==> x <= y)
 define implied-by-equal := (forall x y . x = y ==> x <= y)
 define implies-not-reverse := (forall x y . x <= y ==> ~ y < x)
 define PO-reflexive := (forall x . x <= x)
 define PO-transitive := (forall x y z . x <= y & y <= z ==> x <= z)
 define PO-antisymmetric := (forall x y . x <= y & y <= x ==> x = y)

 define theorems := [<=-definition implied-by-less implied-by-equal
                     implies-not-reverse PO-reflexive
                    PO-antisymmetric PO-transitive]

 define proofs :=
   method (theorem adapt)
     let {adapt := (o adapt SPO.renaming);
          [_ prove chain chain-> _] := (proof-tools adapt SPO.theory);
          [< <=] := (adapt [< <=]);
          irreflexive := Strict-Partial-Order.irreflexive;
          transitive  := Strict-Partial-Order.transitive;
          asymmetric  := (!prove Strict-Partial-Order.asymmetric)}
       match theorem {
         (val-of implied-by-less) =>
           pick-any x y
             (!chain [(x < y) ==> (x < y | x = y)          [alternate]
                               ==> (x <= y)                 [<=-definition]])
       | (val-of implied-by-equal) =>
           pick-any x y
             (!chain [(x = y) ==> (x < y | x = y)          [alternate]
                             ==> (x <= y)                   [<=-definition]])
       | (val-of implies-not-reverse) =>
           pick-any x y
```

```
        assume x<=y := (x <= y)
          (!cases (!chain-> [x<=y ==> (x < y | x = y) [<=-definition]])
            (!chain [(x < y) ==> (~ y < x)             [asymmetric]])
            assume (x = y)
              (!by-contradiction (~ y < x)
                assume (y < x)
                  let {y<y := (!chain-> [(y < x)
                                    ==> (y < y)        [(x = y)]]);
                       -y<y := (!chain-> [true
                                    ==> (~ y < y)      [irreflexive]])}
                    (!absurd y<y -y<y)))
    | (val-of PO-reflexive) =>
      pick-any x
        let {IBE := (!prove implied-by-equal)}
          (!chain-> [(x = x) ==> (x <= x)              [IBE]])
    | (val-of PO-antisymmetric) =>
      (!force (adapt theorem))
    | (val-of PO-transitive) =>
      (!force (adapt theorem))
    }

  (add-theorems SPO.theory |{theorems := proofs}|)
}
```

Exercise 14.19: Fill in the proof of `PO-antisymmetric`. □

Exercise 14.20: Fill in the proof of `PO-transitive`. □

We can check these proofs as follows:

```
(test-proofs [PO-from-SPO.PO-antisymmetric PO-from-SPO.PO-transitive] 'SPO)
```

Proving the other direction, that one can start from a partial order and obtain a corresponding strict partial order, is left as an exercise (Exercise 14.26).

14.9.6 Strict weak order

As mentioned at the beginning of our discussion of ordering relations, we will focus mainly on theories relevant to order-related algorithms, such as sorting and searching algorithms. Partial orders, whether strict or nonstrict, are not sufficient for this purpose (except for weak notions of sorting such as *topological sorting* [58], which we will not discuss here). The strict order relation < on natural numbers, as defined in Chapter 8, is a *total order* relation—that is, the trichotomy property is satisfied—but classical sorting algorithms such as mergesort and quicksort can, with a little care, be made to work with a more general notion of order, making them much more widely applicable than if they required total ordering.

The essential idea is that when sorting a sequence of values we often don't care about the order of some of the values; we are willing to consider them "equivalent" with respect to the ordering. A simple example is case-insensitive comparison of strings, when strings that differ only in case are considered equivalent. Likewise, in ordering personnel records in a database, we might not care about the ordering of records that have the same entry—"assistant manager," say—in a job-classification field; names, salaries, and the like, could all be different, since we only want to group together all persons having the same job classification.

Accordingly, given a strict partial order $<$, we define a relation E as follows: $(x\ E\ y)$ iff neither $(x < y)$ nor $(y < x)$.[6] It turns out that transitivity is the only property that we must assume E satisfies. The other properties of an equivalence relation—reflexivity and symmetry—follow from E's definition and transitivity. We thus define the theory SWO, *strict weak order*, as a refinement of SPO as follows:

```
extend-module SPO {
  declare E: (T) [T T] -> Boolean [100]

  define E-definition := (forall x y . x E y <==> ~ x < y & ~ y < x)

  (add-axioms 'SPO [E-definition])
}

module SWO {
  open SPO

  define E-transitive := (forall x y z . x E y & y E z ==> x E z)

  define theory := (make-theory ['SPO] [E-transitive])
```

Along with the reflexive and symmetry properties of E, we gather together a number of other useful theorems that relate $<$ and E (or in some cases are just about one or the other):

```
  define E-reflexive := (forall x . x E x)
  define E-symmetric := (forall x y . x E y ==> y E x)
  define <-E-transitive-1 := (forall x y z . x < y & y E z ==> x < z)
  define <-E-transitive-2 := (forall x y z . x < y & x E z ==> z < y)
  define not-<-property := (forall x y . ~ x < y ==> y < x | y E x)
  define <-transitive-not-1 := (forall x y z . x < y & ~ z < y ==> x < z)
  define <-transitive-not-2 := (forall x y z . x < y & ~ x < z ==> z < y)
  define <-transitive-not-3 := (forall x y z . ~ y < x & y < z ==> x < z)
  define not-<-is-transitive :=
    (forall x y z . ~ x < y & ~ y < z ==> ~ x < z)

  define <-E-theorems :=
    [E-reflexive E-symmetric <-E-transitive-1 <-E-transitive-2
```

6 It is customary to call such elements *incomparable* with respect to $<$.

```
        not-<-property <-transitive-not-1 <-transitive-not-2
        <-transitive-not-3 not-<-is-transitive]
```

In the proof method we include actual proofs only for `E-reflexive` and `E-symmetric` and leave the others as exercises in Section 14.10.

```
  define sm := (get-renaming 'SPO)
  define irreflexive := (sm Strict-Partial-Order.irreflexive)
  define transitive := (sm Strict-Partial-Order.transitive)
  define asymmetric := (sm Strict-Partial-Order.asymmetric)

  define <-E-proofs :=
   method (theorem adapt)
     let {[_ _ _ chain-> _] := (proof-tools adapt theory);
          E := lambda (x y) (adapt (x E y));
          < := lambda (x y) (adapt (x < y))}
       match theorem {
         (val-of E-reflexive) =>
           pick-any x
             (!chain-> [true
                    ==> (~ x < x)                [irreflexive]
                    ==> (~ x < x & ~ x < x)      [augment]
                    ==> (x E x)                  [E-definition]])
       | (val-of E-symmetric) =>
           pick-any x y
             assume (x E y)
               (!chain-> [(x E y)
                      ==> (~ x < y & ~ y < x)    [E-definition]
                      ==> (~ y < x & ~ x < y)    [comm]
                      ==> (y E x)                [E-definition]])
       | _ => (!force (adapt theorem))
       }

 (add-theorems theory |{<-E-theorems := <-E-proofs}|)
} # close module SWO
```

14.9.7 A preorder

Suppose in module `SWO` we define a relation <E as follows:

```
extend-module SWO {
  declare <E: (T) [T T] -> Boolean

  define <E-definition := (forall x y . x <E y <==> ~ y < x)

 (add-axioms theory [<E-definition])
}
```

The following exercise shows that `<E` is a *preorder*, as defined on page 668.

Exercise 14.21: Show that `<E` is reflexive and transitive, and evolve `SWO.theory` with these two theorems and their proofs. □

14.9.8 Strict total order

If we have a strict weak order and also have the corresponding trichotomy law, we have a *strict total order*:

```
module STO {
  open SWO

  define strict-trichotomy := (forall x y . ~ x < y & ~ y < x ==> x = y)

  define theory := (make-theory ['SWO] [strict-trichotomy])
```

The only theorem we will concern ourselves with here is one stating that the *E* relation is the same as equality:

```
  define E-iff-equal := (forall x y . x E y <==> x = y)
} # close module STO
```

Exercise 14.22: Define a corresponding proof method containing a single proof, that of `E-iff-equal`. Follow it with an `add-theorems`, then write a test with `test-proofs`. □

14.9.9 Lists over a strict weak order

In Section 8.3 we defined ordered lists on natural numbers in terms of the `(List N)` datatype introduced in Section 3.9 and the natural-number ordering properties introduced in the first part of Chapter 8. Now that we have abstract ordering theories, we can generalize to ordered lists `(List` *S*`)`, where *S* can be any sort that satisfies the strict weak order theory. We choose SWO because it is the minimal theory that allows algorithms on lists such as binary search, merging, and sorting.

As a generalization of the `<=L` binary predicate defined in Section 8.3, we first declare and axiomatize a binary predicate `<EL` that asks if a value of sort *S* is related by < or `E` to the first element of a list of `S` values (which we define as true by default if the list is empty):

```
extend-module SWO {
  declare <EL: (S) [S (List S)] -> Boolean

  module <EL {

   define [L M] := [?L:(List 'S) ?M:(List 'S)]
   define [++ in] := [List.++ List.in]
```

```
  define def := (close [(x <EL nil)
                        (x <EL y::L <==> x <E y)])

  define [empty nonempty] := def

  (add-axioms theory def)
```

Corresponding to theorems proved about <=L in Section 8.3, we have the following:

```
  define left-transitive := (forall L x y . x <E y & y <EL L ==> x <EL L)

  define before-all-implies-before-first :=
    (forall L x . (forall y . y in L ==> x <E y) ==> x <EL L)

  define append := (forall L M x . x <EL L & x <EL M ==> x <EL L ++ M)

  define append-2 := (forall L M x . x <EL L ++ M ==> x <EL L)

  define theorems := [left-transitive before-all-implies-before-first
                      append append-2]
  define proofs :=
   method (theorem adapt)
     let {[_ prove _ chain-> _] := (proof-tools adapt theory);
          [< <E <EL] := (adapt [< <E <EL])}
       match theorem {
         (val-of left-transitive) =>
           datatype-cases theorem {
             nil =>
               pick-any x y
                 assume (x <E y & y <EL nil)
                   (!chain-> [true ==> (x <EL nil)       [empty]])
           | (z :: M) =>
               let {ET := (!prove <E-transitive)}
                 pick-any x y
                   assume (x <E y & y <EL z::M)
                     conclude (x <EL z::M)
                       (!chain-> [(x <E y & y <EL z::M)
                              ==> (x <E y & y <E z)      [nonempty]
                              ==> (x <E z)               [ET]
                              ==> (x <EL z::M)           [nonempty]])}
       | _ => (!force (adapt theorem))
       }

   (add-theorems theory |{theorems := proofs}|)
 } # close module <EL
} # close module SWO
```

Exercise 14.23: Fill in the proof of <EL.before-all-implies-before-first. □

Exercise 14.24: Fill in the proof of `<EL.append`. □

Exercise 14.25: Fill in the proof of `<EL.append-2`. □

We now define a unary predicate, `ordered`, that asks whether the elements of a list are in (nondescending) order:

```
extend-module SWO {

 declare ordered: (S) [(List S)] -> Boolean

 module ordered {
   open <EL

   define def := (close [(ordered nil)
                          (ordered x::L <==> x <EL L & ordered L)])

   define [empty nonempty] := def

   (add-axioms theory def)
```

The following are very simple but still useful lemmas:

```
   define head := (forall L x . ordered x::L ==> x <EL L)
   define tail := (forall L x . ordered x::L ==> ordered L)

   define proofs :=
     method (theorem adapt)
       let {[_ _ chain _ _] := (proof-tools adapt theory);
            [< ordered <EL] := (adapt [< ordered <EL])}
         match theorem {
           (val-of head) =>
             pick-any L x
               (!chain [(ordered x::L)
                    ==> (x <EL L & ordered L)    [nonempty]
                    ==> (x <EL L)                [left-and]])
         | (val-of tail) =>
             pick-any L x
               (!chain [(ordered x::L)
                    ==> (x <EL L & ordered L)    [nonempty]
                    ==> (ordered L)              [right-and]])
         }

   (add-theorems theory |{[head tail] := proofs}|)
```

Finally, these are the main theorems about ordered lists over elements in a strict weak ordering:

```
define first-to-rest-relation :=
  (forall L x y . ordered x::L & y in L ==> x <E y)

define cons := (forall L x . ordered L &
                             (forall y . y in L ==> x <E y)
                             ==> ordered x::L)
define append :=
  (forall L M . ordered L &
                ordered M &
                (forall x y . x in L & y in M ==> x <E y)
                ==> ordered L ++ M)

define append-2 :=
  (forall L M . ordered L ++ M ==> ordered L & ordered M)

define theorems := [first-to-rest-relation cons append append-2]

define ftr-proof :=
  method (theorem adapt)
    let {[_ prove chain chain-> _] := (proof-tools adapt theory);
         [ordered <EL] := (adapt [ordered <EL])}
      match theorem {
        (val-of first-to-rest-relation) =>
          by-induction (adapt theorem) {
            nil => pick-any x y
                     assume (ordered x::nil & y in nil)
                       (!from-complements
                         (x <E y)
                         (y in nil)
                         (!chain-> [true
                                ==> (~ y in nil)     [List.in.empty]]))
        | (L as (z :: M)) =>
          let {IH := (forall x y . ordered x::M &
                                   y in M ==> x <E y);
               transitive := (!prove <EL.left-transitive)}
            pick-any x y
              assume (ordered x::L & y in z::M)
                let {p := (x <E z & z <EL M & ordered M);
                     _ := (!chain-> [(ordered x::z::M)
                                 ==> (x <EL z::M &
                                      ordered z::M) [nonempty]
                                 ==> (x <EL z::M &
                                      z <EL M &
                                      ordered M)    [nonempty]
                                 ==> p              [<EL.nonempty]]);
                     _ := (!chain-> [p
                                 ==> (ordered M)    [prop-taut]]);
                     _ := (!chain->
                            [p
                         ==> (x <E z & z <EL M)     [prop-taut]
                         ==> (x <EL M)              [transitive]
```

```
                ==> (x <EL M &
                     ordered M)                   [augment]
                ==> (ordered x::M)                [nonempty]])}
          (!cases (!chain->
                     [(y in L)
                ==> (y = z | y in M)              [List.in.nonempty]])
            assume (y = z)
              (!chain-> [p ==> (x <E z)           [left-and]
                             ==> (x <E y)          [(y = z)]])
              (!chain [(y in M)
                  ==> (ordered x::M &
                        y in M)                   [augment]
                  ==> (x <E y)                    [IH]]))
      }
  }

(add-theorems theory |{first-to-rest-relation := ftr-proof}|)
} # close module ordered
} # close module SWO
```

Proving the other three theorems is left as Exercise 14.28.

14.9.10 Relational theory diagram

Figure 14.2 depicts the theory hierarchy developed in the preceding sections. As we mentioned in section 14.7.5, a directed arrow means the target node is a refinement of the source node.

14.10 Additional exercises

⋆ ***Exercise 14.26:*** Show that if one has a partial order relation $\leq$ and defines $x < y$ as $x \leq y$ and $x \neq y$, then $<$ is a strict partial order relation. □

Exercise 14.27: Each part of this exercise requires filling in the proof of one of the `SWO.<-E-theorems` (page 672) in method `SWO.<-E-proofs`, that is, replacing the use of `force` with an actual proof. *Hint:* These theorems are generalizations of theorems about the natural numbers, so it may be helpful to review their proofs as given in examples or exercises in Chapter 8 (in the latter case, if you haven't done the exercise already, try doing it before tackling the more general proof necessary here).

In each case, check your proof by creating an `SWO` instance (`no-renaming` is OK), asserting the `SWO` theory axioms, and evaluating the proof method with `test-proofs` (use the transcript beginning on page 671 as a guide).

(a) `SWO.<-E-transitive-1`

(b) `SWO.<-E-transitive-2`

(c) `SWO.not-<-property`

(d) `SWO.<-transitive-not-1`

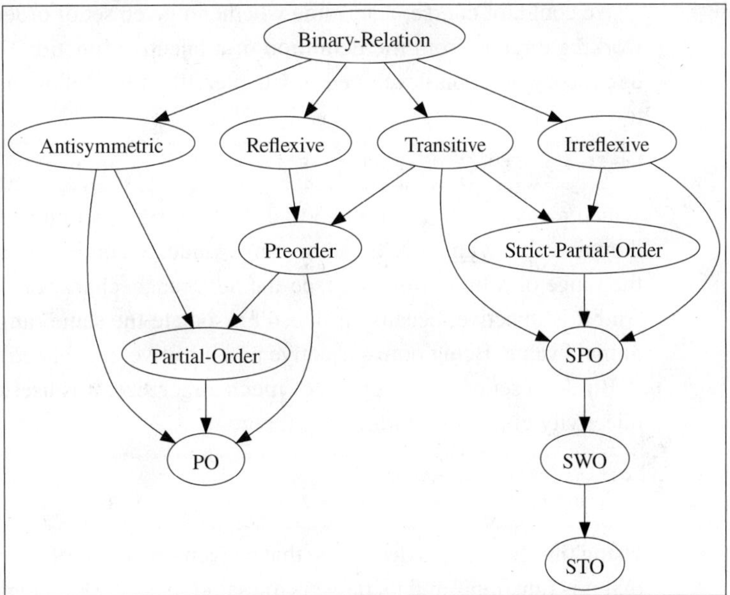

Figure 14.2
Refinement hierarchy of relational theories

(e) `SWO.<-transitive-not-2`

(f) `SWO.<-transitive-not-3`

(g) `SWO.not-<-is-transitive` □

Exercise 14.28: Each part of this exercise requires filling in the proof of one of the `SWO.ordered.theorems` (page 676) in method `SWO.ordered.proofs`. Check your solution with `test-proofs`.

(a) `SWO.cons`

(b) `SWO.append-2`

⋆ (c) `SWO.append` □

⋆⋆ ***Exercise 14.29:*** A function with the same *finite* set D as its domain and range must be a bijection, and therefore a permutation. This exercise formalizes and proves this statement in terms of the finite map and set theories presented in Chapter 10.[7]

7 Finite permutations play important roles in algebraic theory and in analysis of algorithms (such as sorting algorithms), especially in light of their group theory properties. A much studied example is the *symmetric group* of degree n, denoted S_n, which is defined as the set of all permutations on a set of size n. The number of members of S_n is $n!$.

We could, of course, determine whether a given set of ordered pairs is a permutation by working directly from the definition of a bijective function. A simple example, using the Set theory notation developed in Chapter 10, is the following set of ordered pairs of Int terms:

```
define R := ((1 @ 3) ++ (2 @ 1) ++ (3 @ 2) ++ null)
```

which maps 1 to 3, 2 to 1 and 3 to 2. First, R is a function because it doesn't associate any domain value with more than one range value. Second, it is surjective just by definition of the range of a finite function (see the Set.range-characterization theorem, page 522). And it is injective, because it doesn't associate the same range value with more than one domain value. Being both surjective and injective, it is bijective.

But for a set of ordered pairs of much larger size, it is useful to have the simpler test for bijectivity given by the following theorem:

```
define bijective-theorem :=
  (forall R . functional R & range R = dom R ==> bijective R)
```

Using this theorem, to determine that a given set of pairs R is bijective we only have to check that it is functional and its range is the same set as its domain. Furthermore, if we start with a finite map m and compute R as (map->set m), we are guaranteed that R is functional, so the only test required is whether the domain and range of R are the same set.

Enclose all of the following development in an extension of the Set module: **extend-module** Set { ... } and try to take advantage of axioms and theorems from that module.

(a) Define the predicate functional as follows:

```
define [s x y y' rest] :=
  [?s:(Set (Pair 'S 'T)) ?x:'S ?y:'T ?y':'T ?rest:(Set (Pair 'S 'T))]

declare functional: (S, T) [(Set (Pair S T))] -> Boolean

assert* functional-def :=
  [(functional (x @ y ++ rest) <==>
      (x @ y ++ rest) ^1 x = singleton (x @ y) & functional rest)
   (functional null)]
```

Prove:

```
define functional-characterization :=
  (forall R . functional R <==>
              forall x y y' . x @ y in R & x @ y' in R ==> y = y')
```

(Prove each direction of the equivalence as a separate theorem and then combine them using the equiv primitive method.)

(b) Prove the following lemma:

```
define range-stays-smaller :=
  (forall s t .
    card range t < card dom t &
    functional (s \/ t)
    ==> card range (s \/ t) < card dom (s \/ t))
```

(c) Using range-stays-smaller, prove the following lemma:

```
define [x1 x2] := [?x1 ?x2]

define smaller-range :=
  (forall R x1 x2 y .
    functional R &
    x1 @ y in R & x2 @ y in R & x1 =/= x2
    ==> card range R < card dom R)
```

(d) Introduce the following definitions:

```
declare injective: (S, T) [(Set (Pair S T))] -> Boolean

assert* injective-def :=
   (injective R <==>
     forall x x' y . x @ y in R & x' @ y in R ==> x = x')

declare surjective: (S, T) [(Set (Pair S T))] -> Boolean

assert* surjective-def :=
   (surjective R <==>
      forall y . y in range R <==> exists x . x @ y in R)

declare bijective: (S, T) [(Set (Pair S T))] -> Boolean

assert* bijective-def :=
  (bijective R <==> injective R & surjective R)

define injective-theorem :=
  (forall R .  functional R & range R = dom R ==> injective R)

define surjective-theorem := (forall R . surjective R)

define bijective-theorem :=
  (forall R . functional R & range R = dom R ==> bijective R)
```

Then prove the three theorems. □

15 Abstract Algorithms

THE advantages of abstraction and specialization are not limited to the static algebraic or relational structures discussed in the previous chapter. In the realm of computer programming, we can take the same approach to the creation and use of *abstract algorithms and data structures* (also called *generic* algorithms and data structures). At an abstract level, some issues of algorithm and data structure design and implementation may become more difficult to understand and work with, but others may actually become easier! But whatever extra effort is required is then repaid many times over as the abstractions are specialized to various concrete domains. In the case of correctness proofs, replaying the existing abstract-level proofs with appropriate parameters is all that's needed.

In this chapter we illustrate this point with abstractions of algorithms and data structures we treated in Chapters 11 and 12. In Section 15.1, we develop an abstract version of one of the concrete binary search algorithms we studied in Chapter 11. Instead of the natural number domain, our abstract binary search works with any domain satisfying the strict weak order axioms. We naturally choose as the starting point for generalization the more efficient concrete algorithm discussed in Section 11.4—it would be counterproductive to develop an abstract algorithm based on an inefficient concrete algorithm, since the result would only produce inefficient concrete algorithms when specialized back to the original or any other domain. As we will see, the abstract binary search algorithm that we develop is not only more efficient but also more general than would be possible if we had based it on the more "obvious" algorithm that we first considered in Chapter 11.

In Section 15.2, the goal is an abstract but efficient algorithm for raising a value to a power, where the value can be from any monoid, not just for the natural numbers, for which we developed a fast power algorithm in Chapter 12. As noted in Section 12.7, both the embedded and the tail-recursive algorithms developed in the chapter suffer from performing unneeded operations, and thus there is still need for further optimization. In lifting the algorithm to the abstract level, we work out the proofs for a version of the algorithm that eliminates the unneeded operations, essentially adopting the abstract-level solution given (in C++ code) by Stepanov and McJones [93].

15.1 An abstract binary search algorithm

We begin by recalling the definition of the natural-number binary search function from Section 11.4.

```
declare binary-search: [N (BinTree N)] -> (BinTree N)

module binary-search {
define [x y L R T] := [?x:N ?y:N ?L:(BinTree N) ?R:(BinTree N)
                       ?T:(BinTree N)]
```

```
assert axioms :=
  (fun
    [(binary-search x (node L y R)) =
      [(binary-search x L)   when (x < y)
       (binary-search x R)   when (y < x)
       (node L y R)          when (~ x < y & ~ y < x)]
     (binary-search x null) = null])

define [go-left go-right at-root empty] := axioms
```

Here are the two main correctness properties we proved about this algorithm (recall that the `BST` predicate tests whether a binary tree is a binary search tree):

```
define found :=
  (forall T .
    BST T ==>
      forall x L y R .
        (binary-search x T) = (node L y R) ==> x = y & x in T)

define not-found :=
  (forall T .
    BST T ==> forall x . (binary-search x T) = null ==> ~ x in T)
```

As we begin to examine how to generalize this algorithm to an abstract domain, one question to ask is whether the correctness properties, as stated, might be stronger than necessary. In this case, in the `found` property, do we really need to require exact equality, `(x = y)`? All we can conclude at the `at-root` exit from the function is that `(~ x < y & ~ y < x)`, which is not enough to guarantee that `(x = y)` unless the domain is totally ordered, as is the case for the natural numbers. But it could be useful to search a tree when we only want to find a value that is equivalent, in some sense, to the search key rather than exactly equal to it. As mentioned in Section 14.9.6, a simple example is case-insensitive comparison of strings (treating strings that differ only in case as equivalent). Looking back at all of the ordering theories we studied in Chapter 14, it appears that the one that has the most general requirements for supporting algorithms based on such equivalences is the SWO (strict weak order) theory: SWO *defines* an equivalence `(x E y)` based on any strict partial order <, precisely as `(~ x < y & ~ y < x)`, and requires it to be transitive—nothing more is needed to yield a useful equivalence.

We therefore choose to try to generalize the above binary-search algorithm to any SWO domain, but we recognize that we need to "loosen up" the correctness properties. In `found`, we must replace the exact equality relation with the weaker `E` equivalence. And correspondingly, we must replace our definition of the `in` relation (which is actually the `BinTree.in` relation) with a weaker one: instead of defining

```
(x in (node L y R) <==> x = y | x in L | x in R)
```

we will define

```
(x in (node L y R) <==> x E y | x in L | x in R).
```

Here, then, is the abstracted algorithm:

```
extend-module SWO {
 open BST

 declare binary-search: (S) [S (BinTree S)] -> (BinTree S)

 module binary-search {

  define axioms :=
    (fun
     [(binary-search x (node L y R)) =
         [(binary-search x L)    when (x < y)
          (binary-search x R)    when (y < x)
          (node L y R)           when (~ x < y & ~ y < x)]
      (binary-search x null) = null])

  define [go-left go-right at-root empty] := axioms

  (add-axioms theory axioms)      # Here theory is SWO.theory
```

The BST module opened here should not be the one introduced for the natural numbers but a new one, described in the next section, that provides an appropriately general definition of a binary search tree.

The theorems to be proved are then as follows (expressed in terms of convenient property procedures):

```
  define (found-property T) :=
    (forall x L y R .
       (binary-search x T) = (node L y R) ==> x E y & x in T)

  define found := (forall T . BST T ==> found-property T)

  define (not-found-property T) :=
    (forall x . (binary-search x T) = null ==> ~ x in T)

  define not-found := (forall T . BST T ==> not-found-property T)

 } # binary-search
} # close module SWO
```

To be able to prove these general theorems, we start in the next section with a generalization of binary search trees and prove some lemmas about the new `in` relation. But before we do that, let's try evaluating a few `binary-search` calls, just as we did with the natural-number version in Chapter 11. For evaluation to be possible, we must adapt an abstract algorithm to a particular instance. The sample `tree1` in Figure 8.1 (page 464) contains `Int` values, so we create an instance capable of searching it, as follows:

```
declare bs-int: [Int (BinTree Int)] -> (BinTree Int)
define swo1 := (renaming |{SWO.binary-search := bs-int, SWO.< := Real.<}|)
assert (swo1 SWO.binary-search.axioms)

define tree1 :=
  (node (node null
              2
              (node (node null
                          3
                          null)
                    5
                    null))
        7
        (node (node null
                    11
                    null)
              13
              null))

> (eval (bs-int 5 null))

Term: null:(BinTree Int)

> (eval (bs-int 5 tree1))

Term: (node (node null:(BinTree Int)
                  3
                  null:(BinTree Int))
            5
            null:(BinTree Int))

> (eval (bs-int 13 tree1))

Term: (node (node null:(BinTree Int)
                  11
                  null:(BinTree Int))
            13
            null:(BinTree Int))

> (eval (bs-int 9 tree1))

Term: null:(BinTree Int)
```

Note that we did not need input expansion or output transformation for these `eval` calls, as the `Int` terms in the tree are directly accepted by this instance of `binary-search`.

15.1.1 Abstract-level binary search trees

At the abstract level of SWO relations, we define binary search trees as follows. Note that we now directly give a nonconstructive definition, although we could also take a similar approach to that taken in Section 8.4 (whereby a constructive definition is given recursively and then `empty` and `nonempty` are derived as characterization results). In module `BST` we first define `BST.in`, with, as we noted earlier, comparison using `E` rather than `=`. We then use this new membership predicate in defining `BST`.

```
extend-module SWO {
 open BinTree

 declare BST: (S) [(BinTree S)] -> Boolean

 module BST {

  declare in: (S) [S (BinTree S)] -> Boolean

  module in {

    define def := (close [(~ x in null)
                          (x in (node L y R) <==> x E y | x in L | x in R)])

   define [empty nonempty] := def

   (add-axioms theory def)
  } # in

  define empty := (BST null)
  define nonempty :=
   (forall L y R .
     BST (node L y R) <==>
       BST L & (forall x . x in L ==> x <E y) &
       BST R & (forall z . z in R ==> y <E z))

  (add-axioms theory [empty nonempty])
 } # BST
} # SWO
```

In defining `BST` we also now replace `N.<=` with the abstract relation `<E`, the ordering relation we defined in Section 14.9.7. We repeat its definition here:

```
extend-module SWO {

  declare <E: (S) [S S] -> Boolean

  define <E-definition := (forall x y . x <E y <==> ~ y < x)
```

```
  (add-axioms theory [<E-definition])
}
```

Note Exercise 14.21, which showed that <E is reflexive and transitive (the requirements for a preorder; see page 668).

It will be handy to have abstract versions of the in.root, in.left, and in.right lemmas of Section 8.4:

```
extend-module SWO {
  extend-module BST {
    extend-module in {

      define root  := (forall x L y R . x E y  ==> x in (node L y R))
      define left  := (forall x L y R . x in L ==> x in (node L y R))
      define right := (forall x L y R . x in R ==> x in (node L y R))

    } # in
  } # BST
} # SWO
```

Exercise 15.1: Define an abstract method that derives these three lemmas and test it. □

15.1.2 Abstract-level binary search correctness theorems

Using the proofs in Chapter 11 of the concrete-level binary search correctness theorems as a guide, we can now prove the corresponding abstract-level correctness theorems stated on page 685. Here is an almost-complete proof of SWO.binary-search.not-found, which involves finding a contradiction in each of several different cases. To save space, we introduce bsearch as an abbreviation for binary-search.

```
extend-module SWO {
extend-module binary-search {

define tree-axioms := (datatype-axioms "BinTree")
define bsearch := binary-search

define not-found-proof :=
  method (theorem adapt)
    let {[_ _ _ chain-> _] := (proof-tools adapt theory);
         [< <E E BST bsearch] := (adapt [< <E E BST bsearch])}
      by-induction (adapt theorem) {
        null =>
          assume (BST null)
            conclude (not-found-property null)
              pick-any x
                assume ((bsearch x null) = null)
                  (!chain-> [true ==> (~ x in null)      [BST.in.empty]])
```

```
| (T as (node L y R)) =>
  let {[p1 p2] := [(not-found-property L) (not-found-property R)];
       [IH1 IH2] := [(BST L ==> p1) (BST R ==> p2)]}
    assume (BST T)
      conclude (not-found-property T)
        let {smaller-in-left := (forall x . x in L ==> x <E y);
             larger-in-right := (forall z . z in R ==> y <E z);
             _ := (!chain->
                    [(BST T)
                 ==> (BST L &
                      smaller-in-left &
                      BST R &
                      larger-in-right)            [BST.nonempty]]);
             _ := (!chain->
                    [(BST L)
                 ==> (not-found-property L)       [IH1]]);
             _ := (!chain->
                    [(BST R)
                 ==> (not-found-property R)       [IH2]])}
          pick-any x
            assume failed-search :=
                     ((bsearch x T) = null)
              (!by-contradiction (~ x in T)
                assume (x in T)
                  let {C := (!chain->
                              [(x in T)
                           ==> (x E y  |
                                x in L |
                                x in R)           [BST.in.nonempty]])}
                    (!two-cases
                      assume (x < y)
                        let {_ :=
                              (!chain->
                                [(bsearch x L)
                               = (bsearch x T)    [go-left]
                               = null             [failed-search]
                             ==> (~ x in L)       [p1]])}
                          (!cases C
                            assume (x E y)
                              (!absurd
                                (x < y)
                                (!chain->
                                  [(x E y)
                               ==> (~ x < y)      [E-definition]]))
                            assume (x in L)
                              (!absurd
                                (x in L)
                                (~ x in L))
                            assume (x in R)
```

```
                    (!absurd
                      (x < y)
                      (!chain->
                        [(x in R)
                     ==> (y <E x)          [larger-in-right]
                     ==> (~ x < y)         [<E-definition]])))
              assume (~ x < y)
                (!two-cases
                  assume (y < x)
                    let {_ :=
                          (!chain->
                           [(bsearch x R)
                          = (bsearch x T) [go-right]
                          = null          [failed-search]
                        ==> (~ x in R)    [p2]])}
                      (!force false)
                  assume (~ y < x)
                    (!absurd
                      (!chain->
                        [null
                       = (bsearch x T)    [failed-search]
                       = T                [at-root]])
                      (!chain->
                        [true
                     ==> (null =/= T)     [tree-axioms]])))))
     } # by-induction

  (add-theorems theory |{[not-found] := not-found-proof}|)

  } # close module binary-search
} # close module SWO
```

Exercise 15.2: In method `SWO.binary-search.not-found-proof`, complete the proof by replacing the use of `(!force false)`: Do another case analysis on sentence `C`, establishing a contradiction in each case. □

* ***Exercise 15.3:*** Prove `SWO.binary-search.found`. □

15.2 An abstract fast-power algorithm

15.2.1 Raising to a power in a monoid

The monoid abstract structure (Section 14.2) provides a suitable abstract basis for defining the operation of raising a value to a nonnegative power.

```
load "nat-power"
load "nat-half"
load "strong-induction"

extend-module Monoid {

  declare +*: (S) [S N] -> S [400]

  define [x y m n] := [?x:'T ?y:'T ?m:N ?n:N]

  module Power {

    define def := (close [(x +* zero = <0>)
                          (x +* (S n) = x + x +* n)])

    define [right-zero right-nonzero] := def

    (add-axioms theory def)

  }
}
```

Here, we have written the monoid operation as +, as we did in Chapter 14 in defining `Monoid` as a theory. Thus, in the definition above and in the following development of properties of `+*` and of an abstract fast-power algorithm, the notation may appear to suggest that we are computing nx rather than x^n. And indeed we are, if + is the addition operation on, say, natural numbers. Remember, though, that we can just as well work with multiplication (for any multiplication operator that obeys the monoid laws), and then we obtain the more familiar notion of power, x^n. Recall `MM` (multiplicative monoid), a shared theory we defined in Chapter 14 with:

```
module MM {

  declare <1>: (S) [] -> S

  define theory :=
    (adapt-theory 'Monoid |{Semigroup.+ := MSG.*, Monoid.<0> := <1>}|)
}
```

We can rename `+*` as `**`, obtaining a multiplicative-monoid counterpart of the `**` operation on natural numbers that we worked with in Chapter 12. We return to this point in Section 15.2.3.

From the defining axioms in `Power` we obtain the following familiar properties—all of which, except `left-neutral`, are used in the proof of correctness of the abstract fast power algorithm to follow. We give the proof of only one, `right-one`, leaving the others as exercises.

```
extend-module Monoid {
 extend-module Power {
  define [+' *'] := [N.+ N.*]

  define right-plus    := (forall m n x . x ** (m +' n) = x ** m + x ** n)
  define left-neutral  := (forall n . <0> ** n = <0>)
  define right-one     := (forall x . x ** N.one = x)
  define right-two     := (forall x . x ** N.two = x + x)
  define right-times   := (forall n m x . x ** (m *' n) = (x ** m) ** n)

  define theorems :=
    [right-plus left-neutral right-one right-two right-times]

  define proofs :=
    method (theorem adapt)
      let {[get prove chain chain-> chain<-] := (proof-tools adapt theory);
           [+ <0> **] := (adapt [+ <0> **])}
        match theorem {
          (val-of right-one) =>
            pick-any x:(sort-of <0>)
              (!chain [(x ** N.one)
                   --> (x ** (S zero))    [N.one-definition]
                   --> (x + x ** zero)    [right-nonzero]
                   --> (x + <0>)          [right-zero]
                   --> x                  [right-identity]])
        | _ => (!force (adapt theorem))
        }

  (add-theorems theory |{theorems := proofs}|)
 } # close module Power
} # close module Monoid
```

Exercise 15.4: Prove `Monoid.Power.right-plus`. □

Exercise 15.5: Prove `Monoid.Power.left-neutral`. □

Exercise 15.6: Prove `Monoid.Power.right-two`. □

⋆ ***Exercise 15.7:*** Prove `Monoid.Power.right-times`. □

The following property does not hold in all monoids, but does hold when we have commutativity of + (i.e., in an Abelian monoid).

```
extend-module Abelian-Monoid {

  define ** := Monoid.**

  define [x y n] := [?x:'T ?y:'T ?n:N]
```

```
  define Power-left-times := (forall n x y . (x + y) ** n = x ** n + y ** n)
}
```

While this property is not needed for proving correctness of the abstract fast power algorithm (and thus we do not need to situate the algorithm less abstractly on an Abelian monoid), it is certainly useful when we do have commutativity.

⋆ ***Exercise 15.8:*** Write a method for the proof of `Power-left-times` in the `Abelian-Monoid` module, and write a test of the proof. □

15.2.2 A monoid version of fast-power

We now proceed to develop a monoid version of the fast power algorithm we studied in Chapter 12. Recall that we actually presented two versions, one with embedded recursion and one that was tail-recursive, but we noted that both do unnecessary operations. If we were to promote this defect to the abstract level, we would suffer from the resulting inefficiency with every concrete instance of the algorithm that we create. Thus, although it takes considerable care to craft an algorithm that avoids the unneeded operations, the effort is well justified.

An elegant solution is given by Stepanov and McJones in their book *Elements of Programming* [93]. The following code, extracted from pages 41–42, is written in a small subset of C++.

```
template<typename I, typename Op>
    requires(Integer(I) && BinaryOperation(Op))
Domain(Op) power_accumulate_positive(Domain(Op) r,
                                     Domain(Op) x, I n,
                                     Op op)
{
    // Precondition: associative(op) ∧ positive(n)
    while (true) {
      if (odd(n)) {
          r = op(r, x);
          if (one(n)) return r;
      }
      x = op(x, x);
      n = half_nonnegative(n);
    }
}

template<typename I, typename Op>
    requires(Integer(I) && BinaryOperation(Op))
Domain(Op) power(Domain(Op) x, I n, Op op)
{
    // Precondition: associative(op) ∧ positive(n)
    while (even(n)) {
        x = op(x, x);
```

```
        n = half_nonnegative(n);
    }
    n = half_nonnegative(n);
    if (zero(n)) return x;
    return power_accumulate_positive(x, op(x, x), n, op);
}

template<typename I, typename Op>
    requires(Integer(I) && BinaryOperation(Op))
Domain(Op) power(Domain(Op) x, I n, Op op, Domain(Op) id)
{
    // Precondition: associative(op) ∧ ¬negative(n)
    if (zero(n)) return id;
    return power(x, n, op);
}
```

As far as C++ compilers are concerned, template functions such as these express only syntactic requirements. For example, the use of op in the body of the function as a binary function places a requirement on any function passed for op to be binary, but there is no compiler-checkable requirement of any semantic property such as associativity. Such requirements are only expressed informally, in comments such as the precondition comment and using requires, which is just a macro that swallows its argument. Thus, the stated semantic requirements are not actually enforced by the C++ compiler. By contrast, in reformulating these function definitions as function-defining axioms associated with a theory—monoid, in this case—we will be able to impose semantic requirements with the same force as syntactic ones.

The other transformation we perform to obtain function-defining axioms is replacing iteration by pure recursion. To do so, we will introduce recursive functions corresponding to each of the three iterative functions that together form the Stepanov-McJones power algorithm:

- pap_1 and pap_2 together implement a recursive version of the C++ iterative function power-accumulate-positive.
- fpp_1 and fpp_2, "fast-power-positive," together implement a recursive version of the C++ iterative function power (first overloading).
- fast-power corresponds to the C++ function power (second overloading).

We now define all three functions, putting the definitions together in one call of fun:

```
extend-module Monoid {

 declare pap_1, pap_2: (S) [S S N] -> S
 declare fpp_1, fpp_2: (S) [S N] -> S
 declare fast-power: (S) [S N] -> S
```

```
module fast-power {
 define [+'   *    half    even    odd    one    two] :=
        [N.+ N.* N.half N.even N.odd N.one N.two]

 define [r a x n] := [?r:'S ?a:'S ?x:'S ?n:N]

 define axioms :=
  (fun
   [(fast-power a n) =
        [<0>                              when (n = zero)       # right-zero
         (fpp_1 a n)                      when (n =/= zero)]    # right-nonzero

    (fpp_1 a n) =
        [(fpp_1 (a + a) (half n))         when (even n)         # fpp-even
         (fpp_2 a (half n))               when (~ even n)]      # fpp-odd
   (fpp_2 a n) =
        [a                                when (n = zero)       # fpp-zero
         (pap_1 a (a + a) n)              when (n =/= zero)]    # fpp-nonzero

   (pap_1 r a n) =
        [(pap_2 (r + a) a n)              when (odd n)          # pap-odd
         (pap_1 r (a + a) (half n))       when (~ odd n)]       # pap-even
   (pap_2 r a n) =
        [r                                when (n = one)        # pap-one
         (pap_1 r (a + a) (half n))       when (n =/= one)]])   # pap-not-zone

 define [right-zero right-nonzero fpp-even fpp-odd fpp-zero fpp-nonzero
         pap-odd pap-even pap-one pap-not-one] := axioms

 (add-axioms theory axioms)
```

In the preceding code, we use the technique of defining a set of mutually recursive functions to achieve the same result as an iterative procedure, each function in the set corresponding to a particular control point in the iterative code. In this case, `pap_1` corresponds to the entry into the while loop in `power_accumulate_positive`, and `pap_2` to the first statement following the `if` statement in the loop; similarly, `fpp_1` and `fpp_2` correspond to control points in the first overloading of `power` in the C++ iterative code.

Note that although there is mutual recursion in these functions, none of the function calls is embedded in other computation. That is, they are all *tail calls* (Section 12.7). This property is required for faithful correspondence to the original iterative code in `power-accumulate-positive`. To achieve it, the argument r in `pap_1` and `pap_2` is used as an accumulator.

A crucial property of the original function is indicated by Stepanov and McJones in appending "positive" to its name: It is intended to be called only with positive values of the exponent argument. And being positive and decreasing by halving, the exponent cannot decrease to 0 without going through 1. But when 1 is reached, the function terminates with the accumulated value as the result. A key idea here is that the 0 case can be taken care of by the other functions, leaving this "workhorse" function to operate more efficiently by not having to test for 0 in its loop.

We formulate correctness properties for pap_1 first with quantifiers in the best position for proof by strong induction, then with all quantifiers at the top, as is more convenient for use in justifying equational chain steps.

```
define pap_1-correctness0 :=
  (forall n . n =/= zero ==> (forall x r . (pap_1 r x n) = r + x ++* n))

define pap_1-correctness :=
  (forall n x r . n =/= zero ==> (pap_1 r x n) = r + x ++* n)
```

Like pap_1, fpp_1 has the precondition $n > 0$. This function first removes any positive power of 2, say 2^k, from n and accumulates (x ++* 2^k) in its first argument. It then halves the resulting n (which is odd) and passes on the computation to fpp_2. If the result is 0, the accumulated value is the answer, as we prove below. If the result is not 0, fpp_2 calls pap_1 to finish the job. Note that the call to pap_1 respects the latter's precondition, $n > 0$.

Function fast-power is thus the only place where there is a check for $n = 0$.

Just as we did with pap_1, we first state a correctness theorem for fpp_1 with quantifiers best positioned for strong induction. We also have a correctness theorem for fpp_2 and, finally, for fast-power.

```
define fpp_1-correctness0 :=
  (forall n .  n =/= zero ==> forall x . (fpp_1 x n) = x ++* n)

define fpp_1-correctness :=
  (forall n x . n =/= zero ==> (fpp_1 x n) = x ++* n)

define fpp_2-correctness :=
  (forall n x . n =/= zero ==> (fpp_2 x n) = x ++* (two * n +' one))

define correctness := (forall n x . (fast-power x n) = x ++* n)

define theorems := [pap_1-correctness0 pap_1-correctness fpp_1-correctness0
                    fpp_1-correctness fpp_2-correctness correctness]
```

As we did with the binary search correctness proofs, we draw the main ideas for the proofs from our correctness proof of the natural number fast-power in Chapter 12.

```
define proofs :=
 method (theorem adapt)
  let {[_ prove chain chain-> _] := (proof-tools adapt theory);
       [+ <0> ++*] := (adapt [+ <0> ++*]);
       [pap_1 pap_2 fpp_1 fpp_2 fast-power] :=
         (adapt [pap_1 pap_2 fpp_1 fpp_2 fast-power]);
```

```
      [not-odd-if-even odd-if-not-even even-def odd-def] :=
          [N.EO.not-odd-if-even N.EO.odd-if-not-even
           N.EO.even-definition N.EO.odd-definition];
      PR1 := (!prove Power.right-one);
      PR2 := (!prove Power.right-two);
      PRT := (!prove Power.right-times);
      PRP := (!prove Power.right-plus)}
 match theorem {
  (val-of pap_1-correctness0) =>
   let {theorem' := (adapt theorem)}
     (!strong-induction.principle theorem'
       method (n)
         assume IH := (strong-induction.hypothesis theorem' n)
          conclude (strong-induction.conclusion theorem' n)
           assume (n =/= zero)
            pick-any x
             (!two-cases
               assume (n = one)
                 pick-any r
                   let {p := (!chain->
                               [(odd S zero)
                           <==> (odd n)              [(n = one)
                                                      N.one-definition]

                           ==>  ((pap_1 r x n) =
                                (pap_2 (r + x)
                                       x
                                       n))           [pap-odd]])}
                     (!combine-equations
                       (!chain
                         [(pap_1 r x n)
                        = (pap_2 (r + x)
                                 x
                                 n)                  [p]
                        = (r + x)                    [pap-one (n = one)]])
                       (!chain
                         [(r + x ** n)
                        = (r + x ** one)             [(n = one)]
                        = (r + x)                    [PR1]]))
               assume (n =/= one)
                 let {p1a :=
                       (half n =/= zero ==>
                         (forall a r .
                           (pap_1 r
                                  a
                                  (half n)) =
                           r + a
                             ** (half n)));
                      p1b :=
                        (forall a r .
                          (pap_1 r
                                 a
```

```
                    (half n)) =
             r + a
               +* (half n));
         p2 := (forall r .
                 (pap_1 r
                        (x + x)
                        (half n)) =
                 (r + x
                    +* (two * half n)));
       _ := (!chain->
             [(n =/= zero)
          ==> (half n N.< n)                    [N.half.less]
          ==> p1a                               [IH]]);
       p3 := (!by-contradiction
               (half n =/= zero)
               assume (half n = zero)
                 (!cases
                   (!chain<-
                      [(n = zero |
                        n = one)
                    <== (half n = zero) [N.half.equal-zero]])
                   assume (n = zero)
                     (!absurd
                       (n = zero)
                       (n =/= zero))
                   assume (n = one)
                     (!absurd
                       (n = one)
                       (n =/= one))));
       _ := (!chain-> [p3 ==> p1b              [p1a]]);
       _ := conclude p2
              pick-any r
                (!chain
                  [(pap_1 r
                          (x + x)
                          (half n))
                 = (r + (x + x)
                      +* (half n))             [p1b]
                 = (r + (x ++ two)
                      +* (half n))             [PR2]
                 = (r + x
                      +* (two *
                              half n))         [PRT]])}
  (!two-cases
    assume (even n)
      pick-any r
        let {p4 := (!chain->
                     [(even n)
                  ==> (~ odd n)                [not-odd-if-even]])}
          (!chain
            [(pap_1 r x n)
```

```
                      = (pap_1 r
                               (x + x)
                               (half n))              [pap-even p4]
                      = (r + x
                           ++ (two * (half n)))       [p2]
                      = (r + x ++ n)                  [even-def]])
             assume (~ even n)
               pick-any r
                 let {p5 := (!chain->
                               [(~ even n)
                            ==> (odd n)               [odd-if-not-even]])}
                   (!chain
                     [(pap_1 r x n)
                     = (pap_2 (r + x)
                               x
                               n)                     [pap-odd p5]
                      = (pap_1 (r + x)
                                (x + x)
                                (half n))             [pap-not-one]
                      = ((r + x)
                          + x
                          ++ (two * half n))          [p2]
                      = (r +
                           (x + x
                              ++ (two * half n)))     [associative]
                      = (r +
                           (x ++ one
                              + x
                              ++ (two * half n)))     [PR1]
                      = (r +
                           (x ++
                             (one +'
                              two * half n)))         [PRP]
                      = (r +
                            (x ++
                              (two * (half n)
                                    +' one)))         [N.Plus.commutative]
                      = (r + x ++ n)                  [odd-def]]))))
  | (val-of pap_1-correctness) =>
      let {PC0 := (!prove pap_1-correctness0)}
        pick-any n x r
          assume (n =/= zero)
            let {p := (!chain->
                         [(n =/= zero) ==>
                            (forall x r .
                               (pap_1 r x n) =
                               r + x ++ n)      [PC0]])}
              (!chain [(pap_1 r x n)
                     = (r + x ++ n) [p]])
  | _ => (!force (adapt theorem))
```

```
  }

(add-theorems theory |{theorems := proofs}|)
} # close module fast-power
} # close module Monoid
```

Exercise 15.9: Fill in the proof of `fpp_2-correctness`. □

⋆ ***Exercise 15.10:*** Fill in the proof of `fpp_1-correctness0`, using strong induction. □

Exercise 15.11: Fill in the proof of `fpp_1-correctness` and `correctness`. □

15.2.3 Multiplicative version of fast power

We easily obtain the multiplicative version of `fast-power` and verify its correctness as follows:

```
declare **: (S) [S N] -> S [400]

define M1 := (renaming |{Monoid.+* := **}|)

assert (M1 (theory-axioms 'MM))

(!prove-property Monoid.fast-power.correctness M1 'MM)
```

Exercise 15.12: On page 595, we noted that although `fast-power-accumulate` is tail-recursive, and therefore equivalent to a loop, it was still potentially inefficient. The problem was that at the end it could do an extra multiplication whose result was not used. For example, we showed that for $n = 13$ it unnecessarily computed $x^8 \cdot x^8 = x^{16}$. Write a trace of the computation of x^{13} with `fast-power` (in its multiplicative version), verifying that it obtains the result x^{13} without computing $x^8 \cdot x^8$. The trace begins as follows:

$$
\begin{aligned}
(\text{fast-power } x \ 13) &= (\text{pap\_1 } x \ 13) \\
&= (\text{pap\_2 } x \ 6) \\
&= \ldots
\end{aligned}
$$

15.2.4 A nonnumeric application

Since list concatenation (`List.join`) has the monoid properties (with `nil` as the neutral element), raising a list L to power n can be interpreted as computing a list consisting of n copies of L. While a `join*` instance of `+*` would do the computation simply by starting with `nil` and repeatedly appending L, we can create an instance of `fast-power` that may do the job faster (with approximately $\log_2 n$ `List.join` operations rather than $n - 1$):

```
declare list-pap_1, list-pap_2: (T) [(List T) (List T) N] -> (List T)
declare list-fpp_1, list-fpp_2: (T) [(List T) N] -> (List T)

declare join*: (T) [(List T) N] -> (List T)
declare fast-join*: (T) [(List T) N] -> (List T)

define M1 :=
  (renaming |{Monoid.++ := join*, Monoid.+ := List.join, Monoid.<0> := nil,
              Monoid.fast-power := fast-join*, Monoid.pap_1 := list-pap_1,
              Monoid.pap_2 := list-pap_2, Monoid.fpp_1 := list-fpp_1,
              Monoid.fpp_2 := list-fpp_2}|)

assert (M1 (theory-axioms Monoid.theory))

(!prove-property Monoid.fast-power.correctness M1 'Monoid)
```

16 Algorithms on Memory Abstractions

IN this chapter, we turn to methods of reasoning about computer memory. Of course, the datatypes we have studied in previous chapters—such as lists and trees—use memory, but in an abstracted and limited way: Memory does not appear explicitly in the interfaces to these structures, nor do they allow updating of the values that they hold. While we could extend such datatypes to allow access to and updating of the underlying memory locations, we will concentrate here on simpler and more basic memory concepts. First, we consider accessing and updating individual memory locations. (Another treatment of these memory issues is presented in Chapter 18, there being a close correspondence between the treatment of the **While** language's semantics of assignment statements and evaluation of variables and the memory axioms we introduce here.)

We next consider accessing and updating a *memory range*, which is a sequence of memory locations. But "sequence" will be defined abstractly in a *memory range theory*, so that, for example, instances include not only a sequence of memory locations with addresses in an arithmetic progression, but also the data fields of a linked list and those of a binary tree traversal. All that is necessary is that the memory be accessible via functions that obey the axioms of the range theory. The theory is based mainly on *iterator* operations as defined in the C++ Standard Template Library (STL). An iterator in the STL is, roughly speaking, an abstraction of a limited form of pointer. Readers who have programmed with STL iterators will find at least some aspects of our treatment quite familiar. However, whereas traditional descriptions of STL iterators in textbooks and reference manuals (and even in the C++ Standard) only present an informal account of their meaning, we are able to define and analyze them here with the rigor of a formulation in logic, supported by Athena's facilities for mechanized reasoning. Among the benefits that derive from both the abstract level of the STL iterator concept itself and our formulation of it as a logical theory is assistance in implementing this concept in other languages—or, more commonly, thoroughly understanding it in C++ or in existing ports to other languages.

16.1 Axioms and theorems for individual memory locations

We axiomatize individual memory location access and updating in terms of the following polymorphic domains and functions:

```
domain (Memory S)

module Memory {
  domain (Change S)
  domain (Loc S)

  declare \: (S) [(Memory S) (Change S)] -> (Memory S)
```

```
declare \\: (S, T) [(Memory S) T] -> T
declare at: (S) [(Memory S) (Loc S)] -> S
```

A (Memory.Loc S) value represents a single location in (Memory S), a memory holding values of sort S. Using the at function, we write (M at a) to denote the value of sort S contained in location a in memory M. The domain (Change S) is introduced for use as the return sort of operations that are executed just for their effect on memory (rather than for computing an output value).[1]

We use the \ operator to provide a memory argument M to a function f in addition to other arguments $a_1 \; \ldots \; a_n$:

$$(M \setminus \; (f\, a_1 \; \ldots \; a_n)).$$

As a special case, this convention allows us to apply the assignment operator <- with just its usual two arguments, the location to be updated and the value to be stored, rather than having memory as a explicit extra argument:

$$(M \setminus a \text{ <- } x).$$

Note that (M \ a <- x) is parsed as (\ M (<- a x)).

The intended meaning of this term is the memory in which the value at location a is x and all other locations have the same value as in memory M. We make these meanings precise with the following axioms of Memory.theory:

```
define [M M' M1 M2 M3 a b c x y] :=
  [?M:(Memory 'S) ?M':(Memory 'S) ?M1:(Memory 'S)
   ?M2:(Memory 'S) ?M3:(Memory 'S)
   ?a:(Loc 'S) ?b:(Loc 'S) ?c:(Loc 'S) ?x:'S ?y:'S]

define equality :=
  (forall M1 M2 . (forall a . M1 at a = M2 at a) <==> M1 = M2)

declare <-: (S) [(Loc S) S] -> (Change S)

module assign {
  define (axioms as [equal unequal]) :=
    (fun [((M \ a <- x) at b) =
          [x                    when (a = b)
           (M at b)             when (a =/= b)]])

}

define theory := (make-theory [] [equality assign.equal assign.unequal])
```

---

1 Operations that both update memory and return a value will be modeled by two separate functions, one to express the memory updating and the other to compute the return value, as discussed more fully in Section 16.5.

The memory equality axiom says we should regard two memories as equal if and only if one holds the same values as the other at all locations.

The second assignment axiom, `Memory.assign.unequal`, is just as important for reasoning about memory as the first: It says that the memory obtained via the assignment is unchanged at all locations other than the one assigned to. This is a simple case of a kind of "memory safety property" that we will see in more complex forms later in this chapter.

The `\\` operator is used similarly to `\`, but instead of returning a memory value, it returns a value of the same sort as its second argument. We will begin making use of this operator in Section 16.3, in modeling operations that return a value in addition to possibly updating memory.

We continue developing the `Memory` module and theory by declaring and defining the following swap operation:

```
declare swap: (S) [(Loc S) (Loc S)] -> (Change S)

module swap {

  define (axioms as [equal1 equal2 unequal]) :=
    (fun [((M \ (swap a b)) at c) =
          [(M at b)                 when (a = c)
           (M at a)                 when (b = c)
           (M at c)                 when (a =/= c & b =/= c)]])

  (add-axioms theory axioms)
 }
} # close module Memory
```

Again note the `swap.unequal` axiom, which says that the memory that results from the `swap` operation is unchanged at all locations besides `a` and `b`.

In dealing with a sequence of memory updating operations, we must be careful to propagate each new memory created to subsequent operations in the sequence.[2] Consider, for example, an "open implementation" of `swap` using a temporary location, `t`, which we might write informally as

```
t <- a; a <- b; b <- t.
```

To express this accurately in our memory theory, we must write:

```
M1 = M \ t <- (M at a) &
M2 = M1 \ a <- (M1 at b) &
M3 = M2 \ b <- (M2 at t)
```

2 Such propagation can be systematized, as in the theoretical *monad* notion (see Wadler [108]), which has been introduced into functional languages such as Haskell, thus enabling simpler reasoning about memory updating and real-world interaction.

Note that the right argument of an assignment requires obtaining the *value* at the location in the propagated memory. What can we say about memory M3? Actually, not as much as we would like. We cannot show that the first two swap axioms are satisfied without the assumption that t is not the same location as a or b. But even with that assumption, we haven't exactly implemented swap, because the third swap axiom fails. The first of the following two theorems is the best we can say about this (not-quite-correct) implementation of swap:

```
extend-module Memory {

 define t := ?t:(Loc 'S)

 define swap-open-implementation :=
   (forall M a b t M1 M2 M3 .
     a =/= t & b =/= t &
     M1 = M \ t <- (M at a) &
     M2 = M1 \ a <- (M1 at b) &
     M3 = M2 \ b <- (M2 at t)
     ==> M3 = (M \ t <- (M at a)) \ (swap a b))

 define swap-implementation :=
   (forall M a b x M1 M2 .
     x = (M at a) &
     M1 = M \ a <- (M at b) &
     M2 = M1 \ b <- x
     ==> M2 = M \ (swap a b))
```

The second theorem shows how we can obtain an implementation that satisfies all three swap axioms: Treat the temporary not as a location in memory but as a separate entity that "doesn't count." That is, of course, essentially what we do when we do the swap in a procedure with a and b as formal parameters and the temporary t as a local variable, as in the following C++ code:

```
template <typename S>
void swap(S& a, S& b) {S t = a; a = b; b = t;}
```

In such cases, the memory occupied by t cannot possibly be in the same location as a or b, and it disappears upon procedure exit (assuming the usual treatment of local variables as pushed onto a stack and popped off when the procedure exits), so that we don't need to consider it a part of memory as seen from outside the procedure. In the logic, instead of treating the temporary as a variable of sort (Loc S) that participates in the memory updates, we simply use a logical variable x of sort S.

We provide a proof of the second theorem and leave the first as an exercise.

```
define proofs :=
  method (theorem adapt)
    let {[_ _ chain _ _] := (proof-tools adapt theory);
         [at \ swap]     := (adapt [at \ swap]);
         [eq uneq]       := [assign.equal assign.unequal]}
      match theorem {
        (val-of swap-open-implementation) => (!force (adapt theorem))
      | (val-of swap-implementation) =>
          pick-any M:(Memory 'S)
                   a:(Memory.Loc 'S)
                   b:(Memory.Loc 'S)
                   x:'S
                   M1:(Memory 'S)
                   M2:(Memory 'S)
            assume h1 := (x = M at a);
                   h2 := (M1 = M \ a <- (M at b));
                   h3 := (M2 = M1 \ b <- x)
              conclude (M2 = M \ (swap a b))
                let {p1 := (!chain
                             [(M2 at a)
                             = ((M1 \ b <- x) at a)         [h3]
                             = ((M1 \ b <- (M at a)) at a)  [h1]]);
                     p2 := conclude (M2 at a = M at b)
                             (!two-cases
                               assume (b = a)
                                 (!chain
                                   [(M2 at a)
                                  = ((M1 \ b <- (M at a))
                                      at a)                 [p1]
                                  = (M at a)                [eq]
                                  = (M at b)                [(b = a)]])
                               assume (b =/= a)
                                 (!chain
                                   [(M2 at a)
                                  = ((M1 \ b <- (M at a))
                                      at a)                 [p1]
                                  = (M1 at a)               [uneq]
                                  = ((M \ a <- (M at b))
                                     at a)                  [h2]
                                  = (M at b)                [eq]]));
                     p3 :=
                       pick-any u
                         conclude (M2 at u =
                                     (M \ (swap a b)) at u)
                           (!three-cases
                             assume (a = u)
                               (!combine-equations
                                 (!chain
                                   [(M2 at u)
                                  = (M2 at a)               [(a = u)]
                                  = (M at b)                [p2]])
```

```
                (!chain
                  [((M \ (swap a b))
                    at u)
                  = ((M \ (swap a b))
                     at a)                          [(a = u)]
                  = (M at b)                        [swap.equal1]]))
            assume (b = u)
              (!combine-equations
                (!chain
                  [(M2 at u)
                  = (M2 at b)                       [(b = u)]
                  = ((M1 \ b <- x) at b)            [h3]
                  = x                               [eq]
                  = (M at a)                        [h1]])
                (!chain
                  [((M \ (swap a b))
                    at u)
                  = ((M \ (swap a b))
                    at b)                           [(b = u)]
                  = (M at a)                        [swap.equal2]]))
            assume (a =/= u & b =/= u)
              (!combine-equations
                (!chain
                  [(M2 at u)
                  = ((M1 \ b <- x) at u)            [h3]
                  = (M1 at u)                       [uneq]
                  = ((M \ a <- (M at b))
                    at u)                           [h2]
                  = (M at u)                        [uneq]])
                (!chain
                  [((M \ (swap a b))
                    at u)
                  = (M at u)                        [swap.unequal]]))))}
        (!chain [M2 = (M \ (swap a b))              [equality]])
    } # match theorem {

(add-theorems theory |{[swap-open-implementation
                        swap-implementation] := proofs}|)
```

* ***Exercise 16.1:*** Fill in the proof of `Memory.swap-open-implementation`. □

Before continuing, we close the `Memory` module.

```
} # close module Memory
```

16.2 Iterators and ranges

To define a memory range—a sequence of memory locations—we first define an *iterator range*. One kind of iterator is an ordinary pointer, as defined in C/C++ and other languages that allow such access to memory. There are other kinds of iterators, but they are required to behave like pointers in the sense that operations like incrementing and dereferencing are defined on them and behave in similar ways to the corresponding operations on pointers. In other words, iterators are an abstraction of pointers, defined by axioms, and this abstraction has many instances besides pointers. By defining algorithms on an iterator range, defined by a pair of iterators, rather than just on a memory range defined by a pair of pointers, we obtain abstract algorithms that have many useful instances.

Consider, for example, an algorithm for counting the number of occurrences of a given value in a range. Such an algorithm could be written in C++ specifically for a memory range of locations holding `double` values where the range is defined by pointers:

```
int count(double* first, double* last, const double& x) {
  int n = 0;
  while (first != last) {
    if (*first == x) ++n;
    ++first;
  }
  return n;
}
```

This function computes the number of occurrences of `x` in the range `*first`, `*(first+1)`, `...`, `*(last-1)`. Note that the location to which `last` points is not included in the range; if `last == first`, the range is considered empty, the while loop does nothing, and `count` returns `0`. If the range is nonempty, the first iteration of the while loop reduces the computation to doing the same thing on a range from `first + 1` to `last`. The length of this range is one less than that of the original range, which suggests that we could reason about the behavior of the algorithm by induction on the length of the range. Alternatively, in Athena, by defining a range as a datatype, we can do induction directly with **`by-induction`** on ranges rather than on the natural number lengths. This is the approach we take below.

The other issue at hand is abstraction away from the specific details of the `double` pointer type, `double*`. The most obvious generalization is to replace `double` by a template type parameter, `T`:

```
template <typename T>
int count(T* first, T* last, const T& x) {
  int n = 0;
  while (first != last) {
    if (*first == x) ++n;
    ++first;
```

```
  }
  return n;
}
```

We have thus replaced the `double*` pointers with `T*` pointers, where `T` can be replaced by any type, but they are still pointers. What is done instead in the STL is generalization of the pointers to iterators:[3]

```
template <typename I> // where I is a Forward Iterator with value type T
int count(I first, I last, const T& x) {
  int n = 0;
  while (first != last) {
    if (*first == x) ++n;
    ++first;
  }
  return n;
}
```

Exactly what is meant by a *Forward Iterator* with *value type* `T` is not defined in the code; C++ is incapable of expressing such requirements except in comments. Both implementers and users of the STL must rely on informal and sometimes ambiguous requirements specifications in manuals, textbooks, and even in the C++ Standard. Although the code for `count` is short and seemingly simple, reasoning about it in its full generality raises nontrivial issues, since its use is not limited to cases where `first` and `last` point to positions in an array. By instantiating with appropriate definitions of `++` and `*`, such algorithms can work on sequences of linked list elements: define `++` to follow a node's "next" link and `*` to return a node's data field. Or they could work on binary trees by defining `++` to advance to the next node in, say, the preorder sequence. Fully understanding iterators and algorithms just based on informal specifications can be difficult, due to the great generality achieved by such abstractions.

By contrast, we are able to define and use such notions rigorously, by axiomatizing iterators and iterator ranges. In particular, we use such a treatment in logic as a basis for proving correctness of abstract algorithms on sequences in the STL.

All of the abstract sequence algorithms we consider take as one their inputs an iterator range. In the STL an iterator range is syntactically just a pair of iterators, as in the `count` algorithm. There is, however, the further semantic requirement that the second iterator of the pair be *reachable* from the first; that is, in the pair `(first, last)`, it must be the case that `last = first +` n where n is a natural number and `first +` n means n-fold application of `++` to `first`. Otherwise, the behavior of the algorithm is undefined. In the development below, we build this requirement into the definition of an iterator range as an inductive

3 In the STL, there is a further generalization, which we do not consider here, to ensure that the type of the counter variable `n` is wide enough to hold the maximum count that could result from the computation, by using an appropriate "difference type" instead of `int`.

datatype of values from which the pair of iterators can be extracted, an approach that also allows inductions to be done in Athena directly with **by-induction**.[4]

16.2.1 Iterator and range axioms and theorems

We begin by defining a polymorphic domain (It X S), in terms of which we then define the polymorphic datatype (Range X S):

```
domain (It X S)

datatype (Range X S) := (stop (It X S)) | (back (Range X S))
assert Range-axioms := (datatype-axioms "Range")
```

The sort parameter S is the "value type" of the iterator or range, and the sort parameter X is included just to allow for different iterators or ranges with the same value type.[5] The base (nonrecursive) Range constructor stop constructs an empty range beginning and ending at the given iterator, while back constructs a range that begins "one step back" from where the argument range begins. (This notion is made precise in the axioms below.) The beginning and ending iterators of the range are then extracted with functions start and finish:

```
module Range {

  define theory := (make-theory [] [])

  define [h i i' j j' k k' r r' r1 r2 r3] :=
      [?h:(It 'X1 'S1) ?i:(It 'X2 'S2) ?i':(It 'X3 'S3)
       ?j:(It 'X4 'S4) ?j':(It 'X5 'S5) ?k:(It 'X6 'S6) ?k:(It 'X7 'S7)
       ?r:(Range 'X8 'S8) ?r':(Range 'X9 'S9) ?r'':(Range 'X10 'S10)
       ?r1:(Range 'X11 'S11) ?r2:(Range 'X12 'S12) ?r3:(Range 'X13 'S13)]

  declare start: (X, S) [(Range X S)] -> (It X S)

  module start {

    define of-stop := (forall i . start stop i = i)
    define injective := (forall r r' . start r = start r' ==> r = r')

    (add-axioms theory [of-stop injective])
  }

  declare finish: (X, S) [(Range X S)] -> (It X S)
```

4 An alternative approach is to define a predicate *valid-range* on pairs of iterators to mean that the second is reachable from the first; to assume *valid-range* as a precondition in sentences about algorithm correctness; and to do induction with a range-induction method that transforms sentences about the range into ones about the length of the range; and, finally, uses **by-induction** to prove them. For details, see [106].

5 Think of vector<int>::iterator and list<int>::iterator. These are two different iterator types that have the same value type. We can model this distinction simply by using any two distinct sorts, arbitrarily introduced, say with domains X1, X2, to obtain two different iterator sorts (It X1 Int) and (It X2 Int).

```
module finish {

  define of-stop := (forall i . finish stop i = i)
  define of-back := (forall r . finish back r = finish r)

  (add-axioms theory [of-stop of-back])
}
```

The first theorems we wish to prove both state, in different ways, that the back constructor "makes a difference." We give the proof of the first, which is by contradiction:

```
define nonempty-back := (forall r . start back r =/= finish back r)
define back-not-same := (forall r . back r =/= r)

define proofs :=
 method (theorem adapt)
   let {[_ prove chain chain-> _] := (proof-tools adapt theory)}
     match theorem {
       (val-of nonempty-back) =>
         pick-any r
           (!by-contradiction (start back r =/= finish back r)
             assume hyp := (start back r = finish back r)
               (!absurd
                 (!chain-> [(start back r)
                           = (finish back r)           [hyp]
                           = (finish r)                [finish.of-back]
                           = (start stop finish r)     [start.of-stop]
                         ==> (back r = stop finish r)  [start.injective]])
                 (!chain-> [true
                         ==> (stop finish r =/= back r)  [Range-axioms]
                         ==> (back r =/= stop finish r)  [sym]])))
     | (val-of back-not-same) => (!force adapt theorem)
     }

(add-theorems theory |{[nonempty-back back-not-same] := proofs}|)
```

Exercise 16.2: Fill in the proof of Range.back-not-same. *Hint:* Use induction. □

The algorithms on ranges that we will consider are based on ones in the STL that are defined directly in terms of pairs of iterators, rather than the Range datatype. To express our versions more closely to their STL counterparts, we need a way of combining two iterators to form a range. Since not every pair of iterators forms a range, there is the possibility of error, which we handle via *option values* (Section 10.2). We thus define the following function in terms of the Option datatype:

```
declare range: (X, S) [(It X S) (It X S)] -> (Option (Range X S))

module range {

  define collapse := (forall r . (range (start r) (finish r)) = SOME r)
  define injective :=
    (forall i j i' j' . (range i j) = (range i' j') ==> i = i' & j = j')
  define start-back :=
    (forall i j r . (range i j) = SOME back r ==> i = start back r)

  (add-axioms theory [collapse injective start-back])
}
```

Whenever we use range on iterators i and j, we will make the assumption that (exists r . (range i j) = SOME r).

Two other useful functions on ranges are empty and length. We declare and define these functions as follows:

```
declare empty: (X, S) [(Range X S)] -> Boolean

module empty {

  define of-stop := (forall i . empty stop i)
  define of-back := (forall r . ~ empty back r)

  (add-axioms theory [of-stop of-back])
}

declare length: (X, S) [(Range X S)] -> N

module length {

  define of-stop :=  (forall j . length stop j = zero)
  define of-back :=  (forall r . length back r = S length r)

  (add-axioms theory [of-stop of-back])
}
```

Another characterization of an empty range is in terms of range:

```
define empty-range := (forall i . (range i i) = SOME stop i)

define proof :=
  method (theorem adapt)
    let {[get prove chain chain-> chain<-] := (proof-tools adapt theory)}
      match theorem {
        (val-of empty-range) =>
          pick-any i
            (!chain
              [(range i i)
```

```
          = (range (start stop i)
                   (finish stop i))  [start.of-stop finish.of-stop]
        = (SOME stop i)              [range.collapse]])
  }

(add-theorems theory |{empty-range := proof}|)
```

We also have the converse:

```
define empty-range1 :=
  (forall h i j . (range i j) = SOME stop h ==> i = j)

define proof :=
  method (theorem adapt)
    let {[get prove chain chain-> chain<-] := (proof-tools adapt theory)}
    match theorem {
    | (val-of empty-range1) =>
       pick-any h:(It 'X 'S) i:(It 'X 'S) j:(It 'X 'S)
         assume hyp := ((range i j) = SOME stop h)
           conclude (i = j)
             let {EL := (!prove empty-range);
                  _ := (!chain->
                        [(range i j)
                       = (SOME stop h)          [hyp]
                       = (range h h)            [EL]
                     ==> (i = h & j = h)        [range.injective]])}
               (!chain [i = h                   [(i = h)]
                          = j                   [(j = h)]])
    }

(add-theorems theory |{empty-range1 := proof}|)
```

Exercise 16.3: Prove the following:

```
define nonempty-back1 :=
  (forall i j r . (range i j) = SOME back r ==> i =/= j)
```

using nonempty-back as a lemma. □

Exercise 16.4: The following theorems provide additional relations between the `stop` and `back` constructors and the `length` function:

```
define zero-length :=
  (forall r . length r = zero ==> exists i . r = stop i)
define nonzero-length :=
  (forall r . length r =/= zero ==> exists r' . r = back r')
```

Prove them. □

Another useful function on ranges is the membership function Range.in:

```
declare in: (X, S) [(It X S) (Range X S)] -> Boolean

module in {

  define of-stop := (forall i j . ~ i in stop j)
  define of-back :=
    (forall i r . i in back r <==> i = start back r | i in r)

  (add-axioms theory [of-stop of-back])
```

We conclude this subsection with proofs of two simple theorems about range membership (the second being the contrapositive of the first):

```
  define range-expand := (forall i r . i in r ==> i in back r)

  define range-reduce := (forall i r . ~ i in back r ==> ~ i in r)

  define proofs :=
    method (theorem adapt)
      let {[_ prove chain _ _] := (proof-tools adapt theory)}
        match theorem {
          (val-of range-expand) =>
            pick-any i:(It 'X 'S) r:(Range 'X 'S)
              (!chain
                [(i in r)
               ==> (i = start back r | i in r)    [alternate]
               ==> (i in back r)                  [of-back]])
        | (val-of range-reduce) =>
            pick-any i r
              let {RE := (!prove range-expand)}
                (!chain [(~ i in back r)
                     ==> (~ i in r)               [RE]])
        }

  (add-theorems theory |{[range-expand range-reduce] := proofs}|)
 } # close module in
} # close module Range
```

16.2.2 Trivial iterator: The base of a hierarchy of iterator theories

In the STL, a hierarchy of iterator requirements is defined: *trivial*, *forward*, *bidirectional*, and *random-access* iterators, each including the requirements of the previous. As in the algebraic structures case, such a hierarchy allows properties—including correctness properties of algorithms—to be stated and proved at the most general level possible, and therefore to be most broadly applied. Here we present these requirements in a series of theory refinements.

The only operation assumed on trivial iterators is *dereferencing*. Dereferencing an iterator produces a memory location:[6]

```
module Trivial-Iterator {
  open Range, Memory

  define theory := (make-theory ['Range 'Memory] [])

  define v := ?v

  declare deref: (X, S) [(It X S)] -> (Memory.Loc S)

  module deref {

    define injective := (forall i j . deref i = deref j ==> i = j)

    (add-axioms theory [injective])

  }
```

Using `deref`, it is useful to introduce a variant of `Range.in`. Instead of asking whether a given iterator i is in the set of iterators in a given range r,

$$(i \; \texttt{*in} \; r)$$

asks whether the memory location to which i points is in the set of memory locations to which the iterators in r point.

```
  declare *in: (X, Y, S) [(It X S) (Range Y S)] -> Boolean

  module *in {

    define of-stop := (forall i k . ~ i *in stop k)
    define of-back :=
      (forall i r . i *in back r
                  <==> deref i = deref start back r | i *in r)

    define first-not-in-rest := (forall r . ~ start back r *in r)

    (add-axioms theory [of-stop of-back first-not-in-rest])
}
```

Exercise 16.5: Prove the `*in` counterparts of the two `in` theorems proved at the end the previous subsection, namely

---

6 In C++, the dereference operator is written `*`. Note that there may be a further implicit operation of retrieving the contents of the location. For example, corresponding to the C++ assignment `*i = *j`, we would write `(M \ (deref i) <- (M at (deref j))`.

```
define range-expand := (forall i r . i *in r ==> i *in back r)
define range-reduce := (forall i r . ~ i *in back r ==> ~ i *in r)
```

Put the theorems and proofs within module *in. □

The following collect function will be useful as a specification tool; that is, we will use it in specifying correctness properties of algorithms. This function gathers up the values in a range and returns them in a list.

```
  declare collect: (S, X) [(Memory S) (Range X S)] -> (List S)
  define ++ := List.join

  module collect {

   define (axioms as [of-stop of-back]) :=
     (fun
       [(collect M stop _) = nil
        (collect M back r) = ((M at deref start back r) :: (collect M r))])

   (add-axioms theory axioms)
```

Since collect is intended just as a specification tool rather than as a practical programming tool, we have included a memory value as one of its arguments, and we have not bothered to make it tail-recursive.

The following theorem confirms that memory updates outside of the given range do not affect the list of values collected from the range.

```
define (unchanged-prop r) :=
  (forall M i v .
    ~ i *in r ==> (collect (M \ (deref i) <- v) r) = (collect M r))

define unchanged := (forall r . unchanged-prop r)

define proof :=
  method (theorem adapt)
    let {[_ prove chain chain-> _] := (proof-tools adapt theory);
         [deref *in] := (adapt [deref *in])}
      match theorem {
        (val-of unchanged) =>
          by-induction (adapt theorem) {
            (stop h:(It 'Y 'S)) =>
              pick-any M:(Memory 'S)
                       i:(It 'X 'S)
                       v:'S
                assume (~ i *in stop h)
                  let {M1 := (M \ (deref i) <- v)}
                    (!chain [(collect M1 (stop h))
                             = nil:(List 'S)              [of-stop]
                             = (collect M (stop h))       [of-stop]])
```

```
        | (r as (back r':(Range 'Y 'S))) =>
            pick-any M:(Memory 'S)
                     i:(It 'X 'S)
                     v:'S
              let {IH := (unchanged-prop r');
                   k' := (start r);
                   M1 := (M \ (deref i) <- v)}
                assume hyp := (~ i *in r)
                  let {p1 := (!chain->
                                [hyp
                           ==> (~ (deref i = deref k' |
                                   i *in r'))              [*in.of-back]
                           ==> (deref i =/= deref k' &
                                 ~ i *in r')               [dm]
                           ==> (deref i =/= deref k')      [left-and]]);
                       rr := (!prove *in.range-reduce);
                       p2 := (!chain->
                                [hyp
                            ==> (~ i *in r')               [rr]
                            ==> ((collect M1 r') =
                                  (collect M r'))          [IH]])}
                    (!chain
                      [(collect M1 r)
                     = ((M1 at deref k') ::
                         (collect M1 r'))                  [of-back]
                     = ((M at deref k')  ::
                         (collect M1 r'))                  [p1 assign.unequal]
                     = ((M at deref k')  ::
                         (collect M r'))                   [p2]
                     = (collect M r)                       [of-back]])
      }
    }

    (add-theorems theory |{unchanged := proof}|)
  } # close module collect
} # close module Trivial-Iterator
```

16.2.3 Forward iterators

In the `collect` specification tool there is an implicit forward movement through the range. For writing functions to be used in practical programming, it is useful to allow more direct expression of this forward movement with an additional iterator operation, `successor`, for advancing to the next position in a range:

```
module Forward-Iterator {
  open Trivial-Iterator
```

```
define theory := (make-theory ['Trivial-Iterator] [])

declare successor: (X, S) [(It X S)] -> (It X S)

module successor {

  define of-start := (forall r . successor start back r = start r)

  define injective := (forall i j . successor i = successor j ==> i = j)

  define deref-of :=
    (forall i r . deref successor i = deref start r
                  ==> deref i = deref start back r)

  (add-axioms theory [of-start injective deref-of])
}
```

Here are a couple of theorems relating successor to other operations:

```
define start-shift :=
  (forall i r . successor i = start r ==> i = start back r)

define range-back :=
  (forall i j r . (range (successor i) j) = SOME r
                <==> (range i j) = SOME back r)
```

We give the proof of the first and leave the second as an exercise.

```
define proofs :=
  method (theorem adapt)
    let {[_ prove chain chain-> _] := (proof-tools adapt theory);
         [deref *in successor] := (adapt [deref *in successor])}
      match theorem {
        (val-of start-shift) =>
          pick-any i:(It 'X 'S) r:(Range 'X 'S)
            assume I := (successor i = start r)
              (!chain->
                [(successor i)
                 = (start r)                           [I]
                 = (successor start back r)            [successor.of-start]
               ==> (i = start back r)                  [successor.injective]])
      | (val-of range-back) =>
          (!force (adapt theorem))
      }

(add-theorems theory |{[start-shift range-back] := proofs}|)
```

Exercise 16.6: Fill in the proof of `Forward-Iterator.range-back`. □

The following theorem does not mention successor in its statement, but successor can be used in the proof, which is by induction.

```
define (finish-not-*in-prop r) :=
  (forall i j k . (range i j) = SOME r & k *in r ==> k =/= j)

define finish-not-*in := (forall r . finish-not-*in-prop r)

define proof :=
  method (theorem adapt)
    let {[_ prove chain chain-> chain<-] := (proof-tools adapt theory);
         [deref *in successor] := (adapt [deref *in successor])}
      match theorem {
        (val-of finish-not-*in) =>
          by-induction (adapt theorem) {
            (stop h) =>
              pick-any i j k
                assume ((range i j) = SOME stop h & k *in stop h)
                  (!from-complements (k =/= j)
                    (k *in stop h)
                    (!chain-> [true ==> (~ k *in stop h) [*in.of-stop]]))
          | (r as (back r':(Range 'X 'S))) =>
              (!force (finish-not-*in-prop r))
          }
      }

(add-theorems theory |{finish-not-*in := proof}|)
```

⋆ ***Exercise 16.7:*** Fill in the proof of the inductive step in the above proof. □

Exercise 16.8: In Forward-Iterator theory, we can say more about the Range.in and Range.*in functions. First, for Range.in, we have:

```
define range-shift1 :=
  (forall r i . (successor i) in r ==> i in back r)

define range-shift2 :=
  (forall i r . ~ i in back r ==> ~ (successor i) in r)
```

Define a method that proves these theorems and evolve Forward-Iterator.theory to include the theorems and proofs. Note that range-shift2 is just the contrapositive of range-shift1. □

Exercise 16.9: State the corresponding range-shift theorems *in.range-shift1 and *in.range-shift2 with *in in place of in, define a method that proves the theorems, and evolve Forward-Iterator.theory to include the theorems and proofs. □

```
} # close module Forward-Iterator
```

16.3 Range count algorithm

The first algorithm on iterator ranges we consider is `Forward-Iterator.count`. Analogously to `List.count` (see Exercise 8.29 on page 467), this function computes the number of occurrences of a given value in a given range. Note that this function only needs to read values in the memory range; it does not need to update them. For efficiency, we define it in terms of a tail-recursive function `count1` that keeps track of the count in an extra accumulator argument, `A`. In order to associate a memory value with `count` and `count1` terms, we use the `Memory.\\` operator defined in Section 16.1. This is similar to our use of `Memory.\`, but instead of returning a memory value, `Memory.\\` returns a value of the same sort as its second argument.

```
extend-module Forward-Iterator {

  define collect := Trivial-Iterator.collect

  declare count1: (S, X) [S (It X S) (It X S) N] -> N

  declare count: (S, X) [S (It X S) (It X S)] -> N

  module count {

    define A := ?A:N

    define (axioms as [if-empty if-equal if-unequal definition]) :=
     (fun
      [(M \\ (count1 x i j A)) =
       [A                                        when (i = j)
        (M \\ (count1 x (successor i) j (S A))) when (i =/= j &
                                                       M at deref i = x)
        (M \\ (count1 x (successor i) j A))     when (i =/= j &
                                                       M at deref i =/= x)]

       (M \\ (count x i j)) = (M \\ (count1 x i j zero))])

    (add-axioms theory axioms)
```

We express the correctness of these functions in terms of the `collect` function, introduced on page 717 for just such a purpose, together with the `List.count` function.

```
define count' := List.count
overload + N.+

define (correctness1-prop r) :=
```

```
      (forall M x i j A .
        (range i j) = SOME r ==>
        M \\ (count1 x i j A) = (count' x (collect M r)) + A)

define correctness1 := (forall r . correctness1-prop r)

define correctness :=
  (forall r M x i j .
    (range i j) = SOME r ==>
    M \\ (count x i j) = (count' x (collect M r)))
```

Here are partial proofs, to be completed in the exercises that follow.

```
define proofs :=
  method (theorem adapt)
   let {[_ prove chain chain-> _] := (proof-tools adapt theory);
        [deref successor] := (adapt [deref successor])}
    match theorem {
      (val-of correctness1) =>
        by-induction (adapt theorem) {
          (stop h:(It 'X 'S)) =>
            pick-any M:(Memory 'S) x:'S i:(It 'X 'S) j:(It 'X 'S) A:N
              assume hyp := ((range i j) = (SOME stop h))
                let {er1 := (!prove empty-range1);
                     _ := (!chain->
                            [hyp ==> (i = j)         [er1]])}
                 (!force ((M \\ (count1 x i j A)) =
                            (count' x'
                                    (collect
                                      M
                                      (stop h)))
                          + A))
      | (r as (back r':(Range 'X 'S))) =>
         let {ind-hyp := (correctness1-prop r')}
           pick-any M:(Memory 'S)
                    x:'S
                    i:(It 'X 'S)
                    j:(It 'X 'S)
                    A:N
             assume hyp := ((range i j) = SOME r)
               let {goal := (M \\ (count1 x i j A) =
                              (count' x
                                      (collect M r))
                              + A);
                    nb1 := (!prove nonempty-back1);
                    lb := (!prove range-back);
                    ii := (!chain->
                            [hyp ==> (i =/= j)      [nb1]]);
                    iii := (!chain->
                           [hyp
```

```
                       ==> ((range
                              (successor i) j) =
                            SOME r')                  [lb]]);
                 iv := conclude (i = start r)
                         (!chain->
                           [(range i j)
                          = (SOME r)                  [hyp]
                          = (range (start r)
                                    (finish r))       [range.collapse]
                        ==> (i = start r &
                              j = finish r)           [range.injective]
                        ==> (i = start r)             [left-and]])}
          (!two-cases
            assume case1 := (M at deref i = x)
              (!force goal)
            assume case2 := (M at deref i =/= x)
              conclude goal
                let {_ := (!sym case2)}
                  (!combine-equations
                    (!chain
                      [(M \\ (count1 x i j A))
                       = (M \\ (count1
                                  x
                                  (successor i)
                                  j
                                  A))                 [if-unequal]
                       = ((count'
                            x
                            (collect M r')) + A)    [iii ind-hyp]])
                    (!chain
                      [((count' x
                                 (collect M r))
                         + A)
                       = ((count'
                            x
                            (M at deref i) ::
                            (collect M r')) + A)
                                                      [iv collect.of-back]
                       = ((count'
                            x
                            (collect M r')) + A)
                                                      [case2
                                                       List.count.same]])))
      } # by-induction
    | (val-of correctness) => (!force (adapt theorem))
    } # match theorem

  (add-theorems theory |{[correctness1 correctness] := proofs}|)
 } # close module count
} # close module Forward-Iterator
```

Exercise 16.10: In the proof of `Forward-Iterator.count.correctness1`, fill in the proof of the basis case. □

Exercise 16.11: In the proof of `Forward-Iterator.count.correctness1`, fill in the proof of `goal` in `case1` of the inductive step. □

Exercise 16.12: Fill in the proof of `Forward-Iterator.count.correctness`. □

16.4 Range replace algorithm

We next consider an algorithm on iterator ranges that updates the memory locations in the range. This `Forward-Iterator.replace` function is analogous to the `List.replace` function introduced in Exercise 11.9. The memory returned is a copy of the initial memory except that, within the given range, all occurrences of a given value `x` are replaced by another given value `y`.

```
extend-module Forward-Iterator {

  declare replace: (S, X) [(It X S) (It X S) S S] -> (Memory.Change S)

  module replace {

  define (axioms as [if-empty if-equal if-unequal]) :=
    (fun
     [(M \ (replace i j x y)) =
         [M                         when (i = j)
          ((M \ (deref i) <- y) \ (replace (successor i) j x y))
                                    when (i =/= j & M at deref i = x)
          (M \ (replace (successor i) j x y))
                                    when (i =/= j & M at deref i =/= x)]])

  (add-axioms theory axioms)
```

As with `Forward-Iterator.count`, we specify correctness in terms of the corresponding list function on the list obtained from the range with `collect`:

```
define replace' := List.replace
define q := ?q:(It 'Z 'S)

define (correctness-prop r) :=
  (forall M M' i j x y .
    (range i j) = SOME r &
    M' = (M \ (replace i j x y))
    ==> (collect M' r) = (replace' (collect M r) x y) &
        forall q . ~ q *in r ==> M' at deref q = M at deref q)
```

```
define correctness := (forall r . correctness-prop r)
```

Note that this states not only what happens to memory in the range; we also require that memory outside the range is unchanged. In fact, including this memory safety property as part of correctness is essential to being able to carry out the induction argument; without it, we would not be assured that the recursive call on the shorter range,

`(M \ (replace (successor i) j x y)),`

could not change the contents of the location `(deref i)` already processed.

Again, we give a proof outline with some parts left as exercises.

```
define proof :=
  method (theorem adapt)
    let {[_ prove chain chain-> _] := (proof-tools adapt theory);
         [deref *in successor] := (adapt [deref *in successor]);
          replace-eq := List.replace.equal}
      match theorem {
       (val-of correctness) =>
         by-induction (adapt theorem) {
          (stop h:(It 'X 'S)) =>
            pick-any M: (Memory 'S)
                     M':(Memory 'S)
                     i: (It 'X 'S)
                     j: (It 'X 'S)
                     x:'S
                     y:'S
              assume hyp1 := ((range i j) = SOME stop h);
                     hyp2 := (M' = M \ (replace i j x y))
                let {el1 := (!prove empty-range1);
                     _ := (!chain-> [hyp1 ==> (i = j)       [el1]]);
                     _ := conclude (M' = M)
                            (!chain
                              [M'
                             = (M \ (replace i j x y))       [hyp2]
                             = M                             [(i = j)
                                                              if-empty]]);
                     p1 := (!force
                             ((collect M' stop h) =
                              (replace'
                                (collect M stop h) x y)));
                     p2 := conclude
                             (forall k .
                                ~ k *in stop h ==>
                                  M' at deref k =
                                  M at deref k)
                             pick-any k:(It 'Z 'S)
                               assume (~ k *in stop h)
                                 (!chain
```

```
                                [(M' at deref k)
                               = (M at deref k)            [(M' = M)]])}
            (!both p1 p2)
      | (r as (back r':(Range 'X 'S))) =>
          let {ind-hyp := (correctness-prop r')}
            pick-any M:(Memory 'S)
                     M':(Memory 'S)
                     i:(It 'X 'S)
                     j:(It 'X 'S)
                     x:'S
                     y:'S
              assume hyp1 := ((range i j) = SOME r);
                     hyp2 := (M' = M \ (replace i j x y))
                let {p1 := ((collect M' r) =
                           (replace'
                              (collect M r) x y));
                     p2 := (forall h .
                              ~ h *in r ==>
                                 M' at deref h =
                                 M at deref h);
                     nb1 := (!prove nonempty-back1);
                     i=/=j := (!chain->
                                [hyp1
                             ==> (i =/= j)                [nb1]]);
                     lb := (!prove range-back);
                     p3 := (!chain->
                              [hyp1
                           ==> ((range (successor i) j)
                                 = SOME r')               [lb]]);

                     p4 := conclude (i = start r)
                             (!chain->
                              [(range i j)
                             = (SOME r)                   [hyp1 hyp2]
                             = (range (start r)
                                      (finish r))         [range.collapse]
                           ==> (i = start r &
                                 j = finish r)            [range.injective]
                           ==> (i = start r)              [left-and]]);
                     fnir := (!prove
                               *in.first-not-in-rest);
                     rr := (!prove *in.range-reduce)}
             conclude (p1 & p2)
               (!two-cases
                 assume (M at deref i = x)
                   let {M1 := (M \ (deref i) <- y);
                        q1 :=
                         (!chain
                           [M'
                          = (M \
                              (replace i j x y))      [hyp2]
```

```
                        = (M1 \
                           (replace
                            (successor i) j x y))  [if-equal]]);
                      (q2 as (and q2a q2b)) :=
                        (!chain->
                         [q1
                     ==> (p3 & q1) [augment]
                     ==> ((collect M' r') =
                          (replace'
                            (collect M1 r') x y) &
                          forall h .
                            ~ h *in r' ==>
                              M' at deref h =
                              M1 at deref h)       [ind-hyp]]);
                      q3 := (!chain->
                             [true
                          ==> (~ start r *in r')   [fnir]
                          ==> (~ i *in r')         [p4]]);
                      _ := (!sym
                             (M at deref i = x));
                      cu := (!prove
                              collect.unchanged);
                      _ := conclude p1
                            (!combine-equations
                             (!chain
                              [(collect M' r)
                             = ((M' at deref i) ::
                                  (collect M' r'))   [p4
                                                      collect.of-back]
                             = ((M1 at deref i) ::
                                  (replace'
                                    (collect M1 r')
                                     x y))           [q2a q2b]
                             = (y ::
                                  (replace'
                                    (collect M1 r')
                                      x y))          [assign.equal]
                             = (y ::
                                  (replace'
                                    (collect M r')
                                      x y))          [cu]])
                             (!chain
                              [(replace'
                                  (collect M r) x y)
                             = (replace'
                                  ((M at deref i) ::
                                    (collect M r'))
                                     x y)            [p4
                                                      collect.of-back]
                             = (y ::
```

```
                                      (replace'
                                        (collect M r')
                                          x y))          [replace-eq]]));
                              _ := (!force p2)}
                           (!both p1 p2)
                        assume (M at deref i =/= x)
                          let {M1 := M;
                              q1 := (!chain
                                      [M'
                                       = (M \ (replace
                                                 i j x y))    [hyp2]
                                       = (M \ (replace
                                               (successor i)
                                                 j x y))
                                                              [if-unequal]]);
                              (q2 as (and q2a q2b)) :=
                                (!chain->
                                  [q1
                               ==> (p3 & q1)                  [augment]
                               ==> ((collect M' r') =
                                      (replace'
                                        (collect M r')
                                          x y) &
                                    forall h .
                                      ~ h *in r' ==>
                                      M' at deref h =
                                      M at deref h)           [ind-hyp]]);
                              q3 := (!chain->
                                      [true
                                   ==> (~ start r *in r')  [fnir]
                                   ==> (~ i *in r')         [p4]]);
                              _ := (!sym
                                      (M at deref i =/= x));
                              _ := (!force p1);
                              _ := (!force p2)}
                           (!both p1 p2))
        } # by-induction
    } # match theorem

  (add-theorems theory |{correctness := proof}|)
  } # close module replace
} # close module Forward-Iterator
```

Exercise 16.13: Fill in the proof of p1 in the basis case. □

Exercise 16.14: Fill in the proof of p2 in the first case of the inductive step. □

⋆ ***Exercise 16.15:*** Give proofs for p1 and p2 in the second case of the inductive step. □

16.5 Range copy algorithm

The third algorithm we consider on forward iterator ranges both updates memory and returns an iterator value. To express this in the logic, we introduce two separate functions, one to express the memory updating and the other to compute the iterator return value.

```
extend-module Forward-Iterator {

  declare copy-memory: (S, X, Y) [(It X S) (It X S) (It Y S)] -> (Change S)
  declare copy: (S, X, Y) [(It X S) (It X S) (It Y S)] -> (It Y S)

  module copy-memory {

    define (axioms as [empty nonempty]) :=
      (fun
        [(M \ (copy-memory i j k)) =
           [M                                          when (i = j)
            ((M \ (deref k) <- (M at (deref i)))
                   \ (copy-memory (successor i)
                                  j
                                  (successor k)))    when (i =/= j)]])

    (add-axioms theory axioms)
  }

  module copy {

    define (axioms as [empty nonempty]) :=
      (fun
        [(M \\ (copy i j k)) =
           [k                                        when (i = j)
            ((M \ (deref k) <- (M at (deref i)))
                   \\ (copy (successor i)
                            j
                            (successor k)))        when (i =/= j)]])

    (add-axioms theory axioms)
```

Recall that the `Memory.\\` operator returns a value of the same sort as its second argument, which in this case is the return sort of copy, namely `(It X S)`.

Once again, we specify correctness in terms of `collect`, but instead of relying on a corresponding copy function on lists (which wouldn't be very useful), we just use equality of lists returned by `collect` applied to the original memory range and the final one.

```
define (correctness-prop r) :=
  (forall i j M k M' k' .
    (range i j) = SOME r &
```

```
  ~ k *in r &
  M' = M \ (copy-memory i j k) &
  k' = M \\ (copy i j k)
  ==> exists r1 .
        (range k k') = SOME r1 &
        (collect M' r1) = (collect M r) &
        forall h . ~ h *in r1 ==> M' at deref h = M at deref h)

define correctness := (forall r . correctness-prop r)
```

Note the precondition (`~ k *in r`); without this restriction on `k`, the starting point of the destination range, values in the source range would be overwritten before they were read. See Figure 16.1.

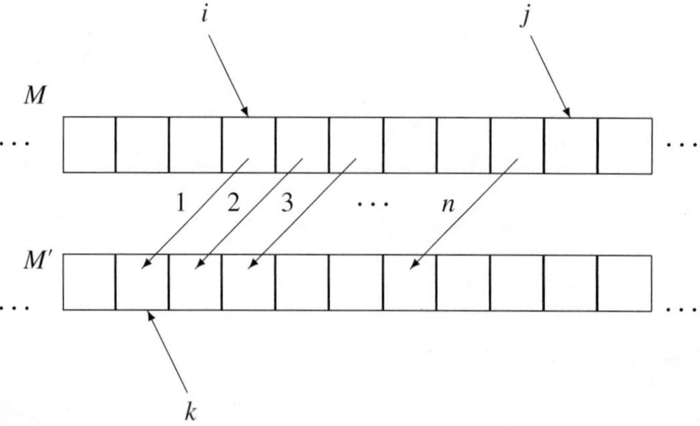

Figure 16.1

Illustrating $M' = M \setminus$ `(copy-memory i j k)`. The numbers labeling the arrows indicate the order in which the memory locations are copied.

Note also that we again have a safety property as part of the requirements: Memory locations outside the destination range are unchanged (but this doesn't mean that the source range isn't changed; it will be if it overlaps with the destination range).

```
define proof :=
 method (theorem adapt)
  let {[_ prove chain chain-> _] := (proof-tools adapt theory);
       [deref *in successor] := (adapt [deref *in successor]);
       [cp-mem-empty cp-mem-nonempty] :=
       [copy-memory.empty copy-memory.nonempty]}
    match theorem {
```

```
(val-of correctness) =>
 by-induction (adapt theorem) {
  (stop q:(It 'X 'S)) =>
   pick-any i:(It 'X 'S)
            j:(It 'X 'S)
            M:(Memory 'S)
            k:(It 'Y 'S)
            M':(Memory 'S)
            k':(It 'Y 'S)
     assume hyp := ((range i j) = SOME stop q &
                    ~ k *in stop q &
                    M' = M \ (copy-memory i j k) &
                    k' = M \\ (copy i j k))
       conclude goal := (exists r1 .
                          (range k k') = SOME r1 &
                          (collect M' r1) = (collect M stop q) &
                          forall h .
                            ~ h *in r1 ==>
                              M' at deref h =
                              M at deref h)
         let {er1 := (!prove empty-range1);
              _ := conclude (i = j)
                     (!chain->
                       [(range i j)
                        = (SOME stop q)              [hyp]
                     ==> (i = j)                     [er1]]);
              _ := conclude (M' = M)
                     (!chain->
                       [M'
                      = (M \ (copy-memory i j k))    [hyp]
                      = M                            [cp-mem-empty]]);
              _ := conclude (k' = k)
                    (!chain->
                      [k'
                      = (M \\ (copy i j k))          [hyp]
                      = k                            [empty]]);
              _ := (!chain [(start stop k)
                           = k                       [start.of-stop]
                           = (finish stop k)         [finish.of-stop]]);
              protected := pick-any h
                             assume (~ h *in stop k)
                               (!chain
                                [(M' at deref h)
                               = (M at deref h)      [(M' = M)]]);
              er := (!prove empty-range);
              p := (!both
                    (!chain [(range k k')
                           = (range k k)             [(k' = k)]
                           = (SOME stop k)           [er]])
                    (!both (!chain
```

```
                          [(collect M' stop k)
                          = nil:(List 'S)          [collect.of-stop]
                          = (collect M stop q)     [collect.of-stop]])
                         protected))}
        (!chain-> [p ==> goal                      [existence]])
  | (r as (back r':(Range 'X 'S))) =>
    let {ind-hyp := (correctness-prop r')}
     pick-any i:(It 'X 'S)
              j:(It 'X 'S)
              M:(Memory 'S)
              k:(It 'Y 'S)
              M':(Memory 'S)
              k':(It 'Y 'S)
      let {M1 := (M \ deref k <- M at deref i);
           hyp1 := ((range i j) = SOME r);
           hyp2 := (~ k *in r);
           hyp3 := (M' = M \ (copy-memory i j k));
           hyp4 := (k' = M \\ (copy i j k))}
        assume (hyp1 & hyp2 & hyp3 & hyp4)
         conclude goal := (exists r1 .
                           (range k k') =
                           SOME r1             &
                           (collect M' r1) =
                           (collect M r)       &
                           forall h .
                             ~ h *in r1 ==>
                               M' at deref h =
                               M at deref h)
           let {(q as (and q1 q2)) :=
                  (!chain->
                   [(range i j)
                  = (SOME r)                       [hyp1]
                  = (range (start r)
                           (finish r))             [range.collapse]
                ==> (i = (start r) &
                     j = (finish r))               [range.injective]]);
                nb := (!prove nonempty-back);
                _ := (!chain->
                      [true
                   ==> (start r =/= finish r)      [nb]
                   ==> (i =/= j)                   [q1 q2]]);
                rr := (!prove
                        *in.range-reduce);
                cu := (!prove
                        collect.unchanged);
                q3 := (!chain->
                        [hyp2
                     ==> (~ k *in r')              [rr]
                     ==> ((collect M1 r') =
                          (collect M r'))          [cu]]);
```

```
q4 := conclude
        (M' = M1 \
              (copy-memory
                (successor i)
                j
                (successor k)))
        (!chain
          [M'
         = (M \
             (copy-memory i j k))  [hyp3]
         = (M1 \
              (copy-memory
                (successor i)
                j
               (successor k)))     [cp-mem-nonempty]]);
q5 := conclude
        (k' = M1 \\
              (copy
                (successor i)
                j
                (successor k)))
        (!chain
          [k'
         = (M \\ (copy i j k))     [hyp4]
         = (M1 \\
              (copy
                (successor i)
                j
                (successor k)))    [nonempty]]);
lb := (!prove range-back);
hyp1' := (!chain->
           [hyp1
        ==> ((range
                (successor i) j)
               = SOME r')          [lb]]);
rs2 := (!prove
          *in.range-shift2);
q6 := (!chain->
         [hyp2
     ==> (~ (successor k) *in r')  [rs2]
     ==> (hyp1' &
          ~ (successor k) *in r' &
          q4 &
          q5)                      [augment]
     ==> (exists r1 .
           (range
             (successor k) k') =
           SOME r1         &
          (collect M' r1) =
          (collect M1 r') &
```

```
                          forall h .
                           ~ h *in r1 ==>
                             M' at deref h =
                             M1 at deref h)          [ind-hyp]])}
                  pick-witness r1 for q6 q6-w
                    (!force goal)
        }
      } # match theorem { ...

   (add-theorems theory |{correctness := proof}|)
  } # close module copy
} # close module Forward-Iterator
```

** ***Exercise 16.16:*** Finish the proof of Forward-Iterator.copy.correctness.

16.6 Range copy-backward algorithm

The next iterator theory refinement we consider is Bidirectional-Iterator, in which we axiomatize a predecessor function that enables iterator movement in the direction opposite to that of successor:

```
module Bidirectional-Iterator {
  open Forward-Iterator

  declare predecessor: (X, S) [(It X S)] -> (It X S)

  module predecessor {

    define of-start :=
      (forall r .  predecessor start r = start back r)

    define of-successor :=
      (forall i . predecessor successor i = i)
  }

  define theory :=
    (make-theory ['Forward-Iterator]
                 [predecessor.of-start predecessor.of-successor])
```

The following theorem is simple to prove:

```
define successor-of-predecessor :=
  (forall i . successor predecessor i = i)

define proof :=
  method (theorem adapt)
```

```
    let {[_ _ chain _ _] := (proof-tools adapt theory);
         [successor predecessor] := (adapt [successor predecessor])}
      match theorem {
        (val-of successor-of-predecessor) =>
          pick-any i:(It 'X 'S)
            (!chain
              [(successor predecessor i)
             = (successor predecessor start stop i)  [start.of-stop]
             = (successor start back stop i)         [predecessor.of-start]
             = (start stop i)                        [successor.of-start]
             = i                                     [start.of-stop]])
      }

  (add-theorems theory |{successor-of-predecessor := proof}|)
} # close module Bidirectional-Iterator
```

The STL copy algorithm, which uses forward iterators, has a counterpart `copy_backward` that uses bidirectional iterators. In general, algorithms on bidirectional iterators can use both `predecessor` and `successor`, but `copy_backward` only uses `predecessor`, to progress through the source range from its end to its beginning. Why is such an algorithm needed? For the answer, recall that one of the preconditions for the `copy.correctness` theorem is

```
(~ k *in r),
```

where `r` is the source range and `k` is the start of the destination range. Without that restriction, as previously noted, memory locations in the source range would be overwritten before being read. The correctness theorem that we will state and prove for our `copy-backward` function has a similar restriction,

```
(~ (predecessor k) *in r),
```

but in this case, `k` is the *end* of the destination range. See Figure 16.2. (The reason for `(predecessor k)` instead of just `k` is a technical point that will be explained shortly.) With `copy-backward` there is no problem if the *beginning* of the destination range lies within the source range, and thus it can be used in the cases where `copy` would fail. (Either `copy` or `copy-backward` works equally well when there is no overlap between the source and destination ranges.)

To prove theorems about algorithms on bidirectional iterator ranges requires a different approach from the one we have taken with the three algorithms we have considered on forward iterator ranges. Whereas it was natural to set up inductions based on the `Range` datatype and use the axioms for `successor`, the same approach no longer works when an algorithm uses `predecessor` to progress through a range. For an alternative approach, there are several choices.

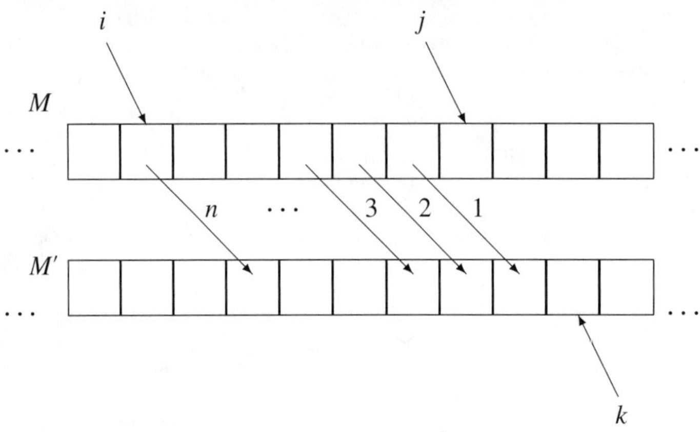

Figure 16.2
Illustrating $M' = M \setminus$ `(copy-memory-backward` i j k`)`. The numbers labeling the arrows indicate the backward order in which the memory locations are copied. Compare with Figure 16.1, where the order of copying is in the forward direction.

1. Define a new datatype `Reverse-Range` with constructors `halt` and `ahead`, say, and instead of the axioms for `successor`, define analogous axioms for `predecessor`. For example, instead of

```
(forall r . successor start back r = start r)
```

   we would have

```
(forall r' . predecessor start ahead r' = start r')
```

   While this would work for reasoning about an algorithm like `copy-backward` that only uses `predecessor`, it is not suitable for bidirectional range algorithms that iterate in both directions.

2. Instead of trying to set up inductions based on a range datatype, do natural number inductions based on range length. This would work for `copy-backward` and more generally for algorithms that iterate in both directions, like the STL's `reverse` algorithm (which positions iterators at both ends of a range and moves them in lock step toward the center, swapping the referenced memory locations as it goes). We pass over this choice for now, but we will use it in the next section to reason about a binary search algorithm on ranges (where in fact we do a strong induction, since the algorithm halves the length of the range at each step).

3. Introduce *reverse iterator adapters* like those in the STL: `(reverse-iterator` i`)` is an iterator such that applying `successor` to it applies `predecessor` to i, and applying `predecessor` to it applies `successor` to i. Then we can implement `copy-backward` simply by applying `copy` to appropriately defined reverse iterators. This is the approach we take here.

16.7 Adapters: Reverse-iterator and reverse-range

So our first task is to define the reverse iterator adapters, for which it is useful to also define reverse range adapters (which have no counterpart in the STL).

```
extend-module Bidirectional-Iterator {

  module Reversing {
    define reverse := List.reverse

    declare reverse-iterator: (X, S) [(It X S)] -> (It (It X S) S)
    declare reverse-range:    (X, S) [(Range X S)] -> (Range (It X S) S)
    declare base-iterator:    (X, S) [(It (It X S) S)] -> (It X S)
    declare base-range:       (X, S) [(Range (It X S) S)] -> (Range X S)

    define deref-reverse :=
      (forall i . deref reverse-iterator i = deref predecessor i)

    define *reverse-in :=
      (forall i r . (reverse-iterator i) *in (reverse-range r) <==>
                    (predecessor i) *in r)

    define base-reverse-range :=
      (forall r . base-range reverse-range r = r)

    define reverse-base-range :=
      (forall r . reverse-range base-range r = r)

    define reverse-of-range :=
      (forall i j r .
        (range (reverse-iterator j) (reverse-iterator i)) = SOME r
        <==> (range i j) = SOME base-range r)

    define reverse-base :=
      (forall i . reverse-iterator base-iterator i = i)

    define collect-reverse-stop :=
      (forall M i . (collect M (reverse-range stop i)) = nil)

    define collect-reverse-back :=
      (forall M r .
        (collect M reverse-range back r) =
```

```
      (collect M reverse-range r) ++ ((M at deref start back r) :: nil))

    (add-axioms theory [deref-reverse *reverse-in base-reverse-range
                        reverse-base-range reverse-of-range reverse-base
                        collect-reverse-stop collect-reverse-back])
```

In deref-reverse, note the offset using predecessor. This offset is necessary because of the asymmetric way in which an iterator pair (i,j) delimits a range. Let n be the length of the range, and let $i+k$ denote k-fold application of successor to i and $i-k$ denote k-fold application of predecessor to i. Then the iterators in the range are

$$i,\ i+1,\ i+2,\ \ldots,\ i+n-1=j-1.$$

If $i'=$(reverse-iterator i) and $j'=$(reverse-iterator j), then the iterators in the reverse range are

$$j',\ j'+1,\ j'+2,\ \ldots,\ j'+n-1,$$

and

$$\begin{aligned}
(\texttt{deref}\ j') &= (\texttt{deref}\ (j-1)) \\
(\texttt{deref}\ (j'+1)) &= (\texttt{deref}\ (j-2)) \\
(\texttt{deref}\ (j'+2)) &= (\texttt{deref}\ (j-3)) \\
&\cdots \\
(\texttt{deref}\ (j'+n-1)) &= (\texttt{deref}\ (j-n)) = (\texttt{deref}\ i).
\end{aligned}$$

See also Figure 16.2.

The following theorems will be useful in correctness proofs of algorithms that use these adapters.

```
define reverse-range-reverse :=
  (forall i j r .
    (range (reverse-iterator j)
           (reverse-iterator i)) = SOME (reverse-range r)
    <==> (range i j) = SOME r)

define collect-reverse :=
  (forall r M .
    (collect M reverse-range r) = reverse (collect M r))

define collect-reverse-corollary :=
  (forall M r M' r' .
    (collect M' r') = (collect M reverse-range r)
    ==> (collect M' base-range r') = (collect M r))

define proofs :=
 method (theorem adapt)
  let {[_ _ chain chain-> _] := (proof-tools adapt theory);
```

```
      [deref *in successor predecessor] :=
        (adapt [deref *in successor predecessor]);
      [crs reverse-empty] := [collect-reverse-stop List.reverse.empty]}
   match theorem {
     (val-of reverse-range-reverse) =>
       pick-any i:(It 'X 'S)
                j:(It 'X 'S)
                r:(Range 'X 'S)
         (!chain
           [((range (reverse-iterator j)
                    (reverse-iterator i)) =
             SOME (reverse-range r))
       <==> ((range i j) =
              SOME (base-range (reverse-range r)))    [reverse-of-range]
       <==> ((range i j) = SOME r)                    [base-reverse-range]])

   | (val-of collect-reverse) =>
       by-induction (adapt theorem) {
         (stop i) =>
           pick-any M
             (!combine-equations
               (!chain->
                 [(collect M (reverse-range stop i))
                  = nil                                [crs]])
               (!chain
                 [(reverse (collect M stop i))
                  = (reverse nil)                      [collect.of-stop]
                  = nil                                [reverse-empty]]))
       | (r as (back r')) =>
           (!force (forall M .
                     (collect M reverse-range r) =
                     reverse (collect M r)))
       }
   | (val-of collect-reverse-corollary) =>
       (!force (adapt theorem))
   }

  (add-theorems theory |{[reverse-range-reverse collect-reverse
                          collect-reverse-corollary] := proofs}|)

 } # close module Reversing
} # close module Bidirectional-Iterator
```

Exercise 16.17: Fill in the proof of the inductive step in the proof of `Bidirectional-Iterator.Reversing.collect-reverse`. □

Exercise 16.18: Fill in the proof of `Bidirectional-Iterator.Reversing.collect-reverse-corollary`. □

16.8 Implementing copy-backward

As with copy, STL's copy_backward function both changes memory and returns a value, in this case an iterator pointing to the beginning of the destination range. We therefore model it with two functions:

```
extend-module Bidirectional-Iterator {
  open Reversing

  declare copy-memory-backward: (S, X, Y) [(It X S) (It X S) (It Y S)] ->
                                          (Change S)
  declare copy-backward: (S, X, Y) [(It X S) (It X S) (It Y S)] -> (It Y S)
```

These functions are then defined in terms of copy and copy-memory on adapted iterators, as follows:

```
  module copy-memory-backward {
    define [def] :=
      (fun
        [(M \ (copy-memory-backward i j k)) =
                (M \ (copy-memory (reverse-iterator j)
                                  (reverse-iterator i)
                                  (reverse-iterator k)))])
  }

  module copy-backward {
    define [def] :=
      (fun
        [(M \\ (copy-backward i j k)) =
                 (base-iterator (M \\ (copy (reverse-iterator j)
                                            (reverse-iterator i)
                                            (reverse-iterator k))))])
  }

  (add-axioms theory [copy-memory-backward.def copy-backward.def])
```

The correctness theorem is then:

```
extend-module copy-backward {

  define correctness :=
    (forall r i j M k M' k' .
      (range i j) = (SOME r) &
      ~ (predecessor k) *in r &
      M' = (M \ (copy-memory-backward i j k)) &
      k' = (M \\ (copy-backward i j k))
      ==> exists r' .
            (range k' k) = SOME r' &
```

```
          (collect M' r') = (collect M r) &
          forall h . ~ h *in r' ==> M' at deref h = M at deref h)
```

We do not need induction, since we can apply the correctness theorem for copy.

```
define proof :=
  method (theorem adapt)
    let {[_ prove chain chain-> _] := (proof-tools adapt theory);
         [deref *in successor predecessor] :=
           (adapt [deref *in successor predecessor]);
         [cp-mem-bk-df cp-bk-df rev-of-rng] :=
           [copy-memory-backward.def copy-backward.def reverse-of-range]}
      match theorem {
        (val-of correctness) =>
         pick-any r:(Range 'X 'S)
                  i:(It 'X 'S)
                  j:(It 'X 'S)
                  M:(Memory 'S)
                  k:(It 'Y 'S)
                  M':(Memory 'S)
                  k':(It 'Y 'S)
           let {h1 := ((range i j) = SOME r);
                h2 := (~ predecessor k *in r);
                h3 := (M' = M \ (copy-memory-backward i j k));
                h4 := (k' = M \\ (copy-backward i j k));
                goal := (exists r' .
                           (range k' k) = SOME r' &
                           (collect M' r') = (collect M r) &
                           forall h . ~ h *in r' ==>
                             M' at deref h = M at deref h)}
             assume (h1 & h2 & h3 & h4)
               let {ri := reverse-iterator;
                    rlr := (!prove reverse-range-reverse);
                    p1 := (!chain->
                            [h1
                         ==> ((range (ri j)
                                     (ri i)) =
                                SOME reverse-range r)        [rlr]]);
                    p2 := (!chain->
                            [h2
                         ==> (~ (ri k) *in
                                 reverse-range r)            [*reverse-in]]);
                    p3 := (!chain
                            [M'
                           = (M \ (copy-memory-backward
                                     i j k))                 [h3]
                           = (M \ (copy-memory (ri j)
                                               (ri i)
                                               (ri k)))      [cp-mem-bk-df]]);
```

```
          p4 := (!chain
                  [(ri k')
                  = (ri (M \\ (copy-backward
                                i j k)))        [h4]
                  = (ri (base-iterator
                          (M \\ (copy (ri j)
                                      (ri i)
                                      (ri k)))))  [cp-bk-df]

           = (M \\ (copy (ri j)
                         (ri i)
                         (ri k)))                [reverse-base]]);
          cc := (!prove copy.correctness);
          p5 := (!chain->
                  [(p1 & p2 & p3 & p4)
              ==> (exists r' .
                    (range
                      (ri k)
                      (ri k')) = SOME r' &
                    (collect
                      M' r') =
                    (collect
                      M reverse-range r) &
                    forall h .
                      ~ h *in r' ==>
                        M' at deref h =
                        M at deref h)            [cc]])}
      pick-witness r' for p5
        let {p5-w1 := ((range (ri k)
                              (ri k')) =
                        SOME r');
             p5-w2 := ((collect M' r') =
                        (collect
                          M reverse-range r));
             p5-w3 := (forall h .
                         ~ h *in r' ==>
                           M' at deref h =
                           M at deref h);
             q1 := (!chain->
                     [p5-w1
                  ==> ((range k' k) =
                        SOME base-range r')       [rev-of-rng]]);
             crc := (!prove
                      collect-reverse-corollary);
             q2 := (!chain->
                     [p5-w2
                   ==> ((collect
                           M' base-range r') =
                         (collect M r))           [crc]]);
             q3 := (!force
```

```
                    (forall h .
                      ~ h *in base-range r' ==>
                      M' at deref h =
                      M at deref h))}
          (!chain->
            [(q1 & q2 & q3) ==> goal               [existence]])
  }

  (add-theorems theory |{correctness := proof}|)
 } # close module copy-backward
} # close module Bidirectional-Iterator
```

Exercise 16.19: Finish the proof of `copy-backward.correctness`. □

16.9 Random-access iterators

`Random-Access-Iterator` is the final iterator theory refinement we consider. As with STL random-access iterators, we introduce functions that enable movement of an iterator an arbitrary distance away from its current position. Of course, such functions could be programmed using repeated `successor` or `predecessor` function calls, but that would mean taking computing time linear in the distance traveled. What makes "random access" appealing is the assumption that the "big jumps" that the new functions perform are done in constant time. We do not deal with computing time in our specifications and proofs here, but one must keep in mind this crucial issue in determining which algorithms to associate with which iterator theory refinement. A classic case is binary search on a range, which can be much faster than linear search but only if jumping to the middle of a range is a constant time operation.[7] The algorithm example we study in the next section is `lower_bound`, one of several forms of binary search found in the STL.

Let's get started with the functions and axioms for random-access iterators:

```
module Random-Access-Iterator {
  open Bidirectional-Iterator
  overload + N.+

  declare I+N: (X, S) [(It X S) N] -> (It X S) [+]
  declare I-N: (X, S) [(It X S) N] -> (It X S) [-]
  declare I-I: (X, S) [(It X S) (It X S)] -> N [-]

  define [m n] := [?m:N ?n:N]

  define I+0 :=   (forall i . i + zero = i)
```

7 This statement is an oversimplification; see Exercise 16.34.

```
define I+pos := (forall i n . i + S n = (successor i) + n)
define I-0 :=   (forall i . i - zero = i)
define I-pos := (forall i n . i - S n = predecessor (i - n))
define I-I :=   (forall i j n . i - j = n <==> j = i - n)

define theory :=
  (make-theory ['Bidirectional-Iterator] [I+0 I+pos I-0 I-pos I-I])
```

16.9.1 Relationships among iterator functions

We introduce theorems about random access iterators in three groups. The first provides additional relationships between the various iterator functions.

```
  define I-I-self :=             (forall i . i - i = zero)
  define I+N-cancellation := (forall n i . (i + n) - n = i)
  define I-I-cancellation := (forall n i . (i + n) - i = n)
  define successor-in :=         (forall n i . successor (i + n) =
                                               (successor i) + n)

define I-M-N := (forall n m i . (i - m) - n = i - (m + n))
```

We give a couple of the proofs and leave the others as exercises.

```
define proofs :=
  method (theorem adapt)
   let {[_ prove chain chain-> _] := (proof-tools adapt theory);
        [successor predecessor I+N I-N I-I] :=
          (adapt [successor predecessor I+N I-N I-I])}
      match theorem {
        (val-of I-I-self) =>
          pick-any i:(It 'X 'S)
            (!chain->
              [(i - i = zero) <== (i = i - zero)            [I-I]
                              <== (i = i)                   [I-0]])
      | (val-of I+N-cancellation) =>
          (!force (adapt theorem))
      | (val-of I-I-cancellation) =>
          let {IC := (!prove I+N-cancellation)}
            pick-any n i:(It 'X 'S)
              (!chain->
                [(i = i)
             ==> (i = (i + n) - n)                          [IC]
             ==> ((i + n) - i = n)                          [I-I]])
    | (val-of successor-in) => (!force (adapt theorem))
    | (val-of I-M-N) =>          (!force (adapt theorem))
    }
```

```
  (add-theorems theory |{[I-I-self I+N-cancellation I-I-cancellation
                          successor-in I-M-N] := proofs}|)

} # close module Random-Access-Iterator
```

Exercise 16.20: Fill in the proof of `I+N-cancellation`. □

Exercise 16.21: Fill in the proof of `successor-in`. □

Exercise 16.22: Fill in the proof of `I-M-N`. □

16.9.2 New properties of the length function

The second group of `Random-Access-Iterator` theorems provides new properties of the `length` function on ranges that can now be stated in terms of random-access operations.

```
extend-module Random-Access-Iterator {
  overload <= N.<=

  define length1 := (forall r . start r = (finish r) - (length r))

  define length2 := (forall r . length r = (finish r) - (start r))

  define length3 :=
    (forall i j r . (range i j) = SOME r ==> length r = j - i)

  define length4 :=
    (forall i j n r r' r'' .
      (range i j) = SOME r &
      (range i i + n) = SOME r' &
      (range i + n j) = SOME r''
      ==> length r = (length r') + (length r''))

  define (contained-range-prop n) :=
    (forall r i j k .
      (range i j) = SOME r &
      k = i + n &
      n <= length r
      ==> exists r' . (range i k) = SOME r')

  define contained-range := (forall n . contained-range-prop n)

  define (collect-split-range-prop n) :=
    (forall i j r .
      (range i j) = SOME r & n <= length r
      ==> exists r' r'' .
            (range i i + n) = SOME r'  &
            (range i + n j) = SOME r'' &
```

```
      forall M .
        (collect M r) = (collect M r') ++ (collect M r''))

  define collect-split-range := (forall n . collect-split-range-prop n)
```

We give proofs of only the first three theorems, leaving the others as exercises.

```
define [n0 r0 r0'] := [?n0 ?r0 ?r0']

define proofs :=
  method (theorem adapt)
    let {[_ prove chain chain-> _] := (proof-tools adapt theory);
         [successor predecessor I+N I-N I-I] :=
           (adapt [successor predecessor I+N I-N I-I]);
         [pred-of-start range-inj] := [predecessor.of-start range.injective]}
      match theorem {
        (val-of length1) =>
          by-induction (adapt theorem) {
            (stop j) =>
              conclude (start stop j = (finish stop j) -
                                        (length stop j))
                (!combine-equations
                  (!chain [(start stop j) = j                    [start.of-stop]])
                  (!chain [((finish stop j) -
                             (length stop j))
                           = (j - zero)                          [finish.of-stop
                                                                  length.of-stop]
                           = j                                   [I-0]]))
          | (r as (back r')) =>
             conclude (start r = (finish r) - (length r))
               let {ind-hyp := (start r' = (finish r') -
                                            length r')}
                 (!combine-equations
                   (!chain [(start r)
                            = (predecessor start r')             [pred-of-start]
                            = (predecessor
                                ((finish r') - length r'))       [ind-hyp]])
                   (!chain [((finish r) - length r)
                            = ((finish r') - S length r')        [finish.of-back
                                                                  length.of-back]
                            = (predecessor
                                 ((finish r') - length r'))      [I-pos]]))
          } # by-induction (adapt length1) {...
      | (val-of length2) =>
          pick-any r:(Range 'X 'S)
            let {rl1 := (!prove length1)}
              (!chain->
                [true
              ==> (start r = ((finish r) - length r))           [rl1]
              ==> ((finish r) - (start r) = length r)           [I-I]
              ==> (length r = (finish r) - start r)             [sym]])
```

```
      | (val-of length3) =>
          pick-any i:(It 'X 'S)
                   j:(It 'X 'S)
                   r:(Range 'X 'S)
            assume hyp := ((range i j) = SOME r)
              let {p := (!chain->
                          [(range (start r) (finish r))
                          = (SOME r)                               [range.collapse]
                          = (range i j)                            [hyp]
                        ==> (start r = i & finish r = j)           [range-inj]]);
                   rl2 := (!prove length2)}
                (!chain->
                  [(length r)
                 = ((finish r) - start r)                          [rl2]
                 = (j - i)                                         [p]])
      | (val-of length4) =>
          (!force (adapt theorem))
      | (val-of contained-range) =>
          (!force (adapt theorem))
      | (val-of collect-split-range) =>
          (!force (adapt theorem))
      }
(add-theorems theory |{[length1 length2 length3] := proofs}|)

} # close module Random-Access-Iterator
```

Exercise 16.23: Fill in the proof of `length4`. □

⋆ ***Exercise 16.24:*** Fill in the proof of `contained-range`. □

⋆⋆ ***Exercise 16.25:*** Fill in the proof of `collect-split-range`. □

16.9.3 Theorems about collecting locations

The third and final group of `Random-Access-Iterator` theorems is based on a `collect-locs` function, similar to `collect` but gathering into a list the locations in a range rather than the values in those locations.

```
extend-module Random-Access-Iterator {
define in := List.in

declare collect-locs: (S, X) [(Range X S)] -> (List (Memory.Loc S))

module collect-locs {

define [< <=] := [N.< N.<=]
```

```
define (axioms as [of-stop of-back]) :=
  (fun
    [(collect-locs stop h) = nil
     (collect-locs back r) = (deref start back r :: (collect-locs r))])

(add-axioms theory axioms)
```

The first theorem, split-range, shows how a range can be split into two parts whose respective lists of locations can be joined to give the list of locations in the whole range.

```
define (split-range-prop n) :=
  (forall i j r .
    (range i j) = SOME r & n <= length r
    ==>
    exists r' r'' .
      (range i i + n) = SOME r' &
      (range i + n j) = SOME r'' &
      (collect-locs r) = (collect-locs r') ++ (collect-locs r''))

define split-range := (forall n . split-range-prop n)
```

The next theorem, *in-relation, gives a new characterization of the *in predicate introduced on page 716.

```
define *in-relation :=
  (forall r i . i *in r <==> deref i in collect-locs r)
```

Finally, we have three additional theorems concerning the *in predicate.

```
define all-*in :=
  (forall n i j r .
    (range i j) = SOME r & n < length r ==> (i + n) *in r)

define *in-whole-range :=
  (forall n i j k r r' .
    (range i j) = SOME r &
    n < length r &
    (range i i + n) = SOME r' &
    (k *in r' | k = i + n)
    ==> k *in r)

define *in-whole-range-2 :=
  (forall n i j k r r' .
    (range i j) = SOME r &
    n <= length r &
    (range i + n j) = SOME r' &
    (k *in r' | k = j)
    ==> k *in r | k = j)
```

The first three of these theorems are used as lemmas in the proofs of the last two, which in turn play a crucial role in the proof of correctness of the binary search function on ranges to be presented in Section 16.10. We present the proof of `split-range` in full, but leave the other proofs as exercises.

```
define [succ pred cl] := [successor predecessor collect-locs]

define split-range-proof :=
 method (theorem adapt)
  let {[_ prove chain chain-> _] := (proof-tools adapt theory);
       [succ *in pred I+N I-N I-I] :=
         (adapt [succ *in pred I+N I-N I-I]);
       [DAO join-l-empty join-l-nempty <=-inj range-sb] :=
           [(datatype-axioms "Option")
            List.join.left-empty List.join.left-nonempty
            N.Less=.injective range.start-back]}
    match theorem {
      (val-of split-range) =>
        by-induction (adapt theorem) {
          zero =>
            pick-any i:(It 'X 'S)
                     j:(It 'X 'S)
                     r:(Range 'X 'S)
              assume h1 := ((range i j) = SOME r);
                     h2 := (zero <= length r)
                let {goal := (exists r' r'' .
                                (range i i + zero) =
                                   SOME r'              &
                                (range i + zero j) =
                                   SOME r''             &
                                (collect-locs r) =
                                  (collect-locs r') ++
                                  (collect-locs r''));
                     el := (!prove empty-range);
                     p1 := (!chain
                              [(range i i + zero)
                             = (range i i)                      [I+0]
                             = (SOME stop i)                    [el]]);
                     p2 := (!chain
                              [(range i + zero j)
                             = (range i j)                      [I+0]
                             = (SOME r)                         [h1]]);
                     p3 := (!chain->
                              [((collect-locs stop i) ++
                                (collect-locs r))
                             = (nil ++ (collect-locs r))        [of-stop]
                             = (collect-locs r)                 [join-l-empty]
                         ==>  (collect-locs r =
                               (collect-locs stop i) ++
                               (collect-locs r))                [sym]])}
                (!chain->
```

```
              [(p1 & p2 & p3) ==> goal                   [existence]])
   | (n as (S m)) =>
       pick-any i:(It 'X 'S)
                j:(It 'X 'S)
                r:(Range 'X 'S)
         assume h1 := ((range i j) = SOME r);
                h2 := (n <= length r)
           let {goal := (exists r' r'' .
                           (range i i + n) = SOME r'   &
                           (range (i + n) j) = SOME r'' &
                           (collect-locs r) =
                             (collect-locs r') ++
                             (collect-locs r''));
                ind-hyp := (split-range-prop m);
                 p1 := (!chain->
                          [h2
                        ==> (exists n0 . length r = S n0)  [N.Less=.S4]])}
             pick-witness n0 for p1 p1-w
               let {nl := (!prove nonzero-length);
                    p2 := (!chain->
                             [true
                          ==> (S n0 =/= zero)              [N.S-not-zero]
                          ==> (length r =/= zero)          [p1-w]
                          ==> (exists r0 .
                                 r = back r0)              [nl]])}
                 pick-witness r0 for p2 p2-w
                   let {lb := (!prove range-back);
                        p3 := (!chain->
                                 [(range i j)
                                = (SOME r)                 [h1]
                                = (SOME back r0)           [p2-w]
                              ==> ((range (succ i) j) =
                                     SOME r0)              [lb]]);
                        p4 := (!chain->
                                 [h2
                              ==> (n <= length back r0)    [p2-w]
                              ==> (n <= S length r0)       [length.of-back]
                              ==> (m <= length r0)         [<=-inj]]);
                        p5 := (!chain->
                                 [(p3 & p4)
                              ==> (exists r0' r'' .
                                     (range (succ i)
                                            (succ i)
                                            + m) =
                                       SOME r0'      &
                                     (range (succ i) + m
                                            j) =
                                      SOME r''       &
                                     (cl r0) =
                                       (cl r0') ++
                                       (cl r''))           [ind-hyp]])}
```

```
pick-witnesses r0' r'' for p5 p5-w
  let {p5-w1 := ((range
                   (succ i)
                   (succ i) + m)
                 = SOME r0');
       p5-w2 := ((range
                   (succ i) + m
                   j) =
                 SOME r'');
       p5-w3 := ((cl r0) =
                  (cl r0') ++
                  (cl r''));
       q1 := (!chain->
               [p5-w1
             ==> ((range
                  (succ i)
                    i + n) =
                 SOME r0')             [I+pos]
             ==> ((range i
                         i + n) =
                 SOME back r0')        [lb]]);
       q2 := (!chain->
               [p5-w2
             ==> ((range
                    i + n j) =
                  SOME r'')            [I+pos]]);
       q3 := let {q3-1 :=
                   (!chain->
                     [q1
                  ==> (i = start
                           back r0')
                                       [range-sb]]);
                   q3-2 :=
                    (!chain->
                      [(range i j)
                   = (SOME r)          [h1]
                   = (SOME
                       back r0)        [p2-w]
                ==> (i =
                     start
                       back r0)        [range-sb]]);
                   q3-3 :=
                    (!chain
                      [(start
                         back r0)
                        = i            [q3-2]
                        = (start
                            back r0')  [q3-1]])}
                 (!chain
                  [(cl r)
                 = (cl back r0)        [p2-w]
```

```
                                        = ((deref
                                             start
                                             back r0) ::
                                           (cl r0))           [of-back]
                                        = ((deref
                                             start
                                             back r0) ::
                                           ((cl r0') ++
                                            (cl r'')))        [p5-w3]
                                        = (((deref
                                              start
                                              back r0) ::
                                            (cl r0')) ++
                                            (cl r''))         [join-l-nempty]
                                        = (((deref
                                              start
                                              back r0') ::
                                             (cl r0')) ++
                                             (cl r''))        [q3-3]
                                        = ((cl back r0') ++
                                           (cl r''))          [of-back]])}
                          (!chain->
                            [(q1 & q2 & q3) ==> goal          [existence]])
      } # by-induction
    } # match theorem

define (make-trivial-proof-method desired-theorem) :=
  method (theorem adapt)
    let {[get prove chain chain-> chain<-] := (proof-tools adapt theory);
         [successor *in predecessor I+N I-N I-I] :=
           (adapt [successor *in predecessor I+N I-N I-I]);
         DAO := (datatype-axioms "Option")}
      match theorem {
        (val-of desired-theorem) => (!force (adapt desired-theorem))
      }

define *in-relation-proof  :=
  (make-trivial-proof-method *in-relation)
define all-*in-proof :=
  (make-trivial-proof-method all-*in)
define *in-whole-range-proof :=
  (make-trivial-proof-method *in-whole-range)
define *in-whole-range-2-proof :=
  (make-trivial-proof-method *in-whole-range-2)

(add-theorems theory |{split-range       := split-range-proof,
                       *in-relation      := *in-relation-proof,
                       all-*in           := all-*in-proof,
                       *in-whole-range   := *in-whole-range-proof,
                       *in-whole-range-2 := *in-whole-range-2-proof}|)
```

```
} # close module collect-locs
} # close module Random-Access-Iterator
```

Exercise 16.26: Provide a proper definition for `*in-relation-proof`. □

⋆ ***Exercise 16.27:*** Provide a proper definition for `all-*in`. □

⋆ ***Exercise 16.28:*** Provide a proper definition of `*in-whole-range-proof`. □

⋆ ***Exercise 16.29:*** Provide a proper definition of `*in-whole-range-2-proof`. □

16.9.4 Ordered range

For a binary search algorithm to be applicable on a range, the values in the range must be in order. We specify this ordering requirement here by taking advantage of the ordered-list evolution of the strict weak order theory presented in Section 14.9.9:

```
module Ordered-Range {
  open SWO, Random-Access-Iterator

  declare ordered: (S, X) [(Memory S) (Range X S)] -> Boolean

  define ordered' := SWO.ordered

  define def := (forall r M . (ordered M r) <==> ordered' (collect M r))

  define theory := (make-theory ['SWO 'Random-Access-Iterator] [def])
}
```

Although this specification of ordered ranges does not require random access, we have included `Random-Access-Iterator` along with `SWO` as a theory that `Ordered-Range` refines, because the statement and proof of one the following theorems about ordered ranges does require it:

```
extend-module Ordered-Range {

  define ordered-rest-range :=
    (forall M r . (ordered M back r) ==> (ordered M r))

  define ordered-empty-range := (forall M i . (ordered M stop i))

  define ordered-subranges :=
    (forall M r i j n . (range i j) = SOME r &
                        (ordered M r) &
                        n <= length r
                        ==> exists r' r'' .
                              (range i i + n) = SOME r' &
```

```
                    (range i + n j) = SOME r'' &
                    (ordered M r') &
                    (ordered M r''))
```

We give the proof of the first two theorems, leaving the third as an exercise.

```
define proofs :=
  method (theorem adapt)
    let {[_ prove chain chain-> chain<-] := (proof-tools adapt theory);
         [deref <EL I+N I-N I-I ordered ordered'] :=
            (adapt [deref <EL I+N I-N I-I ordered ordered'])}
      match theorem {
        (val-of ordered-rest-range) =>
          pick-any M r
            (!chain
              [(ordered M (back r))
            ==> (ordered' (collect M (back r)))      [def]
            ==> (ordered'
                  ((M at deref start back r)
                   :: (collect M r)))                [collect.of-back]
            ==> ((M at deref start back r)
                  <EL (collect M r)          &
                 (ordered' (collect M r)))           [SWO.ordered.nonempty]
            ==> (ordered' (collect M r))             [right-and]
            ==> (ordered M r)                        [def]])
      | (val-of ordered-empty-range) =>
          pick-any M i
            (!chain->
              [true
            ==> (ordered' nil)                       [SWO.ordered.empty]
            ==> (ordered' (collect M stop i))        [collect.of-stop]
            ==> (ordered M stop i)                   [def]])
      | (val-of ordered-subranges) =>
          (!force (adapt theorem))
    }

 (add-theorems theory |{[ordered-rest-range ordered-empty-range
                          ordered-subranges] := proofs}|)
} # close module Ordered-Range
```

Exercise 16.30: Fill in the proof of `ordered-subranges`. □

16.10 A binary search algorithm

While binary search on a range can be programmed in only a few lines of code, there are two issues that make the task challenging, especially if the goal is an abstract algorithm.

First, an algorithm that checks for equality between the searched-for value and the value at the midpoint of the range may be the most obvious way to program it, but in fact the resulting algorithm is not as efficient on average as one that only does strictly-less-than comparisons all the way down until the range is narrowed to a single value. We have already dealt with this issue with binary search on a binary search tree, in Chapter 15, and we must keep in mind this efficiency issue here too. The `lower-bound` abstract algorithm we present (which, again, is based closely on one in the STL) reduces the average number of comparisons it does by approximately 25 percent relative to the more obvious algorithm, and this is an important advantage in specializations of the algorithm for which comparison operations are expensive.

The second consideration is how to specify precisely what the abstract algorithm does, in a way that captures its full generality without becoming too difficult to understand—or to reason about in the correctness proof. If we look at how binary search on a range is described in expositions of the STL, we typically find something like the following:

> *The* `lower-bound` *algorithm returns an iterator* `i` *referring to the first position into which the value can be inserted while maintaining the sorted ordering.*

This is concise, general, and—at least in some contexts—a useful way to think about binary search on a range, especially if the purpose of calling the algorithm is actually to find where to insert the value into the range in a subsequent program step. But in other contexts, this kind of specification in terms of a hypothetical use of a state-changing operation like range-insertion is unsuitable, such as when we are doing a binary search only to determine whether or not the search value is present in the range. (Note that although `lower_bound` does not answer this question by itself, one only has to compare the value at the returned iterator position for equality with the search value.)

The following definition of our `lower-bound` function is tailored to addressing the first consideration; like the binary search algorithm on binary search trees developed in Section 15.1, it does no strict-equality comparisons at all:

```
extend-module Ordered-Range {

  declare lower-bound: (S, X) [(It X S) (It X S) S] -> (It X S)

  module lower-bound {

    define half := N.half

    define (axioms as [empty go-right go-left]) :=
      (fun
        [(M \\ (lower-bound i j x)) =
          let {mid := (i + half (j - i))}
            [i                                  when (i = j)
```

```
          (M \\ (lower-bound
                    (successor mid) j x))       when (i =/= j &
                                                      M at deref mid < x)

          (M \\ (lower-bound i mid x))          when (i =/= j &
                                                     ~ M at deref mid < x)]])

    (add-axioms theory axioms)
```

Turning to the second issue, we have to be careful in specifying the algorithm to properly account for the boundary cases at the ends of the range. The sentence position-found below is the main correctness result for lower-bound. It is expressed as the proposition that every range r has the property position-found-prop. That property is itself expressed in terms of two auxiliary procedures that construct the respective hypothesis and conclusion, respectively.

```
define (position-found-condition r M i j x k) :=
   ((range i j) = SOME r &
    (ordered M r) &
    k = M \\ (lower-bound i j x))

define (position-found-conclusion r M i j x k) :=
   ((k *in r | k = j) &
    (k =/= i ==> M at deref predecessor k < x) &
    (k =/= j ==> x <E M at deref k))

define (position-found-prop r) :=
   (forall M i j x k .
     (range i j) = SOME r &
     (ordered M r) &
     k = M \\ (lower-bound i j x)
     ==> (k *in r | k = j) &
         (k =/= i ==> M at deref predecessor k < x) &
         (k =/= j ==> x <E M at deref k))

 define position-found := (forall r . position-found-prop r)
```

For the proof, we will need to apply an induction hypothesis for ranges that are half the length of the given range, so strong induction is required. The *measure-induction* approach described in Section 12.9 allows us to do this without artificially introducing a natural number quantifier. Instead, we call upon the measure-induction method and use Range.length as the measure. We begin with a proof sketch, expressed as a higher-order binary method make-proof, that sets up the top-level structure of the measure-induction proof. The two arguments to make-proof are themselves methods, which handle the two cases of the main **datatype-cases** deduction in the body of the output method: one case for stop and one for back. Each of these two case handlers takes a number of arguments capturing useful values, namely, the tools returned by the call to `proof-tools`, the various adapted function symbols, the main universally quantified variable `r:(Range 'X 'S)` introduced by the **`pick-any`** on line 16, various other variables introduced by the subsequent **`pick-any`** on line 20, as well as the pattern variable from the corresponding branch of the **`datatype-cases`** deduction on line 29: `i0` in the `stop` case and `r0` in the `back` case. To get things started, we define trivial versions of these two handlers that simply **`force`** the needed result in each case.

```
1  define <' := N.<
2  overload * N.*
3
4  define (make-proof stop-case-handler back-case-handler) :=
5    method (theorem adapt)
6      let {tools := (proof-tools adapt theory);
7           symbols := (adapt [< <E ordered deref *in
8                              successor predecessor
9                              I+N I-N I-I])}
10       match theorem {
11         (val-of position-found) =>
12           (!strong-induction.measure-induction
13             (adapt theorem)
14             length
15             pick-any r:(Range 'X 'S)
16               assume ind-hyp := (forall r' .
17                                   length r' <' length r
18                                    ==> position-found-prop r')
19                 pick-any M:(Memory 'S)
20                          i:(It 'X 'S)
21                          j:(It 'X 'S)
22                          x:'S
23                          k:(It 'X 'S)
24                   assume hyp := (position-found-condition r M i j x k)
25                     let {uvars := [M i j x k];
26                          goal :=  lambda (r)
27                                     (position-found-conclusion r M i j x k)}
28                       datatype-cases (goal r) on r {
29                         (st as (stop i0:(It 'X 'S))) =>
30                           conclude (goal st)
31                             (!stop-case-handler tools symbols r uvars i0)
32                       | (bk as (back r0:(Range 'X 'S))) =>
33                           conclude (goal bk)
34                             (!back-case-handler tools symbols r uvars r0)
35                       } # datatype-cases
36           ) # strong-induction.measure-induction
37       } # match theorem
38
39 define (trivial-stop-case-handler tools fsyms r uvars i0) :=
40   match uvars {
41     [M i j x k] => (!force (position-found-conclusion (stop i0) M i j x k))
42   }
```

```
43
44 define (trivial-back-case-handler tools fsyms r uvars r0) :=
45   match uvars {
46     [M i j x k] => (!force (position-found-conclusion (back r0) M i j x k))
47   }
48
49 (add-theorems theory |{position-found :=
50                        (make-proof trivial-stop-case-handler
51                                    trivial-back-case-handler)}|)
52
53  } # close module lower-bound
54 } # close module Ordered-Range
```

⋆ ***Exercise 16.31:*** Give a proper definition of stop-case-handler. □

For the (back r0) case, we give a definition of back-case-handler but leave two parts of it as exercises. These parts are themselves methods to be defined. Their initial definition simply uses force. For illustration purposes, we put these methods inside cells so that later on we can simply alter the contents of these cells (by inserting into them proper definitions of the relevant methods) without having to reinvoke add-theorems with updated proofs.

```
define case1-handler-cell :=
  (cell method (tools fsyms r r1 r2 r3 uvars r0 case1 goal)
        (!force goal))

define case2-handler-cell :=
 (cell method (tools fsyms r r1 r2 r3 uvars r0 case2 goal)
        (!force goal))

define (back-case-handler tools syms r uvars r0) :=
  let {[<:(OP 2) <E:(OP 2) ordered:(OP 2) deref:(OP 1)
        *in:(OP 2) successor:(OP 1) predecessor:(OP 1)
        I+N:(OP 2) I-N:(OP 2) I-I:(OP 2)] := syms;
       [M i j x k] := uvars;
       [parity +=-cancellation] := [N.parity N.Plus.=-cancellation];
       [_ prove chain chain-> _] := tools}
    conclude goal := (position-found-conclusion (back r0) M i j x k)
      let {hyp1 := ((range i j) = SOME r);
           hyp2 := (ordered M r);
           hyp3 := (k = M \\ (lower-bound i j x));
           nb := (!prove nonempty-back1);
           i=/=j := (!chain->
                      [(range i j)
                     = (SOME r)                        [hyp1]
                     = (SOME back r0)                  [(r = back r0)]
                   ==> (i =/= j)                       [nb]]);
           (and p2 p3) := (!chain->
                            [((range i j) = SOME r)
```

```
                  ==> ((range i j) =
                       (range start r
                              finish r))     [range.collapse]
                  ==> (i = start r &
                       j = finish r)         [range.injective]]);
     rl2 := (!prove length2);
     n := (length r);
     p4 := (!chain
              [n = ((finish r) - (start r))    [rl2]
                 = (j - i)                     [p2 p3]]);
     p4' := (!by-contradiction (n =/= zero)
               assume (n = zero)
                 (!absurd
                   (!chain
                     [(S length r0)
                      = (length back r0)       [length.of-back]
                      = n                      [(r = back r0)]
                      = zero                   [(n = zero)]])
                   (!chain->
                     [true
                    ==> (S length r0 =/= zero)  [N.S-not-zero]])));
     p5 := (!chain->
              [(n =/= zero)
           ==> (half length r <' n)            [N.half.less]
           ==> (half (j - i) <' n)             [p4]]);
     p6 := (!chain->
              [p5
           ==> (half (j - i) <= n)             [N.Less=.Implied-by-<]]);
     mid := (i + half (j - i));
     os := (!prove ordered-subranges);
     p7 := (!chain->
              [(hyp1 & hyp2 & p6)
           ==> (exists r1 r2 .
                 (range i mid) = SOME r1 &
                 (range mid j) = SOME r2 &
                 (ordered M r1) &
                 (ordered M r2))               [os]])}
  pick-witnesses r1 r2 for p7 p7-w
    let {p7-w1 := ((range i mid) = SOME r1);
         p7-w2 := ((range mid j) = SOME r2);
         p7-w3 := (ordered M r1);
         p7-w4 := (ordered M r2);
         iic := (!prove I-I-cancellation);
         rl3 := (!prove length3);
         p8 := (!chain
                  [p7-w1
               ==> (length r1 = mid - i)      [rl3]]);
         p9 := (!chain
                  [p7-w2
               ==> (length r2 = j - mid)      [rl3]]);
         q1 := (!chain [(length r1)
```

```
                  = (mid - i)              [p8]
                  = (half (j - i))         [iic]
                  = (half n)               [p4]]);
       rl4 := (!prove length4);
        q2 := (!chain->
                [(hyp1 & p7-w1 & p7-w2)
             ==> (n = (length r1) +
                      (length r2))         [rl4]]);
       q3 := (!chain
              [n
              = (N.two * (half n)
                     + (parity n))         [N.parity.half-case]
              = (((half n) + (half n))
                     + (parity n))         [N.Times.two-times]
              = (((half n) + (parity n))
                     + (half n))           [N.Plus.associative
                                            N.Plus.commutative]]);
       q4 := (!chain->
              [((length r2) + (half n))
              = ((half n) + (length r2))   [N.Plus.commutative]
              = ((length r1) +
                 (length r2))              [q1]
              = n                          [q2]
              = (((half n) + (parity n))
                   + (half n))             [q3]
            ==> (length r2 = (half n) +
                             (parity n))   [+=-cancellation]]);
       nzl := (!prove nonzero-length);
       q5 := (!chain->
              [(n =/= zero)
            ==> ((half n) + (parity n)
                    =/= zero)              [N.parity.plus-half]
            ==> (length r2 =/= zero)       [q4]
            ==> (exists r3 . r2 = back r3) [nzl]])}
    pick-witness r3 for q5 q5-w
    # We now have (r2 = back r3)
      (!two-cases
        assume case1 := ((M at (deref mid)) < x)
          (!(ref case1-handler-cell)
            tools syms r r1 r2 r3 uvars r0 case1 goal)
        assume case2 := (~ (M at (deref mid)) < x)
          (!(ref case2-handler-cell)
            tools syms r r1 r2 r3 uvars r0 case2 goal))
```

⋆ ***Exercise 16.32:*** Insert a proper method definition into `case1-handler-cell`. □

⋆ ***Exercise 16.33:*** Insert a proper method definition into `case2-handler-cell`. □

Exercise 16.34: If random-access iterators are not available, such as with a linked list for which only forward iterators are possible, it takes linear rather than constant time to compute the midpoint of a range. In some circumstances it might nevertheless be advantageous to use binary search rather than a linear search algorithm. Explain. □

16.11 Summary and suggestions for continued study

In this chapter, we have written formal specifications for a small number of abstract algorithms on memory abstractions, presented implementations expressed as recursively defined functions, and developed proofs of their correctness with respect to their specifications (with some details to be filled in by the reader). The memory range abstractions and algorithms we have chosen are inspired by those in the C++ Standard Template Library, but they are only a small sample of that library's collection of abstract (or generic) algorithms and data structures. A natural continuation of this line of study would be to select other algorithms from the collection and attempt to write specifications, implementations, and proofs using the same basic approach as we have illustrated in this chapter. The easiest cases would be algorithms that are very closely related to ones studied in this chapter. For example, `count_if` differs from `count` only in having an extra predicate argument to use instead of an equality test, and similarly for `replace_if`. In these cases, only relatively simple modifications of the developments of the algorithm studied in this chapter would be necessary. A somewhat more interesting and challenging project would be to develop a parameterized version of the correctness proof that would apply to both the equality and predicate versions.

Another case of closely related algorithms is the `lower_bound`, `upper_bound` pair of binary search algorithms. In this case, adapting the development we have given for `lower_bound` would require careful attention to the details of the formal specification of `upper_bound`, particularly the boundary conditions, which would in turn require many small adjustments in the proof of correctness. For this case, and probably for many of the STL algorithms that are not closely related to any of the ones studied in this chapter, the effort required would be nontrivial, constituting at least a small project and perhaps a major one in some cases.

Some guidance about the difficulty of such projects can be derived based on a classification of STL algorithms such as the one given by Musser et al. [76]:

- *Nonmutating Sequence Algorithms* do not modify the memory ranges on which they operate. The `count` algorithm falls in this category. As we have seen with `count`, it is unnecessary to model changes to memory, and there are thus no memory safety issues to deal with in the specification or proof of correctness. Some of the simpler algorithms in this category are `find`, `adjacent_find`, and `mismatch`, but also included is `search` for finding an occurrence of one given sequence in another (a generalization of string search), the best algorithms for which are quite sophisticated. Even with the `lower_bound` binary search algorithm we have studied, there are many subtle details in specifying and reasoning about it. (Binary search algorithms, although they are nonmutating, are classified instead as *Sorting Related Algorithms* [76].)

- *Mutating Sequence Algorithms* do modify memory ranges, and thus the memory safety issues raised must be dealt with in specifications and proofs, as we have done in this chapter with `replace`, `copy`, and `copy-backward`. Some of the simpler algorithms in this category are `fill`, `generate`, and `transform`, with `unique`, `rotate` and `partition` probably being somewhat more difficult.
- *Generalized Numeric Algorithms* includes both nonmutating (`accumulate` and `inner_product`) and mutating algorithms (`partial_sum` and `adjacent_difference`) that operate on numeric sequences. Although they bring into play some of the algebraic theories that we studied in Chapter 14, they are probably no more difficult than some of the simpler algorithms in the preceding categories.
- Some of the *Sorting-Related Algorithms* and *Set Operations on Sorted Structures* may present the greatest challenges. We have already seen that even a nonmutating algorithm, `lower_bound`, can require attention to many details even in its specification and a proof that is many times longer than its code. The mutating `sort` and `merge` algorithms may be even bigger projects in terms of proof difficulty, though they will be simpler to specify than `lower_bound` because they do not have any special boundary conditions. Specifying and reasoning about any of the algorithms in this category will have greatest generality if the strict weak order relation (Section 14.9.6) is imposed on sequence values rather than total ordering.

Of course, it's a big world of algorithms and data structures, of which the STL represents only a tiny part. The advantage of working within this subset is that they are relatively simple but still of great importance, because of their generality of application and their widespread use in practice. But we hope that readers will also be motivated to go beyond the STL and apply the experience gained with the examples and exercises in this book to tackle specification and proof projects drawn from elsewhere.

Finally, note that all of the proofs in this chapter were written entirely in Athena and without recourse to any automated theorem proving (apart from the very limited form of proof search performed by the various chaining methods). While this approach has advantages, especially for instructive purposes, it also has limitations: Proofs of nontrivial results tend to grow significantly in size and complexity. While abstraction and structure can control that growth to a considerable extent, ultimately we need more reasoning automation if we wish to take larger steps, as we indeed often do in more challenging modeling and verification projects. The ATP appendix on the book's web site describes Athena's facilities for automated reasoning, including theorem proving and model building.

VI PROOFS ABOUT PROGRAMMING LANGUAGES

In this final part of the book we illustrate a number of techniques for representing and reasoning about programming languages. We begin in Chapter 17 by introducing an expression language, restricted to numeric expressions defined by a datatype, and specify the semantics of these expressions by a simple recursive interpreter. We then introduce a machine language that manipulates a stack and specify its operational semantics through a simple virtual machine. Next, we define a compiler that translates numeric expressions into the machine language and prove that the compiler is correct. Finally, we extend the compiler with explicit error handling and redo the correctness proof accordingly.

In Chapter 18 we study a richer programming language, the **While** language. Although still very simple compared to real programming languages, it contains some of the most fundamental and essential features of imperative languages, providing sufficient context for discussion of various important issues in programming language syntax and semantics.

17 A Correctness Proof for a Toy Compiler

IN this chapter we introduce a simple language for numeric expressions and specify its semantics by a recursive interpreter. We then introduce a stack-based machine language, define a compiler that translates numeric expressions into programs for that machine, and prove that the compiler is correct. The initial version of the compiler, presented in Section 17.1, does not specify what happens when there is an error, but in Section 17.2 we explicitly add and reason about error handling.

The languages studied in this chapter are tiny, not even having variables, but they are sufficiently interesting to illuminate a number of techniques that are useful in the study of programming languages and even more broadly. The compilation algorithm for expressions that we study here is similar to the way in which real-world compilers produce machine code for expressions.

17.1 Interpreting and compiling numeric expressions

17.1.1 Representation and notation

Let us define a datatype for representing simple numeric expressions. To keep things really simple, we only consider ground expressions, that is, expressions without variables. So an expression here will be either a constant natural number, such as (the unary Peano representations of) 4 or 58, or else it will be the sum, difference, product, or quotient of two expressions. Accordingly, the abstract grammar[1] of such expressions can be informally described as follows:

$$e ::= n \mid e_1 + e_2 \mid e_1 - e_2 \mid e_1 \cdot e_2 \mid e_1 / e_2$$

where n ranges over the natural numbers. The inductive datatype `Exp` below can be used to represent expressions of this form. We will put all the work of this chapter inside a module called `TC` (for "`Toy Compiler`"):

```
module TC {

  datatype Exp := (const N)
                | (sum Exp Exp)
                | (diff Exp Exp)
                | (prod Exp Exp)
                | (quot Exp Exp)

  ...
```

1 Refer to the next chapter for more extensive remarks on abstract syntax.

The constructor const builds simple (constant) natural-number expressions; the constructor sum builds sums, and so on. Thus, for instance, the expression $1 + 2$ is captured by the term

```
(sum (const (S zero))
     (const (S (S zero))))
```

while $1 + (2 \cdot 3)$ is represented by

```
(sum (const (S zero))
     (prod (const (S (S zero)))
           (const (S (S (S zero))))))
```

If we use infix notation for the binary operators and omit parentheses around consecutive applications of unary operators, then these two terms can be written as:

```
((const S zero) sum (const S S zero))
```

and

```
((const S zero) sum (const S S zero) prod (const S S S zero)).
```

We specify a high precedence level for const, to ensure that it binds tighter than the other constructors (whose default precedence level is 110):

```
set-precedence const 350
```

Thus, the preceding terms representing $1 + 2$ and $1 + (2 \cdot 3)$ can now be written as:

```
(const S zero sum const S S zero)                                  (17.1)
```

and

```
(const S zero sum const S S zero prod const S S S zero).           (17.2)
```

It is also convenient to make the expression constructorsaccept integer numeral arguments as well as natural numbers. Here is one way to do that:

```
define (int->const x) :=
  check {(integer-numeral? x) => (const int->nat x)
       | else => x}

expand-input const [int->nat]
expand-input sum, diff, prod, quot [int->const int->const]
```

Intuitively, the first **expand-input** directive forces const to accept integer numerals as inputs, which will be converted to natural numbers by the procedure int->nat. The second directive does the same for the four binary constructors of Exp, and for both argument positions of each, but using int->const rather than int->nat. Thus, (17.1) and (17.2) can now be written quite simply as follows, respectively:

```
(1 sum 2)
```

and

```
(1 sum 2 prod 3).
```

Going further still, let us overload the usual numeric function symbols (+, etc.) to do double duty as constructors of Exp, and then also as the corresponding operators on natural numbers:

```
overload (+ sum) (- diff) (* prod) (/ quot)
overload (+ N.+) (- N.-) (* N.*) (/ N./)
```

and let us define a few variables ranging over Exp and N for future reference:

```
define [e e' e1 e2 e3 e4] := [?e:Exp ?e':Exp ···]

define [n n' n1 n2 n3 n4] := [?n:N ?n':N ?n1:N ···]
```

The sorts of these variables can now serve to disambiguate uses of the overloaded symbols:

```
> (e1 + e2)

Term: (sum ?e1:Exp ?e2:Exp)

> (n1 * n2)

Term: (N.* ?n1:N ?n2:N)
```

17.1.2 Defining the interpreter

Next, we introduce an interpreter that takes as input an expression *e* and produces a natural number denoting the value of *e*:

```
declare I: [Exp] -> N [350]
```

For example, the application of I to the expression

```
(1 sum 2)
```

should represent the result 3, or more precisely, the natural number

```
(S (S (S zero))).
```

Note that we have given I a precedence level higher than the precedence levels of the numeric operators N.+, N.-, etc. Thus, for instance,

```
(I e1 + I e2)
```

is parsed as the application of `N.+` to `(I e1)` and `(I e2)`.

We can now specify the behavior of the interpreter `I` as follows:

```
assert* I-def :=
  [(I const n = n)
   (I (e1 + e2) = I e1 + I e2)
   (I (e1 - e2) = I e1 - I e2)
   (I (e1 * e2) = I e1 * I e2)
   (I (e1 / e2) = I e1 / I e2)]
```

If we view `I` as a specification of the meaning of these expressions, then the first equation can be understood as saying that the meaning of a constant expression (`const` n) is n. The remaining equations are all similar in their structure, and are paradigms of recursive compositionality: They specify that the meaning of a complex expression is obtained by applying the corresponding numeric operation to the meanings of the two immediate subexpressions. We emphasize the overloaded use of these symbols. The occurrence of + on the left resolves to `sum`, whereas the occurrence of + on the right resolves to `N.+`, and so forth.

As usual, we can turn to `eval` to test the definition. It will be clearer to see evaluation results in integer notation, so we first issue a **`transform-output`** directive. (The role of the second transformer, from list terms to Athena lists, will become clearer when we come to discuss the compiler, which produces lists of instructions as its output. Refer to Exercise 3.17 for a discussion of `clist->alist` and its converse, `alist->clist`.)

```
transform-output eval [nat->int (clist->alist nat->int)]

> let {2+3  := (2 sum 3);
       10-6 := (10 diff 6);
       10/2 := (10 quot 2);
       9*9  := (9 prod 9)}
    (print "\nValue of 2+3:\t" (eval I 2+3) "\nand 10-6:\t" (eval I 10-6)
           "\nand 10/2:\t" (eval I 10/2) "\nand 9*9:\t" (eval I 9*9) "\n")

Value of 2+3:    5
and 10-6:        4
and 10/2:        5
and 9*9:         81

Unit: ()

> (eval I (4 quot 0))

Unable to reduce the term:

(TC.I (TC.quot (TC.const (S (S (S (S zero)))))
               (TC.const zero)))
```

```
to a normal form.

Unit: ()
```

Our definition of `I` seems to have captured our intentions.

17.1.3 An instruction set and a virtual machine

We are now ready to introduce a lower-level "machine language" into which we will be compiling the numeric expressions (ground `Exp` terms). This machine language will consist of a few simple commands for manipulating a stack, namely:

- pushing a natural number on top of the stack; and
- adding, subtracting, multiplying, or dividing the top two elements of the stack, and replacing them with the result of the operation.

The datatype `Command` represents these possibilities:

```
datatype Command := (push N) | add | sub | mult | div
```

As with `const`, it will be convenient to use integer numerals with `push`, in addition to natural numbers. We also set its precedence level to 350:

```
expand-input push [int->nat]
set-precedence push 350
```

A program in this machine language is simply a list of commands. We define an abbreviation for this sort:

```
define-sort Program := (List Command)
```

We also introduce a sort `Stack` as an abbreviation for lists of natural numbers[2] and some appropriately named variables for these two sorts:

```
define-sort Stack := (List N)

define [cmd cmd' prog prog1 prog2 stack stack1 stack2] :=
       [?cmd:Command ··· ?prog:Program ··· ?stack:Stack ···]
```

We can now define the operation of the abstract machine by means of a binary function `exec` that takes as inputs (a) a program (i.e., a list of commands) and (b) a stack of natural numbers; and executes the program with respect to the given stack, eventually producing a natural number as the result (barring cases of error, which will be discussed later):

2 A directive of the form **`define-sort`** *I* := *S* simply introduces *I* as a name for the (possibly complex) sort *S*.

```
declare exec: [Program Stack] -> N [wrt 101 [(alist->clist id)
                                              (alist->clist int->nat)]]
```

We have expanded the inputs of exec so that it can work directly with Athena (square-bracket) lists on both argument positions, and so that any elements of the second list argument that happen to be integer numerals are automatically converted to natural numbers. We have also set its precedence level to 101, and we have introduced the name `wrt` as a more infix-friendly alias for it, so we can write, for instance, `(prog wrt stack)` interchangeably with `(exec prog stack)`. Because we will be dealing with lists quite a bit, we also expand the input range of the "consing" constructor `::` (the reflexive constructor for the `List` datatype), so that it can accept Athena lists in the second argument position, and we define `++` as `List.join`.

```
expand-input :: [id (alist->clist int->nat)]
define ++ := List.join
```

The definition of exec will constitute a sort of (small-step[3]) operational semantics for our machine language. Again, we define it with a number of axioms using structural recursion on the first argument—the input list of commands:

```
assert* exec-def :=
    [([] wrt n::_ = n)
     ((push n)::rest wrt stack = rest wrt n::stack)
     (add::rest  wrt n1::n2::stack = rest wrt (n1 + n2)::stack)
     (sub::rest  wrt n1::n2::stack = rest wrt (n1 - n2)::stack)
     (mult::rest wrt n1::n2::stack = rest wrt (n1 * n2)::stack)
     (div::rest  wrt n1::n2::stack = rest wrt (n1 / n2)::stack)]
```

We again use eval to test the definition of exec:

```
# Execute an empty program on a nonempty stack:

> (eval [] wrt [99])

Term: 99

# Execute an addition instruction on a stack with at least two numbers:

> (eval [add] wrt [2 3])

Term: 5

# First add 2 and 3, then multiply the result by 6:

> (eval [add mult] wrt [2 3 6])
```

3 See the next chapter for additional discussion of small-step vs. big-step operational semantics.

```
Term: 30

# Add 2 and 3, multiply by 6, then divide by 10:

> (eval [add mult div] wrt [2 3 6 10])

Term: 3

# Execute a binary instruction with too few stack operands:

> (eval [add] wrt [2])

Unable to reduce the term:

(TC.exec (:: TC.add
             nil)
         (:: (S (S zero))
             nil))

to a normal form.

Unit: ()

# Divide by zero:

> (eval [div] wrt [8 0])

Unable to reduce the term:

(TC.exec (:: TC.div
             nil)
         (:: (S (S (S (S (S (S (S (S zero))))))))
             (:: zero
                 nil)))

to a normal form.

# And a longer example:

> define input-program := [(push 1) (push 2) add (push 2) mult mult]

List input-program defined.

> define input-stack := [4]

List input-stack defined.

> (eval input-program wrt input-stack))

Term: 24
```

Let us now define a virtual machine that executes a given program on the empty stack:

```
declare run-vm: [Program] -> N [[(alist->clist id)]]

assert* vm-def := [(run-vm prog = prog wrt [])]
```

And again using eval:

```
> (eval run-vm [(push 2) (push 3) add])

Term: 5
```

17.1.4 Compiling numeric expressions

We are finally ready to define the compiler. It takes an arithmetic expression and returns a program in the machine language of the previous section that will (hopefully) compute the value of the expression:

```
declare compile: [Exp] -> Program [380]
```

As usual, the definition of the compiler will be given by structural recursion, with constant natural numbers constituting the basis case.

```
assert* compiler-def :=
  [(compile const n = (push n)::[])
   (compile (e1 + e2) = compile e2 ++ compile e1 ++ [add])
   (compile (e1 - e2) = compile e2 ++ compile e1 ++ [sub])
   (compile (e1 * e2) = compile e2 ++ compile e1 ++ [mult])
   (compile (e1 / e2) = compile e2 ++ compile e1 ++ [div])]
```

For an expression of the form `(const` n`)`, the compiled program is just the one-element list `[(push` n`)]`. For a compound expression `(`e_1 *op* e_2`)`, the idea is to recursively compile e_1 and e_2, join the obtained instruction lists in the right order (so that the "program" for e_2 ends up above the program for e_1, which means that later on its result will get pushed on the stack *below* the result of e_1's program), and then finally append the proper machine instruction for *op*. We can test the compiler on a number of different inputs:

```
> (eval compile const 5)

List: [(TC.push (S (S (S (S (S zero))))))]

> (eval compile (2 sum 3))

List: [(TC.push (S (S (S zero))))          # push 3
       (TC.push (S (S zero)))              # push 2
       TC.add]                             # add
```

```
> (eval compile (2 sum 3 prod 5))

List: [(TC.push (S (S (S (S (S zero))))))  # push 5
       (TC.push (S (S (S zero))))          # push 3
       TC.mult                             # multiply
       (TC.push (S (S zero)))              # push 2
       TC.add]                             # add
```

17.1.5 Correctness

Is this simple compiler correct? What does correctness mean in this context? As with all compilers, correctness here means that the semantics of the target program are the same, in some appropriate sense, as the semantics of the source. In this case, the semantics of the source expression are given by the interpreter `I`, while the semantics of the compiled program are given by `exec`. To say that the two semantics are the same means that if we apply the virtual machine to the compiled program we will obtain the same result that we get by applying the interpreter `I` to the source expression:

```
define compiler-correctness := (forall e . run-vm compile e = I e)
```

We can express this condition graphically as follows:

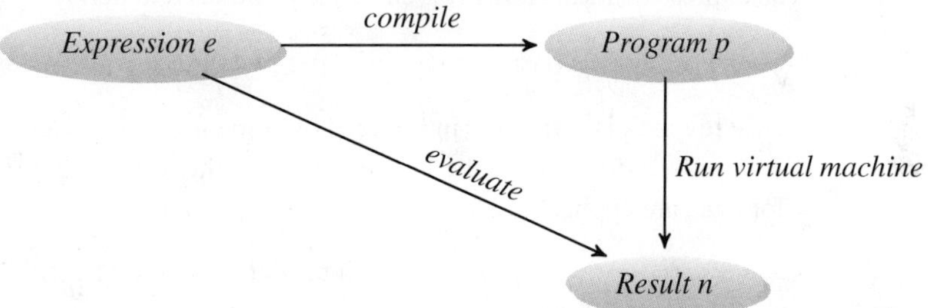

A diagram of this form is known as a *commuting diagram*. In such a diagram, every directed arc represents a function (or a *morphism*, in alternative terminology). The idea is that the function compositions represented by any two directed paths from the same starting point to the same destination are identical functions: They produce the same outputs when applied to the same inputs.

So `compiler-correctness` is the property that we ultimately need to prove. Since this property is universally quantified over all expressions, which are inductively generated, we decide to go about it by way of structural induction. But a bit of experimentation will reveal that, as stated, the property is not quite suitable for induction. The semantics of our machine language are given in terms of `exec`, which takes a stack as an explicit second argument, and we need to take that into account. We thus define the property to be proven inductively as follows:

```
define (correctness e) := (forall stack . compile e wrt stack = I e)
```

If we manage to prove

$$\texttt{(forall e . correctness e)} \tag{17.3}$$

by induction, we can then easily derive the desired result, `compiler-correctness`, by specializing the stack argument with the empty list.

But before we go on and try to *prove* (17.3), let us first test it to make sure that it doesn't fail for any obvious cases that we might have overlooked:

```
define conjecture := (forall e . correctness e)

> (falsify conjecture 30)

Term: 'failure
```

With our confidence in the compiler's correctness thus bolstered, let us go on and try to prove it. Unfortunately, when we do so we soon realize that the present definition of correctness will not quite do; we need to *strengthen the inductive property*. The problem with this definition arises on the inductive steps. Suppose, for example, that we are inducting on an expression of the form (e_1 `sum` e_2), so we need to derive

$$\texttt{(compile (}e_1\texttt{ sum }e_2\texttt{) wrt stack = I (}e_1\texttt{ sum }e_2\texttt{))} \tag{17.4}$$

(where `stack` is a variable introduced by some preceding **`pick-any`**). The only option here is to reduce the term on the left-hand side to the term on the right-hand side. No axioms for `exec` are applicable, so our only course of action is to expand the subterm

$$\texttt{(compile (}e_1\texttt{ sum }e_2\texttt{))},$$

which will transform the left-hand side of (17.4) into:

$$\texttt{(compile }e_1\texttt{ ++ compile }e_2\texttt{ ++ [add] wrt stack).} \tag{17.5}$$

At this point, however, we are not able to exploit our inductive hypotheses about e_1 and e_2. The problem is that, in (17.5), the inductive subterms (`compile` e_1) and (`compile` e_2) are directly embedded inside `List.join` superterms,[4] whereas our inductive property, `correctness`, requires them to be embedded directly inside `exec` (`wrt`). Thus, (17.5) is irreducible and we cannot get any further in deriving (17.4).

As we have seen before, oftentimes the first attempt at an inductive proof does not go through, but when we *generalize* the property we are trying to prove—a technique known

4 Recall that we are using ++ as an abbreviation for `List.join`.

as *strengthening* the inductive property—then the proof succeeds, and we can obtain the initially desired property by simple specialization. If this still seems counterintuitive, keep in mind that the stronger the property we are trying to prove, the stronger the inductive hypothesis; so, by strengthening the inductive property, we give ourselves more ammunition during the inductive step. In this case, we will generalize the inductive property so that `List.join` (++) appears directly in it:

```
define (correctness e) :=
  (forall prog stack . compile e ++ prog wrt stack = prog wrt I e::stack)
```

Crucially, this is a generalization of our initial correctness property because the latter can be readily obtained from this by substituting the empty list of commands for the universally quantified `prog`.

Let us now go ahead and prove that every expression has this property:

`(forall e . correctness e).`

As usual, we proceed by structural induction. The basis step occurs when the expression is of the form `(const` n`)`. We express the step as a method parameterized over n:

```
define (basis-step n) :=
  pick-any p:Program stack:Stack
    (!chain [((compile const n) ++ p wrt stack)
            = ([(push n)] ++ p wrt stack)        [compiler-def]
            = ((push n)::p wrt stack)            [List.join.left-singleton]
            = (p wrt n::stack)                   [exec-def]
            = (p wrt I const n :: stack)         [I-def]])
```

`List.join.left-singleton` is the following property:

`(forall x L . [x] ++ L = x::L).`

We can test the basis step by itself:

```
> (!basis-step zero)

Theorem: (forall ?p:(List TC.Command)
           (forall ?stack:(List N)
             (= (TC.exec (List.join (TC.compile (TC.const zero))
                                     ?p:(List TC.Command))
                         ?stack:(List N))
                (TC.exec ?p:(List TC.Command)
                         (:: (TC.I (TC.const zero))
                             ?stack:(List N))))))
```

We continue with the four inductive cases, when the expression is either a sum, a difference, a product, or a quotient. Instead of writing four separate inductive subproofs, one for each case, we will write a single method that takes the overall expression as input, determines its form (whether it is a sum, product, etc.), and then acts accordingly. This is possible because the reasoning is essentially the same in all four cases, as we might have anticipated in advance (e.g., by looking at the `exec` axioms for the four commands, which shows that they all have the exact same structure). We will package this reusable bit of reasoning into a unary method called `inductive-step`. We first need a simple procedure that takes a binary expression constructor *exp-op* and produces the corresponding machine-language command and the corresponding operator on natural numbers:

```
define (exp-op->cmd-and-num-op exp-op) :=
  match exp-op {
    sum =>  [add N.+]
  | diff => [sub N.-]
  | prod => [mult N.*]
  | quot => [div N./]
  }
```

We can now define the `inductive-step` method as follows. To save space, we abbreviate `compile` as `C`. The **OP** annotations ensure that the corresponding names can be used in infix with the arities specified in the annotations. For example, the annotation `num-op:(OP 2)` allows `num-op` to be subsequently used as a binary function symbol.[5]

```
define (inductive-step e) :=
 match e {
  ((some-symbol exp-op:(OP 2)) e1:Exp e2:Exp) =>
    let {[cmd num-op:(OP 2)] := (exp-op->cmd-and-num-op exp-op);
         C:(OP 1) := compile;
         # Two inductive hypotheses, one for each sub-expression,
         # can be assumed to be in the a.b. when this method is called.
         [ih1 ih2] := [(correctness e1) (correctness e2)]}
      pick-any p:Program stack:Stack
       (!chain
         [(C (e1 exp-op e2) ++ p wrt stack)
        = (C e2 ++ C e1 ++ [cmd] ++ p wrt stack)  [compiler-def
                                                    List.join.Associative]
        = ([cmd] ++ p wrt (I e1)::(I e2)::stack)  [ih1 ih2]
        = (cmd::p wrt (I e1)::(I e2)::stack)      [List.join.left-singleton]
        = (p wrt (I e1 num-op I e2)::stack)       [exec-def]
        = (p wrt (I e)::stack)                    [I-def]])
 }
```

`List.join.Associative` is simply the associativity property for `List.join`:

```
(forall L1 L2 L3 . (L1 ++ L2) ++ L3 = L1 ++ (L2 ++ L3)).
```

5 Such "infix annotations" are occasionally needed because Athena has no global information available about locally introduced names like `num-op`. So if we want to use such a local name in infix, we need to attach an **OP** annotation to it (with an arity of either 1 or 2).

Let's go through the reasoning line by line. The first order of business is to decompose the input expression, which must be of the form (e_1 *exp-op* e_2). We then use

exp-op->cmd-and-num-op

to obtain the machine-language command *cmd* and natural-number operator *num-op* that correspond to the constructor *exp-op*; and we give the names ih1 and ih2 to the inductive hypotheses for the subexpressions e_1 and e_2, respectively. Then, on the **pick-any** line, we consider an arbitrary program (i.e., command list) p and an arbitrary stack and proceed to derive the desired identity by chaining. The first step goes through by the definition of compile and the associativity of list concatenation (++).[6] Observe that this step will go through only if cmd is the correct analogue of exp-op. The second step goes through owing to the inductive hypotheses. (We are assuming, of course, that the method will only be invoked in assumption bases that contain these inductive hypotheses.) The third step simply replaces the subterm [cmd] ++ p with cmd::p, which follows readily from the cited property of List.join. The fourth step follows from the definition of exec. Note here, too, that this will go through only if num-op correctly corresponds to exp-op: N.+ to sum, and so forth. Finally, the fifth step goes through by the definition of the interpreter.

A benefit of factoring out all the inductive work into a separate method is that we can test its reasoning independently:

```
set-flag print-var-sorts "off"

> pick-any e1:Exp e2:Exp
    assume (correctness e1)
      assume (correctness e2)
        (!inductive-step (e1 + e2))

Theorem: (forall ?e1:Exp
           (forall ?e2:Exp
             (if (forall ?prog:(List Command)
                   (forall ?stack:(List N)
                     (= (exec (List.join (compile ?e1)
                                         ?prog)
                              ?stack)
                        (exec ?prog
                              (:: (I ?e1)
                                  ?stack)))))
                 (if (forall ?prog:(List Command)
                       (forall ?stack:(List N)
                         (= (exec (List.join (compile ?e2)
                                             ?prog)
                                  ?stack)
                            (exec ?prog
                                  (:: (I ?e2)
```

6 Convince yourself that both are necessary for the step, and that nothing else is needed.

```
                                    ?stack)))))
                        (forall ?p:(List Command)
                          (forall ?stack:(List N)
                            (= (exec (List.join (compile (sum ?e1 ?e2))
                                                        ?p)
                                      ?stack)
                               (exec ?p
                                      (:: (I (sum ?e1 ?e2))
                                          ?stack)))))))))
```

The overall inductive proof now takes the following form:

```
> define main-correctness-theorem :=
    by-induction (forall e . correctness e) {
      (const n) =>           (!basis-step n)
    | (e as (sum _ _)) =>  (!inductive-step e)
    | (e as (diff _ _)) => (!inductive-step e)
    | (e as (prod _ _)) => (!inductive-step e)
    | (e as (quot _ _)) => (!inductive-step e)
    }

Theorem: (forall ?e:TC.Exp
           (forall ?prog:(List TC.Command)
             (forall ?stack:(List N)
               (= (exec (List.join (compile ?e)
                                    ?prog)
                         ?stack)
                  (exec ?prog
                         (:: (I ?e)
                             ?stack))))))

Sentence main-correctness-theorem defined.
```

The top-level result that we have been after can now be derived in a few lines:

```
conclude compiler-correctness
  pick-any e
    (!chain
      [(run-vm compile e)
     = (compile e wrt [])              [vm-def]
     = (compile e ++ [] wrt [])        [List.join.right-empty]
     = ([] wrt (I e)::[])              [main-correctness-theorem]
     = (I e)                           [exec-def]])

Theorem: (forall ?e:TC.Exp
           (= (TC.run-vm (TC.compile ?e))
              (TC.I ?e)))
```

As you might guess from its name, `List.join.right-empty` is the following property:

`(forall L . L ++ [] = L).`

All three of the proofs we have given could be shortened significantly. In general, a chaining proof can be shortened in the following ways: by reducing the number of steps in it, or by reducing the amount of justification provided for each step, or both. For instance, the basis step could be carried out in just two steps, and without citing any justification whatsoever, in which case Athena's rewrite engine tries to discover the appropriate justifications on its own, by examining the entire assumption base:

```
define basis-step :=
  method (n)
    pick-any p:Program stack:Stack
      (!chain [(compile const n ++ p wrt stack)
             = (p wrt n::stack)
             = (p wrt I const n :: stack)])
```

The inductive step, too, could be abbreviated considerably—here, in three steps:

```
define inductive-step-shorter :=
 method (exp)
  match exp {
    ((some-symbol exp-op:(OP 2)) e1:Exp e2:Exp) =>
      let {[cmd num-op:(OP 2)] := (exp-op->cmd-and-num-op exp-op);
           [ih1 ih2] := [(correctness e1) (correctness e2)]}
        pick-any p:Program stack:Stack
         (!chain
           [(compile (e1 exp-op e2) ++ p wrt stack)
          = (compile e2 ++
              compile e1 ++
                [cmd] ++ p wrt stack)                [compiler-def
                                                      List.join.Associative]

          = ([cmd] ++ p wrt (I e1)::(I e2)::stack)   [ih1 ih2]

          = (p wrt (I exp)::stack)                   [List.join.left-singleton
                                                      exec-def I-def]])
      }
```

Indeed, the body of the method could be written in just one step, and without any justification:

```
define inductive-step :=
  method (exp)
    match exp {
      ((some-symbol exp-op:(OP 2)) e1:Exp e2:Exp) =>
        let {[cmd num-op:(OP 2)] := (exp-op->cmd-and-num-op exp-op);
             [ih1 ih2] := [(correctness e1) (correctness e2)]}
          pick-any p:Program stack:Stack
            (!chain [(compile (e1 exp-op e2) ++ p wrt stack)
                   = (p wrt (I exp)::stack)])
    }
```
(17.6)

Chains that take large steps without citing any justification, such as the one above, rely on nondeterministic proof search, so one and the same call to `chain` might succeed some times but not others, due to a certain amount of randomness in the search. However, if it succeeds at all, it is typically after no more than two or three attempts, each of which will be fast (no more than a few seconds). Moreover, when the attempt succeeds, we can capture the discovered proof via a call to `find-eqn-proof`, as described in footnote 21, and then, if desired, we can readily transcribe that into a more detailed `chain` proof that does not need to perform any search. For instance, suppose we place a call to `find-eqn-proof` in the body of (17.6) as follows:

```
define inductive-step :=
  method (exp)
    match exp {
      ((some-symbol exp-op:(OP 2)) e1:Exp e2:Exp) =>
        let {[cmd num-op:(OP 2)] := (exp-op->cmd-and-num-op exp-op);
             [ih1 ih2] := [(correctness e1) (correctness e2)]}
          pick-any p:Program stack:Stack
            let {left :=  (compile (e1 exp-op e2) ++ p wrt stack);
                 right := (p wrt (I exp)::stack);
                 proof := (find-eqn-proof left right (ab));
                 _     := (print "\nFound proof:\n" proof "\n")}
              (!chain [left = right])
    }
```

Then if and when the method works, we will not only get a theorem from it, but the output will also present the discovered proof in a detailed format that could be easily reformulated as a larger `chain` proof. In this case, the automatically discovered proof will essentially be the same as the one we manually presented in our first definition of `inductive-step`.

The question of what is the proper level of detail for a chaining proof, or for any proof for that matter, does not have a straightforward answer. It involves a classic computer science tradeoff: time vs. space. Longer proofs are more tedious to write but can be checked very efficiently because they involve little or no search. Shorter proofs, by contrast, are easier to write but harder to check, as the verification of each step might require a good deal of search. Where to draw the line depends on various factors, such as the level of sophistication of the intended audience and, in the case of mechanical proofs, the difficulty of the problem and the abilities of our proof-search algorithms and hardware.

In this book we try to strike a middle ground, though we generally prefer somewhat longer proofs for a number of reasons. First, readability. The last version of the `inductive-step` method, for instance, does not provide much information at all. In fact, it leaves us in the dark as to why the identity holds. We typically want to depict at least the key steps of a derivation, and that is why we prefer the original version of `inductive-step`; and we also want to pinpoint the sentences in the assumption base that justify each step. Of course, what exactly constitutes the "key steps" of a derivation can be somewhat subjective, and we must be careful not to tilt the balance too heavily in the opposite direction—too much detail can clutter a proof. Also, when it comes to justifying an individual step of the derivation, we generally do not require the list of cited sentences to contain *all and only* the information that is required for that step. Oftentimes we list a superset of that information and leave it to Athena to filter out the superfluous stuff. For instance, instead of specifying a particular interpreter axiom, we may simply cite `I-def` in general—the list of all defining interpreter equations. This saves us the effort of revisiting earlier definitions, recalling what specific names we gave to what sentences, and figuring out exactly which of these sentences are pertinent to the step.

Another reason why we favor slightly longer proofs is efficiency. As already noted, shorter proofs require search, and when the assumption base contains hundreds or thousands of members, such search may not be expedient. It might not be feasible at all. Finally, we prefer longer proofs in this textbook for instructive reasons: Descending to a lower level of detail and explicitly breaking down a large proof step into a number of smaller steps can be a valuable learning experience. This too, however, must not be taken too far. In most cases, individual steps need not be expressed at the level of primitive methods such as introduction and elimination methods for logical connectives or equality. Roughly speaking, the granularity level that we have been advocating is attained when each step of the chain is justified by a fairly small set of structurally similar sentences or methods; the smaller that set the better, though it need not be the smallest possible.

17.2 Handling errors explicitly

The foregoing specification did not deal with errors, neither in the interpreter nor in the compiler. What happens if the interpreter is used to evaluate a division by `zero`? What happens if the empty program is applied on the empty stack, or if a binary command is executed on a stack with only one member (or no members!), or if the virtual machine encounters a division by zero? Our specification so far has not provided answers to these questions. That can be a legitimate way of dealing with errors: Just don't say anything about them. This technique is known as *underspecification*, and it results in a theory with so-called *loose semantics*, which essentially means that the theory has many different models, and specifically that there are very many different functions that satisfy the demands made by the theory. Consider, for instance, our specification of the interpreter `I`. That specification does not determine a *unique* function from `Exp` to `N`. It does not tell us, for instance, the value of the term

```
(I (0 quot 0)).
```
(17.7)

As it stands, there are many different functions from `Exp` to `N` that satisfy the given axioms,[7] and it is in that sense that the specification is loose. From an operational perspective, you can observe this by trying to evaluate a term such as (17.7):

```
> (eval I (0 quot 0))

Unable to reduce the term:

(TC.I (TC.quot (TC.const zero)
               (TC.const zero)))

to a normal form.

Unit: ()
```

Athena cannot find any information in the assumption base that can be used to obtain a canonical natural number from this input term.

In situations in which we want to state exactly when errors arise and how they are to be handled, and to be able to reason about them, underspecification is not the right approach. Such situations are not uncommon, especially when we are defining machines. We then want our specification to handle errors explicitly, and thus to be a theory with so-called tight—rather than loose—semantics. In other words, we want our theory to have a unique model (up to isomorphism). In this case, for instance, it should pick out unique functions for `I` and for `exec`.

There are various ways of dealing with errors explicitly. One elegant approach is to use subsorting. While Athena supports subsorting, we will not pursue that approach here. Instead, we will handle errors using the `Option` datatype (Section 10.2). In the case of the interpreter `I`, for instance, the new signature will be:

```
declare I': [Exp] -> (Option N)
```

(We prime the name of the interpreter to distinguish it from the original version.) Intuitively, this says that `I'` takes an expression and *might* return a natural number, if all goes well; otherwise it will return `NONE`. Options are pervasive in functional programming, and their use for handling anomalous situations is more realistic (and perhaps more intuitive) than other approaches.

17.2.1 Extending the compiler with error handling

We begin by changing the signature of the interpreter as discussed above:

```
declare I': [Exp] -> (Option N)
```

7 For instance, one function might map (17.7) to 2, while another might map it to 10^{83}.

We now need to reformulate the definition of I in accordance with this new signature. The equation for the const constructor is simple enough:

```
(I' const n = SOME n)
```

The treatment of the remaining constructors is somewhat complicated in that we must distinguish between recursive calls that fail and those that succeed. Consider, for instance, how to go about obtaining the value of a term of the form:[8]

$$(\texttt{I'}\ e_1\ +\ e_2). \qquad (17.8)$$

In the underspecification approach, the algorithm was simple: Recursively obtain the values of the two subexpressions e_1 and e_2 and then return their sum. Now, however, we must keep in mind that I returns a natural number *option*, and hence that the recursive invocations of I on either of the two subexpressions might fail (i.e., they might return NONE). So the algorithm now is this: If the recursive evaluation of *both* subexpressions succeeds, that is, both e_1 and e_2 return values of the form (SOME n_1) and (SOME n_2), respectively, then we return (SOME (n_1 + n_2)). But if either recursive call fails, then we fail as well. So we have two axioms for sums, one for success (a "positive" axiom), and one for failure (a "negative" axiom). The positive axiom is this:

```
(I' e1 = SOME n1 & I' e2 = SOME n2 ==> I' e1 + e2 = SOME n1 + n2)
```

And the negative axiom is this:

```
(I' e1 = NONE | I' e2 = NONE ==> I' e1 sum e2 = NONE)
```

The treatment of subtractions and products is similar. Division must be handled a little differently. The positive axiom for division must ensure not only that the recursive calls to the two subexpressions succeed, but also that the value returned by the second subexpression (the divisor operand) is nonzero:

```
(I' e1 = SOME n1 & I' e2 = SOME n2 & n2 =/= zero ==>
   I' e1 / e2 = SOME n1 / n2)
```

Symmetrically, the negative axiom must fail if either of the two subexpressions fails or if the divisor yields a zero value:

```
(I' e1 = NONE | I' e2 = NONE | I' e2 = SOME zero ==> I' e1 / e2 = NONE)
```

Thus we arrive at the following definition:

```
assert* I'-def :=
  [(I' const n = SOME n)
```

8 Note, by the way, that we have not declared a specific precedence for it, so I' gets the default precedence, 110. Consequently, a term like (I' e_1 + e_2) is parsed as (I' (e_1 + e_2)).

```
  (I' e1 = SOME n1 & I' e2 = SOME n2 ==> I' e1 + e2 = SOME n1 + n2)
  (I' e1 = SOME n1 & I' e2 = SOME n2 ==> I' e1 - e2 = SOME n1 - n2)
  (I' e1 = SOME n1 & I' e2 = SOME n2 ==> I' e1 * e2 = SOME n1 * n2)
  (I' e1 = SOME n1 & I' e2 = SOME n2 & n2 =/= zero ==>
                                          I' e1 / e2 = SOME n1 / n2)

  (I' e1 = NONE | I' e2 = NONE ==> I' e1 + e2 = NONE)
  (I' e1 = NONE | I' e2 = NONE ==> I' e1 - e2 = NONE)
  (I' e1 = NONE | I' e2 = NONE ==> I' e1 * e2 = NONE)
  (I' e1 = NONE | I' e2 = NONE | I' e2 = SOME zero ==>
                                          I' e1 / e2 = NONE)]
```

We go on to test the new interpreter:

```
> (eval I' 2 sum 3)

Term: (SOME 5)

> (eval I' 3 prod 5)

Term: (SOME 15)

> (eval I' 11 sum 3 prod 3)

Term: (SOME 20)

> (eval I' 18 quot 6)

Term: (SOME 3)

> (eval I' 4 quot 0)

Term: NONE:(Option N)

> (eval I' 4 quot (2 diff 2))

Term: NONE:(Option N)
```

Let us now proceed to reformulate the semantics of exec, using a primed version of the original symbol. It will now return a natural number option:

```
declare exec': [Program Stack] -> (Option N)
               [wrt' 101 [(alist->clist id)
                          (alist->clist int->nat)]]
```

Consider first the case when the program to be executed is empty. We can distinguish two subcases: The stack is also empty, or it is not. The first subcase now leads to an error (NONE result), while the second is handled as before but wrapped inside a SOME constructor:

```
define exec-axiom-empty-1 := (nil wrt' nil = NONE)

define exec-axiom-empty-2 := (nil wrt' n::_ = SOME n)
```

The treatment of push is the same as before:

```
define exec-axiom-push :=
  ((push n)::prog wrt' stack = prog wrt' n::stack)
```

We continue with the binary commands. We will handle division last, so let's consider the other three commands first. When can their execution go wrong? Only if the stack does not have enough operands. So we'll assert a negative axiom for each such command, and one positive axiom:

```
define neg-op-axioms :=
  (flatten (map lambda (op) [(op::_ wrt' [] = NONE)
                             (op::_ wrt' [_] = NONE)]
                [add sub mult]))
```

On the positive side we have:

```
define pos-op-axioms :=
  [(add::prog wrt'  n1::n2::stack = prog wrt' (n1 + n2)::stack)
   (sub::prog wrt'  n1::n2::stack = prog wrt' (n1 - n2)::stack)
   (mult::prog wrt' n1::n2::stack = prog wrt' (n1 * n2)::stack)]
```

Finally, for the division operation we have:

```
define pos-div-axiom :=
  (n2 =/= zero ==> div::prog wrt' n1::n2::stack = prog wrt' (n1 / n2)::stack)

define neg-div-axiom :=
  (n2 = zero ==> div::prog wrt' _::n2::stack = NONE)
```

We collect all relevant equations in a single list and assert them:

```
assert* exec'-def :=
  (join [exec-axiom-empty-1 exec-axiom-empty-2 exec-axiom-push]
        pos-op-axioms neg-op-axioms [pos-div-axiom neg-div-axiom])
```

And we are done!

The virtual machine is defined as before, except that its signature is now different:

```
declare run-vm': [Program] -> (Option N) [[(alist->clist id)]]

assert* vm-def' := (run-vm' prog = prog wrt' [])
```

We can now test the new semantics:

```
> (eval [] wrt' [5])

Term: (SOME 5)

> (eval [(push 3)] wrt' [])

Term: (SOME 3)

> (eval [(push 3) (push 4) add] wrt' [])

Term: (SOME 7)

> (eval run-vm' [(push 1) add])

Term: NONE:(Option N)

> (eval run-vm' [(push 0) (push 4) div])

Term: NONE:(Option N)
```

The compiler has the same signature and definition as before, which we do not repeat here. The previous correctness property needs to be slightly modified to ensure that the expression being compiled has *some* value:

```
define (correctness' e) :=
  (forall n prog stack .
    I' e = SOME n ==> compile e ++ prog wrt' stack = prog wrt' n::stack)
```

Let's first check to see whether we might have missed anything obvious:

```
define conjecture := (forall e . correctness' e)

> (falsify conjecture 10)

Term: 'failure
```

Because correctness is now conditional, we will often find ourselves assuming a hypothesis of the form

$$(\texttt{I'}\ e\ \texttt{=}\ \texttt{SOME}\ n). \tag{17.9}$$

When e is a complex (nonconstant) expression, there are certain useful conclusions we can derive from (17.9). Suppose, for instance, that e is of the form $(e_1\ \texttt{sum}\ e_2)$. Then, from the assumption $(\texttt{I'}\ e_1\ \texttt{sum}\ e_2\ \texttt{=}\ \texttt{SOME}\ n)$, we should be able to conclude that there exist natural numbers n_1 and n_2 such that:

1. $(\texttt{I'}\ e_1\ \texttt{=}\ \texttt{SOME}\ n_1)$;
2. $(\texttt{I'}\ e_2\ \texttt{=}\ \texttt{SOME}\ n_2)$; and
3. $(n = n_1 + n_2)$.

In other words, if (I' e_1 sum e_2 = SOME n) holds, then the recursive evaluations for the two subexpressions e_1 and e_2 must have succeeded, and moreover, the results returned by them must be related to n in the appropriate way. The reasoning required to derive this conclusion can be packaged into a generic unary method, get-lemma, that takes any premise of the form

$$(\texttt{I'}\ e_1\ op\ e_2 = \texttt{SOME}\ n),$$

where op is either sum, diff, or prod (we will handle quot with a separate method); and returns the preceding existential quantification, which appears in more detail as the result of the application of **force** below:

```
define get-lemma :=
  method (premise)
    match premise {
      (= (I' (exp-op:(OP 2) e1 e2))
         (SOME n)) =>
        let {[_ num-op:(OP 2)] := (exp-op->cmd-and-num-op exp-op)}
          (!force (exists n1 n2 .
                    I' e1 = SOME n1 &
                    I' e2 = SOME n2 &
                    n = n1 num-op n2))
    }
```

We leave it as an exercise to replace the use of **force** with a proper deduction. Keep in mind that the input premise will be in the assumption base at the time when get-lemma is invoked.

Exercise 17.1: Implement get-lemma. □

We define a similar method, get-div-lemma, to obtain a similar result for divisions. Here the input premise will be of the form (I' e_1 div e_2 = SOME n), and the method will conclude that there exist natural numbers n_1 and n_2 such that:

1. (I' e_1 = SOME n_1);
2. (I' e_2 = SOME n_2);
3. (n_2 =/= zero); and
4. $n = n_1$ / n_2.

So the body of the conclusion here is slightly more complicated in that it includes an additional conjunct: that n_2 is nonzero. Here is the skeleton of this method:

```
define get-div-lemma :=
  method (premise)
    match premise {
      ((I' (e1 quot e2)) = (SOME n)) =>
```

```
    (!force (exists n1 n2 .
              I' e1 = SOME ?n1 &
              I' e2 = SOME ?n2 &
              n2 =/= zero &
              n = n1 / n2))
  }
```

The detailed implementation is also left as an exercise.

Exercise 17.2: Implement `get-div-lemma`. □

Note that both of these properties can and should be independently developed and tested. For example, in the case of `get-lemma` we might write something like this:

```
define (test-get-lemma op:(OP 2)) :=
  let {goal := (I' e1 op e2 = SOME n ==>
                 exists n1 n2 . I' e1 = SOME n1 &
                                I' e2 = SOME n2 &
                                 n = n1 op n2);
       result := assume h := (I' e1 op e2 = SOME n)
                   (!get-lemma h)}
   check {(result equal? goal) =>
              let {_ := (print "\nTest successful!\n")} true
        | else =>
              let {_ := (print "\nTest failed!\n")} false
         }

> (test-get-lemma +)

Test successful!

Term: true

> (test-get-lemma -)

Test successful!

Term: true

> (test-get-lemma *)

Test successful!

Term: true
```

Likewise for `get-div-lemma`.

Let us now proceed with the inductive proof of the desired property:

```
(forall e . correctness' e).
```

As before, the basis case treats expressions of the form (const *k*) and it is parameterized over *k*:

```
define basis-step :=
  method (k)
    pick-any n:N prog:Program stack:Stack
      assume hyp := (I' const k = SOME n)
        let {k=n := (!chain->
                       [(SOME k) <-- (I' const k)  [I'-def]
                                 = (SOME n)        [hyp]
                     ==> (k = n)                   [option-results]])}
          (!chain
            [(compile const k ++ prog wrt' stack)
           = ([(push k)] ++ prog wrt' stack)       [compiler-def]
           = (push k :: prog wrt' stack)           [List.join.left-singleton]
           = (prog wrt' k::stack)                  [exec'-def]
           = (prog wrt' n::stack)                  [k=n]])
```

The inductive-step method for the first three expression constructors is defined as follows:

```
define istep :=
  method (exp)
    match exp {
     ((some-symbol op:(OP 2)) e1 e2) =>
        pick-any n:N prog:Program stack:Stack
          assume hyp := (I' e1 op e2 = SOME n)
             let {[cmd num-op:(OP 2)] := (exp-op->cmd-and-num-op op);
                  [ih1 ih2] := [(correctness' e1) (correctness' e2)];
                  lemma := (!chain->
                              [hyp ==> (exists n1 n2 .
                                           I' e1 = SOME n1 &
                                           I' e2 = SOME n2 &
                                           n = n1 num-op n2)
                                 [get-lemma]])
                }
             pick-witnesses n1 n2 for lemma spec-lemma
               (!chain
                 [(compile (e1 op e2) ++ prog wrt' stack)

                = ((compile e2 ++ compile e1 ++ [cmd]) ++ prog wrt' stack)
                    [compiler-def]

                = (compile e2 ++ compile e1 ++ [cmd] ++ prog wrt' stack)
                    [List.join.Associative]

                = ([cmd] ++ prog wrt' n1::n2::stack)
                    [ih1 ih2]

                = (cmd::prog wrt' n1::n2 ::stack)
                    [List.join.left-singleton]
```

```
                = (prog wrt' (n1 num-op n2)::stack)
                    [exec'-def]

                = (prog wrt' (n :: stack))
                    [spec-lemma]])
    }
```

The inductive step for quotients is similar:

```
define istep-div :=
  method (exp)
    match exp {
     (quot e1 e2) =>
       pick-any n:N prog:Program stack:Stack
         assume hyp := (I' e1 quot e2 = SOME n)
           let {[ih1 ih2] := [(correctness' e1) (correctness' e2)];
                lemma := (!chain->
                           [hyp ==> (exists n1 n2 .
                                      I' e1 = SOME n1 &
                                      I' e2 = SOME n2 &
                                      n2 =/= zero &
                                      n = n1 / n2)
                            [get-div-lemma]])}
            pick-witnesses n1 n2 for lemma spec-lemma
              (!chain
                [((compile (e1 / e2)) ++ prog wrt' stack)

                 = ((compile e2 ++ compile e1 ++ [div]) ++ prog wrt' stack)
                     [compiler-def]

                 = (compile e2 ++ compile e1 ++ [div] ++ prog wrt' stack)
                     [List.join.Associative]

                 = ([div] ++ prog wrt' n1::n2::stack)
                     [ih1 ih2]

                 = (div::prog wrt' n1::n2::stack)
                     [List.join.left-singleton]

                 = (prog wrt' (n1 N./ n2)::stack)
                     [exec'-def]

                 = (prog wrt' n::stack)
                     [spec-lemma]])
    }
```

The final proof now becomes:

```
conclude main-correctness-theorem' := (forall e . correctness' e)
  by-induction main-correctness-theorem' {
    (const k) => (!basis-step k)
```

```
  | (e as (_ sum _))  => (!istep e)
  | (e as (_ diff _)) => (!istep e)
  | (e as (_ diff _)) => (!istep e)
  | (e as (_ prod _)) => (!istep e)
  | (e as (_ quot _)) => (!istep-div e)
  }
```

Once again, the top-level result that we want can now be derived in a few lines:

```
conclude compiler-correctness' :=
  (forall e n . I' e = SOME n ==> run-vm' compile e = SOME n)
    pick-any e:Exp n:N
      assume hyp := (I' e = SOME n)
        (!chain
          [(run-vm' compile e)
        = (compile e wrt' [])           [vm-def']
        = (compile e ++ [] wrt' [])     [List.join.right-empty]
        = ([] wrt' n::[])               [main-correctness-theorem']
        = (SOME n)                      [exec'-def]])

Theorem: (forall ?e:Exp
           (forall ?n:N
             (if (= (TC.I' ?e:Exp)
                    (SOME ?n:N))
                 (= (run-vm' (TC.compile ?e:Exp))
                    (SOME ?n:N)))))
```

As before, these proofs could be shortened. For instance, the basis step could be expressed as follows:

```
define basis-step :=
  method (k)
    pick-any n:N prog:Program stack:Stack
      assume hyp := (I' const k = SOME n)
        let {k=n := (!chain-> [(SOME k) <-- (I' const k)
                                        = (SOME n)
                               ==> (k = n) ])}
          (!chain [(compile const k ++ prog wrt' stack)
                 = (prog wrt' n::stack)])
```

Likewise, the main inductive step could be written with only two rewriting steps as:

```
define istep :=
  method (exp)
    match exp {
     ((some-symbol op:(OP 2)) e1 e2) =>
        pick-any n:N prog:Program stack:Stack
          assume hyp := (I' e1 op e2 = SOME n)
            let {[_ num-op:(OP 2)] := (exp-op->cmd-and-num-op op);
                 lemma := (!chain->
```

```
                              [hyp ==> (exists n1 n2 .
                                          I' e1 = SOME n1 &
                                          I' e2 = SOME n2 &
                                          n = n1 num-op n2)
                             [get-lemma]])}
              pick-witnesses n1 n2 for lemma
                (!chain
                  [(compile (e1 op e2) ++ prog wrt' stack)
                  = (prog wrt' (n1 num-op n2)::stack)
                  = (prog wrt' n::stack)])
      }
```

Indeed, we could shrink the body even further, down to one single step:

```
  (!chain [(compile (e1 op e2) ++ prog wrt' stack)
         = (prog wrt' n::stack)])
```

Similarly, the body of istep-div could be whittled down to one or two chaining steps, essentially leaving it to Athena to discover the proof. As we discussed earlier, however, such drastic reductions are not always desirable (and certainly not always feasible).

The correctness result we have established for the error-handling version of the compiler is only partial. It says only that *if* the evaluation of an expression *e* produces *some* natural number *n*, then running the virtual machine on the program obtained by compiling *e* will result in the same number *n*. It does not, however, say anything about the relationship between the interpreter and the virtual machine for those expressions *e* whose evaluation fails. If the evaluation of *e* produces NONE, then what happens if we compile *e* and run the virtual machine on the output program? Will we also get NONE, or might we end up with some number on the top of the stack? The result we have so far does not address that question. To do so, we must also prove the following:

```
(forall e . I' e = NONE ==> run-vm' compile e = NONE).        (17.10)
```

We leave that part as an exercise. The main inductive property to be derived is this (we have called it correctness-conv, for the converse direction):

```
define (correctness-conv e) :=
  (forall prog stack .
    I' e = NONE ==> compile e ++ prog wrt' stack = NONE)
```

Let's put to the test the conjecture that every expression has this property:

```
define conjecture := (forall e . correctness-conv e)

> (falsify conjecture 30)

Term: 'failure
```

Once we have shown that every expression *e* has the property correctness-conv, let us call this main-correctness-theorem-conv, then (17.10) can be derived as follows:

```
conclude (forall e . I' e = NONE ==> run-vm' compile e = NONE)
 pick-any e:Exp
   assume (I' e = NONE)
     (!chain
       [(run-vm' compile e)
      = (compile e wrt' [])            [vm-def']
      = (compile e ++ [] wrt' [])  [join-right-empty]
      = NONE                           [main-correctness-theorem-conv]])
```

Now main-correctness-theorem-conv is again derived by induction on the structure of expressions, with the proof following the same outline as the inductive proof of the first direction of the correctness result. We write one method for the basis step, parameterized by the natural number that is the argument to const, and one method for the three inductive-step cases. Division is again best handled by a different method.

The basis step is somewhat different, in that the main assumption of the property we are now trying to prove, namely, that the evaluation of the expression produces NONE, is incompatible with constant expressions, as these always produce some natural number, never NONE. Therefore, the assumption yields an inconsistency with the definition of the interpreter and the result follows vacuously:

```
define (basis-step-conv k) :=
  pick-any prog:Program stack:Stack
    assume hyp := (I' const k = NONE)
      let {_ := (!absurd
                  (!chain-> [true
                         ==> (I' const k = SOME k)    [I'-def]])
                  (!chain-> [hyp
                         ==> (I' const k =/= SOME k) [option-results]]))}
        (!from-false (compile const k ++ prog wrt' stack = NONE))
```

The two inductive methods (one for the division and one for the remaining three expression constructors) are left as exercises. It will help to have a parameterized lemma, encoded as a method analogous to the get-lemma method, to draw some useful conclusions from the assumption that the evaluation of an expression produces NONE. Let us call this one get-lemma-conv:

```
define get-lemma-conv :=
  method (premise)
    match premise {
      (= (I' (_ e1 e2)) NONE) =>
        (!force (I' e2 = NONE | I' e2 =/= NONE & I' e1 = NONE))
    }
```

The conclusion we draw here is that either the evaluation of the right subexpression e2 fails, or else the evaluation of e2 does not fail but that of the left subexpression (e1) does. It is left as an exercise to replace the use of **force** with a proper deduction.

Exercise 17.3: Define get-lemma-conv. □

A similar lemma will come in handy specifically for division: If we know that the evaluation of (e1 quot e2) fails, then we may conclude one of the following: Either the evaluation of e1 or e2 fails, or else the value of the right subexpression, e2, is zero:

```
define get-lemma-conv-div :=
  method (premise)
    match premise {
      (- (I' (quot e1 e2)) NONE) =>
        (!force (I' e1 = NONE | I' e2 = NONE | I' e2 = SOME zero))
    }
```

Exercise 17.4: Define get-lemma-conv-div. □

After these two lemma-producing methods have been defined, the remaining two exercises ask you to write the methods istep-conv and istep-conv-div that appear in the inductive proof of main-correctness-theorem-conv below:

```
conclude main-correctness-theorem-conv := (forall e . correctness-conv e)
  by-induction main-correctness-theorem-conv {
    (const k) => (!basis-step-conv k)
  | (e as (_ sum _))  => (!istep-conv e)
  | (e as (_ diff _)) => (!istep-conv e)
  | (e as (_ prod _)) => (!istep-conv e)
  | (e as (_ quot _)) => (!istep-conv-div e)
  }
```

Exercise 17.5: Define istep-conv. □

Exercise 17.6: Define istep-conv-div. □

Exercise 17.7: While the second version of the compiler, with the explicit error handling, is a more realistic model, it is still fairly abstract in that it does not distinguish between different errors. All errors result in the same output—NONE. Modify the specification so that it can model different error messages. □

Exercise 17.8: In all of the inductive proofs shown in this chapter, we have defined a unary procedure correctness that takes an input expression and builds a sentence that expresses the desired property for that expression. The inductive proofs themselves then took the following form:

```
by-induction (forall e . correctness e) {
  ...
}
```

As discussed before (Section 3.8), this is an approach that is often useful with inductive proofs over some sort S. We define the inductive property P as a unary procedure that takes any term t of sort S as input and builds the sentence that expresses the property for t. The overall proof then takes the form:

```
by-induction (forall ?x:S . P ?x) {
  ...
}
```

There is one subtle point that must be kept in mind when defining inductive properties in this way. Inductive properties are often themselves quantified (look at the various definitions of `correctness`, for example), and hence they are typically defined as follows:

```
define (P t) :=
  (forall ?x ?y ...
     body)
```

where, presumably, the input parameter `t` occurs free inside the *body*. But what happens if we apply P to a term that has `?x` or `?y` among its variables? Those variables might be accidentally captured. To guard against this possibility, it is best to write such a procedure as follows:

```
define (P t) :=
  let {x := (fresh-var ...);
       y := (fresh-var ...);
       ...}
    (forall x y ...
  body')
```

where *body*′ uses `x` instead of `?x`, `y` instead of `?y`, and so on. This ensures that the bound variables of the output sentence will be distinct from any variable occurrences inside the input term. You will notice, however, that we did not do that in the various definitions of `correctness` in this chapter. The reason is that it was not necessary—variable capture could not possibly occur here. Explain why. □

17.3 Chapter notes

In this chapter we have presented a couple of different models and correctness proofs for an expression compiler. Our treatment was inspired by a CafeObj [31] tutorial on the subject, which itself seems to go back to earlier proofs in collaboration with Goguen [41], which in turn go back to Burstall's seminal work on the subject [16]; the general approach is standard. For instance, the algorithm given in Section 2.3 of *Compiler Design* [114] for translating expressions into machine code is virtually identical to the algorithm formalized here. While both versions of the compiler we have considered are quite simple, the specifications and proofs we have presented reinforce a number of important points that have already emerged in previous chapters, and which are of key importance in more challenging projects. They can be summarized as follows:

- *Computable definitions (and executable specifications in general)*: the ability to compute with the logical content of sentences is a powerful tool for theory development. It allows us to mechanically test our definitions and conjectures.
- *Computation in the service of notation*: Dynamic overloading, precedence levels, input expansions, output transformations, and even simple static scoping combined with the ability to name symbols and even variables[9] can greatly enhance the readability of specifications and proofs. Likewise for the use of procedures for building sentences with similar structure, defining the inductive predicate for a proof by structural induction, and so on.
- *Assuming more to prove more*: We have again demonstrated the usefulness of induction strengthening.
- *Proof abstraction*: Using methods to package up recurring bits of reasoning is invaluable for structured proof engineering.
- *Fundamental datatypes*: Here we have used options to model errors and lists to model programs as sequences of instructions. A vast range of interesting systems and processes can be represented using the basic structures we have studied: numbers, options, ordered pairs, lists, sets, and maps.

9 For instance, locally renaming a function symbol *f* for the scope of a given proof or proof method.

18 A Simple Imperative Programming Language

We start this chapter by introducing a simple imperative programming language, the **While** language, formally encoding its abstract syntax and natural semantics in Athena.[1] We then proceed to prove some nontrivial results about the language, including a proof showing that the formal semantics are deterministic.

This chapter serves a number of goals. First, because this material is relatively more complicated, it constitutes a sort of stress test for techniques that we have already introduced in earlier chapters. We will see that these techniques scale well; they can be used to develop well-structured proofs of nontrivial results. Second, it introduces some new techniques of its own. The natural semantics of **While** are expressed by means of an inference system. Proofs in that system have a recursive tree-like structure, and they are explicitly represented by an inductive datatype. Reasoning about the semantics often requires induction on the shape of such proof trees. Induction of this form goes beyond simple structural induction on the abstract syntax trees of the language (indeed, the determinism result for the semantics cannot be obtained by structural induction on **While** ASTs, for reasons we shall explain in due course; induction on proof shapes is necessary). The methodology introduced in this chapter has wide applicability because inference systems are routinely used to define formal languages and type systems in computer science, and induction over proofs is an important general technique for reasoning about such languages and systems. Note that the proofs in question are object-level proofs (in the inference system being encoded, which will be the object system), and our inductive Athena proofs over them will be metaproofs, that is, proofs about the (encoded) proofs.

Moreover, because natural semantics are more intuitively represented by relations rather than functions, evaluation by rewriting is not directly applicable and therefore we cannot straightforwardly "execute" our specifications in order to test them, at least not by simple rewriting, as we have done previously. This provides a good opportunity to illustrate the use of logic programming for similar evaluation purposes. We will see that logic programming is in certain respects more flexible than rewriting. For instance, it will allow us not only to compute the semantics of **While** programs, but also to simultaneously compute the object-level proofs that derive the corresponding evaluation judgments.

18.1 A simple imperative language

We begin by defining the abstract syntax of the **While** language. There are five syntactic categories:

1. nonnegative *numerals*;

1 With a few minor differences, our presentation of the syntax and semantics of **While** follows the discussion of Nielson and Nielson [77].

2. *variables*;
3. *expressions*;
4. *Boolean conditions*; and
5. *commands*.

We use the letters n, x, e, b, and c (possibly with subscripts, etc.) as variables ranging over these five domains, respectively. Numerals and variables are primitive, while the other three domains are defined by the following abstract grammars:

$$e ::= n \mid x \mid e_1 \textbf{ sum } e_2 \mid e_1 \textbf{ prod } e_2 \mid e_1 \textbf{ diff } e_2 \tag{18.1}$$

$$b ::= e_1 \textbf{ eq } e_2 \mid e_1 \textbf{ leq } e_2 \mid \textbf{neg } b \mid b_1 \textbf{ conj } b_2 \tag{18.2}$$

$$c ::= \textbf{skip} \mid x \mathrel{\textbf{:=}} e \mid c_1 \textbf{ ; } c_2 \mid \textbf{if } b \textbf{ then } c_1 \textbf{ else } c_2 \mid \textbf{while } b \textbf{ do } c \tag{18.3}$$

We informally explain each grammar. (18.1) says that an expression is either a numeral n, or a variable x, or the sum, product, or difference of two expressions. (18.2) says that a Boolean condition is either an identity between two numeric expressions, of the form e_1 **eq** e_2; or an inequality of the form e_1 **leq** e_2 (**leq** here should be read as "less than or equal to"); or a negation of a Boolean condition (**neg** b); or a conjunction of two Boolean conditions (b_1 **conj** b_2). Finally, a command is either a **skip** statement, which, as we will see, does nothing; or an *assignment* of the form x **:=** e, intended to assign to the variable x the value of the expression e; or a *sequence* c_1 **;** c_2, intended to compose the effects of the two commands c_1 and c_2 by executing them sequentially, in the given order; or a conditional statement of the form

$$\textbf{if } b \textbf{ then } c_1 \textbf{ else } c_2,$$

which will execute either c_1 or c_2, depending on whether or not the Boolean condition b holds; or, finally, a loop of the form

$$\textbf{while } b \textbf{ do } c,$$

which will keep executing the body c as long as the condition b holds.

It is worthwhile at this point to say a few things about abstract grammars. There are, essentially, two different ways of looking at syntactic objects such as expressions and commands. One way is through the lens of a concrete syntax. We can view expressions, for instance, as linear strings of symbols drawn from a certain alphabet. In that case we would use a context-free grammar to specify the exact membership of the set of all expressions. Such a grammar would tell us whether or not any given string is an expression.

That viewpoint is too low-level for doing formal semantics. For the purpose of specifying the formal *meaning* of expressions, for instance, we want to be able to say something along the following lines: The value of an expression e that represents the addition of two subexpressions e_1 and e_2 is the sum of the values of e_1 and e_2. Here we are not—and should not be—concerned with how such expressions are represented as strings of symbols. It is irrelevant whether "an expression that represents the addition of two subexpressions e_1 and e_2" is concretely written down as "(e_1 **sum** e_2)" or as "($e_1 + e_2$)" or as "$e_1 \ + \ e_2$" or as "(+ e_1 e_2)" or "[+ e_1 e_2]" or "+(e_1, e_2)," and so on. All we care about is the *abstract tree structure* of the expression, namely, that there is an addition operator at the root of the tree,[2] and the two subtrees corresponding to e_1 and e_2 on the left and right sides, respectively:

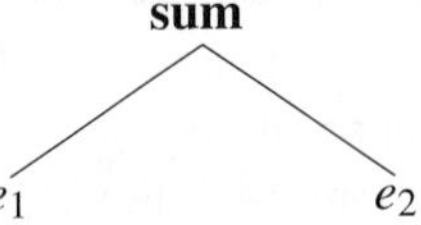

When we treat syntactic objects as strings of symbols, we are concerned with concrete syntax. When we treat them as two-dimensional tree structures, we are dealing with abstract syntax. These two viewpoints converge in parsing theory and practice: A parser is a program that takes a string as input and either rejects it as ill-formed or else outputs a "parse tree" that captures the string's abstract syntactic structure. We will implement such a parser for the **While** language in the exercises.

Although in formal semantics we are almost always concerned exclusively with abstract syntax, it would be cumbersome to represent and manipulate abstract syntax trees with two-dimensional pictures all the time. Instead, we use a linear notation specified by "abstract grammars" such as (18.1)—(18.3). Such grammars are very much like concrete context-free grammars: There are nonterminals and terminals, and a set of productions for the nonterminals of interest. But despite the similarities, it is generally not feasible to treat abstract grammars as conventional context-free grammars. For one thing, nonterminals corresponding to primitive syntactic categories such as numbers or variables are usually left unspecified (no productions are given for them), because their details are a matter of concrete rather than abstract syntax. Moreover, when understood as conventional context-free grammars, abstract grammars are typically ambiguous and cannot be used for parsing purposes—at least not by themselves.[3] For instance, (18.1) does not tell us whether

$$a \ \textbf{sum} \ b \ \textbf{prod} \ c$$

should be parsed as the sum of a and the product of b and c, or as the product of the sum of a and b on the one hand and c on the other. But with a set of appropriate conventions, such

2 More precisely, a function symbol that is to be interpreted as representing addition.

3 There are parsing algorithms that can handle ambiguous context-free grammars. Such algorithms produce a set—or ranked list—of abstract syntax trees for a given input string. That is actually par for the course in parsing natural languages such as English or Chinese. However, for *formal* (artificial) languages, such as programming languages, we usually require unambiguous grammars for clarity (as we typically want a program to have a unique meaning, derived from a unique abstract syntax tree), and also for increased parsing efficiency.

as precedence levels and/or parenthesization, we can usually turn an abstract grammar into an unambiguous concrete grammar without much difficulty.[4]

Abstract syntax trees (ASTs) can be naturally represented by *terms*, owing to the fact that terms are essentially tree structures (recall the discussion of Section 3.3). For example, the abstract syntax tree depicted above could be represented by—or indeed identified with—a term of the form (**sum** t_1 t_2), where the subterms t_1 and t_2 are the term representations of the subexpressions e_1 and e_2, respectively. That is the main idea behind the use of algebraic datatypes for defining sets of abstract syntax trees: The constructors of the datatype represent the main abstract syntactic operators (addition on expressions, sequencing on commands, etc.), and canonical terms built from those constructors represent abstract syntax trees under the usual understanding of terms as trees.

In the case of **While**, the three abstract syntactic categories of interest can be defined by the following datatypes:

```
datatype Exp := (num N)
              | (var Ide)
              | (sum Exp Exp)
              | (diff Exp Exp)
              | (prod Exp Exp)

datatype BCond := (eq Exp Exp)
                | (leq Exp Exp)
                | (neg BCond)
                | (conj BCond BCond)

datatype Cmd := skip
              | (asgn Ide Exp)
              | (sequence Cmd Cmd)
              | (cond BCond Cmd Cmd)
              | (while-loop BCond Cmd)
```

The correspondence between these datatypes and the abstract grammars (18.1)—(18.3) is straightforward. Terms of the form (num t), where t is a canonical natural number, represent

4 The easiest way to obtain an unambiguous context-free grammar from *any* abstract grammar is to fully parenthesize the right-hand side of every production of the abstract grammar that contains a nonterminal. For example, in the case of expressions, we would get:

$$e ::= n \mid x \mid (e_1 \ \textbf{sum} \ e_2) \mid (e_1 \ \textbf{prod} \ e_2) \mid (e_1 \ \textbf{diff} \ e_2) \qquad (18.4)$$

Interpreted as a conventional context-free grammar, (18.4) is unambiguous: Every string in the language defined by (18.4) has a unique parse tree (assuming reasonable specifications for n and x). Strictly speaking, the set of strings defined by (18.4) is not the same set of strings that we would obtain by treating (18.1) as a conventional grammar, but that is not the objective here anyway. The point, rather, is to find some unambiguous grammar that provides an acceptable linear notation for the abstract tree structures described by (18.1), and (18.4) fits that bill. Of course, we might want to tweak the concrete grammar to make the linear notation more palatable, say, to avoid the excessive parenthesization required by (18.4). That can be done by precedence levels and other similar conventions. But parenthesizing the right-hand sides of the abstract grammar's productions always provides us with an unambiguous context-free grammar that we can use as a starting point for developing a concrete syntax to our liking. Parenthesizing, by the way, does not necessarily mean using the usual pair of parentheses, (and). Square brackets would also do, as would begin-end pairs, or indeed any other two distinct matching terminals.

numerals; terms of the form (var x), where x is a meta-identifier, represent variables. Terms of the form (sum t_1 t_2), where t_1 and t_2 are expression terms, represent additions; and so on.

Note that the free-generation axioms for datatype constructors are crucial for ensuring that such definitions can be understood as grammars. If an identity such as

skip = (asgn s t)

held, for some s and t, then it would make no sense to treat asgn and skip as *syntactic* operators. If assignment commands are to be understood as pieces of syntax, whether abstract or concrete, then they must be guaranteed to be distinct from skip commands, and indeed distinct from all other commands produced by other constructors. Further, two assignment commands must be identical, as syntactic objects, iff their respective variables and expressions are identical:

(asgn x_1 e_1) = (asgn x_2 e_2) iff $x_1 = x_2$ and $e_1 = e_2$.

Likewise for conditional commands, while loops, and so on. Moreover, it is understood that there are no commands other than those that can be built from the given constructors in a finite number of steps. But these requirements are precisely the free-algebra constraints that must hold for all datatypes. As was mentioned before (Section 2.7), these constraints can always be satisfied by requiring that the set defined by the datatype is the *smallest* set that satisfies the recursive equations implicit in the datatype definition. And that requirement, in turn, obtains in virtue of the syntactic form of the datatype definitions, along with some set-theoretic results ensuring that certain functions have least fixed points.

Given that Athena terms have unique string representations as parenthesized s-expressions,[5] this scheme also provides us with an unambiguous linear notation for abstract syntax trees. We will use this notation extensively for specifying the functions and relations of our formal semantics. For instance, the function V that gives the meaning (value) of an expression with respect to a given state will be defined as follows for sums:

```
(forall e1 e2 s . e1 sum e2 wrt s = (e1 wrt s) + (e2 wrt s))
```

where e1 and e2 are variables ranging over Exp; s is a variable ranging over states; + is natural-number addition, that is, N.+;[6] and we write (e wrt s) ("the value of e with respect to state s") as an infix shorthand for (V e s). This notation works well for defining

5 Essentially, the strings that represent Athena terms, like all s-expressions, are obtained through the parenthesization device discussed in footnote 4, with the additional prefix convention that the operator always appears to the left. This is what makes them somewhat bulky to use on certain occasions, as we will shortly illustrate in the main text.

6 We will overload + to make it represent either natural-number addition *or* the constructor sum, so in fact this equation will be expressible as

```
(forall e1 e2 s . e1 + e2 wrt s = (e1 wrt s) + (e2 wrt s)),
```

where the occurrence of + on the left side is an alias for sum while that on the right stands for N.+.

semantic functions because the abstract-syntax terms that appear in such definitions contain various universally quantified variables designating arbitrary subtrees (e.g., the term `(e1 sum e2)` in the preceding quantification), and they are therefore short and easy to understand. (Essentially, they are patterns.)

However, this notation is too cumbersome to use for the purpose of representing *ground* ASTs denoting *particular* expressions, commands, and so on. Consider, for instance, a simple composition of two commands that we might informally write as follows:

`x := 0;` **while** `x <= y` **do** `x := x + 1` (18.5)

The AST representing this command is the following rather unwieldy term:

```
(sequence (asgn 'x
                (num zero))
          (while-loop (leq (var 'x)
                           (var 'y))
                (asgn 'x
                      (sum (var 'x)
                           (num (S zero))))))
```

We will need to deal with particular (ground) commands, expressions, and so on, in order to test our formal semantics. Hence, in order to represent such objects more succinctly and in more familiar notation, we will use parsers that can accept string inputs such as (18.5) and automatically produce the corresponding AST terms as output. We will use three such parsers: `exp-parser`, `bcond-parser`, and `cmd-parser`, for expressions, Boolean conditions, and commands, respectively. In what follows we briefly explain the set of strings accepted by each of these parsers.

The simplest (atomic) sort of expressions will be:

1. nonnegative numerals (where a numeral is either `0` or a nonempty sequence of digits in the `0`, ..., `9` range that does not begin with `0`); and
2. variables, where a variable is any nonempty sequence of printable characters that does not begin with a digit and does not contain any of the following eleven lexically significant characters: `; = < + - * : ( ) & ~`.

More complicated (compound) expressions will be sums, differences, and products, which will be written as e_1 `+` e_2, e_1 `-` e_2, and e_1 `*` e_2, respectively. Products bind tighter than sums and differences, which bind equally tightly. All three associate to the right. Precedence levels can be overridden by the use of parentheses. Some examples:

```
> (exp-parser "1")

Term: (num (S zero))

> (exp-parser "x")
```

```
Term: (var 'x)

> (exp-parser "x + 1")

Term: (sum (var 'x)
          (num (S zero)))

> (exp-parser "x + y * z")

Term: (sum (var 'x)
          (prod (var 'y)
                (var 'z)))

> (exp-parser "(x + y) * z")

Term: (prod (sum (var 'x)
                (var 'y))
           (var 'z))
```

Atomic Boolean conditions (identities and inequalities) are written as e_1 = e_2 and e_1 <= e_2, respectively. Conjunctions are written as b_1 & b_2 and associate to the right. Negations must be written in the form ~ (b), that is, the scope of a negation must be explicitly marked with parentheses. We allow for the syntax (e_1 != e_2) as an abbreviation for the negation of (e_1 = e_2). Some examples:

```
> (bcond-parser "x = 1")

Term: (equal (var 'x)
            (num (S zero)))

> (bcond-parser "x <= 2 * y & ~(z = 0)")

Term: (conj (leq (var 'x)
                (prod (num (S (S zero)))
                      (var 'y)))
           (neg (eq (var 'z)
                    (num zero))))

> (bcond-parser "x != 1")

Term: (neg (eq (var 'x)
              (num (S zero))))

> (bcond-parser "~ (x <= y & z = 1) & w = w & 1 <= foo")

Term: (conj (neg (conj (leq (var 'x)
                            (var 'y))
                       (eq (var 'z)
                           (num (S zero)))))
```

```
          (conj (eq (var 'w)
                    (var 'w))
                (leq (num (S zero))
                     (var 'foo))))
```

Finally, we explain how `cmd-parser` works. Assignments are written as x := e; skip commands as `skip`; the composition of c_1 and c_2 is written as c_1; c_2; conditionals are written as `if` b `then` c_1 `else` c_2; and while loops are written as `while` b `do` c. Sequences bind less tightly than while loops or conditionals, but that can be overridden by the use of `begin-end` pairs. Some examples:

```
define parse := cmd-parser

> (parse "x := 1")

Term: (asgn 'x
            (num (S zero)))

> (parse "x := 1; y := x * 2")

Term: (sequence (asgn 'x
                      (num (S zero)))
                (asgn 'y
                      (prod (var 'x)
                            (num (S (S zero))))))

> (parse "x := 0; while x <= y do x := x + 1; z := z + x")

Term: (sequence (asgn 'x
                      (num zero))
                (sequence (while-loop (leq (var 'x)
                                           (var 'y))
                                      (asgn 'x
                                            (sum (var 'x)
                                                 (num (S zero)))))
                          (asgn 'z
                                (sum (var 'z)
                                     (var 'x)))))

> (parse "x := 0; while x <= y do begin x := x + 1; z := z + x end")

Term: (sequence (asgn 'x
                      (num zero))
                (while-loop (leq (var 'x)
                                 (var 'y))
                            (sequence (asgn 'x
                                            (sum (var 'x)
```

```
                    (num (S zero))))
           (asgn 'z
                 (sum (var 'z)
                      (var 'x))))))
```

We impose two last requirements on these parsers that will come handy. First, they must be able to handle Athena variables (of the form ?*I*) *inside* the strings that they are parsing. These will essentially stand for arbitrary subexpressions (or subcommands, etc.), and will appear as the corresponding variables in the output terms. For example:

```
> (exp-parser "x + 2")        # as already specified

Term: (sum (var 'x)
           (num (S (S zero))))

> (exp-parser "x + ?FOO")     # note the occurrence of ?FOO

Term: (sum (var 'x)
           ?FOO:Exp)

> (cmd-parser "x := 1; ?C") # note the occurrence of ?C

Term: (sequence (asgn 'x
                      (num (S zero)))
                ?C:Cmd)
```

Finally, we require of each of these parsers that if its input *x* is *not* a string, then *x* is returned unchanged. That will make these parsers suitable as input expanders for function symbols.

Exercise 18.1: Implement the three parsers. □

A few more notational conventions before we turn to semantics. First, some variable abbreviations:

```
define [n n' n1 n2 n3 m m'] := [?n:N ?n':N ···]
define [e e' e1 e2 e3] := [?e:Exp ?e':Exp ···]
define [x y z x' y' z'] := [?x ?y ···]
define [cmd cmd' cmd1 cmd2 cmd3 body body' c c' c1 c2 c3 c1' c2' c3'] :=
       [?cmd:Cmd ?cmd':Cmd ···]
```

We define ^ as an alias for sequence and <- as an alias for asgn:

```
define [^ <-] := [sequence asgn]
```

Thus, we can write an assignment of the form (asgn *x* *e*) in infix as (*x* <- *e*):

```
> (x <- e)

Term: (asgn ?x:Ide ?e:Exp)
```

We also define a few function symbol abbreviations:

```
define [+ - * <= null ++ in subset singleton \/ /\ \] :=
       [N.+ N.- N.* N.<= Set.null DMap.++ Set.in Set.subset
        Set.singleton Set.\/ Set./\ Set.\]
```

and we overload +, -, *, and <= so that they can be used in place of `sum`, `diff`, `prod`, and `leq`, respectively:

```
overload (+ sum) (- diff) (* prod) (<= leq) (/\ conj)
```

Further, we make the `num` constructor accept integer numerals (which will be converted to natural numbers), in addition to natural numbers, and give a high precedence level to the unary expression constructors `num` and `var`:

```
expand-input num [int->nat]
set-precedence num 350
set-precedence var 350
```

Thus, we can now write expressions like this:

```
> (e1 + num 3 * e2)

Term: (sum ?e1:Exp
           (prod (num (S (S (S zero))))
                 ?e2:Exp))
```

We also define some convenient infix abbreviations for the Boolean-condition constructors:

```
define [== --] := [eq neg]

set-precedence <= 110
set-precedence == 110
```

18.2 Semantics of expressions

The meaning or value of an expression e will be a natural number n. That value can only be obtained if we are given values for the variables that occur in e. There is no way to tell the value of $x+1$, for instance, unless we are first given a value for the variable x. If we are told that x is 2, then we can compute the value of $x+1$ to be 3; if x is 99, then the value of $x+1$ will be 100; and so on. This means that the value of an expression can only be given relative to, or with respect to, the values of its variables.

A *state* is a data structure that provides precisely that type of information: It tells us the values of certain variables. We will represent states as finite maps from variables to values (i.e., to natural numbers). In particular, a state will be an object of sort (DMap Ide N) with a default value of zero, meaning that if a key is not explicitly bound in a state then its value will be zero by default (see Section 10.4.3 for a discussion of these maps and their theory). We define a few abbreviations for variables ranging over this State sort:

```
define-sort State := (DMap.DMap Ide N)

define [s s' s'' s1 s1' s2 s2' s3 is is'] := [?s:State ?s':State ···]
```

It will be convenient to have a procedure state that takes a list of pair bindings in Athena-list (square-bracket) notation and produces the corresponding map, converting integer numerals to natural numbers along the way:

```
define (state L) :=
  match L {
    (some-list _) => (DMap.alist->dmap-general [zero L] int->nat)
  | _ => L}
```

For example:

```
> (state [])

Term: (DMap.empty-map zero)

> (state [['a --> 1] ['b --> 2]])

Term: (DMap.update (pair 'a
                         (S zero))
                   (DMap.update (pair 'b
                                      (S (S zero)))
                                (DMap.empty-map zero)))
```

The meaning of an expression e can now be given by a binary function V that takes e as its first input and a state s as a second input:

```
declare V: [Exp State] -> N
           [150 [exp-parser
                 lambda (x) (DMap.alist->dmap-general x int->nat)]]
```

We also introduce wrt ("with respect to") as an infix variant of V, writing (e wrt s) interchangeably with (V e s);[7] and we define at as DMap.at:

```
define [wrt at] := [V DMap.at]
```

7 We read (e wrt s) as "the value of expression e with respect to state s."

The output of eval is transformed so that natural numbers are converted to integers and states are converted to lists of pairs:

```
transform-output eval
   [nat->int lambda (m) (DMap.dmap->alist-general m nat->int)]
```

Because we have listed exp-parser as an input expander for the first argument of V, we can, for example, write:

```
> ("x + 1" wrt state [['x --> 2]])

Term: (V (sum (var 'x)
              (num (S zero)))
         (DMap.update (pair 'x
                            (S (S zero)))
                      (DMap.empty-map zero)))
```

We define this evaluation function by the following axioms, one for each constructor of Exp:

```
assert* V-def :=
  [(num n wrt _ = n)
   (var x wrt s = s at x)
   (e1 + e2 wrt s = (e1 wrt s) + (e2 wrt s))
   (e1 - e2 wrt s = (e1 wrt s) - (e2 wrt s))
   (e1 * e2 wrt s = (e1 wrt s) * (e2 wrt s))]
```

Observe the overloaded use of numeric operators such as +: On the left side they represent constructors of the Exp datatype, while on the right side they represent operations on N.

The first axiom essentially says that numeric expressions are self-evaluating. The second says that the value of a variable relative to a given state is the result obtained by applying that state to the variable. The third one says that the value of an expression of the form (sum e_1 e_2) with respect to a state s is the numeric sum of the values of e_1 and e_2 with respect to s. And likewise for the other two axioms. Let us test the definition:

```
define sample-state := (state [['x --> 2] ['y --> 3]])

> (eval "x + 1" wrt sample-state)

Term: 3

> (eval "x * 3" wrt sample-state)

Term: 6

> (eval "x + 2 * y" wrt sample-state)

Term: 8
```

```
> (eval "(x - 1) + (y - 1)" wrt sample-state)

Term: 3

> (eval "2 * counter" wrt sample-state)

Term: 0
```

The last output is `0` because `sample-state` does not contain a binding for the key `'counter`, so it gets mapped to `zero` by default, which is then multiplied by two to yield a final value of `zero`.

Next, let us define the set of variables that occur in a given expression:

```
declare evars: [Exp] -> (Set.Set Ide) [105 [exp-parser]]
```

(Recall that, for any sort *S*, `(Set.Set` *S*`)` is the sort of finite sets of elements drawn from *S*; see Section 10.3.) In conventional mathematical notation, we would define `evars` as follows:

$$\begin{aligned} evars(n) &= \emptyset; \\ evars(x) &= \{x\}; \\ evars(e_1 + e_2) &= evars(e_1) \cup evars(e_2); \\ evars(e_1 - e_2) &= evars(e_1) \cup evars(e_2); \\ evars(e_1 * e_2) &= evars(e_1) \cup evars(e_2); \end{aligned}$$

where all the variables are understood to be universally quantified. The definition is quite similar in Athena:

```
assert* evars-def :=
  [(evars num _ = null)
   (evars var x = singleton x)
   (evars e1 + e2 = (evars e1) \/ (evars e2))
   (evars e1 - e2 = (evars e1) \/ (evars e2))
   (evars e1 * e2 = (evars e1) \/ (evars e2))]

transform-output eval [Set.set->lst]

> (eval evars "x + y")

List: ['x 'y]
```

We introduce a similar function for Boolean conditions:

```
declare bcond-vars: [BCond] -> (Set.Set Ide) [105 [bcond-parser]]
```

and leave its definition as an exercise.

Exercise 18.2: Define bcond-vars and test your definition. □

Let us now prove that if two states s_1 and s_2 agree on the variables of an expression e, meaning that each such variable is mapped to identical numbers by the two states, then the value of e is the same with respect to the two states:

```
define var-agreement :=
  (forall e s1 s2 . (DMap.agree-on s1 s2 evars e) ==> e wrt s1 = e wrt s2)

> var-agreement

Sentence: (forall ?e:Exp
            (forall ?s1:(DMap.DMap Ide N)
              (forall ?s2:(DMap.DMap Ide N)
                (if (DMap.agree-on ?s1:(DMap.DMap Ide N)
                                   ?s2:(DMap.DMap Ide N)
                                   (evars ?e:Exp))
                    (= (V ?e:Exp
                          ?s1:(DMap.DMap Ide N))
                       (V ?e:Exp
                          ?s2:(DMap.DMap Ide N)))))))
```

Recall that two maps agree on a set of keys K iff every key in K is mapped to the same value by the two maps (that is the content of the theorem DMap.agreement-characterization).

This goal is universally quantified over expressions, and its body involves the function V, which is defined by structural recursion on expressions. This is a strong hint that the proof requires structural induction over expressions. Let us begin by defining the relevant inductive property as a unary procedure:

```
define (var-agreement-property e) :=
  (forall s1 s2 . (DMap.agree-on s1 s2 evars e) ==> e wrt s1 = e wrt s2)
```

The goal now can be formulated as

```
(forall e . var-agreement-property e)
```

The skeleton of the inductive proof appears below. The two basis steps, for numerals and variables, are handled inline. The remaining three (inductive) cases are relegated to a method that handles the inductive steps:

```
1  by-induction var-agreement {
2    (num n) =>
3      pick-any s1:State s2:State
4        assume hyp := (DMap.agree-on s1 s2 evars num n)
5          (!chain [(num n wrt s1) --> n                   [V-def]
6                                  <-- (num n wrt s2) [V-def]])
7  | (e as (var x)) =>
8      pick-any s1:State s2:State
9        assume hyp := (DMap.agree-on s1 s2 evars e)
```

```
10        (!chain<- [(e wrt s1 = e wrt s2)
11                   <== (s1 at x = s2 at x) [V-def]
12                   <== (x in evars e)       [DMap.agreement-characterization]
13                   <== (x in singleton x)  [evars-def]
14                   <== true                 [Set.singleton-lemma]])
15  | (sum e1 e2)  => (!istep + e1 e2)
16  | (diff e1 e2) => (!istep - e1 e2)
17  | (prod e1 e2) => (!istep * e1 e2)
18  }
```

Being optimistic, we are assuming here that a single method, istep, will be able to handle all three inductive cases. That is not an unrealistic assumption; we have seen before (e.g., in Chapter 17) that many inductive subproofs have similar structures and their reasoning can be abstracted into one single master method. For now we will treat istep as a black box, using **force** to make it produce the right result:

```
define (istep op:(OP 2) e1 e2) :=
  (!force (var-agreement-property (e1 op e2)))
```

We will shortly provide a proper implementation of istep, but using **force** calls in this manner—and then gradually removing them—is a convenient top-down methodology.

How did we arrive at this proof skeleton? The overall form of the proof is of course dictated by the syntax of the **by-induction** construct, but what about the two basis steps? Also, why did we decide to make istep ternary rather than unary? And why are we passing the overloaded operators as arguments to istep, instead of the original constructors? For example, on line 15, why are we passing + rather than sum as the first argument to istep? We briefly take up these questions in what follows.

Let us consider the basis step for numerals first, on lines 2–6. The first two steps of this proof, on lines 3 and 4, are determined purely by the syntactic structure of the corresponding subgoal: We need to prove that for *all* states s1 and s2, *if* s1 and s2 agree on the variables of (num n), *then* (num n wrt s1) is identical to (num n wrt s2). Thus, we start by picking two arbitrary states s1 and s2, on line 2, and assuming that the two agree on the variables of (num n). We must now proceed to derive the identity (num n wrt s1 = num n wrt s2). As usual, we do that by chaining. Both chain steps follow directly from the definition of V.

Consider now the basis step for variables, on lines 7–14. As was the case with the previous basis step, the first two moves, on lines 8 and 9, are dictated by the syntactic structure of the current subgoal. We need to show that for all states s1 and s2, if the two agree on the variables of the expression (var x) (which we are abbreviating as e via the **as** pattern mechanism), then the value of (var x) wrt s1 is identical to the value of (var x) wrt s2. So we consider arbitrary s1 and s2, assume that the two agree on the expression variables of (var x), and proceed to derive the identity (var x wrt s1 = var x at s2) by chaining. We do this in a backward direction, ultimately reducing the goal to Set.singleton-lemma, namely the result

(forall x . x in singleton x),

which of course reduces to true because it is unconditional. The first backward step of the chain is determined by the definition of V. The second step is perhaps the crucial one, noting that the current goal (s1 at x = s2 at x) would in fact follow from (x in evars e) by virtue of DMap.agreement-characterization, given that we are currently operating under the assumption that s1 and s2 agree on the set (evars e). Once we have taken that step, the next backward step, from (x in evars e) to (x in singleton x), follows directly from the definition of evars, with the final step supported by Set.singleton-lemma.

Let us turn to istep, first considering the questions that were raised earlier. While we could make istep unary, in which case the call on line 12 would be (!istep (sum e1 e2)) instead of (!istep + e1 e2), and likewise for the other two calls, that would require istep to start by taking apart the input term in order to extract its individual components. Passing the individual components as three separate arguments saves us the effort of that initial pattern matching. As to why we are passing + instead of sum, and so on, the answer is that + is overloaded and hence more flexible. Suppose we passed sum instead. Then if the body istep needed to use the corresponding operator on natural numbers, Plus, we would need to write additional code to obtain Plus from sum, Times from prod, and so on. But overloaded operators can be used in both capacities, as required by the context, so there is no need for such additional code. This will become clearer when we define istep below.

Recall that istep takes op, e1, and e2 as inputs and has to derive

(var-agreement-property (e1 op e2)),

so it must prove that for all states s1 and s2, if the two agree on the variables of (e1 op e2), then the values of (e1 op e2) under s_1 and s_2 are identical:

(e1 op e2 at s1 = e1 op e2 at s2),

where, of course, istep has access to the corresponding inductive hypotheses about e1 and e2, namely, (var-agreement-property e1) and (var-agreement-property e2). Let us proceed syntactically:

```
define (istep op:(OP 2) e1 e2) :=
  pick-any s1:State s2:State
    assume hyp := (DMap.agree-on s1 s2 evars e1 op e2)
      let {ih1 := (var-agreement-property e1);
           ih2 := (var-agreement-property e2)}
      # ih1 and ih2 will be in the a.b. when istep is called
        conclude ((e1 op e2) wrt s1 = (e1 op e2) wrt s2)
```

```
        ⋮
```

How do we go about filling in the dots? Since the goal is an identity, we use chaining. The first step is simply to apply the recursive definition of V. V will thus be driven inward to the subexpressions e1 and e2, and at that point the inductive hypotheses will hopefully become applicable. While we're at it, we abbreviate (e1 op e2) as e:

```
1  define (istep op:(OP 2) e1 e2) :=
2    pick-any s1:State s2:State
3      assume hyp := (DMap.agree-on s1 s2 evars e1 op e2)
4        let {ih1 := (var-agreement-property e1);
5             ih2 := (var-agreement-property e2);
6             e := (e1 op e2)}
7        # ih1 and ih2 will be in the a.b. when istep is called
8          conclude (e wrt s1 = e wrt s2)
9            (!chain [(e wrt s1)
10                   = ((e1 wrt s1) op (e2 wrt s1)) [V-def]
11                   = ((e1 wrt s2) op (e2 wrt s2)) [ih1 ih2]
12                   = (e wrt s2)                   [V-def]])
```

(Here you can observe the aforementioned advantage of passing the overloaded op as a separate argument: op is used in two different capacities in the body of this method.)

Unfortunately, this version does not work. The first and last steps are correct, but the second step, from line 10 to line 11, does not go through. Keep in mind that the inductive hypotheses are conditional (since var-agreement-property is conditional). Thus, to apply the inductive hypothesis ih1, for example, in order to replace (e1 wrt s1) by (e1 wrt s2), *the antecedent of the inductive hypothesis must hold*, meaning that the sentence

```
(DMap.agree-on s1 s2 evars e1)
```

must be in the assumption base. Likewise for applying the second inductive hypothesis. Accordingly, the body of the method must take the following form:

```
define (istep op:(OP 2) e1 e2) :=
  pick-any s1:State s2:State
    assume hyp := (DMap.agree-on s1 s2 evars e1 op e2)
      let {ih1 := (var-agreement-property e1);
           ih2 := (var-agreement-property e2);
           e := (e1 op e2);
           _ := ··· # somehow derive (DMap.agree-on s1 s2 evars e1);
           _ := ··· # somehow derive (DMap.agree-on s1 s2 evars e2)
          }
        conclude (e wrt s1 = e wrt s2)
          (!chain [(e wrt s1)
                 = ((e1 wrt s1) op (e2 wrt s1)) [V-def]
                 = ((e1 wrt s2) op (e2 wrt s2)) [ih1 ih2]
                 = (e wrt s2)                   [V-def]])
```

So how do we derive these two lemmas? We already have a tool for doing just that,

`DMap.downward-agreement-lemma`,

namely:

```
(forall B A m1 m2 .
   (agree-on m1 m2 A) & B subset A ==> agree-on m1 m2 B)
```

The variables of e_1 and e_2 are a subset of the variables of

$$e \equiv (e_1 \;\texttt{op}\; e_2),$$

and therefore, if s_1 and s_2 agree on the variables of e, they must also agree on the variables of e_1 and e_2, by the downward agreement lemma. Of course, we still need to prove that "the variables of e_1 and e_2 are a subset of the variables of $e \equiv (e_1 \;\texttt{op}\; e_2)$." So let us assume the existence of a method `var-subset` that takes the same three arguments that `istep` takes, `op`, `e1`, and `e2`, and derives the following conjunction:

`(evars e1 subset evars e & evars e2 subset evars e)`,

where `e` is the compound expression `(e1 op e2)`. With that method at our disposal, `istep` can be defined as follows:

```
define (istep op:(OP 2) e1 e2) :=
  pick-any s1:State s2:State
    assume hyp := (DMap.agree-on s1 s2 evars e1 op e2)
      let {ih1 := (var-agreement-property e1);
           ih2 := (var-agreement-property e2);
           dal := DMap.downward-agreement-lemma;
           e := (e1 op e2);
           _ := conclude (evars e1 subset evars e & evars e2 subset evars e)
                  (!var-subset op e1 e2);
           _ := (!chain-> [hyp
                      ==> (hyp & evars e1 subset evars e) [augment]
                      ==> (DMap.agree-on s1 s2 evars e1)  [dal]]);
           _ := (!chain-> [hyp
                      ==> (hyp & evars e2 subset evars e) [augment]
                      ==> (DMap.agree-on s1 s2 evars e2)  [dal]])
           }
        conclude (e wrt s1 = e wrt s2)
          (!chain [(e wrt s1)
                 = ((e1 wrt s1) op (e2 wrt s1)) [V-def]
                 = ((e1 wrt s2) op (e2 wrt s2)) [ih1 ih2]
                 = (e wrt s2)                   [V-def]])
```

The definition of `var-subset` is left as an exercise.

Exercise 18.3: Implement var-subset. □

Similar to V, we need a function that gives us the value of a Boolean condition in a given state. We thus define a binary predicate true-in that tells us whether or not a Boolean condition holds in a given state:

```
declare true-in: [BCond State] -> Boolean
                 [105 [bcond-parser
                       lambda (L) (DMap.alist->dmap-general L int->nat)]]
```

Its definition uses structural recursion:

```
assert* true-in-def :=
  [(e1 == e2 true-in s <==> e1 wrt s = e2 wrt s)
   (e1 <= e2 true-in s <==> e1 wrt s <= e2 wrt s)
   (-- b true-in s      <==> ~ b true-in s)
   (b1 /\ b2 true-in s <==> b1 true-in s & b2 true-in s)]
```

The definition should be fairly straightforward to read. The first axiom says that an equality between two expressions holds in a given state iff the expressions have identical values in that state. Likewise, an inequality between two expressions holds in a state iff the value of the first expression in that state is less than or equal to the value of the second expression in that state. The third axiom states that a negation is true in a state iff the negated condition is not true in that state, while the last one gives the semantics of conjunctions: A conjunction of two Boolean conditions is true in a state iff both conditions are true in that state. Some tests:

```
> (eval "x + 10 = 12" true-in sample-state)

Term: true

> (eval "x + 10 != 15 - 3" true-in sample-state)

Term: false
```

18.3 Semantics of commands

As is the case in most imperative languages, the purpose of a **While** program[8] is to change states—specifically, to transform starting states to desired final states. Suppose, for example, that we write a **While** program to compute the factorial of x, putting the result in the

8 We use the terms "program" and "command" synonymously, as a **While** program is just a (potentially large) command.

"output" variable y:

```
1  y := 1;
2  while (x != 0) do
3    begin
4      y := y * x;
5      x := x - 1;
6    end
```
(18.6)

Starting from any state that assigns a value n to variable x, this program succeeds if and only if it terminates in a state in which the value of y is $n!$.

In general, the fundamental semantic judgments in this language are of this form:

Starting from so-and-so state, executing this program terminates and results in such-and-such state. (18.7)

In more mathematical parlance, the relation that we are interested in for semantic purposes is a binary relation, let us call it *yields*, which holds between starting configurations on the one hand and final states on the other:

$$\mathit{yields} \subseteq \mathit{Configuration} \times \mathit{State},$$

where a configuration is simply a pair consisting of a program to be executed and the initial state in which execution is to be commenced:

$$\mathit{Configuration} = \mathit{Cmd} \times \mathit{State}.$$

To specify the formal semantics of **While** is nothing more and nothing less than to specify the precise extension of the *yields* relation—in short, to define *yields*.[9]

Writing $[c, s]$ for a configuration pair consisting of command c and state s, and using infix notation for binary relations, a semantic judgment of the form (18.7) may be more compactly notated as:

$$[c, s] \ \mathit{yields} \ s'. \tag{18.8}$$

Using the symbol $\hookrightarrow$ as a more intuitive alias for *yields*, (18.8) may be written as

$$[c, s] \hookrightarrow s'. \tag{18.9}$$

As mentioned, the task of a formal semantics for **While** is to rigorously define *yields*; that is, to tell us, for any given c, s, and s', whether or not $[c, s] \hookrightarrow s'$ holds.

This is what we will now proceed to do. The definition of *yields* will be given by an inference system: a collection of axioms and inference rules[10] that establish judgments of

9 At least that is the case for so-called "natural" formal semantics. Formal semantics given in other styles may have different mathematical purposes.

10 Strictly speaking, axioms are inference rules with no premises, so it is not necessary to treat them separately. Nevertheless, it is customary to distinguish the two and we will generally follow that practice here. But when we use the term "rule" without qualification, it should be understood as inference rule *or* axiom.

the form $[c,s] \hookrightarrow s'$. It will then be understood that $[c,s] \hookrightarrow s'$ holds *if and only if the judgment* $[c,s] \hookrightarrow s'$ *can be formally derived using the given inference rules.*

We will first present the inference rules in conventional mathematical notation. Later we will encode them in Athena. We start with the axiom that gives the semantics of skip commands:

$$\frac{}{[\mathbf{skip}, s] \hookrightarrow s} \quad \texttt{[skip-axiom]}$$

Axioms are always written in this format, which consists of three parts:

1. A horizontal line with nothing above it. The fact that there is nothing above the line signifies that there are no premises to be derived in order to apply the axiom.
2. An axiom name (in this case `skip-axiom`) enclosed within square brackets and appearing to the right of the horizontal line. This is just a label that we give to the axiom so that we can refer to it.
3. The *body* or conclusion of the axiom, which appears directly below the horizontal line, and in this particular inference system is always of the form $[c,s] \hookrightarrow s'$.

Although we refer to it as an axiom in the interest of simplicity, `skip-axiom` is really an axiom *schema*, because s is a variable that ranges over all states. Accordingly, `skip-axiom` has infinitely many *instances*, as many as can be obtained by replacing s by a particular ground state. Read as a schema, the axiom essentially says that **skip** does nothing: Executing **skip** in any given state s results in s unchanged.

We continue with the axiom that gives the semantics of assignments:

$$\frac{}{[x := e, s] \hookrightarrow [x, V(e,s)] +\!\!+\, s} \quad \texttt{[asgn-axiom]}$$

This is also an axiom schema that can be instantiated by assigning particular values to x, e, and s. Here we write $V(e,s)$ for the value of expression e in state s,[11] and $[x, V(e,s)] +\!\!+\, s$ for the state obtained by prepending the pair $[x, V(e,s)]$ to the state s.[12] The axiom says that executing an assignment of the form $x := e$ in a state s will result in the state $[x, V(e,s)] +\!\!+\, s$. Recall that s and $[x, V(e,s)] +\!\!+\, s$ agree everywhere, except perhaps on x.

The semantics of compositions are given by the following inference rule:

$$\frac{[c_1, s] \hookrightarrow s' \quad [c_2, s'] \hookrightarrow s''}{[c_1 \,;\, c_2, s] \hookrightarrow s''} \quad \texttt{[seq-rule]}$$

---

11 We have formally defined this function in Athena by structural recursion. $V(e,s)$ is just a more customary mathematical notation that we use here to denote the application of V to e and s.

12 More precisely, applying the constructor `DMap.++` to: (a) the pair of x and $V(e,s)$; and (b) the map s.

Inference rules are written and named just like axioms, the only difference being that they might have one or more *premises* above the horizontal line, in addition to the body (or conclusion) of the rule below the line. Intuitively, the operational meaning of an inference rule with k premises is this: If k conclusions have been derived that match the rule's premises, then the corresponding body of the inference rule may also be derived.

We speak of matching because `seq-rule`, like the two axioms we have already seen, is really a schema with as many instances as can be obtained by (consistently) replacing the variables c_1, c_2, s, s', and s'' by particular commands and states.[13] The rule is perhaps best read backward: To obtain the state that results from executing a sequence $c_1\,;c_2$ in s, first execute c_1 in s to get some state s', and then execute c_2 in s' to get the final state for the entire composition.

Occasionally, an axiom or inference rule might have certain *side conditions*. These are provisos that further delimit the applicability of the axiom or rule. The stronger the provisos, the fewer the instances of the axiom or rule. Side conditions, when there are any, are written farther below the body of the rule. An example is provided by `if-true-rule` and `if-false-rule`, which together specify the semantics of conditional commands:

$$\frac{[c_1, s] \hookrightarrow s'}{[\textbf{if } b \textbf{ then } c_1 \textbf{ else } c_2, s] \hookrightarrow s'} \qquad \texttt{[if-true-rule]}$$
$$\text{provided that } T(b,s)$$

$$\frac{[c_2, s] \hookrightarrow s'}{[\textbf{if } b \textbf{ then } c_1 \textbf{ else } c_2, s] \hookrightarrow s'} \qquad \texttt{[if-false-rule]}$$
$$\text{provided that } \neg\, T(b,s)$$

Consider `if-true-rule` first. The side condition here dictates that $T(b,s)$ must hold, where we write $T(b,s)$ to indicate that condition b is true in state s.[14] Thus, the rule essentially says that if b is true in state s and the execution of c_1 in s yields s', then the execution of **if** b **then** c_1 **else** c_2 in s yields s'. Read backward: If b is true in s, then to execute a conditional statement **if** b **then** c_1 **else** c_2 in s, we execute c_1 in s. The second rule, `if-false-rule`, is read likewise, except that it requires the condition b to fail in s. Since any condition b will either hold or not hold in any given state s, these two rules are exhaustive for conditional commands.

13 The result of carrying out such a replacement can be viewed as a triple of judgments $[j_1, j_2, j_3]$, where the first two judgments are instances of the two rule premises, while j_3 is an instance of the conclusion. More generally, an $n+1$-element list $L = [j_1, \ldots, j_n, j_{n+1}]$ of particular semantic judgments is said to be an instance of an inference rule with n premises iff L can be obtained from the $(n+1)$-element list comprising the rule's premises and conclusion by consistently substituting particular syntactic objects and states in place of the variables that occur in the premises and conclusion.

14 Again, we formally defined this predicate with structural recursion earlier (where we called it `true-in`); here we are just using $T(b,s)$ as a piece of more conventional mathematical notation for the same thing.

There are also two rules for **while** loops, one for when the loop condition is true in the starting state and one for when it is false. The first one has no premises (although it has a side condition), and it is therefore an axiom:

$$\frac{}{[\textbf{while}\ b\ \textbf{do}\ c, s] \hookrightarrow s} \quad \texttt{[while-axiom]}$$

provided that $\neg T(b, s)$

This axiom says that if b does not hold in s, then the loop does nothing: Its body is not executed and the resulting state is just the starting state s unchanged. By contrast, we have the following inference rule for the case when b is true in s:

$$\frac{[c, s] \hookrightarrow s' \quad [\textbf{while}\ b\ \textbf{do}\ c, s'] \hookrightarrow s''}{[\textbf{while}\ b\ \textbf{do}\ c, s] \hookrightarrow s''} \quad \texttt{[while-rule]}$$

provided that $T(b, s)$

According to this rule, when b is true in s, the final state is obtained in two stages, as indicated by the two premises: first by executing the body of the loop, c, thereby obtaining an intermediate state s'; and then executing the entire loop again, but with s' as our starting state.

Figure 18.1 summarizes the axioms and rules of this inference system. Observe that there is exactly one rule for each syntactic operator that builds commands (essentially, for each constructor of the `Cmd` datatype), except for **while** and **if**, each of which has a pair of rules, one covering the case where the relevant Boolean condition is true and one for when it is false.

A *proof* in this inference system is a tree τ,[15] every node of which contains two things:

1. a judgment of the form $[c, s] \hookrightarrow s'$; and
2. the name R of an axiom or inference rule.

We say that τ successfully derives its *conclusion*, namely, the judgment at the root of τ, iff for every node u in τ that contains a judgment j and a rule label R, j follows from the judgments $j_1, \ldots, j_n$ at the children of u (where we will have $n = 0$ if u is a leaf) by virtue of the rule named by R. More precisely, the $(n+1)$-element list $[j_1, \ldots, j_n, j]$ must be an instance of that rule.[16]

As a simple example, here is a proof that derives the conclusion

$$[c, s] \hookrightarrow [[x, 1], [y, 2]],$$

where c is the command $x := x + 1\ ;\ y := 2 * x$ and s is the state $[[x, 0], [y, 0]]$:

15 We are overloading the term "proof" here, which is also used throughout the book to designate Athena deductions. In what follows, the context will always make clear whether we are talking about an Athena proof or about a proof in the restricted sense of this particular inference system.

16 When $n = 0$, we regard the one-element list $[j]$ as an instance of an axiom iff j is an instance of that axiom.

$$\frac{}{[\mathbf{skip}, s] \hookrightarrow s} \quad \texttt{[skip-axiom]}$$

$$\frac{}{[x := e, s] \hookrightarrow [x, V(e,s)] +\!+\, s} \quad \texttt{[asgn-axiom]}$$

$$\frac{[c_1, s] \hookrightarrow s' \quad [c_2, s'] \hookrightarrow s''}{[c_1 \,; c_2, s] \hookrightarrow s''} \quad \texttt{[seq-rule]}$$

$$\frac{[c_1, s] \hookrightarrow s'}{[\mathbf{if}\ b\ \mathbf{then}\ c_1\ \mathbf{else}\ c_2, s] \hookrightarrow s'} \quad \texttt{[if-true-rule]}$$
provided that $T(b,s)$

$$\frac{[c_2, s] \hookrightarrow s'}{[\mathbf{if}\ b\ \mathbf{then}\ c_1\ \mathbf{else}\ c_2, s] \hookrightarrow s'} \quad \texttt{[if-false-rule]}$$
provided that $\neg T(b,s)$

$$\frac{[c, s] \hookrightarrow s' \quad [\mathbf{while}\ b\ \mathbf{do}\ c, s'] \hookrightarrow s''}{[\mathbf{while}\ b\ \mathbf{do}\ c, s] \hookrightarrow s''} \quad \texttt{[while-true-rule]}$$
provided that $T(b,s)$

$$\frac{}{[\mathbf{while}\ b\ \mathbf{do}\ c, s] \hookrightarrow s} \quad \texttt{[while-axiom]}$$
provided that $\neg T(b,s)$

Figure 18.1
Inference system defining the formal semantics of **While** commands.

$[c, s] \hookrightarrow [[x, 1], [y, 2]]$
[seq-rule]

$[x := x + 1, s] \hookrightarrow [[x, 1], [y, 0]]$
[asgn-axiom]

$[y := 2 * x, [[x, 1], [y, 0]]] \hookrightarrow [[x, 1], [y, 2]]$
[asgn-axiom]

As can be verified, the conclusion at the root of this proof tree follows from the two leaves through [seq-rule], as required. The leaves, in turn, are instances of [asgn-axiom], as is also readily confirmed.

Let us start to encode this inference system in Athena. We start by introducing a datatype that captures the fundamental semantic judgments of this system:

```
datatype Judgment := (yields (Pair Cmd State) State)
```

So judgments here are of the form

(yields (pair c s) s')

for an arbitrary command c and states s and s', corresponding to the judgments of the form (18.8). To mimic the graphical symbol $\hookrightarrow$, we could introduce something like the following as an alias for yields:

```
define -->> := yields
```

That would allow us to write a judgment in the following form:

((pair c s) -->> s').

But it would be nicer still if we could pass a two-element list [c s] as the left argument of -->>, instead of a term of the form (pair c s), so that we could write judgments in the form

([c s] -->> s'),

which are almost visually identical to the "conventional" judgments of the form (18.9).

As we have done before, the trick here is to make -->> accept a two-element list on its left side and automatically convert it to the corresponding pair. (Of course, it should also be able to accept regular pair terms.) That is again achieved with **expand-input**:

```
expand-input yields [lambda (x) (lst->pair-general x cmd-parser state) state]
define -->> := yields
```

(recall the definition of lst->pair-general from page 494). Accordingly, we can now write:

```
> ([c s] -->> s')

Term: (yields (pair ?c:Cmd
                    ?s:(DMap.DMap Ide N))
              ?s':(DMap.DMap Ide N))

> ((pair c s) -->> s')
```

```
Term: (yields (pair ?c:Cmd
                    ?s:(DMap.DMap Ide N))
              ?s':(DMap.DMap Ide N))

> (c pair s -->> s')

Term: (yields (pair ?c:Cmd
                    ?s:(DMap.DMap Ide N))
              ?s':(DMap.DMap Ide N))
```

All three expressions are equivalent, in that they produce the same output, but we will generally prefer the first alternative.

Next, we specify the structure of the proofs of this inference system:

```
datatype Proof := asgn-axiom
                | skip-axiom
                | (seq-rule Proof Proof)
                | (if-true-rule Proof)
                | (if-false-rule Proof)
                | while-axiom
                | (while-rule Proof Proof)
```

Observe that there is exactly one constructor for each inference rule of the system. Constructors representing axioms are irreflexive, in fact nullary constants, whereas the constructors corresponding to proper inference rules are reflexive, meaning that they take proof objects as their arguments. Indeed, each such rule constructor receives exactly as many arguments as there are premises. This reflects the fact that axioms have no premises and thus serve as the leaves of proof trees, whereas inference rules have premises and their constructors therefore require proofs of those premises as their children (arguments). We introduce some convenient shorthands for variables ranging over `Proof` and `Judgment`:

```
define [proof proof' proof1 proof2 p p' p1 p2 p1' p2' p3 p4] :=
       [?proof:Proof ···]

define [judgment j j' j1 j2] := [?judgment:Judgment ··· ?j2:Judgment]
```

Although we refer to them as proofs in the interest of simplicity, objects of sort `Proof` are really *proof shapes*. A canonical term of this sort represents the tree structure of a proof in the sense in which proofs were defined above (page 821). More precisely, a `Proof` object is what we get from a regular proof tree by eliding all judgments from the nodes and keeping only the names of the inference rules and axioms. Consider, for example, the proof tree depicted earlier:

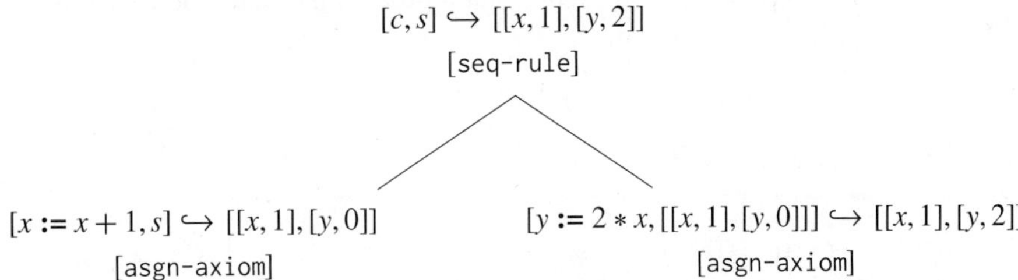

If we strike out the judgments from the nodes, what we are left with is this tree:

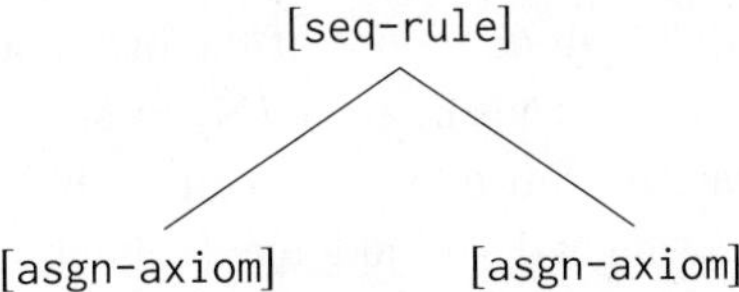

which is represented by the canonical `Proof` term

`(seq-rule asgn-axiom asgn-axiom).`

To put the meat back on such proofs, we must connect them with judgments, and that is done by giving the operational meaning of proofs, essentially by transcribing the content of Figure 18.1 in Athena. To do that, we first introduce a binary relation `derives` that holds between a proof and a judgment:

```
declare derives: [Proof Judgment] -> Boolean
```

In the vein of continuing to use conventional notation, we introduce the following alias for `derives`:

```
define |- := derives
```

So we can now say that a given proof derives a given judgment as follows:

```
> (proof |- judgment)

Term: (derives ?proof:Proof ?judgment:Judgment)
```

Let us proceed to specify exactly what it is that proofs do, meaning exactly what judgments they derive and under what circumstances. That is, for each axiom and inference rule R, we will assert an axiom of the form

R derives a judgment of the form `[c s] -->>` s' iff $\cdots$.

More precisely, for each constructor *proof-con* of the Proof datatype, we will give exactly one axiom of the form

(*proof-con* $\cdots$) derives a judgment of the form [*c* *s*] -->> *s'* iff $\cdots$.

We start with skip-axiom:

```
assert* skip-semantics :=
  (skip-axiom |- [c s] -->> s' <==> c = skip & s' = s)
```

This says that a proof of the form skip-axiom derives a judgment [c s] -->> s' iff the command c is none other than skip and the resulting state s' is none other than the initial state s. Note that both directions of the biconditional are important. It is the right-to-left direction that corresponds more closely to skip-axiom as shown in Figure 18.1, but if we did not also have the left-to-right direction, then we would be leaving open the possibility that skip-axiom might derive judgments about commands other than skip, or that the final state might be something other than the initial state. In a similar vein, here is the meaning of asgn-axiom:

```
assert* asgn-semantics :=
  (asgn-axiom |- [c s] -->> s' <==>
     exists x e . c = x <- e & s' =  [x (e wrt s)] ++ s)
```

And our first reflexive constructor:

```
assert* sequence-semantics :=
  ((seq-rule p1 p2) |- [c s] -->> s' <==>
     exists c1 c2 s'' . c = c1 ^ c2
                      & p1 |- [c1 s] -->> s''
                      & p2 |- [c2 s''] -->> s')
```

Observe the recursive constraints on the subproofs p1 and p2: They must themselves derive appropriate judgments if (seq-rule p1 p2) is to do its work. Here s'' is an "intermediate state" such that

p1 |- [c1 s] -->> s'' and p2 |- [c2 s''] -->> s'.

We continue with the remaining rules:

```
assert* if-true-semantics :=
 (if-true-rule p |- [c s] -->> s' <==>
   exists b c1 c2 .
     c = (cond b c1 c2) & b true-in s & p |- [c1 s] -->> s')

assert* if-false-semantics :=
 (if-false-rule p |- [c s] -->> s' <==>
  exists b c1 c2 .
     c = (cond b c1 c2) & ~ b true-in s & p |- [c2 s] -->> s')

assert* while-false-semantics :=
```

```
  (while-axiom |- [c s] -->> s' <==>
     exists b body . c = (while-loop b body) & ~ b true-in s & s' = s)

assert* while-true-semantics :=
  ((while-rule p1 p2) |- [c s] -->> s' <==>
     exists b body s'' . c = (while-loop b body) &
                          b true-in s &
                          p1 |- [body s] -->> s'' &
                          p2 |- [(while-loop b body) s''] -->> s')
```

And we are done! For ease of reference, we collect these axioms into a single list that we call semantics:

```
define semantics :=
  [skip-semantics asgn-semantics sequence-semantics if-true-semantics
   if-false-semantics while-true-semantics while-false-semantics]
```

Finally, we say that a judgment is *provable* iff there is some proof that derives it:

```
declare provable: [Judgment] -> Boolean

assert* provable-def :=
  [(provable judgment <==> exists proof . proof |- judgment)]
```

Before we proceed, we will have to assert the free-generation axioms for each of the datatypes we have introduced so far that ought to be freely generated (thus not counting maps). We will refer to them collectively as structural-axioms:

```
assert structural-axioms :=
  (join (datatype-axioms "Exp")
        (datatype-axioms "BCond")
        (datatype-axioms "Cmd")
        (datatype-axioms "Judgment")
        (datatype-axioms "Proof"))
```

Note that this is a fairly long list:

```
> (length structural-axioms)

Term: 70
```

18.4 ⋆ Testing the semantics

The semantics of proof trees have been formulated using biconditionals and existential quantifications. As we have already indicated, this is necessary in order to ensure, for instance, that skip-axiom can *only* be used to derive judgments about skip commands and not, say, about assignments or loops. And it is also necessary in order to "complete" the inference rules of the system in both directions, so that we can be assured, for instance, that if a proof of the form (seq-rule p_1 p_2) derives a judgment about a certain composition (c_1 ^ c_2), then p_1 derives the appropriate judgment about c_1 and p_2 about c_2.

Consequently, these axioms have somewhat complicated logical structure. They are certainly not universally quantified conditional equations, which means that we cannot use evaluation based on rewriting to "test" our definitions or to falsify conjectures in the way we have done before. Nevertheless, the semantics do have constructive content (unsurprisingly, given that they are computational semantics), and that content can be readily mined in the setting of logic programming. Indeed, using the tools described in Appendix B, that content can be extracted completely automatically by analyzing the assumption base, so there is no need to explicitly specify a logic program—we can just pose a query and let `Prolog.auto-solve` try to figure out how to answer it.

As discussed in Section B.4, `auto-solve` will not only extract relevant Horn clauses that are shallowly embedded in the assumption base, but will also perform a deeper translation of all relevant (conditional) equations into relational form. This means that we can freely mix paradigms when we develop our theories, combining equational with nonequational axioms such as the ones we used to specify the natural semantics of **While** commands; `auto-solve` should be able to treat these uniformly. For instance, let us test the semantics of expressions and Boolean conditions, which were defined using equations:

```
(Prolog.add-transformer nat->int)

> (Prolog.auto-solve "x + 3" wrt sample-state = ?RESULT)

List: [true [[?RESULT --> 5]]]
```

More interestingly, in some cases logic programming can even find for us the *expressions* (or subexpressions) that we need in order to achieve a certain result. To take a simple example, given a state that maps `'x` to 2, what expression (from the abstract grammar of all expressions) do we need to add to `(var 'x)` in order to obtain 4?

```
> (Prolog.auto-solve ?E1 + var 'x wrt state [['x --> 2]] = S S S S zero)


List: [true [[?E1 --> (num (S (S zero)))]]]
```

In fact, we can ask for a number of different solutions. When we ask for three such expressions, `auto-solve` obliges with the following answers:

```
define goal := (?E1 + var 'x wrt state [['x --> 2]] = S S S S zero)

> (Prolog.auto-solve-N goal 3)

List: [[[?E1 --> (num (S (S zero)))]]
       [[?E1 --> (var 'x)]]
```

```
      [[?E1 --> (prod (num (S (S zero)))
                      (num (S zero)))]]]
```

The second suggestion is particularly interesting because it is not a constant expression—the system has discovered that, in the given state, one can add (var 'x) to itself in order to produce the sum 4.

Let us try testing a simple Boolean condition:

```
define sample-state := (state [['x --> 2] ['y --> 3]])

define goal := ("5 + x <= y" true-in sample-state)

> (Prolog.auto-solve goal)

List: [false {}]
```

If we negate the goal we see that it becomes true, as it should:

```
> (Prolog.auto-solve "~ 5 + x <= y" true-in sample-state)

List: [true {}]
```

And a more interesting goal involving Boolean conditions:

```
> (Prolog.auto-solve num 5 <= ?E true-in sample-state)

List: [true {?E:Exp --> (num (S (S (S (S (S (S ?v53781:N)))))))}]
```

Let us now move on to the bread and butter of the **While** language—commands. We start with a very simple (the simplest possible) command: skip.

```
(Prolog.add-transformer
   lambda (m) (DMap.dmap->alist-canonical-general m nat->int))

> (Prolog.auto-solve skip-axiom |- [skip sample-state] -->> ?RESULT)

List: [true [[?RESULT --> [zero [['x --> 2] ['y --> 3]]]]]]
```

Thus, we see that executing the trivial command skip in the starting state sample-state has no effect, in that the final state is identical to the starting state, namely

[zero [['x --> 2] ['y --> 3]]].

In this example we explicitly specified the relevant proof object ourselves on the left of the |- symbol, namely skip-axiom. But Prolog could also find this for us automatically:

```
> (Prolog.auto-solve ?PROOF |- [skip sample-state] -->> ?STATE)

List: [true [[?STATE --> [zero [['x --> 2] ['y --> 3]]]]
             [?PROOF --> skip-axiom]]]
```

Here is a simple program for swapping the values of two variables x and y:

```
define swap := "z := x; x := y; y := z"

define goal :=
  (?PROOF |-  [swap (state [['x --> 1] ['y --> 2]])] -->> ?RESULT)

> (Prolog.auto-solve goal)

List: [true [[?RESULT --> [zero [['y --> 1] ['x --> 2] ['z --> 1]]]]
             [?PROOF --> (seq-rule asgn-axiom
                                   (seq-rule asgn-axiom
                                             asgn-axiom))]]]
```

Indeed, we can execute this code completely symbolically by introducing two natural number constants x-val and y-val for the values of the x and y variables respectively, and verifying that these are swapped in the output state:

```
declare x-val, y-val: N

define goal :=
  (?PROOF |-  [swap (state [['x --> x-val] ['y --> y-val]])] -->> ?RESULT)

> (Prolog.auto-solve goal)

List: [true [[?RESULT --> [zero [['y --> x-val]
                                 ['x --> y-val]
                                 ['z --> x-val]]]]
             [?PROOF --> (seq-rule asgn-axiom
                                   (seq-rule asgn-axiom
                                             asgn-axiom))]]]
```

To test conditionals, consider a simple program that sets y to twice the value of x *if* x is positive, and sets it to 1 otherwise:

```
define prog := "if (x = 0) then y := 1 else y := 2 * x"

define goal :=
  (?PROOF |-  [prog (state [['x --> 0]])] -->> ?RESULT)

> (Prolog.auto-solve goal)

List: [true [[?RESULT --> [zero [['y --> 1] ['x --> 0]]]]
             [?PROOF --> (if-true-rule asgn-axiom)]]]
```

```
> (Prolog.auto-solve
     ?PROOF |-  [prog (state [['x --> 4]])] -->> ?RESULT)

List: [true [[?RESULT --> [zero [['y --> 8] ['x --> 4]]]]
             [?PROOF --> (if-false-rule asgn-axiom)]]]
```

Let us now consider a more interesting example, involving looping—a program for computing the factorial function:

```
define factorial-program :=
   "y := 1; while (x !=  0) do begin y := y * x; x := x - 1 end"

define goal :=
  (?PROOF |-  [factorial-program (state [['x --> 3]])] -->> ?RESULT)

> (Prolog.auto-solve goal)

List: [true [[?RESULT --> [zero [['x --> 0] ['y --> 6]]]]
             [?PROOF -->
                (seq-rule
                  asgn-axiom
                  (while-rule
                    (seq-rule asgn-axiom asgn-axiom)
                    (while-rule
                      (seq-rule asgn-axiom asgn-axiom)
                      (while-rule
                        (seq-rule asgn-axiom asgn-axiom)
                        while-axiom))))]]]
```

Observe that the discovered proof object essentially tracks an unfolding of the program as it goes through the various loop iterations. Note, in particular, that the number of applications of `while-rule` in the computed proof object is equal to the number of times the body of the **while** loop is executed at run time, which in this case is identical to the value of the input x:

```
define goal :=
  (?PROOF |-  [factorial-program (state [['x --> 5]])] -->> ?RESULT)

> (Prolog.auto-solve goal)

List: [true [[?RESULT --> [zero [['x --> 0] ['y --> 120]]]]
             [?PROOF -->
               (seq-rule asgn-axiom
                 (while-rule
                   (seq-rule asgn-axiom asgn-axiom)
                   (while-rule
                     (seq-rule asgn-axiom asgn-axiom)
                     (while-rule
```

```
                      (seq-rule asgn-axiom asgn-axiom)
                      (while-rule
                        (seq-rule asgn-axiom asgn-axiom)
                        (while-rule
                           (seq-rule asgn-axiom asgn-axiom)
                           while-axiom)))))))]]]
```

Let us now compute a function of two arguments. The following program raises x to the nth power and stores the result in y:

```
y := 1;
while (n != 0) do
  begin
     y := y * x;
     n := n -1;
  end
```

(18.10)

We check that it produces the right result:

```
define pow :=
 "y := 1; while (n != 0) do begin y := y * x; n := n - 1 end"

define goal :=
  (?PROOF |-  [pow (state [['x --> 2] ['n --> 3]])] -->> ?RESULT)

> (Prolog.auto-solve goal)

List: [true [[?RESULT --> [zero [['n --> 0] ['y --> 8] ['x --> 2]]]]
             [?PROOF --> (seq-rule
                            asgn-axiom
                            (while-rule
                              (seq-rule asgn-axiom asgn-axiom)
                              (while-rule
                                 (seq-rule asgn-axiom asgn-axiom)
                                 (while-rule
                                    (seq-rule asgn-axiom asgn-axiom)
                                    while-axiom))))]]]
```

Finally, here is a program that loops forever:

```
define loop := "while (1 = 1) do skip"

define goal := (?PROOF |-  [loop sample-state] -->> ?RESULT)

> (Prolog.auto-solve goal)
      ...
     # Infinite loop ultimately failing
     # with an out-of-stack error message from Prolog
```

18.5 Some lemmas

In this section we derive some useful lemmas from the formal semantics.

While `Prolog.auto-solve` automatically extracts Horn clauses from the assumption base as needed to answer its input queries, we will see that, for the natural semantics of **While** at least, it is also useful to extract Horn clauses more explicitly and permanently—to *prove* them. Starting from the strong semantics as we have stated them, it takes little effort to derive the unidirectional inference rules of Figure 18.1 as lemmas. These lemmas will come in handy in some of the forthcoming proofs.

Let us begin with `skip`. The lemma that we want to derive is essentially [`skip-axiom`] of Figure 18.1:

```
define skip-horn-clause := (forall s . skip-axiom |- [skip s] -->> s)
```

How can we deduce this result from `skip-semantics`? Recall the content of that axiom:

```
(forall cmd s s' .
  skip-axiom |- [cmd s] -->> s' <==> cmd = skip & s' = s)
```

Since `skip-horn-clause` aims to prove `(skip-axiom |- [skip s] -->> s)` for arbitrary `s`, we need to use the biconditional of `skip-semantics` in a right-to-left direction, so we need to establish the right-hand side of that biconditional for the appropriate values of `cmd`, `s`, and `s'`. Chaining can do that in one step:

```
conclude skip-horn-clause
  pick-any s:State
    (!chain-> [(skip = skip & s = s)
           ==> (skip-axiom |- [skip s] -->> s) [skip-semantics]])
```

(Recall from page 414 that `chain->` will automatically derive the first element of the chain, `(skip = skip & s = s)`, since it is a conjunction of two identities of the form $(t = t)$.)

We continue with `asgn-axiom`. Here we want to derive the following lemma, which captures the [`asgn-axiom`] of Figure 18.1:

```
define asgn-horn-clause :=
  (forall x e s . asgn-axiom |- [(x <- e) s] -->> [x (V e s)] ++ s)
```

The idea is similar to the proof of `skip-horn-clause`: We derive the lemma from the biconditional of `asgn-semantics`, except that now there is an extra step required to introduce the necessary existential quantification:

```
conclude asgn-horn-clause
  pick-any x e s:State
    (!chain->
      [(x <- e = x <- e & [x (e wrt s)] ++ s = [x (e wrt s)] ++ s)
```

```
  ==> (exists x' e' .
        (x <- e) = (x' <- e') &
        [x (e wrt s)] ++ s = [x' (e' wrt s)] ++ s)    [existence]
  ==> (asgn-axiom |-
        [(x <- e) s] -->> [x (e wrt s)] ++ s)          [asgn-semantics]])
```

We state the remaining Horn clauses here and leave their proofs as exercises:

```
define seq-horn-clause :=
  (forall cmd1 cmd2 s s' s'' p1 p2 .
    p1 |- [cmd1 s] -->> s' & p2 |- [cmd2 s'] -->> s'' ==>
      (seq-rule p1 p2) |- [(cmd1 ^ cmd2) s] -->> s'')

define if-true-horn-clause :=
  (forall b cmd1 cmd2 p s s' .
    b true-in s & p |- [cmd1 s] -->> s' ==>
       (if-true-rule p) |- [(cond b cmd1 cmd2) s] -->> s')

define if-false-horn-clause :=
  (forall b cmd1 cmd2 p s s' .
    ~ b true-in s & p |- [cmd2 s] -->> s' ==>
        (if-false-rule p) |- [(cond b cmd1 cmd2) s] -->> s')

define while-true-horn-clause :=
  (forall b body p1 p2 s s' s'' .
    b true-in s &
    p1 |- [body s] -->> s' &
    p2 |- [(while-loop b body) s'] -->> s''
  ==> (while-rule p1 p2) |- [(while-loop b body) s] -->> s'')

define while-false-horn-clause :=
  (forall b body s .
    ~ b true-in s ==>
      while-axiom |- [(while-loop b body) s] -->> s)
```

We gather all the Horn clauses into a single list for ease of reference:

```
define semantics-horn-clauses :=
  [skip-horn-clause asgn-horn-clause seq-horn-clause
   if-false-horn-clause if-false-horn-clause
   while-false-horn-clause while-true-horn-clause]
```

Exercise 18.4: Derive the remaining Horn clause lemmas. □

We refer to the lemmas that are formulated and proven in the remainder of this section as "exclusivity lemmas" because each of them states that, roughly, for any given command c of **While** there is *only* one inference rule that can derive semantic judgments about c. For instance, given some composition $(c_1 \wedge c_2)$, only proofs of the form `(seq-rule ···)`

are capable of deriving judgments about it; only `skip-axiom` can derive judgments about `skip` commands; and so on. These are not particularly glamorous or interesting lemmas. But they are quite useful, as we will see later. Moreover, they provide a good opportunity to demonstrate, once again, the benefits of using methods to automate recurring patterns of reasoning.

Let us start with the exclusivity lemma for compositions:

```
define skip-excl :=
  (forall proof s s' .
     proof |- [skip s] -->> s' ==> proof = skip-axiom & s' = s)
```

This sentence says that if any proof tree derives a judgment of the form `[skip s] -->> s'`, then that proof tree must be the leaf `skip-axiom`, and further, the state `s'` must be identical to `s`. This ought to follow from our semantics, and indeed we will prove that it does. For now, we simply list all the exclusivity lemmas:

```
define skip-excl :=
  (forall proof s s' .
     proof |- [skip s] -->> s' ==> proof = skip-axiom & s' = s)

define asgn-excl :=
  (forall proof x e s s' .
    proof |- [(x <- e) s] -->> s' ==> proof = asgn-axiom &
                                      s' =  [x (V e s)] ++ s)


define seq-excl :=
  (forall p c1 c2 s s' .
    p |- [(c1 ^ c2) s] -->> s' ==>
      exists p1 p2 s'' . p = (seq-rule p1 p2) &
                         p1 |-  [c1 s] -->> s'' &
                         p2 |-  [c2 s''] -->> s')

define if-true-excl :=
 (forall p b c1 c2 s s' .
   p |- [(cond b c1 c2) s] -->> s' & b true-in s ==>
     exists p' . p = if-true-rule p' & p' |- [c1 s] -->> s')

define if-false-excl :=
 (forall p b c1 c2 s s' .
   p |- [(cond b c1 c2) s] -->> s' & ~ b true-in s ==>
     exists p' . p = if-false-rule p' & p' |- [c2 s] -->> s')

define while-true-excl :=
 (forall p b body s s' .
   p |- [(while-loop b body) s] -->> s' & b true-in s ==>
     exists p1 p2 s'' . p = (while-rule p1 p2) &
                        p1 |- [body s] -->> s'' &
                        p2 |- [(while-loop b body) s''] -->> s')
```

```
define while-false-excl :=
 (forall p b body s s' .
   p |- [(while-loop b body) s] -->> s' & ~ b true-in s ==>
     p = while-axiom & s' = s)

define excl-lemmas :=
  [skip-excl asgn-excl seq-excl if-true-excl if-false-excl
   while-true-excl while-false-excl]
```

So how do we go about proving these lemmas? Let's consider the case of compositions first, seq-excl. We need to show that for *any* proof p, *if* p derives a judgment of the form [(c_1 ^ c_2) s] -->> s', *then* p must be of the form (seq-rule p_1 p_2), where p_1 and p_2 themselves derive appropriate judgments. This goal is universally quantified over proofs, so we must show that the aforementioned conditional holds no matter what p is (i.e., no matter what the *form* of p is); in other words, no matter what the top constructor of p is. So we have to consider each possible proof constructor and establish the conditional separately in each case. One way to do that is to use **datatype-cases**. So the desired proof now takes the following skeletal form:

```
datatype-cases seq-excl {
  skip-axiom => ...
| asgn-axiom => ...
| (seq-rule p1 p2) => ...
| (if-true-rule p) => ...
| (if-false-rule p) =>
| (while-rule p1 p2) => ...
| while-axiom => ... }
```

Let us take up the case of skip-axiom first. We need to show that *if* skip-axiom derives a judgment of the form [(c_1 ^ c_2) s] -->> s', *then* skip-axiom must be of the form (seq-rule p_1 p_2), where p_1 and p_2 themselves derive appropriate judgments. How can we establish such a conditional? The answer is by contradiction. We are assuming that skip-axiom derives a judgment of the form [(c_1 ^ c_2) s] -->> s':

$$\texttt{skip-axiom |- [(}c_1 \wedge c_2\texttt{)}\ s\texttt{] -->> } s'. \tag{18.11}$$

But that's inconsistent with skip-semantics! Recall:

```
assert* skip-semantics :=
 (skip-axiom |- [cmd s] -->> s' <==> cmd = skip & s' = s)
```

So our assumption (18.11) can be combined with skip-semantics to derive the absurd conclusion:

$$(c_1 \wedge c_2) = \texttt{skip}.$$

This conclusion is absurd on account of the free-generation axioms of the Cmd datatype (or less precisely, from what we earlier called structural-axioms). Accordingly, we can start to fill that part of the overall proof skeleton as follows:

```
datatype-cases seq-excl {
  skip-axiom =>
    pick-any cmd1 cmd2 s:State s':State
      assume hyp := (skip-axiom |- [(cmd1 ^ cmd2) s] -->> s') {
        goal :=  (exists p1 p2 is . skip-axiom = (seq-rule p1 p2) &
                                   p1 |-  [cmd1 s] -->> is &
                                   p2 |-  [cmd2 is] -->> s');
        (!chain-> [hyp ==> (cmd1 ^ cmd2 = skip) [semantics]
                       ==> goal                 [free-gen]])}
| asgn-axiom => ...
| (seq-rule p1 p2) => ...
| (if-true-rule p) => ...
| (if-false-rule p) =>
| (while-rule p1 p2) => ...
| while-axiom => ... }
```

where, on first approximation, free-gen is a binary method that takes a premise of the form

`(cmd1 ^ cmd2 = skip)`

as its first argument (or more generally, a premise of the form $((c \cdots) = (c' \cdots))$ where c and c' are *distinct* constructors of Cmd); and an arbitrary goal as its second argument; and derives that goal. (We deliberately defer further discussion of free-gen for now.) Let us continue by filling in the case of if-true-rule, which is quite similar:

```
datatype-cases seq-excl {
  skip-axiom =>
    pick-any c1 c2 s:State s':State
      assume hyp := (skip-axiom |- [(c1 ^ c2) s] -->> s')
        let {goal :=  (exists p1 p2 s'' . skip-axiom = (seq-rule p1 p2) &
                                          p1 |-  [c1 s] -->> s'' &
                                          p2 |-  [c2 s''] -->> s')}
          (!chain-> [hyp ==> (c1 ^ c2 = skip) [semantics]
                         ==> goal              [free-gen]])

| (if-true-rule p) =>
    pick-any c1 c2 s:State s':State
      assume hyp := ((if-true-rule p) |-  [(c1 ^ c2) s] -->> s')
        let {goal := (exists p1 p2 s'' .
                        (if-true-rule p) = (seq-rule p1 p2) &
                        p1 |-  [c1 s] -->> s'' &
                        p2 |-  [c2 s''] -->> s')}
      (!chain-> [hyp
              ==> (exists b cmd1 cmd2 .
                     c1 ^ c2 = (cond b cmd1 cmd2) &
```

```
                  b true-in s &
                  p |- [cmd1 s] -->> s')  [semantics]
             ==> goal                       [free-gen]])

| asgn-axiom => ...
| (seq-rule p1 p2) => ...
| (if-false-rule p) => ...
| (while-rule p1 p2) => ...
| while-axiom =>...
}
```

Two points emerge when we examine the two subproofs in this skeleton:

- The `free-gen` method needs to be somewhat more sophisticated. Its input premise need not be a simple premise of the form

$$((c \; \cdots) = (c' \; \cdots)), \tag{18.12}$$

  that is, an identity between two terms built with distinct `Cmd` constructors c and c'. Instead, as the case of `if-true-rule` shows, the input premise of `free-gen` could be an existential quantification, only one conjunct of which is an identity of the form (18.12). However, we specify that the existential quantification in question might have zero variables, which means that `free-gen` should also be able to handle equations of the form (18.12). Of course, such an existential quantification is inconsistent with the free-generation axioms of `Cmd`, so `free-gen` should be able to derive its second argument by contradiction, in the way outlined above. Therefore, the skeleton of `free-gen` can be expressed as follows:

```
define (free-gen premise goal) :=
  match premise {
    (exists (some-list evars) body) =>
       # granting that body contains a conjunct of the form
       #                ((c ...) = (c' ...))
       # for two distinct constructors c and c' of Cmd,
       # derive goal.
       ...
  }
```

  We leave the complete definition of this method as an exercise.

- For both subproofs shown above, we defined the appropriate `goal` that was derived in the subsequent chain. These goals have a common structure which can be abstracted into a procedure:

```
define (make-goal proof c1 c2 s s') :=
    (exists p1 p2 s'' . proof = (seq-rule p1 p2) &
                        p1 |- [c1 s] -->> s'' &
                        p2 |- [c2 s''] -->> s')
```

Exercise 18.5: Implement free-gen. □

With free-gen under our belt, most remaining cases become fairly routine, as they are quite similar to the two cases shown above: By assuming that a proof constructor *other* than seq-rule derives a judgment of the form

$$[(c_1 \wedge c_2)\ s] \dashrightarrow\!\!> s',$$

we get a contradiction that allows us to derive our goal vacuously. The only interesting case, which does not involve a contradiction, is the case of seq-rule. In that case the result can be derived from the semantics and structural-axioms of our theory. Here is the proof in its entirety:

```
datatype-cases seq-excl {
  asgn-axiom =>
    pick-any c1 c2 s s'
      assume hyp := (asgn-axiom |- [(c1 ^ c2) s] -->> s')
        let {goal := (make-goal asgn-axiom c1 c2 s s')}
          (!chain-> [hyp ==> (exists x e . c1 ^ c2 = x <- e &
                                         s' = [x (V e s)] ++ s)
                             [semantics]
                     ==> goal [free-gen]])
| skip-axiom =>
     pick-any c1 c2 s:State s':State
       assume hyp := (skip-axiom |- [(c1 ^ c2) s] -->> s')
         let {goal := (make-goal skip-axiom c1 c2 s s')}
           (!chain-> [hyp ==> (c1 ^ c2 = skip & s' = s) [semantics]
                          ==> goal [free-gen]])
| (seq-rule p1 p2) =>
     pick-any c1 c2 s:State s':State
       assume hyp := ((seq-rule p1 p2) |- [(c1 ^ c2) s] -->> s')
         let {goal := (make-goal (seq-rule p1 p2) c1 c2 s s');
              L := (!chain->
                     [hyp ==>  (exists cmd1 cmd2 s'' .
                                  c1 ^ c2 = cmd1 ^ cmd2 &
                                  p1 |- [cmd1 s] -->> s'' &
                                  p2 |- [cmd2 s'']  -->> s')
                                            [semantics]])}
           pick-witnesses c1' c2' s'' for L
             (!chain-> [(p1 |- [c1' s] -->> s'' &
                         p2 |-  [c2' s'']  -->> s')
                    ==> (p1 |- [c1 s] -->> s'' &
                         p2 |-  [c2 s'']  -->> s')
                           [(c1 = c1' & c2 = c2') <==
                            (c1 ^ c2 = c1' ^ c2')     [structural-axioms]]
                    ==> ((seq-rule p1 p2) = (seq-rule p1 p2) &
                         p1 |- [c1 s] -->> s'' &
                         p2 |-  [c2 s'']  -->> s')   [augment]
```

```
                ==> goal [existence]])
| (if-true-rule p) =>
    pick-any c1 c2 s:State s':State
      assume hyp := ((if-true-rule p) |-  [(c1 ^ c2) s] -->> s')
        let {goal := (make-goal (if-true-rule p) c1 c2 s s')}
          (!chain->
             [hyp ==> (exists b cmd1 cmd2 .
                        c1 ^ c2 = (cond b cmd1 cmd2) &
                        b true-in s &
                        p |- [cmd1 s] -->> s') [semantics]
                  ==> goal [free-gen]])
| (if-false-rule p) =>
    pick-any c1 c2 s:State s':State
      assume hyp := (if-false-rule p |- [(c1 ^ c2) s] -->> s')
        let {goal := (make-goal (if-false-rule p) c1 c2 s s')}
          (!chain->
             [hyp ==> (exists b cmd1 cmd2 .
                        c1 ^ c2 = (cond b cmd1 cmd2) &
                        ~ b true-in s &
                        p |- [cmd2 s] -->> s') [semantics]
                 ==> goal [free-gen]])
| (while-rule p1 p2) =>
    pick-any c1 c2 s:State s':State
      assume hyp := ((while-rule p1 p2) |- [(c1 ^ c2) s] -->> s')
        let {goal := (make-goal (while-rule p1 p2) c1 c2 s s')}
          (!chain-> [hyp
                 ==> (exists b body s'' .
                       c1 ^ c2 = (while-loop b body) &
                       b true-in s &
                       p1 |- [body s] -->> s'' &
                       p2 |- [(while-loop b body) s''] -->> s')
                                                    [semantics]
                ==> goal [free-gen]])
| while-axiom =>
    pick-any c1 c2 s:State s':State
      assume hyp := (while-axiom |- [(c1 ^ c2) s] -->> s')
        let {goal := (make-goal while-axiom c1 c2 s s')}
          (!chain-> [hyp ==> (exists b body .
                               c1 ^ c2 = (while-loop b body) &
                              ~ b true-in s & s' = s) [semantics]
                         ==> goal [free-gen]])
}
```

This is a rather long and tedious method. We have written 77 lines of code, not counting some auxiliary methods and procedures, and have only managed to prove one of the seven exclusivity lemmas. If we need to write another 80–100 lines of code for each of the remaining lemmas, we are looking at about 800 lines of code just for the exclusivity lemmas. There has to be a better way.

Once again, method abstraction can come to our aid. The inelegance of the above proof stems from the fact that there is a lot of unnecessary repetition of the same pattern of reasoning: All cases other than `seq-rule` do essentially the same thing. Abstracting that pattern into a separate method, if it can be done, would save us a great deal of effort and also make our proof more intelligible. In fact, when we think about the situation a bit more carefully, we see that there is common structure not just in the proof of this particular exclusivity lemma, but in the proofs of all exclusivity lemmas. Each such lemma will have to be proved by **datatype-cases**, and the proof will have to proceed along similar lines: There will be one `Proof` constructor for which we will do some honest work, while for the remaining constructors we will be using `free-gen` to derive the relevant goal by contradiction. Is there any way we can automate this pattern across all different exclusivity lemmas?

The answer is affirmative. The key is to think of the reasoning for the various branches of **datatype-cases** as an independent unit that can be passed around and applied; specifically, as a unary method that takes a given proof object $(c \cdots)$ as its argument and acts accordingly, depending on c. To achieve this decoupling, we formulate a binary method

`auto-dtc-for-proofs`

that takes a goal as its first argument, universally quantified over all proofs, and a method M as its second argument, and proceeds to prove that goal by **datatype-cases**, using M in each case:

```
define (auto-dtc-for-proofs goal M) :=
  datatype-cases goal {
    asgn-axiom           => (!M asgn-axiom)
  | skip-axiom           => (!M skip-axiom)
  | (seq-rule p1 p2)     => (!M (seq-rule p1 p2))
  | (if-true-rule p)     => (!M (if-true-rule p))
  | (if-false-rule p)    => (!M (if-false-rule p))
  | while-axiom          => (!M while-axiom)
  | (while-rule p1 p2)   => (!M (while-rule p1 p2))
}
```

We can now prove an exclusivity lemma, say, the one for `skip`, in a few lines as follows:

```
(!auto-dtc-for-proofs skip-excl
  method (proof)
    pick-any s:State s':State
      assume hyp := (proof |- [skip s] -->> s')
        let {goal := (proof = skip-axiom & s' = s)}
        match proof {
          skip-axiom =>
            (!chain-> [hyp ==> (s' = s)              [semantics]
                           ==> goal                  [augment]])
        | _ => (!chain-> [hyp
                     ==> (fire-semantic-axiom hyp) [semantics]
```

```
                    ==> goal                    [free-gen]])
     })
```

where fire-semantic-axiom is defined as follows:

```
define (fire-semantic-axiom hyp) :=
  (find-first
    semantics
    lambda (axiom)
      match axiom {
        (forall (some-list _) (left <==> right)) =>
           match (match-sentences hyp left) {
             (some-sub sub) => (sub right)
           | _ => false
           }
      })
```

The generic utility find-first is a binary procedure that takes a list L and a unary procedure f and returns the first (leftmost) element x of L such that $(f\ \ x)$ is a value other than false. An error occurs if L has no such x:

```
define (find-first L f) :=
  match L {
    (list-of x rest) => match (f x) {
                          false => (find-first rest f)
                        | res => res
                        }
  }
```

Note that we have to use something like fire-semantic-axiom in the proof of skip-excl, because we want to handle *all* cases other than skip-axiom in one fell swoop, so we need to retrieve and use the proper semantic axiom automatically, depending on the case at hand. Thus, for instance:

```
> (fire-semantic-axiom (asgn-axiom |- [c s] -->> s'))

Sentence: (exists ?v70422:Ide
            (exists ?v70423:Exp
              (and (= ?c:Cmd
                      (asgn ?v70422:Ide ?v70423:Exp))
                   (= ?s':(DMap.DMap Ide N)
                      (DMap.update (pair ?v70422:Ide
                                         (V ?v70423:Exp
                                            ?s:(DMap.DMap Ide N)))
                                   ?s:(DMap.DMap Ide N))))))

> (fire-semantic-axiom ((if-true-rule p) |- [c s] -->> s'))
```

```
Sentence: (exists ?v70424:BCond
            (exists ?v70425:Cmd
              (exists ?v70426:Cmd
                (and (= ?c:Cmd
                        (cond ?v70424:BCond ?v70425:Cmd ?v70426:Cmd))
                     (and (true-in ?v70424:BCond
                                   ?s:(DMap.DMap Ide N))
                          (derives ?p:Proof
                                   (yields (pair ?v70425:Cmd
                                                 ?s:(DMap.DMap Ide N))
                                           ?s':(DMap.DMap Ide N))))))))
```

Using this approach we can prove seq-excl as follows:

```
(!auto-dtc-for-proofs seq-excl
  method (proof)
    pick-any c1 c2 s s'
      assume hyp := (proof |- [(c1 ^ c2) s] -->> s')
        let {goal := (make-goal proof c1 c2 s s')}
        match proof {
          (seq-rule p1 p2) =>
            let {L := (!chain->
                        [hyp
                     ==> (exists cmd1 cmd2 s'' .
                           c1 ^ c2 = cmd1 ^ cmd2 &
                           p1 |- [cmd1 s] -->> s'' &
                           p2 |- [cmd2 s''] -->> s') [semantics]])}
            pick-witnesses c1' c2' s'' for L
              (!chain->
                 [(p1 |- [c1' s] -->> s'' & p2 |- [c2' s''] -->> s')
              ==> (p1 |- [c1 s] -->> s'' & p2 |-  [c2 s''] -->> s')
                    [(c1 = c1' & c2 = c2') <==
                     (c1 ^ c2 = c1' ^ c2') [structural-axioms]]
              ==> ((seq-rule p1 p2) = (seq-rule p1 p2) &
                   p1 |- [c1 s] -->> s'' &
                   p2 |- [c2 s''] -->> s')          [augment]
              ==> goal [existence]])
        | _ => (!chain->
                 [hyp ==> (fire-semantic-axiom hyp) [semantics]
                      ==> goal                      [free-gen]])
    })
```

So we have not only eliminated the redundancies from our first proof attempt, thereby going from 77 to 26 lines, but we have found a general template that can be used to prove all exclusivity lemmas with significantly less effort. The proofs of the other exclusivity lemmas are left as exercises.

Exercise 18.6: Prove the remaining exclusivity lemmas. □

18.6 Reasoning about the language

In this section we prove some results about the semantics of the **While** language. Two of these results make use of a notion of *program equivalence*. We say that two programs (commands) c_1 and c_2 are equivalent iff executing c_1 and c_2 in the same initial state results in the same final state. We can define this notion more precisely in Athena as follows:

```
declare equivalent: [Cmd Cmd] -> Boolean [[cmd-parser cmd-parser]]

assert equivalent-def :=
  (forall cmd1 cmd2 . cmd1 equivalent cmd2 <==>
      (forall s s' . provable [cmd1 s] -->> s' <==>
                     provable [cmd2 s] -->> s'))
```

We overload == so that we can use it as a more intuitive mnemonic for `equivalent`:

```
overload == equivalent

set-precedence == 105
```

The first result that we will prove is the associativity of composition:

```
define assoc-theorem :=
  (forall c1 c2 c3 . c1 ^ (c2 ^ c3) == (c1 ^ c2) ^ c3)
```

Assuming that the result is true, how do we prove it? Consider arbitrary commands c_1, c_2, and c_3, and let `left` and `right` refer to the compositions

$$(c_1 \wedge (c_2 \wedge c_3))$$

and $((c_1 \wedge c_2) \wedge c_3)$, respectively. Consider, further, two arbitrary states s and s'. Now, equivalence is a bidirectional notion, so there are two parts that we need to establish. First, in one direction, we need to show that *if* the judgment `[left` s`] -->>` s' is derivable, *then* the judgment `[right` s`] -->>` s' is also derivable. And conversely, in the other direction, we need to show that if the judgment `[right` s`] -->>` s' is derivable, then `[left` s`] -->>` s' is also derivable. It is therefore natural to break up the proof effort into two proof methods that will do the brunt of the work: `assoc-dir-1` will take a premise of the form

$$[(c_1 \wedge (c_2 \wedge c_3))\ s] \dashrightarrow\!\!> s' \tag{18.13}$$

and will derive

$$[((c_1 \wedge c_2) \wedge c_3)\ s] \dashrightarrow\!\!> s'. \tag{18.14}$$

The second method, `assoc-dir-2`, will work in the reverse direction: It will take a premise of the form (18.14) and will derive (18.13). Equipped with these two methods, `assoc-theorem` can be derived as follows:

```
conclude assoc-theorem
  pick-any c1 c2 c3
    let {left := (c1 ^ (c2 ^ c3));
         right := ((c1 ^ c2) ^ c3);
         eqv := pick-any s:State s':State
                  let {imp1 := assume hyp1 := (provable [left s] -->> s')
                                 (!assoc-dir-1 hyp1);
                       imp2 := assume hyp2 := (provable [right s] -->> s')
                                 (!assoc-dir-2 hyp2)}
                     (!equiv imp1 imp2)}
      (!chain-> [eqv ==> (left == right) [equivalent-def]])
```

All we now have to do is implement the two methods. Here is the first one:

```
define assoc-dir-1 :=
  (method (premise)
    match premise {
      (provable (bind judgment
                  (yields (pair (bind left (sequence c1
                                                     (sequence c2 c3)))
                                start-s)
                          final-s))) => {
        right := ((c1 ^ c2) ^ c3);
        conclude goal := (provable [right start-s] -->> final-s)
          pick-witness p for
            (!chain-> [premise ==> (exists p . p |- judgment)
                                       [provable-def]]) {
              L1 := (!chain->
                       [(p |- judgment)
                    ==> (exists p1 p2 s' .
                           p = (seq-rule p1 p2) &
                           p1 |- [c1 start-s] -->> s' &
                           p2 |- [(c2 ^ c3) s'] -->> final-s)
                                                         [seq-excl]]);
              pick-witnesses p1 p2 s' for L1 {
                L2 := (!chain->
                         [(p2 |- [(c2 ^ c3) s'] -->> final-s)
                      ==> (exists p3 p4 s'' .
                             p2 = (seq-rule p3 p4) &
                             p3 |- [c2 s'] -->> s'' &
                             p4 |- [c3 s''] -->> final-s) [seq-excl]]);
                pick-witnesses p3 p4 s'' for L2
                  (!chain->
                     [(p1 |- [c1 start-s] -->> s')
                  ==> (p1 |- [c1 start-s] -->> s' &
                       p3 |- [c2 s'] -->> s'')   [augment]
                  ==> ((seq-rule p1 p3) |-
                          [(c1 ^ c2) start-s] -->> s'')
                                                  [seq-horn-clause]
                  ==> ((seq-rule p1 p3) |-
                         [(c1 ^ c2) start-s] -->> s'' &
```

```
                p4 |- [c3  s''] -->> final-s) [augment]
          ==> ((seq-rule (seq-rule p1 p3) p4) |-
                 [right start-s] -->> final-s) [seq-horn-clause]
          ==> (exists p .
                 p |- [right start-s] -->> final-s) [existence]
          ==> goal [provable-def]])
    }
   }}})
```

The definition of the converse method, assoc-dir-2, is left as an exercise.

Exercise 18.7: Implement assoc-dir-2. □

Our next result will be the determinism of the semantics:

```
define deterministic-semantics :=
  (forall c s s' s'' .
    provable [c s] -->> s' & provable [c s] -->> s''  ==> s' = s'')
```

We will prove this result in a form in which the definition of provable has been expanded:

```
define determinism :=
  (forall proof1 proof2 c s s' s'' .
     proof1 |- [c s] -->> s' &
     proof2 |- [c s] -->> s'' ==> s'' = s')
```

Deriving deterministic-semantics from determinism is straightforward.

How do we go about proving determinism? The goal is universally quantified over all proofs, so one thought would be to use **datatype-cases**. However, if we tried that avenue we would soon realize that it will not work. We need a more powerful mechanism—induction. So we will proceed instead by structural induction on proof objects.

We begin by defining the relevant inductive property as a unary procedure:

```
define det-property :=
  lambda (proof1)
    (forall proof2 c s s' s'' .
      proof1 |- [c s] -->> s' &
      proof2 |- [c s] -->> s'' ==> s'' = s')
```

Thus, determinism can be understood as the sentence

```
(forall proof . det-property proof).
```

Our inductive proof inlines the basis cases and separates out the inductive subproofs in different methods that can be independently tested:

```
by-induction (forall proof . det-property proof) {
  skip-axiom =>
   pick-any proof2 c s:State s':State s'':State
```

```
     assume hyp1 := (skip-axiom |- [c s] -->> s');
            hyp2 := (proof2 |- [c s] -->> s'')
      (!chain->
        [hyp2 ==> (proof2 |- [skip s] -->> s'')
                                   [(c = skip) <== hyp1 [semantics]]
              ==> (s'' = s)        [skip-excl]
                       = s'        [(s' = s) <== hyp1  [semantics]]])
| asgn-axiom =>
     pick-any proof2 c s:State s':State s'':State
       assume hyp1 := (asgn-axiom |- [c s] -->> s');
              hyp2 := (proof2 |- [c s] -->> s'')
         let {L1 := (!chain->
                      [hyp1 ==>
                        (exists x e . c = x <- e &
                                      s' = [x (V e s)] ++ s) [semantics]])}
           pick-witnesses x e for L1
             (!chain->
               [hyp2
           ==> (proof2 |- [(x <- e) s] -->> s'') [(c = x <- e)]
           ==> (s'' = [x (V e s)] ++ s)          [asgn-excl]
                   = s'                          [(s' = [x (V e s)] ++ s)]])
| (seq-rule p1 p2) =>    (!determinism-sequence-case (seq-rule p1 p2))
| (if-true-rule p) =>    (!determinism-if-true-case (if-true-rule p))
| (if-false-rule p) =>   (!determinism-if-false-case (if-false-rule p))
| (while-rule p1 p2) =>  (!determinism-while-true-case (while-rule p1 p2))
| while-axiom =>         (!determinism-while-axiom-case)
}
```

We present the methods for compositions and the "if-true" cases below, and leave the rest as exercises.

```
define determinism-sequence-case :=
  method (seq-proof)
    match seq-proof {
      (seq-rule p1 p2) =>
        pick-any other-proof:Proof c:Cmd s:State s':State s'':State
          assume hyp1 := (seq-proof    |- [c s] -->> s') ;
                 hyp2 := (other-proof |- [c s] -->> s'')
           let {ih1 := (det-property p1);
                ih2 := (det-property p2);
                L1 := (!chain->
                        [hyp1
                     ==> (exists c1 c2 is .
                            c = c1 ^ c2 &
                            p1 |- [c1 s] -->> is &
                            p2 |-  [c2 is]  -->> s')        [semantics]])}
             pick-witnesses c1 c2 is for L1
               let {L2 := (!chain->
                            [hyp2
                         ==> (other-proof |- [(c1 ^ c2) s] -->> s'')
                                                             [(c = c1 ^ c2)]
```

```
                         ==> (exists proof1 proof2 s1 .
                               other-proof = (seq-rule proof1 proof2) &
                               proof1 |- [c1 s] -->> s1 &
                               proof2 |- [c2 s1] -->> s'') [seq-excl]])}
             pick-witnesses proof1 proof2 s1 for L2
               (!chain->
                 [(p2 |- [c2 is] -->> s' &
                   proof2 |- [c2 s1] -->> s'')
                ==> (p2 |- [c2 is] -->> s' &
                     proof2 |- [c2 is] -->> s'')
                       [(s1 = is) <== (p1 |- [c1 s] -->> is &
                                       proof1 |- [c1 s] -->> s1) [ih1]]
                ==> (s'' = s')  [ih2]])
   }
```

We can test the method independently as follows:

```
set-flag print-var-sorts "off"

> pick-any p1:Proof p2:Proof
    assume (det-property p1)
      assume (det-property p2)
        let {composition := (seq-rule p1 p2)}
          conclude (det-property composition)
            (!determinism-sequence-case composition)

Theorem: (forall ?p1:Proof
           (forall ?p2:Proof
             (if (forall ?proof2:Proof
                   (forall ?c:Cmd
                     (forall ?s:(DMap.DMap Ide N)
                       (forall ?s':(DMap.DMap Ide N)
                         (forall ?s'':(DMap.DMap Ide N)
                           (if (and (derives ?p1
                                             (yields (pair ?c
                                                           ?s)
                                                     ?s'))
                                    (derives ?proof2
                                             (yields (pair ?c
                                                           ?s)
                                                     ?s'')))
                               (= ?s''
                                  ?s')))))))
                 (if (forall ?proof2:Proof
                       (forall ?c:Cmd
                         (forall ?s:(DMap.DMap Ide N)
                           (forall ?s':(DMap.DMap Ide N)
                             (forall ?s'':(DMap.DMap Ide N)
                               (if (and (derives ?p2
                                                 (yields (pair ?c
                                                               ?s)
```

```
                                                      ?s'))
                                        (derives ?proof2
                                                 (yields (pair ?c
                                                               ?s)
                                                         ?s'')))
                                   (= ?s''
                                      ?s')))))))
                    (forall ?other-proof:Proof
                      (forall ?c:Cmd
                        (forall ?s:(DMap.DMap Ide N)
                          (forall ?s':(DMap.DMap Ide N)
                            (forall ?s'':(DMap.DMap Ide N)
                              (if (and (derives (seq-rule ?p1 ?p2)
                                                (yields (pair ?c
                                                              ?s)
                                                        ?s'))
                                       (derives ?other-proof
                                                (yields (pair ?c
                                                              ?s)
                                                        ?s'')))
                                  (= ?s''
                                     ?s'))))))))))
```

Finally, here is determinism-if-true-case:

```
define determinism-if-true-case :=
  method (if-true-proof)
    match if-true-proof {
      (if-true-rule p) =>
        pick-any other-proof:Proof c:Cmd s:State s':State s'':State
          assume hyp1 := (if-true-proof |- [c s] -->> s');
                 hyp2 := (other-proof   |- [c s] -->> s'')
            let {ih := (det-property p);
                 L1 := (!chain->
                        [hyp1 ==> (exists b c1 c2 .
                                   c = (cond b c1 c2) &
                                   b true-in s &
                                   p |- [c1 s] -->> s') [semantics]])}
              pick-witnesses b c1 c2 for L1
               let {L2 := (!chain->
                           [hyp2
                      ==> (other-proof |-
                             [(cond b c1 c2) s] -->> s'')
                                                  [(c = (cond b c1 c2))]
                      ==> (other-proof |-
                            [(cond b c1 c2) s] -->> s'' &
                             b true-in s)         [augment]
                      ==> (exists proof' .
                             other-proof = if-true-rule proof' &
                             proof' |- [c1 s] -->> s'') [if-true-excl]])}
                 pick-witness proof' for L2
```

```
              (!chain->
                [(p |- [c1 s] -->> s' &
                  proof' |- [c1 s] -->> s'')
             ==> (s'' = s') [ih]])
    }
```

We can test this method as we tested the previous one:

```
> pick-any p:Proof
    assume (det-property p)
      conclude (det-property if-true-rule p)
        (!determinism-if-true-case (if-true-rule p))

Theorem: (forall ?p:Proof
           (if (forall ?proof2:Proof
                 (forall ?c:Cmd
                   (forall ?s:(DMap.DMap Ide N)
                     (forall ?s':(DMap.DMap Ide N)
                       (forall ?s'':(DMap.DMap Ide N)
                         (if (and (derives ?p
                                            (yields (pair ?c
                                                          ?s)
                                                    ?s'))
                                  (derives ?proof2
                                            (yields (pair ?c
                                                          ?s)
                                                    ?s'')))
                             (= ?s''
                                ?s')))))))
               (forall ?proof2:Proof
                 (forall ?c:Cmd
                   (forall ?s:(DMap.DMap Ide N)
                     (forall ?s':(DMap.DMap Ide N)
                       (forall ?s'':(DMap.DMap Ide N)
                         (if (and (derives (if-true-rule ?p)
                                            (yields (pair ?c
                                                          ?s)
                                                    ?s'))
                                  (derives ?proof2
                                            (yields (pair ?c
                                                          ?s)
                                                    ?s'')))
                             (= ?s''
                                ?s'))))))))
```

Exercise 18.8: Implement the remaining three methods. □

Exercise 18.9: Recall the formal specification of the semantics of `seq-rule`:

```
assert* sequence-semantics :=
  ((seq-rule p1 p2) |- [c s] -->> s' <==>
    exists c1 c2 is .
       c = c1 ^ c2 &
       p1 |- [c1 s] -->> is & p2 |- [c2 is] -->> s')
```

Why did we existentially quantify over the intermediate state is, instead of universally quantifying over it? That is, why not express the above in the following simpler format:

```
assert* sequence-semantics :=
  ((seq-rule p1 p2) |- [c s] -->> s' <==>
    exists c1 c2 .
       c = c1 ^ c2 &
       p1 |- [c1 s] -->> is & p2 |- [c2 is] -->> s')
```

What, if anything, would go wrong? □

Exercise 18.10: Prove the following:[17]

```
(forall ?B ?C .
   "while ?B do ?C" equivalent "if ?B then
                                  begin
                                    ?C;
                                      while ?B do ?C
                                  end
                                else skip")
```

In more conventional notation, prove that for all b and c,

while b **do** c

and

if b **then begin** c ; **while** b **do** c **end else skip**

have identical semantics. □

A common pattern has emerged in many of these recent proofs. Starting with a premise of the form

$$p \vdash [c\ s] \hookrightarrow s', \tag{18.15}$$

we deploy either a semantics axiom or an exclusivity lemma, depending on whether p or c is structured (a complex term rather than a variable or constant), and infer the existence of appropriate subproofs of p and/or subcommands of c. After picking witnesses for these existentially quantified statements, some of the instantiated premises might themselves be

17 Note that because equivalent is input-expanded to convert strings to Cmd terms via cmd-parser, and because our parser understands Athena variables inside expressions, commands, etc., this quantified sentence is actually natively understood by Athena.

of the form (18.15), and we then repeat the process for those premises. We keep going until we have "unrolled" the structure of all premises that can be unpacked in this manner.

Consider, for instance, the definition of the `assoc-dir-1` method (page 845). Starting with a premise of the form

$$p \vdash \texttt{[(}c_1\texttt{;(}c_2\texttt{;}c_3\texttt{))}\ s_1\texttt{]} \hookrightarrow s_2$$

and using `seq-exlc`, we infer the following existence statement:

$$\exists\ p_1, p_2, s' \,.\, p = \texttt{(seq-rule}\ p_1\ p_2\texttt{)}\ \&\ p_1 \vdash \texttt{[}c_1\ s_1\texttt{]} \hookrightarrow s'\ \&\ p_2 \vdash \texttt{[(}c_2\texttt{;}c_3\texttt{)}\ s'\texttt{]} \hookrightarrow s_2.$$

After picking witnesses for this statement, we repeat the process for the instantiated premise

$$p_2 \vdash \texttt{[(}c_2\texttt{;}c_3\texttt{)}\ s'\texttt{]} \hookrightarrow s_2,$$

thereby inferring:

$$\exists\ p_3, p_4, s'' \,.\, p_2 = \texttt{(seq-rule}\ p_3\ p_4\texttt{)}\ \&\ p_3 \vdash \texttt{[}c_2\ s'\texttt{]} \hookrightarrow s''\ \&\ p_4 \vdash \texttt{[}c_3\ s''\texttt{]} \hookrightarrow s_2.$$

And we then proceed again to pick witnesses for this new conclusion.

Repeatedly manipulating such existential quantifiers, especially in this rote fashion, can become cumbersome. Can we somehow automate this process so that we never have to explicitly deal with any existential quantifiers, relegating all such manipulation (deriving existential quantifications, picking witnesses for them, etc.) to the computer? Using recursion and higher-order methods (proof continuations), the answer is affirmative. We can define a method `unpack` that takes a premise of the form (18.15) and recursively unpacks it as described above, keeping track of newly instantiated premises, until no such premises can be further unpacked. At that point, all of the intermediate conclusions can be passed on to a proof continuation that is given as the second argument to `unpack`. Using such a method, many of these proofs can be made both shorter and simpler. The details are explored in the following exercise.

Exercise 18.11: ⋆ **[Eliminating existential quantifiers]** Implement the `unpack` method and use it to reformulate some of the recent proofs and/or methods. □

18.7 Chapter notes

The **While** language and much of our development of the semantics we have given here follows the presentation of Nielson and Nielson [77]. This style of semantics, originally introduced by Kahn [54], is known as *natural semantics* or *big-step operational semantics*, because the core semantic judgments relate starting configurations to final ones (by taking "big steps"). In contradistinction we have *small-step operational semantics*, or structural operational semantics (often abbreviated as SOS), mainly associated with Plotkin [82].[18] In that style, an initial state is gradually transformed (over the course of many small steps) into a final state. Both are species of operational semantics, which aim to specify not only the results of computations but also the way in which the computations are performed. In contrast to operational semantics we have *denotational* semantics [96], [86], which focus on specifying the results of computations rather than the manner in which these are performed (the "what" versus the "how"). Denotational semantics are always compositional, meaning that the denotation (meaning) of a syntactic object is defined as a function of the denotations (meanings) of its proper parts. The relevant equations are typically expressed in the λ-calculus, and the fixed-point theory of complete partial orders is used to capture the meaning of iteration constructs such as loops.

Axiomatic semantics, introduced by Floyd [37] and Hoare [47], is a third type of formalism for mathematically specifying the meanings of programs. Unlike both operational and denotational semantics, it is not based on the notion of a program state. Rather, it is based on relationships between assertions that hold prior to a program's execution and assertions that hold after its execution. It is the most heavily proof-oriented of all three formalisms, based on inference rules for deriving final assertions from initial ones. These are typically expressed in first-order logic, but other formalisms—such as temporal logics—are also used. The field of program verification has been based largely on axiomatic semantics.

A somewhat challenging aspect of the particular natural semantics studied in this chapter is that they are *not* compositional. More specifically, it is the inference rule `while-rule` that is not compositional, meaning that, unlike the other rules, the syntactic objects that appear in the antecedent of that rule (above the horizontal line) are not proper parts of the syntactic objects that appear in the consequent (below the line). The entire command in question (**while** b **do** c) appears in both antecedent and consequent. That is why simple structural induction on commands does not suffice for analyzing these semantics. And that is why the proof system defined by the semantic rules had to be explicitly encoded and we had to painstakingly prove things at the (meta) Athena level about the encoded (object) proofs. When the semantics are compositional, as is often the case, that is not necessary and the resulting proofs tend to be simpler.

Even then, however, we have to somehow "complete" the semantics by ensuring that the transition relation is the smallest relation that contains the given inference rules. We can always do that manually by explicitly resorting to biconditionals, possibly along with existential quantification, as we did in this chapter. But it is also possible to state the inference rules simply as Horn clauses, and even list them in ML-style order (whereby a defining pattern is only assumed to be applicable if none of the previous patterns are), and then invoke a procedure that will automatically turn the clauses into biconditionals, possibly

18 Although similar ideas predated that work by almost twenty years, such as Landin's seminal work on the SECD machine from 1964 [63].

with existential quantifiers on their right-hand sides. Inductive proofs on such automatically completed relations could then proceed in a natural style that tracks the given rules and respects the order in which these were listed.

Appendices

Appendices

Athena Reference

THIS APPENDIX provides a compact description of the syntax and operational semantics of Athena's core language constructs.

A.1 Syntax

As explained in Chapter 2, there are two main syntactic categories in Athena: *expressions* (E) and *deductions* (D). A *phrase* F is either an expression or a deduction:

$$F ::= E \mid D.$$

The syntax of expressions is specified in Figure A.1, while that of deductions is shown in Figure A.2. Figure A.4 depicts the syntax of patterns. In the rest of this section we describe the syntax of character constants (C), string constants (T), and the sorts (S) used for annotations. ASCII characters are generated by the following grammar:

C	::=	‘*printable*	(e.g., `‘A`)
	|	‘*code*	(e.g., `‘\73`)
	|	‘\\^*control-character*	(e.g., `‘\^C`)
	|	‘*escape-character*	(e.g., `‘\n`)
escape-character	::=	`"` | `\` | `a` | `b` | `n` | `r` | `f` | `t` | `v`	
control-character	::=	`A` | ⋯ | `Z` | `@` | `[` | `]` | `/` | `^` | `_` | `‘\blank`	

where *printable* is any printable ASCII character other than the backslash \ (i.e., any character with ASCII code from 32 through 91 or 93 through 127, inclusive); and *code* is any sequence of one, two, or three digits: `9`, `03`, `124`, and so on. Since any ASCII character with code c can be expressed as ‘\\c, the other three ways of representing characters are strictly speaking redundant, but convenient nevertheless. Escape characters have standard meanings; for example, `\n` is the newline character, `\t` is the tab, and so forth. An Athena string T is simply a list of characters. String constants can be directly input between quotes:

T ::= "C^*" (e.g., `"\nHello world!"`)

where C here is just as described by the foregoing grammar for characters, except that the opening quote mark ‘ is omitted.[1] Thus, control characters and escape characters can be directly embedded inside string constants. As a notational convention, for any nonterminal X, we write X^* for a sequence of zero or more strings generated by X, and X^+ for a sequence of one or more such strings.

Finally, the following grammar gives the syntax of sorts:

1 If included, it will count as a separate character.

E	$::=$	I	Identifiers
	$\mid$	C	Characters
	$\mid$	T	String constants
	$\mid$	$()$	Unit
	$\mid$	$?I{:}S$	Term variables (annotated)
	$\mid$	$?I$	Term variables (unannotated)
	$\mid$	$'I$	Meta-identifiers
	$\mid$	$\mathtt{check}\ \{F_1 \Rightarrow E_1 \mid \cdots \mid F_n \Rightarrow E_n\}$	Conditional expressions
	$\mid$	$\mathtt{lambda}\ (I^*)\ E$	Procedures
	$\mid$	$(E\ F^*)$	Applications
	$\mid$	$[F^*]$	Lists
	$\mid$	$\mathtt{method}\ (I^*)\ D$	Methods
	$\mid$	$\mathtt{let}\ \{\pi_1 := F_1;\ \cdots;\ \pi_n := F_n\}\ E$	Let expressions
	$\mid$	$\mathtt{letrec}\ \{I_1 := E_1;\ \cdots;\ I_n := E_n\}\ E$	Recursive let expressions
	$\mid$	$\mathtt{match}\ F\ \{\pi_1 \Rightarrow E_1 \mid \cdots \mid \pi_n \Rightarrow E_n\}$	Match expressions
	$\mid$	$\mathtt{try}\ \{E_1 \mid \cdots \mid E_n\}$	Try expressions
	$\mid$	$\mathtt{cell}\ F$	Cells
	$\mid$	$\mathtt{set!}\ E\ F$	Assignments
	$\mid$	$\mathtt{ref}\ E$	References
	$\mid$	$\mathtt{while}\ F\ E$	Loops
	$\mid$	$\mathtt{make\text{-}vector}\ E\ F$	Vector creation
	$\mid$	$\mathtt{vector\text{-}sub}\ E_1\ E_2$	Vector access
	$\mid$	$\mathtt{vector\text{-}set!}\ E_1\ E_2\ F$	Vector assignment
	$\mid$	$(\mathtt{seq}\ F^*)$	Sequences
	$\mid$	$(\&\&\ F^*)$	Short-circuit Boolean "and"
	$\mid$	$(\Vert\ F^*)$	Short-circuit Boolean "or"

Figure A.1
Syntax of Athena expressions

S	$::=$	$'I$	(Sort variables, e.g., `'S5`)
	$\mid$	I	(Constant sort constructors, e.g., `Boolean` or `Int`)
	$\mid$	$(I\ S_1 \cdots S_n)$	(Compound sorts, e.g., `(List Int)`)

Note that the patterns that may appear in a clause of a **`by-induction`** or **`datatype-cases`** proof are of a restricted form; their syntax is described by the following simple grammar:

$$\pi ::= I \mid I{:}S \mid (I\ \pi^+) \mid (I\ \mathtt{as}\ \pi)$$

where S ranges over sorts.

A.2 Values

Figure A.3 shows the types of values that Athena phrases denote. Some brief remarks on each type follow.

```
D ::= (apply-method E F_1 ··· F_n)                              Method calls
    | (!E F_1 ··· F_n)                                          Method calls (usual syntax)
    | conclude F D                                              Conclusion-annotated deductions
    | assume F D                                                Hypothetical deductions
    | assume I := F D                                           Named hypothetical deductions
    | assume I_1 := F_1; ··· ;I_{n+1} := F_{n+1} D              Conjunctive named hypothetical deductions
    | suppose-absurd F D                                        Proofs by contradiction
    | generalize-over E D                                       Universal generalizations
    | pick-any I D                                              Universal generalizations
    | pick-any I:S D                                            Universal generalizations
    | with-witness E F D                                        Existential instantiations
    | pick-witness I for F D                                    Existential instantiations
    | pick-witnesses I_1 I_2 ··· I_n for D                      Existential instantiations
    | by-induction F {π_1 => D_1 | ··· | π_n => D_n}            Structural induction
    | datatype-cases F {π_1 => D_1 | ··· | π_n => D_n}          Structural case analysis
    | check {F_1 => D_1 | ··· | F_n => D_n}                     check deductions
    | match F {π_1 => D_1 | ··· | π_n => D_n}                   match deductions
    | let {π_1 := F_1; ··· ; π_n := F_n} D                      let deductions
    | letrec {I_1 := E_1; ··· ; I_n := E_n} D                   letrec deductions
    | try {D_1 | ··· | D_n}                                     try deductions
```

Figure A.2
Syntax of Athena deductions

1. The unit value, denoted by the expression `()`, is a single special object, distinct from all other values. It is used primarily as the result of expressions with side effects.
2. Function symbols are discussed in Section 2.2; they are datatype or structure constructors, or else symbols introduced with **`declare`**. This means that symbols such as `false` and `N.+` are actual values, possible results of computations.
3. Terms and sentences are as explained in Chapter 2.
4. The sentential connectives are the five operators used to build compound sentences: `not`, `and`, `or`, `if`, and `iff`. In this group of values we also have the quantifiers `forall` and `exists`. These are values, so they too can be the results of computations.
5. A list of values is just that: a finite list $[V_1 \cdots V_n]$, $n \geq 0$, possibly empty.
6. A *function value* is the mathematical object denoted by an Athena procedure. Specifically, a function value is a computable ternary function that takes as arguments (1) a list of values $[V_1 \cdots V_m]$, $m \geq 0$; (2) an assumption base β; and (3) a store σ; and returns an ordered pair (V, σ') consisting of a value V and a store σ', or else halts in error or diverges. Function values are thus higher-order: Some of the V_i inputs might themselves be function values, and the output returned by applying a function value may itself be a function value.

The unit value	Function symbols	Sentential connectives and quantifiers
Terms	Sentences	Lists of values
Substitutions	Function values	Method values
Cells	Vectors	Characters
Tables	Maps	

Figure A.3
Types of Athena values

7. A *method value* is the deductive analogue of a function value; it is the mathematical object denoted by an Athena method. Specifically, a method value is a computable ternary function that takes as arguments (1) a list of values $[V_1 \cdots V_m]$, $m \geq 0$; (2) an assumption base β; and (3) a store σ; and returns an ordered pair (p, σ') consisting of a sentence p and a store σ' (or else generates an error or diverges). Note that a method value may take other method values as inputs, but it may not produce them as results. (Function values, by contrast, may produce method values as results.)
8. Characters are individual ASCII symbols.
9. A substitution is a computable function from term variables to terms that is the identity almost everywhere; substitutions are discussed in Section 2.14.8.
10. Tables and maps are type-tagged finite functions from values to values. So while a table and a map might abstractly represent the same underlying function from values to values, they are guaranteed to be distinct entities.
11. Finally we have cells and vectors, which act as storage containers that can hold arbitrary values (possibly other cells and/or vectors). For mathematical purposes, a cell may be associated with a unique natural number l, and we may then think of the computer's memory as an infinite list: cell 0, cell 1, cell 2, cell 3, and so on. All of these cells are initially unassigned. Vector values are modeled as sequences of cells.

These families of values are not quite pairwise disjoint; there is some overlap between (a) terms and function symbols, and (b) terms and sentences. First, a constant symbol such as `zero` or `peter` counts both as a symbol and as a term. And second, a term of sort `Boolean` counts both as a term and as a sentence. In each of these two cases, a value of one type may be coerced into the other type, as required by the context. It is fine, for instance, to pass a term t of sort `Boolean` to a procedure that expects a sentence; t will just be treated as a sentence. Athena performs such conversions automatically.

Finally, a word about value equality. Any two values of the same type may be compared for equality (e.g., with the primitive binary procedure `equal?`). Equality for sentences is alpha-equivalence. For terms it is syntactic identity modulo sort renaming. A cell is only identical to itself. If you think about cells as locations indexed by natural numbers, that makes sense: Locations i and j are the same iff $i = j$. Accordingly, different cells may have identical contents. Similar remarks apply to vectors. Two substitutions are equal iff they have identical supports, and assign the same term to each variable in their supports. Equality is not decidable for function and method values, and an error will occur if one attempts to apply `equal?` to such values. Equality for characters is obvious: Two characters are the same iff they have the same code. Two lists of values $[V_1 \cdots V_n]$ and $[V_1' \cdots V_m']$ are identical iff $n = m$ and V_i is equal to V_i' for $i = 1, \ldots, n$. Two tables (or two maps) are identical iff they have the exact same domain and range; that is, they map the same keys to the same values. The unit value is only identical to itself. Finally, identity on function symbols, sentential connectives, and quantifiers is clear: A symbol or a sentential connective or quantifier is identical only to itself. Values of different types are considered distinct by default, unless one of the values can be converted to the type of the other and the two can then be determined to be identical on the basis of the above conventions.

A.3 Operational semantics

In this section we specify in detail the result of evaluating any phrase F in a given environment ρ, assumption base β, store σ, and symbol set γ. We clarify these parameters below:

1. An *environment* ρ is a computable function that maps any given identifier I either to a value V or to a special *unbound* token.
2. An *assumption base* β is a finite set of sentences.
3. A *store* σ is a computable function that maps any natural number (representing a memory location) to a value (the location's contents) or to a special *unassigned* token. Infinitely many numbers must be unassigned in any given store.
4. A *symbol set* γ is a collection of function symbols and their respective signatures,[2] along with a collection of sort constructors and their arities; γ also includes information on whether a given sort constructor is a datatype or structure, and if so, which function symbols are its constructors.

The result of evaluating a phrase F with respect to given ρ, β, σ, and γ, is one of three things:

1. a pair (V, σ') consisting of a value V and a store σ', where V is the output of the evaluation and σ' reflects any side effects accumulated during the evaluation; or

2 See page 24 for a discussion of signatures.

2. a pair consisting of an error message and a store σ', indicating the occurrence of an error during the computation (a store σ' is still necessary to reflect side effects accrued prior to the error); or
3. nontermination.

The specification of the evaluation process below is given in English, but without ambiguity. It can serve as the basis for implementing a core Athena interpreter.

Since the evaluation of most expressions leaves the store unaffected, if we do not explicitly specify what the new store is then it should be assumed that it is the same as that in which the phrase was evaluated: $\sigma' = \sigma$. Also, unless we explicitly say otherwise, we will generally assume that the evaluation of a phrase F is immediately halted if the evaluation of a subphrase of F produces an error message and a store σ'; in such cases the result of evaluating F becomes that same error message and σ'.

The evaluation algorithm proceeds by a case analysis of the syntactic structure of the given phrase. We begin with expressions, but first a piece of notation. Consider any function f from a set A to a set B. When $a_1, \ldots, a_n$ are distinct elements of A and $b_i \in B$, $i = 1, \ldots, n$, we write $f[a_1 \mapsto b_1, \ldots, a_n \mapsto b_n]$ for that function from A to B which maps a_i to b_i and every other $x \in A$ to $f(x)$.

- **Identifiers**: If the given expression is an identifier I, and if I is bound in the given environment ρ to some value V, then the output value is V; it is an error if I is unbound in ρ.[3]
- **Unit**: The value of the expression `()` is always the unit value.
- **Term variables, meta-identifiers**, and **characters** are always self-evaluating, regardless of ρ, β, σ, and γ.
- **String constants**: A string constant results in a list of the characters that constitute the string, in the given order.
- **check expressions**: To evaluate an expression of the form

  $$\texttt{check } \{F_1 \texttt{ => } E_1 \mid \cdots \mid F_n \texttt{ => } E_n\}$$

  in ρ, β, σ, and γ, we first evaluate F_1 in ρ, β, σ, and γ, possibly resulting in a value V_1 and store σ_1.[4] If V_1 is the constant term `true`, then the result of the entire **check** expression is the result obtained by evaluating E_1 in ρ, β, σ_1, and γ. Otherwise, if V_1 is

3 Note that when a function symbol f is first introduced, the name f is automatically bound to that function symbol. The name f can later perhaps be redefined (bound to another value), but the underlying function symbol will continue to exist in the relevant symbol set (and can always be retrieved, for example, by the `string->symbol` procedure). Numerals (such as 5 or `3.14`) can be understood as always being bound to the corresponding numeric terms.

4 In accordance with the convention made above, if the evaluation of F_1 in ρ, β, σ, and γ generates an error and some store σ_1, we simply return that as the result of the entire **check** expression.

false, the process continues with the second clause, taking into account any side effects that the evaluation of F_1 might have engendered: The phrase F_2 is evaluated in ρ, β, σ_1, and γ, possibly resulting in a value V_2 and a store σ_2. If V_2 is true, then the final result is that of evaluating E_2 in ρ, β, σ_2, and γ. Otherwise, if V_2 is false, F_3 is evaluated in ρ, β, σ_2, and γ; its value is compared to true, and so forth. Clearly, evaluation might diverge if some F_i or E_i diverges. It is an error if there are no alternatives (i.e., if $n = 0$), in which case we return an appropriate error message and the store σ as the result of the evaluation; or if no alternative succeeds (i.e., no F_i ever produces true), in which case we return an error message and the store σ_n; or if some F_i results in a value other than true or false, in which case we return an error message and the store σ_i. If F_n is the keyword **else** and no F_j produced true, $j < n$, then the final result is that of evaluating E_n in ρ, β, σ_{n-1}, and γ.[5]

- **Procedures**: The value of **lambda** $(I_1 \cdots I_n)$ E in ρ, β, σ, and γ is a function value that takes a list of n values $V_1, \ldots, V_n$ along with an assumption base β' and a store σ' as arguments, and produces the result of evaluating the body E in $\rho[I_1 \mapsto V_1, \ldots, I_n \mapsto V_n]$, β', σ', and γ. Note that the environment and symbol set are statically determined, whereas the store and the assumption base are dynamic. The evaluation of procedures always terminates successfully.
- **Methods**: The value of **method** $(I_1 \cdots I_n)$ D in ρ, β, σ, and γ is a method value that takes a list of n values $V_1, \ldots, V_n$ along with an assumption base β' and store σ' as arguments, and produces the result of evaluating the deduction D in $\rho[I_1 \mapsto V_1, \ldots, I_n \mapsto V_n]$, β', σ', and γ. Here too, the environment and symbol set are statically determined while the store and the assumption base are dynamic. The evaluation of methods always terminates successfully.
- **Applications**: The value of an expression of the form

$$(E \; F_1 \cdots F_n)$$

in ρ, β, σ, and γ is obtained as follows. First, E is evaluated in ρ, β, σ, and γ, possibly resulting in a value V and store σ'. We then proceed with a case analysis of V:

1. If V is a function value ϕ, we evaluate the arguments $F_1, \ldots, F_n$ sequentially, starting with F_1 in ρ, β, σ', and γ, to obtain values $V_1, \ldots, V_n$ and a store σ_n. This sequential evaluation threads the store from the evaluation of F_i to that of F_{i+1}. Specifically, first we evaluate F_1 in ρ, β, σ', and γ to obtain a value V_1 and a store σ_1; then we evaluate F_2 in ρ, β, σ_1, and γ to obtain a value V_2 and store σ_2; we continue in this fashion until we evaluate F_n in ρ, β, σ_{n-1}, and γ, resulting in a value V_n and store σ_n. The final result is that of applying ϕ to the list $[V_1 \cdots V_n]$, β, and σ_n.

5 The keyword **else** may only appear in the position of F_n.

2. If V is a function symbol f of arity n, we sequentially evaluate $F_1, \ldots, F_n$, starting with F_1 in ρ, β, σ and γ, to obtain values $V_1, \ldots, V_n$ and a store σ_n. If each V_i is a term t_i, then the resulting value is the term $(f \ t_1 \cdots t_n)$, provided that this term is well sorted. If it is not, we return an error message and the store σ_n as the result. (We also return an error message and the store σ_n if some V_i is not a term.)
3. If V is a sentential connective $\circ$, we evaluate $F_1, \ldots, F_n$ sequentially to obtain values $V_1, \ldots, V_n$ and a store σ_n. If each V_i is a sentence p_i, then the resulting value is the sentence $(\circ \ p_1 \cdots p_n)$, provided that the latter is well sorted, and the resulting store is σ_n. If the sentence is not well sorted or if some V_i is not a sentence, we return an error message and the store σ_n.
4. If V is a quantifier Q, we check to make sure that $n > 1$; if not, we return an error message and the store σ'. We then evaluate $F_1, \ldots, F_n$ sequentially to obtain values $V_1, \ldots, V_n$ and a store σ_n. The first $n-1$ values must all be term variables $x_1, \ldots, x_{n-1}$, while the last value V_n must be a sentence p. If not, we halt with an error message and σ_n; otherwise the resulting value is the quantified sentence:

   $$(Q \ x_1 \ (Q \ x_2 \ \cdots \ (Q \ x_n \ p) \ \cdots)),$$

   provided that it is well sorted, while the resulting store is σ_n. If the above is not a well-sorted sentence, we return an error message and the store σ_n.[6]
5. If V is a substitution θ, we check whether $n = 1$. If not, we return an error message along with the store σ' and halt. Otherwise we evaluate F_1 in ρ, β, σ', and γ to obtain a value V_1 and a store σ_1. Then:

   - If V_1 is a term t_1 (or sentence p_1), the output is the term $\theta(t_1)$ (or sentence $\theta(p_1)$, respectively) and σ_1, provided that $\theta(t_1)$ (respectively, $\theta(p_1)$) is well sorted; if it is not, we return an error message and σ_1.[7]
   - If V_1 is a list `[`$V_1 \cdots V_n$`]` where each V_i is a term or sentence, then the output is the list $[\theta(V_1) \cdots \theta(V_n)]$ and the store σ_1, provided that each $\theta(V_i)$ is well sorted. If one is not, of if V_1 is not such a list, we return an error message and the store σ_1.
   - If V_1 is neither a term nor a list of terms, we halt with an error message and the store σ_1.

6. If V is a map, we check whether $n = 1$. If not, like before, we return an error message along with the store σ' and halt. Otherwise we evaluate F_1 in ρ, β, σ', and γ to obtain a value V_1 and a store σ_1. Then, if V_1 is a key in the map V, the output is the value that the map prescribes for that key, along with σ_1. If V_1 is not a key in that map, we return an error message and σ_1.

---

6 Note that applications such as `(forall` x `.` p`)` are syntax sugar for `(forall` x p`)`. The latter is the more fundamental syntactic construct.

7 We write $\theta(t)$ (respectively, $\theta(p)$) for the term obtained by applying the substitution θ to the term t (respectively, sentence p).

7. If V is neither a function value, nor a function symbol, sentential connective, quantifier, substitution, or map, then we halt with an error message and σ'.

- **`match` expressions**: To evaluate `match` F $\{\pi_1$ `=>` E_1 `|` $\cdots$ `|` π_n `=>` $E_n\}$ in ρ, β, σ, and γ, we first evaluate the discriminant F in ρ, β, σ, and γ, possibly obtaining a value V and store σ'. We then go through the patterns $\pi_1, \ldots, \pi_n$ sequentially, trying to find a pattern that is matched by V in ρ and γ (using the algorithm of Section A.4). If no such pattern is found, we halt with an error message and σ'. Otherwise, let π_j be the first matching pattern and let $\{I_1 \mapsto V_1, \ldots, I_k \mapsto V_k\}$ and τ be the finite set of bindings and sort valuation, respectively, returned by the matching algorithm. The output then becomes the result of evaluating E_j in $\rho[I_1 \mapsto V_1, \ldots, I_k \mapsto V_k]$, β, σ', and γ. If that result is a term or sentence, then τ is applied to it before returning.
- **`let` expressions**: For semantic purposes, an expression of the form

  $$\texttt{let}\ \{\pi_1\ \texttt{:=}\ F_1;\ \cdots;\ \pi_n\ \texttt{:=}\ F_n\}\ E \tag{1}$$

  is treated as syntax sugar. The desugaring proceeds by induction on n. When $n = 0$, (1) reduces to E. When $n > 0$, (1) is desugared into `match` F_1 $\{\pi_1$ `=>` $E'\}$, where E' is the result of desugaring

  $$\texttt{let}\ \{\pi_2\ \texttt{:=}\ F_2;\ \cdots;\ \pi_n\ \texttt{:=}\ F_n\}\ E.$$

- **`letrec` expressions**: The result of evaluating

  $$\texttt{letrec}\ \{I_1\ \texttt{:=}\ E_1;\ \cdots;\ I_n\ \texttt{:=}\ E_n\}\ E$$

  in ρ, β, σ, and γ is that of evaluating the following expression in ρ, β, σ, and γ :

  ```
  let {I1 := cell (); ···; In := cell ();
       _ := set! I1 E1';
          ···;
       _ := set! In En'}
    E'
  ```

  where each $E_j{}'$ is obtained from E_j by replacing every free occurrence of I_j[8] by `ref` I_j, $j = 1, \ldots, n$; and E' is likewise obtained from E. This desugaring is the classic way to "tie the knot," and it is often used to implement recursion when the implementation language supports state. The various E_i are usually `lambda` or `method` expressions.
- **`try` expressions**: The `try` construct implements backtracking: To evaluate

  $$\texttt{try}\ \{E_1\ \texttt{|}\ \cdots\ \texttt{|}\ E_n\}$$

  in ρ, β, σ, and γ, we first evaluate E_1 in ρ, β, σ, and γ. If that results in a value V_1 and store σ_1, we return V_1 and σ_1. But if it results in an error message and a store σ_1, we go

8 We have not given a precise definition of when an identifier occurs free inside a phrase, but an intuitive understanding will suffice for present purposes.

on to evaluate E_2 in ρ, β, σ_1, and γ; and so forth. If the evaluation of every E_i fails, we return an error message and σ_n.

- **Cells**: To evaluate an expression of the form `cell` F in ρ, β, σ, and γ, we first evaluate F in ρ, β, σ, and γ to obtain some value V and store σ'. We then let l be the smallest natural number such that $\sigma'(l)$ is unassigned, and we return as output the cell represented by the location l along with the store $\sigma'[l \mapsto V]$.
- **References**: To evaluate an expression of the form `ref` E in ρ, β, σ, and γ, we evaluate E in ρ, β, σ, and γ, possibly obtaining a value V and store σ'. If the value V is *not* a memory location (cell), we return an error message and σ'. If it is a memory location l, we return the value $\sigma'(l)$ and the store σ'.[9]
- **Assignments**: To evaluate an expression of the form `set!` E F in ρ, β, σ, and γ, we first evaluate E in ρ, β, σ, and γ, possibly obtaining a value V_1 and store σ_1. If V_1 is not a memory location (cell), we halt with an error message and σ_1 as the result. If it is a cell l, we proceed to evaluate F in ρ, β, σ_1, and γ, possibly obtaining a value V_2 and store σ_2. We then return the unit value and the store $\sigma_2[l \mapsto V_2]$.
- `while` **loops**: To evaluate an expression of the form `while` F E in ρ, β, σ, and γ:
  1. We evaluate F in ρ, β, σ, and γ, possibly obtaining a value V_1 and store σ_1.
  2. If V_1 is the constant term `false`, we return the unit value and store σ_1. Otherwise, if V_1 is `true`, we evaluate E in ρ, β, σ_1, and γ to obtain some value V_2 and store σ_2. We then continue with the first step, only now F is evaluated in ρ, β, σ_2, and γ (rather than ρ, β, σ, and γ). If V_1 is neither `true` nor `false`, we return an error message and σ_1.

- **Short-circuit Boolean operations**: Expressions of the `&&` and `||` form `(&&` $F_1 \cdots F_n$`)` and `(||` $F_1 \cdots F_n$`)` are evaluated as explained in Section 2.16, making sure to thread the store in the obvious manner (the store resulting from the evaluation of F_1 in ρ, β, σ, and γ becomes the store in which the evaluation of F_2 takes place, if F_2 is evaluated at all; and so on).
- **Vector creation**: To evaluate an expression `make-vector` E F in ρ, β, σ, and γ, we start by evaluating E in ρ, β, σ, and γ to obtain a value V_1 and a store σ_1. If V_1 is not a nonnegative integer constant, we halt with an error message and σ_1. If it is a nonnegative integer constant n, we proceed to evaluate F in ρ, β, σ_1, and γ, possibly obtaining a value V_2 and store σ_2. We then let $l_1, \ldots, l_n$ be the smallest n natural numbers such that $\sigma_2(l_i)$ is unassigned for each $i = 1, \ldots, n$, and we return as output the list of memory locations $[l_1 \cdots l_n]$ along with the store $\sigma_2[l_1 \mapsto V_2, \ldots, l_n \mapsto V_2]$.
- **Vector access**: To evaluate an expression `vector-sub` E_1 E_2 in ρ, β, σ, and γ, we first evaluate E_1 in ρ, β, σ, and γ, obtaining a value V_1 and store σ_1. If V_1 is not a list

9 It is an error if cell l is unassigned, although this could never happen in our semantics.

of memory locations previously created by `make-vector`,[10] we output an error message and σ_1. If V_1 is a list of memory locations $[l_1 \cdots l_n]$ previously created by `make-vector`, $n \geq 0$, we proceed to evaluate E_2 in ρ, β, σ_1, and γ, obtaining a value V_2 and store σ_2. If V_2 is not a nonnegative integer constant, we halt with an error message and σ_2. If it is a nonnegative integer constant i, we return the value $\sigma_2(l_{i+1})$ and σ_2, provided that the index i is between 0 and $n-1$ (if it is not, we halt with an error message and σ_2).

- **Vector assignment**: To evaluate an expression `vector-set!` E_1 E_2 F in ρ, β, σ, and γ, we first evaluate E_1 in ρ, β, σ, and γ, obtaining from it some value V_1 and store σ_1. If V_1 is not a list of memory locations previously created by `make-vector`, we output an error message and σ_1. If V_1 is a list of memory locations $[l_1 \cdots l_n]$ previously created by `make-vector`, $n \geq 0$, we proceed to evaluate E_2 in ρ, β, σ_1, and γ, obtaining a value V_2 and store σ_2. If V_2 is not a nonnegative integer constant between 0 and $n-1$, we halt with an error message and σ_2. Otherwise, if it is a some such constant i, we proceed to evaluate F in ρ, β, σ_2, and γ, obtaining from it some value V_3 and store σ_3. We then return the unit value along with the store $\sigma_3[l_{i+1} \mapsto V_3]$.

We continue with the evaluation of deductions, which again proceeds by a case analysis of syntactic structure:

- **Method calls**: To evaluate a deduction `(!`E $F_1 \cdots F_n$`)` in ρ, β, σ, and γ, we first evaluate E in ρ, β, σ, and γ, obtaining a value V and store σ' from it. If V is not a method value, we halt with an error message and σ'. Otherwise, if V is a method value M, we go on to evaluate the arguments $F_1, \ldots, F_n$, in that turn. Letting $\sigma_0 = \sigma'$ and $\beta' = \emptyset$, we evaluate each F_i in ρ, β, σ_{i-1}, and γ, obtaining from it a value V_i and store σ_i. In addition, if the argument F_i is a deduction whose value V_i is a conclusion (sentence) p_i, we add p_i to β'. When all arguments have been evaluated, we apply M to the list of values $[V_1 \cdots V_n]$, $\beta \cup \beta'$, and σ_n. Note that the assumption base in which M is applied will include the conclusion of every argument phrase F_i that is a deduction.
- **Conclusion-annotated deductions**: To evaluate a deduction `conclude` F D in ρ, β, σ, and γ, we first evaluate F in ρ, β, σ, and γ, to obtain a value V and a store σ'. If V is not a sentence, we halt with an error message and σ'. Otherwise, if V is a sentence p, we proceed to evaluate the body D in ρ, β, σ', and γ, obtaining from it a conclusion q and store σ''. If p and q are alpha-equivalent, we halt with p and σ'' as the output; otherwise we halt with an appropriate error message and σ''.
- **Hypothetical deductions**: To evaluate `assume` F D in ρ, β, σ, and γ, we first evaluate F in ρ, β, σ, and γ, producing a value V and store σ'. If V is not a sentence, we halt with an error message and σ'. Otherwise, if V is a sentence p, we proceed to evaluate the body D

10 Strictly speaking this requires that we tag lists of cells created by `make-vector` to distinguish them from lists of cells that may have been created by other means, but this need not detain us here.

in ρ, $\beta \cup \{p\}$, σ', and γ, obtaining from it some sentence q and store σ''. We then return the conditional (*p* ==> *q*) and σ'' as the result.

- **Named hypothetical deductions**: A deduction `assume` *I* := *F D* is treated as syntax sugar for the following:

```
let {I := F}
  assume I
    D
```

- **Conjunctive named hypothetical deductions**: A deduction of the form

$$\texttt{assume}\ I_1 := F_1;\ \cdots\ ;I_{n+1} := F_{n+1}\ D$$

for $n > 0$ is treated as syntax sugar for the following:

```
let {I1 := F1; ··· ;In+1 := Fn+1}
  assume (I1 & I2 & ··· & In+1)
    D
```

- **Proof by contradiction**: To evaluate `suppose-absurd` *F D* in ρ, β, σ, and γ, we begin by evaluating *F* in ρ, β, σ, and γ, producing a value *V* and store σ'. If *V* is not a sentence, we halt with an error message and σ'. If *V* is a sentence *p*, we evaluate the body *D* in ρ, $\beta \cup \{p\}$, σ', and γ, obtaining from it some sentence q and store σ''. If q is the constant `false`, we return (~ *p*) and σ'' as the result; if q is not `false`, we output an error message and σ''.
- **Universal generalizations**: To evaluate a deduction `generalize-over` *E D* in ρ, β, σ, and γ, we first evaluate *E* in ρ, β, σ, and γ, obtaining some value *V* and store σ'. If *V* is a variable *x* (say, `?p:Person`), we first check to see that *x* does not occur free in β.[11] If it does, we halt with an error message and σ'. If *x* does not occur free in β, we proceed to evaluate the body *D* in ρ, β, σ', and γ, obtaining a conclusion *p* and store σ''. We then return the sentence (`forall` *x p*) and σ'', provided that (`forall` *x p*) is well sorted (if it is not, we output an error message and σ'').
- **`pick-any` universal generalizations**: A deduction of the form `pick-any` *I D* is treated as syntax sugar for:

```
let {I := (fresh-var)}
  generalize-over I
    D
```

11 A variable ?*I*:*S* is said to occur free in an assumption base β iff there is some $p \in \beta$ such that p contains a free occurrence of some variable of the form ?*I*:*S*′, where *S* is an instance of *S*′. Thus, if $\beta = \{$`(P ?z:'S3)`$\}$, then `?z:Person` is considered to occur free in β, as `Person` is an instance of `'S3`.

And a deduction of the form **pick-any** $I : S\ D$ is treated as syntax sugar for

```
let {I := (fresh-var "S")}
  generalize-over I
    D
```

- **with-witness deductions**: To evaluate **with-witness** $E\ F\ D$ in ρ, β, σ, and γ, we first evaluate E in ρ, β, σ, and γ, to obtain a value V and store σ'. If V is not an Athena variable, or if it is a variable that occurs free in β, we halt with an error message and σ'. Otherwise, if V is a variable w that does not occur free in β, we proceed to evaluate F in ρ, β, σ', and γ, obtaining a value V' and store σ''. If V' is not an existential quantification, we halt with an error message and σ''. Otherwise V' must be an existential quantification $p =$ `(exists y . q)`. If the phrase F is a deduction, let $\beta' = \beta \cup \{p\}$; otherwise let $\beta' = \beta$. Now, let q' be the sentence obtained from the body q by replacing every free occurrence of y by the variable w, provided that the result is well sorted (if not, we halt with an error message and σ''). We then evaluate the deduction D in ρ, $\beta' \cup \{q'\}$, σ'', and γ, obtaining some sentence r and store $\widehat{\sigma}$. If the variable w occurs free in r, we halt with an error message and $\widehat{\sigma}$, otherwise we return r and $\widehat{\sigma}$.
- **pick-witness deductions**: A deduction **pick-witness** I **for** $F\ D$ is treated as syntax sugar for

```
let {I := (fresh-var)}
  with-witness I F D
```

A deduction **pick-witness** I **for** $F\ I_2\ D$ is treated as syntax sugar for

```
let {I1 := (fresh-var);
     equant := F}
  match equant {
    (exists (some-var v) body) =>
      let {I2 := (replace-var v I1 body)}
        with-witness I1 equant D
  }
```

(Recall that `(replace-var v t p)` produces the sentence obtained from p by replacing every free occurrence of variable v with the term t, provided that the result is well sorted.)
- **check deductions**: A deduction **check** $\{F_1$ `=>` D_1 `|` $\cdots$ `|` F_n `=>` $D_n\}$ is evaluated by the same algorithm used for **check** expressions, save for the obvious difference that, barring error or divergence, the output value will always be a sentence produced by some D_i.

- **`match` deductions**: To evaluate `match` F $\{\pi_1$ `=>` D_1 `|` $\cdots$ `|` π_n `=>` $D_n\}$ in ρ, β, σ, and γ, we first evaluate the discriminant F in ρ, β, σ, and γ, obtaining a value V and store σ'. If the discriminant is a deduction, then V must be a sentence p, and we let $\beta' = \beta \cup \{p\}$. If, by contrast, F is an expression rather than a deduction, we let $\beta' = \beta$. We then go through the patterns $\pi_1, \ldots, \pi_n$ sequentially, trying to find a pattern that is matched by V in ρ and γ. If no such pattern is found, we halt with an error message and σ'. Otherwise, let π_j be the first matching pattern and let $\{I_1 \mapsto V_1, \ldots, I_k \mapsto V_k\}$ and τ be the set of bindings and sort valuation returned by the matching algorithm, respectively. The output then becomes the result of applying τ to the conclusion obtained by evaluating D_j in $\rho[I_1 \mapsto V_1, \ldots, I_k \mapsto V_k]$, β', σ', and γ. Note that if the discriminant is a deduction, then its conclusion is available as a lemma inside D_j.

- **`let` deductions**: A deduction of the form

$$\texttt{let } \{\pi_1 \texttt{ := } F_1\texttt{;} \cdots \texttt{;} \pi_n \texttt{ := } F_n\} \; D \tag{2}$$

is treated as syntax sugar, with the desugaring proceeding by induction on n. When $n = 0$, (2) reduces to D. When $n > 0$, (2) is desugared into `match` F_1 $\{\pi_1$ `=>` $D'\}$, where D' is the result of desugaring `let` $\{\pi_2$ `:=` F_2`;` $\cdots$`;` π_n `:=` $F_n\}$ D.

- **`letrec` deductions**: A deduction of the form `letrec` $\{I_1$ `:=` E_1`;` $\cdots$`;` I_n `:=` $E_n\}$ D is desugared into the following:

```
let {I1 := (cell ());
       ...
     In := (cell ());
     _ := (set! I1 E1');
       ...
     _ := (set! In En')}
  D'
```

where each $E_j{}'$ is obtained from E_j by replacing every free occurrence of I_j by `(ref` I_j`)`, $j = 1, \ldots, n$; and D' is likewise obtained from D.

- **`try` deductions**: To evaluate `try` $\{D_1$ `|` $\cdots$ `|` $D_n\}$ in ρ, β, σ, and γ, we first evaluate D_1 in ρ, β, σ, and γ. If that results in a sentence p and store σ_1, we return p and σ_1 as the result. But if it results in an error message and a store σ_1, we go on to evaluate D_2 in ρ, β, σ_1, and γ; and so forth. If the evaluation of every D_i fails, we return an error message and σ_n.

- **Structural induction**: To evaluate a deduction of the form

$$\texttt{by-induction } F \; \{\pi_1 \texttt{ => } D_1 \texttt{ | } \cdots \texttt{ | } \pi_n \texttt{ => } D_n\} \tag{3}$$

in ρ, β, σ, and γ, we start by evaluating F in ρ, β, σ, and γ, obtaining a value V and store σ_1. If V is not a universally quantified sentence of the form

$$\texttt{(forall } I\texttt{:}S_D \; p\texttt{)} \tag{4}$$

for some inductively generated sort S_D in γ,[12] we halt with an error message and σ_1. Otherwise, we check to make sure that the patterns $\pi_1, \ldots, \pi_n$ are proper patterns for S_D (in accordance with γ) and that they are jointly exhaustive.[13] We will now explain in detail what it means for the patterns to be "proper" for S_D in the context of γ. Recall that a pattern π_i here is not an arbitrary Athena pattern (as can appear, e.g., inside a `match` clause), but is instead an element of a more restricted class of patterns described by the following grammar:

$$\pi ::= I \mid I{:}S \mid (I\ \pi^+) \mid (I\ \texttt{as}\ \pi) \tag{5}$$

where S ranges over sorts. An identifier I inside such a pattern is said to be a *pattern variable* iff it is not the name of a function symbol in γ. It is an error if there are any duplicate pattern variable occurrences in one of the patterns.

To each given pattern π we assign a term t_π, defined by structural recursion on the syntax of π as explained below. The algorithm takes as input not only the pattern π but also a finite function M mapping pattern variables to terms (mostly term variables), and it outputs not only the term t_π but also an extension of the mapping M. Initially the algorithm is invoked with $M = \emptyset$. The algorithm is this:

1. If π is an identifier I, then if I is the name of a constant symbol c in γ, we let t_π be c and return M unchanged. Otherwise I must be a pattern variable, and in that case we let t_π be a *fresh* variable x and we return x and $M[I \mapsto x]$. Note that the sort of this fresh Athena variable will be some fresh sort variable (e.g., `'T145`).
2. If π is of the form $I{:}S$ for some sort S in γ, we proceed as above except that now, if I is a pattern variable, the fresh variable x will be of sort S (rather than some fresh sort variable, as before). Also, if I is not a pattern variable, we need to ensure that $I{:}S$ is a well-sorted term.
3. If π is of the form $(I\ \pi_1 \cdots \pi_k)$ for some $k > 0$, we apply the algorithm recursively to $(\pi_1, M_0), (\pi_2, M_1), \ldots, (\pi_k, M_{k-1})$, starting with $M_0 = M$, obtaining outputs $(t_{\pi_1}, M_1), \ldots, (t_{\pi_k}, M_k)$. Thus, each output map M_i becomes the input map to the next call. Finally, if I is the name of a constructor of S_D (according to the information in γ), we output $((I\ t_{\pi_1} \cdots t_{\pi_k}), M_k)$, provided that $(I\ t_{\pi_1} \cdots t_{\pi_k})$ is a well-sorted term (an error is generated if it is not).
4. Finally, if π is of the form $(I\ \texttt{as}\ \pi)$, we apply the algorithm recursively to (π, M) to obtain an output (t_π, M_1) and we then return $(t_\pi, M_1[I \mapsto t_\pi])$.

---

12 By an inductively generated sort we mean a sort $(S\ S_1 \cdots S_k)$, $k \geq 0$, whose outer sort constructor S is the name of a datatype or structure. Thus, e.g., both `N` and `(List 'T35)` are datatype sorts.

13 Joint exhaustiveness means that every canonical term of sort S_D matches some such pattern. There is an algorithm for deciding that condition, though we need not describe it here.

A pattern π_i in (3), $i \in \{1,\ldots,n\}$, is considered proper for S_D in the context of γ provided that the term t_{π_i} obtained by applying the above algorithm to $(\pi_i, \emptyset)$ is a legal term of sort S_D.

Note that the final sort checking of a term t_{π_i} might refine the sorts of the fresh variables that appear in the mapping M produced by the algorithm. Suppose, for example, that the pattern is `(:: head tail)`. Here `::` is a function symbol (a constructor of the datatype `List`), while `head` and `tail` are pattern variables. Since both `head` and `tail` are unannotated, they will be mapped to fresh variables of completely unconstrained sorts, say to `?v10:'T145` and `?v11:'T147`, respectively. Later, sort inference will refine these sorts by realizing that `'T147` must be `(List 'T145)`. We assume that the sorts of the various fresh variables that appear in the map returned by the algorithm have been properly updated in this fashion to reflect constraints inferred by sort checking. In this case, for instance, the map M might assign the fresh variables `?v10:'T145` and `?v11:(List 'T145)` to `head` and `tail`, respectively: $M = \{\ldots,$ `head` $\mapsto$ `?v10:'T145`, `tail` $\mapsto$ `?v11:(List 'T145)`, $\ldots\}$.

Once we have gone through each pattern π_i and computed the corresponding term t_{π_i} and mapping M_i and ensured that each π_i is proper for S_D in the context of γ, we proceed to do the following for each clause π_i `=>` D_i, $i = 1,\ldots,n$. First, we compute all the appropriate inductive hypotheses for the clause. Specifically, for each fresh variable x in the range of M_i whose sort is an instance of the datatype sort S_D, we generate an inductive hypothesis, obtained from the body p by replacing every free occurrence of $I{:}S_D$ by x. Let β_i consist of β augmented with all the inductive hypotheses thus generated, and let ρ_i be the environment ρ augmented with the bindings of M_i. We then evaluate D_i in ρ_i, β_i, σ_i, and γ, producing some conclusion p_i and store σ_{i+1} (this new store will become the one in which the deduction of the next clause will be evaluated). We must now check that p_i is of the right form; if it is not, we will halt with an error message and σ_{i+1}. We say that p_i is of the right form if it is alpha-equivalent to the sentence we obtain from p by replacing every free occurrence of $I{:}S_D$ by t_{π_i}. If p_i is indeed of the right form, we continue with the next clause. If this process is successfully carried out for every clause, we return as output the universal generalization (4) along with the store σ_{n+1}.

The generation of inductive hypotheses is somewhat more sophisticated than just described, allowing, in particular, for nested inductive proofs. We give a brief illustrating example. Suppose we have a datatype describing the abstract syntax of λ-calculus expressions as follows:

```
datatype Exp :=
   (var Ide)                    # variables
 | (abs Ide Exp)                # abstractions
 | (app Exp (List Exp))         # applications
```

The interesting point for our purposes here is that the reflexive constructor App takes a *list* of expressions as its second argument. Should we ever need to perform an induction on the structure of that list in the midst of performing an outer structural induction on Exp, appropriate inductive hypotheses should be generated not only for the tail of that list, but also for the head of the list on the basis of the outer induction. Suppose, for example, that we have some unary procedure exp-property that takes an arbitrary expression e and builds some sentence expressing a proposition about e, and we are interested in proving (forall e . exp-property e) (where e is a variable ranging over Exp). We proceed by structural induction:

```
by-induction (forall e . exp-property e) {
  (var x) => D1
| (abs x e) => D2
| (app proc args) => D3
}
```

Consider now D_3, the proof in the third clause. It is not uncommon in such cases to proceed by induction on the structure of the list args, that is, to show that for all such lists L, we have (exp-property (app proc L)). Our goal will then follow simply by specializing L with args:

```
by-induction (forall e . exp-property e) {
  (var x) => D1
| (abs x e) => D2
| (app proc args) =>
    let {list-property := lambda (L)
                            (exp-property (app proc L));
         lemma := by-induction (forall L (list-property L)) {
                    nil => D3
                  | (:: head:Exp tail:(List Exp)) => D4
                  }
        }
      (!uspec lemma args)
}
```

Here, inside D_4 we will not only get an appropriate inductive hypothesis for tail (namely, (list-property tail)),[14] but we will also get one for head, namely (exp-property head). Athena knows that it is appropriate to generate this outer inductive hypothesis for head because it is keeping track of the nesting of the various structural inductions and their respective goals, as well as the sorts of the corresponding universally quantified variables, and realizes that, in the given context, head will in fact be a "smaller" expression than the outer (app proc args), since head will essentially be a

14 Strictly speaking, of course, when we say "tail" we are referring to the fresh term variable to which tail will be bound.

proper part of `args`, which is itself a proper part of `(app proc args)`.[15] Inductive proofs can be nested to an arbitrarily deep level (not just lexically, i.e., as determined by the nesting of the proofs in the actual text, but dynamically as well, via method calls), and for every clause of every such proof, appropriate inductive hypotheses will be generated for every enclosing inductive proof.

- **Structural case analysis**: A deduction of the form

$$\texttt{datatype-cases}\ F\ \{\pi_1\ \texttt{=>}\ D_1\ \texttt{|}\ \cdots\ \texttt{|}\ \pi_n\ \texttt{=>}\ D_n\} \quad (6)$$

is evaluated (in some ρ, β, σ, and γ) exactly like a structural induction deduction of the form (3), except that no inductive hypotheses are generated for the various cases.

There is a variant of the **datatype-cases** construct that is occasionally convenient, namely:

$$\texttt{datatype-cases}\ F\ \texttt{on}\ E\ \{\pi_1\ \texttt{=>}\ D_1\ \texttt{|}\ \cdots\ \texttt{|}\ \pi_n\ \texttt{=>}\ D_n\}.$$

Here, the evaluation of F in ρ, β, σ, and γ must produce a sentence p, along with a store σ_1. Further, the evaluation of E in ρ, β, σ_1, and γ must produce a variable $x{:}S$, where S is a datatype, and where p possibly contains free occurrences of $x{:}S$. The idea now is that we want to derive p, and we will proceed by a constructor case analysis on $x{:}S$. That is, because we know that $x{:}S$ is of sort S, and because we know that every value of S is obtainable by some constructor application, we essentially reason that $x{:}S$ must be of the form of one of the n listed patterns, $\pi_1, \ldots, \pi_n$. Assuming that these patterns jointly exhaust S (a condition that is checked after F and E are evaluated), we must then show that each deduction D_i derives the desired goal p but with every free occurrence of $x{:}S$ replaced by the term t_{π_i}, where t_{π_i} is obtained from π_i exactly as described above for **by-induction** proofs. There are no inductive hypotheses in this case, but each D_i is evaluated in β augmented with

15 Note, however, that the generation of outer induction hypotheses depends crucially on the form of the inner goal. While in a case such as shown here it would indeed be valid to generate an outer inductive hypothesis for `head`, that might not be so if `list-property` were defined differently. Athena will generate an outer inductive hypothesis only if it can ensure that it is appropriate to do so by conservatively matching the current (inner) goal to the outer goal. Specifically, an outer inductive hypothesis will be generated only if it can be shown that the current goal is an instance of the outer goal via some substitution of the form $\{x_1 \mapsto t_1, \ldots, x_n \mapsto t_n\}$ such that (a) each x_i is a variable in the outer inductive pattern and (b) each t_i is the current inductive pattern. In this example, the current goal (inside D_4) is `(list-property (:: head tail))`, that is, `(exp-property (app proc (:: head tail)))`. The prior goal (in the current stack of inductive clauses) is `(exp-property (app proc args))`. Thus, the current goal matches the prior goal under the substitution {`args` ↦ `(:: head tail)`}, where `args` is a variable in the outer inductive pattern and `(:: head tail)` is the current (inner) inductive pattern. Since `head` is a variable in the current inductive pattern of the same sort as the universally generalized variable in the outer clause, and since the substitution conditions (a) and (b) are satisfied, an outer inductive hypothesis will be generated in this case. We do not define here the notion of one sentence matching another sentence under some substitution, but see page 97 for a related discussion (alternatively, consult the definition of the procedure `match-sentences` in `lib/basic/util.ath`).

the identity $(x{:}S = t_{\pi_i})$. If the evaluation of each D_i in the corresponding environment (obtained from ρ and π_i as in a **by-induction** proof), the aforementioned assumption base, the appropriate store, and γ, results in the correct sentence (namely, p with every free occurrence of $x{:}S$ replaced by t_{π_i}), then p is finally produced as a result.

A.4 Pattern matching

Athena's pattern language is shown in Figure A.4. These patterns make it easy to take apart complicated structures. Disregarding parentheses and square brackets, a pattern is made up of (a) keywords such as **list-of** and **some-quant**, and (b) three kinds of identifiers:

1. The sentential connectives and quantifiers: `not`, `and`, `or`, `if`, `iff` (or their infix counterparts), along with `forall` and `exists`.
2. Function symbols, such as `nil`, `true`, and so on.
3. All remaining identifiers, which serve as *pattern variables*.

Function symbols along with sentential connectives and quantifiers are pattern constants, in that they can only be matched by the corresponding values. Pattern variables, on the other hand, can be matched by arbitrary values.

Pattern matching works as follows: a value V is matched against a pattern π, with respect to some environment ρ and symbol set γ, and results either in failure (in which case we say that V did not match the pattern); or in an environment (finite set of bindings) $\rho' = \{I_1 \mapsto V_1, \ldots, I_n \mapsto V_n\}$ that assigns values to the pattern variables of π, along with a sort valuation, that is, a finite function from sort variables to sorts. Suppose, for example, that V is the term `(S S zero)` and π is the pattern `(S n)`. Here we have a successful match, resulting in the environment {`n` $\mapsto$ `(S zero)`} (and the empty sort valuation).

In what follows we describe in detail an algorithm that takes as input: (a) a value V, (b) a pattern π, (c) an environment ρ, and (d) a symbol set γ; and outputs either (i) a failure token, indicating that V does not match π with respect to ρ and γ, or else (ii) an environment ρ' that represents a successful match, along with a sort valuation τ'. Both the output environment ρ' and the output sort valuation τ' are built up in stages, and it is convenient to express the algorithm so that it takes ρ' and τ' as two additional inputs. Initially the algorithm is called with $\rho' = \emptyset$ and $\tau' = \emptyset$.

The sort valuation is needed primarily in order to handle polymorphism. For instance, we would like to ensure that a term such as

```
(pair ?x:'S ?y:'T)
```

is successfully matched against a pattern like `(pair left:Ide right:Real)`. Intuitively, `(pair ?x:'S ?y:'T)` represents infinitely many terms, as many as can be obtained by consistently replacing the sort variables `'S` and `'T` by ground sorts. One of these infinitely many terms is the term `(pair ?x:Ide ?y:Real)`, obtained by replacing `'S` by `Ide` and `'T` by

π	::=	I	(Pattern variables, symbols, connectives, quantifers)
	|	I:S	(Sort-annotated pattern variables or constant symbols)
	|	?I:S	(Athena variables)
	|	'I	(Meta-identifiers)
	|	C	(Character constants)
	|	T	(String constants)
	|	()	(Unit pattern)
	|	_	(Wildcard pattern)
	|	(bind I π)	(Named patterns, same as (I as π))
	|	(val-of I)	(val-of patterns)
	|	(list-of π_1 π_2)	(list-of patterns)
	|	(split π_1 π_2)	(split patterns)
	|	[$\pi_1 \cdots \pi_n$]	(Fixed-length list patterns)
	|	($\pi_1 \cdots \pi_n$)	(Compound patterns)
	|	(π where E)	(Where patterns)
	|	(some-var J)	
	|	(some-sent-con J)	
	|	(some-quant J)	
	|	(some-term J)	
	|	(some-atom J)	
	|	(some-sentence J)	
	|	(some-list J)	
	|	(some-cell J)	
	|	(some-vector J)	
	|	(some-proc J)	
	|	(some-method J)	
	|	(some-symbol J)	
	|	(some-table J)	
	|	(some-map J)	
	|	(some-sub J)	
	|	(some-char J)	
J	::=	I | _	

Figure A.4
Syntax of Athena patterns

Real; and that term does indeed match the given pattern, under {left $\mapsto$?x:Ide, right $\mapsto$?y:Real}; our pattern-matching algorithm should infer this automatically. Roughly, when matching a term or sentence against a pattern, our algorithm will first try to unify the sorts of the various components of the term (or sentence) with the sort constraints expressed in the pattern. If that succeeds, a match may be obtained; otherwise the match will fail. The incrementally built sort valuation is used primarily for that purpose.

In summary, then, the algorithm takes as input: (a) a value V; (b) a pattern π; (c) an environment ρ; (d) a symbol set γ; (e) the auxiliary environment ρ' that is to be incrementally built up; and (f) a sort valuation τ.[16] The empty set $\emptyset$ is given as the value of the last two arguments when the algorithm is first invoked. The algorithm will either fail or it will produce a pair (ρ'', τ') consisting of an environment ρ'' and a sort valuation τ' that extend ρ' and τ, respectively.[17] The algorithm proceeds by a case analysis of the structure of π:

- Case 1: π is the wildcard pattern _. In that case return (ρ', τ).
- Case 2: π is a term variable of sort S (such as `?x:Boolean`). In that case, if V is a term variable of sort T such that $\tau(S)$ and $\tau(T)$ are unifiable under a most general unifier τ', return (ρ', τ''), where τ'' is the composition of τ and τ'; otherwise fail.
- Case 3: π is a meta-identifier (such as `'foo`), or a character (such as `‘A`), or a string (such as `"Hello world!"`), or the unit value `()`. Then if V is the exact same value (`'foo`, `‘A`, etc.), return (ρ', τ); otherwise fail.
- Case 4: π is an identifier I, possibly annotated with a sort S. Then if I is a sentential connective or quantifier, the match succeeds iff V is the corresponding value and there is no sort annotation attached to I, in which case we simply return ρ' and τ unchanged. Otherwise we consult γ to see whether I is a function symbol.

  1. If it is, we check to see if there is a sort annotation S:

     - If there is, we check whether the value V is a constant term t whose root is I. If it is not, we fail. If it is, let S_t be the sort of t. We then check to see whether $\tau(S_t)$[18] and $\tau(S)$ can be unified under some most general unifier τ'. If not, we fail, otherwise we return ρ' unchanged along with the composition of τ' and τ.
     - If there is not, we check to see if the value V is that exact same function symbol (I).[19] If it is not, we fail; otherwise we return ρ' and τ unchanged.

  2. If it is not, that means that I is a pattern variable. We then check to see if ρ' already assigns a value V_I to I:

     - Suppose that it does. Then we again check whether there is a sort annotation S:

---

16 The arguments ρ and γ are only used for lookups, so these two values could be held fixed throughout. An inner matching algorithm that would do all the work could then take four arguments only: (a), (b), (e), and (f). But for simplicity we describe a single six-input algorithm here.

17 To be perfectly complete and precise, we would also need to pass two additional inputs to the matching algorithm: an assumption base and a store. These are needed to evaluate expressions in **`where`** patterns. And we would also need to modify the output of the pattern matching algorithm to include an output store, that which might be produced as a side effect of evaluating such expressions. However, because these are needed only for **`where`** patterns, we omit them from the overall specification in order to avoid further complicating the description of the algorithm. The changes that would be needed to arrive at a working implementation are straightforward.

18 For any sort S and sort valuation τ, $\tau(S)$ denotes the sort obtained from S by replacing every occurrence of a sort variable in S by the unique sort that τ assigns to that variable (if a variable is not in the domain of τ, then it is returned unchanged).

19 Note that a constant term (such as `zero` or `nil`) can be coerced into a function symbol and vice versa.

* If there is not, then we fail if V and V_I are not identical; otherwise, if the two values are the same, we return ρ' and τ unchanged.
* If there is a sort annotation S, we proceed as follows. If either V or V_I is not a term value, we fail. Otherwise both V and V_I are term values, call them t and t' respectively, with corresponding sorts S_t and $S_{t'}$. Then if the sorts $\tau(S)$ and $\tau(S_t)$ are not unifiable, we fail. If they are unifiable under some τ', then let t_1 and $t_1{}'$ be the terms obtained from t and t', respectively, by applying the composition of τ' and τ to their sorts.[20] If these two terms are identical, then we return ρ' extended with the assignment that maps I to t_1, along with the aforementioned composition. If t_1 and $t_1{}'$ are not identical, we fail.

- Suppose that it does not. We again check to see whether there is a sort annotation:

  * If not, then we return (a) ρ' augmented with $\{I \mapsto V\}$; and (b) τ unchanged.
  * If I is annotated with a sort S, then we check to see if the value V is a term t, with some sort S_t. If it is not a term, we fail. Otherwise, we check whether $\tau(S_t)$ and $\tau(S)$ are unifiable under some τ'. If not, we fail. Otherwise, we return (a) ρ' augmented with $\{I \mapsto t'\}$, where t' is the term obtained by applying the composition of τ' and τ to t; and (b) the said composition.

- Case 5: π is a list pattern of the form `[`$\pi_1 \cdots \pi_n$`]`. In that case, if V is a list of values $[V_1 \cdots V_n]$, we return the result of trying to sequentially match $V_1, \ldots, V_n$ against the patterns $\pi_1, \ldots, \pi_n$ in ρ, γ, ρ', and τ (see the paragraph at the end of this section on how to sequentially match a number of values $V_1, \ldots, V_n$ against patterns $\pi_1, \ldots, \pi_n$). Otherwise we fail.
- Case 6: π is a list pattern of the form `(`**`list-of`** π_1 π_2`)`. In that case, if V is a nonempty list of values $[V_1 \cdots V_n]$, $n > 0$, we return the result of trying to sequentially match the values $V_1, [V_2 \cdots V_n]$ against the patterns π_1, π_2 (again, in ρ, γ, ρ', and τ). Otherwise we fail.
- Case 7: π is a list pattern of the form `(`**`split`** π_1 π_2`)`. In that case, if V is not a list of values, we fail. Otherwise, we determine whether V can be expressed as the concatenation of two lists L_1 and L_2 such that L_1 is the smallest prefix of V for which the values L_1, L_2 sequentially match the patterns π_1, π_2 under ρ, γ, ρ', and τ, producing a result (ρ'', τ'). If so, we return that result. If no such decomposition of V exists, we fail.
- Case 8: π is of the form `(`**`val-of`** I`)`. We then check to see if I is bound in ρ. If it is not, we fail. If it is bound to some value V', we check whether the two values V and V' are

20 To apply a sort valuation to a term means to apply the sort valuation to every sort annotation in that term.

identical. If the values do not admit equality testing, or if they do but are not identical, we fail, else we return ρ' and τ unchanged.

- Case 9: π is of the form (**bind** I π'), or equivalently, (I **as** π'). In that case we match V against π' in ρ, γ, ρ', and τ. If that fails, we fail. Otherwise, if it produces a result (ρ'', τ'), we return $(\rho''[I \mapsto V], \tau')$.

- Case 10: π is of the form (π' **where** E). We then match V against π' in ρ, γ, ρ', and τ. If that fails, we fail. Otherwise, if we get a result (ρ'', τ'), we evaluate the expression E in the environment ρ augmented with the bindings in ρ'', γ, and some appropriate assumption base and state. If that evaluation results in `true`, we return (ρ'', τ'), otherwise we fail.

- Case 11: π is of the form $(\pi_1\ \pi_2\ \cdots\ \pi_{n+1})$, for $n > 0$. As explained in Section 2.11, patterns of this form are used for decomposing terms and sentences. Accordingly, we distinguish the following cases:

  - (i) V is a term t of the form $(f\ t_1 \cdots t_m)$, $m \geq 0$. We then distinguish two subcases:

    1. If $n = 2$, π_1 is a quantifier pattern,[21] and π_2 is a list pattern, then:

       * If t is term of sort `Boolean`, so that it can be treated as a sentence p, try to match it as a sentence against π (in ρ, γ, ρ', and τ), using the algorithm given below (under case (ii)).
       * If t is not a term of sort `Boolean`, fail.

    2. Otherwise, we distinguish the following subcases:

       * $n = 1$ and π_2 is a list pattern. In that case we try to sequentially match the values f, `[`$t_1 \cdots t_m$`]` against the patterns π_1, π_2 (in ρ, γ, ρ', and τ).
       * Otherwise, if $n > 1$ or π_2 is not a list pattern, we try to sequentially match the values $f, t_1, \ldots, t_m$ against the patterns $\pi_1, \pi_2, \ldots, \pi_{n+1}$ (in ρ, γ, ρ', and τ).

  - (ii) V is a sentence p. We then distinguish two subcases, on the basis of the pattern:

    1. $n = 2$, π_1 is a quantifier pattern, and π_2 is a list pattern. In that case, we first express p in the form

    $$(Q\ x_1 \cdots x_k\ p') \tag{7}$$

    for some quantifier Q, $k \geq 0$, and a sentence p' that is *not* of the form $(Q\ y_1 \cdots y_m\ p'')$, $m > 0$. Note that this can always be done, for any sentence p, although the identity of Q will not be uniquely determined if $k = 0$, that is, if p is not an actual quantified sentence (a condition that might affect the first of the

21 That is, one of the two pattern constants `forall`, `exists`; or a pattern of the form (**some-quant** J); or else a pattern of the form (**bind** I π), where π is a quantifier pattern.

steps below). After we express p in the form (7), we do the following, in sequential order:

* If $k = 0$, then if the quantifier pattern π_1 is of the form (`some-quant` I) or (`bind` I π') for some quantifier pattern π', fail. Otherwise, let $\rho'' = \rho'$ and $\tau' = \tau$, and continue.
* If $k > 0$ then match Q against the pattern π_1 in ρ, γ, ρ', and τ, and let (ρ'', τ') be the output of that match. (If the match fails, we fail.)
* Match the largest possible prefix [$x_1 \cdots x_i$], $i \leq k$, against π_2 in ρ, γ, ρ'' and τ', and let (ρ''', τ'') be the result.
* Return the result of matching the sentence (Q $x_{i+1} \cdots x_k$ p') against π_3 in ρ, γ, ρ''', and τ''. (If $k - 0$ then this sentence will simply be p'.)

2. Otherwise we distinguish the following additional subcases, obtained by analyzing the structure of p:

* If p is an atomic sentence t, we match t as a term against π (in ρ, γ, ρ', and τ).
* If p is a compound sentence of the form ($\circ$ $p_1 \cdots p_k$), for some sentential connective $\circ$ and $k > 0$, then:

  (a) If $n = 1$ and π_2 is a list pattern, we sequentially match $\circ$, [$p_1 \cdots p_k$] against the patterns π_1, π_2 in ρ, γ, ρ', and τ.

  (b) Otherwise, we sequentially match $\circ, p_1, \ldots, p_k$ against the patterns $\pi_1, \pi_2, \ldots, \pi_{n+1}$ (in ρ, γ, ρ', and τ).

* Finally, if p is a quantified sentence of the form (Q x p'), we sequentially match the values Q, x, p' against the patterns $\pi_1, \ldots, \pi_{n+1}$.

• (iii) If V is neither a term of the form (f $t_1 \cdots t_m$) nor a sentence, we fail.

• Case 12: π is a filter pattern, that is, of the form (`some-`$\cdots$ J). Suppose first that it is of the form (a) (`some-var` I) or (b) (`some-var _`). In that case we check to see whether V is some term variable. If it is not, we fail. If it is, then, in case (b), we return ρ' and τ unchanged. In case (a), we return $\rho'[I \mapsto V]$ and τ. When π is of the form (a) (`some-atom` I) or (b) (`some-atom _`), we check to see whether V is an atomic sentence (i.e., a term of sort `Boolean`), and if so, we return $(\rho'[I \mapsto V], \tau)$ in case (a) and (ρ', τ) in case (b). Likewise for the remaining filter patterns. For instance, if π is of the form (a) (`some-sent-con` I) or (b) (`some-sent-con _`), we check whether V is one of the five sentential connectives, and if so, we return $(\rho'[I \mapsto V], \tau)$ in case (a) and (ρ', τ) in case (b).

Finally, to *sequentially match* a number of values $V_1, \ldots, V_n$ against a number of patterns $\pi_1, \ldots, \pi_m$ in given ρ, γ, ρ', and τ, we do the following: First, if $n \neq m$, we fail. Otherwise, if $n = 0$, we return (ρ', τ). Finally, if $n = m$ and $n > 0$, we match V_1 against π_1 in ρ, γ, ρ', and τ, resulting in some output pair $({\rho_1}', \tau_1)$, and then we proceed to sequentially match (recursively) $V_2, \ldots, V_n$ against $\pi_2, \ldots, \pi_n$ in ρ, γ, ${\rho_1}'$, and τ_1.

A.5 Selectors

A constructor profile $(c\ S_1 \cdots S_n)$ in the definition of a datatype or structure can optionally have one or more *selectors* attached to various argument positions. To introduce a selector by the name of g for the ith argument position of c, simply write the profile as: $(c\ S_1 \cdots\ g{:}S_i\ \cdots S_n)$. For instance, for the natural numbers we might have:

```
datatype N := zero | (succ pred:N)
```

This introduces the selector `pred` as a function from natural numbers to natural numbers such that (`pred succ` n `=` n) for all n:`N`. It is generally a good idea to introduce selectors.[22] When a constructor has arity greater than 1, we can introduce a selector for every argument position. For instance, in the definition of natural-number lists below, we introduce `head` as a selector for the first argument position of `nat-cons` and `tail` as a selector for the second argument position:

```
datatype Nat-List := nat-nil | (nat-cons head:N tail:Nat-List)
```

Axioms specifying the semantics of selectors are automatically generated and can be obtained by applying the procedure `selector-axioms` to the name of the datatype:

```
> (selector-axioms "Nat-List")

List: [
(forall ?v561:N
  (forall ?v562:Nat-List
    (= (head (nat-cons ?v561 ?v562))
       ?v561)))

(forall ?v561:N
  (forall ?v562:Nat-List
    (= (tail (nat-cons ?v561 ?v562))
       ?v562)))
]
```

22 And if the datatype is to be used in SMT solving (see the "Automated Theorem Proving" appendix available on the book's website), then every constructor *must* have a selector attached to every argument position.

The behavior of head or tail on nat-nil is unspecified. In general, the behavior of any selector of a constructor c when applied to a term built by a different constructor c' is unspecified.

However, it is possible to provide a total specification for selectors by instructing Athena to treat selectors as functions from the datatype at hand to *optional* values of the corresponding sorts. This is done by turning on the flag option-valued-selectors, which is off by default. For instance:

```
set-flag option-valued-selectors "on"

datatype Nat-List-2 := nat-nil-2 | (nat-cons-2 head-2:N tail-2:Nat-List-2);;

> (selector-axioms "Nat-List-2")

List: [
(forall ?v617:N
  (forall ?v618:Nat-List-2
    (= (head-2 (nat-cons-2 ?v617 ?v618))
       (SOME ?v617))))

(forall ?v617:N
  (forall ?v618:Nat-List-2
    (= (tail-2 (nat-cons-2 ?v617 ?v618))
       (SOME ?v618))))

(= (head-2 nat-nil-2)
   NONE)

(= (tail-2 nat-nil-2)
   NONE)
]
```

A.6 Prefix syntax

In this book we have used mostly infix syntax,both for issuing directives (such as **declare** and **define**) and for writing phrases. However, Athena also supports a fully prefix syntax based on s-expressions. An advantage of s-expressions is that they lend themselves exceptionally well to indentation, which can serve to clarify the structure of complex syntactic constructions. (That is why Athena always displays output in fully indented prefix.) Below we present s-expression variants of all major syntax forms (shown on the left side in the infix form in which they have been used in this book).

declare I: [$S_1 \cdots S_n$] -> S	(declare I (-> ($S_1 \cdots S_n$) S))
declare I: ($I_1,\ldots,I_k$) [$S_1 \cdots S_n$] -> S	(declare I (($I_1 \cdots I_k$) -> ($S_1 \cdots S_n$) S))
declare $I_1,\ldots,I_m$: ($I_1{}',\ldots,I_k{}'$) [$S_1 \cdots S_n$] -> S	(declare ($I_1 \cdots I_m$) (($I_1{}' \cdots I_k{}'$) -> ($S_1 \cdots S_n$) S))
assert $E_1,\ldots,E_n$	(assert $E_1 \cdots E_n$)
assert I := E	(assert I := E)
set-flag I "on"/"off"	(set-flag I "on"/"off")
overload I_1 I_2	(overload I_1 I_2)
set-precedence I E	(set-precedence I E)
set-precedence ($I_1 \cdots I_n$) E	(set-precedence ($I_1 \cdots I_n$) E)
lambda ($I_1 \cdots I_n$) E	(lambda ($I_1 \cdots I_n$) E)
method ($I_1 \cdots I_n$) D	(method ($I_1 \cdots I_n$) D)
cell F	(cell F)
set! E F	(set! E F)
ref E	(ref E)
while F E	(while F E)
make-vector E F	(make-vector E F)
vector-sub E_1 E_2	(vector-sub E_1 E_2)
vector-set! E_1 E_2 F	(vector-set! E_1 E_2 F)
check {F_1 => $F_1{}'$ | $\cdots$ | F_n => $F_n{}'$}	(check (F_1 $F_1{}'$) $\cdots$ (F_n $F_n{}'$))
match F {π_1 => F_1 | $\cdots$ | π_n => F_n}	(match F (π_1 F_1) $\cdots$ (π_n F_n))
let {π_1 := F_1; $\cdots$; π_n := F_n} F	(let ((π_1 F_1) $\cdots$ (π_n F_n)) F)
letrec {I_1 := E_1; $\cdots$; I_n := E_n} F	(letrec ((I_1 E_1) $\cdots$ (I_n E_n)) F)
try {F_1 | $\cdots$ | F_n}	(try F_1 $\cdots$ F_n)
assume F D	(assume F D)
assume I := F D	(assume (I F) D)
pick-any I D	(pick-any I D)
pick-witness I for F D	(pick-witness I F D)
pick-witnesses $I_1 \cdots I_n$ for F D	(pick-witnesses ($I_1 \cdots I_n$) F D)

B Logic Programming and Prolog

THIS APPENDIX contains a brief discussion of logic programming and Prolog. We present an Athena implementation of a Prolog interpreter for definite clauses, and we also discuss Athena's integration with external Prolog systems such as SWI Prolog. We provide a rudimentary overview of the central concepts of logic programming, at least as they manifest themselves in Athena, but due to space limitations the treatment here cannot be exhaustive. For in-depth introductions to these subjects, consult a textbook [18], [78], [60], [65], [95].

B.1 Basics of logic programming

The key notion in logic programming is that of a *definite Horn clause*.

In the present context a definite Horn clause can be understood as an Athena sentence of the following kinds:

1. A *fact* of the form

   $$\texttt{(forall } x_1 \cdots x_n \texttt{ . } p\texttt{)},$$

   where $n \geq 0$ and p is an atomic sentence whose free variables are exactly $x_1, \ldots, x_n$; or

2. a *rule*, that is, a conditional of the form

   $$\texttt{(forall } x_1 \cdots x_n \texttt{ . } p_1 \texttt{ \& } \cdots \texttt{ \& } p_k \texttt{ ==> } p_{k+1}\texttt{)},$$

   where $k > 0$ and $x_1, \ldots, x_n$ are all and only the free variables that occur in the various p_i. We refer to p_{k+1} as the conclusion or *head* of the rule and to $p_1, \ldots, p_k$ as the *antecedents*. (We may also collectively refer to the entire conjunction of $p_1, \ldots, p_k$ as *the* antecedent of the rule. This is also known as the *body* of the rule in logic programming terminology.)

Facts can also be viewed as rules with the trivial body `true`, and we will occasionally treat them as such, so the distinction between facts and rules is not theoretically essential.

A (definite) *logic program* is simply a finite set of definite Horn clauses, that is, a finite collection of facts and rules. A (definite) Prolog program, by contrast, is a *list* of definite Horn clauses—a logic program where the order in which the clauses are listed is significant. We will see shortly why clause order is significant in Prolog. In addition to clause order, the order in which one lists the atoms of a rule's body is also significant in Prolog (whereas in the world of pure logic that order is immaterial, given that conjunction is commutative and associative).

The rules and facts of a logic program collectively express a declarative body of information: They state *what is* (or what might be) the case in a given domain. The programs are called definite because, viewed as theories (as sets of first-order logic sentences), they always have unique minimal models, namely, the so-called *least Herbrand models*. Roughly speaking, the least Herbrand model of a logic program $\mathcal{P}$ is the set of sentences obtained by starting with the facts in $\mathcal{P}$ and then adding all possible consequences thereof, possibly infinitely many, as dictated by the rules in $\mathcal{P}$, where a sentence counts as "a possible consequence" only if it involves ground terms that can be formed with symbols that occur in the program.[1] For example, suppose we have a domain `D`, a constant `a` of sort `D`, a unary function symbol `f` from `D` to `D`, and a unary relation symbol `P` on `D` (a function symbol from `D` to `Boolean`). And suppose that $\mathcal{P}$ consists of the sole rule

```
(forall x . P x ==> P f x)
```
(1)

and the sole fact `(P a)`. Then the least Herbrand model of this program would be the following infinite set of sentences:

{`(P a)`, `(P (f a))`, `(P (f (f a)))`, `(P (f (f (f a))))`, ...}.

The model is said to be the "least" or "smallest" Herbrand model in that it can be shown that it is contained in (it is a subset of) *every* Herbrand model of the program, that is, it is contained in every set of facts over the Herbrand universe that satisfies the program. If we allowed disjunction in rule heads this property (the existence of a unique smallest model) would no longer hold.

As a stylistic point, a rule such as (1) is typically written in the reverse direction, as `(forall x . P f x <== P x)`, and the quantification is also implicit, with the tacit understanding that all variables in the rule are universally quantified. The backward direction is more in keeping with the *operational meaning* of such a rule, which can be loosely put as follows:[2] *To solve a goal of the same form as the head of the rule, solve the subgoals comprising the body of the rule.* To figure out whether a goal is "of the same form" as the head of a rule, we use unification. If successful, this produces a substitution θ, and we then recursively proceed to solve the subgoals obtained by applying θ to the atoms in the body of the rule. This process, which will be made fully precise in Section B.3, is called *backchaining* on the rule, and can be viewed as the inverse of modus ponens: Instead of proceeding from the antecedent(s) to the conclusion, we go from the conclusion to the antecedents. This algorithm must eventually be grounded (it must terminate) in the facts of the program, or otherwise we might keep backchaining indefinitely. As mentioned earlier, a fact can be regarded as a rule with the trivial body `true`, so backchaining on a fact immediately terminates successfully, assuming that the goal is of the same form as—unifies with—the (head of the) fact.

---

1 The set of all such ground terms is called the *Herbrand universe* of $\mathcal{P}$. To ensure that this set is nonempty, it is typically assumed that the program contains at least one constant symbol.

2 As opposed to their declarative meaning, which is the usual meaning as dictated by the standard Tarskian semantics of first-order logic (as discussed in Section 5.6).

Definite programs can be viewed as *positive*, insofar as they only express affirmative information. They cannot express negative rules such as "a man is a person who is not a woman." It is possible to extend logic programming to allow for negative clauses of that form, called *general clauses*. Neither the semantics nor the computational interpretation of such extensions are entirely satisfactory, and we do not discuss them here. Nevertheless, negation is often useful in clause bodies. For instance, many conditional equations have negative conditions, and it is useful to encode these directly when we translate such equations into clausal form, as we do in Section B.4. Full-blown Prolog implementations do allow for negated subgoals via a mechanism known as (finite) *negation as failure*, and since Athena is integrated with such systems, one can work with general clauses via the existing translation, as described in Section B.4.

So how do we compute with a logic program $\mathcal{P}$? By issuing *queries*, which are then automatically answered by a logic program engine. A simple query is just an atomic sentence, such as `(father-of adam peter)` or `(parent-of ?X peter)`. If the query is ground, like `(father-of adam peter)`, then it is basically asking whether the ground atom in question holds, so the answer will be a simple yes or no, respectively indicating that the ground atom does or does not follow from $\mathcal{P}$. If the query contains variables, like

`(parent-of ?X peter),` (2)

then it can be understood as asking whether there are any values for these variables that make the atom true according to the program. For instance, is there a value of `?X` that makes (2) true according to `person-database`, the small program introduced in the first example of the next section? A positive answer in this case will be a *substitution* from the query variables to appropriate terms, such as

`{?X:Person` $\longrightarrow$ `seth},`

that renders the query a logical consequence of $\mathcal{P}$.[3] This is largely where the power of logic programming stems from: the ability to not merely confirm or deny, but to also *discover* objects that satisfy a given query.[4] As the queries become more complex, this ability to discover objects that may be related in arbitrarily complicated ways becomes increasingly useful. By contrast, a negative answer for a nonground query would indicate that *no* substitution instance of the query follows logically from $\mathcal{P}$.

A query need not be a simple atom. It may be conjunctive, typically represented by a *list* of atoms. An example might be

```
[(parent-of adam ?X) (parent-of ?X ?Y)].
```

3 More precisely, the sentence obtained by applying the substitution to the query is a logical consequence of $\mathcal{P}$.

4 Such queries might look familiar to readers who have been exposed to query languages like SQL or SPARQL. There is indeed a very close connection between logic programming and database (or knowledge base) queries, which is studied in the field known as *deductive databases* [105].

A positive answer here will again be a substitution which, when applied to the query atoms, will yield instances that are logical consequences of $\mathcal{P}$. In this case a substitution will give us a value for ?X, representing a child of adam, *and* a value for ?Y, representing a child of ?X. Thus, in one fell swoop we will retrieve both a child and a grandchild of adam. As before, a negative answer will indicate that there is no (joint) ground instance of the query atoms that follows logically from $\mathcal{P}$.

There may well be multiple substitutions (indeed, potentially infinitely many) resulting in ground logical consequences of $\mathcal{P}$. Prolog engines are able to incrementally produce as many of these as desired, as is the interpreter we implement in Section B.3.

In addition to the possibilities already mentioned, there is another alternative for what might happen when a user poses a query: nontermination. The query engine might get into an infinite loop. This is where the aforementioned issue of clause and body orderings comes into the picture. Naive ways of expressing clauses and programs can easily send a Prolog system into an infinite loop. It takes some experience to write logic programs in a way that avoids such pitfalls.

B.2 Examples

Let us look at some examples to get a feeling for what we can do with logic programming. Queries will be answered here by the interpreter that we present in the next subsection. We will look at two domains, some genealogical relationships and a few arithmetic operations (addition and multiplication) on the natural numbers. We start with the former:

```
domain Person

declare father-of, mother-of, parent-of, grandparent-of, ancestor-of:
        [Person Person] -> Boolean

declare adam, eve, seth, peter, mary, paul, joe: Person

define facts :=
   [(adam father-of seth)
    (eve mother-of seth)
    (seth father-of peter)
    (peter father-of paul)
    (mary mother-of paul)
    (paul father-of joe)]

define rules :=
 (close [(x parent-of y <== x father-of y)
         (x parent-of y <== x mother-of y)
         (x grandparent-of z <== x parent-of y & y parent-of z)
         (x ancestor-of y <== x parent-of y)
         (x ancestor-of z <== x ancestor-of y & y parent-of z)])
```

```
define person-database := (facts joined-with rules)
```

We now answer a few queries against this program/database. The main tool we use to answer queries is the binary procedure `Prolog_Interpreter.solve`, which takes a query (either an atom or a list thereof) and a list of definite clauses and returns either a substitution θ such that the θ-instance of the query[5] is a logical consequence of the given clauses; or `false` if no such substitution can be found. If the query is a ground fact that holds,[6] or a conjunction or list of such facts, then the empty substitution is returned.

We also use the ternary procedure `Prolog_Interpreter.solve-N`, whose first two arguments are the same as those of `Prolog_Interpreter.solve`, and whose third argument is a nonnegative integer N indicating the desired number of solutions (substitutions). If at least N satisfying substitutions can be found, they are returned in a list. If fewer than N substitutions can be found, they are also returned in a list. If no satisfying substitutions can be found, then the empty list is returned. Some examples:

```
define (solve q) := (Prolog_Interpreter.solve q person-database)

define (solve-N q N) := (Prolog_Interpreter.solve-N q person-database N)

set-precedence solve 80

> (solve paul father-of joe)

Substitution: {}  # An empty substitution indicates a fact

> (solve paul mother-of joe)

Term: false       # false means the query cannot be proved/solved

> (solve ?X father-of peter))

Substitution: {?X:Person --> seth}

> (solve seth father-of ?Y)

Substitution: {?Y:Person --> peter}

> (solve ?X father-of ?Y)

Substitution:
{?Y:Person --> seth
?X:Person --> adam}
```

5 Meaning the sentence we obtain by applying θ to the conjunction of all the query atoms.

6 A sentence "holds" in this context iff it is a member of the input program/database, *not* if it is in the assumption base.

```
> (solve-N (?X father-of ?Y) 10)

List: [
{?X:Person --> adam
?Y:Person --> seth}

{?X:Person --> seth
?Y:Person --> peter}

{?X:Person --> peter
?Y:Person --> paul}

{?X:Person --> paul
?Y:Person --> joe}]
```

Note that the last result gave us four substitutions listing all pairs of fathers and sons in the database.

We continue with some more interesting queries that essentially require reasoning (using the rules):

```
> (solve eve grandparent-of ?X)

Substitution: {?X:Person --> peter}

> (solve eve ancestor-of ?X)

Substitution: {?X:Person --> seth}

> (solve-N (eve ancestor-of ?X) 10)

    ...
```

The last query sends the system into an infinite loop. The reason is the last of the given rules, namely

```
(x ancestor-of z <== x ancestor-of y & y parent-of z).
```

The problem here is that we have listed `(x ancestor-of y)`, a recursive subgoal (recursive in the sense that its predicate is the same as the predicate in the head of the rule), as the *first* subgoal in the body of the rule, which means that backchaining on this rule for certain goals will lead to an infinite regress. This will become clearer in the next subsection. Once we reformulate the rule as

```
(x ancestor-of z <== x parent-of y & y ancestor-of z)
```

and reload, we get the following complete list of all of Eve's ancestors:

```
> (solve-N (eve ancestor-of ?X) 10)

List: [{?X:Person --> seth}
       {?X:Person --> peter}
       {?X:Person --> paul}
       {?X:Person --> joe}]
```

This was an example of a case where the order of the subgoals in the body of a rule is important. In the next section we will see an example illustrating the importance of rule ordering.

Let us now give some definite clauses for addition and multiplication on the natural numbers. First, we introduce ternary predicates `plus-p` and `times-p` such that

(plus-p x y z)

holds iff z is the sum of x and y, and likewise for `times-p`:

```
declare plus-p, times-p: [N N N] -> Boolean [[int->nat int->nat int->nat]]
```

We now define `plus-p` by the following definite clauses:

```
define plus-clauses :=
  (close [(plus-p x zero x)
          ((plus-p x (S y) (S z)) <== (plus-p x y z))])
```

Compare this relational definition with a functional definition of natural-number addition in the style given in the text:

```
assert* Plus-def := [(x + zero = x)
                     (x + S y = S (x + y))]
```

The similarities here are deeper than the differences. If we read `plus-clauses` backward, then the first clause says that every instance of `(plus-p x zero x)` holds: For any `x`, the sum of `x` and `zero` is `x`, which is precisely the content of the first equation in the functional definition. The second clause says that to prove a goal of the form `(plus-p x (S y) (S z))`, we first need to prove the subgoal `(plus-p x y z)`. Thus, if the first operand to the addition is any number `x` and the second operand is nonzero (of the form `(S y)`), then we know that the result will be nonzero, of the form `(S z)`, and the precise value of `z` can be computed by backchaining and solving the subgoal `(plus-p x y z)`.

```
> (Prolog_Interpreter.solve (plus-p 1 2 ?RESULT) plus-clauses)

Substitution: {?RESULT:N --> (S (S (S zero)))}
```

To increase readability, let us print out substitutions so that natural numbers are written as integers:

```
define (transform-sub sub) :=
  (map lambda (v) [v --> (nat->int (sub v))]
       (supp sub))

define (solve q program) :=
  match (Prolog_Interpreter.solve q program) {
    (some-sub sub) => (transform-sub sub)
  | r => r}

define (solve-N q program N) :=
  (map transform-sub (Prolog_Interpreter.solve-N q program N))
```

The following queries begin to illustrate the flexibility of logic programming:

```
> (solve (plus-p 3 ?Y 10) plus-clauses)

List: [[?Y --> 7]]
```

This shows that logic programming is not unidirectional, meaning that it need not always go from fixed inputs to an output. We can just as well fix the output and some inputs and solve for other inputs, or fix only the output and solve for any inputs (as does the next query). Queries are viewed essentially as constraints, and the logic programming engine attempts to find any or all values satisfying the constraints. In this particular example we used the addition clauses to essentially perform subtraction: We asked what number `?Y` we must add to 3 in order to obtain 10, and we got the answer 7, which is of course the result of subtracting 3 from 10.

The following query produces all ways in which `?X` and `?Y` can add up to 3:

```
> (solve-N (plus-p ?X ?Y 3) plus-clauses 10)

List: [[[?Y --> 0] [?X --> 3]]
       [[?Y --> 1] [?X --> 2]]
       [[?Y --> 2] [?X --> 1]]
       [[?Y --> 3] [?X --> 0]]]
```

We can likewise define multiplication as follows:

```
define times-clauses :=
   (close [(times-p x zero zero)
           ((times-p x (S y) z) <== (times-p x y w) & (plus-p x w z))])

define num-clauses := (plus-clauses joined-with times-clauses)

> (solve (times-p 2 3 ?RES) num-clauses)

List: [[?RES --> 6]]
```

Reading the clauses backward, the first says that every goal of the form

```
(times-p x zero zero)
```

holds, which is another way of saying that the product of any number with zero is zero. The second clause says that to solve a goal of the form `(times-p x (S y) z)`, we first need to solve `(times-p x y w)`, and then solve `(plus-p x w z)`. In other words, letting `w` be the product of `x` and `y`, the product `z` of `x` and `(S y)` is `(x + w)`, that is, `(x + (x * y))`, which is precisely how an equational formulation would express this constraint. As before, however, we can not only compute products with these clauses, but quotients too! For example, the following computes 12 divided by 4:

```
> (solve (times-p 4 ?X 12) num-clauses)

List: [[?X --> 3]]
```

B.3 Implementing a Prolog interpreter

A nondeterministic algorithm $\mathcal{A}$ for answering a list of queries $L = [q_1 \cdots q_n]$ (or "solving the goals $q_1, \ldots, q_n$") with respect to a definite logic program $\mathcal{P}$ can be formulated as follows. The algorithm takes as input not just the list of goals L and the program $\mathcal{P}$, but also a substitution θ. The initial value of θ will be the empty substitution $\{\}$. As $\mathcal{A}$ progresses, this substitution will become increasingly specialized. The algorithm, which will either return a substitution or else fail, is simple:

1. If L is empty ($n = 0$), return θ.
2. Otherwise let θ' be the substitution obtained by executing q_1 against $\mathcal{P}$ and with respect to θ, and apply the algorithm recursively to $[(\theta' \; q_2) \; \cdots \; (\theta' \; q_n)]$, $\mathcal{P}$, and θ'.

The only undefined operation here is that of "executing a query q against $\mathcal{P}$ and with respect to θ," an operation that is supposed to return a substitution θ'. This is actually where the nondeterminism comes in. We perform this operation by first choosing a (freshly renamed) rule R in $\mathcal{P}$ whose head unifies with $(\theta \; q)$ under some substitution σ (if there is no such rule we raise an exception and the whole algorithm halts in failure); and then recursively applying $\mathcal{A}$ to (a) the list of atoms comprising the body of R; (b) $\mathcal{P}$, and (c) the composition of σ and θ. If R is a fact, then this second step can be omitted and we can simply return the composition of σ and θ. Note that freshly renaming the rules (alpha-renaming them, to be more precise) is important. Without such renaming we could get incorrect results.

There are two sources of nondeterminism here: the choice of R and the choice of how to list the subgoals (atoms) in the body of R. By somehow making all the "right choices" along the way, this algorithm is guaranteed to arrive at a solution if one exists. In other words this algorithm is both *sound*, meaning that if it produces a substitution θ then all θ-instances of the input queries are logical consequences of $\mathcal{P}$; and *complete*, meaning that if there exists such a substitution at all, the algorithm will return one. Unfortunately, deterministic versions such as the one we are about to formulate (the same algorithm used by Prolog systems) are incomplete, as satisfying substitutions might only exist in parts of the search space that will never be explored by a fixed search strategy.

To arrive at a deterministic algorithm, we arrange all possible computations of $\mathcal{A}$ in one big search tree, called an *SLD tree*, and fix a strategy for exploring that tree. The nodes of the tree will be pairs (L,θ) consisting of a list of goals (atoms) $L = [q_1 \cdots q_n]$ and a substitution θ.

Given a pair (L,θ), the SLD tree for (L,θ) is built recursively: The root of the tree is (L,θ), and its children are defined as follows. If L is empty then there are no children—the node is a leaf. Otherwise L is of the form $[q_1 \cdots q_n]$ for $n > 0$. Let $R_1,\ldots,R_k$ be all and only those (freshly renamed) rules in $\mathcal{P}$ whose head unifies with q_1 under some substitution θ_i, where $R_1,\ldots,R_k$ are listed according to the order in which they appear in $\mathcal{P}$.[7] (If $k = 0$, i.e., if there are no such matching rules, then the node is again a leaf—there are no children.) For $i = 1,\ldots,k$, let σ_i be the composition of θ_i with θ. Then the node (L,θ) has exactly k children, namely

$$(body(R_1,\sigma_1,L),\sigma_1),\ldots,(body(R_k,\sigma_k,L),\sigma_k),$$

in that order, with $body(R_i,\sigma_i,L)$ defined as

$$[(\sigma_i\ A_1^i),\ldots,(\sigma_i\ A_{m_i}^i),(\sigma_i\ q_2),\ldots,(\sigma_i\ q_n)],$$

where $A_1^i,\ldots,A_{m_i}^i$ are the atoms that appear in the body (antecedent) of R_i, in the same left-to-right order in which they appear in R_i. We then proceed in the same fashion to construct the children of each of these children, and so on. This process need not terminate if the tree is infinite.

The `solve-all` algorithm given below works by building/exploring the SLD tree for the pair consisting of the input list of queries and the empty substitution in a depth-first manner, maintaining at all times what is essentially a stack of nodes to explore,

$$[[L_1\ \theta_1]\ \cdots\ [L_m\ \theta_m]],$$

and starting off with the one-element stack

$$[[L\ \{\}]],$$

where L is the input list of queries. The main loop, implemented by the recursive procedure `search`, keeps popping the leftmost pair off the stack, building its children, and pushing these back on the stack. Note that the stack is encoded as the (only) argument of `search`.

7 Recall that $\mathcal{P}$ is a list, not a set.

By using higher-order procedures—essentially continuations—this algorithm not only returns a substitution if there is one, but also returns a *thunk*, that is, a nullary procedure that can be invoked later to obtain yet more satisfying substitutions, if any exist. That thunk will continue exploring the SLD tree at the point where the procedure left off last time, namely, at the point where the last satisfying substitution was found. Calling that thunk will again result in a pair consisting of a new substitution and yet another thunk, and so on. When there are no more substitutions to be found, a thunk will return the empty list [] rather than a substitution-thunk pair. In this way it is possible to eventually obtain *every* satisfying substitution that is reachable at all by a depth-first strategy, even if there are infinitely many such substitutions.

```
module Prolog_Interpreter {

define (restrict sub L) :=
  let {V := (vars* L)}
    (make-sub (list-zip V (sub V)))

define (conclusion-of P) :=
   match P {
    (forall (some-list _) (_ ==> p)) => p
  | (forall (some-list _) p) => p
  }

define (match-conclusion-of p goal) :=
  let {q := (rename p)}
    match (goal unified-with conclusion-of q) {
      (some-sub sub) => [q sub]
    | _ => false
    }

define (get-matches goal db) :=
  (filter (map lambda (rule) (match-conclusion-of rule goal) db)
          (unequal-to false))

define (subgoals-of clause) :=
  match clause {
    (forall (some-list _) (p ==> _)) => (get-conjuncts p)
  | _ => []
  }

define (make-new-goals previous-sub previous-goals) :=
  lambda (pair)
    match pair {
      [clause sub] =>
        let {new-goals := (sub (join (subgoals-of clause) previous-goals))}
          [new-goals (compose-subs sub previous-sub)]
    }

define (solve-all queries db) :=
```

```
letrec {search := lambda (L)
                   match L {
                     [] => []
                   | (list-of [[] sub] rest) =>
                       [(restrict sub queries) lambda () (search rest)]
                   | (list-of [(list-of goal more-goals) sub] rest) =>
                       match (get-matches goal db) {
                         [] => (search rest)
                       | rule-matches =>
                           let {L' := (map (make-new-goals sub more-goals)
                                           rule-matches)}
                             (search (join L' rest))
                       }}}
  (search [[queries empty-sub]])

} # module Prolog_interpreter
```

Given solve-all, it is easy to implement a solve-N procedure that will attempt to retrieve N solutions, if N solutions exist, otherwise producing the maximum possible number $K < N$ of solutions:

```
1  define (solve-N queries program N) :=
2    let {queries := match queries {
3                      (some-sent p) => [p]
4                    | _ => queries}}
5      letrec {loop :=
6                lambda (i res thunk)
7                  check {(greater? i N) => (rev res)
8                        | else => match (thunk) {
9                                    [] => (rev res)
10                                 | [sub thunk'] => (loop (i plus 1)
11                                                          (add sub res)
12                                                          thunk')
13                                 }
14                 }
15             }
16       match (solve-with-thunks queries program) {
17         [] => []
18       | [sub rest] => (loop 2 [sub] rest)
19       }
```

Note that the code on lines 2–4 ensures that the procedure works when the first input argument is either a list of atoms or a single atom by itself. We can likewise implement a solve procedure that either returns a single substitution, if one can be found, or else returns false:

```
define (solve queries program) :=
  match (solve-N queries program 1) {
    [] => false
  | (list-of sub _) => sub
  }
```

Both solve and solve-N are parts of the Prolog_Interpreter module. Let us test the system on our earlier example:

```
domain D
declare a: D
declare f: [D] -> D
declare P: [D] -> Boolean
define clauses := (close [(P a)
                          (P f x <== P x)])

open Prolog_Interpreter

> (solve (P a) clauses)

Substitution: {}

> (solve (P f ?X) clauses)

Substitution: {?X:D --> a}

> (solve-N (P f ?X) clauses 2)

List: [{?X:D --> a}
       {?X:D --> (f a)}]

> (solve-N (P f ?X) clauses 10)

List: [{?X:D --> a}
       {?X:D --> (f a)}
       {?X:D --> (f (f a))}
       {?X:D --> (f (f (f a)))}
       {?X:D --> (f (f (f (f a))))}
       {?X:D --> (f (f (f (f (f a)))))}
       {?X:D --> (f (f (f (f (f (f a))))))}
       {?X:D --> (f (f (f (f (f (f (f a)))))))}
       {?X:D --> (f (f (f (f (f (f (f (f a))))))))}
       {?X:D --> (f (f (f (f (f (f (f (f (f a)))))))))}]
```

It should be clear by now why rule ordering (as well as the ordering of rule bodies) is important. Rules are tried in the textual order in which they are listed in a program (list of clauses), so a recursive rule might send our depth-first strategy into an infinite loop if it is listed before other rules for the same predicate. For example:

```
define clauses' := (close [(P f x <== P x)
                           (P a)])

> (solve (P f ?X) clauses')

      ... # infinite loop...
```

B.4 Integration with external Prolog systems

The Prolog interpreter we have presented is adequate for small problems and useful for instructive purposes, but it has two serious drawbacks. First, it doesn't handle negation. Second, and more important, the interpretation layer incurs a severe efficiency penalty. The performance of this interpreter on realistic problems would be unacceptable, as would indeed be the performance of any Prolog interpreter.

The first issue, negation, is not too serious; it would not be difficult to modify our interpreter so as to implement finite negation as failure. But efficiency is a more pressing concern. To get Prolog search to run fast, Prolog must be compiled into machine code.

For these reasons, Athena is integrated with two state-of-the-art Prolog systems, SWI Prolog and XSB, that can execute Prolog queries much more efficiently. The integration is seamless in that both queries and programs are expressed in Athena notation, as is the output. The integration with XSB is tighter, in that XSB can be started when Athena starts and thereafter the two programs will communicate at the process level. This has some efficiency advantages, as new knowledge (facts or rules) can be added incrementally. XSB is also a *tabled* Prolog system, which enables a whole class of interesting applications that are not otherwise possible. However, integration with XSB is only available on Linux, and not by default, so in what follows we only describe the connection to SWI Prolog. That connection works on a per-query basis: To solve a given query (or list of queries) with respect to a given list of clauses, Athena prepares an equivalent Prolog program and query, invokes SWI Prolog from scratch on these inputs, and then translates the output back into Athena notation.[8] Because this approach involves some file IO and starting the `swipl` process from scratch, there is a fixed runtime cost associated with it.

The main API is captured by three procedures in the module `Prolog`:

1. `(solve` *goal program*`)`;
2. `(solve-all` *goal program*`)`;
3. `(solve-N` *goal program N*`)`.

We describe each in turn.

In a call of the form `(solve` *goal program*`)`, *goal* is either a single query (atom) or a list of queries, and *program* is a list of (general) Horn clauses. Disjunctions of literals may also appear in the bodies of these clauses. A clause may be explicitly universally quantified, but need not be. If it is not quantified, it is tacitly assumed that all variables in it are universally quantified. In fact, *program* may contain completely arbitrary sentences, for instance, one may give the entire assumption base as the value of *program*, with a call of the

8 SWI Prolog must be locally installed on your machine and must be runnable from the command line with the command `swipl`.

form (solve *goal* (ab)). The implementation of Prolog will automatically try to extract general Horn clauses from the given sentences, and will simply ignore those sentences that it cannot convert into Horn clauses.

The output is always a two-element list whose first element is true or false, respectively indicating success or failure. When the first element of the output list is true, the second element is a substitution, which, when applied to the query goals, results in sentences that follow logically from the clauses. Some examples:

```
domain D
declare a,b,c: D
declare P: [D] -> Boolean
declare R: [D D] -> Boolean

define program := [(P a) (P b) (a R b) (b R c)]

> (Prolog.solve (P a) program)

List: [true {}]

> (Prolog.solve (P ?X) program)

List: [true {?X:D --> a}]

> (Prolog.solve (R ?X ?Y) program)

List: [true
       {?Y:D --> c
        ?X:D --> b}]

> (Prolog.solve (R ?X ?X) program)

List: [false {}]
```

The arguments of solve-all are the same as those of solve, but the result is a *list of all* satisfying substitutions (assuming there is a finite number of them). Failure here is signified by an empty output list, which means that no substitutions could satisfy the goal(s).

```
> (Prolog.solve-all (P ?X) program)

List: [{?X:D --> a} {?X:D --> b}]

> (Prolog.solve-all (R ?X ?X) program)

List: []
```

Note that the goal(s) given to solve-all must contain at least one variable. If the goal(s) are all ground, use solve instead.

Finally, solve-N takes an extra argument at the end, a positive integer *N* indicating the number of desired solutions. However, this is implemented by obtaining one substitution, then calling Prolog again with an explicit added constraint that rules out the first substitution; then calling Prolog again with yet another constraint that rules out the first two substitutions; and so on. This is not particularly efficient. It may be more expedient to compute a list of all solutions (using solve-all) and then simply take *N* elements from that list. However, if there are infinitely many solutions discoverable by depth-first search, then solve-all will diverge, so in that case solve-N is a more viable alternative.

There are timed versions of the above three procedures that can take a nonnegative integer *M* as a last argument, which will then force Prolog to quit after *S* seconds have gone by, if it is not done by then:

1. Prolog.solve-with-time-limit;
2. Prolog.solve-all-with-time-limit; and
3. Prolog.solve-N-with-time-limit.

(For the last procedure, the time limit applies to each of the *N* separate calls, so if you want all solving to last no more than *M* seconds, give a time limit of *M* divided by *N*.)

The equality symbol, when it appears in clauses, has the usual Prolog semantics (unifiability). There are a number of predefined Prolog symbols introduced in the Prolog module that can also be used in queries:

- The binary polymorphic symbols Prolog.== and Prolog.\== represent strict term identity and inequality, respectively.
- The binary symbols Prolog.arith-eq and Prolog.arith-uneq represent numeric Prolog identity and inequality; they are polymorphic in both argument positions.
- The ternary polymorphic symbols setof, findall, and bagof represent the corresponding Prolog symbols. The results are collected into Athena lists. For example:

```
> (Prolog.solve (Prolog.findall ?X (P ?X) ?Result) program)

List: [true
       {?Result:(List D) --> (a :: b :: nil:(List D))
        ?X:D --> ?v53752:D}]
```

  As in Prolog, the first argument can be an arbitrarily complex term containing variables to be solved for:

```
> (Prolog.solve (Prolog.findall (?X :: ?Y :: nil)
                                (?X R ?Y)
                                ?Result)
                program)

List: [true
```

```
{?Result:(List (List D)) --> (:: (:: a (:: b nil:(List D)))
                                 (:: (:: b (:: c nil:(List D)))
                                     nil:(List (List D))))
          ?Y:D --> ?v1663:D
          ?X:D --> ?v1662:D}]
```

Here the results were assembled into a list of two lists, essentially the pairs `[a b]` and `[b c]`, that is, all and only the pairs of ground terms for which `R` holds according to the given program.[9]

- The `Prolog` module also introduces the symbols `cut`, `fail`, `call`, and `once`, whose use should be clear to experienced Prolog programmers. They can be used to provide greater control over the Prolog search. The `call` predicate is particularly useful for metaprogramming.
- The `Prolog.write` symbol, with signature `[Ide] -> Boolean`, can be used to perform output from inside Prolog. This can be useful for debugging purposes.

We stress that the inputs and outputs of these procedures are given entirely at the Athena level. Thus, for instance, syntactic idiosyncrasies of Prolog are immaterial: Any legal Athena variables and function symbols may appear inside clauses and queries, no matter how complicated, even if they are not legal Prolog syntax. To further facilitate output, one can specify "term transformers" that are automatically applied to resulting substitutions in order to print their content in a more customized (and presumably more readable) format. This is done by invoking the unary procedure `Prolog.add-transformer` as follows:

`(Prolog.add-transformer` *f*`)`,

where *f* is a unary procedure that takes a term and produces a term. For instance, to print natural numbers as integer numerals we may say:

```
(Prolog.add-transformer nat->int).
```

Multiple transformers can be specified in this way, and all of them will be applied to any substitution returned by any of the search procedures we have discussed so far.[10] Because a transformer may change the sort of a term, if there are any transformers specified then the returned results are no longer substitutions but rather lists of triples of the form

`[`*v* `-->` *t*`]`,

where *v* is a query variable and *t* is a transformed term.

9 More precisely, according to the completion of the given program.

10 However, the order in which these are applied is not guaranteed. If that is important then it might be better to specify one single transformer that performs all the necessary conversions in the desired order.

Moreover, as mentioned earlier, the "program" that gets passed to these procedures may contain arbitrary Athena sentences. Athena will try to extract appropriate Horn clauses from any sentences given as clauses.

B.5 Automating the handling of equations

The procedures described in the previous section do not properly address equations, simple or conditional. Consider, for instance, the defining equations for addition on natural numbers:

```
assert* Plus-def := [(x + zero = x)
                     (x + S y = S (x + y))]
```

These are universally quantified identities, so the naive approach of these procedures will translate them as facts, with the identity symbol = as the main predicate. But to properly capture the semantics of such equations in the setting of logic programming, we need new predicate symbols and new clauses for those symbols, and some of these clauses (the ones that correspond to recursive equations) will be proper rules rather than facts. In this particular case, for example, we would need to introduce a new ternary relation symbol `plus-p`, as we did earlier, and introduce new clauses that capture the computational content of the above equations:

```
define plus-clauses :=
  (close [(plus-p x zero x)
          ((plus-p x (S y) (S z)) <== (plus-p x y z))])
```

As the equations get more complex, this can get cumbersome. Accordingly, the `Prolog` module provides a unary procedure, `auto-solve`, that attempts to automate this process. This procedure is unary so it only takes a goal, not a program; it will attempt to automatically construct the right logic program for the given goal, by examining the assumption base and the form of the goal. Specifically, the procedure takes a goal like

```
(? N.+ 3 = ?Result)
```

and automatically tries to:

1. Determine the (transitive closure of all the) defining equations of all function symbols that appear in the goal.
2. Introduce predicate symbols for all these symbols. Typically, the predicate symbol introduced for a function symbol *f* is *f*_p.[11]
3. Translate all defining equations and conditional equations for all of these symbols into general Prolog clauses.

11 That can differ if there is an already existing symbol by that name.

4. Include any other (nonequational) clauses that may be necessary for some of these symbols.
5. Transform the query itself into relational form.
6. Solve the transformed query with respect to the Prolog program produced the previous steps.
7. Output the results.

For example, assuming we have defined `N.+` and `N.*` with the usual recursive equations:

```
define goal := (2 N.+ 3 = ?sum-of-2-and-3)

> (Prolog.auto-solve goal)

List: [true {?sum-of-2-and-3:N --> (S (S (S (S (S zero)))))}]
```

If we wish to display natural numbers as integer numerals:

```
(Prolog.add-transformer nat->int)

> (Prolog.auto-solve goal)

List: [true [[?sum-of-2-and-3 --> 5]]]
```

There is also a unary `auto-solve-all` and a binary `auto-solve-N` procedure, whose second argument is the desired number of solutions:

```
define goal := (2 N.* ?X = ?Y)

> (Prolog.auto-solve-N goal 3)

List: [[[?Y --> 0] [?X --> 0]]
       [[?Y --> 2] [?X --> 1]]
       [[?Y --> 4] [?X --> 2]]]
```

We can inspect the functional-to-relational transformation by asking to see the constructed clausal program for a given goal by using the procedure `Prolog.defining-clauses`:

```
define goal := (2 N.* ?X = ?Y)

> (Prolog.defining-clauses goal)

List: [
(n_+_p ?n zero ?n)

(if (n_+_p ?n ?m ?v784)
    (n_+_p ?n
           (S ?m)
           (S ?v784)))
```

```
(n_*_p ?x zero zero)

(if (and (n_*_p ?x ?y ?v785)
         (n_+_p ?v785 ?x ?v786))
    (n_*_p ?x
           (S ?y)
           ?v786))
]
```

Here, n_+_p and n_*_p are the ternary predicate symbols that were automatically introduced for N.+ and N.*, respectively.

As we have seen, of course, order is important, both for clauses and for body literals. The implementation of auto-solve and its two sibling procedures applies a few simple heuristics in order to determine the best possible ordering for clauses and goal literals, but the heuristics are not perfect. If these procedures fail to produce the expected results, it might be useful to look at the clause list produced for a given goal, using Prolog.defining-clauses, possibly make some changes to it, and then apply Prolog.solve (or solve-all, etc.) to the given goal and the modified clause list.

We can do the same thing for a goal; that is, we can inspect the clauses that are automatically produced for it (via auto-solve, etc.), with the unary procedure Prolog.query-clauses:

```
define goal := (2 N.* ?X = ?Y)

> (Prolog.query-clauses goal)

List: [
(n_*_p (S (S zero))
       ?X
       ?v785)

(= ?v785 ?Y)
]
```

Again, by applying these two procedures to a given goal, we can obtain and possibly rearrange clause lists for both the extracted program and the goal itself, and then use Prolog.solve, etc., to solve with respect to both.

References

[1] H. Abelson and G. J. Sussman. *Structure and Interpretation of Computer Programs*. MIT Press, Cambridge MA, 1985.

[2] J. Alama, T. Heskes, D. Kühlwein, E. Tsivtsivadze, and J. Urban. Premise selection for mathematics by corpus analysis and kernel methods. *J. Automated Reasoning*, 52(2):191–213, 2014.

[3] Aristotle. *The Basic Works of Aristotle*. Modern Library, 2001. Edited by R. McCeon.

[4] K. Arkoudas. Simplifying proofs in Fitch-style natural deduction systems. *J. Automated Reasoning*, 34(3):239–294, 2005.

[5] K. Arkoudas, C. Chadha, and J. Chiang. Sophisticated access control via SMT and logical frameworks. *ACM Transactions on Information Systems Security*, 16(4):17:1–17:31, April 2014.

[6] K. Arkoudas, R. Chadha, and J. C. Chiang. An application of formal methods to cognitive radios. In *Proceedings of DIFTS 2011 (Design and Implementation of Formal Tools and Systems), a workshop of FMCAD 2011 (Formal Methods in Computer Aided Design)*, pages 3–13, Austin TX, 2011.

[7] K. Arkoudas, R. Chadha, and J. C. Chiang. An efficient policy engine for dynamic spectrum access. In *Proceedings of CogArt 2011, the Fourth International Conference on Cognitive Radios*, Barcelona, Spain, October 2011.

[8] E. W. Beth. Semantic entailment and formal derivability. *Mededlingen der Koninklijke Nederlandse Akademie van Wetenschappen*, 18(13):309–342, 1955.

[9] A. Biere, M. Heule, H. van Maaren, and T. Walsh. *Handbook of Satisfiability: Volume 185 of Frontiers in Artificial Intelligence and Applications*. IOS Press, Amsterdam, The Netherlands, 2009.

[10] G. Birkhoff. On the structure of abstract algebras. *Mathematical Proceedings of the Cambridge Philosophical Society*, 31:433–454, Oct 1935.

[11] G. Boole. *An Investigation of the Laws of Thought*. Dover, New York, 1958.

[12] A. R. Bradley and Z. Manna. *The Calculus of Computation: Decision Procedures with Applications to Verification*. Springer-Verlag, Secaucus, NJ, USA, 2007.

[13] M. D. S. Braine and D. P. O'Brien. The theory of mental-propositional logic: Description and illustration. In *Mental Logic*, pages 79–89. Lawrence Erlbaum Associates, 1998.

[14] I. Bratko. *PROLOG Programming for Artificial Intelligence*. Addison-Wesley Longman, Boston MA, 2nd edition, 1990.

[15] L. Burkholder. The halting problem. *ACM SIGACT News*, 18(3):48–60, 1987.

[16] R. M. Burstall. Proving properties of programs by structural induction. *Computer Journal*, 1969.

[17] J. Caldwell. Structural induction principles for functional programmers. In *Proceedings Second Workshop on Trends in Functional Programming In Education, TFPIE 2013, Provo, UT, May 2013*, pages 16–26, 2013.

[18] W. F. Clocksin and C. S. Mellish. *Programming in Prolog*. Springer Verlag, New York, 4th edition, 1994.

[19] S. A. Cook. The complexity of theorem-proving procedures. In *Proceedings of the Third Annual ACM Symposium on Theory of Computing*, STOC '71, pages 151–158, 1971.

[20] I. M. Copi. *Symbolic Logic*. Macmillan Publishing Co., New York, 5th edition, 1979 [1954].

[21] J. F. Couchot and S. Lescuyer. Handling polymorphism in automated deduction. In *Proceedings of the 21st International Conference on Automated Deduction (CADE-2007)*, volume 4603 of *Lecture Notes in Computer Science*, pages 263–278. Springer, 2007.

[22] J. Cussens. Bayesian network learning by compiling to weighted MAX-SAT. In *UAI 2008, Proceedings of the 24th Conference in Uncertainty in Artificial Intelligence, Helsinki, Finland, July 9–12, 2008*, pages 105–112, 2008.

[23] A. Darwiche. *Modeling and Reasoning with Bayesian Networks*. Cambridge University Press, New York, 2009.

[24] M. Davis. Mathematical logic and the origin of modern computers. In R. Herken, editor, *The Universal Turing Machine: A Half-Century Survey*, pages 135–158. Oxford University Press, 2nd edition, 1995.

[25] M. Davis. SAT: Past and future. In *Proceedings of the Tenth International Conference on Theory and Applications of Satisfiability Testing (SAT 2007)*, volume 4501 of *Lecture Notes in Computer Science (LNCS)*, pages 1–2. Springer, 2007.

[26] M. D. Davis, R. Sigal, and E. J. Weyuker. *Computability, Complexity, and Languages*. Morgan Kaufmann, San Diego CA, 2nd edition, 1994.

[27] T. B. de la Tour. Minimizing the number of clauses by renaming. In *Proceedings of the Tenth International Conference on Automated Deduction (CADE 1990)*, volume 449 of *Lecture Notes in Artificial Intelligence (LNAI)*, pages 558–572. Springer, 1990.

[28] L. de Moura and N. Bjørner. Satisfiability modulo theories: Introduction and applications. *Communications of the ACM*, 54(9):69–77, September 2011.

[29] L. de Moura, B. Dutertre, and N. Shankar. A tutorial on satisfiability modulo theories. In *Computer Aided Verification*, volume 4590 of *LNCS*, pages 20–36. Springer, 2007.

[30] N. Dershowitz and J.-P. Jouannaud. Rewrite systems. In J. van Leeuwen, editor, *Handbook of Theoretical Computer Science*, volume B, pages 243–320. Elsevier, Amsterdam, 1990.

[31] R. Diaconescu, K. Futatsugi, and K. Ogata. CafeOBJ: Logical foundations and methodologies. *Computers and Artificial Intelligence*, 22(3-4):257–283, 2003.

[32] J. A. Díez. A hundred years of numbers: A historical introduction to measurement theory (1887–1990). *Studies in History and Philosophy of Science*, 28(1):167–185, 1997.

[33] A. Diller. *Z: An Introduction to Formal Methods*. Wiley & Sons, 1994.

[34] H.-D. Ebbinghaus, J. Flum, and W. Thomas. *Mathematical Logic*. Springer-Verlag, Berlin, 2nd edition, 1994.

[35] M. Felleisen and D. Friedman. *The Little MLer*. MIT Press, Cambridge MA, 1997.

[36] F. B. Fitch. *Symbolic Logic: An Introduction*. The Ronald Press Co., New York, 1952.

[37] R. W. Floyd. Assigning meanings to programs. In *Proceedings of American Mathematical Society Symposia in Applied Mathematics*, volume 19, pages 19–31, 1967.

[38] A. A. Fraenkel, Y. Bar-Hillel, and A. Levy. *Foundations of Set Theory*. North-Holland, Amsterdam, 2nd edition, 1973.

[39] G. Frege. *The Foundations of Arithmetic: A Logico-Mathematical Enquiry into the Concept of Number*. Northwestern University Press, Evanston IL, 1968 [1884]. Translated by J. L. Austin.

[40] D. P. Friedman and M. Felleisen. *The Little Schemer*. MIT Press, Cambridge MA, 4th edition, 1996.

[41] K. Futatsugi, J. A. Goguen, and K. Ogata. Verifying design with proof scores. In *Verified Software: Theories, Tools, Experiments, First IFIP TC 2/WG 2.3 Conference, VSTTE 2005, Zurich, Switzerland, October 10-13, 2005, Revised Selected Papers and Discussions*, pages 277–290, 2005.

[42] G. Gentzen. *The Collected Papers of Gerhard Gentzen*. North-Holland, Amsterdam, The Netherlands, 1969. English translations of Gentzen's papers, edited and introduced by M. E. Szabo.

[43] G. P. Goodwin and P. N. Johnson-Laird. Transitive and pseudo-transitive inferences. *Cognition*, 108:320–352, 2008.

[44] J. Y. Halpern, R. Harper, N. Immerman, P. G. Kolaitis, M. Y. Vardi, and V. Vianu. On the unusual effectiveness of logic in computer science. *Bulletin of Symbolic Logic*, 7:213–236, 6 2001.

[45] L. Henkin. The completeness of the first-order functional calculus. *J. Symbolic Logic*, 14(3):159–166, 09 1949.

[46] P. M. Hill and J. W. Lloyd. *The Gödel Programming Language*. MIT Press, Cambridge MA, 1994.

[47] C. A. R. Hoare. An axiomatic basis for computer programming. *Acta Informatica*, 1:271–281, 1972.

[48] K. Hoder and A. Voronkov. Sine qua non for large theory reasoning. In N. Björner and V. Sofronie-Stokkermans, editors, *Automated Deduction, CADE-23*, volume 6803 of *Lecture Notes in Computer Science*, pages 299–314. Springer Berlin Heidelberg, 2011.

[49] A. D. Irvine and H. Deutsch. Russell's paradox. In Edward N. Zalta, editor, *The Stanford Encyclopedia of Philosophy*. Winter 2014 edition, 2014.

[50] J. Jacky. *The Way of Z: Practical Programming with Formal Methods*. Cambridge University Press, New York, 1996.

[51] S. Jáskowski. On the rules of suppositions in formal logic. *Studia Logica*, 1, 1934.

[52] T. Jech. *Set Theory: The Third Millennium Edition*. Springer Monographs in Mathematics. Springer Berlin Heidelberg, 3rd edition, 2006.

[53] P. N. Johnson-Laird and F. Savary. Illusory inferences: a novel class of erroneous deductions. *Cognition*, 71:191–229, 1999.

[54] G. Kahn. Natural semantics. In *Proceedings of Theoretical Aspects of Computer Science*, Passau, Germany, February 1987.

[55] D. Kalish and R. Montague. *Logic: Techniques of Formal Reasoning*. Harcourt Brace Jovanovich, New York, 1964. Second edition in 1980, with G. Mar.

[56] S. C. Kleene. *Mathematical Logic*. John Wiley & Sons, 1967.

[57] D. Knuth. *Seminumerical Algorithms*, volume 2 of *The Art Of Computer Programming*. Addison Wesley, Reading MA, 1969.

[58] D. Knuth. *Fundamental Algorithms*, volume 1 of *The Art Of Computer Programming*. Addison Wesley, Reading MA, 2nd edition, 1973.

[59] D. E. Knuth and P. B. Bendix. Simple word problems in universal algebras. In J. Leech, editor, *Computational Problems in Abstract Algebra*, pages 266–297. Pergamon Press, Oxford, 1970.

[60] R. Kowalski. *Logic for Problem Solving*. North-Holland, Amsterdam, 1979.

[61] D. Kroening and O. Strichman. *Decision Procedures: An Algorithmic Point of View*. Springer, 1st edition, 2008.

[62] K. Kuratowski. Sur la notion de l'ordre dans la théorie des ensembles. *Fundamenta Mathematica*, 2(1):161–171, 1921.

[63] Peter J. Landin. The mechanical evaluation of expressions. *Computer Journal*, 6(4):308–320, January 1964.

[64] E. J. Lemmon. *Beginning Logic*. Hackett Publishing Company, 1978 [1965].

[65] J. W. Lloyd. *Foundations of Logic Programming*. Springer Verlag, Berlin, Germany, 1984.

[66] Z. Manna and R. Waldinger. *The Logical Basis for Computer Programming*. Addison-Wesley, Reading MA, 1985.

[67] M. Manzano. *Extensions of First-Order Logic*. Cambridge Tracts in Theoretical Computer Science. Cambridge University Press, 1996.

[68] V. W. Marek. *Introduction to Mathematics of Satisfiability*. Chapman & Hall/CRC Press, New York, 2009.

[69] D. A. McAllester and K. Arkoudas. Walther recursion. In *Proceedings of the 13th Conference on Automated Deduction (CADE 13)*, pages 643–657. Springer, 1996.

[70] John McCarthy. Towards a mathematical science of computation. In C. M. Popplewell, editor, *Proceedings of the 1962 International Congress on Information Processing*, pages 21–28. North Holland, Amsterdam, 1963.

[71] K. Meinke and J. V. Tucker. Universal algebra. In S. Abramsky and T. S. E. Maibaum, editors, *Handbook of Logic in Computer Science (Vol. 1)*, pages 189–368. Oxford University Press, Inc., New York, 1992.

[72] K. Meinke and J. V. Tucker. *Many-sorted Logic and Its Applications*. Wiley Professional Computing. Wiley, 1993.

[73] E. Mendelson. *Introduction to Mathematical Logic*. Chapman & Hall, New York, 4th edition, 1994 [1964].

[74] Robin Milner. A theory of type polymorphism in programming. *J. Computer and System Sciences*, 17:348–375, 1978.

[75] D. R. Musser. On proving inductive properties of abstract data types. In *Proceedings of the 7th ACM SIGPLAN-SIGACT Symposium on Principles of Programming Languages*, POPL '80, pages 154–162, 1980.

[76] D. R. Musser, G. J. Derge, and A. Saini. *STL Tutorial and Reference Guide: C++ Programming with the Standard Template Library*. Addison-Wesley, Reading MA, 2001.

[77] H. R. Nielson and F. Nielson. *Semantics with Applications*. Wiley Professional Computing, New York, 1992. Available from `http://www.daimi.au.dk/~bra8130/Wiley_book/wiley.html`.

[78] U. Nillson and J. Maluszynski. *Logic, Programming and Prolog*. John Wiley and Sons, New York, 2nd edition, 1995.

[79] N. J. Nilsson. *Artificial Intelligence: A New Synthesis*. Morgan Kaufmann, San Francisco, 1998.

[80] F. J. Pelletier. Seventy-five problems for testing automatic theorem provers. *J. Automated Reasoning*, 2(2):191–216, 1986.

[81] D. A. Plaisted and S. Greenbaum. A structure-preserving clause form translation. *J. Symbolic Computation*, 2(3):293–304, September 1986.

[82] G. D. Plotkin. A structural approach to operational semantics. Research Report DAIMI FN-19, Computer Science Department, Aarhus University, Aarhus, Denmark, September 1981.

[83] F. D. Portoraro. Strategic construction of Fitch-style proofs. *Studia Logica*, 60:45–66, 1998.

[84] L. J. Rips. *The Psychology of Proof*. MIT Press, Cambridge MA, 1994.

[85] A. Robinson and A. Voronkov, editors. *Handbook of Automated Reasoning*. Elsevier Science Publishers B. V., Amsterdam, The Netherlands, 2001.

[86] D. A. Schmidt. *Denotational Semantics*. Allyn and Bacon, Boston MA, 1986.

[87] R. Shostak. On the sup-inf method for proving Presburger formulas. *J. ACM*, 24(4):529–543, October 1977.

[88] R. Shostak. Deciding linear inequalities by computing loop residues. *J. ACM*, 28(4):769–779, October 1981.

[89] W. Sieg and J. Byrnes. Normal natural deduction proofs (in classical logic). *Studia Logica*, 60:67–106, 1998.

[90] W. Sieg and R. Scheines. Searching for proofs (in sentential logic). In L. Burkholder, editor, *Philosophy and the Computer*, pages 137–159. Westview Press, Boulder CO, 1992.

[91] R. M. Smullyan. *First-Order Logic*. Dover, New York, 1995 [1968].

[92] K. E. Stanovich. Rational and irrational thought: The thinking that IQ tests miss. *Scientific American*, November–December 2009.

[93] A. Stepanov and P. McJones. *Elements of Programming*. Pearson Education Inc., New York, 2009.

[94] A. Stepanov and D. Rose. *From Mathematics to Generic Programming*. Addison-Wesley, Reading MA, 2015.

[95] L. Sterling and E. Shapiro. *The Art of Prolog*. MIT Press, Cambridge MA, 2nd edition, 1994.

[96] J. E. Stoy. *Denotational Semantics: The Scott-Stratchey Approach to Programming Language Theory*. MIT Press, Cambridge MA, 1981.

[97] O. Strichman, S. A. Seshia, and R. E. Bryant. Deciding separation formulas with SAT. In Ed Brinksma and Kim Guldstrand Larsen, editors, *Computer Aided Verification*, volume 2404 of *Lecture Notes in Computer Science*, pages 209–222. Springer Berlin Heidelberg, 2002.

[98] P. Suppes. A set of independent axioms for extensive quantities. *Portugaliae Mathematica*, 10(4):163–172, 1951.

[99] P. Suppes. *Introduction to Logic*. Van Nostrand, Princeton, NJ, 1957.

[100] P. Suppes. *Axiomatic Set Theory*. Dover Books on Mathematics Series. Dover, New York, 1960.

[101] A. Tarski. Sur les ensembles définissables de nombres réels. *Fundamenta Mathematicae*, 17(1):210–239, 1931.

[102] A. Tarski. *Introduction to Logic and to the Methodology of the Deductive Sciences*. Oxford Logic Guides. Oxford University Press, 1994 [1936].

[103] G. S. Tseitin. On the complexity of derivation in propositional calculus. In A. O. Silenko, editor, *Studies in Constructive Mathematics and Mathematical Logic, part II*, pages 115–125. Consultants Bureau, 1970.

[104] F. A. Turbak and D. K. Gifford. *Design Concepts in Programming Languages*. MIT Press, Cambridge MA, 2008.

[105] J. D. Ullman. *Principles of Database and Knowledge-Base Systems: Volume II: The New Technologies*. W. H. Freeman & Co., New York, 1990.

[106] A. Vargun. Code carrying theory. PhD thesis, Rensselaer Polytechnic Institute, 2006.

[107] A. Vargun. Termination checking without using an ordering relation. In *Proceedings of the 11th IASTED International Conference on Software Engineering and Applications*, SEA '07, pages 130–135, 2007.

[108] P. Wadler. Comprehending monads. *Mathematical Structures in Computer Science*, 2:461–493, 1992.

[109] H. Wang. Logic of many-sorted theories. *J. Symbolic Logic*, 17:105–116, 6 1952.

[110] P. C. Wason and P. N. Johnson-Laird. *Psychology of Reasoning*. Harvard University Press, Cambridge MA, 1972.

[111] C. Weidenbach. Combining superposition, sorts, and splitting. In A. Robinson and A. Voronkov, editors, *Handbook of Automated Reasoning*, volume 2. North-Holland, 2001.

[112] A. N. Whitehead and B. Russell. *Principia Mathematica*, volume 1. Cambridge University Press, 1963 [1910].

[113] E. Wigner. The unreasonable effectiveness of mathematics in the natural sciences. *Communications in Pure and Applied Mathematics*, 13(1), February 1960.

[114] R. Wilhelm and D. Maurer. *Compiler Design*. Addison-Wesley, Reading MA, 1995.

Glossary

abstract algorithm An algorithm expressed as a function on an abstract structure, as described in Chapter 15. Also called a *generic algorithm*. (In some cases we use two functions to express an abstract algorithm, one to return a value and another to model memory updates, as described in Chapter 16.) 1, 14, 599, 625, 627, 683, 685, 709, 710, 754, 761

abstract structure A collection of polymorphic operators (function symbols) and theories (axioms defining the operators and theorems with proofs) that can be specialized to many different structures through adaptation with operator mappings, as described in Chapter 14. In Athena, we express an abstract structure with a module containing declarations of its operators and axioms defining them, along with theory refinement and theory evolutions. 13, 475, 642, 690

abstract-level proof A proof that can be specialized in different ways to prove concrete (or even other abstract-level) theorems, such as a proof about group operations that can be adapted to prove theorems about corresponding integer operations. Also called a generic proof. In this book, abstract-level proofs are encapsulated in a method whose definition follows certain conventions outlined in Section 14.4. 14, 637, 641, 683

alpha-equivalence Two sentences are alpha-equivalent iff each can be obtained from the other by consistently renaming bound variables. Athena treats alpha-equivalent sentences as identical for deductive purposes. 39

assumption base The set of axioms, assumptions, and theorems available at any given time for use in deductions. 21, 42, 911, 915

axiom A sentence of first-order logic that is taken as given (i.e., one that is neither the result of a deduction, nor a temporary hypothesis introduced by conditional reasoning). In Athena, `assert` p causes a sentence p to be treated as an axiom by inserting it into the assumption base. 42, 911

bound variable occurrence An occurrence of a variable x in a sentence p is bound iff it is within a subsentence of p of the form $(Q\,x\ .\ p')$, for some quantifier Q; otherwise it is a free variable occurrence. See also *variable capture*. 39, 913, 918

canonical term A ground term that contains only constructors and possibly numeric literals (integer or real) and/or meta-identifier literals. It represents an irreducibly simple term. 140

complement Complementation is an operation defined on sentences: The complement of a sentence p, denoted by $\overline{p}$, is defined as q if p is a negation of the form `(~ q)`, and as `(~ p)` otherwise (if p is not a negation). 205, 914

constructor A function symbol introduced by the definition of a datatype or structure D. A constructor c with profile $(c\ S_1 \cdots S_n)$ can be understood as a particularly simple function that takes n values $v_1, \ldots, v_n$ of sorts $S_1, \ldots, S_n$ and builds an element of D, $(c\ v_1 \cdots v_n)$. Datatype constructors are injective, every element of the datatype is the in the image of some constructor, and applications of distinct constructors build distinct elements of D. The first and third requirements may be relaxed for structures, which can be viewed as datatypes with a coarser identity relation. 45, 911

continuation-passing style A style of writing methods in which one of the arguments is itself a method K (of arbitrary arity), called a proof continuation and representing "the rest of the proof." The final theorem is whatever result is produced by applying K to interim values. A similar style is possible for procedures. Continuation-passing style for regular computations is a powerful and well-known technique in functional programming. 235

deduction Deductions form one of the two main syntactic categories in Athena, the other being expressions. The syntax of deductions is specified in Appendix A.2. A deduction D represents a proof, that is, a logical argument, and therefore its evaluation can only result in a value of one type: a sentence. That sentence is viewed as the *conclusion* of D and is guaranteed to be a logical consequence of the assumption base in which D was evaluated. Contrast this with expressions, which can freely produce values of any type whatsoever, including arbitrary sentences that are in no way constrained by the assumption base. 912

elimination method A proof method that eliminates the main sentential connective or quantifier from a given premise. An example is Athena's `left-and` primitive method for the `and` connective, which takes a conjunction in the assumption base and detaches its left component. 64

expression Expressions form one of the two main syntactic categories in Athena, the other being deductions. The syntax of expressions is specified in Figure A.1. An expression E represents an arbitrary *computation*; its evaluation can result in a value of any type: a term, a sentence, a vector, a hash table, the unit value, a list, and so on. 912, 915, 916

free algebra An algebra in which constructor terms represent distinct elements unless they are syntactically identical, and there are no values in the algebra other than those generated by the constructors in a finite number of steps. An Athena **datatype** is always interpreted as a free algebra. 46

free variable occurrence An occurrence of a variable x in a sentence p is free iff it is not in a subsentence of p of the form $(Q\ x\ .\ p')$ for some quantifier Q; otherwise it is a bound variable occurrence. See also *variable capture*. 38, 101, 911

free-generation axioms *No confusion* and *no junk* axioms for the constructors of a **datatype**, namely, axioms stating that distinct constructors terms represent distinct elements and that every element is in the image of some constructor. For a datatype named D, a list of these axioms can be produced by the expression `(datatype-axioms "D")`. 46

heterogeneous list A list of values that are not necessarily of the same type. Athena's built-in lists, like those of Scheme or Python, are heterogeneous. Do not confuse Athena's built-in lists, which can hold arbitrary Athena *values* of any *type*, with list *terms* of sort `(List S)`. The latter can only contain terms as members, not arbitrary values, and all of those terms must be of the same sort, S. 19

higher-order language A programming language in which procedures are first-class values that can be passed as arguments to, and returned from, other procedures. Athena's conventional programming language is fully higher-order, and its deductive layer has a method counterpart to procedures that is partially higher-order: A method can accept procedures and methods as arguments, but can only result in a sentence. 19

homomorphic mapping A function f from one domain D to another domain D' such that all relations between elements of D hold for the corresponding (via f) elements of D'. Also called a *homomorphism*. The mapping might involve a single operator (e.g, an additive or multiplicative homomorphism), or all of the operators of an algebraic structure, for instance, a monoid or group homomorphism. D and D' can be the same domain. 16, 481, 628, 651

Horn clause A Horn clause is a universally quantified conditional whose antecedent is a conjunction of one or more atoms and whose consequent is an atom. (By allowing for a sole `true` atom in the antecedent, we can represent facts as Horn clauses.) There are several variations on this theme depending on the context (for example, we may allow for literals in the antecedent, instead of plain atoms only). 885

infix notation Syntax for terms and sentences in which binary operators (function symbols or sentential constructors) are written between their two operands, and an operand appearing between two adjacent operators is associated with the operator with higher precedence (or in case of ties, by a left- or right-associativity declaration). Unary operators also have a precedence that is compared with that of neighboring unary or binary operators. The associations that would be determined by such precedence or associativity rules can always be overridden by use of parentheses. Athena accepts input in infix notation but always outputs results in prefix notation. 24, 916

injective function A function f such that $f(a_1,\ldots,a_n) = f(b_1,\ldots,b_n)$ implies $a_1 = b_1,\ldots,a_n = b_n$. 47

interpretation A specification of the meanings of the primitive syntactic components of a sentence or set of sentences. Once an interpretation is fixed, we can determine whether it satisfies a sentence in accordance with Tarski's recursive definition of the satisfaction relation. In sentential logic, an interpretation is simple: It is an assignment of truth or falsity to each atomic sentence. Interpretations of first-order logic sentences are more complex: Among other things, we must specify the carriers of the various sorts and the functions corresponding to the various symbols. 239, 917

introduction method A proof method that introduces a sentential connective or quantifier, such as Athena's primitive `both` method, which forms the conjunction of two sentences in the assumption base. More generally, we speak of introduction (and elimination) constructs or mechanisms. Most of these are primitive methods, but some (such as **`pick-any`** and **`pick-witness`**) are special syntactic forms. 64

lambda calculus A mathematical model of computation whose extremely simple syntax and semantics belie its usefulness for exploring many essential features of programming languages and computation in general. Its syntax consists only of expressions, including λ abstractions (simple function definitions of the form $(\lambda x \,.\, e)$, where free occurrences of x in expression e are said to be bound to λx in $(\lambda x \,.\, e)$), and whose operational semantics consists of two rules for reduction of an expression to another expression: (1) α-renaming, which simply renames bound variables; and (2) β-reduction, which applies a λ abstraction $(\lambda x \,.\, e_1)$ to an expression e_2 by replacing all free occurrences of x in e_1 by e_2 (after first doing α reduction if necessary to avoid variable capture). Scheme, ML, and other (partially) functional languages derive the core of their syntax and semantics from the λ-calculus, or at least from (possibly typed) variants that allow λ abstractions to have multiple variables and constants in expressions representing externally defined values (including functions). Athena's procedural language follows in this line, and its deductive language shares some aspects. 26, 916

literal A sentence of first-order logic that is either atomic (a `Boolean` term) or else a negation of an atomic sentence. Literals are closed under the complement operation. 249

logic programming A computation paradigm based on the notion of Horn clauses and unification. A program consists of a set of Horn clauses and computation is performed by posing and answering queries, where a query is a list of atoms, possibly containing variables. The output is either a substitution to be applied to these variables that makes the resulting query atoms logical consequences of the program, or else failure, indicating that no instances of the queries follow from the program. Various extensions of this basic paradigm can handle more expressive programs and queries, for example, accommodating negation (both in queries and in Horn clauses) or disjunctions in the conclusions of Horn clauses. 885

meta-identifier A meta-identifier is a term of sort `Ide`. There are infinitely many built-in constant literals of that sort, of the form `'`I for every identifier I. These are also all and only the elements of the carrier of that sort. Two such terms `'`I_1 and `'`I_2 are distinct iff I_1 and I_2 are distinct. Meta-identifiers are useful for encoding symbolic information in a way that can be reasoned about (quantified over, etc.), and also for representing variables of an object language whose syntax is being encoded by a datatype. 61, 911

method An Athena (proof) method is similar to a procedure in conventional programming languages, except that it must always return a sentence (if it doesn't halt in error or loop forever), and is much more carefully circumscribed in its behavior, constrained by the assumption base. A user-defined method must have a deduction D as its body. It performs inferences (which are carried out when D is evaluated) by calling other methods, either primitive methods or other user-defined methods, or by evaluating special syntax forms for deduction, such as **assume** or **pick-any**. If a method application succeeds, it returns a sentence that is treated as a theorem and is typically inserted into the assumption base. 912–914

method call (or application) A method call is a deduction of the form (**apply-method** E $F_1 \cdots F_n$), or equivalently and more commonly, (!E $F_1 \cdots F_n$). Such a call is evaluated by applying whatever method M is denoted by expression E to the values of the phrases $F_1 \cdots F_n$—the *arguments* to the method. M can be either a primitive method or a user-defined one. 62

natural deduction A style of deductive reasoning characterized primarily by the notion of conditional proof, whereby a conditional $p \Rightarrow q$ is derived by assuming p as a hypothesis and proceeding to give a subproof that derives q. That subproof constitutes the scope of the hypothesis p and is visually apparent by the lexical structure of the overall deduction. The scope of one hypothesis can thus be nested inside the scope of an outer hypothesis, and so on. Assumption scope can be captured either by visual devices, for example, by enclosing subproofs inside boxes, as was done by the inventor of the method, Jáskowski, or else by way of block structure as that is understood in contemporary programming languages. The latter is the approach taken in Athena. A secondary characteristic of natural deduction is the use of introduction and elimination inference rules for logical connectives and quantifiers. 315

overloading Layering a new meaning on a unary or binary operator (function symbol, sentential connective, or procedure), in such a way that the context determines which of the old or new meanings is in effect. In Athena, a form of overloading is possible with the top-level directive **`overload`**. The context used to select among two or more meanings of an overloaded operator is usually fixed by the sorts of the operator's arguments, but it can be arbitrarily complex in the general case. 86

precedence Rules of infix notation by which it is determined, for a unary or binary operator in a term or sentence, which are its operands. In Athena, precedence levels are positive integer values, predefined but reassignable for some unary and binary operators, and assignable for newly declared operators within their declaration or by using the **`set-precedence`** directive. An operator's precedence may be queried with the `get-precedence` procedure. Precedence does not matter for terms written entirely in prefix notation or in fully-parenthesized infix notation. 32, 913

prefix notation Concrete syntax in which all operators (function symbols, sentential connectives, procedures, etc.) and their operands are written within parentheses and with the operator placed before the operands. Advantages of prefix notation are that no appeal to precedence or associativity rules is required, and it is easy to automatically indent for output. Athena allows infix notation for input, but always outputs its results in prefix. 24, 914, 916

premise In traditional logic terminology, a premise (also known as *premiss*) is a sentence that is taken as a working assumption for the purposes of a logical argument or a given stretch of discourse. In this book we also use the term "premise" to indicate an argument of a method M that must be in the assumption base when M is called (i.e., an argument whose presence in the assumption base is a precondition of M). 21, 912

procedure In Athena, a procedure is either primitive (built-in) or else it is defined by a lambda abstraction whose body is an arbitrary expression. A non-primitive procedure performs computations by calling other procedures (either primitive ones or other library or user-defined procedures) and/or evaluating built-in syntax forms (such as **`match`** or **`check`**). See also *lambda calculus*. 20

SAT problem The satisfiability (SAT) problem for sentential logic is this: Given a sentence p, determine whether or not p is satisfiable, that is, whether or not there exists an interpretation that makes it true. If there a satisfying interpretation, produce one (or more). The problem can be likewise posed for first-order logic (and indeed for many other logics). For sentential logic this problem is decidable but theoretically intractable, although programs that solve it (so-called "SAT solvers") perform remarkably well in practice. 246

sentence In first-order logic, a syntactic class of expressions built up inductively from terms, quantifiers, and sentential constructors. 19, 911, 914, 918

sort In Athena, a sort is a class of terms. See also *type*. 24

substitution A substitution is a finite function from term variables to terms. Substitutions form a native type of values in Athena. A substitution θ can be applied to any term t or sentence p; these applications are written as if θ were a procedure: (θ t) and (θ p). The application (θ t) produces the term obtained from t by replacing every occurrence of a variable v that is in the domain of θ by whatever term θ assigns to v; while (θ p) is the sentence obtained from p by safely replacing every free occurrence of a variable v that is in the domain of θ by whatever term θ assigns to v, possibly alpha-renaming along the way to avoid variable capture. In conventional notation we write $\theta(t)$ and $\theta(p)$ instead of (θ t) and (θ p). 94

tail recursion A recursive procedure call that occurs last in execution order (as opposed to *embedded recursion*, where a recursive call is followed by additional computation). Important for space and time efficiency, as tail recursion can be replaced by a loop, whereas embedded recursion requires manipulating the run-time stack. 93

tautology A sentence that is true under all interpretations. 196

term Terms are symbolic structures intended to represent elements of a given set. A term is essentially a tree whose every node contains either a *function symbol* or a *variable*. A variable is of the form ?I:S where I is any Athena identifier and S is a *sort*, for instance, `?flag:Boolean` or `?counter:Int`. Terms are regular denotable values in Athena, so we can name them. For instance, we can (and frequently do) define a variable with a directive like `define L := ?L:(List 'S)`. 19, 914

term matching Term matching is the process of determining whether a term s *matches* a term t, which is the case iff there exists a substitution θ whose application to t yields s. The primitive

Athena procedure `match-terms` can be used to determine whether or not a term matches another, and if so, to produce a matching substitution. 96

theorem A sentence that is derived from and entered into the assumption base by way of a deduction D. 42, 911, 915

theory evolution Adding axioms and/or theorem/proof pairs to the theory of an abstract structure. In Athena, theory evolution is performed with the procedures `add-axioms` and `add-theorems` (defined in module `ST` and available at the top level), as described in Section 14.3. 911

theory refinement Defining the theory of a new abstract structure as a combination of existing theories and additional axioms. In Athena, theory refinement is performed with `theory`, a procedure (defined in module `ST`) that organizes axioms and theorems in a *structured theory* data structure, as described in Section 14.2. The refinement relation between theories constitutes a *theory refinement hierarchy*, such as the hierarchies of algebraic theories in Figure 14.1 on p. 653 and of relational theories in Figure 14.2 on p. 679. 631, 911

type Although *type* is often used synonymously with *sort* in the literature, in Athena there is a distinction: Types are used to classify Athena values (as enumerated at the end of Chapter 2 and summarized in Figure A.3), while sorts are used to classify terms. Terms are only one type of value; there are several more types. There is no upper limit on how many sorts there are. There are some primitive sorts (such as `Boolean` and `Int`) but the user can introduce arbitrarily many, via **`domain`** declarations and **`datatype`** or **`structure`** definitions. 912, 917

unification Unification is a process that determines whether two (or more) terms are *unifiable*, and if so, produces a unifying substitution for them. Two terms s and t are unifiable iff there is a substitution θ that renders them identical, so that applying θ to s and t yields the same result. Such a substitution is called a *unifier* of s and t. Unification can be performed in Athena with the primitive procedure `unify`, which is guaranteed to produce the most general possible unifier, if one exists at all. 98, 915

universal algebra The field of study of *algebraic structures* (or simply *algebras*) in general. An algebraic structure is a set together with a collection of operations on it. Chapter 14 deals with some of the most commonly studied algebraic structures. 46

variable capture A phenomenon that occurs when replacing a free variable occurrence by a term t that contains another variable that thereby becomes a bound occurrence in the resulting sentence. For example, substituting y for x in the body of the theorem $(\forall\, x{:}\texttt{Int}\,.\,\exists\, y{:}\texttt{Int}\,.\, x < y)$ results in capture of y and yields the false sentence $(\exists\, y{:}\texttt{Int}\,.\, y < y)$. Variable capture can be avoided by first renaming all bound variables of a sentence to be distinct from any free variables that occur in the term. Athena accomplishes this (in primitive methods such as `uspec` and in other similar contexts) by renaming bound variables with fresh variables. 39, 911, 913, 914, 917

Index

! shorthand for **apply-method**, 62
*
  in MSG, 647
**, 130, 579, 589
  in FMap, 545
*in
  in Trivial-Iterator, 716, 748
+, -, *, /, % predefined binary function symbols for Int and Real arithmetic, 59
+
  in N, 434
  in Semigroup, 632
  in Z, 476
  in ZPS, 485
+*
  in Monoid, 690
++, 141
  in FMap, 535
  in Set, 501
-
  in DMap, 553
  in FMap, 537
  in Group, 632
  in N, 451
  in Set, 511
  in Z, 476
--
  in Set, 524
-->, 91, 534
-->>, 823
/
  in MG, 649
  in N, 603
:: (List constructor), 140
<, >, <=, >= predefined binary function symbols for Int and Real comparison, 60
<
  in N, 439
  in SPO, 669
  in SWO, 686
<-, 807
  in Memory, 704
<->
  in FMap, 548
<0>
  in Identity, 632
<1>
  in MM, 648
<=
  in N, 447
  in PO, 669
<==, 413, 422
<==>, 34
<=L
  in List, 460
<E
  in SWO, 673, 687
<EL
  in SWO, 674
=, 26
=/=, 35
==, 844
==>, 12, 34
>
  in SPO, 669
>=
  in PO, 669
&, 12, 34
&&, 99, 866
&&*, 239
|, 34
||, 99, 866
||*, 239
_
  for dummy variables, 101
  wildcard pattern, 79
|^
  in DMap, 557
  in FMap, 543
@, 479, 493
\
  in Memory, 703
\\
  in Memory, 703, 721, 729
^
  in Set, 527
^1
  in Set, 527
^^
  in FMap, 548
;; end-of-input marker, 23
%
  in N, 603
~, 12, 34
λ-calculus, 20, 62
^, 807

ab, 43
Abelian group, 644–646, 649
Abelian monoid, 645, 692

abs, 94
abstract algorithm, 14, 599, 627, 683, 710, 754
abstract data type, 559
abstract grammar, 767, 800
abstract syntax, 118, 767, 799
  terms as a linear notation for, 803
abstract syntax tree, 171, 227, 799, 801–803
abstract-level proof, 14, 641, 683
abstraction/specialization, 13–14, 683
absurd, 12, 191, 202, 204, 276, 281
accumulator, 93, 594, 695, 721
adapt-theory, 647–649, 669, 691
adapter, 630
add, 29, 771
  in Map, 92
add-axioms, 632, 641
add-theorems, 632, 635, 637, 641, 642, 646
add-transformer
  in Prolog, 828, 901
additive homomorphism, 481, 484, 628, 651
ADT, *see* abstract data type
agree-on
  in DMap, 556
  in FMap, 545
agreement-characterization, 814
algebraic datatype, 8, 57, 493
algebraic formulations of ADTs, 559
algorithm requirements specification, 574–575
alist->dmap
  in DMap, 550
alist->dmap-general
  in DMap, 550
alist->map
  in FMap, 534
alist->pair
  in FMap, 534
alist->set
  in Set, 500
all-interpretations, 246
all-primes, 622
alpha-convertible sentences, 39, 325
alpha-equiv?, 322
alpha-equivalence, 39, 64, 322, 416
alpha-renaming, 39, 325, 893
alternate, 448
analytic tactics, *see* backward proof tactics
and, sentential constructor, 34, 191
  polyadic, 236
and-comm, 214
and-cong, 233
and-intro, 237
AND/OR graph, 285, 316
antecedent, 35
antecedent, 346
Antisymmetric, 668
apply
  in DMap, 552
  in FMap, 536
apply-method, 62
apply-module, 437
apply-to-key-values
  in Map, 93
Aristotle, 393
arity, 24, 122, 163, 237, 577, 629
as pattern, 137
asgn, 802
asgn-axiom, 819
asgn-excl, 835
asgn-semantics, 826
assert, 7, 43, 115
assert*, 7, 44, 176
associative arrays, 559
associativity, 4
assume, 195, 397, 867
assumption base, 7, 19, 21, 42–44, 861
  augmenting, 68, 69
assumption scope, 195, 315
AST, *see* abstract syntax tree
at, 809
  in DMap, 552
  in FMap, 536
  in Function, 654
  in Memory, 703
Athena, 19–22
  interacting with, 22–23
atom, 34
atoms, 247
augment, 398
auto-assert-dt-axioms, 48
auto-dtc-for-proofs, 841
auto-solve
  in Prolog, 828, 833, 902
automated proof, 15, 424, 473
automated testing, 17, 168
automated theorem prover (ATP), 15
axiom, 818
axiom schema, 121, 819
axiomatic semantics, 853

backjumping, 259
backtracking expression/deduction, 73
backward chaining, 414
backward proof tactics, 274–276, 354–355
base theory, 647

basis case, 8
basis-step, 777, 791
batch mode, 22
BCond, 802
bcond-parser, 804
bcond-vars, 811
BCP, *see* Boolean constraint propagation
bdn, 214
Bendix, Peter, 189
Beth, Evert, 317
bicond-def, 225
bicond-def', 226
biconditional, 206
  similarity to conjunctions, 207
biconditional definition (method), 225
biconditional elimination, 206
biconditional introduction, 207
bidirectional double negation, 214
Bidirectional-Iterator, 734
  copy-backward, 740
  copy-memory-backward, 740
  predecessor, 734
big-step operational semantics, 852
bijective
  in Function, 655
binary relation, 521
  composition, 525
  converse, 524
  domain, 521
  image, 528
  range, 521
  restriction, 527
binary search algorithm
  in C++ code, 566
  in ML code, 566
  interface design, 567
  on a binary search tree over N, 563–577, 683
  on a binary search tree over a strict weak order, 685
  on a range, 754
  optimization of, 575–576
binary search tree, 468–469
binary tree, 463–469
binary tree membership, 463
Binary-Relation, 664
  R', 664
  R, 664
binary-search
  in BinTree, 563
  in SWO, 685
BinTree, 463, 563
  binary-search, 563
  BST, 468, 569
  count, 467
  in, 463
  inorder, 464
Birkhoff, Garrett, 189
body of a quantification, 35
Boolean constraint propagation, 250, 310
both, 64, 193
Bottenbruch, Hermann, 623
bound variable occurrence, 38
bound-var-occs, 322
Bratko, Ivan, 317
BST
  in BinTree, 468, 569
built-in constants, 25
Burstall, Rod, 798
by-contradiction, 202
by-induction, 136, 870
Byrnes, John, 316

cancellation property, 450–452, 459
canonical term, 140
card
  in Set, 529
carrier, 372
case analysis, 10–12, 199, 440
  based on conditional definitions, 456
cases, 199
  for polyadic disjunctions, 237
CC
  in Set, 525
cell, 90, 866
chain, 116, 123, 491
chain->, 411
chain-first, 414
chain-help, 641
chain-last, 411
chain<-, 414
chaining, *see* equality chaining, implication chaining, equivalence chaining
  nested inside a justification item, 421–423
Change
  in Memory, 703
character, 90
  syntax, 857
check, 71, 869
children (of an application), 28
children, 28
choose-cong-method, 234
choose-lit, 253
claim, 63
clause, 249

clear
  in HashTable, 92
clear-assumption-base, 44
closed branch (in a semantic tableau), 295
Cmd, 802, 821, 837
cmd-parser, 804
CNF, *see* conjunctive normal form
CNF conversion, 249
cnf, 310
cnf-core, 255
cnf?, 310
collect
  in Trivial-Iterator, 717, 721
collect-locs
  in Random-Access-Iterator, 747
combine-equations, 132
comm, 214
Command, 771
commutation, 213
commutative, 9
Commutative-Ring, 649
Commutative-Ring-With-Identity, 649
commuting diagram, 775
compile, 774
compiler-correctness, 775
compiler-correctness', 793
compiler-def, 774
complement, 205
complementation, 205
complements?, 216
completeness, 384
completion procedure, 189
compose
  in FMap, 547
composed-with
  in Set, 525
composing proofs, *see* proof composition
compositional semantics, 239, 853
compositionality, 770, 853
compound (or complex) sort, 51
conclude, 70
conclusion
  in strong-induction, 606
conclusion-annotated deduction, 70, 867
concrete syntax, 118, 800
  vs. abstract syntax, 800
cond, 802
cond-def, 221
cond-def', 222
conditional branching, 71, 72
conditional deduction, 195, 867
  assumption, 195
  assumption scope, 195
  body, 195
  hypothesis, 195
  subdeduction, 195
conditional definition, 221
conditional elimination, 194
conditional equation, 125
conditional introduction, 195
conditional negation, 224
conditional proof, *see* conditional deduction
conditional rewriting, 172–178
conflict learning, 259
confluence, 180, 189
congruence relation, 232
conj, 802
conj-elim, 237
conj-intro, 237
conjecture
  counterexample to, 168
  falsification, *see* falsification
conjunct, 35
conjunction, 192
  commutativity, 213
  distributing over disjunctions, 219
  elimination, 192
  introduction, 193
conjunctive named hypothetical deductions, 868
conjunctive normal form, 227, 310
conjunctive query (in logic programming), 887
connective
  sentential, 34
consequence, 243, 373
consequent, 35
consing, 141
const, 767
constant symbol, 25
constructive definition, 468, 491
  vs. nonconstructive definition, 502–505
constructor, 8
  constant, 45
  datatype, 45
  irreflexive, 46
  nullary, 45
  reflexive, 46
  sentential, 34
context-free grammar, 800
continuation-passing style, 235, 302
contra-pos, 213
contractum, 126
contradiction tactic, 283
contrapositive, 213

conv
  in Set, 524
converters, 158–162
copy
  in Forward-Iterator, 729
copy-backward
  in Bidirectional-Iterator, 740
copy-memory
  in Forward-Iterator, 729
copy-memory-backward
  in Bidirectional-Iterator, 740
count
  in BinTree, 467
  in Forward-Iterator, 721
  in List, 467, 721
CPC
  in Set, 520
CPC-2
  in Set, 520
CPS, *see* continuation-passing style
creator, 559
cycles in goal graphs, 284

data type, 491
datatype, 24, 44–49
  admissible definition, 57
  polymorphic, 57–58
datatype, 44
datatype-axioms, 47, 57
datatype-cases, 442, 836, 846
datatypes, 48
Davis, Martin, 249
DC
  in Set, 511
De Bruijn arguments, 311
De Morgan's laws, 217
declare, 24, 45, 52, 167–168
deduction, 20, 62–78
  prefix syntax, 882
  syntax, 857
  vs. expression, 20, 77–78
deduction sequence, *see* proof sequence
dedup, 236, 500
default map, 809
default
  in DMap, 552
define, 40, 74, 77, 79
define-sort, 771
defining-clauses
  in Prolog, 903
definite Horn clause, 885
definite logic program, 885
definition, 40
denotational semantics, 853
deref
  in Trivial-Iterator, 716
derive-cnf, 310
derive-dnf, 310
derive-nnf, 310
derives, 825
Dershowitz, Nachum, 189
determinism, 846
deterministic semantics, 799, 846
deterministic-semantics, 846
Dewey path, 118
dictionaries, 559
dictionary, 533
diff, 767, 802
  in Set, 508
DIMACS format, 256
directive, 85–86
discrete mathematics, 491
discriminant, 72, 78, 84
disjointness, 180, 616
disjunct, 35
disjunction
  commutativity, 213
  distributing over conjunctions, 219
  elimination, 199
  introduction, 201
disjunctive normal form, 227, 310
disjunctive syllogism, 215
dist, 221
distributivity, 219, 459
div, 60, 771
divisibility, 608–609, 611–615
division algorithm (theorem), 605–608
dm, 218
dm-1, 217
dm-2, 217
dm-3, 217
dm-4, 217
dm', 219
DMap, 549
  -, 553
  |^, 557
  agree-on, 556
  alist->dmap, 550
  alist->dmap-general, 550
  apply, 552
  at, 552
  default, 552
  dmap->alist, 550
  dmap->alist-canonical-general, 551

dmap->alist-general, 550
dmap->set, 555
dom, 553
empty-map, 549
remove, 553
remove-from, 551
restricted-to, 557
size, 553
update, 549
dmap->alist
in DMap, 550
dmap->alist-canonical-general
in DMap, 551
dmap->alist-general
in DMap, 550
dmap->set
in DMap, 555
dn, 65, 202, 214
dn*, 227
DNF, *see* disjunctive normal form
dnf, 310
dnf?, 310
dom
in DMap, 553
in FMap, 539
in Set, 521
dom-characterization
in FMap, 539
domain, 23–27
domains, 24
DOMC
in Set, 522
double negation, 202
bidirectional, 214
downward-agreement-lemma, 816
DPLL procedure, 249
as constraint solving, 249
dpll, 250, 252
dpll0, 252
draw-all-theories, 653
draw-theory, 653
drinker's principle, 391
drop, 108
dsyl, 216
dsyl-1, 216
dummy variable, 101
dynamic assumption scoping, 76
dynamic scoping, 98
dynamic term construction, 100

E
in SPO, 672

Ebbinghaus, Heinz-Dieter, 393
egen, 331
egen*, 343
egen-cong, 351
either, 201
for polyadic disjunctions, 237
elaborated ground instance, 383
elaboration of a sentence, 382
elimination method, 64
else, 72
embedded recursion, 592
empty
in Range, 713
empty-map
in DMap, 549
in FMap, 533
empty-sub, 95
entailment relation, 243, 373
environment, *see* lexical environment
EOF, 23
eq, 802
equal?, 322
equal?, 60
equality axioms, 120
equality chaining, 3–8, 117
initial examples, 116–117, 128–135
logic behind, 120–128
mixing with equivalence steps, 418
mixing with implication steps, 418
equality predicate, 26
equational chaining, *see* equality chaining
equational proof, 113
equiv, 207
equivalence chaining, 6, 415–420, 448
mixing with equality steps, 418, 419
mixing with implication steps, 417
structural, 415
with alpha-equivalence, 416
equivalent, 844
equivalent-def, 844
Euclid, 315, 623
Euclid
is-common-divisor, 617
euclid
in N, 616
Euclid's algorithm, 603, 616–623
Eudoxus of Cnidus, 623
eval, 151, 153
evaluation, 21
evaluation judgment, 799
evaluation of ground terms, 150–152
evars, 811

even, 587
ex falso quodlibet, 215
ex-middle, 213
excl-lemmas, 836
excluded middle, 200, 213
exclusivity lemmas, 834
exec, 771
exec-def, 772
exec', 786
exec'-def, 787
exhaustiveness, 179, 189, 616
existence, 399
existential elimination, *see* existential instantiation
existential generalization, 331
existential instantiation, 332
  body, 333
existential premise, 333
existential quantification, 331–338
  elimination, 332
  introduction, 331
  similarity to disjunction, 336
existential specialization, *see* existential instantiation
exists, 35
Exp, 767
exp-op->cmd-and-num-op, 778
exp-parser, 804
expand-input, 162, 475, 501, 535, 551, 568, 768, 771, 772, 808, 823
exponentiation algorithm
  applied to lists, 700
  in C++ code, 693
  multiplicative version, 700
  on N, 579–601
  on a monoid, 693–700
export*, 309
expression, 20, 62–78
  prefix syntax, 882
  sequence, 93
  syntax, 857
  vs. deduction, 20, 77–78
expression compiler, 774–797
  correctness, 775–781
  with error handling, 784–796
extend-module, 433
extend-sub, 95
extensionality principle, 500
extraction tactics, 279

fact (in logic programming), 885
fail, 102
false, 191
false-elim, 192
false-elimination tactic, 284
falsification, 168–172
  examples, 169–171, 183, 515, 517, 523, 532, 543, 547, 776, 788
falsify, 135, 169, 491, 794
fast-power, 580, 590
fcong, 122, 123
filter, 108
find-element, 247
find-eqn-proof, 178, 782
find-first, 842
find-first-element, 234, 299
finish
  in Range, 711
fire, 346
fire-semantic-axiom, 842
First Substitution Theorem, 123
first-order logic, 319
  entailment, 373
  heuristics, 353–372
  satisfaction relation, 373
  semantics, 372–385
  tautology, 374
Fitch, Frederic, 316
Fitch-style natural deduction, 316
flatten, 109
Floyd, Robert W., 853
FMap, 533
  **, 545
  ++, 535
  -, 537
  <->, 548
  |^, 543
  ^^, 548
  agree-on, 545
  alist->map, 534
  alist->pair, 534
  apply, 536
  at, 536
  compose, 547
  dom, 539
  dom-characterization, 539
  empty-map, 533
  identity-characterization, 542
  iterate, 548
  Map, 533
  map->alist, 534
  map->set, 540
  map->set-characterization, 542
  map-identity, 541

o, 547
o', 549
override, 545
range, 542
range-characterization, 542
remove, 537
remove-correctness, 538
set->map, 541
size, 539
update, 533
foldl, 108
in Map, 93
foldr, 108
for-each, 108
for-some, 108
forall, 35
force, 207–209
formal semantics, 800
forward Horn clause inference, 344
forward proof tactics, 276–282, 355–371
Forward-Iterator, 718
copy, 729
copy-memory, 729
count, 721
replace, 724
successor, 718
Fraenkel, Abraham, 558
free algebra, 46, 379, 475
free variable, 97, 116
computation, 101
replacement of, 101
free variable occurrence, 38, 322
free-gen, 837, 839
free-generation axioms, 46, 803
free-var-occs, 322
free-vars, 101
freely generated, 499
freely generated datatype, 46
Frege, Gottlob, 315, 393
fresh variable, 327
fresh-var, 100
from-complements, 206, 215
from-false, 206, 215
from-to, 108
Fun
in Function, 654
fun, 425, 563, 577
fun->perm
in Permutation, 659
fun-def-conds, 181
fun-def-conds-d, 181
fun-def-conds-e, 181
function symbol, 19, 22–27, 859
admissible declaration, 52
arity, 24
associativity, 33
built-in, 26, 59
declaration, 24
input sorts, 24
numeric, 59
output sort, 24
polymorphic, 52–56
precedence, 32
range, 24
function theory, 654–658
function value, 859
Function, 654
at, 654
bijective, 655
Fun, 654
identity, 654
injective, 655
o, 654
surjective, 655
functional, 680
fv, 101

general clause, 887
generalize-over, 330, 868
generalized disjunction tactic, 283
generic algorithm, 14, *see* abstract algorithm
generic programming, 14
Gentzen, Gerhard, 315
get-ab, 43
get-assoc, 33
get-conjuncts, 109
get-div-lemma, 789
get-lemma, 789
get-precedence, 32, 34
get-property, 636
get-symbol-map, 647
Gilbert random graph model, 267
goal tree, 272, 284
Gödel, Kurt, 393
Goguen, Joseph, 797
graph coloring, 261
as a SAT problem, 263
graph-coloring, 265
greater?, 60
ground sort, *see* sort, monomorphic
ground term, 30, 113
grounding, 380
group theory, 627–641
permutation example, 653–664

refinements, 644–653
Group, 632
-, 632
U-, 632
guard, 172, 424

half, 584
halting problem, 388
hash table, 91
HashTable, 91
clear, 92
lookup, 91
remove, 92
size, 92
table->list, 92
table->string, 92
head, 29
Henkin, Leon, 246
Herbrand universe, 886
heterogeneous list, 19, 29
heuristics
for first-order logic proofs, 353
for sentential logic proofs, 272
Hilbert systems, 312
Hilbert, David, 315, 393
Hindley, Roger, 394
Hindley-Milner polymorphism, 394
Hoare, Tony, 853
hold?, 43
holds?, 43
homomorphic mapping, 481, 628
Horn clause, 344, 828, 833
Horn rule, *see* Horn clause
hsyl, 215
hypothesis matching (heuristic), 290
hypothesis scope, *see* assumption scope
hypothesis
in strong-induction, 606
hypothetical deduction, *see* conditional deduction
hypothetical proof, *see* conditional deduction
hypothetical syllogism, 215

I, 769
I+N
in Random-Access-Iterator, 743
I-def, 770
I-def', 785
I-I
in Random-Access-Iterator, 743
I-N
in Random-Access-Iterator, 743
I', 784
IC
in Set, 510
id, 163
Ide, 61
identifier (in Athena), 21, 24, 104, 862
identity symbol, *see* equality predicate
Identity, 632
<0>, 632
identity
in Function, 654
in Permutation, 658
identity-characterization
in FMap, 542
idn, 214
if, sentential constructor, 34, 191
if-chain, 248
if-cong, 233
if-false-excl, 835
if-false-rule, 820
if-false-semantics, 826
if-then-else operator, *see* ite
if-true-excl, 835
if-true-rule, 820
if-true-semantics, 826
iff, sentential constructor, 34, 191
iff-comm, 214
image
in Set, 528
IMGC
in Set, 529
imperative formulations of ADT operations, 559
implication chaining, 395–402
backward, 413–415
mixing with equality steps, 418, 419
mixing with equivalence steps, 417
structural, 411
import, 229
import*, 229
importation, 228
in
in BinTree, 463
in List, 460
in Range, 715
in Set, 501
in-all
in Set, 531
in-all-characterization
in Set, 531
in-lemma-2
in Set, 507
in-lemma-3
in Set, 507

in-lemma-4
  in Set, 507
indirect proof, 202
indirect tactic, 283
induction, *see* mathematical induction
induction hypothesis, 8, 136
induction step, 8, 136, 146
induction strengthening, 156, 776–777
induction variable
  choosing an, 445
induction*, 183
inductive datatype, 767
inductive property, 776, 797
inductive-step, 778
inductively generated, 44, 499
inference block, 67
inference rule, 818
  graphical presentation, 820
  side conditions, 820
inference system, 799
infix notation, 24, 31, 118
injective function, 47
injective
  in Function, 655
inorder tree walk, 464
inorder
  in BinTree, 464
input expansion, 153, 162–168
insert
  in Set, 499
instance (of a term), 96
instance, 115, 122, 128, 173
Int, 59
int->nat, 159, 189
integer
  mappings between representations, 480–483
  pair representation, 478–480
  signed representation, 475–478
integer addition, 476
  associativity and commutativity, 483–484
integer numeral, 25
integers, 59
integral domain theory, 652
Integral-Domain, 652
interactive mode, 22
interpretation, 239
  appropriate (for a given sentence), 241
  for first-order logic, 372
  of algebraic datatypes, 379
  of primitive sorts, 384
interpreter, 767–771
  with error handling, 783–784
intersection
  in Set, 508
intersection-subset-theorem
  in Set, 511
introduction and elimination methods, 64, 192
introduction method, 64
Inv
  in MG, 649
irreflexive constructor, 146
Irreflexive, 665
irreflexivity, 12
is-common-divisor
  in Euclid, 617
istep, 791, 813
It, 711
ite, 424
ite*, 425
iterate
  in FMap, 548
iterator, 703
  bidirectional, 715, 734–743
    copy-backward algorithm, 734–743
  dereferencing of, 716
  forward, 710, 715, 718–721
    copy algorithm, 729–734
    count algorithm, 721–724
    replace algorithm, 724–728
  random access, 715, 743–761
    lower-bound algorithm, 754–761
  reachable, 710
  trivial, 715–718
iterator range, 709
ith, 119

Jáskowski, Stanisław, 315
Jech, Thomas, 558
Joaunnaud, Jean-Pierre, 189
join, 29, 90, 141
  in List, 147
Judgment, 823, 824
justifier, 117

Kahn, Gilles, 852
key-values
  in Map, 93
keys
  in Map, 93
Kleene, Stephen, 393
Knuth, Donald, 189, 623

lambda expression
  body, 74

  formal parameters, 74
lambda abstraction, 26
lambda calculus, 20, 62
lambda, 74
Landin, Peter J., 853
last-val, 102
leaf node, 117
least Herbrand model, 886
left-and, 64, 192
left-assoc, 33, 35
left-either, 201
left-iff, 206
length, 29
  in Range, 713
leq, 802
less?, 26, 60
let, 69, 865, 870
letrec, 93, 865, 870
lexical environment, 21, 861
  augmenting/extending, 79
lexical scoping, *see* static scoping
list (of Athena values), 29, 82, 104, 859
  head, 29
  join, 29
  length, 29
  rev, 29
  tail, 29
list equation
  examples, 140–149
list membership, 460
List, 147
  <=L, 460
  count, 467, 721
  in, 460
  join, 147
  ordered, 460, 461
  replace, 577, 724
list-replace, 109
lists (as terms), 147
literal, 249
literal?, 298
load, 22
Loc
  in Memory, 703
Logemann, George, 249
logic program, 828
logic programming, 799, 828, 885–888
logical psychologism, 315
lookup
  in HashTable, 91
loose semantics, 783
Loveland, Donald, 249
lower-bound
  in Random-Access-Iterator, 755
lst->pair, 494
lst->pair-general, 494, 823

machine language, 767, 771
make
  in Map, 92
make-graph, 262
make-random-graph, 267
make-random-sentence, 309
make-sub, 95
make-term, 100, 161
make-theory, 632
make-vector, 91, 866
many-sorted logic, 393
Manzano, Maria, 393
map, 92, 860
map (theory), 533–558
  applying, 535
  composition, 545, 546
  default, 549
  domain, 539
  lookup, 535
  overriding, 545
  range, 542
  range restriction, 544
  restriction, 543
Map, 92
  add, 92
  apply-to-key-values, 93
  in FMap, 533
  foldl, 93
  key-values, 93
  keys, 93
  make, 92
  map-to-key-values, 93
  map-to-values, 93
  remove, 92
  size, 92
  values, 93
map, 74, 108
map->alist
  in FMap, 534
map->set
  in FMap, 540
map->set-characterization
  in FMap, 542
map-identity
  in FMap, 541
map-to-key-values
  in Map, 93

map-to-values
  in Map, 93
match, 72, 78, 865, 870
  clause, 72
  discriminant, 72
match-sentences, 97
match-terms, 97, 128
mathematical induction, 8–10
  for binary trees, 466
  for lists of natural numbers, 146
  for lists over sort *S*, 148
  for natural numbers, 136, 440
  for natural numbers (variant), 584
  measure-induction, 185, 541, 597–599, 756
  natural number examples, 135–140
  ordinary, 8, 185
  strong, 8, 10, 185, 581–583
  strong (variant), 600
Mauchly, John, 623
McCarthy, John, 560
McJones, Paul, 683, 693–695
measure-induction
  in strong-induction, 541, 598
  strong-induction, 756
Meinke, Karl, 394
memory range, 703
memory safety property, 705, 725, 730, 762
Memory, 703
  <-, 704
  \, 703
  \\, 703, 721, 729
  at, 703
  Change, 703
  Loc, 703
  swap, 705
Mendelson, Elliott, 393
meta-identifier, 61–62
metavariable, 28
method, 62, 75
  anonymous, 75
  defining, 74–226
  higher-order, 229
  recursive, 226–236
method abstraction, 841
method call, 62, 867
method value, 860
MG, 649
  /, 649
  Inv, 649
Milner, Robin, 394
MiniSat, 259
minus, 26, 60
MM, 648
  <1>, 648
mod, 60
model checking, 168
module, 6
  extending, 433
  indentation conventions, 437
  introduction, 429–431
  nested, 429
  scope, 429
module->string, 430
module-domain, 436
module-size, 436
modus ponens, 3, 194
modus tollens, 194
monad, 705
monoid, 638, 683, 690
Monoid, 632
  ++, 690
monotonicity, 196
morphism, 775
mp, 194
MSG, 647
  *, 647
mt, 194, 202
mult, 771
multi-sorted logic, *see* many-sorted logic
multiplicative theory, 646–649
mutator, 559
mutual recursion, 94
mutually recursive datatypes, 48

N, 46, 113
  +, 434
  -, 451
  /, 603
  <, 439
  <=, 447
  %, 603
  euclid, 616
name (of a variable), 27
named hypothetical deductions, 868
natural number
  equality properties, 113–189
  ordering properties, 439–473
natural semantics, 799, 852
NC
  in Set, 502
NC-2
  in Set, 506
NC-3
  in Set, 506

neg, 802
neg-cond, 224
negate
  in Z, 476
negation as failure, 887
negation elimination, 202
negation normal form, 227, 309
negative polarity, 268
nested chaining rules, 407
nested method calls, 68
Nielson, Flemming, 852
Nielson, Hanne Ries, 852
nil (List constructor), 140
NNF, *see* negation normal form
nnf, 309
nnf?, 309
no-confusion condition, 46
no-junk condition, 46
no-renaming, 643
No-Zero-Divisors, 652
non-empty
  in Set, 506
noncompositional semantics, 853
NONE, 497
nonground (or general) term, 113
normal forms in sentential logic, 227
not, sentential constructor, 34, 191
not-cong, 232
notational conventions, 103–105
null
  in Set, 499
null-characterization
  in Set, 502
nullable types, 558
nullary procedure, 43
num, 802
numeric equations, 113–116
numeric function symbols, 59
numeric procedures, built-in, 60

o
  in FMap, 547
  in Function, 654
  in Permutation, 658
  in Set, 525
o'
  in FMap, 549
observer, 559
OC
  in Set, 526
odd, 587
OP, 483, 778
open, 432
  transitivity, 435
open branch (in a semantic tableau), 295
operational semantics, 772, 853
  big-step, 852
  of Athena phrases, 861–875
  small-step, 772, 852
opt-axioms, 498
option, 497–499
option value, 497, 712, 784
Option, 497, 712, 784
option-results, 499
option-val, 497
option-valued-selectors, 882
or, sentential constructor, 34, 191
  polyadic, 237
or-comm, 214
ordered list, 459–463
ordered pair, 493–497
ordered range, 753–754
ordered
  in List, 460, 461
  in Ordered-Range, 753
  in SWO, 676
Ordered-Range
  ordered, 753
ordering property
  asymmetry, 12, 444–447, 666
  combining with arithmetic, 450–451
  irreflexivity, 439
  nonstrict, 447–450
  reflexivity, 447
  strict, 439–447
  transitivity, 444–447
  trichotomy, 443–444, 447
output transformation, 153, 165–168
overload, 86, 424, 465, 467, 476, 477, 479–481, 535, 551, 721, 743, 745, 757, 769, 808, 844
overloading, 9, 24, 86–89, 114
override
  in FMap, 545

Pair, 493
pair, 493
pair->lst, 494
pair->lst-general, 494
pair-converter, 496
pair-left, 493
pair-right, 493
paired-with
  in Set, 518
parsing, 801

partial deduction, 272
pattern variable, 871, 875
patterns
  for lists, 82
  for quantified sentences, 81
  for terms, 80
  infix notation, 80
  inside **let** phrases, 102
  matching, 29, 72–73, 78–84, 875–881
Peano, Giuseppe, 393
Perm
  in Permutation, 658
perm->fun
  in Permutation, 659
permutation theory, 658–664
Permutation, 658
  fun->perm, 659
  identity, 658
  o, 658
  Perm, 658
  perm->fun, 659
phrase, 20, 62, 857
pick-all-witnesses, 387
pick-any, 327, 868
pick-witness, 334, 869
pick-witnesses, 335
pigeonhole principle (logical encoding), 313
Plotkin, Gordon, 853
Plus, 113
plus, 21, 60
PO, 669
  <=, 669
  >=, 669
polarities, 269
polarity, 268
poly?, 55, 56
polyadic sentential constructor, 35
polymorphic domain, 50
polymorphic sort, 51
polymorphic term, *see* term, polymorphic
polymorphism, 50–61
Portoraro, Frederic, 316
POSC
  in Set, 531
position
  of a node in a tree, 118
positions-and-subterms, 119, 127
positive polarity, 268
positive-negative polarity, 269
power series, 484–487
powerset, 530–533
powerset
  in Set, 530
precedence, 32, 603, 769
predecessor
  in Bidirectional-Iterator, 734
predicate logic, *see* first-order logic
predicate symbol, *see* relation symbol
prefix notation, 24, 118
premise, 42, 820
Preorder, 668
prime number, 621
prime, 621
primitive goal, 276
primitive method, 102, 312
primitive recursion, 540
primitive-method, 103, 312
principle
  in strong-induction, 590, 596, 600
principle2
  in strong-induction, 600
print-instance-check, 637
private, 430
procedure, 20
  anonymous, 74
  defining, 74
procedure call, 62
  embedded vs. tail, 592
  tail, 695
prod, 767, 802
producer, 559
program equivalence, 844
Program, 771
programming, 89–98
Prolog interpreter implementation, 893–897
Prolog program, 885
Prolog
  add-transformer, 828, 901
  auto-solve, 828, 833, 902
  defining-clauses, 903
  solve, 898
  solve-all, 898
  solve-all-with-time-limit, 900
  solve-N, 898
  solve-N-with-time-limit, 900
  solve-with-time-limit, 900
Prolog_Interpreter
  solve, 889
  solve-all, 894
  solve-N, 889
proof by contradiction, 12–13, 202–206, 440, 868
proof composition, 65–67

proof conclusion, 20, 42
proof continuation, 235
proof continuation-passing style, 235
proof heuristics, 272, 316, 456, 605
proof method, 3
proof sequence, 65–67
proof shortening, 781–783, 793–794
proof simplification, 286
proof spec, 272
proof strategies, 282–294, 371–372
  vs. proof tactics, 274
proof tactics, 272
proof terms, 353
proof tree, 799, 821
  shape, 799, 824
`Proof`, 824
`proof-error`, 102
`proof-tools`, 635, 640
`prop-taut`, 294
proper function definition, 179–185
proper sentence matching, 359
proper subset, 507
`proper-subset`
  in `Set`, 507
`proper-subset-lemma`
  in `Set`, 507
property procedure, 131
propositional logic, *see* sentential logic
`provable`, 827
`provable-def`, 827
`prove-antecedent`, 346
`prove-equiv`, 231
`prove-property`, 636
proving negations directly (heuristic), 292
`PSC`
  in `Set`, 507
pure literal propagation, 250
`pure-literal`, 253
`push`, 771
Putnam, Hilary, 249
`PWC`
  in `Set`, 519

`qn`, 350
qualified name, 429
`quant-body`, 400
`quant-dist`, 351, 388
`quant-swap`, 350
quantification, 320
quantified logic, *see* first-order logic
quantified sentence, 320
  body, 320
  variable, 35
quantifier, 35, 320
quantifier distribution, 351, 388
quantifier negation, 348
quantifier swapping, 350
query (in logic programming), 887
`quot`, 767
quotient and remainder functions, 603–605
`qvars-of`, 400

`R'`
  `Binary-Relation`, 664
`R`
  `Binary-Relation`, 664
`RANC`
  in `Set`, 522
random sentence, 309
`Random-Access-Iterator`, 743
  `collect-locs`, 747
  `I+N`, 743
  `I-I`, 743
  `I-N`, 743
  `lower-bound`, 755
`Range`, 711
  `empty`, 713
  `finish`, 711
  `in`, 715
  `length`, 713
  `range`, 712
  `start`, 711
`range`
  in `FMap`, 542
  in `Range`, 712
  in `Set`, 521
`range-characterization`
  in `FMap`, 542
`RC`
  in `Set`, 512
`rcong`, 122
`rd`, 109, 377
real numbers, 59
real numeral, 25
`Real`, 59
reasoning by cases, *see* case analysis
reasoning by symmetry, 513
recursion, 93
  well-founded, 184
recursive methods, 226
redex, 126
reductio ad absurdum, 202
reduction to a common term, 7
`ref`, 90, 866

ref-equiv, 232
reflex, 122
Reflexive, 668
register-sort-stream, 172
reiteration tactic, 282
relation, 521–529
relation symbol, 25
remove
  in DMap, 553
  in FMap, 537
  in HashTable, 92
  in Map, 92
  in Set, 511
remove-correctness
  in FMap, 538
remove-from
  in DMap, 551
renaming, 629
replace
  in Forward-Iterator, 724
  in List, 577, 724
replace-subsentence, 120, 188
replace-subterm, 119, 188
replace-var, 101
replace-var-by-term, 323
replacement rule, 229
replacement rule for first-order logic, 351
replacement tactics, 282
resolution, 216, 310
resolve, 310
restrict
  in Set, 527
restrict1
  in Set, 527
restricted-to
  in DMap, 557
retract, 44
rev, 29, 90
reverse iterator adapter, 737–743
reverse, 152
rewrite rule, 125, 125
  conditional, 172
  constructor-based, 125
rewriting, 97, 799
Rewriting Theorem, 126
right-and, 64, 192
right-assoc, 33
right-either, 201
right-iff, 206
ring theory, 649–652
Ring, 649
Ring-With-Identity, 649
Ring-With-No-Zero-Divisors, 652
root symbol, 28
root, 28
Rose, Daniel, 623
RSC
  in Set, 528
rule (a sentence used as a justifier), 402
  biconditional, 404
  nested, 407
rule (in logic programming), 885
  body, 885
  head, 885
  operational meaning, 886
run-vm, 774
run-vm', 787
Russell's paradox, 390
Russell, Bertrand, 315, 393, 558

SAT, *see* sentential logic satisfiability problem
SAT solver, 249, 256, 259
SAT solving, 15, 246–268, 316, 473
sat, 375
sat-solve, 259
sat?, 247
satisfiability, 243, 316
satisfiability modulo theories, *see* SMT solving
satisfiable, 243
SC
  in Set, 504
Scheines, Richard, 316
Scheme, 19
scope of a quantified variable, 35
scope of **pick-any** variable, 327
SECD machine, 853
Second Substitution Theorem, 124
selector, 493, 881–882
selector-axioms, 881
semantic judgment, 818
semantic tableaux, 294
semantics
  of sentential logic, *see* sentential logic semantics
semantics, 827
semantics-horn-clauses, 834
Semigroup, 632, 641
  +, 632
sent->tree, 321
sent-rename, 323
sentence, 19, 34–40, 42
  as justifier, 402–409
  as tree-structured object, 117, 119
  atomic, 34

  Boolean combinations, 34
  matching, 97
  quantified, 35
sentence matching, 359
sentential connective, 12, 34, 191, 859
sentential constructor, 34
sentential logic, 191
  automated theorem proving, 294–304
  entailment, 243
  heuristics, 272–294
  Hilbert system, 312
  satisfaction relation, 241
  satisfiability problem, 245
  semantics, 238–246
  tautology, 241
`Sentential-Semantic-Tableaux`, 298
`seq-excl`, 835
`seq-horn-clause`, 834
`seq-rule`, 819
`sequence`, 802
`sequence-semantics`, 826
sequent systems, 315
set (finite), 499–518
  cardinality, 529–530
  cartesian product, 518–521
  identity relation, 505–506
  membership relation, 501–502
  subset relation, 502–505, 506–507
  union, intersection, difference, 508–518
set theory, 558
`Set`, 499
  `++`, 501
  `-`, 511
  `--`, 524
  `^`, 527
  `^1`, 527
  `alist->set`, 500
  `card`, 529
  `CC`, 525
  `composed-with`, 525
  `conv`, 524
  `CPC`, 520
  `CPC-2`, 520
  `DC`, 511
  `diff`, 508
  `dom`, 521
  `DOMC`, 522
  `IC`, 510
  `image`, 528
  `IMGC`, 529
  `in`, 501
  `in-all`, 531
  `in-all-characterization`, 531
  `in-lemma-2`, 507
  `in-lemma-3`, 507
  `in-lemma-4`, 507
  `insert`, 499
  `intersection`, 508
  `intersection-subset-theorem`, 511
  `NC`, 502
  `NC-2`, 506
  `NC-3`, 506
  `non-empty`, 506
  `null`, 499
  `null-characterization`, 502
  `o`, 525
  `OC`, 526
  `paired-with`, 518
  `POSC`, 531
  `powerset`, 530
  `proper-subset`, 507
  `proper-subset-lemma`, 507
  `PSC`, 507
  `PWC`, 519
  `RANC`, 522
  `range`, 521
  `RC`, 512
  `remove`, 511
  `restrict`, 527
  `restrict1`, 527
  `RSC`, 528
  `SC`, 504
  `Set`, 499
  in `Set`, 499
  `set->alist`, 500
  `set-identity`, 505
  `set-identity-intro`, 505
  `set-identity-intro-direct`, 505
  `SIC`, 505
  `singleton`, 507
  `singleton-characterization`, 507
  `subset`, 502
  `subset-antisymmetry`, 506
  `subset-characterization`, 504
  `subset-def`, 503
  `subset-intro`, 504
  `subset-reflexivity`, 506
  `subset-transitivity`, 506
  `UC`, 509
  `union`, 508
  `X`, 518

set->alist
  in Set, 500
set->map
  in FMap, 541
set-flag
  auto-assert-dt-axioms, 48
  option-valued-selectors, 882
  print-qvar-sorts, 85
  print-var-sorts, 32, 85
set-identity
  in Set, 505
set-identity-intro
  in Set, 505
set-identity-intro-direct
  in Set, 505
set-precedence, 32, 768, 771, 808, 844
set!, 90, 866
short-circuit evaluation, 99, 239
Shostak, Robert, 473
show-assumption-base, 43
SIC
  in Set, 505
side effect, 559
Sieg, Wilfried, 316
signature, 24
simple term, 28
Single Rewriting Step, 126
single-sorted logic, 393
singleton
  in Set, 507
singleton-characterization
  in Set, 507
size
  in DMap, 553
  in FMap, 539
  in HashTable, 92
  in Map, 92
skip, 802
skip-axiom, 819
skip-excl, 835
skip-semantics, 826
SLD tree, 894
small-step operational semantics, 852
SMT solving, 15, 424, 473, 549
Smullyan, Raymond, 317
solve
  in Prolog, 898
  in Prolog_Interpreter, 889
solve-all
  in Prolog, 898
  in Prolog_Interpreter, 894
solve-all-with-time-limit
  in Prolog, 900
solve-N
  in Prolog, 898
  in Prolog_Interpreter, 889
solve-N-with-time-limit
  in Prolog, 900
solve-with-time-limit
  in Prolog, 900
SOME, 497
sort, 24
  (of a variable), 27
  annotation, 36, 54
  checking, 30
  constructor, 50
  definition, 50
  error, 30
  matching, 51
  monomorphic, 51
  parameter, 50
  syntax, 857
  unification, 51
  valuation, 51, 875
  variable, 50, 61
  vs. type, 31, 105
sort parameter, 147
SOS, *see* structural operational semantics
soundness, 238, 384
specification, 559
SPO, 669
  <, 669
  >, 669
  E, 672
square, 589
square-bracket list notation, 42, 493, 500, 550
ST, 632
stack (for machine-language semantics), 771
Stack, 771
Standard Template Library (STL), 14, 703, 755, 761 763
start
  in Range, 711
state, 559, 809
State, 809
state, 809
static scoping, 41, 76, 98–99
  in modules, 435
Stepanov, Alexander, 623, 683, 693–695
STO, 674
store, 21, 861

Strichman, Ofer, 473
strict partial order, 446
strict total order, 674
strict weak order, 459, 671–674, 683, 762
  list example, 674–679
Strict-Partial-Order, 666
string, 90
  syntax, 857
string->symbol, 89, 166
string->var, 311
strong induction hypothesis, 10
strong-induction
  conclusion, 606
  hypothesis, 606
  measure-induction, 541, 598, 756
  principle, 590, 596, 600
  principle2, 600
structural induction, 46, 870
structural operational semantics, 853
structural-axioms, 827
structure, 49–50, 499
structure, 49, 475
structure-axioms, 49, 57
structured theory, 627
  dynamic evolution of, 641–642
  proof method conventions, 635–641
  refinement, 631–635
  shared (or adapted), 647
sub, 771
subsentence, 38
  immediate, 38
  proper, 38
subsentence, 120, 188
subsentence-node, 120, 188
subset
  in Set, 502
subset-antisymmetry
  in Set, 506
subset-characterization
  in Set, 504
subset-def
  in Set, 503
subset-intro
  in Set, 504
subset-reflexivity
  in Set, 506
subset-transitivity
  in Set, 506
substitution, 94, 860
  applying, 95
  composition, 96
  extending, 95
  support of, 94
subterm, 30
  immediate, 30
  proper, 30
subterm, 119
subterm-node, 119
subtraction, 451–459
successor function, 8, 113
successor
  in Forward-Iterator, 718
sufficient completeness, 189
sum, 767, 802
superior, 631
supp, 95
Suppes, Patrick, 393, 491, 558
surjective
  in Function, 655
swap, 495
  in Memory, 705
SWI Prolog, 898
SWO, 672
  <, 686
  <E, 673, 687
  <EL, 674
  binary-search, 685
  ordered, 676
sym, 122
symbol, *see* function symbol
symbol instance, 381
symbol map, 629
symbol set, 21, 861
synthetic tactics, *see* forward proof tactics

table, 91, 860
table->list
  in HashTable, 92
table->string
  in HashTable, 92
tactic, *see* proof tactics
tactic ranking, 286
tail, 29
tail-recursive definition, 592
take, 108
Tarski, Alfred, 393
taut?, 247
tautology, 196, 241
TC, 767
term, 19, 27–34
  as tree-structured object, 117
  canonical, 140

children, 28
complex, 28
denotation of, 19, 373
general, 113
ground, 113
ill-sorted, 30, 508
instance, 96
matching, 96
polymorphic, 54
root, 28
simple, 28
unification, 96
variable replacement, 101, 323
term rewriting, 123, 125
term->tree, 321
termination, 184, 189, 616
terms and sentences as trees, 117–120
test-all-proofs, 644
test-proofs, 643
testing abstract proofs, 642–644
theorem, 42, 63
theorem proving, 294
theory, 628
theory diagram
algebraic, 652
relational, 678
theory of measurement, 391
theory refinement hierarchy, 632
theory-axioms, 632
theory-name, 632
theory-superiors, 632
theory-theorems, 632
thunk, 895
tight semantics, 784
Times, 129
times, 26, 60
Top, 436
top-down proof development, 152–158
vs. bottom-up, 154
total function specification, 179
total ordering, 443
tran, 122
transform, 230, 351
transform-output, 165, 508, 530, 535, 539, 568, 579, 603, 770, 810, 811
Transitive, 665
transitivity, 12, 395
tree-leaves, 322
Trivial-Iterator, 716
*in, 716, 748
collect, 717, 721
deref, 716
true, 191
true-in, 817
true-in-def, 817
true-intro, 63
true-intro, 191
try, 73, 865, 870
Tseitin variable, 255
Tucker, John, 394
tv, 375
two-cases, 200
type, 31, 104
vs. sort, 31, 105
type theory, 559

U-
in Group, 632
UC
in Set, 509
UFV (universal free variables), 356
ugen-cong, 351
underspecification, 180, 604, 783
unequal, 35
unequal-to, 300
unification, 96, 98
unify, 98
union
in Set, 508
unit clause, 250
unit clause propagation, 250
unit value, 27, 859
unit-propagation, 253
universal algebra, 46, 394
universal free variables, 356
universal instantiation, 323
universal quantification, 323–331
elimination, 323
introduction, 326
similarity to conjunction, 326
universal sentence occurrence, 355
universal specialization, 323
update
in DMap, 549
in FMap, 533
uspec, 173, 323
uspec*, 342

V, 809
V-def, 810
value (of an Athena phrase), 21, 858–861
equality, 860
hashable, 91
type, 31, 104

value type, 710
values
  in Map, 93
var, 802
var-agreement, 812
var-occs, 322
variable, 19, 27
  capture, 39, 116, 325, 797
  dummy, 101
  free, 44
  fresh, 100
  polymorphic, 53
  replacement, 101, 323
variable assignment, 373
variable occurrence, 321
  bound, 38, 321
  free, 38, 321
vector, 91
vector-set!, 91, 867
vector-sub, 91, 866
virtual machine, 771–774
vm-def, 774
vm-def', 787
vocabulary, 372

Wang, Hao, 393
where, 84
While language, 799
  associativity of composition, 844
  exclusivity lemmas, 834
  parser, 801
  state, 809
while, 93, 866
while-axiom, 821
while-false-excl, 836
while-false-semantics, 826
while-loop, 802
while-rule, 821, 853
while-true-excl, 836
while-true-semantics, 827
Whitehead, Alfred, 315
wildcard pattern, 79
with, 401
with-witness, 337, 869
witness hypothesis, 333
witness variable, 333
wrt, 240, 771, 809
wrt', 786

X
  in Set, 518

yields, 818
yields, 823
  $\hookrightarrow$, 818

Z, 559
Z, 475
  +, 476
  -, 476
  negate, 476
Zermelo, Ernst, 558
zero-order sentence, 191
  polarity, 268
  random, 309
  satisfiable, 243
  unsatisfiable, 243
zip, 108
ZPS, 485
  +, 485